McGRAW-HILL

CONCISE

ENCYCLOPEDIA OF

EARTH

SCIENCE

McGraw-Hill

New York   Chicago   San Francisco   Lisbon   London   Madrid   Mexico City
Milan   New Delhi   San Juan   Seoul   Singapore   Sydney   Toronto

The **McGraw·Hill** Companies

Library of Congress Cataloging in Publication Data

McGraw-Hill concise encyclopedia of earth science.
    p.   cm.
  Includes bibliographical references and index.
  ISBN 0-07-143954-4
    1. Earth sciences—Encyclopedias. I. Title: Concise encyclopedia of earth
science.

  QE5.M363   2004
  550′.3—dc22   2004061052

1234567890   DOC/DOC 0109876 5

ISBN 0-07-143954-4

This book was printed on acid-free paper.

It was set in Helvetica Black and Souvenir by TechBooks, Fairfax, Virginia.

The book was printed and bound by RR Donnelley, The Lakeside Press.

CONTENTS

EDITORIAL STAFF

EDITING, DESIGN, AND PRODUCTION STAFF

CONSULTING EDITORS

PREFACE

For more than four decades, the *McGraw-Hill Encyclopedia of Science & Technology* has been an indispensable scientific reference work for a broad range of readers, from students to professionals and interested general readers. Found in many thousands of libraries around the world, its 20 volumes authoritatively cover every major field of science. However, the needs of many readers will also be served by a concise work covering a specific scientific or technical discipline in a handy, portable format. For this reason, the editors of the *Encyclopedia* have produced this series of paperback editions, each devoted to a major field of science or engineering.

The articles in this *McGraw-Hill Concise Encyclopedia of Earth Science* cover all the principal topics of this field. Each one is a condensed version of the parent article that retains its authoritativeness and clarity of presentation, providing the reader with essential knowledge in Earth Science without extensive detail. The initials of the authors are at the end of the articles; their full names and affiliations are listed in the back of the book.

The reader will find 900 alphabetically arranged entries, many illustrated with images or diagrams. Most include cross references to other articles for background reading or further study. Dual measurement units (U.S. Customary and International System) are used throughout. The Appendix includes useful information complementing the articles. Finally, the Index provides quick access to specific information in the articles.

This concise reference will fill the need for accurate, current scientific and technical information in a convenient, economical format. It can serve as the starting point for research by anyone seriously interested in Earth Science, even professionals seeking information outside their own specialty. It should prove to be a much used and much trusted addition to the reader's bookshelf.

MARK D. LICKER
Publisher

ORGANIZATION OF THE ENCYCLOPEDIA

Alphabetization. The approximately 900 article titles are sequenced on a word-by-word basis, not letter by letter. Hyphenated words are treated as separate words. In occasional inverted article titles, the comma provides a full stop. The index is alphabetized on the same principles. Readers can turn directly to the pages for much of their research. Examples of sequencing are:

Air temperature	**Earthquake**
Airglow	**Sea breeze**
Earth, heat flow in	**Sea-floor imaging**
Earth crust	**Seamount and guyot**

Cross references. Virtually every article has cross references set in CAPITALS AND SMALL CAPITALS. These references offer the user the option of turning to other articles in the volume for related information.

Measurement units. Since some readers prefer the U.S. Customary System while others require the International System of Units (SI), measurements in the Encyclopedia are given in dual units.

Contributors. The authorship of each article is specified at its conclusion, in the form of the contributor's initials for brevity. The contributor's full name and affiliation may be found in the "Contributors" section at the back of the volume.

Appendix. Every user should explore the variety of succinct information supplied by the Appendix, which includes conversion factors, measurement tables, fundamental constants, and a biographical listing of scientists. Users wishing to go beyond the scope of this Encyclopedia will find recommended books and journals listed in the "Bibliographies" section; the titles are grouped by subject area.

Index. The 4400-entry index offers the reader the time-saving convenience of being able to quickly locate specific information in the text, rather than approaching the Encyclopedia via article titles only. This elaborate breakdown of the volume's contents assures both the general reader and the professional of efficient use of the *McGraw-Hill Concise Encyclopedia of Earth Science.*

A

Acanthite   The mineral name applied to the monoclinic form of silver sulfide (Ag_2S) that is stable at room temperature. Two different crystalline structures are assumed in succession as temperature is increased. At a temperature of about 176–178°C (349–352°F), depending on stoichiometry, acanthite transforms rapidly to argentite, which is body-centered cubic, space group *Im 3m*. The cell contains two Ag_2S units and has a lattice constant that ranges from 0.4860 nm at 186°C (367°F) to 0.4889 nm at 325°C (617°F). At 586–620°C (1087–1148°F), again depending on stoichiometry, argentite transforms to a face-centered cubic structure. The details of the atomic arrangement in this phase are not known.

Acanthite is an important silver ore. Notable deposits in the United States include Butte, Montana; Aspen and Leadville, Colorado; and the Comstock Lode in Nevada. The mineral is also common in Mexican mines at Guanajuato. Other major deposits occur in Bolivia, Peru, and Chile. Especially fine crystals have been found in Saxony and in the Harz Mountains in Germany. Silver sulfide is encountered in everyday life as the black tarnish that develops on silver objects.   [B.J.Wu.]

Acid rain   Precipitation that incorporates anthropogenic acids and acidic materials. The deposition of acidic materials on the Earth's surface occurs in both wet and dry forms as rain, snow, fog, dry particles, and gases. Although 30% or more of the total deposition may be dry, very little information that is specific to this dry form is available. In contrast, there is a large and expanding body of information related to the wet form: acid rain or acid precipitation. Acid precipitation, strictly defined, contains a greater concentration of hydrogen (H^+) than of hydroxyl (OH^-) ions, resulting in a solution pH less than 7. Under this definition, nearly all precipitation is acidic. The phenomenon of acid deposition, however, is generally regarded as being anthropogenic, that is, resulting from human activity.

Theoretically, the natural acidity of precipitation corresponds to a pH of 5.6, which represents the pH of pure water in equilibrium with atmospheric concentrations of carbon dioxide. Atmospheric moisture, however, is not pure, and its interaction with ammonia, oxides of nitrogen and sulfur, and windblown dust results in a pH between 4.9 and 6.5 for most "natural" precipitation. The distribution and magnitude of precipitation pH in the United States (see illustration) suggest the impact of anthropogenic rather than natural causes. The areas of highest precipitation acidity (lowest pH) correspond to areas within and downwind of heavy industrialization and urbanization where emissions of sulfur and nitrogen oxides are high. It is with these emissions that the most acidic precipitation is thought to originate.

The transport of acidic substances and their precursors, chemical reactions, and deposition are controlled by atmospheric processes. In general, it is convenient to distinguish between physical and chemical processes, but it must be realized that both

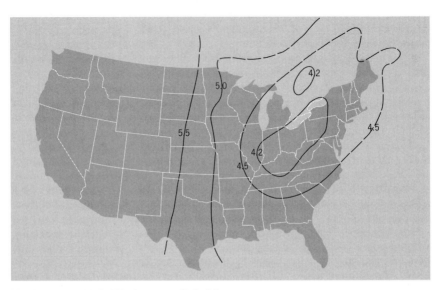

Distribution of rainfall pH in the eastern United States.

types may be operating simultaneously in complicated and interdependent ways. The physical processes of transport by atmospheric winds and the formation of clouds and precipitation strongly influence the patterns and rates of acidic deposition, while chemical reactions govern the forms of the compounds deposited.

There are a number of chemical pathways by which the primary pollutants, sulfur dioxide (SO_2) from industry, nitric oxide (NO) from both industry and automobiles, and reactive hydrocarbons mostly from trees, are transformed into acid-producing compounds. Some of these pathways exist solely in the gas phase, while others involve the aqueous phase afforded by the cloud and precipitation. As a general rule, the volatile primary pollutants must first be oxidized to more stable compounds before they are efficiently removed from the atmosphere. Ironically, the most effective oxidizing agents, hydrogen peroxide (H_2O_2) and ozone (O_3), arise from photochemical reactions involving the primary pollutants themselves. *See* AIR POLLUTION.

The effect of acid deposition on a particular ecosystem depends largely on its acid sensitivity, its acid neutralization capability, the concentration and composition of acid reaction products, and the amount of acid added to the system. As an example, the major factors influencing the impact of acidic deposition on lakes and streams are (1) the amount of acid deposited; (2) the pathway and travel time from the point of deposition to the lake or stream; (3) the buffering characteristics of the soil through which the acidic solution moves; (4) the nature and amount of acid reaction products in soil drainage and from sediments; and (5) the buffering capacity of the lake or stream.

Acid precipitation may injure trees directly or indirectly through the soil. Foliar effects have been studied extensively, and it is generally accepted that visible damage occurs only after prolonged exposure to precipitation of pH 3 or less (for example, acid fog or clouds). Measurable effects on forest ecosystems will then more likely result indirectly through soil processes than directly through exposure of the forest canopy. Many important declines in the condition of forest trees have been reported in Europe and North America during the period of increasing precipitation acidity. These cases include injury to white pine in the eastern United States, red spruce in the Appalachian

Mountains of eastern North America, and many economically important species in central Europe. Since forest trees are continuously stressed by competition for light, water, and nutrients; by disease organisms; by extremes in climate; and by atmospheric pollutants, establishing acid deposition as the cause of these declines is made more difficult. Each of these sources of stress, singly or in combination, produces similar injury. However, a large body of information indicates that accelerated soil acidification resulting from acid deposition is an important predisposing stress that in combination with other stresses has resulted in increased decline and mortality of sensitive tree species and widespread reduction in tree growth. *See* TERRESTRIAL ECOSYSTEM.

Acidic deposition impacts aquatic ecosystems by harming individual organisms and by disrupting flows of energy and materials through the ecosystem. The effect of acid deposition is commonly assessed by studying aquatic invertebrates and fish. Aquatic invertebrates live in the sediments of lakes and streams and are vitally important to the cycling of energy and material in aquatic ecosystems. These small organisms break down large particulate organic matter for further degradation by microorganisms, and they are an important food source for fish, aquatic birds, and predatory invertebrates.

Currently, there are concerns that acid deposition is causing the loss of fish species, through physiological damage and by reproductive impairment. While fish die from acidification, their numbers and diversity are more likely to decline from a failure to reproduce.The effects of acid deposition on individuals in turn elicit changes in the composition and abundance of communities of aquatic organisms. The degree of change depends on the severity of acidification, and the interaction of other factors, such as metal concentrations and the buffering capacity of the water. The pattern most characteristic of aquatic communities in acidified waters is a loss of species diversity, and an increase in the abundance of a few, acid-tolerant taxa.

Community-level effects may occur indirectly, as a result of changes in the food supply and in predator-prey relations. Reduction in the quality and amount of periphyton may decrease the number of herbivorous invertebrates, which may in turn reduce the number of organisms (predatory invertebrates and fish) that feed upon herbivorous invertebrates. The disappearance of fish may result in profound changes in plant and invertebrate communities. Dominant fish species function as keystone predators, controlling the size distribution, diversity, and numbers of invertebrates. Their reduction alters the interaction within and among different levels of the food web and the stability of the ecosystem as a whole.

The impact of acid deposition on terrestrial and aquatic ecosystems is not uniform. While increases in acid deposition may stress some ecosystems and reduce their stability and productivity, others may be unaffected. The degree and nature of the impact depend on the acid input load, organismal susceptibility, and buffering capacity of the particular ecosystem. *See* BIOGEOCHEMISTRY.   [R.R.Sc./D.Lam./H.B.Pi./D.Ge.]

Aeronautical meteorology   The branch of meteorology that deals with atmospheric effects on the operation of vehicles in the atmosphere, including winged aircraft, lighter-than-air devices such as dirigibles, rockets, missiles, and projectiles. The air which supports flight or is traversed on the way to outer space contains many potential hazards. Poor visibility caused by fog, snow, dust, and rain is a major cause of aircraft accidents and the principal cause of flight cancellations or delays.

The weather conditions of ceiling and visibility required by regulations for crewed aircraft during landing or takeoff are determined by electronic and visual aids operated by the airport and installed in the aircraft. The accurate forecasting of terminal conditions is critical to flight economy, and to safety where sophisticated landing aids are not available. Improved prediction methods are under continuing investigation and development, and are based on mesoscale and microscale meteorological analyses, electronic computer calculations, radar observations of precipitation areas, and observations of fog trends. *See* MESOMETEOROLOGY; MICROMETEOROLOGY.

Atmospheric turbulence is principally represented in vertical currents and their departures from steady, horizontal airflow. When encountered by an aircraft, turbulence produces abrupt excursions in aircraft position, sometimes resulting in discomfort or injury to passengers, and sometimes even structural damage or failure. Major origins of turbulence are (1) mechanical, caused by irregular terrain below the flow of air; (2) thermal, associated with vertical currents produced by heating of air in contact with the Earth's surface; (3) thunderstorms and other convective clouds; (4) mountain wave, a regime of disturbed airflow leeward of mountains or hills, often comprising both smooth and breaking waves formed when stable air is forced to ascend over the mountains; and (5) wind shear, usually variations of horizontal wind in the vertical direction, occurring along air-mass boundaries, temperature inversions (including the tropopause), and in and near the jet stream.

While encounters with strong turbulence anywhere in the atmosphere represent substantial inconvenience, encounters with rapid changes in wind speed and direction at low altitude can be catastrophic. Generally, wind shear is most dangerous when encountered below 1000 ft (300 m) above the ground, where it is identified as low-altitude wind shear. Intense convective microbursts, downdrafts usually associated with thunderstorms, have caused many aircraft accidents often resulting in a great loss of life. The downdraft emanating from convective clouds, when nearing the Earth's surface, spreads horizontally as outrushing rain-cooled air. When entering a microburst outflow, an aircraft first meets a headwind that produces increased performance by way of increased airspeed over the wings. Then within about 5 s, the aircraft encounters a downdraft and then a tailwind with decreased performance. A large proportion of microburst accidents, both after takeoff and on approach to landing, are caused by this performance decrease, which can result in rapid descent. *See* THUNDERSTORM.

Turbulence and low-altitude wind shear can readily be detected by a special type of weather radar, termed Doppler radar. By measuring the phase shift of radiation backscattered by hydrometeors and other targets in the atmosphere, both turbulence and wind shear can be clearly identified. It is anticipated that Doppler radars located at airports, combined with more thorough pilot training regarding the need to avoid microburst wind shear, will provide desired protection from this dangerous aviation weather phenomenon.

Since an aircraft's speed is given by a propulsive component plus the speed of the air current bearing the aircraft, there are aiding or retarding effects depending on wind direction in relation to the track flown. Wind direction and speed vary only moderately from day to day and from winter to summer in certain parts of the world, but fluctuations of the vector wind at middle and high latitudes in the troposphere and lower stratosphere can exceed 200 knots (100 mph). The role of the aeronautical meteorologist is to provide accurate forecasts of the wind and temperature field, in space and time, through the operational ranges of each aircraft involved. For civil jet-powered aircraft, the optimum flight plan must always represent a compromise among wind, temperature, and turbulence conditions. *See* UPPER-ATMOSPHERE DYNAMICS; WIND.

The jet stream is a meandering, shifting current of relatively swift wind flow which is embedded in the general westerly circulation at upper levels. Sometimes girdling the globe at middle and subtropical latitudes, where the strongest jets are found, this band of strong winds, generally 180–300 mi (300–500 km) in width, has great operational significance for aircraft flying at cruising levels of 4–9 mi (6–15 km). The jet stream challenges the forecaster and the flight planner to utilize tailwinds to the greatest extent possible on downwind flights and to avoid retarding headwinds as much as practicable on upwind flights. As with considerations of wind and temperature, altitude and horizontal coordinates are considered in flight planning for jet-stream conditions. Turbulence in the vicinity of the jet stream is also a forecasting problem. *See* JET STREAM.

An electrical discharge or lightning strike to or from an aircraft is experienced as a blinding flash and a muffled explosive sound. Atmospheric conditions favorable for lightning strikes follow a consistent pattern, characterized by solid clouds or enough clouds for the aircraft to be flying intermittently on instruments; active precipitation of an icy character; and ambient air temperature near or below 32°F (0°C). Saint Elmo's fire, radio static, and choppy air often precede the strike. However, the charge separation processes necessary for the production of strong electrical fields is destroyed by strong turbulence. Thus turbulence and lightning usually do not coexist in the same space. *See* ATMOSPHERIC ELECTRICITY; LIGHTNING.

Modern aircraft operation finds icing to be a major factor in the safe flight. Icing usually occurs when the air temperature is near or below freezing (32°F or 0°C) and the relative humidity is 80% or more. Clear ice is most likely to form when the air temperature is between 32 and −4°F (0 and −20°C) and the liquid water content of the air is high (large drops or many small drops). As these drops impinge on the skin of an aircraft, the surface temperature of which is 32°F (0°C) or less, the water freezes into a hard, high-density solid. When the liquid water content is small and when snow or ice pellets may also be present, the resulting rime ice formation is composed of drops and encapsulated air, producing an ice that is less dense and opaque in appearance. Accurate forecasts and accurate delineation of freezing conditions are essential for safe aircraft operations.   [J.T.Le.; J.M.]

Aeronomy   The study of the chemistry and physics of the regions above the tropopause or upper part of the atmosphere. The region of the atmosphere below the tropopause is the site of most of the weather phenomena that so directly affect all life on the planet; this region has primarily been the domain of meteorology.

The chemical and physical properties of the atmosphere and the changes that result from external and internal forces impact all altitudes and global distributions of atoms, molecules, ions, and electrons, both in composition and in density. Dynamical effects are seen in vertical and horizontal atmospheric motion, and energy is transferred through radiation, chemistry, conduction, convection, and wave propagation.

The atmosphere of the Earth is separated into regions defined by the variation of temperature with height. In the middle atmosphere, that region of the atmosphere between the tropopause and the mesopause (10–100 km or 6–60 mi), the temperature varies from 250 K (−9.7°F) at the tropopause to 300 K (80°F) at the stratopause and back down to 200 K (−100°F) at the mesopause (see illustration). These temperatures are average values, and they vary with season and heat and winds due to the effect of the Sun on the atmosphere. Over this same height interval the atmospheric density varies by over five orders of magnitude. Although there is a constant mean molecular weight over this region, that is, a constant relative abundance of the major atmospheric constituents of molecular oxygen and nitrogen, there are a number of minor constituents that have a profound influence on the biosphere and an increasing influence on the change in the atmosphere below the tropopause associated with the general topic of global change. These constituents, called the greenhouse gases (water vapor, ozone, carbon dioxide, methane, chlorine compounds, nitrogen oxides, chlorofluorocarbons, and others), are all within the middle atmosphere. *See* MESOSPHERE; STRATOSPHERE; TROPOSPHERE.

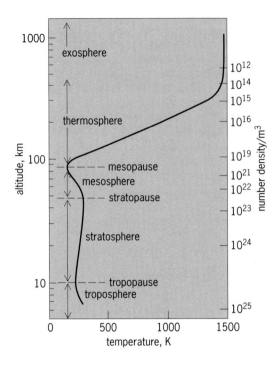

Typical atmospheric temperature variation with altitude and constituent number density (number of atoms and molecules per cubic meter) at high solar activity. $^\circ$F = (K $\times$ 1.8) $-$ 459.67. 1 km = 0.6 mi.

Understanding the aeronomy of the middle atmosphere requires the study of the physical motion of the atmosphere. The particular composition and chemistry at any given time, location, or altitude depends on how the various constituents are transported from one region to another. Thus, global circulation models for both horizontal and vertical motions are needed to completely specify the chemical state of the atmosphere. In understanding the dynamics of the middle atmosphere, internal gravity waves and acoustic gravity waves play significant roles at different altitudes, depending on whether the particle motion associated with the wave is purely transverse to the direction of propagation or has some longitudinal component. These waves originate primarily in meteorological events such as wind shears, turbulent storms, and weather fronts; and their magnitude can also depend on orographic features on the Earth's surface. *See* MIDDLE-ATMOSPHERE DYNAMICS.

The upper atmosphere is that region above the middle atmosphere that extends from roughly 100 km (60 mi) to the limit of the detectable atmosphere of the planet. This region is characterized by an increasing temperature until it reaches a constant exospheric temperature. There is a slow transition from the region of constant mean molecular weight associated with the middle atmosphere to that of almost pure atomic hydrogen at high altitudes of the exosphere. This is also the region of transition between transport dominated by collision and diffusion, and transport influenced by plasma convection in the magnetic field. The neutral density varies by over ten orders of magnitude from one end to the other and is dominated by molecular processes in the high-density region and an increasing importance of atomic, electron, and ion processes as the density decreases with altitude.

As the Sun sets on the atmosphere, dramatic changes due to the loss of the solar radiation occur. The excitation of the atmosphere declines, and chemical reactions between the constituents become more and more dominant as the night progresses. When observed with very sensitive instruments, the night sky appears to glow at various colors of light. Prominent green and red atomic emissions due to oxygen at 555.7 and 630 nm, respectively, and yellow sodium light at 589 nm appear, while molecular bands of the hydroxyl radical and molecular oxygen even further in the red collectively contribute most of the total intensity of the nightglow spectrum.

The aurora that appears in the southern and northern polar regions is the optical manifestation of the energy loss of energetic particles precipitating into the atmosphere. The region of highest probability of occurrence is called the auroral oval. At high altitudes, electrons and ions present in the magnetosphere are accelerated along magnetic field lines into the atmosphere at high polar latitudes. *See* AURORA; MAGNETOSPHERE.

[G.J.R.]

Africa   A continent that straddles the Equator, extending between 37°N and 35°S. It is the second largest continent, exceeded by Eurasia. The area, shared by 55 countries, is 11,700,000 mi^2 (30,300,00 km^2), approximately 20% of the world's total land area. Despite its large area, it has a simple geological structure, a compact shape with a smooth outline, and a symmetrical distribution of climate and vegetation.

Africa has few inlets or natural harbors and a small number of offshore islands that are largely volcanic in origin. Madagascar is the largest island, with an area of 250,000 mi^2 (650,000 km^2).

Africa is primarily a high interior plateau bounded by steep escarpments. These features show evidence of the giant faults created during the drift of neighboring continents. The surface of the plateau ranges from 4000–5000 ft (1200–1500 m) in the south to about 1000 ft (300 m) in the Sahara. These differences in elevation are particularly apparent in the Great Escarpment region in southern Africa, where the land suddenly drops from 5000 ft (1500 m) to a narrow coastal belt. Although most of the continent is classified as plateau, not all of its surface is flat. Rather, most of its physiographic features have been differentially shaped by processes such as folding, faulting, volcanism, erosion, and deposition. *See* ESCARPMENT; FAULT AND FAULT STRUCTURES; PLATEAU.

The rift valley system is one of the most striking features of the African landscape. Sliding blocks have created wide valleys 20–50 mi (30–80 km) wide bounded by steep walls of variable depth and height. Within the eastern and western branches of the system, there is a large but shallow depression occupied by Lake Victoria. *See* RIFT VALLEY.

Several volcanic features are associated with the rift valley system. The most extensive of these are the great basalt highlands that bound either side of the rift system in Ethiopia. These mountains rise over 10,000 ft (3000 m), with the highest peak, Ras Dashan, reaching 15,158 ft (4500 m). There are also several volcanic cones, including the most renowned at Mount Elgon (14,175 ft; 4321 m); Mount Kenya (17,040 ft; 5194 m); and Mount Kilimanjaro, reaching its highest point at Mount Kibo (19,320 ft; 5889 m). Mounts Kenya and Kilimanjaro are permanently snowcapped. *See* BASALT.

Since the Equator transects the continent, the climatic conditions in the Northern Hemisphere are mirrored in the Southern Hemisphere. Nearly three-quarters of the continent lies within the tropics and therefore has high temperatures throughout the year. Frost is uncommon except in mountainous areas or some desert areas where nighttime temperatures occasionally drop below freezing. These desert areas also record some of the world's highest daytime temperatures, including an unconfirmed record of 136.4°F (58°C) at Azizia, Tripoli. *See* EQUATOR.

Africa can be classified into broad regions based on the climatic conditions and their associated vegetation and soil types. The tropical rainforest climate starts at the Equator and extends toward western Africa. The region has rainfall up to 200 in. (500 cm) per year and continuously high temperatures averaging 79°F (26°C). The eastern equatorial region does not experience these conditions because of the highlands and the presence of strong seasonal winds that originate from southern Asia.

The areal extent of the African rainforest region (originally 18%) has dwindled to less than 7% as a result of high rates of deforestation. Despite these reductions, the region is still one of the most diverse ecological zones in the continent.

Extensive savanna grasslands are found along the Sudanian zone of West Africa, within the Zambezian region and the Somalia-Masai plains. Large areas such as the Serengeti plains in the Somalia-Masai plains are home to a diverse range of wild animals.

The tropical steppe forms a transition zone between the humid areas and the deserts. This includes the area bordering the south of the Sahara that is known as the Sahel, the margins of the Kalahari basin, and the Karoo grasslands in the south.

The structural evolution of the continent has much to do with the drainage patterns. Originally, most of the rivers did not drain into the oceans, and many flowed into the large structural basins of the continent. However, as the continental drift occurred and coasts became more defined, the rivers were forced to change courses, and flow over the escarpments in order to reach the sea. Several outlets were formed, including deep canyons, waterfalls, cataracts, and rapids as the rivers carved out new drainage patterns across the landscape. Most of the rivers continue to flow through or receive some of their drainage from the basins, but about 48% of them now have a direct access into the surrounding oceans. The major rivers are the Nile, Congo (Zaire), Niger, and Zambezi. *See* RIVER.

The tremendous diversity in wildlife continues to be one of the primary attractions of this continent. Africa is one of the few remaining places where one can view game fauna in a natural setting. There is a tremendous diversity in species, including birds, reptiles, and large mammals. Wildlife are concentrated in central and eastern Africa because of the different types of vegetation which provide a wide range of habitats.

Africa is not a densely populated continent. With an estimated population of 743,000,000, its average density is 64 per square mile (26 per square kilometer). However, some areas have large concentrations, including the Nile valley, the coastal areas of northern and western Africa, the highland and volcanic regions of eastern Africa, and parts of southern Africa. These are mostly areas of economic or political significance.   [F.L.M.]

Agate   A variety of chalcedonic quartz that is distinguished by the presence of color banding in curved or irregular patterns (see illustration). Most agate used for ornamental purposes is composed of two or more tones or intensities of brownish-red, often interlayered with white, but it is also commonly composed of various shades of gray and white. Since agate is relatively porous, it can be dyed permanently in red, green, blue, and a variety of other colors.

The term agate is also used with prefixes to describe certain types of chalcedony in which banding is not evident. Moss agate is a milky or almost transparent chalcedony containing dark inclusions in a dendritic pattern. Iris agate exhibits an iridescent color effect. Fortification, or landscape, agate is translucent and contains inclusions that give it an appearance reminiscent of familiar natural scenes. Banded agate is distinguished from onyx by the fact that its banding is curved or irregular, in contrast to the straight, parallel layers of onyx. The properties of agate are those of chalcedony: refractive

Section of polished agate showing the characteristic banding. (*Field Museum of Natural History, Chicago*)

indices of 1.535 and 1.539, a hardness of $6\frac{1}{2}$ to 7, and a specific gravity of about 2.60. *See* CHALCEDONY; GEM; QUARTZ.   [R.T.L.]

Agricultural meteorology
A branch of meteorology that examines the effects and impacts of weather and climate on crops, rangeland, livestock, and various agricultural operations. The branch of agricultural meteorology dealing with atmospheric-biospheric processes occurring at small spatial scales and over relatively short time periods is known as micrometeorology, sometimes called crop micrometeorology for managed vegetative ecosystems and animal biometeorology for livestock operations. The branch that studies the processes and impacts of climatic factors over larger time and spatial scales is often referred to as agricultural climatology. *See* CLIMATOLOGY; MICROMETEOROLOGY.

Agricultural meteorology, or agrometeorology, addresses topics that often require an understanding of biological, physical, and social sciences. It studies processes that occur from the soil depths where the deepest plant roots grow to the atmospheric levels where seeds, spores, pollen, and insects may be found. Agricultural meteorologists characteristically interact with scientists from many disciplines.

Agricultural meteorologists collect and interpret weather and climate data needed to understand the interactions between vegetation and animals and their atmospheric environments. The climatic information developed by agricultural meteorologists is valuable in making proper decisions for managing resources consumed by agriculture, for optimizing agricultural production, and for adopting farming practices to minimize any adverse effects of agriculture on the environment. Such information is vital to ensure the economic and environmental sustainability of agriculture now and in the future. *See* WEATHER OBSERVATIONS.

Agricultural meteorologists also quantify, evaluate, and provide information on the impact and consequences of climate variability and change on agriculture. Increasingly, agricultural meteorologists assist policy makers in developing strategies to deal with climatic events such as floods, hail, or droughts and climatic changes such as global warming and climate variability.

Agricultural meteorologists are involved in many aspects of agriculture, ranging from the production of agronomic and horticultural crops, trees, and livestock to the final delivery of agricultural products to market. They study the energy and mass exchange processes of heat, carbon dioxide, water vapor, and trace gases such as methane, nitrous oxide, and ammonia, within the biosphere on spatial scales ranging from a leaf to a watershed and even to a continent. They study, for example, the photosynthesis, productivity, and water use of individual leaves, whole plants, and fields. They also examine climatic processes at time scales ranging from less than a second to more than a decade.   [B.L.B.]

Air   A predominantly mechanical mixture of a variety of individual gases enveloping the terrestrial globe to form the Earth's atmosphere. In this sense air is one of the three basic components, air, water, and land (atmosphere, hydrosphere, and lithosphere), that interblend to form the life zone at the face of the Earth. *See* ATMOSPHERE.   [C.V.C.]

Air mass   In meteorology, an extensive body of the atmosphere which is relatively homogeneous horizontally. An air mass may be followed on the weather map as an entity in its day-to-day movement in the general circulation of the atmosphere. The expressions air mass analysis and frontal analysis are applied to the analysis of weather maps in terms of the prevailing air masses and of the zones of transition and interaction (fronts) which separate them.

The relative horizontal homogeneity of an air mass stands in contrast to sharper horizontal changes in a frontal zone. The horizontal extent of important air masses is reckoned in millions of square miles. In the vertical dimension an air mass extends at most to the top of the troposphere, and frequently is restricted to the lower half or less of the troposphere. *See* FRONT; METEOROLOGY; WEATHER MAP.

The occurrence of air masses as they appear on the daily weather maps depends upon the existence of air-mass source regions, areas of the Earth's surface which are sufficiently uniform that the overlying atmosphere acquires similar characteristics throughout the region. *See* ATMOSPHERIC GENERAL CIRCULATION.

The thermodynamic properties of air mass determine not only the general character of the weather in the extensive area that it covers, but also to some extent the severity of the weather activity in the frontal zone of interaction between air masses. Those properties which determine the primary weather characteristics of an air mass are defined by the vertical distribution of water vapor and heat (temperature). On the vertical distribution of water vapor depend the presence or absence of condensation forms and, if present, the elevation and thickness of fog or cloud layers. On the vertical distribution of temperature depend the relative warmth or coldness of the air mass and, more importantly, the vertical gradient of temperature, known as the lapse rate. The lapse rate determines the stability or instability of the air mass for thermal convection and consequently, the stratiform or convective cellular structure of the cloud forms and precipitation. The most unstable moist air mass is characterized by severe turbulence and heavy showers or thundershowers. In the most stable air mass there is observed an actual increase (inversion) of temperature with increase of height at low elevations. *See* TEMPERATURE INVERSION.   [H.C.Wi.; E.Ke.]

Air pollution   The presence in the atmospheric environment of natural and artificial substances that affect human health or well-being, or the well-being of any other specific organism. Pragmatically, air pollution also applies to situations where contaminants impact structures and artifacts or esthetic sensibilities (such as visibility or smell). Most artificial impurities are injected into the atmosphere at or near the Earth's surface. The lower atmosphere (troposphere) cleanses itself of some of these pollutants in a few hours or days as the larger particles settle to the surface and soluble gases and particles encounter precipitation or are removed through contact with surface objects.

Unfortunately, removal of some pollutants (for example, sulfates and nitrates) by precipitation and dry deposition results in acid deposition, which may cause serious environmental damage. Also, mixing of the pollutants into the upper atmosphere may dilute the concentrations near the Earth's surface, but can cause long-term changes in the chemistry of the upper atmosphere, including the ozone layer. *See* ATMOSPHERE; TROPOSPHERE.

Types of sources. Sources may be characterized in a number of ways. First, a distinction may be made between natural and anthropogenic sources. Another frequent classification is in terms of stationary (power plants, incinerators, industrial operations, and space heating) and moving (motor vehicles, ships, aircraft, and rockets) sources. Another classification describes sources as point (a single stack), line (a line of stacks), or area (city).

Different types of pollution are conveniently specified in various ways: gaseous, such as carbon monoxide, or particulate, such as smoke, pesticides, and aerosol sprays; inorganic, such as hydrogen fluoride, or organic, such as mercaptans; oxidizing substances, such as ozone, or reducing substances, such as oxides of sulfur and oxide s of nitrogen; radioactive substances, such as iodine-131; inert substances, such as pollen or fly ash; or thermal pollution, such as the heat produced by nuclear power plants.

Air contaminants are produced in many ways and come from many sources; it is difficult to identify all the various producers. Also, for some pollutants such as carbon dioxide and methane, the natural emissions sometimes far exceed the anthropogenic emissions.

Both anthropogenic and natural emissions are variable from year to year, depending on fuel usage, industrial development, and climate. In some countries where pollution control regulations have been implemented, emissions have been significantly reduced. For example, in the United States sulfur dioxide emissions dropped by about 30% between 1970 and 1992, and carbon monoxide (CO) emissions were cut by over 30% in the same period. However, in some developing countries emissions continually rise as more cars are put on the road and more industrial facilities and power plants are constructed. In dry regions, natural emissions of nitrogen oxides (NO_x), carbon dioxide (CO_2), and hydrocarbons can be greatly increased during a season with high rainfall and above-average vegetation growth.

The anthropogenic component of most estimates of the methane budget is about two-thirds. Ruminant production and emissions from rice paddies are regarded as anthropogenic because they result from human agricultural activities. The perturbations to carbon dioxide since the industrial revolution are also principally the result of human activities. These emissions have not yet equilibrated with the rest of the carbon cycle and so have had a profound effect on atmospheric levels, even though emissions from fossil fuel combustion are dwarfed by natural emissions.

Effects. The major concern with air pollution relates to its effects on humans. Since most people spend most of their time indoors, there has been increased interest in air-pollution concentrations in homes, workplaces, and shopping areas. Much of the early information on health effects came from occupational health studies completed prior to the implementation of general air-quality standards.

Air pollution principally injures the respiratory system, and health effects can be studied through three approaches, clinical, epidemiological, and toxicological. Clinical studies use human subjects in controlled laboratory conditions, epidemiological studies assess human subjects (health records) in real-world conditions, and toxicological studies are conducted on animals or simple cellular systems. Of course, epidemiological studies are the most closely related to actual conditions, but they are the most difficult to interpret because of the lack of control and the subsequent problems with statistical analysis. Another difficulty arises because of differences in response among different people. For example, elderly asthmatics are likely to be more strongly affected by sulfur dioxide than the teenage members of a hiking club.

Damage to vegetation by air pollution is of many kinds. Sulfur dioxide may damage field crops such as alfalfa and trees such as pines, especially during the growing season (Fig. 1). Both hydrogen fluoride (HF) and nitrogen dioxide (NO_2) in high concentrations have been shown to be harmful to citrus trees and ornamental plants, which are of economic, importance in central Florida. Ozone and ethylene are other contaminants that cause damage to certain kinds of vegetation.

Air pollution can affect the dynamics of the atmosphere through changes in long-wave and shortwave radiation processes. Particles can absorb or reflect incoming short-wave solar radiation, keeping it from the Earth's surface during the day. Greenhouse gases can absorb long-wave radiation emitted by the Earth's surface and atmosphere.

Carbon dioxide, methane, fluorocarbons, nitrous oxides, ozone, and water vapor are important greenhouse gases. These represent a class of gases that selectively absorb long-wave radiation. This effect warms the temperature of the Earth's atmosphere and surface higher than would be found in the absence of an atmosphere (the greenhouse effect). Because the amount of greenhouse gases in the atmosphere is rising, there is a possibility that the temperature of the atmosphere will gradually rise, possibly resulting in a general warming of the global climate over a time period of several generations. *See* GREENHOUSE EFFECT.

Researchers are also concerned with pollution of the stratosphere (10–50 km or 6–30 mi above the Earth's surface) by aircraft and by broad surface sources. The stratosphere is important, because it contains the ozone layer, which absorbs part of the Sun's short-wave radiation and keeps it from reaching the surface. If the ozone layer is significantly depleted, an increase in skin cancer in humans is expected. Each 1% loss of ozone is estimated to increase the skin cancer rate 3–6%. *See* STRATOSPHERE.

Visibility is reduced as concentrations of aerosols or particles increase. The particles do not just affect visibility by themselves but also act as condensation nuclei for cloud or haze formation. In each of the three serious air-pollution episodes discussed above, smog (smoke and fog) were present with greatly reduced visibility.

Chemistry. Air pollution can be divided into primary and secondary compounds, where primary pollutants are emitted directly from sources (for example, carbon monoxide, sulfur dioxide) and secondary pollutants are produced by chemical reactions between other pollutants and atmospheric gases and particles (for example, sulfates, ozone). Most of the chemical transformations are best described as oxidation processes. In many cases these secondary pollutants can have significant environmental effects, such as acid rain and smog.

Smog is the best-known example of secondary pollutants formed by photochemical processes, as a result of primary emissions of nitric oxide (NO) and reactive hydrocarbons from anthropogenic sources such as transportation and industry as well as natural sources. Energy from the Sun causes the formation of nitrogen dioxide, ozone (O_3), and peroxyacetalnitrate, which cause eye irritation and plant damage.

It has been shown that when emissions of sulfur dioxide and nitrogen oxide from tall power plant and other industrial stacks are carried over great distances and combined with emissions from other areas, acidic compounds can be formed by complex chemical reactions. In the absence of anthropogenic pollution sources, the average pH of rain is around 5.6 (slightly acidic). In the eastern United States, acid rain with a pH less than 5.0 has been measured and consists of about 65% dilute sulfuric acid, 30% dilute nitric acid, and 5% other acids. [S.R.H.; P.J.S.]

Air pressure   The force per unit area that the air exerts on any surface in contact with it, arising from the collisions of the air molecules with the surface. It is equal and opposite to the pressure of the surface against the air, which for atmospheric air in normal motion approximately balances the weight of the atmosphere above, about 15 pounds per square inch (psi) at sea level. It is the same in all directions and is the force that balances the weight of the column of mercury in the Torricellian barometer, commonly used for its measurement. *See* BAROMETER.

The units of pressure traditionally used in meteorology are based on the bar, defined as equal to 1,000,000 dynes/cm². One bar equals 1000 millibars or 100 centibars.

In the meter-kilogram-second or International System of Units (SI), the unit of force, the pascal (Pa), is equal to 1 newton/m². One millibar equals 100 pascals. The normal pressure at sea level is 1013.25 millibars or 101.325 kilopascals.

Also widely used in practice are units based on the height of the mercury barometer under standard conditions, expressed commonly in millimeters or in inches. The standard atmosphere (760 mmHg) is also used as a unit, mainly in engineering, where large pressures are encountered. The following equivalents show the conversions between the commonly used units of pressure, where $(mmHg)_n$ and $(in. Hg)_n$ denote the millimeter and inch of mercury, respectively, under standard (normal) conditions, and where $(kg)_n$ and $(lb)_n$ denote the weight of a standard kilogram and pound mass, respectively, under standard gravity.

$$1 \text{ kPa} = 10 \text{ millibars} = 1000 \text{ N/m}^2$$
$$= 7.50062 \ (mmHg)_n$$
$$= 0.295300 \ (in. \ Hg)_n$$
$$1 \text{ millibar} = 100 \text{ Pa} = 1000 \text{ dynes/cm}^2$$
$$= 0.750062 \ (mmHg)_n$$
$$= 0.0295300 \ (in. \ Hg)_n$$
$$1 \text{ atm} = 101.325 \text{ kPa} = 1013.25 \text{ millibars}$$
$$= 760 \ (mmHg)_n = 29.9213 \ (in. \ Hg)_n$$
$$= 14.6959 \ (lb)_n/in.^2$$
$$= 1.03323 \ (kg)_n/cm^2$$
$$1 \ (mmHg)_n = 1 \text{ torr} = 0.03937008 \ (in. \ Hg)_n$$
$$= 1.333224 \text{ millibars}$$
$$= 133.3224 \text{ Pa}$$
$$1 \ (in. \ Hg)_n = 33.8639 \text{ millibars}$$
$$= 25.4 \ (mmHg)_n$$
$$= 3.38639 \text{ kPa}$$

Because of the almost exact balancing of the weight of the overlying atmosphere by the air pressure, the latter decreases with height. A standard equation is used in practice to calculate the vertical distribution of pressure with height above sea level. The temperature distribution in a standard atmosphere, based on mean values in middle latitudes, has been defined by international agreement. The use of the standard atmosphere yields a definite relation between pressure and height. This relation is used in all altimeters which are basically barometers of the aneroid type. The difference between the height estimated from the pressure and the actual height is often considerable; but since the same standard relationship is used in all altimeters, the difference is the same for all altimeters at the same location, and so causes no difficulty in determining the relative position of aircraft. Mountains, however, have a fixed height, and accidents have been caused by the difference between the actual and standard atmosphere.

In addition to the large variation with height, atmospheric pressure varies in the horizontal and with time. The variations of air pressure at sea level, estimated in the case of observations over land by correcting for the height of the ground surface, are routinely plotted on a map and analyzed, resulting in the familiar weather map representation with its isobars showing highs and lows. The movement of the main features of the sea-level pressure distribution, typically from west to east, produces characteristic fluctuations of the pressure at a fixed point, varying by a few percent within a few days. Smaller-scale variations of sea-level pressure, too small to appear on the ordinary weather map, are also present. These are associated with various forms of atmospheric motion, such as small-scale wave motion and turbulence. Relatively large variations are found in and near thunderstorms, the most intense being the low-pressure region in a tornado. The pressure drop within a tornado can be a large fraction of an atmosphere, and is the principal cause of the explosion of buildings over which a tornado passes. *See* WEATHER MAP.

It is a general rule that in middle latitudes at localities below 1000 m (3280 ft) in height above sea level, the air pressure on the continents tends to be slightly higher in winter than in spring, summer, and autumn; whereas at considerably greater heights on the continents and on the ocean surface, the reverse is true.

The practical importance of air pressure lies in its relation to the wind and weather. It is because of these relationships that pressure is a basic parameter in weather forecasting, as is evident from its appearance on the ordinary weather map.

The large-scale variations of pressure at sea level shown on a weather map are associated with characteristic patterns of vertical motion of the air, which in turn affect the weather. Descent of air in a high heats the air and dries it by adiabatic compression, giving clear skies, while the ascent of air in a low cools it and causes it to condense and produce cloudy and rainy weather. These processes at low levels, accompanied by others at higher levels, usually combine to justify the clear-cloudy-rainy marking on the household barometer.   [R.J.D.; E.Ke.]

Air temperature   The temperature of the atmosphere represents the average kinetic energy of the molecular motion in a small region, defined in terms of a standard or calibrated thermometer in thermal equilibrium with the air. Many different types of thermometer are used for the measurement of air temperature, the most common depending on the expansion of mercury with temperature, the variation of electrical resistance with temperature, or the thermoelectric effect (thermocouple).

The temperature of a given small mass of air varies with time because of heat added or subtracted from it, and also because of work done during changes of volume.

The rate at which the temperature changes at a particular point, that is, as measured by a fixed thermometer, depends on the movement of air as well as physical processes such as absorption and emission of radiation, heat conduction, and changes of phase of water involving latent heat of condensation and freezing. The large changes of air temperature from day to day are mainly due to the horizontal movement of air, bringing relatively cold or warm air masses to a particular point, as the large-scale pressure-wind systems move across the weather map. *See* AIR MASS; AIR PRESSURE.

Temperatures near the surface are read at one or more fixed times daily, and the day's extremes are obtained from special maximum and minimum thermometers, or from the trace (thermogram) of a continuously recording instrument (thermograph). The average of these two extremes, technically the midrange, is considered in the United States to be the day's average temperature. The true daily mean, obtained from a thermogram, is closely approximated by the mean of 24 hourly readings, but may differ from the midrange by 1 or $2°F$ (0.6 or $1°C$), on the average. In many countries temperatures are read daily at three or four fixed times, so that their weighted mean closely approximates the true daily mean.

Averages of daily maximum and minimum temperature for a single month for many years give mean daily maximum and minimum temperatures for that month. The average of these values is the mean monthly temperature, while their difference is the mean daily range for that month. Monthly means, averaged through the year, give the mean annual temperature; the mean annual range is the difference between the hottest and coldest mean monthly values. The hottest and coldest temperatures in a month are the monthly extremes; their averages over a period of years give the mean monthly maximum and minimum (used extensively in Canada), while the absolute extremes for the month (or year) are the hottest and coldest temperatures ever observed. The interdiurnal range or variability for a month is the average of the successive differences, regardless of sign, in daily temperatures.

Over the oceans the mean daily, interdiurnal, and annual ranges are slight, because water absorbs the insolation and distributes the heat through a thick layer. In tropical regions the interdiurnal and annual ranges over the land are small also, because the annual variation in insolation is relatively small. The daily range also is small in humid tropical regions, but may be large (up to $40°F$ or $22°C$) in deserts. Interdiurnal and annual ranges increase generally with latitude, and also with distance from the ocean; the mean annual range defines continentality. The daily range depends on aridity, altitude, and noon Sun elevation. *See* ATMOSPHERE; INSOLATION.   [R.J.D.]

Airglow   Visible, infrared, and ultraviolet emissions from the atoms and molecules in the atmosphere above 30 km (20 mi), generally in layers, and mostly between 70 and 300 km (45 and 200 mi). The airglow, together with the ionosphere, is found in the uppermost parts of the atmosphere that absorb the incoming energetic radiations from the Sun. While the airglow consists of spectral features similar to those of the aurora, it is mostly uniform over the sky; and it is caused by the absorption of solar ultraviolet and x-radiations, rather than energetic particles. *See* AURORA.

The daytime airglow (dayglow) is caused mainly by fluorescence processes as molecules and atoms are photodissociated and photoionized. The photoelectrons that are produced in the ionization processes are a further source of airglow in their collisions with other atoms and molecules.

Twilight offers an opportunity to observe resonant scattering of sunlight on layers such as those of the alkali atoms sodium, lithium, and potassium. As the Earth's shadow scans through the layers, the changes of intensity allow their heights (near 90 km or 55 mi) to be measured. *See* ALKALI EMISSIONS.

The nighttime airglow (nightglow) is predominantly due to recombination emissions. The ionospheric plasma recombines near the bottom of the F region (150–200 km or 90–120 mi) where the densities and thus collision frequencies are higher, producing bright atomic oxygen (O) spectral lines in the red (at 630 and 636 nanometers) and a weaker green (558-nm) line.   [B.A.T.]

Albite   A sodium-rich plagioclase feldspar mineral whose composition extends over the range $Ab_{100}An_0$ to $Ab_{90}An_{10}$, where Ab (=albite) is $NaAlSi_3O_8$ and An (=anorthite) is $CaAl_2Si_2O_8$. Albite occurs in crustal igneous rocks as a major component of pegmatites and granites, in association with quartz, mica (usually muscovite), and potassium feldspar (orthoclase or microcline). Sodium and potassium feldspars usually occur as distinct mineral grains, sizes varying from millimeter to meter scale. They are frequently intergrown; if the intergrowth is visually observable in a hand specimen, the composite material is known as macroperthite; if visible only in a microscope, microperthite; and if submicroscopic in scale, cryptoperthite. In metamorphic rocks albite is found in granitic gneisses, and it may be the principal component of arkose, a feldspar-dominant, sedimentary rock. Cleavelandite, a platy variety, is sometimes found in lithium-rich pegmatites. *See* ARKOSE; FELDSPAR; GNEISS; IGNEOUS ROCKS; PEG-MATITE; PERTHITE.   [P.H.R.]

Alkali emissions   Light emissions in the upper atmosphere from elemental lithium, potassium, and especially sodium. These alkali metals are present in the upper atmosphere at altitudes from about 50 to 62 mi (80 to 100 km) and are very efficient in resonant scattering of sunlight. The vertical column contents (number of atoms per square meter) of the alkali atoms are easily deduced from their respective emission intensities. First detected with ground-based spectrographs, the emissions were observed mainly at twilight since they tend to be overwhelmed by intense scattered sunlight present in the daytime. A chemiluminescent process gives rise to so-called nightglow emissions at the same wavelengths. The development of lidars (laser radars) that are tuned to the resonance lines have enabled accurate resolution of the concentrations of these elements versus altitude for any time of the day. *See* AERONOMY; AIRGLOW.

There is little doubt that the origin of these metals is meteoritic ablation. Rocket-borne mass spectrometers have found that meteoritic ions are prevalent above the peaks of neutral atoms with a composition similar to that found in carbonaceous chondrites, a common form of meteorites. The ratio of the concentrations of ions to neutral atoms rises rapidly with altitude above 55 mi (90 km).

Sodium (Na) is the most abundant alkali metal in meteorites. The sodium D-lines, a doublet at 589 and 589.6 nanometers, were first detected in the nightglow in the late 1920s and at twilight a decade later. The nominal peak concentration of sodium is 3×10^9 atoms m^{-3} near 55 mi (90 km) where the total gas concentration of the atmosphere is 7×10^{19} atoms (or molecules) m^{-3}.

Potassium (K) is 15 times less abundant than sodium in meteorites. The ratio of the potassium and sodium column contents in the mesosphere ranges from 1/10 to 1/100. The potassium doublet at 767 and 770 nm is estimated to have a nightglow intensity near the night sky background, 50 times smaller than the typical intensity of the sodium nightglow. The potassium nightglow has never been detected.

Lithium (Li) is 35 times less abundant in meteorites than potassium, and 500 times less abundant than sodium. Nevertheless, the lithium emission at 671 nm has been observed at twilight by spectrometers and at night by lidars. The lithium nightglow emission is undetectable.   [W.Sw.]

Altitudinal vegetation zones   Intergrading regions on mountain slopes characterized by specific plant life forms or species composition, and determined by complex environmental gradients. Along an altitudinal transect of a mountain, there are sequential changes in the physiognomy (growth form) of the plants and in the species composition of the communities.

Such life zones are associated with temperature gradients present along mountain slopes. Research on patterns of altitudinal zonation has centered on the response of species and groups of species to a complex of environmental gradients. Measurements of a species along a gradient, for example, the number of individuals, biomass, or ground coverage, generally form a bell-shaped curve. Peak response of a species occurs under optimum conditions and falls off at both ends of the gradient. The unique response of each species is determined by its physiological, reproductive, growth, and genetic characteristics. Zones of vegetation along mountain slopes are formed by intergrading combinations of species that differ in their tolerance to environmental conditions. Zones are usually indistinct entities rather than discrete groupings of species. However, under some conditions of localized disjunctions, very steep sections of gradients, or competitive exclusion, discontinuities in the vegetation can create discrete communities. Vegetation zones are often defined by the distributions of species having the dominant growth form, most frequently trees.

Altitudinal vegetation zonation, therefore, is an expression of the response of individual species to environmental conditions. Plants along an altitudinal transect are exposed, not to a single environmental gradient, but to a complex of gradients, the most important of which are solar radiation, temperature, and precipitation. Although these major environmental gradients exist in most mountain ranges of the world, the gradients along a single altitudinal transect are not always smooth because of topographic and climatic variability.

The solar energy received by mountain surfaces increases with altitude, associated with decreases in air density and the amount of dust and water vapor. An overcast sky is more efficient at reducing short-wave energy reaching low elevations and can increase the difference in energy input to 160%. However, more frequent clouds over high elevations relative to sunnier lower slopes commonly reduces this difference. Vegetation patterns are also strongly influenced by the decline in air temperature with increasing altitude, called the adiabatic lapse rate. Lapse rates are generally between $1.8°F$ to $3.6°F$ per 1000 ft ($1°C$ to $2°C$ per 300 m), but vary with the amount of moisture present; wet air has a lower lapse rate. Thus, plants occurring at higher elevations generally experience cooler temperatures and shorter growing periods than low-elevation plants. Variation in the temperature gradient can be caused by differences in slope, aspect, radiation input, clouds, and air drainage patterns. The precipitation gradient in most mountains is the reverse of the temperature gradient: precipitation increases with altitude. *See* Air Temperature; Precipitation (meteorology).

General changes in vegetation with increases in altitude include reduction in plant size, slower growth rates, lower production, communities composed of fewer species, and less interspecific competition. However, many regional exceptions to these trends exist.

Characteristics of vegetation zones also vary with latitude. Mountains at higher latitudes have predominantly seasonal climates, with major temperature and radiation extremes between summer and winter. Equatorial and tropical mountains have a strong diurnal pattern of temperature and radiation input with little seasonal variation. The upper altitudinal limit of trees, and the maximum elevation of plant growth generally, decreases with distance from the Equator, with the exception of a depression near the Equator.   [J.S.C.]

Alunite   A mineral of composition $KAl_3(SO_4)_2(OH)_6$. Alunite occurs in white to gray rhombohedral crystals or in fine-grained, compact masses. Alunite is produced by sulfurous vapors on acid volcanic rocks and also by sulfated meteoric waters affecting aluminous rocks. Alunite is used as a source of potash or for making alum. Alum has been manufactured from the well-known alunite deposits at Tolfa, near Civita Vecchia, Italy, since the mid-15th century. In the United States alunite is widespread in the West.
                                                                                    [E.C.T.C.]

Amber   Most commonly, a generic name for all fossil resins, although it has been restricted by some to refer only to succinite, the mineralogical species of fossil resin making up most of the Baltic Coast deposits. Resins generally are complex mixtures of mono-, sesqui-, di-, and triterpenoids; however, some resins contain aromatic phenols. Among the plants, primarily trees, that produce copious amounts of resin that may fossilize to become amber, two-thirds are tropical or subtropical.

Although ambers occur throughout the world in deposits from Carboniferous to Pleistocene in age, they have been reported most commonly from Cretaceous and Tertiary strata and often are associated with coal or lignites. Amber may contain beautifully preserved insects, spiders, flowers, leaves, and even small animals. The most extensively studied deposits are those from the Baltic Coast, Alaska, Canada, Burma, Dominican Republic, and Mexico.

When amber is used for jewelry, it usually is transparent yellow, reddish-brown, or "amber" color. Translucent or semitranslucent amber is used for pipe stems, decorating small boxes, and a variety of ornamental purposes. The specific gravity varies from 1.05 to 1.10, and hardness from 1 to 3 on Mohs scale.

At one time, chemical studies of amber were mineralogically oriented because the purpose was to describe and classify amber as a semiprecious gem. However, phytochemical studies comparing fossil and present-day resins are providing information regarding the botanical origins of ambers.

The predominantly tropical or subtropical occurrence of amber-producing plants through geologic time has led to evolutionary studies of the natural purpose of resins and their possible defensive role for trees against injury and disease inflicted by the high diversity of insects and fungi in tropical environments. See GEM; MINERALOGY.   [J.H.L.]

Amblygonite   A lithium aluminum phosphate mineral of basic formula $LiAl(PO_4F)$. The structure of amblygonite consists of phosphate (PO_4) groups of tetrahedra and AlO_6 groups of octahedra. Each PO_4 tetrahedron is connected to an AlO_6 octahedron. Amblygonite crystallizes in the triclinic system. Its color is commonly white or gray with tints of blue, green, and yellow. Amblygonite is transparent to translucent and has a vitreous to pearly luster.

The best-known occurrences of amblygonite are in Montebras, France; the Black Hills of South Dakota; the White Picacho District in Arizona; pegmatite districts in Maine; the Tanco pegmatite in Manitoba, Canada; and Portland, Connecticut. While amblygonite has been mined as an ore of lithium, it is not a major ore. See PHOSPHATE MINERALS.   [C.K.S.]

Amethyst   The transparent purple to violet variety of the mineral quartz. Amethyst is rare in the deep colors that characterize fine quality. It is usually colored unevenly and is often heated slightly in an effort to distribute the color more evenly. Heating at higher temperatures usually changes it to yellow or brown (rarely green), and further heating removes all color. The principal sources are Brazil, Arizona, Uruguay, and Russia. See GEM; QUARTZ.   [R.T.L.]

Amino acid dating   Determination of the relative or absolute age of materials or objects by measurement of the degree of racemization of the amino acids present. With the exception of glycine, the amino acids found in proteins can exist in two isomeric forms called D- and L-enantiomers. Although the enantiomers of an amino acid rotate plane-polarized light in equal but opposite directions (the D form rotates it to the right and the L form to the left), their other chemical and physical properties are identical. It was discovered by L. Pasteur around 1850 that only L-amino acids are generally found in living organisms, but scientists still have not formulated a convincing reason to explain why life is based on only L-amino acids.

Under conditions of chemical equilibrium, equal amounts of both enantiomers are present ($D/L = 1.0$); this is called a racemic mixture. Living organisms maintain a state of disequilibrium through a system of enzymes that selectively utilize only the L-enantiomers. Once a protein has been synthesized and isolated from active metabolic processes, the L-amino acids are subject to a racemization reaction that converts them into a racemic mixture. Since racemization is a chemical process, the extent of racemization is dependent not only on the time that has elapsed since the L-amino acids were synthesized but also on the exposure temperature: the higher the temperature, the faster the rate of racemization. The rate of racemization is also different for most of the various amino acids.

A variety of analytical procedures can be used to separate amino acid enantiomers; gas chromatography and high-performance liquid chromatography are the most widely used. Since these techniques have sensitivities in the parts per billion range, only a few hundred milligrams of sample material are normally required. Samples are first hydrolyzed in hydrochloric acid to break down the proteins into free amino acids, which are then isolated by cation-exchange chromatography.

Since the late 1960s, the geochemical and biological significance of amino acid racemization has been extensively investigated. Geochemical uses of amino acid racemization include the dating of fossils or, in the case of known age specimens, the determination of their temperature history. Fossil types such as bones, teeth, and shells have been studied, and racemization has been found to be particularly useful for dating specimens that were difficult to date by other methods. Racemization has also been observed in the metabolically inert tissues of living mammals. Racemization can be studied in certain organisms and used to assess the biological age of a variety of mammalian species; in addition, it may be important in determining the biological lifetime of certain proteins. Fossils have been found to contain both D- and L-amino acids, and the extent of racemization generally increases with geologic age. *See* ARCHEOLOGICAL CHEMISTRY; GEOCHRONOMETRY; RADIOCARBON DATING; ROCK AGE DETERMINATION.

[J.L.Ba.]

Amphibole   A group of common ferromagnesian silicate minerals that occur as major or minor constituents in a wide variety of rocks. The crystal structure of the amphiboles is very flexible and, as a result, the amphiboles show a larger range of chemical composition than any other group of minerals. The structural and chemical complexity of the amphiboles reveals considerable information on the geological processes that have affected the rocks in which they occur. *See* MINERAL.

A general formula for amphiboles may be written as $A_{0-1}B_2C_5T_8O_{22}W_2$, where

$$A = \text{Na, K, Ca}$$
$$B = \text{Ca, Na, Mn}^{2+}, \text{Fe}^{2+}, \text{Li, Mg}$$
$$C = \text{Mg, Fe}^{2+}, \text{Al, Fe}^{3+}, \text{Ti}^{4+}, \text{Mn, Li}$$
$$T = \text{Si, Al, Ti}^{4+}$$
$$O = \text{oxygen}$$
$$W = \text{OH, F, O}^{2-}, \text{Cl}$$

and the chemical species are written in order of their importance. Amphiboles are divided into four main groups, according to the type of chemical species in the B group:

$B = (Fe^{2+}, Mg, Mn^{2+}, Li)_2$ Iron-magnesium-manganese amphiboles

$B = Ca_2$ Calcic amphiboles

$B = NaCa$ Sodic-calcic amphiboles

$B = Na_2$ Sodic amphiboles

Amphiboles can have orthorhombic and monoclinic symmetries; these can be distinguished either by x-ray crystallography or by the optical properties of the mineral in polarized light.

Monoclinic amphiboles are by far the most common. The characteristic feature of the amphibole structure is the chain of corner-sharing tetrahedrally coordinated groups. Inspection of the structure down the y axis shows that the amphibole structure consists of sheets of tetrahedra interleaved with octahedrally coordinated C-group cations. In the monoclinic structure, the tetrahedral layers all stack in the x direction with the same sense of displacement (along the z direction) relative to the underlying layer, and hence the x axis is inclined to the z axis, producing a monoclinic structure. In the orthorhombic structure, the displacement of the tetrahedral layers reverses every third layer, and hence the x axis is orthogonal to the z axis, producing an orthorhombic structure.

Amphiboles do not show the complete range of possible compositions suggested by the general chemical formula and common idealized compositions described above. In particular, there is not a continuous range of chemical composition between the four main amphibole groups: the iron-magnesium-manganese amphiboles, the calcic amphiboles, the sodic-calcic amphiboles, and the alkali amphiboles. This lack of so-called solid solution is a result of the structure not being able to accommodate two types of cations (positively charged atoms) of very different size (or charge) at the same set of sites in the structure of a single crystal.

The degree to which two amphiboles are immiscible often varies as a function of temperature and pressure; at high temperatures or pressures, miscibility is usually enhanced. The immiscible region, which is known as a miscibility gap, is narrow at high temperature but widens at lower temperature. When the amphibole composition is within the miscibility gap, a single amphibole is no longer stable. It is here that the process of exsolution (or unmixing) occurs. Coexisting amphiboles and exsolution textures are very informative about temperatures of crystallization and cooling history, particularly in metamorphic rocks. See SOLID SOLUTION.

Amphiboles are common minerals in many types of igneous rocks, and the composition of the amphibole reflects the silica content of the rock. Calcic amphiboles, particularly pargasite, are characteristic of ultramafic and metabasaltic rocks and are usually quite rich in magnesium. Titanium-rich hornblendes and kaersutites occur in intermediate rocks and are often strongly oxidized—an unusual feature in amphiboles from any other environment. Acidic rocks, particularly granites, contain a wide range of amphiboles, from hastingsite to riebeckite and arfvedsonite; these are often iron-rich and can contain significant amounts of more unusual elements such as Li, Zn, and Mn. Amphiboles usually weather very easily and hence are not important in sedimentary rocks, although they can be significant components of soil. Iron-rich alkali amphiboles can form at essentially ambient conditions in the sedimentary environment, but this occurrence is rare. Amphiboles are common and important rock-forming minerals in many types of metamorphic rocks. They are particularly abundant in rocks of basaltic composition at most grades of metamorphism.

Amphiboles are economically important as commercial asbestos minerals and as semiprecious gem materials. World asbestos production is dominated by the serpentine-group mineral chrysotile; but the amphibole minerals anthophyllite, cummingtonite-grunerite (amosite), actinolite, and riebeckite (crocidolite) are also important, particularly in Australia and South Africa. Some amphiboles with attractive physical properties are marketed as semiprecious gem material. Most important is nephrite, a dense compact form of fibrous actinolite that is a principal variety of jade. Fibrous riebeckite is marketed as one of the less common varieties of tiger's eye. Iridescent gedrite and gem-quality pargasite are used as semiprecious gems in contemporary jewelry. *See* GEM; JADE; MINERALOGY. [F.C.Ha.]

Amphibolite A class of metamorphic rocks with one of the amphibole minerals as the dominant constituent. Most of the amphibolites are dark green to black crystalline rocks that occur as extensive layers widely distributed in mountain belts and deeply eroded shield areas of the continental crust. Amphibolite is the main country rock that has been intruded by the large granite masses found in most mountain ranges, with small and large masses of amphibolite present also as inclusions in granites.

Amphibolites are the products of regional metamorphism and crustal deformation of older materials of appropriate composition. The features of the original rock are obliterated; thus it is difficult and sometimes impossible to determine the premetamorphic rock. Apparent differences in the formation of the bulk composition are used to classify amphibolites as ortho or para. Compositional relations between the minor elements titanium, chromium, and nickel have been used to distinguish the ortho from para amphibolites in some occurrences. *See* METAMORPHIC ROCKS. [G.W.DeV.]

Analcime A mineral with a framework structure in which all the aluminosilicate tetrahedral vertices are linked, thus allying it to the feldspars, feldspathoids, and zeolites. Its formula is $Na(H_2O)[AlSi_2O_6]$; in this sense it is a tectosilicate.

The analcime structure type includes several other mineral species. These include high-temperature leucite, pollucite, and wairakite. Crystals are most frequently trapezohedra, and rarely the mineral is massive granular. Hardness is 5–$5^1/_2$ on Mohs scale; specific gravity is 2.27.

Analcime most frequently occurs as a low-temperature mineral in vesicular cavities in basalts, where it is associated with zeolites (particularly natrolite), datolite, prehnite, and calcite. Small grains are frequent constituents of sedimentary rocks and muds in oceanic basins associated with volcanic sources. *See* FELDSPAR; FELDSPATHOID; LEUCITE ROCK; ZEOLITE. [P.B.M.]

Andalusite A nesosilicate mineral, composition Al_2SiO_5, crystallizing in the orthorhombic system. It occurs commonly in large, nearly square prismatic crystals. There is poor prismatic cleavage; the luster is vitreous and the color red, reddish-brown, olive-green, or bluish. Transparent crystals may show strong dichroism, appearing red in one direction and green in another in transmitted light. The specific gravity is 3.1–3.2; hardness is 7.5 on Mohs scale, but may be less on the surface because of alteration. *See* SILICATE MINERALS.

Andalusite was first described in Andalusia, Spain, and was named after this locality. It is found abundantly in the White Mountains near Laws, California, where for many years it was mined for manufacture of spark plugs and other highly refractive porcelain. Chiastolite, in crystals largely altered to mica, is found in Lancaster and Sterling, Massachusetts. Water-worn pebbles of gem quality are found at Minas Gerais, Brazil. [C.S.Hu.]

Andesine   A plagioclase feldspar with composition $Ab_{70}An_{30}$ to $Ab_{50}An_{50}$ (Ab = $NaAlSi_3O_8$; An = $CaAl_2Si_2O_8$). Andesine occurs primarily in igneous rocks, often in a glassy matrix as small, chemically zoned, lathlike crystals known as microlites. The rock types may be called andesinites (if dominantly feldspar), andesites, andesitic basalts (or olivine-bearing andesites, as in Hawaiian lava flows), or pyroxene-, hornblende- or biotite-andesites (all are volcanic). *See* ANDESITE; FELDSPAR; IGNEOUS ROCKS.

The symmetry of andesine is triclinic, hardness on the Mohs scale 6, specific gravity 2.69, melting point $\sim 1210°C$ ($2210°F$). If quenched at very high temperatures, andesine has an albitelike structure, with aluminum (Al) and silicon (Si) essentially disordered in the structural framework of the crystals. But, in the course of cooling, most natural andesines develop an Al-Si ordered structure called e-plagioclases. *See* ALBITE.

Calcic andesines and labradorites ($Ab_{55}An_{45}$-$Ab_{40}An_{60}$) may exsolve into two distinctly intergrown lamellar phases whose regularity of stacking produces beautiful interference colors like those in the feathers of a peacock. Polished specimens of this material are called spectrolite in the gem trade, and at some localities (notably eastern Finland) crystals up to 10 in. (25 cm) are mined by hand. Smaller crystals are made into cabochons for jewelry. They may be abundant enough in the host rock to be valued as a decorative stone. *See* GEM; LABRADORITE.   [P.H.R.]

Andesite   A typical volcanic rock erupted from a volcano associated with convergent plate boundaries. The process of subduction, which defines convergent plate boundaries, pushes oceanic lithosphere beneath either oceanic lithosphere or continental lithosphere. Andesites are the principal rocks forming the volcanoes of the "ring of fire," the arcuate chains of volcanoes which rim the Pacific Ocean basin. The Marianas and Izu-Bonin islands, the islands of Japan, the Aleutian Islands, the Cascades Range of the northwest United States, the Andes mountain chain of South America, and the Taupo Volcanic Zone of New Zealand are andesitic. *See* LITHOSPHERE; PLATE TECTONICS; VOLCANO.

Andesites are mostly dark-colored vesicular volcanic rocks which are typically porphyritic (containing larger crystals set in a fine groundmass). Phenocrysts (the larger crystals) comprise plagioclase; calcium-rich, calcium-poor pyroxene; and iron-titanium oxides set in a fine-grained, frequently glassy, groundmass. Some andesites contain phenocrysts of olivine, and some contain amphibole and biotite; these latter rocks generally contain more potassium. The porphyritic nature of andesites is derived from a complicated history of magmatic crystallization and evolution as the melts rise toward the surface from deep in the Earth. Phenocryst minerals commonly are strongly zoned and show evidence for disequilibrium during growth, consistent with an origin involving crystal fractionation and mixing processes. Andesites are readily classified in terms of their silicon dioxide (SiO_2) content, between 53 and 63 wt %, and potassium oxide (K_2O) content at a given SiO_2 content. They can also be readily discriminated on a total alkali versus SiO_2 diagram. Most andesite volcanoes erupt lavas and tephras (volcanic ash) which range in composition from basaltic andesite to dacite. Eruptions are often explosive, reflecting the relatively high water and gas content of the magmas. Pyroclastic flows are a particular feature of andesite-type volcanism and are among the most dangerous of volcanic hazards. *See* BASALT; LAVA; PYROCLASTIC ROCKS.   [J.Ga.]

Anglesite   A mineral with the chemical composition $PbSO_4$. Anglesite occurs in white or gray, orthorhombic, tabular or prismatic crystals or compact masses. It is a common secondary mineral, usually formed by the oxidation of galena. Fracture is conchoidal and luster is adamantine. Hardness is 2.5–3 on Mohs scale and specific gravity is 6.38. The mineral does not occur in large enough quantity to be mined as an ore of lead, and is therefore of no particular commercial value. Fine exceptional crystals of anglesite have been found throughout the world.   [E.C.T.C.]

Anhydrite   A mineral with the chemical composition $CaSO_4$. Anhydrite occurs commonly in white and grayish granular masses, rarely in large, orthorhombic crystals. Fracture is uneven and luster is pearly to vitreous. Hardness is 3–3.5 on Mohs scale and specific gravity is 2.98. Anhydrite is an important rock-forming mineral and occurs in association with gypsum, limestone, dolomite, and salt beds. Under natural conditions, anhydrite hydrates slowly, but readily, to gypsum. It is not used as widely as gypsum. Anhydrite is of worldwide distribution. Large deposits occur in the Carlsbad district, Eddy County, New Mexico, and in salt-dome areas in Texas and Louisiana. *See* GYPSUM.
[E.C.T.C.]

Ankerite   The carbonate mineral $Ca(Fe,Mg)(CO_3)_2$, also commonly containing some manganese. The mineral has hexagonal (rhombohedral) symmetry and has the cation-ordered structure of dolomite. The name is applied only to those species in which at least 20% of the magnesium positions are occupied by iron or manganese; species containing less iron are termed ferroan dolomites. The pure compound, $CaFe(CO_3)_2$, has never been found in nature and has never been synthesized as an ordered compound. *See* DOLOMITE.

Ankerite is commonly white to light brown, its specific gravity is about 3, and its hardness is about 4 on Mohs scale. *See* CARBONATE MINERALS.   [A.M.G.]

Anorthite   The calcium-rich plagioclase feldspar with composition Ab_0An_{100} to $Ab_{10}An_{90}$ (Ab = $NaAlSi_2O_8$; An = $CaAl_2Si_2O_8$), occurring in olivine-rich igneous rocks and rare volcanic ejecta (for example, at Mount Vesuvius, Italy, and Miyakejima, Japan). Hardness is 6 on Mohs scale, specific gravity 2.76, melting point 1550°C (2822°F). The crystal structure of Ab_0An_{100} consists of an infinite three-dimensional array of corner-sharing $[AlO_4]$ and $[SiO_4]$ tetrahedra, alternately linked together in a framework of $[Al_2Si_2O_8]_\infty^{2-}$ composition in which charge-balancing calcium (Ca^{2+}) cations occupy four distinct, irregular cavities. Natural anorthite has no commercial uses, but the synthetic material $[CaO \cdot Al_2O_3 \cdot 2SiO_2$ (known as CAS_2)] is important in the ceramic industry and in certain composite materials with high-temperature applications. *See* FELDSPAR.   [P.H.R.]

Anorthoclase   The name usually given to alkali feldspars which have a chemical composition ranging from $Or_{40}Ab_{60}$ to $Or_{10}Ab_{90}$ ± up to approximately 20 mole % An (Or, Ab, An = $KAlSi_3O_8$, $NaAlSi_3O_8$, $CaAl_2Si_2O_8$) and which deviate in one way or another from monoclinic symmetry tending toward triclinic symmetry. When found in nature, they usually do not consist of a single phase but are composed of two or more kinds of K- and Na-rich domains mostly of submicroscopic size. In addition, they are frequently polysynthetically twinned after either or both of the albite and pericline laws. It appears that they originally grew as the monoclinic monalbite phase, inverting and unmixing in the course of cooling during geological times. They are typically found in lavas or high-temperature rocks. *See* FELDSPAR; IGNEOUS ROCKS.   [F.H.L.]

Anorthosite   A rock composed of 90 vol % or more of plagioclase feldspar. Strictly, the rock is composed entirely of crystals discernible with the eye, but some finely crystalline examples from the Moon have been called anorthosite or anorthositic breccia. Scientists have been fascinated with anorthosites because they are spectacular rocks (dark varieties are quarried and polished for ornamental use); valuable deposits of iron and titanium ore are associated with anorthosites; and the massif anorthosites appear to have been produced during a unique episode of ancient Earth history (about $1-2 \times 10^9$ years ago).

Pure anorthosite has less than 10% of dark minerals—generally some combination of pyroxene, olivine, and oxides of iron and titanium; amphibole and biotite are rare, as are the light minerals apatite, zircon, scapolite, and calcite. Rocks with less than 90% but more than 78% of plagioclase are modified anorthosites (such as gabbroic anorthosite), and rocks with 78–65% of plagioclase are anorthositic (such as anorthositic gabbro). *See* GABBRO.

The structure, texture, and mineralogy vary with type of occurrence. One type of occurrence is as layers (up to several meters thick) interstratified with layers rich in pyroxene or olivine. The second type of occurrence is the massifs type and can have an area up to $11,600 \text{ mi}^2$ $(30,000 \text{ km}^2)$. Commonly, the massifs are domical in shape and weakly layered. Possibly there is a third group of anorthosite occurrences: extremely ancient bodies of layered rock in which the layers of anorthosite contain calcium-rich plagioclase and the adjacent layers are rich in chromite and amphibole in addition to pyroxene. There are only a few examples of these apparently igneous complexes, in Greenland, southern Africa, and India. However, they appear to be terrestrial counterparts of lunar anorthosites.

By comparison with terrestrial occurrences, most lunar anorthosites are very fine grained, although one rock has crystals up to a centimeter long. Much of the fine grain size results from comminution by meteorite impact, and some of it probably results from rapid crystallization of impact melts. *See* ANDESINE; IGNEOUS ROCKS; LABRADORITE; METAMORPHISM.   [A.T.A.]

Anoxic zones   Oxygen-depleted regions in marine environments. The dynamic steady state between oxygen supply and consumption determines the oxygen concentration. In regions where the rate of consumption equals the rate of supply, seawater becomes devoid of oxygen and thus anoxic. In the open ocean, the only large regions which approach anoxic conditions are between 165 and 3300 ft (50 and 1000 m) deep in the equatorial Pacific and between 330 and 3300 ft (100 and 1000 m) in the northern Arabian Sea and the Bay of Bengal in the Indian Ocean. The Pacific region consists of vast tongues extending from Central America and Peru nearly to the middle of the ocean in some places. In parts of this zone, oxygen concentrations become very low, 15 μmol/liter (atmospheric saturation is 200–300 μmol/liter). Pore waters of marine sediments are sometimes anoxic a short distance below the sediment-water interface. The degree of oxygen consumption in sediment pore waters depends upon the amount of organic matter reaching the sediments and the rate of bioturbation (mixing of the surface sediment by benthic animals). In shallow regions (continental shelf and slope), pore waters are anoxic immediately below the sediment-water interface; in relatively rapid sedimentation-rate areas of the deep sea, the pore waters are usually anoxic within a few centimeters of the interface; and in pore waters of slowly accumulating deep-sea sediments, oxygen may never become totally depleted. *See* MARINE SEDIMENTS.

Restricted basins (areas where water becomes temporarily trapped) are often either permanently or intermittently anoxic. Classic examples are the Black Sea, the Carioca Trench off the coast of Venezuela, and fiords which occupy the Norwegian and British Columbia coasts. Lakes which receive a large amount of nutrient inflow (either from natural or human-produced sources) are often anoxic during the period of summer stratification. *See* BLACK SEA; FIORD.

The chemistry of many elements dissolved in seawater (particularly the trace elements) is vastly changed by the presence or absence of oxygen. Since large areas of the ocean water mass are in contact with oxygen-depleted pore waters, the potential exists for anoxic conditions to have a marked effect on the chemistry of the sea. *See* SEAWATER; SEAWATER FERTILITY.   [S.R.E.]

Antarctic Circle   An imaginary line that delimits the northern boundary of Antarctica. It is a distinctive parallel of latitude at approximately 66°30' south. Thus it is located about 4590 mi (7345 km) south of the Equator and about 1630 mi (2620 km) north of the south geographic pole.

All of Earth's surface south of the Antarctic Circle experiences one or more days when the Sun remains above the horizon for at least 24 h. The Sun is at its most southerly position on or about December 21 (slightly variable from year to year). This date is known as the summer solstice in the Southern Hemisphere and as the winter solstice in the Northern Hemisphere. At this time, because Earth is tilted on its axis, the circle of illumination reaches 23.50° to the far side of the South Pole and stops short 23.50° to the near side of the North Pole.

The longest period of continuous sunshine at the Antarctic Circle is 24 h, and the highest altitude of the noon Sun is 47° above the horizon at the time of the summer solstice. The long days preceding and following the solstice allow a season of about 5 months of almost continuous daylight.

Six months after the summer solstice, the winter solstice (Southern Hemisphere terminology) occurs on or about June 21 (slightly variable from year to year). On this date the Sun remains below the horizon for 24 h everywhere south of the Antarctic Circle; thus the circle of illumination reaches 23.50° to the far side of the North Pole and stops short 23.50° to the near side of the South Pole. *See* ARCTIC OCEAN.   [T.L.M.]

Antarctic Ocean   The Antarctic Ocean, sometimes called the Southern Ocean, is the watery belt surrounding Antarctica. It includes the great polar embayments of the Weddell Sea and Ross Sea, and the deep circumpolar belt of ocean between 50 and 60°S and the southern fringes of the warmer oceans to the north. Its northern boundary is often taken as 30°S (see illustration). The Antarctic is a cold ocean, covered by sea ice during the winter from Antarctica's coast northward to approximately 60°S.

The remoteness of the Antarctic Ocean severely hampers the ability to observe its full character. The sparse data collected and the more recent addition of data obtained from satellite-borne sensors have led to an appreciation of the unique role that this ocean plays in the Earth's ocean and climate. Between 50 and 60°S there is the greatest of all ocean currents, the Antarctic Circumpolar Current sweeping seawater from west to east, blending waters of the Pacific, Atlantic, and Indian oceans. Observed within this current is the sinking of cool (approximately 4°C; 39.2°F), low-salinity waters to depths of near 1 km (0.6 mi), which then spreads along the base of the warm upper ocean waters or thermocline of more hospitable ocean environments. The cold polar atmosphere spreading northward from Antarctica removes great amount of heat from the ocean, heat which is carried to the sea surface from ocean depths, brought into the Antarctic Ocean from warmer parts of the ocean. At some sites along the margins of Antarctica, there is rapid descent of cold (near the freezing point of seawater, −1.9°C; 28.6°F) dense water, within thin convective plumes. This water reaches the sea floor, where it spreads northward, chilling the lower 2 km (1.2 mi) of the global ocean, even well north of the Equator.

The major flow is the Antarctic Circumpolar Current, or West Wind Drift (see illustration). Along the Antarctic coast is the westward-flowing East Wind Drift. The strongest

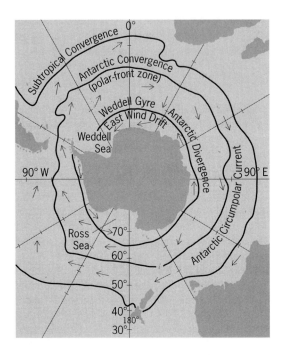

Direction of the surface circulation and major surface boundaries of the Antarctic Ocean.

currents are in the vicinity of the polar front zone and restricted passages such as the Drake Passage, and over deep breaks in the meridionally oriented submarine ridge systems.

The extreme cold of the polar regions causes an extensive ice field to form over the southern regions of the Antarctic Ocean. The extent of the ice is seasonal in that during the October-to-March period the area decreases, and it increases during the remaining months. The seasonal difference in the volume of sea ice is estimated as 2.3×10^{19} grams (8.1×10^{17} oz). Satellite photographs reveal that the sea ice field is not uniform, but has many large polynyas (areas of water). The sea ice plays an important role in the heat balance since it reflects much more solar radiation (and therefore heat) into space than would be the case for a water surface. The polynyas would therefore be of special interest in radiation and heat-balance studies. In addition to the ice formed at sea, the ice calving at the coast of Antarctica introduces icebergs into the ocean at a rate of approximately 1×10^{18} g/year (3.5×10^{12} oz/year). *See* HEAT BALANCE, TERRESTRIAL ATMOSPHERIC; ICEBERG; SEA ICE.

Glacial (fresh-water) ice and the ocean meet along the shores of Antarctica. This occurs not only at the northern face of the ice sheet but also at hundreds of meters depth along the bases of floating ice shelves. Ocean-glacial ice interaction is believed to be a major factor in controlling Antarctica's glacial ice mass balance and stability.   [A.L.G.]

Antarctica   The coldest, windiest, and driest continent, overlying the South Pole. The lowest temperature ever measured on Earth was recorded at the Russian Antarctic station of Vostok at $-89.2°C$ ($-128.5°F$) in July 1983. Katabatic (cold, gravitational) winds with velocities up to 50 km/h (30 mi/h) sweep down to the coast and occasionally turn into blizzards with 150 km/h (nearly 100 mi/h) wind velocities. Antarctica's interior is a cold desert with only a few centimeters of water-equivalent precipitation, while the coastal areas average 30 cm (12 in.).

Antarctica's area is about 14 million square kilometers (5.4 million square miles), which is larger than the contiguous 48 United States and Mexico together. It is the third smallest continent, after Australia and Europe. About 98% of it is buried under a thick ice sheet, which in places is 4 km (13,000 ft) thick, making it the highest continent, with an average elevation of over 2 km (6500 ft).

Although most of Antarctica is covered by ice, some mountains rise more than 3 km (almost 10,000 ft) above the ice sheet. The largest of these ranges is the Transantarctic Mountains separating east from west Antarctica, and the highest peak in Antarctica is Mount Vinson, 5140 m (16,850 ft), in the Ellsworth Mountains. Other mountain ranges, such as the Gamburtsev Mountains in East Antarctica, are completely buried, but isolated peaks called nunataks frequently thrust through the ice around the coast.

The Antarctic ice sheet is the largest remnant of previous ice age glaciations. It has probably been in place for the last 20 million years and perhaps up to 50 million years. It is the largest reservoir of fresh water on Earth, with a volume of about 25 million cubic kilometers (6 million cubic miles). Glaciers flow out from this ice sheet and feed into floating ice shelves along 30% of the Antarctic coastline. The two biggest ice shelves are the Ross and Filchner-Ronne. These shelves may calve off numerous large tabular icebergs, with thicknesses of several hundred meters, towering as high as 70–80 m (250 ft) above the sea surface. *See* GLACIOLOGY.

Year-round life on land in Antarctica is sparse and primitive. North of the Antarctic Peninsula a complete cover of vegetation, including moss carpets and only two species of native vascular plants, may occur in some places. For the rest of Antarctica, only lichen, patches of algae in melting snow, and occasional microorganisms occur. In summer, however, numerous migrating birds nest and breed in rocks and cliffs on the continental margins, to disappear north again at the beginning of winter. South of the Antarctic Convergence, 43 species of flying birds breed annually. They include petrels, skuas, and terns, cormorants, and gulls. Several species of land birds occur on the subantarctic islands. The largest and best-known of the Antarctic petrels are the albatrosses, which breed in tussock grass on islands north of the pack ice. With a wing span of 3 m (10 ft), they roam freely over the westerly wind belt of the Southern Ocean. [G.We.]

Anthophyllite   A magnesium-rich orthorhombic amphibole with perfect {210} cleavage and a color which varies from white to various shades of green and brown. It is a comparatively rare metamorphic mineral which occurs as slender prismatic needles, in fibrous masses, and sometimes in asbestiform masses. Anthophyllite may occur together with calcite, magnesite, dolomite, quartz, tremolite, talc, or enstatite in metacarbonate rocks; with plagioclase, quartz, orthopyroxene, garnet, staurolite, chlorite, or spinel in cordierite-anthophyllite rocks; and with quartz and hematite in metamorphosed iron formations, and with talc, olivine, chlorite, or spinel in metamorphosed ultrabasic rocks. Anthophyllite is distinguished from other amphiboles by optical examination or by x-ray diffraction, and from other minerals by its two cleavage directions at approximately 126° and 54°.

Anthophyllite has the general formula

$$(Mg, Fe^{2+})_{7-x}Al_x(Al_xSi_{8-x})O_{22}(OH, F, Cl)_2$$

with $x < 1.0$. For aluminum-poor varieties, up to about 40% of the Mg may be replaced by Fe^{2+}; higher iron contents result in the formation of the monoclinic amphibole cummingtonite. Increasing the aluminum content in anthophyllite beyond $x = 1.0$ results in the formation of the orthohombic amphibole gedrite; aluminous anthrophyllite can accommodate more Fe^{2+} than Al-poor varieties. *See* AMPHIBOLE; CUMMINGTONITE.

[J.V.C.]

Anticline   A fold in layered rocks in which the strata are inclined down and away from the axes. The simplest anticlines (see illustration) are symmetrical, but in more

Diagram relating anticlinal
structure to topography.

highly deformed regions they may be asymmetrical, overturned, or recumbent. Most anticlines are elongate with axes that plunge toward the extremities of the fold, but some have no distinct trend; the latter are called domes. Generally, the stratigraphically older rocks are found toward the center of curvature of an anticline, but in more complex structures these simple relations need not hold. Under such circumstances, it is sometimes convenient to recognize two types of anticlines. Stratigraphic anticlines are those folds, regardless of their observed forms, that are inferred from stratigraphic information to have been anticlines originally. Structural anticlines are those that have forms of anticlines, regardless of their original form. *See* SYNCLINE.   [P.H.O.]

Apatite   The most abundant and widespread of the phosphate minerals, crystallizing in the hexagonal system. The apatite structure type includes no less than 10 mineral species and has the general formula $X_5(YO_4)_3Z$, where X is usually Ca^{2+} or Pb^{2+}, Y is P^{5+} or As^{5+}, and Z is F^-, Cl^-, or $(OH)^-$. The apatite series takes $X = Ca$, whereas the pyromorphite series includes those members with $X = Pb$. Three end members form a complete solid-solution series involving the halide and hydroxyl anions. These are fluorapatite, $Ca_5(PO_4)F_3$; chlorapatite, $Ca_5(PO_4)_3Cl$; and hydroxyapatite, $Ca_5(PO_4)_3(OH)$. Thus, the general series can be written $Ca_5(PO_4)_3(F,Cl,OH)$, the fluoride member being the most frequent and often simply called apatite.

The apatite isomorphous series of minerals occurs as grains, blebs, or short to long hexagonal prisms terminated by pyramids, dipyramids, and the basal pinacoid. The minerals are transparent to opaque, and can be asparagus-green (asparagus stone), grayish-green, greenish-yellow, gray, brown, brownish-red, and more rarely violet, pink, or colorless. Apatites are brittle, with hardness 5 on Mohs scale, and specific gravity 3.1–3.2; they are also botryoidal, fibrous, and earthy.

Apatite occurs in nearly every rock type as an accessory mineral. It often crystallizes in regional and contact metamorphic rocks, especially in limestone and associated with chondrodite and phlogopite. It is very common in basic to ultrabasic rocks; enormous masses occur associated with nephelinesyenites in the Kola Peninsula, Russia, and constitute valuable ores which also contain rare-earth elements. Large beds of oolitic, pulverulent, and compact fine-grained carbonate-apatites occur as phosphate rock, phosphorites, or collophanes. Extensive deposits of this kind occur in the United States in Montana and Florida and in North Africa. The material is mined for fertilizer and for the manufacture of elemental phosphorus. *See* PYROMORPHITE.   [P.B.M.]

Aplite   A fine-grained, sugary-textured rock, generally of granitic composition; also any body composed of such rock. This light-colored rock consists chiefly of quartz, microcline, or orthoclase perthite and sodic plagioclase, with small amounts of muscovite, biotite, or hornblende and traces of tourmaline, garnet, fluorite, and topaz. Much quartz and potash feldspar may be micrographically intergrown in cuneiform fashion. See GRANITE; IGNEOUS ROCKS.

Aplites may form dikes, veins, or stringers, generally not more than a few feet thick, with sharp or gradational walls. Some show banding parallel to their margins. Aplites usually occur within bodies of granite and more rarely in the country rock surrounding granite. They are commonly associated with pegmatites and may cut or be cut by pegmatites. See PEGMATITE.   [C.A.C.]

Apophyllite   A hydrous calcium-potassium silicate containing fluorine. The composition is variable but approximates to $KFCa_4(Si_2O_5)_4 \cdot 8H_2O$. It resembles the zeolites, with which it is sometimes classified, but differs from most zeolites in having no aluminum. It exfoliates (swells) when heated, losing water, and is named from this characteristic; the water can be reabsorbed. It is essentially white, with a vitreous luster, but may show shades of green, yellow, or red. The symmetry is tetragonal, and the crystal structure contains sheets of linked SiO_4 groups; this accounts for the perfect basal cleavage of the mineral. It occurs as a secondary mineral in cavities in basic igneous rocks, commonly in association with zeolites. The specific gravity is about 2.3–2.4, the hardness 4.5–5 on Mohs scale, the mean refractive index about 1.535, and the birefringence 0.002. See SILICATE MINERALS; ZEOLITE.   [G.W.Br.]

Aquifer   A subsurface zone that yields economically important amounts of water to wells. The term is synonymous with water-bearing formation. An aquifer may be porous rock, unconsolidated gravel, fractured rock, or cavernous limestone.

Aquifers are important reservoirs storing large amounts of water relatively free from evaporation loss or pollution. If the annual withdrawal from an aquifer regularly exceeds the replenishment from rainfall or seepage from streams, the water stored in the aquifer will be depleted. This mining of groundwater results in increased pumping costs and sometimes pollution from sea water or adjacent saline aquifers. Lowering the piezometric pressure in an unconsolidated artesian aquifer by overpumping may cause the aquifer and confining layers of silt or clay to be compressed under the weight of the overlying material. The resulting subsidence of the ground surface may cause structural damage to buildings, altered drainage paths, increased flooding, damage to wells, and other problems. See ARTESIAN SYSTEMS.   [R.K.Li.]

Aragonite   One of three naturally occurring mineral forms of calcium carbonate ($CaCO_3$). The other forms (or polymorphs) are the abundant mineral calcite and the relatively rare mineral vaterite. Still other forms of calcium carbonate are known, but only as products of laboratory experiments. The name aragonite comes from Aragon, a province in Spain where especially fine specimens occur. See CALCITE.

Aragonite has an orthorhombic crystal structure in which layers of calcium (Ca) atoms alternate with layers of offset carbonate (CO_3) groups. A common crystallographic feature of aragonite is twinning, in which regions of crystal are misoriented as though they were mirror images of each other. This can give rise to a pseudohexagonal symmetry which is readily identified in large crystals (see illustration). Aragonite crystals are usually colorless or white if seen individually; however, aggregates of small crystals may exhibit different colors. Most aragonites are nearly pure calcium carbonate; however, small amounts of strontium (Sr) and less commonly barium (Ba) and lead (Pb) may be present as impurities.

Aragonite. Pseudohexagonally twinned specimen from Girgenti, Sicily. (*American Museum of Natural History specimens*)

5 cm

At the low temperatures and pressures found near the Earth's surface, aragonite is metastable and should invert spontaneously to calcite, which is stable at these conditions. This, in part, explains why calcite is far more abundant than aragonite. However, at low temperatures the transformation of aragonite to calcite effectively occurs only in the presence of water, and aragonite may persist for long periods of geologic time if isolated from water. Increased temperature also promotes the transformation to calcite. Despite being metastable, aragonite rather than calcite is sometimes the favored precipitate from certain solutions, such as seawater, in which magnesium (Mg) ions inhibit precipitation of calcite.

Aragonite occurs most abundantly as the hard skeletal material of certain freshwater and marine invertebrate organisms, including pelecypods, gastropods, and some corals. The accumulated debris from these skeletal remains can be thick and extensive, usually at the shallow sea floor, and with time may transform into limestone. Most limestones, however, contain calcite and little or no aragonite. The transformation of the aragonite to calcite is an important step in forming limestone and proceeds by the dissolution of aragonite followed by the precipitation of calcite in the presence of water. This process may take more than 100,000 years. *See* LIMESTONE.

Other occurrences of aragonite include cave deposits (often in unusual shapes) and weathering products of calcium-rich rocks.   [R.J.Re.]

Archean   A period of geologic time from about 3.8 to 2.5 billion years ago (Ga). During the Archean Eon a large percentage of the Earth's continental crust formed, plate tectonics began, very warm climates and oceans existed, and life appeared on Earth in the form of unicellular organisms.

	1	PROTEROZOIC (ALGONKIAN)	LATE
PRECAMBRIAN	2		MIDDLE
	3	ARCHEAN (ARCHEOZOIC)	EARLY
	4		

The occurrence of rock assemblages typical of arcs, oceanic plateaus, and oceanic islands and the presence of accretionary orogens in the very earliest vestiges of the geologic record at 4–3.5 Ga strongly supports some sort of plate tectonics operating on the Earth by this time. By 3 Ga, cratons, passive margins, and continental rifts were also widespread. Although plate tectonics appears to have occurred since 4 Ga, there are geochemical differences between Archean and younger rocks that indicate that Archean tectonic regimes must have differed in some respects from modern ones. The degree that Archean plate tectonics differed from modern plate tectonics is unknown; however, these differences are important in terms of the evolution of the Earth. *See* PLATE TECTONICS.

The oldest rocks occur as small, highly deformed terranes tectonically incorporated within Archean crustal provinces. Although the oldest known igneous rocks on Earth are the 4 Ga Acasta gneisses of northwest Canada, the oldest minerals are detrital zircons (zircons in sediments) from the 3 Ga Mount Narryer quartzites in western Australia.

The oldest isotopically dated rocks on Earth are the Acasta gneisses, which are a heterogeneous assemblage of highly deformed granitic rocks, tectonically interleaved with mafic and ultramafic rocks, and metasediments. Uranium-lead zircon ages from the granitic components of these gneisses range from 4 to 3.6 Ga, and thus it would appear that this early crustal segment evolved over about 400 million years and developed a full range in composition of igneous rocks. The chemical compositions of Acasta mafic rocks are very much like less deformed Archean greenstones representing various oceanic tectonic settings. *See* DATING METHODS; ROCK AGE DETERMINATION.

The largest and best-preserved fragment of early Archean continental crust is the Itsaq Gneiss Complex in southwest Greenland. In this area, three terranes, each with its own tectonic and magmatic history, collided about 2.7 Ga, forming the continental nucleus of Greenland. Although any single terrane records less than 500 million years of precollisional history, collectively the terranes record over 1 billion years of history before their amalgamation.

The Archean is known for its reserves of iron, copper, zinc, nickel, and gold. Some of the world's largest copper-zinc deposits occur as massive sulfide beds associated with submarine volcanics in Archean greenstones in Canada and western Australia.

The Earth's first atmosphere was probably composed chiefly of gases such as helium and hydrogen inherited from the solar nebula from which the solar system formed, as well as from the asteroidlike bodies that collided to form Earth. As Earth heated up from core formation, it released gases and formed a secondary atmosphere composed chiefly of CO_2, methane (CH_4), nitrogen (N_2), and water (H_2O). In support of this view, the surviving rock record includes carbonates that reflect a carbon dioxide-rich atmosphere; and also one or more greenhouse gases (carbon dioxide, methane) must have been present to prevent the surface of the Earth from freezing over.

There are three lines of evidence for life in the Archean: (1) fossil stromatolites, which are laminated structures deposited by microorganisms; (2) fossils of cells or cellular tissue; and (3) carbonaceous matter identifiable from its carbon isotopic composition as a product of biologic activity. Some of the oldest fossil stromatolites occur in the 3.5 Ga Barberton greenstone in southern Africa and in the 3.5 Ga Pilbara greenstone in western Australia. [K.C.C.]

Archeological chemistry The application of chemical techniques to the study of archeological finds, natural or anthropogenic, in order to ascertain their composition or age. Traditional chemical analysis uses wet methods, in which a sample is brought into solution and its components are assayed by precipitation or titration. These methods were applied to ancient coins as early as the late eighteenth century.

The obvious need to minimize damage to an irreplaceable object spurred the development of microchemical techniques. Modern analysis relies on instrumental methods that require only very small samples or are entirely nondestructive. Although these methods rely on physical phenomena rather than chemical transformation, all procedures that are capable of the qualitative and quantitative determination of the atomic or molecular composition of the object under study are usually included under the broad heading of archeological chemistry.

Various analytical methods are utilized in archeological chemistry, including optical emission spectography, atomic absorption spectroscopy, inductively coupled plasma, neutron activation analysis, x-ray fluorescence spectrometry, electron microprobe analysis, proton-induced x-ray emission, Auger electron spectroscopy, and x-ray photoelectron spectroscopy. It should be noted that these methods of analysis are not competing but complementary. The choice of method depends on the nature of the object, on the elements to be determined, and on the accuracy required.

Organic materials constitute only a small portion of archeological finds, but since they include such basic necessities as food, drink, and clothing, they have the potential of revealing much about past life. Because they consist of covalently bound, complex, and sensitive molecules, their study requires special methods of analysis. Organic archeometry is the newest and most rapidly expanding field of archeological chemistry. Organic dyes have long been determined qualitatively and quantitatively by absorption spectroscopy in the visible and ultraviolet ranges. The extension into the infrared range allows not only the identification of organic materials by visual or computer-aided comparison of infrared spectra ("fingerprinting") but also some structural interpretation. Since organic residues typically consist of mixtures of dozens or even hundreds of individual compounds, the progress of organic archeometry has crucially depended on the development of chromatographic separation procedures. These include column chromatography, paper and thin-layer chromatography, gas chromatography, with or without prior pyrolysis, and liquid chromatography. All of these techniques not only separate mixtures into individual components but permit their identification if the rate at which they travel through the chromatographic substrate, the retention time, can be matched to those of authentic reference compounds.

Another method that is gaining use in organic archeometry is nuclear magnetic resonance spectrometry (NMR), which detects a limited number of atomic nuclei, among them ordinary hydrogen, carbon-13, nitrogen-15, fluorine-19, and phosphorus-31, by their simultaneous interaction with an external magnetic field and a radio-frequency field.

The determination of the chemical composition of an archeological find is not an end in itself, but provides the archeologist with factual evidence not otherwise obtainable and touching on many aspects of early human life. The changing elemental composition of coins detects progressive debasement and reveals economic history and fiscal policy. The metals added to copper to make bronze and brass outline the history and spread of technology. The foodstuffs consumed are indicators of the advent and progress of agriculture and animal husbandry. Together, all these paint a picture of prehistoric social, cultural, and economic stratification. The composition of an object also offers clues to its geographic origin, which may be far from the excavation site. This provides evidence of trade and exchange in commodities and raw materials.

While the most widely used methods for dating archeological material—radioactive decay, thermoluminescence, and archeomagnetism—deal with physical processes, three depend on the progress of conventional chemical reactions. (1) Amino acid dating uses the rate of racemization of optically active organic molecules. (2) Hydration dating measures the thickness of the weathering layer produced by the action of water on natural and artificial glass, including obsidian and flint. (3) NFU dating of bone relies on the loss of nitrogen (N) from the organic collagen component and on the uptake of fluorine (F) and uranium (U) by the inorganic hydroxyapatite component. Like all nonnuclear chemical reactions, these changes are a function not only of time but also of temperature, of acidity and, in the case of fluorine and uranium uptake, of the concentrations of these elements in the surrounding soil. Chemical methods cannot produce absolute dates unless these other variables are known or can be estimated reasonably closely. They are, however, useful in establishing relative ages of finds within a single site in which the depositional characteristics are likely to have been uniform. See AMINO ACID DATING; ARCHEOLOGICAL CHRONOLOGY; DATING METHODS; PALEOMAGNETISM; RADIOCARBON DATING.   [C.W.Be.]

Archeological chronology   The establishment of the temporal sequence of human cultures. Prior to the discovery of nuclear and chemical dating methods, which provide an absolute time scale, archeologists used stratigraphy, lithic and ceramic typology, seriation, index fossils, and a limited range of chemical techniques to establish relative chronologies for cultural remains. The major chemical techniques used in dating of bones in relative sequence have been labeled F-U-N (for fluorine-uranium-nitrogen). In a single site or environment, bones of the same age usually absorb the same amount of fluorine and uranium while losing the same amount of nitrogen.

Since the mid-twentieth century numerous absolute (sometimes called chronometric) dating techniques have been devised by natural scientists. Some of these techniques can give results in calendar years, whereas others yield dates which are expressed in years but which cannot always be correlated precisely with the calendar. The major methods for absolute dating of archeological materials used include radiocarbon dating, dendrochronology, thermoluminescence dating, hydration dating, racemization, potassium-argon dating, lead-210 dating, and archeomagnetism. Many other techniques, for example, fission track dating, have been used to establish archeological chronology, and a host of physical and, to a lesser degree, chemical dating methods have been investigated. See DATING METHODS; DENDROCHRONOLOGY; FISSION TRACK DATING; PALEOMAGNETISM; RADIOCARBON DATING; ROCK AGE DETERMINATION.   [G.R.]

Archeology   The scientific study of past material culture. The initial objective of archeology is to construct cultural chronologies, attempting to order past material culture into meaningful temporal segments. The intermediate objective is to breathe life into these chronologies by reconstructing past lifeways. The ultimate objective of contemporary archeology is to determine the cultural processes that underlie human behavior, both past and present.

The material culture of the past is of infinite variety. The scientific study of this evidence is such a broad task that there is no such thing as any single archeological method, although over the past century archeologists have evolved what can be termed an overall archeological approach. By constant confirmation, the archeologist often attempts to establish synchronism with what has already been established historically.

Archeologists use a number of types in order to categorize similar artifacts. Most common is the temporal type, a principle similar to the index fossil concept used by the geologist. A temporal type can be any kind of archeological artifact or feature, but ideally it is some object of common use in which the form is subject to change, due to either the whim of fashion or technological improvement. One example is the simple flint arrowhead with side barbs and central tang. It is typical of the British Bronze Age and was not in fashion earlier or later. Ceramic types have been established by archeologists working around the world, and a thoroughly tested ceramic chronology is invaluable as a temporal ordering device, no matter where the archeologist is working. The nature of the artifact employed as a temporal type is irrelevant, and its use may not even be known. Archeologists also establish other kinds of types. Functional types attempt to group artifacts on the basis of known or presumed functions. Technological types, divisions which reflect the mode of manufacture, are particularly helpful when studying stone tool manufacture.

The concept of culture is used in two different ways by contemporary archeologists. When dealing with cultural chronologies, the archeologist most commonly uses a modal or shared view of culture. It is this normative collection of shared ideas which causes artifacts to change in systematic ways through time, and temporal types can be established on the basis of this shared culture. When attempting to reconstruct lifeways, however, the archeologist can no longer rely on the shared aspects of culture. When transcending temporal associations, contemporary archeologists tend to view culture systematically, as people's extrasomatic (that is, learned) method of dealing with the social and cultural environment.

The principles of stratigraphy are applied to archeology in terms of the law of superposition, which states that, all else being equal, older deposits will tend to be buried beneath younger ones. Mere stratigraphic equivalence, however, does not necessarily indicate contemporaneity, as there can be misleading mixtures of successive occupational debris on one surface. Archeologists must therefore study the processes of cultural deposition in order to recognize the difference between intact and disturbed strata.

Contemporary excavation must be conducted with a plan, a firm research design that attempts to provide answers to definite questions. Archeology is one of the few sciences which destroys its own data in the process of generating them. Archeologists must therefore be extremely careful to make the appropriate observations at the time of excavation.

The task of deciphering meaning from past material culture is so complex that the archeologist is often required to borrow from allied disciplines in the physical and natural sciences, including geology, climatology, paleobotany, paleontology, mineralogy, physics, chemistry, and anthropology. The archeologist must have some understanding of all these sciences to extract from sites and materials every possible piece of information which may lead to a better understanding of prehistory. The archeologist must be able to record and publish every minor fact for the benefit of colleagues and successors, because the writing of prehistory requires the synthesis of all archeological discovery and interpretation. *See* ARCHEOLOGICAL CHEMISTRY.   [D.H.Th.]

Arctic and subarctic islands   Defined primarily by climatic rather than latitudinal criteria, arctic islands are those in the Norhern Hemisphere where the mean temperature of the warmest month does not exceed 50°F (10°C) and that of the coldest is not above 32°F (0°C). Subarctic islands are those in the Northern Hemisphere where the mean temperature of the warmest month is over 50°F (10°C) for less than 4 months and that of the coldest is less than 32°F (0°C). Such islands generally are in high latitudes. Distribution of land and sea masses, ocean currents, and atmospheric circulation greatly modifies the effect of latitude so that it is often misleading to use location relative to the Arctic Circle as a significant criterion of arctic or subarctic. The largest proportion by area of the islands lies in the Western Hemisphere, primarily in Greenland and in the Canadian Arctic Archipelago. Within this general description, individual islands vary considerably (see table).

Physiographically, the islands include all the varied major landforms found elsewhere in the world, from rugged mountains over 8000 ft (2500 m) high, through plateaus and hills, to level plains only recently emerged from the sea. All have been glaciated except

Size of larger arctic and subarctic Islands*

Name	Area mi²	Area km²
Aleutian Is.		
Unimak I.	15,500	40,100
Unalaska I.	10,800	28,000
St. Lawrence I.	18,200	47,100
Nunivak I.	16,000	41,400
Kodiak I.	37,400	96,900
Canadian Arctic		
Archipelago	500,000	1,295,000
Baffin I.	196,000	507,000
Ellesmere I.	76,000	197,000
Victoria	84,000	217,000
Banks	27,000	70,000
Devon	21,000	55,000
Axel Heiberg	17,000	43,000
Melville	16,000	42,000
Southhampton	16,000	42,000
Prince of Wales	13,000	33,033
Newfoundland	42,734	109,000
Greenland	840,000	2,176,000
Iceland	39,961	102,000
Svalbard (archipelago)	24,100	62,000
Vest-Spitsbergen	15,250	39,000
Franz Josef Land (archipelago)	7,000	18,000
Novaya Zemlya (archipelago)	36,000	93,000
Severny I.	21,000	54,000
Yughny I.	15,000	39,000
Severnaya Zemlya (archipelago)	14,000	36,000
New Siberian Is.	12,000	31,000
Wrangel I.	2,000	5,000
Sakhalin I.	27,000	70,000
Kuriloe Is.	6,000	16,000

*Approximate only in some cases because of incomplete mapping.

Sakhalin and some of the islands in the Bering Sea sector. Removal of the weight of ice sheets and the resultant crustal rebound has exposed prominent marine beaches and wave-cut cliffs on many of the islands. These now commonly occur at elevations of over 300 ft (150 m) above sea level.

The general climatic pattern of these islands is set by their location relative to the two semipermanent centers of low pressure over the Aleutian Islands and over Iceland. Most of the precipitation is cyclonic in origin. Because they are marine areas, the islands receive more precipitation than they otherwise would, yet even so this is very light for most of the arctic islands removed from the zone of cyclonic activity. Also, because they are marine areas, the islands, regions of low temperatures by definition, are not regions of extreme low temperatures. In general, the larger the island and the closer its proximity to a continental landmass, the higher are the summer temperatures and the lower its winter temperature. *See* POLAR METEOROLOGY.

The climatic differences between arctic and subarctic islands are reflected in their natural vegetation. The arctic islands are treeless. Natural vegetation consists of the tundra—mosses, sedges, lichens, grasses, and creeping shrubs. Bare ground is often exposed and in some places plant growth may be lacking completely except for a few rock-encrusting lichens. In such places the ground surface may consist of frost-shattered rock fragments, tidal mud flats, boulder-strewn fell fields, or snow patches and ice. Permafrost (permanently frozen ground) occurs throughout the Arctic (and in parts of the subarctic) and is reflected in impeded drainage and patterned ground. *See* PERMAFROST.

The natural vegetation of subarctic islands characteristically is the boreal forest or taiga, composed predominantly of conifers such as spruce, fir, pine, and larch with deciduous trees such as birch, aspen, and willow; the latter are especially common in regrowth of clearings in the forest. Impeded drainage because of permafrost or glaciation gives rise to numerous ponds and muskeg areas. A transitional type of vegetation, the forest-tundra, is recognized on some subarctic islands in sectors where smaller trees are widely spaced and abundant mosses cover the ground.

The typical soils of the subarctic islands are podzols—the grayish-white surface soil beneath the raw humus layer and highly acidic in nature. The tundra soils of the arctic islands really consist only of a dark-brown peaty surface layer over poorly defined thin horizons, and much of the ground cannot properly be termed soil.   [W.C.Wo.]

Arctic Circle   The parallel of latitude approximately $66^1/_2°$ (66.55°) north of the Equator, or $23^1/_2°$ from the North Pole. The Arctic Circle has the same angular distance from the Equator as the inclination of the Earth's axis from the plane of the ecliptic. Thus, when the Earth in its orbit is at the Northern Hemisphere summer solstice, June 21, and the North Pole is tilted $23^1/_2°$ toward the Sun, the Sun's rays extend beyond the pole $23^1/_2°$ to the Arctic Circle that parallel 24 h of sunlight. On this same date the Sun's rays at noon will just reach the horizon at the Antarctic Circle, $66^1/_2°$ south. The highest altitude of the noon Sun at the Arctic Circle is on June 21, when it is 47° above the horizon.

At the Arctic Circle the Sun remains above the horizon continuously only 24 h at the longest period. However, with twilight considered, it remains daylight or twilight continuously for about 5 months. Twilight can be considered to last until the Sun drops 18° below the horizon. *See* MATHEMATICAL GEOGRAPHY.   [V.H.E.]

Arctic Ocean   The north polar ocean lying between North Armerica and Asia, extending over about 386,000 mi^2 (10^6 km^2). It is nearly completely covered by 6–9 ft (2–3 m) of ice in winter, and in summer it becomes substantially open only at its peripheries. Its extent has been variably defined, but it is oceanogaphically appropriate to consider it bounded on the south by a line running from northern Greenland through Smith, Jones, and Lancaster sounds, along northwestern Baffin Island to the Canadian mainland, thence to the Alaskan coast, across Bering Strait, along the Siberian coast to Novaya Zemlya, across to Franz Josef Land and Spitsbergen, and over to northern Greenland. This definition omits the Barents, Norwegian, and Greenland seas and Baffin Bay, which have a pronounced North Atlantic character.

The central polar basin, somewhat triangular in shape, is surrounded by continental shelves which are interrupted only by the deep passage running through Fram Strait. The upper 650 ft (200 m) of the Arctic Ocean, referred to as Surface Water or Arctic Water, is characterized by a significant density stratification produced by the strong increase in salinity downward from the surface. This density stratification is of considerable importance, for it prevents a deep-reaching convection from developing within the Arctic Ocean and also prevents the heat of the underlying warm Atlantic Water from reaching the surface. The relatively low salinity at the surface is maintained against the upward diffusion of salt by the addition of fresh water, principally through river outflow. The upper 100–160 ft (30–50 m) of Surface Water tends to be relatively uniform vertically in temperature and salinity. Except for areas which become ice-free in summer, the water will be near the freezing point. Currents in the upper waters tend to be relatively slow (4 in./s or 10 cm/s or less), and they are similar in both speed and direction to the ice motion. The overall circulation in the upper waters has its ultimate cause in the prevailing wind pattern over the Arctic Ocean.

As in other oceans, the current at any instant can vary greatly from the mean condition. The most spectacular example observed in the Arctic Ocean occurs on an occasional basis in the Canadian Basin, consisting of a high-speed current core. See OCEAN CIRCULATION; SEAWATER.

Below the Surface Water, the temperature increases to a maximum, which over most of the region is about 33°F (0.5°C) and lies between 1000 and 1500 ft (300 and 500 m). The salinity is nearly uniform, and since at low temperatures the density of seawater depends almost solely on salinity, there is virtually no density stratification beneath the upper waters. Significant deviations from the stated temperature occur only in the southern Eurasian Basin closest to Spitsbergen, for it is there that the warm and saline water (called Atlantic Water) which maintains the temperature maximum throughout the Arctic Ocean first enters. This water has its origin in the North Atlantic. Once into the Arctic Ocean it sinks because of its high salinity and moves eastward along the Eurasian continental slope. Beneath the Atlantic Water lies cold, nearly uniform Bottom Water. These two water masses together constitute over 90% of the volume of the Arctic Ocean. The Bottom Water is formed in the Greenland Sea.   [K.A.]

Arenaceous rocks   The arenaceous rocks (arenites) include all those classic rocks whose particle sizes range from 2 to 0.0625 mm, or if silt is included, to 0.0039 mm. Some arenites are composed primarily of carbonate particles, in which case they are called calcarenites and grouped with the limestones. Some oolitic iron ores and glauconite beds are properly classified as arenites. But the vast majority of arenites are commonly called sandstones, and the two words are almost synonymous. See GRAYWACKE; OOLITE; SANDSTONE; SEDIMENTARY ROCKS.   [R.Si.]

Argillaceous rocks   The argillaceous rocks (lutites) include shales, argillites, siltstones, and mudstones; they are clastic sediments whose constituent particles are less than 0.0625 mm (if siltstones are included) or less than 0.0039 mm (if siltstones are excluded). They are the most abundant sedimentary rock type, varying according to different estimates from 44 to 56% of the total sedimentary rock column. Claystone is indurated clay, which consists dominantly of fine material of which at least a major proportion is clay mineral (hydrous aluminum silicates). Shale is a laminated or fissile claystone or siltstone, in general more consolidated than claystone. Mudstone is a claystone that is blocky and massive. The term argillite is used for rocks which are more indurated than claystone or shale but not metamorphosed to slate. All these argillaceous rocks are consolidated equivalents of muds, oozes, silts, and clays. Loess is a finegrained, unconsolidated, wind-blown deposit. The term shale has been used by many authors generically to denote all of these types of rock. See BENTONITE; CLAY; CLAY MINERALS; LOESS; SEDIMENTARY ROCKS; SHALE.   [R.Si.]

Arkose   An arenaceous rock that contains a high proportion of feldspar in addition to quartz and other detrital minerals. Arkose is also known as feldspathic sandstone. Although there is no universal agreement, many geologists consider a minimum of 25% feldspar a requisite for calling sandstone an arkose. Other geologists accept a lower value. Arkoses may contain a high proportion of other nonquartz detritus, such as igneous and metamorphic rock fragments, micas, amphiboles, and pyroxenes. Frequently the accessory heavy mineral suite consists of a variety of species.

Sedimentary structures of arkoses are similar in kind to those of the orthoquartzites. Cross-bedding, the major feature, may be displayed on a huge scale, some cross-bedded units being many feet thick. Arkoses are associated with a variety of clastic rocks, dominantly conglomerates, and reddish-colored shales. Arkoses also are found with basic lava flows. Most arkoses are found in geosynclinal areas, but the thin, reworked, granite wash arkoses can be found on stable continental platforms. *See* GEOSYNCLINE.

The granite-wash arkoses appear to have formed as the result of a transgression of the sea over a land area underlain by granite. The fragmented granite in the soil and mantle rock is incorporated in the basal sediment. In some areas the original granite is changed so slightly that the arkose is called recomposed granite and may be almost indistinguishable from the original granite. Since high relief and climatic extremes generally are associated with orogenic movements, arkoses are usually interpreted as sediments that result from tectonically active regions. *See* ARENACEOUS ROCKS; FELDSPAR; GRAYWACKE; SANDSTONE; SEDIMENTARY ROCKS.   [R.Si.]

Arsenopyrite   A mineral having composition FeAsS and crystallizing in the monoclinic system. Crystals have pseudo-orthorhombic symmetry because of twinning. The Mohs hardness is 5.5–6.0, and the specific gravity is 6.0. The luster is metallic and the color silver-white. Arsenopyrite is the most widespread arsenic-bearing mineral. It is commonly found in veins containing gold (Lead, South Dakota; Deloro, Ontario), tin or tungsten minerals (Bolivia; Cornwall, England), or nickel-cobalt-silver minerals (Cobalt, Ontario; Freiberg, Germany).   [L.Gr.]

Artesian systems   Groundwater conditions formed by water-bearing rocks (aquifers) in which the water is confined above and below by impermeable beds. Because the water table in the intake area of an artesian system is higher than the top of the aquifer in its artesian portion, the water is under sufficient head to cause it to rise in a well above the top of the aquifer. Many of the systems have sufficient head to cause the water to overflow at the surface, at least where the land surface is relatively low. Flowing artesian wells were extremely important during the early days of the development of groundwater from drilled wells, because there was no need for pumping. Their importance has diminished with the decline of head that has occurred in many artesian systems and with the development of efficient pumps and cheap power with which to operate the pumps.   [A.N.S./R.K.Li.]

Asbestos   Any of six naturally occurring minerals characterized by being extremely fibrous (asbestiform), being incombustible, and having high tensile strength. Historically they were utilized in commerce for fire protection; for fiber-reinforcing material in tiles, plastics, and cements; for friction materials; and for thousands of other uses. Currently the vast majority of asbestos used worldwide is chrysotile type, which is used for asbestos cement, friction products, coating and compounds, and roofing products. Because of great concern over the health effects of asbestos, many countries have promulgated strict regulations or bans on its use.

The six naturally occurring minerals exploited commercially for their desirable physical properties, which are in part derived from their asbestiform habit, are chrysotile asbestos—a member of the serpentine mineral group; and anthophyllite asbestos, grunerite asbestos (known historically by the commercial name amosite), riebeckite asbestos (known historically by the commercial name crocidolite), tremolite asbestos, and actinolite asbestos—all members of the amphibole mineral group. Populations of these mineral fibers, however processed, can be demonstrated to be asbestos if the length varies independently of the diameter. The six minerals designated as asbestos also occur in a nonfibrous form.

The three principal diseases associated with exposure to the asbestos minerals are lung cancer; mesothelioma, a rare cancer of the pleural and peritoneal membranes that enclose the chest and abdominal cavities; and asbestosis, a nonmalignant disease characterized by a diffuse interstitial fibrosis of the lung, which causes the lung tissue to become stiff and exchange oxygen poorly. Excessive exposure to all the asbestos fiber types is associated with asbestosis and increased risk of lung cancer. Mesothelioma, a rare tumor accounting for approximately 1 in 10,000 deaths in the general population, can be dramatically increased by exposure to amosite, crocidolite, or tremolite asbestos. These last two fiber types are strongly associated with an increased incidence of nonoccupational mesothelioma and therefore are thought to present a risk at rather low exposures.   [M.Ro.; R.P.N.]

Asia   The largest of the world's continents. With its peninsular extension, commonly called the continent of Europe, it is the major portion of the broad east-west extent of the Northern Hemisphere land masses. In many ways Asia is more a cultural concept than a physical entity. There is no logical physical separation between Asia and Europe, and even Africa is separated from Asia merely by the width of the Suez Canal. For convenience, however, the Eurasian land mass is considered to be divided by the Ural Mountains into Europe in the west and Asia in the east. Thus restricted, Asia has an area of about 17,700,000 mi^2 (45,800,000 km^2), about one-third of the land area of the Earth. In the north, Siberia reaches past the 80th latitude. Southward, India and Sri Lanka (Ceylon) reach nearer than 10°N of the Equator, while the Indonesian islands extend more than 10°S of the Equator. The continental heart of Asia is more than 2000 mi (3200 km) from the nearest ocean. *See* CONTINENT; EUROPE.

Topography.   In the topographic framework of Asia, the great mountain systems are the most impressive features. From the central knot of the mighty Pamirs and Kopet Dagh in the heart of the continent originate chains radiating in several directions. In the Peter the First Range there are such heights as Qullai Ismoili Somoni, 24,584 ft (7493 m), and Lenin Peak, 23,377 ft (7125 m), above sea level. Running westward through Afghanistan is the Hindu Kush, reaching elevations over 20,000 ft (6100 m). The mountain trendline continues, after a jog northwestward, in the Elburz of northern Iran and thence in the Armenian highlands and the Caucasus, each with elevations reaching 18,000 ft (5500 m), decreasing thereafter to the Pontus and Taurus ranges of northern and southern Turkey. In western and southern Iran are the massive Zagros and Makran ranges.

Southeastward from the Pamir knot run the three most imposing mountain chains on Earth: the Karakorum, which continues the line of the Hindu Kush eastward in an arc convex to the north; the Himalaya in an arc convex to the south; and the shorter Trans-Himalaya, or Nyen-chen Tangla, north of the Himalaya, with higher average elevations but peaks of lesser height. In all of these, the average elevations exceed 4 mi (6400 m), with several scores of peaks reaching a height in excess of 25,000 ft (7600 m) above sea level. Everest, 29,141 ft (8882 m), and Kinchinjunga, 28,146 ft (8579 m), lie in the Himalaya, while the peak designated as K2, 28,250 ft (8611 m), rises in the Karakorum.

In eastern Tibet the Himalaya and Nyen-chen Tangla bend sharply toward the south, and the former is cut through by the gorge of the Brahmaputra River. From the bend zone, great ridges divided by deep gorges run south to form the Burma-China frontiers and the mountain backbones of the Malay peninsula and Vietnam. The Nan-ling system of south China diverges eastward to divide the Yang-tzu (Yangtze) from the Hsi (Si) drainage.

From the western Himalaya, the 11,000-ft (3400-m) Sulaiman Range runs south and, together with the Kirthar Range, divides West Pakistan from Afghanistan.

Beginning at heights over 20,000 ft (6100 m) and branching off from the Karakorum south of Kashgar, the Kuen-lun Mountains run eastward across western China. Genetically they form the longest mountain system of China. With their eastward extensions in the 12,000 ft (3700 m) Ch'in-ling and the lesser Ta-pieh mountains and Huai-yang hills, they reach almost to the Pacific. Together with the northeastward arc of the Altyn Tagh and the Nan Shan branching from it, the Kuen-lun forms the northern wall of the Tibetan plateau. Near the eastern end of the Kuen-lun proper lie the Amne Machin Mountains, with peaks up to 25,000 ft (7600 m) in elevation.

Northeastward of the Pamir knot runs the east-west oriented Tien Shan, over 1000 mi (1600 km) long and maintaining heights of 18,000–20,000 ft (5500–6100 m) over much of its length. Roughly parallel and trending east and west is a series of great ranges to its north, with mutual connections in the west. These include the Altai-Sayan, the Tannu Ola, and the Kentei, which form natural boundaries for Outer Mongolia. They continue the systems of young mountains crossing central Asia; farther northeast, they extend further in the Stanovoi Mountains of Eastern Siberia.

The Asian plateaus are in various stages of erosion and thus present a great variety of landscapes. The Tibetan plateau is a prime example. The western half, because of little rainfall, exhibits a rolling topography with relatively slight local relief except where mountain chains cross it; it is a land of internal drainage basins. Average elevations are over 16,000 ft (4900 m). The eastern half is humid or subhumid and is cut by numerous rivers, producing deep canyons and great ridges. In contrast to this is the Mongolian plateau. This plateau consists mostly of vast, rather level plains 3000–5000 ft (900–1500 m) high, surmounted in places by mountains, and containing broad, shallow basins divided by land swells of low elevation.

Other major topographic units of Asia are blocs of hill lands. Most of southern China and much of southeastern Asia comprise hills which may be roughly defined as slope lands with local relief under 1000–1500 ft (300–450 m) although in absolute elevation they may rise many thousands of feet above sea level. Hilly lands are found to predominate in the northern part of the Indian peninsula and along both flanks of the Indian plateau, where they are called ghats. In southern India are the Nilgiri and Cardomom hills, rising to mountainous elevations of 8000 ft (2400 m). Many parts of different plateaus have hilly regions where erosion has produced uneven local relief, as in the Shan or North Vietnam plateau. Hills are prominent features of southwestern Asia, including eastern Mediterranean regions, such as Israel, Syria, and Lebanon.

The most significant topographic units of Asia are the great alluvial plains and river deltas. The gross drainage pattern of Asia is radial; the rivers flow from the highlands in the heart of the continent and run outward in all directions. Only in the south, east, and north sectors of the continent do the rivers reach the sea. Flowing into the peripheral seas of the Pacific are such mighty rivers as the Mekong, the Hsi, the Yang-tzu, the Huai, the Yellow, and the Amur, each building large, heavily populated plains and, with the exception of the Amur, densely settled deltas. The Yellow Plain (North China Plain), with some 125,000 mi^2 (324,000 km^2) of area, and the Yangtzu Plain, with about 75,000 mi^2 (194,000 km^2), are among the most extensive alluvial plains of the Earth. In the shallow South China, East China, and Yellow seas, the deltas of the first five rivers mentioned above are pushing steadily seaward.

Important sectors of Asia, containing some 200,000,000 people, are completely insular. The most important are the Japanese, Philippine, and Indonesian islands and

Taiwan. Almost all of Asia's islands lie in great volcanic arcs bounding large seas off the continent's Pacific coast. At least 160 active volcanoes are found here and in Kamchatka. Few islands lie along the Asiatic coasts of the Indian Ocean, although the Sunda chain of Indonesia has perhaps more of a claim to Indian Ocean frontage than to Pacific frontage. Sri Lanka is the only significant island in the northern part of the Indian Ocean west of Sumatra. In the Persian Gulf off the north coast of Arabia lies the small island Bahrein.

Few islands lie off the alluviated coastlands of northern Siberia. Some moderately large ones are included in the barren and rocky Severnaya Zemlya group, the New Siberian Islands, and Wrangel Island. The Commander Islands and Karaginski Island lie in the Bering Sea only a short distance from the Aleutians.

Climates. Five major climatic types may be distinguished in the Asian region: (1) the monsoonal system of eastern Asia, (2) the monsoonal system of southern Asia, (3) the equatorial regions of southeastern Asia and their extension into the Southern Hemisphere as they are influenced by the Australian monsoon, (4) the winter rainfall areas of southwestern Asia, and (5) the cyclonic and convectional storm systems of central and northern Asia.

Fundamental to understanding the climates of Asia are the vastness of the unbroken landmass and the long latitudinal stretch from the polar realm to south of the Equator. These are responsible for the great temperature and humidity extremes that occur. The greatest ranges of temperatures in the world have been recorded in interior Asia. Continentality, therefore, is the outstanding feature of climates of interior Asia. In coastal and insular areas of east Asia, however, winds moving over the warm, northward-flowing Japan Current and the western Pacific waters moderate the coastland and island climates. *See* MONSOON METEOROLOGY.

The driest portions of Asia include the vast areas of southern Mongolia, Hsin-chiang, former Soviet Central Asia, and southwestern Asia. Except for small, favored mountain areas, most of this region from the Gobi to the Red Sea gets less than 10 in. (25 cm) of precipitation per year. With the exception of southern Arabia, which is subtropical desert, these are mid-latitude desert and dry steppe regions. Favored with higher rainfall are the Yemen Mountains and the coastal mountains of Turkey, together with Lebanon, Syria, and northern Israel. The highlands of Armenia and the Elburz of Iran are favored also with more abundant rainfall, which may range from 25 to 50 in. (64 to 127 cm) or more per year.

The northeastern Siberian mountains and the Arctic coastal lands also receive meager rainfall, less than 8 in. (20 cm), but are not dry because evaporation is low and the water table is high. Most of Siberia has permafrost below a few feet of surface soil, so that rainwater does not filter far down into the earth. Between the arid belt of central Asia and the northeast Siberian low-precipitation zone, the annual rainfall ranges between 10 and 18 in. (25 and 45 cm).

In eastern Asia the precipitation increases in a southeasterly direction from interior Asia to the coast. The annual maximum seldom exceeds 80 in. (203 cm) in the wetter southeast coastal regions, whereas this drops to less than 30 in. (76 cm) in the North China Plain and less than 15 in. (38 cm) at the Great Wall. In some mountainous parts of Japan and Taiwan, the yearly average may be more than 100 in. (254 cm).

In the Indian subcontinent rainfall is heaviest along the western plateau fringe and in East Bengal, where it may average over 100 in. (254 cm) per year. The interior of the peninsula is relatively dry. Northwestern India and Pakistan share the drought of southwestern Asia. With the exception of the extreme north, Ceylon generally has abundant rainfall.

Southeastern Asia has the heaviest rainfall of the entire Asiatic region. The mainland mountains facing the southwest summer monsoon crossing the Bay of Bengal, and parts of the Vietnamese and Laotian cordilleras facing the humidified northeast winter monsoons of eastern Asia, regularly get average rainfalls of 120–150 in. (305–381 cm) or even more. Equally heavy rainfalls occur in the southwestern half of Sumatra, southwestern Java, the northwestern half of Borneo, and the Pacific fringe of the Philippine Islands. With a few small exceptions, southeastern Asia has no areas that are subject to severe drought.

Vegetation. Asia's vegetation belts and zones follow, in general, the climatic patterns from desert lands through tropical to Arctic margins.

A wide belt of tundra made irregular by topography occupies the entire Arctic lowland of Siberia with widths varying from 250 to 500 mi (400 to 800 km) north and south. It is widest in the extreme northeast and it extends southward and inland with higher elevations. The frozen subsoil permits the growth of little more than mosses, lichens, dwarfed trees, and scrub. *See* PERMAFROST.

The largest unbroken expanse of forest in the world is the Siberian taiga, a dominantly coniferous forest of larches, spruce, fir, and pines, with such deciduous trees as birch and aspen occurring intermixed with the conifers or taking over as a secondary growth in burnt-over areas. The width of this belt in Siberia is more than 1000 mi (1600 km) and it stretches about 4000 mi (6400 km) from the Sea of Okhotsk to the Urals.

Various admixtures of coniferous and deciduous trees compose the vegetation of mid-latitude mixed forests. In the west Siberian plain there is a narrow zone of mixed taiga and deciduous forests including oaks, maples, ash, and lindens. This zone, with a width of 50–100 mi (80–160 km), lies somewhat south of the parallel of 60°N and fades into the steppelands that form the great spring-wheat region of Siberia. Mixed midlatitude deciduous and coniferous forest areas of a similar type occupy most of Korea, the northern half of Honshu in Japan, and the hill lands surrounding the Yellow Plain, as well as the Ch'in-ling Mountains. In southern Asia these forests are found chiefly in a narrow belt of mountain land in the outer ranges of the Himalaya. The remaining areas of these mixed forests run from the Elburz Mountains through the Armenian highlands and the Black Sea fringe of Turkey to the Aegean coast, and in southwestern Asia in the Elburz of northern Iran.

From the mixed and deciduous forests of the west Siberian plain southward, an increasingly dry steppeland is encountered. It extends for 400–500 mi (640–800 km) in a belt about 1000 mi (1600 km) long between the Urals and the Altai-Sayan and associated uplands. The northern half of this belt with its higher annual precipitation of 12–16 in. (30–40 cm) is the agricultural heart of the plain. The southern part gradually changes to desert steppe and then to desert along about the 50th parallel. Eastward of Lake Baikal a broadened steppe zone occupies the Trans-Baikal region extending southward to the Gobi Desert of southern Mongolia and eastward to the Great Hsing-an Mountains, where the zone, about 200 mi (320 km) wide, runs southward in Inner Mongolia. The steppe zone in Inner Mongolia widens with the increasing moisture south of the Great Wall to include most of China's loess plateau. Grasses also form the natural vegetation of the Manchurian plain, with tall grass in the eastern portion thinning out to short-grass steppe in the Hsing-an Mountain flanks. The Gobi Desert is flanked by steppelands to its north, east, and south, as well as by mountain steppe zones in the eastern Altai and eastern T'ien Shan.

Mixed evergreen forests appear to be limited mostly to interior southern China and to Japan from the Kwanto Plain southward. South of the Yang-tzu Valley, this forest type extends from the coast at Shanghai to the gorge lands of eastern Tibet. In Asia the characteristic trees of the mixed forest include broad-leafed evergreen trees such as banyans and camphor, and coniferous trees such as pines, cedars, and cypresses, as well as varieties of bamboo.

Tropical and subtropical rainforest is restricted to warm or hot regions of southern and southeastern Asia which get ample rainfall the year round or get so much rain during a large part of the year that a high groundwater table is maintained during the short dry season. The subtropical sectors are found along the southeastern China coast, in Taiwan, and in northern Burma; they merge with the tropical rainforest farther south, where rainfall and temperature increase.

Monsoon tropical deciduous forests comprise the tropical parts of Asia which have a moderately high rainfall but a long dry season (usually in the low-sun period or winter). These forests consist mostly of mixed species, but sometimes a single species becomes dominant as a result of selection from frequent burnings.

A large region of savanna grassland surrounds the Thar Desert of northwestern India and occupies most of the Indus Valley, the Punjab, and the Kathiawar peninsula. Much of the drier interior peninsular Deccan of India also has this as a natural vegetation. Other Asian regions with similar cover are found in Yemen and the region in southeastern Arabia from Oman as far westward as the Qatar peninsula; and similar vegetation extends over the Korat plateau of Thailand, lower Thailand west of Bangkok, southern Cambodia, and small areas in interior Borneo and the Philippines.

Immense areas of central and southwestern Asia have little or no vegetative cover, and bare rock alternates with sand veneering. In places shifting sand dunes are formed. Although the deserts are not necessarily lifeless, the vegetation is so widely spaced that much bare ground is exposed. The tropical desert areas generally receive their meager rainfall in torrential downpours on rare occasions. After such rains numerous herbs may spring to life and flower, while the bunch grass here and there may become green for a short season.   [H.J.Wi.]

Asphalt and asphaltite   Varieties of naturally occurring bitumen. Asphalt is also produced as a petroleum by-product. Both substances are black and largely soluble in carbon disulfide. Asphalts are of variable consistency, ranging from a highly viscous fluid to a solid, whereas asphaltites are all solid. Asphalts fuse readily, but asphaltites fuse only with difficulty. Asphalts may, moreover, occur with or without appreciable percentages of mineral matter, but asphaltites usually have little or no associated mineral matter. *See* BITUMEN; IMPSONITE; WURTZILITE.

Many asphalts occur as viscous impregnations in sandstones, siltstones, and limestones. Most such deposits are thought to be petroleum reservoirs from which volatile constituents have been stripped by exposure of the rock. Relatively pure asphalt occurs in Kern, San Luis Obispo, and Santa Barbara counties, California. Occurrences of asphalt are also known in Kentucky and Oklahoma. Although asphalt seeps have long been known in France, Greece, Russia, Cuba, and other countries, the best known and largest are those of Venezuela and Trinidad.

The asphaltites (gilsonite, grahamite, and glance pitch) were probably derived from a saline lacustrine sapropel and owe their variable properties to differences in environment of deposition. These substances occur on a large scale in the Uinta Basin of northeastern Utah, where they are derived from upper Eocene Green River sediments, most of which are oil shales high in carbonate content. *See* OIL SHALE; SAPROPEL.

                                                                                                   [I.A.B.]

Asphalt is derived from petroleum in commercial quantities by removal of volatile components. It is an inexpensive construction material used primarily as a cementing and waterproofing agent.

Asphalt is composed of hydrocarbons and heterocyclic compounds containing nitrogen, sulfur, and oxygen; its components vary in molecular weight from about 400 to 5000. It is thermoplastic and viscoelastic; at high temperatures or over long loading times it behaves as a viscous fluid, while at low temperatures or short loading times it behaves as an elastic body.

The three distinct types of asphalt made from petroleum residues are straight-run, air-blown, and cracked. Straight-run asphalt, characterized by a nearly viscous flow, is used in the construction of pavement surfaces for roads and airport runways. Air-blown asphalt is resilient and has a viscosity that is less susceptible to temperature change than that of straight-run asphalt. It is used mainly for roofing, pipe coating, paints, underbody coatings, and paper laminates. Cracked asphalt, with limited applications such as dust laying or as an insulation board saturant, has a nearly viscous flow, and its viscosity is more susceptible to temperature change than straight-run asphalt.   [T.K.M.]

Asthenosphere   A layer in the Earth's interior occurring approximately 50 mi (80 km) below the surface and extending to a depth of about 180 mi (300 km); it consists of rocks possessing less mechanical strength than the rocks above or below it. The asthenosphere is a relatively thin layer contained in a much larger region known as the mantle. The mantle is the solid portion of the Earth's interior that is located between the bottom of the Earth's crust (at about 15 mi or 25 km depth) and the top of the liquid outer core (at 1800 mi or 2900 km depth). The layers in the mantle that are above and below the asthenosphere are known as the lithosphere and the mesosphere respectively. The lithosphere is broken into 12 major tectonic plates that possess much greater mechanical strength than the underlying asthenosphere. See LITHOSPHERE.

The thermal structure of the asthenosphere, and indeed the very existence of this layer, is determined by the thermal convection process in the mantle. The convective flow in the mantle transports heat vertically upward from the deep interior, and it drives the observed horizontal motions of the tectonic plates. In the deep mantle, below the lithosphere, the vertical advection of heat by the convective flow is sufficiently rapid to create an adiabatic depth variation of mantle temperature. In the lithosphere the velocities of the vertical flows are much smaller than in the deep mantle; therefore the depth variation of temperature in this region is determined by a balance between the horizontal advection of heat (due to the horizontal flow associated with the tectonic plate motions) and the vertical conduction of heat to the surface. The asthenosphere is, in effect, a layer in which the depth variation of temperature changes from a steep gradient in the lithosphere to a relatively flat gradient in the deep mantle.

Since the mantle flow occurs over geological time scales, the long-term mechanical strength of the mantle rocks may be defined as the amount of stress that must be applied to produce some specified flow velocity. The flow of the solid mantle is made possible by the presence of naturally occurring microscopic defects in the crystal grains that constitute mantle rocks. The movement of these defects, due to thermally generated internal stresses, allows the mantle to creep as though it were a fluid with an extremely high viscosity. The effective viscosity of mantle rocks is a direct measure of their long-term mechanical strength, and it is strongly dependent on the ratio between the temperature (T) and the melting temperature (T_m) of the rocks. An increase of the scaled temperature T/T_m (also called the homologous temperature) produces exponentially large decreases in the effective viscosity of rocks. In the asthenosphere the average mantle temperature is closest to the melting temperature; thus the effective viscosity (that is, mechanical strength) is lower there than above or below the asthenosphere. There is a smooth transition between the zone of reduced mechanical strength in the asthenosphere and the zone of greater strength in the adjoining portions of the mantle. Therefore it is not possible, or meaningful, to specify precise locations for the upper and lower boundaries of the asthenosphere.

The analysis of seismic data (for example, the travel times of seismic waves) has provided the only direct indication of the presence of the asthenosphere. Seismologists usually refer to the asthenosphere as a low-velocity zone on account of the reduction of seismic wave speeds in this layer. *See* SEISMOLOGY.

Seismologists have made considerable progress in the application of tomographic imaging techniques to map the three-dimensional variation of seismic wave speed in the mantle. A tomographic model of the relative perturbations of seismic shear velocity has been constructed; at a depth of 120 mi (200 km), this model indicates that the shear-velocity perturbations range from -2.5 to $+4.5\%$. The coldest (that is, largest negative perturbation of) temperature is found below the continents. This local reduction of mantle temperature, and the corresponding increase of mechanical strength, may be sufficiently great that the concept of the asthenosphere (as a hotter and mechanically weak region) ceases to be valid below the continents. The concept of the asthenosphere is valid below the oceans, and there is an obvious concentration of hotter material below the plate boundaries, which are zones of active spreading (the so-called mid-oceanic ridges). This pattern suggests that the observed spreading at the mid-oceanic ridges is fed, and perhaps partially driven, by the upward ascent of hotter mantle material across the asthenosphere. When the ascent of this hotter material is sufficiently rapid (that is, adiabatic), the material begins to melt (and may thus produce surface eruptions of lava), because the temperature of this ascending material exceeds the local melting temperature. This partial melting can occur in the asthenosphere. *See* ISOSTASY; PLATE TECTONICS.   [A.M.Fo.]

Atlantic Ocean   The large body of sea water separating the continents of North and South America in the west from Europe and Africa in the east and extending south from the Arctic Ocean to the continent of Antarctica. The Atlantic is the second largest ocean water body and in area covers nearly one-fifth of the Earth's surface. The two major divisions, North and South Atlantic oceans, have the Equator as the common boundary. The North Atlantic, because of projecting land areas and island arcs, has numerous subdivisions. These include three large mediterranean-type seas, the Mediterranean Sea, the Gulf of Mexico plus Caribbean Sea, and the Arctic Ocean; two small mediterranean-type seas, the Baltic Sea and Hudson Bay; and four marginal seas, the North Sea, English Channel, Irish Sea, and Gulf of St. Lawrence. Parts of the Atlantic are given special names but lack precise boundaries, such as the Bahama Sea, Irminger Sea, Labrador Sea, and Sargasso Sea.

The mean depth of the Atlantic Ocean is 12,960 ft (3868 m), and its volume is 76,300,000 mi^3 (318,000,000 km^3). Broad shelves with depths less than 660 ft (200 m) are found in the region of the North Sea and the British Isles, on the Grand Banks of Newfoundland, and off the coasts of northeastern South America and Patagonia. The Mid-Atlantic Ridge, which extends from the Arctic Ocean to 55°S, is less than 9800 ft (3000 m) beneath the surface and is characterized by a pronounced relief. It separates the east and west Atlantic troughs, both of which have relatively uniform relief.

Three marked east-west ridges—the Greenland-Scotland Ridge in the North Atlantic and the Walvis and Rio Grande Ridges in the South Atlantic—and several less-conspicuous east-west rises separate the two Atlantic troughs into a series of basins including the West European, Canary, and Angola in the eastern Atlantic and the North American, Brazilian, and Argentine basins in the western Atlantic.

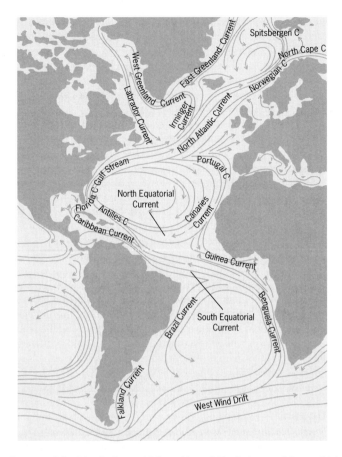

Currents of the Atlantic Ocean. (*Adapted from J. Bartholomew, Advanced Atlas of Modern Geography, McGraw-Hill, 3d ed., 1957*)

Islands in the Atlantic are mostly of volcanic origin. The Bermudas are the northernmost coral reefs, rising from an old submarine volcanic cone. Some islands, such as the British Isles, are continental in character. *See* OCEANIC ISLANDS; WEST INDIES.

The primary circulation of surface winds over the Atlantic Ocean is characterized by a zonal distribution pattern oriented in an east-west direction. The greatest storm frequency, more than 30% in winter, is in the zone of the prevailing westerlies. Air temperatures also follow a zonal pattern of distribution. They are lower in the South Atlantic than in the North Atlantic, and lower in the tropics and subtropics over the eastern Atlantic, than they are in the same latitudes over the western Atlantic. Maximum precipitation occurs in the doldrum zone (80 in. or 2000 mm/year). Precipitation also is relatively great in the zone of westerlies but is low in the trade-wind zones.

Sea ice is formed in the northernmost and southernmost parts of the Atlantic Ocean. From these areas drift ice moves equatorward into neighboring regions where it becomes a hazard to sea traffic and limits fishing. Many icebergs drift southward into the sea lanes of the North Atlantic. Most of these have their origin in the valley glaciers of western Greenland. Icebergs generally drift south of the Grand Banks, and some are known to have drifted southeast of Bermuda. In the South Atlantic large, tabular icebergs separate from the Antarctic ice shelf and drift northward. *See* ICEBERG; SEA ICE.

Surface currents in the Atlantic Ocean flow in much the same direction as the prevailing surface winds (see illustration). Deflections from these directions are caused by the bottom topography and the latitude or increased effect of Coriolis forces. The fairly constant flow of the North and South Equatorial currents is sustained largely by the trade winds. As a result, warm water is piled up along the poleward borders of these currents and on the western sides of the Atlantic Ocean. *See* OCEAN; OCEAN CIRCULATION.

The surface water in certain areas takes on a particularly high density in winter under the influence of climatic conditions. These water masses sink to a depth where the surrounding waters have a corresponding density and then spread out at that level. At the same time they are constantly mixing with the surrounding waters. In this way a multistoried stratification arises. Compared with that of the Indian and Pacific oceans, the deep circulation in the Atlantic Ocean is very vigorous, and the deeper water is therefore rich in oxygen. The abundance of nutrients permits a greater rate of organic production where the nutrient-rich waters nearly reach the surface, as in the Antarctic waters. *See* SEAWATER; SEAWATER FERTILITY.

The semidiurnal tidal form predominates in the Atlantic Ocean. The mean tidal range is about 3.3 ft (1 m) in the open ocean, but it decreases to 6.3 in. (16 cm) off Rio Grande do Sul in southern Brazil and to 3.5 in. (9 cm) off Puerto Rico. Tidal ranges increase beyond broad shelves under favorable physical conditions. The tides of the mediterranean and marginal seas are cooscillations of the tides of the Atlantic Ocean. *See* TIDE.

The Atlantic Ocean, especially the North Atlantic, is by far the most important bearer of the world's sea traffic. Favorable trend include increased transportation capacities for handling bulk goods, regular weather observations for the safety of air and sea traffic by weather ships in selected positions, and the observation and reporting of drifting icebergs by the International Ice Patrol. Communication facilities, including telegraph and telephone cables and radio stations, have been improved and increased in number.

[G.O.D.]

Atmosphere A gaseous layer that envelops the Earth and most other planets in the solar system. Earth, Venus, Mars, Jupiter, Saturn, Uranus, Neptune, and Titan (Saturn's largest satellite) are all known to possess substantial atmospheres that are held by the force of gravity. The structure and properties of the various atmospheres are determined by the interplay of physical and chemical processes. Structural features of Earth's atmosphere detailed below can often be identified in the atmospheres of other planetary bodies.

The composition of the Earth's atmosphere is primarily nitrogen (N_2), oxygen (O_2), and argon (Ar) [see table]. The concentration of water vapor (H_2O) is highly variable, especially near the surface, where volume fractions can vary from nearly 0% to as high as 4% in the tropics. There are many minor constituents or trace gases, such as neon (Ne), helium (He), krypton (Kr), and xenon (Xe), that are inert, and active species such as carbon dioxide (CO_2), methane (CH_4), hydrogen (H_2), nitrous oxide (NO), carbon monoxide (CO), ozone (O_3), and sulfur dioxide (SO_2), that play an important role in radiative and biological processes.

In addition to the gaseous component, the atmosphere suspends many solid and liquid particles. Aerosols are particulates usually less than 1 micrometer in diameter that are created by gas-to-particle reactions or are lifted from the surface by the wind. A portion of these aerosols can become centers of condensation or deposition in the

Composition of the atmosphere*

Molecule	Fraction volume near surface	Vertical distribution
Major constituents		
N_2	7.8084×10^{-1}	Mixed in homosphere; photochemical dissociation high in thermosphere
O_2	2.0946×10^{-1}	Mixed in homosphere; photochemically dissociated in thermosphere, with some dissociation in mesosphere and stratosphere
Ar	9.34×10^{-3}	Mixed in homosphere with diffusive separation increasing above
Important radiative constituents		
CO_2	3.5×10^{-4}	Mixed in homosphere; photochemical dissociation in thermosphere
H_2O	Highly variable	Forms clouds in troposphere; little in stratosphere; photochemical dissociation above mesosphere
O_3	Variable	Small amounts, 10^{-8}, in troposphere; important layer, 10^{-6} to 10^{-5}, in stratosphere; dissociated above
Other constituents		
Ne	1.82×10^{-5}	
He	5.24×10^{-6}	Mixed in homosphere with diffusive separation increasing above
Kr	1.14×10^{-6}	
CH_4	1.15×10^{-6}	Mixed in troposphere; dissociated in upper stratosphere and above
H_2	5×10^{-7}	Mixed in homosphere; product of H_2O photochemical reactions in lower thermosphere, and dissociated above
NO	$\sim 10^{-8}$	Photochemically produced in stratosphere and mesosphere

*Other gases, for example, CO, N_2O, NO_2, and many by-products of atmospheric pollution also exist in small amounts.

growth of water and ice clouds. Cloud droplets and ice crystals are made primarily of water with some trace amounts of particles and dissolved gases. Their diameters range from a few micrometers to about 100 μm. Water or ice particles larger than about 100 μm begin to fall because of gravity and may result in precipitation at the surface. *See* CLOUD PHYSICS; PRECIPITATION (METEOROLOGY).

One of the remarkable properties of the Earth's atmosphere is the large amount of free molecular oxygen in the presence of gases such as nitrogen, methane, water vapor, hydrogen, and others that are capable of being oxidized. The atmosphere is in a highly oxidizing state that is far from chemical equilibrium. This is in sharp contrast to the atmospheres of Venus and Mars, the planets closest to the Earth, which are composed almost entirely of the more oxidized state, carbon dioxide. The chemical disequilibrium on the Earth is maintained by a continuous source of reactive gases derived from biological processes. Life plays a vital role in maintaining the present atmospheric composition. *See* ATMOSPHERIC CHEMISTRY .

The total mass of the Earth's atmosphere is about 5.8×10^{15} tons (5.3×10^{15} metric tons). The vertical distribution of gaseous mass is maintained by a balance between the downward force of gravity and the upward pressure gradient force. The balance is known as the hydrostatic balance or the barometric law. Hence, the declining atmospheric pressure that is measured while ascending in the atmosphere is a result of gravity. The globally averaged pressure at mean sea level is 1013.25 millibars (101,325 pascals).

Below about 60 mi (100 km) in altitude, the atmosphere's composition of major constituents is very uniform. This region is known as the homosphere to distinguish it from the heterosphere above 60 mi (100 km), where the relative amounts of the major constituents change with height. In the homosphere there are sufficient atmospheric motions and a short enough molecular free path to maintain uniformity in composition. Above the boundary between the homosphere and the heterosphere, known as the homopause or turbopause, the mean free path of the individual molecules becomes long enough that gravity is able to partially separate the lighter molecules from the heavier ones. The mean free path is the average distance that a particle will travel before encountering a collision. Hence the average molecular weight of the heterosphere decreases with height as the lighter atoms dominate the composition.

The vertical structure of the atmosphere is in large part determined by the transfer properties of the solar and terrestrial radiation streams. The energy of the smallest unit of radiation, the photon, is directly proportional to its frequency. The type of interaction that occurs between photons and the atmosphere depends on the energy of the photons.

The most energetic of the photons are x-rays and extreme ultraviolet radiation of the eletromagnetic spectrum, which are capable of dissociating and ionizing the gaseous molecules. The less energetic near-ultraviolet photons are able to excite molecules and atoms into higher electronic levels. As a result, most of the ultraviolet and x-ray radiation is attenuated by the upper atmosphere. A cloudless atmosphere, however, is relatively transparent to visible light, where most of the solar energy resides. At the opposite end of the spectrum toward the lower frequencies of radiation is the infrared part, which is capable of inducing various vibrational and rotational motions in triatomic and polyatomic molecules.

In order to maintain an energy balance, the Earth must emit about the same amount of radiation as it absorbs from the Sun. The terrestrial radiation occurs in the infrared part of the spectrum and hence is strongly affected by water vapor, clouds, carbon dioxide, and ozone and other trace gases. The ability of these gases to absorb and emit in the infrared allows them to effectively trap some of the outgoing radiation that is emitted by the surface, creating the so-called greenhouse effect. *See* INSOLATION.

The atmospheric layer that extends from the surface to about 7 mi (11 km) is called the troposphere. The tropopause, which is the top of the troposphere, has an average altitude that varies from about 11 mi (18 km) near the Equator to about 5 mi (8 km) near the Poles. The actual tropopause height varies considerably on time scales from a few days to an entire year. The troposphere contains about 80% of the atmospheric mass and exhibits most of the day-to-day weather fluctuations that are observed from the ground. Temperatures generally decrease with increasing altitude at an average lapse rate of about $17°F/mi$ ($6°C/km$), although this rate varies considerably, depending on time and location. *See* TROPOPAUSE; TROPOSPHERE.

The stratosphere is the atmospheric layer that extends from the tropopause up to the stratopause at about 30 mi (50 km) above the surface. It is characterized by a nearly isothermal layer in the first 6 mi (10 km) overlaid by a layer in which the temperature increases with height to a maximum of about $32°F$ ($0°C$) at the stratopause. The reversal in the temperature lapse rate is a result of direct absorption of solar radiation, mainly by ozone and oxygen at the ultraviolet frequencies. *See* STRATOSPHERE.

The reversal of the temperature lapse rate makes the stratosphere vertically stable. This stability limits the amount of vertical mixing and results in molecular residence times of many months to years. Another consequence of a stable stratosphere is that it acts as a lid on the troposphere, confining the strong vertical overturning and hence most of the surface-based weather phenomena. *See* WEATHER.

The mesosphere is the atmospheric layer extending from the stratopause up to the mesopause at an altitude of about 53 mi (85 km). The mesosphere is characterized by temperatures decreasing with height at a rate of about 12°F/mi (4°C/km). Although the mesosphere has less vertical stability than the stratosphere, it is still more stable than the troposphere and does not experience rapid overturning. The coldest temperatures of the entire atmosphere are encountered at the mesopause, with values as low as −150°F (−100°C). The temperature lapse rate found in the mesosphere is a result of the gradual weakening with height of the direct absorption of solar radiation by ozone. The radiative infrared cooling to space by the carbon dioxide molecules is responsible for the low temperatures near the mesopause. *See* MESOSPHERE.

The thermosphere is found above the mesopause. The thermosphere is characterized by rising temperatures with height up to an altitude of about 190 mi (300 km) and then is nearly isothermal above that. Although there is no clear upper limit to the thermosphere, it is convenient to consider it extending several thousand kilometers. Embedded within the thermosphere is the ionosphere, comprising those atmospheric layers in which the ionized molecules and atoms are dominating the processes.

Molecular species dominate the lower thermosphere, while atomic species are dominant above 190 mi (300 km). The distribution of the constituents is controlled by diffusive equilibrium in which the concentration of each constituent decreases exponentially with height according to its molecular weight. Hence the concentration of the heavier constituents such as nitrogen, oxygen, and carbon dioxide will decrease with height faster than the lighter constituents such as helium and hydrogen. At an altitude of 560 mi (900 km) helium becomes the dominant constituent while hydrogen dominates above 1900 mi (3000 km).

The ionosphere can be defined operationally as that part of the atmosphere that is sufficiently ionized to affect the propagation of radio waves. In the ionosphere, the dominant negative ion is the electron, and the main positive ions include O^+, NO^+, and O_2^+. The ionosphere is classified into four subregions. The D region extends from 40 to 60 mi (60 to 90 km) and contains complex ionic chemistry; most of the ionization is caused by ultraviolet ionization of NO and by galactic cosmic rays. This region is responsible for the daytime absorption of radio waves, which prevents distant propagation of certain frequencies. The E region extends from 60 to 90 mi (90 to 150 km) and is caused primarily by the x-rays from the Sun. The F1 region from 90 to 125 mi (150 to 200 km) is caused by the extreme ultraviolet radiation from the Sun and disappears at night. Finally, the F2 region includes all the ionized particles above 125 mi (200 km), with the peak ion concentrations occurring near 190 mi (300 km). *See* IONOSPHERE.

The exosphere is the atmosphere above 300 mi (500 km) where the probability of interatomic collisions is so low that some of the atoms traveling upward with sufficient velocity can escape the Earth's gravitational field. The dominant escaping atom is hydrogen since it is the lightest constituent. Calculations of the thermal escape of hydrogen (also known as the Jeans escape) yield a value of about 3×10^8 atoms · $cm^{-2} \cdot s^{-1}$. This is a very small amount since at this rate less than 0.5% of the oceans would disappear over the current age of the Earth.

The magnetosphere is the region surrounding the Earth where the movement of ionized gases is dominated by the geomagnetic field. The lower boundary of the magnetosphere, which occurs at an altitude of nearly 75 mi (120 km), can be roughly defined as the height where there are enough neutral atoms that the ion-neutral particle collisions dominate the ion motion. The dynamics of the magnetosphere is dictated in part by its interaction with the plasma of ionized gases that blows away from the Sun, the solar wind. The solar wind interacts with the Earth's magnetic field and severely deforms it, producing a magnetosphere around the Earth. It extends about 40,000 mi (60,000 km) toward the Sun but extends beyond the orbit of the Moon away from the Sun. *See* MAGNETOSPHERE; VAN ALLEN RADIATION.   [G.B.L.]

Atmospheric acoustics   The science of sound in the atmosphere. The atmosphere has a structure that varies in both space and time, and these variations have significant effects on a propagating sound wave. In addition, when sound propagates close to the ground, the type of ground surface has a strong effect.

Atmospheric sound attenuation. As sound propagates in the atmosphere, several interacting mechanism attenuate and change the spectral or temporal characteristics of the sound received at a distance from the source. The attenuation means that sound propagating through the atmosphere decreases in level with increasing distance between source and receiver. The total attenuation, in decibels, can be approximated as the sum of three nominally independent terms, as given in the equation below,

$$A_{total} = A_{div} + A_{air} + A_{env}$$

where A_{div} is the attenuation due to geometrical divergence, A_{air} is the attenuation due to air absorption, and A_{env} is the attenuation due to all other effects and includes the effects of the ground, refraction by a nonhomogeneous atmosphere, and scattering effects due to turbulence.

Sound energy spreads out as it propagates away from its source due to geometrical divergence. At distances that are large compared with the effective size of the sound source, the sound level decreases at the rate of 6 dB for every doubling of distance. The phenomenon of geometrical divergence, and the corresponding decrease in sound level with increasing distance from the source, is the same for all acoustic frequencies. In contrast, the attenuation due to the other two terms in the equation depends on frequency and therefore changes the spectral characteristics of the sound.

Air absorption. Dissipation of acoustic energy in the atmosphere is caused by viscosity, thermal conduction, and molecular relaxation. The last arises because fluctuations in apparent molecular vibrational temperatures lag in phase the fluctuations in translational temperatures. The vibrational temperatures of significance are those characterizing the relative populations of oxygen (O_2) and nitrogen (N_2) molecules. Since collisions with water molecules are much more likely to induce vibrational state changes than are collisions with other oxygen and nitrogen molecules, the sound attenuation varies markedly with absolute humidity.

The total attenuation due to air absorption increases rapidly with frequency. For this reason, applications in atmospheric acoustics are restricted to sound frequencies below a few thousand hertz it the propagation distance exceeds a few hundred meters.

Effects of the ground. When the sound source and receiver are above a large flat ground surface in a homogeneous atmosphere, sound reaches the receiver via two paths. There is the direct path from source to receiver and the path reflected from the ground surface. Most naturally occurring ground surfaces are porous to some degree, and their acoustical property can be represented by an acoustic impedance. The acoustic impedance of the ground is in turn associated with a reflection coefficient that is typically less than unity. In simple terms, the sound field reflected from the ground surface suffers a reduction in amplitude and a phase change.

When the source and receiver are both relatively near the ground and are a large distance apart, the direct and reflected fields become nearly equal and cancel each other.

Refraction of sound. Straight ray paths are rarely achieved outdoors. In the atmosphere, both the wind and temperature vary with height above the ground. The

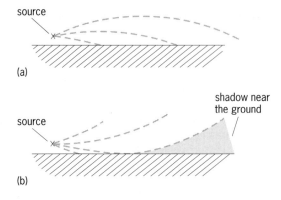

Curved ray paths. (*a*) Refraction downward, during temperature inversion or downwind propagation. (*b*) Refraction upward, during temperature lapse or upwind propagation.

velocity of sound relative to the ground is a function of wind velocity and temperature; hence it also varies with height, causing sound waves to propagate along curved paths.

The speed of the wind decreases with decreasing height above the ground because of drag on the moving air at the surface. Therefore, the speed of sound relative to the ground increases with height during downwind propagation, and ray paths curve downward. For propagation upwind, the sound speed decreases with height, and ray paths curve upward (see illustration). In the case of upward refraction, a shadow boundary forms near the ground beyond which no direct sound can penetrate. Some acoustic energy penetrates into a shadow zone via creeping waves that propagate along the ground and that continually shed diffracted rays into the shadow zones. The dominant feature of shadow-zone reception is the marked decrease in a sound's higher-frequency content. The presence of shadow zones explains why sound is generally less audible upwind of a source.

Refraction by temperature profiles is analogous. During the day, solar radiation heats the Earth's surface, resulting in warmer air near the ground. This condition is called a temperature lapse and is most pronounced on sunny days. A temperature lapse is the common daytime condition during most of the year, and also causes ray paths to curve upward. After sunset there is often radiation cooling of the ground, which produces cooler air near the surface. In summer under clear skies, such temperature inversions begin to form about 2 hours after sunset. Within the temperature inversion, the temperature increases with height, and ray paths curve downward.

The effects of refraction by temperature and wind are additive and produce rather complex sound speed profiles in the atmosphere.

Effects of turbulence. Turbulence in the atmosphere causes the effective sound speed to fluctuate from point to point, so a nominally smooth wave front develops ripples. One result is that the direction of a received ray may fluctuate with time in random manner. Consequently, the amplitude and phase of the sound at a distant point will fluctuate with time. The acoustical fluctuations are clearly audible in the noise from a large aircraft flying overhead. Turbulence in the atmosphere also scatters sound from its original direction.   [G.A.Da.]

Atmospheric chemistry   A scientific discipline concerned with the chemical composition of the Earth's atmosphere. Topics include the emission, transport, and deposition of atmospheric chemical species; the rates and mechanisms of chemical reactions taking place in the atmosphere; and the effects of atmospheric species on human health, the biosphere, and climate.

A useful quantity in atmospheric chemistry is the atmospheric lifetime, defined as the mean time that a molecule resides in the atmosphere before it is removed by chemical reaction or deposition. The atmospheric lifetime measures the time scale on which changes in the production or loss rates of a species may be expected to translate into changes in the species concentration. The atmospheric lifetime can also be compared to the time scales for atmospheric transport to infer the spatial variability of a species in the atmosphere; species with lifetimes longer than a decade tend to be uniformly mixed, while species with shorter lifetimes may have significant gradients reflecting the distributions of their sources and sinks.

The principal constituents of dry air are nitrogen (N_2; 78% by volume), oxygen (O_2; 21%), and argon (Ar; 1%). The atmospheric concentrations of N_2 and Ar are largely determined by the total amounts of N and Ar released from the Earth's interior since the origin of the Earth. The atmospheric concentration of O_2 is regulated by a slow atmosphere-lithosphere cycle involving principally the conversion of O_2 to carbon dioxide (CO_2) by oxidation of organic carbon in sedimentary rocks (weathering), and the photosynthetic conversion of CO_2 to O_2 by marine organisms which precipitate to the bottom of the ocean to form new sediment. This cycle leads to an atmospheric lifetime for O_2 of about 4 million years. *See* BIOSPHERE; LITHOSPHERE.

Water vapor concentrations in the atmosphere range from 3% by volume in wet tropical areas to a few parts per million by volume (ppmv) in the stratosphere. Water vapor, with a mean atmospheric lifetime of 10 days, is supplied to the troposphere by evaporation from the Earth's surface, and it is removed by precipitation. Because of this short lifetime, water vapor concentrations decrease rapidly with altitude, and little water vapor enters the stratosphere. Oxidation of methane represents a major source of water vapor in the stratosphere, comparable to the source contributed by transport from the troposphere.

The most abundant carbon species in the atmosphere is CO_2. It is produced by oxidation of organic carbon in the biosphere and in sediments. The atmospheric concentration of CO_2 is rising, and there is concern that this may cause significant warming of the Earth's surface because of the ability of CO_2 to absorb infrared radiation emitted by the Earth (the greenhouse effect). The total amount of carbon present in the atmosphere is small compared to that present in the other geochemical reservoirs, and therefore it is controlled by exchange with these reservoirs. Equilibration of carbon between the atmosphere, biosphere, soil, and surface ocean reservoirs takes place on a time scale of decades. *See* GREENHOUSE EFFECT.

Methane is the second most abundant carbon species in the atmosphere and an important greenhouse gas. It is emitted by anaerobic decay of biological carbon (for example, in wetlands, landfills, and stomachs of ruminants), by exploitation of natural gas and coal, and by combustion. It has a mean lifetime of 12 years against atmospheric oxidation by the hydroxyl (OH) radical, its principal sink.

Many hydrocarbons other than methane are emitted to the atmosphere from vegetation, soils, combustion, and industrial activities. The emission of isoprene [$H_2C{=}C(CH_3){-}CH{=}CH_2$] from deciduous vegetation is particularly significant. Non-methane hydrocarbons have generally short lifetimes against oxidation by OH (a few hours for isoprene), so that their atmospheric concentrations are low. They are most important in atmospheric chemistry as sinks for OH and as precursors of tropospheric ozone, organic nitrates, and organic aerosols.

Carbon monoxide (CO) is emitted to the atmosphere by combustion, and it is also produced within the atmosphere by oxidation of methane and other hydrocarbons. It is removed from the atmosphere by oxidation by OH, with a mean lifetime of

2 months. Carbon monoxide is the principal sink of OH and hence plays a major role in regulating the oxidizing power of the atmosphere.

Nitrous oxide (N_2O) is of environmental importance as a greenhouse gas and as the stratospheric precursor for the radicals NO and NO_2. The principal sources of N_2O to the atmosphere are microbial processes in soils and the oceans; the main sinks are photolysis and oxidation in the stratosphere, resulting in an atmospheric lifetime for N_2O of about 130 years.

About 90% of total atmospheric ozone (O_3) resides in the stratosphere, where it is produced by photolysis of O_2. The ultraviolet photons ($\lambda < 240$ nm) needed to photolyze O_2 are totally absorbed by ozone and O_2 as solar radiation travels through the stratosphere. As a result, ozone concentrations in the troposphere are much lower than in the stratosphere. *See* STRATOSPHERE; TROPOSPHERE.

Tropospheric ozone plays a central role in atmospheric chemistry by providing the primary source of the strong oxidant OH. It is also an important greenhouse gas. In surface air, ozone is of great concern because of its toxicity to humans and vegetation. Ozone is supplied to the troposphere by slow transport from the stratosphere, and it is also produced within the troposphere by a chain reaction involving oxidation of CO and hydrocarbons by OH in the presence of NO_x. Ozone production by this mechanism is particularly rapid in urban areas, where emissions of NO_x and of reactive hydrocarbons are high.

Sulfuric acid produced in the atmosphere by oxidation of sulfur dioxide (SO_2) is a major component of aerosols in the atmosphere and an important contributor to acid deposition. Sources of SO_2 to the atmosphere include emission from combustion, smelters, and volcanoes, and oxidation of oceanic dimethylsulfide [$(CH_3)_2S$] emitted by phytoplankton. It is estimated that about 75% of total sulfur emission to the atmosphere is anthropogenic. *See* AIR POLLUTION.   [D.J.J.]

Atmospheric electricity   The electrical processes constantly taking place in the lower atmosphere. This activity is of two kinds, the intense local electrification accompanying storms, and the much weaker fair-weather electrical activity over the entire globe, which is produced by the many electrified storms continuously in progress over the Earth. The mechanisms by which storms generate electric charge are unknown, and the role of atmospheric electricity in meteorology has not been determined.

Almost all precipitation-producing storms throughout the year are accompanied by energetic electrical activity. The most intense of these are the thunderstorms, in which the electrification attains values sufficient to produce lightning. Electrical measurements show that most other storms, even though they do not give lightning, are also quite strongly electrified. The electric fields of thunderstorms cause three currents to flow, each of a few amperes: lightning, point discharge from the ground beneath, and conduction in the surrounding air. Because the external field and conductivity are greatest over the top of the cloud, most of the conduction current flows to the ionosphere, the upper, highly conductive layer of the atmosphere. *See* THUNDERSTORM.

Fair-weather measurements, irrespective of place and time, show the invariable presence of a weak negative electric field caused by the estimated several thousand electrified storms continually in progress. Together these storms cause a 2000-A current from the earth to the ionosphere that raises the ionosphere to a positive potential of about 300,000 V with respect to the earth. This potential difference is sufficient to cause a return flow of positive charge to the earth by conduction through the intervening lower atmosphere equal and opposite to the thunderstorm supply current. The fair-weather field is simply the voltage drop produced by the flow of this current through the atmosphere. Because the electrical resistance of the atmosphere decreases with altitude, the field is greatest near the Earth's surface and gradually decreases with attitude until it vanishes at the ionosphere.

No importance is presently attached to fair-weather atmospheric electricity except that according to some theories it is responsible for the initiation of the thunderstorm electrification process. *See* CLOUD PHYSICS; LIGHTNING; SFERICS; STORM DETECTION; TOR-NADO.   [B.V.]

Atmospheric general circulation   The statistical description of atmospheric motions over the Earth, their role in transporting energy, and the transformations among different forms of energy. Through their influence on the pressure distributions that drive the winds, spatial variations of heating and cooling generate air circulations, but these are continually dissipated by friction. While large day-to-day and seasonal changes occur, the mean circulation during a given season tends to be much the same from year to year. Thus, in the long run and for the global atmosphere as a whole, the generation of motions nearly balances the dissipation. The same is true of the long-term balance between solar radiation absorbed and infrared radiation emitted by the Earth-atmosphere system, as evidenced by its relatively constant temperature. Both air and ocean currents, which are mainly driven by the winds, transport heat. Hence the atmospheric and oceanic general circulations form cooperative systems. *See* MARITIME METEOROLOGY.

Owing to the more direct incidence of solar radiation in low latitudes and to reflection from clouds, snow, and ice, which are more extensive at high latitudes, the solar radiation absorbed by the Earth-atmosphere system is about three times as great in the equatorial belt as at the poles, on the annual average. Infrared emission is, however, only about 20% greater at low than at high latitudes. Thus in low latitudes (between about 35°N and 35°S) the Earth-atmosphere system is, on the average, heated by radiation, and in higher latitudes cooled by radiation. The Earth's surface absorbs more radiative heat than it emits, whereas the reverse is true for the atmosphere. Therefore, heat must be transferred generally poleward and upward through processes other than radiation. At the Earth-atmosphere interface, this transfer occurs in the form of turbulent flux of sensible heat and through evapotranspiration (flux of latent heat). In the atmosphere the latent heat is released in connection with condensation of water vapor. *See* CLIMATOLOGY.

Considering the atmosphere alone, the heat gain by condensation and the heat transfer from the Earth's surface exceed the net radiative heat loss in low latitudes. The reverse is true in higher latitudes. The meridional transfer of energy, necessary to balance these heat gains and losses, is accomplished by air currents. These take the form of organized circulations, whose dominant features are notably different in the tropical belt (roughly the half of the Earth between latitudes 30°N and 30°S) and in extratropical latitudes. *See* METEOROLOGY; STORM.

Characteristic circulations over the Northern Hemisphere are shown in the illustration. In the upper troposphere, there are two principal jet-stream systems: the subtropical jet (STJ) near latitude 30°, and the polar-front jet (PFJ), with large-amplitude long waves and superimposed shorter waves associated with cyclone-scale disturbances. The long waves on the polar-front jet move slowly eastward, and the shorter waves move rapidly. At the Earth's surface, northeast and southeast trade winds of the two hemispheres meet at the intertropical convergence zone (ITCZ), in the vicinity of which extensive lines and large clusters of convective clouds are concentrated. Westward-moving waves and vortices form near the intertropical convergence zone and, in summer, within the trades. Heat released by condensation in convective clouds of the intertropical convergence zone, and the mass of air conveyed upward in them,

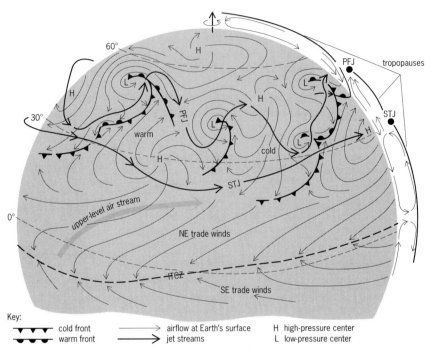

Schematic circulations over the Northern Hemisphere in winter. The intertropical convergence zone (ITCZ) lies entirely north of the Equator in the summer. Eastward acceleration in the upper-level tropical airstream is due to Earth rotation and generates the subtropical jet stream (STJ). The vertical section (right) shows the dominant meridional circulation in the tropics and shows airstreams relative to the polar front in middle latitudes.

drive meridional circulations (right side of the illustration), whose upper-level poleward branches generate the subtropical jet stream at their poleward boundaries.

In extratropical latitudes, the circulation is dominated by cyclones and anticyclones. Cyclones develop mainly on the polar front, where the temperature contrast between polar and tropical air masses is concentrated, in association with upper-level waves on the polar-front jet stream. In winter, cold outbreaks of polar air from the east coasts of continents over the warmer oceans result in intense transfer of heat and water vapor into the atmosphere. Outbreaks penetrating the tropics also represent a sporadic exchange in which polar air becomes transformed into tropical air. Tropical airstreams, poleward on the west sides of the subtropical highs, then supply heat and water vapor to the extratropical disturbances. *See* CYCLONE; FRONT.

The characteristic flow in cyclones takes the form of slantwise descending motions on their west sides and ascent to their east in which extensive clouds and precipitation form. Heat that is released in condensation drives the ascending branch, and the descending branch consists of polar air that has been cooled by radiation in higher latitudes. When viewed relative to the meandering polar-front zone (right side of the illustration), the combined sinking of cold air and ascent of warm air represents a conversion of potential energy into kinetic energy. This process maintains the polar jet stream. The branches of the circulation transfer heat both upward, to balance the radiative heat loss by the atmosphere, and poleward, to balance the radiative heat deficit in high latitudes.   [C.W.N.]

Atmospheric tides   Those oscillations in any or all atmospheric fields whose periods are integral fractions of either lunar or solar days. Oscillations with a period of a day are called diurnal, with a period of a half day semidiurnal, and with a period of a third of a day terdiurnal. The sum of all tidal variations is referred to as the daily variation. As a practical matter, the subject of atmospheric tides is generally restricted to oscillations on a global spatial scale (thus excluding sea breezes). The bulk of attention is devoted to migrating tides, which are those tidal oscillations that depend only on local time.

Atmospheric tides tend to be rather small in the troposphere, although the tidal oscillations in rainfall are surprisingly large. Their importance stems from two primary factors: (1) Tidal oscillations tend to increase in amplitude with height and become major components of the total meteorology above about 50 km (30 mi). (2) The subject has played a prominent role in the intellectual history of meteorology, and it still provides a remarkable example of scientific methodology in an observational science. Tides are unique among meteorological systems in that they have perfectly known periods and relatively well known sources of forcing.

The determination of an oscillation by means of data requires at least two measurements per period. Since most meteorological upper air soundings are taken only twice each day, such data can be used only to marginally determine diurnal oscillations. Occasionally, stations obtain soundings four times per day, which in turn permits determinations of semidiurnal oscillations. Rain gages assign rainfall to specific hours, and averages over many years allow the determination of the daily variation of rainfall. Surface pressure is monitored at a great many stations with effectively (from the point of view of tidal analyses) continuous time resolution. Therefore, surface pressure has traditionally been the field most thoroughly analyzed for tides.

The lunar semidiurnal tide in surface pressure is similar in distribution to the migrating part of the solar semidiurnal tide but only about one-twentieth its strength; maximum lunar semidiurnal surface pressure typically occurs about 1 lunar hour and 13 lunar hours after lunar transit. Clearly, the solar semidiurnal tide dominates the surface pressure. The solar diurnal component is not only smaller but also far more irregular.

Rainfall is commonly observed to have a daily variation. The diurnal component, though often quite large, has a very irregular phase; on the other hand, the solar semidiurnal component is surprisingly uniform, amounting to about 10–20% of the mean daily rainfall in the tropics with maximum rainfall at about 4 A.M. and 4 P.M. local time. Maximum semidiurnal rainfall appears to occur somewhat later in middle latitudes. *See* PRECIPITATION (METEOROLOGY).

Data become more sparse when attempts are made to analyze tides above the surface. Analyses of radiosonde wind measurements have shown that solar semidiurnal oscillations in horizontal wind are primarily in the form of migrating tides. Diurnal oscillations, on the other hand, are significantly affected by regional, nonmigrating components up to at least 20 km (12 mi). Above this height, the diurnal oscillations also tend to be dominated by migrating components. There is a tendency for the diurnal oscillations to have more phase variation with height, especially at low latitudes. As a rough rule of thumb, oscillations in temperature tend to have magnitudes in kelvins comparable to the amplitudes in wind in meters per second. There is also no longer a clear dominance of the semidiurnal oscillations over the diurnal oscillations once the upper-level fields are considered. The amplitude increase with height renders the detection of tidal oscillations at greater altitudes somewhat easier since the tides are becoming a larger feature of the total fields.

While the classical theory of atmospheric tides is adequate for many purposes, recent years have seen a substantial development of theory well beyond the classical theory to include the effects of mean winds, viscosity, and thermal conductivity. See ATMOSPHERE; EARTH TIDES; TIDE.   [R.S.L.]

Atmospheric waves, upper synoptic   Horizontal wavelike oscillations in the pattern of wind flow aloft, usually with reference to the stronger portion of the westerly current in mid-latitudes. The flow is anticyclonically curved in the vicinity of a ridge line in the wave pattern, and is cyclonically curved in the vicinity of a trough line.

Any given hemispheric upper flow pattern may be represented by the superposition of sinusoidal waves of various lengths in the general westerly flow. Analysis of a typical pattern discloses the presence of prominent long waves, of which there are usually three or four around the hemisphere, and of distinctly evident short waves, of about half the length of the long waves.

Typically, each short-wave trough and ridge is associated with a particular cyclone and anticyclone, respectively, in the lower troposphere. The development and intensification of one of these circulations depends in a specific instance upon the details of this association, such as the relative positions and intensities of the upper trough and the low-level cyclone. These circulations produce the rapid day-to-day weather changes which are characteristic of the climate of the mid-latitudes.

The long waves aloft do not generally correspond to a single feature of the circulation pattern at low levels. They are relatively stable, slowly moving features which tend to guide the more rapid motion of the individual short waves and of their concomitant low-level cyclones and anticyclones. Thus, by virtue of their position and amplitude the long waves can exert an indirect influence on the character of the weather over a given region for a period of the order of weeks.

A blocking situation is one in which waves do not progress through the latitude belt of the westerlies. Blocking situations are frequently accompanied by extreme meteorological events; precipitation and cool temperatures persist near upper-level cyclones, and dry, warm weather persists near upper-level anticyclones. A blocking pattern usually consists of a ridge (anticyclone) over a trough (cyclone), a high-amplitude ridge, or flow shaped like an uppercase Greek omega (Ω). Because of the preference for blocking off the west coasts of Europe and North America, it appears that topography must play an important role in blocking. See ATMOSPHERE; JET STREAM; STORM; WIND.   [F.S.; H.B.B.]

Atoll   An annular coral reef, with or without small islets, that surrounds a lagoon without projecting land area. Most atolls are isolated reefs rising from the deep sea, and vary considerably in size. Small rings, usually without islets, may be less than a mile in diameter, but many atolls have a diameter of about 20 mi (32 km) and bear numerous islets.

The reefs of the atoll ring are flat, pavementlike areas, large parts of which, particularly along the seaward margin, may be exposed at times of low tide. The reefs vary in width from narrow ribbons to broad bulging areas more than a mile (1.6 km) across. The structures form a most effective baffle that robs the incoming waves of much of their destructive power, and at the same time brings a constant supply of refreshing sea water with oxygen, food, and nutrient salts to wide expanses of the reef.

Atolls, like other types of coral reefs, require strong light and warm waters and are limited in the existing seas to tropical and near-tropical latitudes. A large percentage of the world's atolls are contained in an area known as the former Darwin Rise that covers much of the central and southwestern Pacific. Atolls are also numerous in parts of the Indian Ocean and a number are found, mostly on continental shelves, in the Caribbean area. *See* OCEANIC ISLANDS; REEF.   [H.S.L.]

Augite A group of monoclinic calcic pyroxenes which have the general chemical formula $(Ca,Mg,Fe)(Mg,Fe)Si_2O_6$, in which calcium is the dominant cation in the first cation position. Monoclinic pyroxene with substantial iron or magnesium in place of calcium is called pigeonite, and has a different crystal structure from augite. Augite is generally considered a combination of the four end members diopside $(CaMgSi_2O_6)$, hedenbergite $(CaFe^{2+}Si_2O_6)$, enstatite $(Mg_2Si_2O_6)$, and ferrosilite $(Fe^{2+}_2Si_2O_6)$, but it almost always has substantial aluminum and minor to substantial amounts of sodium, ferric iron, chromium, and titanium.

Augite occurs in both igneous and metamorphic rocks. It is nearly universal in basalts and gabbros, and occurs somewhat less frequently in less mafic igneous rocks. Magnesium-rich augite is a characteristic mineral in many ultramafic rocks and in rocks of the Earth's mantle. Augite and pigeonite are also rather common constituents of lunar basalts and basaltic meteorites. *See* DIOPSIDE; ECLOGITE; ENSTATITE; PIGEONITE; PYROXENE.   [R.J.Tr.]

Aurora An optical manifestation of a large-scale electrical discharge process which surrounds the Earth. The discharge is powered by the so-called solar wind–magnetosphere generator. The Sun continuously blows out its upper atmosphere, the corona, with a supersonic speed. This fully ionized and magnetized gas flow interacts with the Earth's magnetic field, resulting in a comet-shaped cavity (the magnetosphere) carved around the Earth, while the lines of force of the Earth's magnetic field and of the solar wind magnetic field interconnect. Electric power of as much as 10^{12} W is generated as the solar wind blows across the interconnected field lines near the comet-shaped boundary. A part of the electric current (carried mainly by electrons) thus generated flows between the magnetospheric boundary and an annular, ring-shaped region of the polar upper atmosphere along the lines of force of the Earth's magnetic field. *See* MAGNETOSPHERE.

As these electrons descend toward the Earth, they themselves develop an electrical potential drop of the order of a few kilovolts along the lines of force. As a result, the current-carrying electrons acquire energies of as much as a few kiloelectronvolts, sufficient to ionize and excite a few hundred atoms and molecules before they are stopped by the atmosphere at an altitude of about 60 mi (100 km).

Two ring-shaped glows, one in each hemisphere, are produced by upper atmospheric atoms and molecules which emit their own characteristic light after colliding with the current-carrying electrons. The most common light of the aurora (the greenish-white light) comes from excited oxygen atoms. Excited and ionized molecular nitrogen adds several band emissions. Imaging devices aboard satellites have successfully "photographed" both the northern and southern auroral rings.

From a point on the ground, only a small part of the ring-shaped glow can be observed. It is seen as a curtain-shaped glow, stretching from horizon to horizon across the sky (see illustration). The bottom of the auroral curtain is sharply bounded and is located at about 60 mi (100 km) altitude. The upper boundary diffuses and extends to well above 180 mi (300 km).

The aurora becomes active during geomagnetic storms which occur often about 40 h after an intense solar flare. This is because the efficiency of the solar wind–magnetosphere generator becomes high and variable when the solar wind becomes gusty after a solar flare. During a great magnetic storm, the auroral ring expands from its usual latitude of about 67 to 50° or a little less. It is on such an occasion when

Two curtain-shaped auroras stretching across the sky near Fairbanks, Alaska. (*Courtesy of Lee Snyder*)

the aurora can be seen widely across the continental United States. *See* ATMOSPHERE; GEOMAGNETISM; IONOSPHERE.   [S.I.A.]

Australia   An island continent in the Southern Hemisphere with a total area of 2,941,526 mi² (7,618,552 km²). It is bounded on the west by the Indian Ocean and on the east by the Pacific Ocean and the Tasman Sea. Numerous small and several large islands lie off the coast, including Tasmania and New Zealand. Australia is generally of remarkably low elevation and moderate relief. Three-fourths of the land mass lies between 600 and 1500 ft (180 and 450 m) in the form of a huge plateau. A cross section from east to west shows first a narrow belt of coastal plain, then the steep escarpments of the eastern face of the Great Dividing Range, stretching 1200 mi (1900 km) from the north of Queensland to the south of Victoria. The descent on the western slope of the Dividing Range is gradual until often elevation in the inland basins is below sea level, rising gradually again across the great plateau until the low ranges of western Australia fringing the plateau are reached, and beyond these lies another coastal plain. With the exception of the Gulf of Carpentaria and Cape York peninsula in the north and the Great Australian Bight in the south, there are few striking features in the configuration of the coast. Australia may conveniently be divided into three great structural and landform regions.

The region called East Australian Highlands consists of a narrow plain extending north and south along the eastern coast. Flanking the plain are the series of ranges and tablelands making up the Great Dividing Range. The East Australian Highlands is the best-watered region in Australia, and some of the river systems are of considerable size. On the flanks of the East Australian Highlands are Australia's principal coal deposits—in the vicinity of Sydney and Newcastle and in the Bowen and Ipswich fields in Queensland. Petroliferous basins at Surat (Roma), flanking the divide in Queensland and off the coast of Victoria in Bass Strait, are Australia's most promising deposits of petroleum and natural gas.

The region known as the Interior Lowland Basins comprises a region of sedimentary rocks that occupy one-third of the continent between the western slope of the eastern highlands and the inner eastern margin of the ancient shield which forms the Western Plateau. Little land is over 500 ft (150 m), and some is below sea level. The rivers of the Murray-Darling Basin, draining the western slopes of the Great Dividing Range, have a marked seasonal variation in flow but never dry up in the lower reaches. South Australia's shallow lakes are more often dry expanses of encrusted white salt than bodies of water—the result of low rainfall and high evaporation. In most parts of the region water from deep artesian wells is available.

The region known as the Western Plateau is the largest area, occupying almost three-fifths of the continent, and is a great shield of ancient rocks standing 750–1500 ft (225–450 m) high. Much of it is buried in desert sand, and only a few ridges of ancient mountains (such as the Macdonnel and Musgrave ranges) break the monotony of the plateau surface. Only in the southwestern corner of the continent and along the northwestern coast is rainfall sufficient to support a sclerophyll forest of eucalypts and a monsoon woodland, respectively. In the north, coastal rivers are of considerable size but change from flooded torrents after rains to a succession of water holes in dry seasons.

Tasmania is a small mountainous island lying 150 mi (240 km) southeast of Australia across Bass Strait, with a total area of 26,383 mi^2 (68,332 km^2). The island is structurally similar to the East Australian Highlands. The dominant feature is the central plateau, falling from a general level of 3500 ft (1070 m) in the northwest toward the southeast. A dense eucalyptus forest covers most of the island except along the wetter west coast, where beech forest predominates. The rivers have short, rapid courses with little seasonal variation in flow. *See* NEW ZEALAND.   [K.B.C.]

Authigenic minerals   Minerals that are formed in sediment or a sedimentary rock. Their in-place origin distinguishes them from minerals that are formed elsewhere and transported to the site of deposition (detrital minerals). Authigenic minerals form at the Earth's surface as well as during subsequent burial. The postdepositional processes are referred to as diagenesis, and the resulting minerals are important clues to post-depositional physical and chemical changes in the rock. *See* DIAGENESIS; SEDIMENTARY ROCKS.

Authigenic minerals precipitate from the overlying water column, pore fluids in the sediment, recrystallization or alteration of preexisting minerals, and structural transformation of one mineral to another. The minerals change in an attempt to equilibrate to the physical and chemical conditions present at any given time. Critical factors in their formation are initial mineral assemblage, temperature, pressure, ionic concentration, pH, electron availability, and the fluid flux through the rock.

In sedimentary rocks it is common to find a record of multiple diagenetic events based on the authigenic minerals. For example, in sediments near the surface, meteoric water may displace original marine pore water, resulting in distinct types of cements. Iron oxide can result from oxidizing fluid. Depletion of oxygen by bacteria may result in the formation of iron sulfides, such as pyrite. During burial, the sediments respond to increasing temperature (up to 200°C; 390°F), pressure (up to 2.5 kilobars; 250 megapascals), and fluid movement from compaction-driven waters or influx of water from the basin flanks. As a result, the sedimentary rock may contain authigenic minerals that record a sequence of events ranging from processes occurring near the sediment-water

Common authigenic minerals

Mineral	Formula
Albite	$NaAlSi_3O_8$
Anatase	TiO_2
Anhydrite	$CaSO_4$
Apatite*	$Ca_5(PO_4)_3(F,Cl,OH)$
Aragonite (orthorhombic)	$CaCO_3$
Barite	$BaSO_4$
Boehmite	$AlO(OH)$
Calcite (hexagonal)	$CaCO_3$
Celestite	$SrSO_4$
Clay minerals	
Chlorites*	$(Mg,Fe^{2+},Fe^{3+})_6$- $(Al,Si_3)O_{10}(OH)_8$
Illites*	$K(Al)_2(AlSi_3)O_{10}(OH)_2$
Kaolinite	$Al_2Si_2O_5(OH)_4$
Smectites*	$(Na,0.5Ca)_{0.5}(Al,Mg,Fe)_2$- $(Al,Si_3)O_{10}(OH)_2 \cdot nH_2O$
Dolomite	$CaMg(CO_3)_2$
Gibbsite	$Al(OH)_3$
Glauconite*	$K(Al,Mg,Fe^{2+},Fe^{3+})_2$- $(Al,Si_3)O_{10}(OH)_2$
Goethite	$Fe_2O_3 \cdot n(H_2O)$
Gypsum	$CaSO_4 \cdot 2(H_2O)$
Halite	$NaCl$
Hematite	Fe_2O_3
Leucoxene	TiO_2
Limonite	$FeO(OH) \cdot n(H_2O)$
Opal (amorphous)	$SiO_2 \cdot n(H_2O)$
Orthoclase	$KAlSi_3O_8$
Pyrite (isometric)	FeS_2
Pyrolusite	MnO_2
Quartz	SiO_2
Siderite	$FeCO_3$
Zeolites*	$X_y^{1+,2+}Al_xSi_{1-x}O_z \cdot nH_2O$
Clinoptilolite*	$(Na,0.5Ca,K)_{3.5}\,Al_{3.5}$ $Si_{14.5}O_{36} \cdot nH_2O$
Analcime	$NaAlSi_2O_6 \cdot H_2O$
Laumontite	$CaAl_2Si_4O_{12} \cdot 4H_2O$

*Group of minerals characterized by considerable chemical variation.

interface to those forming during deep burial. Unlike metamorphic rocks, the preexisting (detrital) mineral assemblage is at least partially retained, in part due to the sluggish reaction rates at diagenetic conditions. Early cementation processes often seal up the rock, preventing subsequent diagenetic reactions and preserving the original detrital mineral assemblage.

Authigenic minerals occur in all sedimentary rock and can vary from trace amounts to virtually the total rock (see table). The carbonate minerals calcite, dolomite, and siderite are some of the most common types. They form in a wide range of depositional environments and at varying burial depths. Calcite and dolomite form the principal minerals in limestones and dolostones, respectively, as well as cements in sandstones or shales. Carbonate cements result from recrystallization of detrital carbonates and from dissolution of other calcium, iron, and magnesium minerals with carbon dioxide from organic reactions. Much of the calcite in limestones initially consisted of aragonite or magnesium-rich calcite, whereas most dolomite has been formed by the chemical alteration of calcite. Recrystallization may change aragonite to calcite.

Aragonite (orthorhombic) is a naturally unstable form of calcium carbonate. With the passage of geologic time, aragonite normally inverts to the more stable calcite (hexagonal). The substitution of magnesium for calcium is responsible for the conversion of calcite or aragonite to dolomite, and it has been shown that dedolomitization (replacement of magnesium by calcium) is also possible. *See* ARAGONITE; CALCITE; CARBONATE MINERALS; DOLOMITE.   [J.R.B.]

Avalanche   In general, a large mass of snow, ice, rock, earth, or mud in rapid motion down a slope or over a precipice. In the English language, the term avalanche is reserved almost exclusively for snow avalanche. Minimal requirements for the occurrence of an avalanche are snow and an inclined surface, usually a mountainside. Most avalanches occur on slopes between 30 and 45°.

Two basic types of avalanches are recognized according to snow cover conditions at the point of origin. A loose-snow avalanche originates at a point and propagates downhill by successively dislodging increasing numbers of poorly cohering snow grains, typically gaining width as movement continues downslope. This type of avalanche commonly involves only those snow layers near the surface. The mechanism is analogous to dry sand. The second type, the slab avalanche, occurs when a distinct cohesive snow layer breaks away as a unit and slides because it is poorly anchored to the snow or ground below. A clearly defined gliding surface as well as a lubricating layer may be identifiable at the base of the slab, but the meteorological conditions which create these layers are complex.

In the case of the loose avalanche, release mechanisms are primarily controlled by the angle of repose, while slab releases involve complex strength-stress problems. A release may occur simply as a result of the overloading of a slope during a single snowstorm and involve only snow which accumulated during that specific storm, or it may result from a sequence of meteorological events and involve snow layers comprising numerous precipitation episodes. Most large snow slides are believed to be caused by an unstable layer of ice grains that develop deep in mountain snow. Called depth hoar by students of avalanche dynamics, these crystals owe their formation to heat from earth and rock which are buried by the snow, and which in late autumn are warmer than the surrounding air. Snow nearest the ground vaporizes, causing growth of angular ice grains that exhibit poor bonding qualities. Gravity combining with the weakness of the depth hoar crystals loosens the upper stable layers. Once the stable layers begin to slide, the depth hoar acts in a manner similar to ball beatings to speed the descent of the slide.

Where snow avalanches constitute a hazard, that is, where they directly threaten human activities, various defense methods have evolved. Attempts are made to prevent the avalanche from occurring by artificial supporting structures or reforestation in the zone of origin. The direct impact of an avalanche can be avoided by construction of diversion structures, dams, sheds, or tunnels. Hazardous zones may be temporarily evacuated while avalanches are released artificially, most commonly by explosives. Finally, attempts are made to predict the occurrence of avalanches by studying relationships between meteorological and snow cover factors.   [R.L.A.]

Azurite   A basic carbonate of copper with the chemical formula $Cu_3(OH)_2(CO_3)_2$. Azurite is normally associated with copper ores and often occurs with malachite. Azurite is mono-clinic. It may be massive or may occur in tabular, prismatic, or equant crystals. Invariably blue, azurite was originally used extensively as a pigment. Hardness is $3^1/_2$–4 (Mohs scale) and specific gravity is 3.8. Notable localities for azurite are at Tsumeb, Southwest Africa, and Bisbee, Arizona.   [R.I.Ha.]

B

Baltic Sea A semienclosed brackish sea located in a humic zone, with a positive water balance relative to the adjacent ocean (the North Sea and the North Atlantic). The Baltic is connected to the North Sea by the Great Belt (70% of the water exchange), the Øresund (20% of the water exchange), and the Little Belt. The total area of the Baltic is 147,414 mi^2 (381,705 km^2), its total volume 4982 mi^3 (20,764 km^3), and its average depth 181 ft (55.2 m). The greatest depth is 1510 ft (459 m), in the Landsort Deep.

The topography of the Baltic is characterized by a sequence of basins separated by sills and by two large gulfs, the Gulf of Bothnia (40,100 mi^2 or 104,000 km^2) and the Gulf of Finland (11,400 mi^2 or 29,500 km^2). More than 200 rivers discharge an average of 104 mi^3 (433 km^3) annually from a watershed area of 637,056 mi^2 (1,649,550 km^2). The largest river is the Newa, with 18.5% of the total fresh-water discharge. From December to May, the northern and eastern parts of the Baltic are frequently covered with ice. On the average, the area of maximum ice coverage is 82,646 km^2 (214,000 km^2). The mean maximum surface-water temperature in summer is between 59 and 63°F (15 and 17°C).

As the Baltic stretches from the boreal to the arctic continental climatic zone, there are large differences between summer and winter temperature in the surface waters, ranging from about 68 to 30°F (20 to −1°C) in the Western Baltic and 57 to 32°F (14 to −0.2°C) in the Gulf of Bothnia and the Gulf of Finland.

The salt content of the Baltic waters is characterized by two major water bodies; the brackish surface water and the more saline deep water. Salinities for the surface water range from 8 to 6‰ in the Western and Central Baltic and 6 to 2000 in the Gulf of Bothnia and the Gulf of Finland; salinities for the deep water range from 18 to 13‰ in the Western and Central Baltic and 10 to 4‰ in the Gulf of Bothnia and the Gulf of Finland.

The surface currents of the Baltic are dominated by a general counterclockwise movement and by local and regional wind-driven circulations. A complex system of small- and medium-scale gyres develops especially in the central parts of the Baltic. The currents in the Belt Sea are dominated by the topography; they are due to sea-level differences between the Baltic proper and the North Sea. Tides are of minor importance, ranging between 0.8 and 4.7 in. (2 and 12 cm). Water-level changes of more than 6 ft (2 m) occur occasionally as a result of onshore or offshore winds and the passage of cyclones over the Baltic Sea area. The frequency of longitudinal sea-level oscillations is about 13.5 h. *See* OCEAN CIRCULATION.

The flora and fauna of the Baltic are those of a typical brackish-water community, with considerably reduced numbers of species compared to an oceanic community. The productivity is relatively low compared to other shelf seas. The major commercially exploited species are cod, herring, sprat, flounder, eel, and salmon, and some fresh-water species such as whitefish, pike, perch, and trout. The total annual catch amounts to about 880,000 tons (800,000 metric tons). The Baltic is completely divided into fishery zones, with exclusive fishing rights belonging to the respective countries.

Other than fish the only major resources that have been exploited are sand and gravel in the Western Baltic Sea. It is believed that the deeper layer under the Gotland Basin contains mineral oil, but so far only exploratory drilling has been carried out in the near-coastal regions. Limited amounts of mineral oil have also been located in the Gulf of Kiel.   [K.G.]

Banded iron formation   Banded iron formation is a sedimentary rock that was commonly deposited during the Precambrian. It was probably laid down as a colloidal iron-rich chemical precipitate, but in its present compacted form it consists typically of equal proportions of iron oxides (hematite or magnetite) and silica in the finely crystalline form of quartz known as chert. Its chemical composition is 50% silicon dioxide (SiO_2) and 50% iron oxides (Fe_2O_3 and Fe_3O_4), to give a total iron content of about 30%. Banding is produced by the concentration of these two chemical components into layers about 1–5 cm (1/2–2 in.) thick; typical banded iron formation consists of pale silica-rich cherty bands alternating with black to dark red iron-rich bands (see illustration). These contrasting layers are sharply defined, so that the rock has a striped appearance; banded iron formation is normally a hard, tough rock, highly resistant both to erosion and to breaking with a hammer.

Folded banded iron formation from the Ord Range, Western Australia. The distance between top and bottom of the polished face of the sample is about 15 cm (6 in.). Chert jasper bands alternate with dark magnetite-rich bands. The thin pale layers of irregular thickness are bands of asbestiform amphibole, now replaced by silica, to give the semiprecious material "tiger-eye." (*Photo courtesy of John Blockley*)

The world's iron and steel industry is based almost exclusively on iron ores associated with banded iron formation. Banded iron formation itself may be the primary ore, from which hematite or magnetite is concentrated after crushing. But the main ore now mined globally is high-grade (greater than 60% iron) material that formed within banded iron formation by natural leaching of its silica content.   [A.F.T.]

Barite   An orthorhombic mineral with chemical composition $BaSO_4$. It possesses one perfect cleavage, and two good cleavages, as do the isostructural minerals. The mineral has a specific gravity of approximately 4.5, and is relatively soft, approximately 3 on Mohs scale. The color ranges through white to yellowish, gray, pale blue, or brown, and a thin section is colorless.

Barite is often an accessory mineral in hydrothermal vein systems, but frequently occurs as concretions or cavity fillings in sedimentary limestones. It also occurs as a residual product of limestone weathering and in hot spring deposits. It occasionally occurs as extensive beds in evaporite deposits. Occurrences of barite are extensive. It is found as a vein mineral associated with zinc and lead ores in Derbyshire, England. Large deposits occur at Andalusia, Spain. Commercial residual deposits occur in limestones throughout the Appalachian states such as Georgia, Tennessee, and Kentucky.

Since barite is dense and relatively soft, its principal use is as a weighting agent in rotary well-drilling fluids. It is the major ore of barium salts, used in glass manufacture, as a filler in paint, and, owing to the presence of a heavy metal and inertness, as an absorber of radiation in x-ray examination of the gastrointestinal tract.   [P.B.M.]

Baroclinic field   A distribution of atmospheric pressure and mass such that the specific volume, or density, of air is a function of both pressure and temperature, but not either alone. When the field is baroclinic, solenoids are present, there is a gradient of air temperature on a surface of constant pressure, and there is a vertical shear of the geostrophic wind. Significant development of cyclonic and anticyclonic wind circulations typically occurs only in strongly baroclinic fields. Fronts represent baroclinic fields which are locally very intense. *See* AIR PRESSURE; FRONT; GEOSTROPHIC WIND; SOLENOID (METEOROLOGY); STORM; WIND.   [F.S.; H.B.B.]

Barometer   An absolute pressure gage specifically designed to measure atmospheric pressure. This instrument is a type of manometer with one leg at zero pressure absolute.

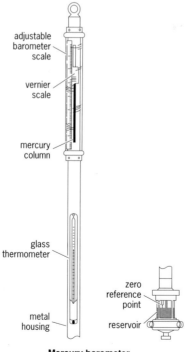

Mercury barometer.

The common meteorological barometer (see illustration) is a liquid-column gage filled with mercury. The top of the column is sealed, and the bottom is open and submerged below the surface of a reservoir of mercury. The atmospheric pressure on the reservoir keeps the mercury at a height proportional to that pressure. An adjustable scale, with a vernier scale, allows a reading of column height. Aneroid barometers using metallic diaphragm elements are usually less accurate, though often more sensitive, devices, and not only indicate pressure but may be used to record it.   [J.H.Z.]

Barotropic field   A distribution of atmospheric pressure and mass such that the specific volume, or density, of air is a function solely of pressure. When the field is barotropic, there are no solenoids, air temperature is constant on a surface of constant pressure, and there is no vertical shear of the geostrophic wind. Significant cyclonic and anticyclonic circulations typically do not develop in barotropic fields. Considerable success has been achieved, paradoxically, in prediction of the flow pattern at middle-tropospheric elevations by methods which are strictly applicable only to barotropic fields, despite the fact that the field in this region is definitely not barotropic. The subtropics, however, are to a large extent barotropic. *See* AIR PRESSURE; BAROCLINIC FIELD; GEOSTROPHIC WIND; SOLENOID (METEOROLOGY); WEATHER FORECASTING AND PREDICTION; WIND.   [F.S.; H.B.B.]

Barrier islands   Elongate, narrow accumulations of sediment which have formed in the shallow coastal zone and are separated from the mainland by some combination of coastal bays and marshes. They are typically several times longer than their width and are interrupted by tidal inlets. Although their origin has been widely discussed, at least three possibilities exist: longshore spit development and subsequent cutting of inlets; drowning of old coastal ridges; and upward shoaling of subtidal sediment accumulations. All three may have occurred; however, the last seems most likely and most prevalent.

Barrier islands must be considered in terms of the adjacent and closely related environments within the coastal system. Beginning offshore and proceeding landward, the sequence of environments crossed is shoreface, beach, dunes, back-island flats or marsh, coastal bay, marsh, and mainland. The barrier island proper consists of the beach, dunes, and back-island flats or marsh; however, of the remaining environments, at least the shoreface is closely integrated with the barrier island in terms of morphology, processes, and sediments. *See* DUNE.

A variety of physical processes exists along the coast. These processes act to shape and maintain the barrier-island system and also to enable the barrier to migrate landward as sea level continues to rise. The most important process in the barrier-island system is the waves, which also give rise to longshore currents. Waves and longshore currents dominate the outer portion of the barrier system, whereas tidal currents are dominant landward of the barrier, although small waves may also be present. Tidal currents are most prominent in and adjacent to the inlets. On the supratidal portion of the barrier island, the wind is the most dominant physical process.

Barrier-island sands represent one of the best sources of oil and gas, with the tight organic-rich source rocks being in the form of the bay and shelf muds and the barrier itself being the reservoir rock. These elongate sand bodies have been sought by exploration geologists for decades. The Tertiary sequences of the Texas Gulf coasts are an example of such barrier systems which have been very productive. *See* COASTAL LANDFORMS.   [R.A.D.]

Basalt   An igneous rock characterized by small grain size (less than about 0.2 in. or 5 mm) and approximately equal proportions of calcium-rich plagioclase feldspar and calcium-rich pyroxene, with less than about 20% by volume of other minerals. Olivine, calcium-poor pyroxene, and iron-titanium oxide minerals are the most prevalent other minerals. Most basalts are dark gray or black, but some are light gray. Various structures and textures of basalts are useful in inferring both their igneous origin and their environment of emplacement. Basalts are the predominant surficial igneous rocks on the Earth, Moon, and probably other bodies in the solar system. Several chemical-mineralogical types of basalts are recognized. The nature of basaltic rocks provides helpful clues about the composition and temperature within the Earth and Moon. The magnetic properties of basalts are responsible in large part for present knowledge of the past behavior of the Earth's magnetic field and of the rate of sea-floor spreading. Some meteorites are basaltic rocks. They differ significantly from lunar basalts and appear to have originated elsewhere in the solar system at a time close to the initial condensation of the solar nebula. *See* IGNEOUS ROCKS.

Basalt erupts out of fissures and cylindrical vents. Repeated or continued extrusion of basalt from cylindrical vents generally builds up a volcano of accumulated lava and tephra around the vent. Fissure eruptions commonly do not build volcanoes, but small cones of tephra may accumulate along the fissures. *See* VOLCANO; VOLCANOLOGY.

Basalts display a variety of structures mostly related to their environments of consolidation. On land, basalt flows form pahoehoe, aa, and block lava, while under water, pillow lava is formed. Basalt also occurs as pumice and bombs. Commonly, basaltic pumice is called scoria to distinguish it from the lighter-colored, more siliceous rhyolitic pumice.

The mineralogy and texture of basalts vary with cooling history and with chemical composition. As basalt crystallizes, both the minerals and the residual melt change in composition because of differences between the composition of the melt and the crystals forming from it. In basalts, because of the rapid cooling, there is little chance for crystals to react with the residual melt after the crystals have formed. Completely

Chemical compositions of basalts, in weight percent

	1*	2	3	4	5	6
SiO_2	49.92	49.20	49.56	51.5	45.90	45.5
TiO_2	1.51	2.32	1.53	1.1	1.80	2.97
Al_2O_3	17.24	11.45	17.88	17.1	15.36	9.69
Fe_2O_3	2.01	1.58	2.78	n.d.	1.22	0.00
Cr_2O_3	0.04	n.d.	n.d.	n.d.	n.d.	0.50
FeO	6.90	10.08	7.26	8.9	8.13	19.7
MnO	0.17	0.18	0.14	n.d.	0.08	0.27
MgO	7.28	13.62	6.97	7.0	13.22	10.9
CaO	11.85	8.84	9.99	9.3	10.71	10.0
Na_2O	2.76	2.04	2.90	4.3	2.29	0.33
K_2O	0.16	0.46	0.73	0.80	0.67	0.06
P_2O_5	0.16	0.23	0.26	n.d.	0.62	0.10
Sum	100.00	100.00	100.00	100.00	100.00	100.02
H_2O	0.4	0.3	n.d.	2	n.d.	0.0
CO_2	0.02	0.01	n.d.	n.d.	n.d.	n.d.
F	0.02	0.03	n.d.	n.d.	n.d.	0.002
Cl	0.02	0.03	n.d.	0.09	0.01	0.0005
S	0.08	0.07	n.d.	0.19	n.d.	0.07

*(1) Average of 10 basalts from oceanic ridges; (2) submarine basalt, Eastern Rift Zone, Kilauea Volcano, Hawaii; (3) average high-alumina basalt of Oregon Plateau; (4) initial melt of Pacaya Volcano, Guatemala; (5) alkali basalt, Hualalai Volcano, Hawaii; (6) average *Apollo 12* lunar basalt.

solid basalts generally preserve a record of their crystallization in the zoned crystals and residual glass.

Most basalts contain minor amounts of chromite, magnetite, ilmenite, apatite, and sulfides in addition to the minerals mentioned above. Magnetite in basalts contains a history of the strength and orientation of the Earth's magnetic field at the time of cooling. Therefore, although magnetite is minor in amount, it is probably the most important mineral in terrestrial basalts, because it enables earth scientists to infer both the magnetic history of the Earth and the rate of the production of basaltic ocean floor at the oceanic ridges.

Basalts occur in all four major tectonic environments: ridges in the sea floor, islands in ocean basins, island arcs and mountainous continental margins, and interiors of continents. The principal environment is the deep sea floor. Significant differences exist in the composition of basalt which relate to different tectonic environments (see table). Chemical analyses of basalts are now used instead of, or together with, the textural criteria as a basis of classification. Geologists customarily recast the chemical analysis into a set of ideal minerals according to a set of rules. The result is called the norm of the rock. In general there is rather close correspondence between the normative minerals and the observed minerals in basaltic rocks. *See* MAGMA; PETROLOGY.   [A.T.A.]

Basin   A low-lying area which is wholly or largely surrounded by higher land. An example is Hudson Bay in northeastern Canada, which was formed by depression beneath the center of a continental ice sheet 18,000 years ago. Another example, the Qattara depression, is 150 mi (240 km) long and the largest of several wind-excavated basins of northern Egypt. Depressions in the ocean floor are also basins, such as the Canary Basin, west of northern Africa, or the Argentine Basin, east of Argentina. These basins occur in regions where cold, dense oceanic crust lies between the topographically elevated ocean ridges and the continental margins. *See* CONTINENTAL MARGIN; MARINE GEOLOGY.

A drainage basin is the entire area drained by a river and its tributaries. Thus, the Mississippi Basin occupies most of the United States between the Rocky Mountains and the Appalachians. Interior drainage basins consist of depressions that drain entirely inward, without outlet to the sea. Examples may be quite small, such as the Salton Sea of southern California or the Dead Sea of central Asia. One of the most remarkable examples of an interior drainage basin is the Chad Basin in northern Africa, the center of which is occupied by Lake Chad. The fresh waters of the lake drain underground to feed oases in the lowlands 450 mi (720 km) to the northeast.

In the geologic sense, a basin is an area in which the continental crust has subsided and the depression has been filled with sediments. Such basins were interior drainage basins at the time of sediment deposition but need not be so today. As these basins subside, the layers of sediment are tilted toward the axis of maximum subsidence. Consequently, when the sedimentary layers are observed in cross section, their geometry is a record of the subsidence of the basin through time and contains clues about the origin of the basin.

The origin of geologic basins is a topic of continuing interest in both applied and basic geological studies. They contain most of the world's hydrocarbon reserves, and they are regarded as some of the best natural laboratories in which to understand the thermal and mechanical processes that operate deep in the interior of the Earth and that shape the Earth's surface.   [G.Bo.; M.Ko.]

Bauxite   A rock mainly comprising minerals that are hydrous aluminum oxides. These minerals are gibbsite, boehmite, and diaspore. The major impurities in bauxite are clay minerals and iron oxides. Bauxite is a weathering product of aluminous rock that results from intense leaching in tropical and subtropical areas, a process called laterization. Bauxite deposits are generally found on plateaus in stable areas where they had sufficient geologic time to form and were protected from erosion.

Bauxite is the primary ore of aluminum. The two types of bauxites that are used commercially as aluminum ores are laterite and karst. Lateritic bauxites constitute more than three-fourths of the world's bauxite resources. Karstic bauxites are formed on a carbonate terrain and are concentrated in sinkholes and solution depressions on the surface of carbonate rocks. See LATERITE.

Bauxite used to produce alumina is called metallurgical grade; approximately 90% of the world's production is for this purpose. Other major uses are in refractories, abrasives, chemicals, and aluminous cements. The compositional requirements are much more rigid for these uses. The alumina content must be higher, and the iron, silica, and titanium contents significantly lower, than for metallurgical-grade bauxite. World resources of bauxite are many tens of billions of tons, so an adequate supply is assured for hundreds of years.   [H.H.Mu.]

Bentonite   The term first applied to a particular, highly colloidal plastic clay found near Fort Benton in the Cretaceous beds of Wyoming. This clay swells to several times its original volume when placed in water and forms thixotropic gels when small amounts are added to water. Later investigations showed that this clay was formed by the alteration of volcanic ash in place; thus, the term bentonite was redefined by geologists to limit it to highly colloidal and plastic clay materials composed largely of montmorillonite clay minerals, and produced by the alteration of volcanic ash in place. The term has been used commercially for any plastic, colloidal, and swelling clays without reference to a particular mode of origin. See CLAY; MONTMORILLONITE.

Bentonites have been found in almost all countries and in rocks of a wide variety of ages. They appear to be most abundant in rocks of Cretaceous age and younger. In the United States, bentonites are mined extensively in Wyoming, Arizona, and Mississippi. England, Germany, Yugoslavia, Russia, Algeria, Japan, and Argentina also produce large tonnages of bentonite. Many bentonites are of great commercial value. They are used in decolorizing oils, in bonding molding sands, in the manufacture of catalysts, in the preparation of oil well drilling muds, and in numerous other relatively minor ways. The properties of a particular bentonite determine its economic use.   [R.E.Gr.; F.M.W.]

Bering Sea   A water body north of the Pacific Ocean, 875,000 mi^2 (2,268,000 km^2) in area, bounded by Siberia, Alaska, and the Aleutian Islands. The Bering Sea is a biologically productive area, with large populations of marine birds and mammals. An active pollock fishery and a developing bottom-fish industry are evidence of its rich biological resources.

The Bering Sea consists of a large, deep basin in the southwest portion, where depths as great as 9900 ft (3000 m) are encountered. To the north and east, an extremely wide, shallow continental shelf extends north to the Bering Strait. The two major regions are separated by a shelf break, the position of which coincides with the southernmost extent of sea ice in a cold season. Ice is a prominent feature of the Bering Sea shelf during the cold months. Coastal ice begins to form in late October, and by February coastal ice is found in the Aleutians. The sea ice may extend as far south as 58°N. Thus, the ice edge in the eastern Bering Sea advances and retreats seasonally over a distance as great as 600 mi (1000 km). Ice-free conditions can be expected throughout the entire region by early July. See SEA ICE.

The main water connections with the Pacific are in the west of the Aleutian Islands, the 6600-ft-deep (2000-m) pass between Attu and Komandorskiye Islands and the 14,000-ft-deep (4400-m) pass between the Komandorskiyes and Kamchatka. Aleutian passes also serve to exchange water. The Bering Sea connection with the Arctic Ocean (Chukchi Sea) is the Bering Strait, 53 mi (85 km) wide and 30 mi (45 m) deep.

Tides in the Bering Sea are semidiurnal, with a strong diurnal inequality typical of North Pacific tides. Three water masses are associated with Bering sea water— Western Subarctic, Bering Sea, and the Alaskan Stream. The general circulation of the Bering Sea is counterclockwise, with many small eddies superimposed on the large-scale pattern. The currents in the Bering Sea are generally a few centimeters per second except along the continental slope, the coast of Kamchatka, and in certain eddies, where somewhat higher values have been found. See OCEAN CIRCULATION; TIDE.   [V.A.]

Beryl   The most common beryllium mineral. Beryl, $Al_2[Be_3Si_6O_{18}]$, crystallizes in the hexagonal system. The crystal structure consists of six-membered rings of corner-sharing silicon-oxygen (SiO_4) tetrahedra cross-linked by corner-sharing beryllium-oxygen (BeO_4) tetrahedra to make a three-dimensional honeycomb structure; aluminum-oxygen (AlO_6) octahedra lie between the Si_6O_{18} rings. Beryl has a vitreous luster and is typically white to bluish- or yellowish-green, but it can also be shades of yellow, blue, and pink. Its hardness is 7.5–8 on Mohs scale; it has an imperfect basal cleavage and a specific gravity of 2.7–2.9 (increasing with alkali content). Weakly colored varieties can be confused with quartz or apatite.

Beryl is a minor accessory mineral in many natural environments, most commonly in granites and associated hydrothermally altered rocks. Granitic pegmatites constitute the major source of beryl (used for beryllium and gemstones); rarely, single crystals weigh many tons. Alkali-rich beryl occurs in complex pegmatites which contain abundant rare-element minerals such as spodumene, lepidolite, and tourmaline. Alkali-poor beryl occurs in mineralogically simple pegmatites, tin and tungsten deposits, and hydrothermal veins. The gem varieties of beryl, aquamarine (blue), emerald (deep green), and morganite (pink to red), are produced from pegmatites (aquamarine, morganite, some emerald), veins (some aquamarine, most emerald), and, rarely, rhyolites (ruby-red morganite). See BERYLLIUM MINERALS; PEGMATITE; SILICATE MINERALS.   [M.D.B.]

Beryllium minerals   Minerals containing beryllium as an essential component. Over 50 beryllium minerals have been identified, even though beryllium is a scarce element in the Earth's crust. The unusual combination of low charge (+2) and small ionic radius (0.035 nanometer) of the beryllium ion accounts for this diverse group of minerals and their occurrence in many natural environments.

Nearly all beryllium minerals can be included in one of three groups: compositionally simple oxides and silicates with or without aluminum; sodium- and calcium-bearing silicates; and phosphates and borates. The first group is by far the most abundant; it contains beryl, the most common beryllium mineral, plus the common minerals phenakite, bertrandite, chrysoberyl, and euclase. Of this group, only beryl shows a wide compositional variation.

The beryllium minerals have many structural characteristics similar to the major rock-forming silicate minerals, but are distinguished by containing large quantities of tetrahedrally coordinated beryllium ion (Be^{2+}) in place of, or in addition to, tetrahedrally coordinated aluminum ion (Al^{3+}) and silicon ion (Si^{4+}). See SILICATE MINERALS.

Beryllium minerals occur in many geological environments, where they are generally associated with felsic (abundant feldspar $\pm$ quartz) igneous rocks and related metasomatically altered rocks.

Beryl and bertrandite, mined from granitic pegmatites and altered volcanic rocks, are the principal ores of beryllium; deposits of chrysoberyl and phenakite may become economically significant in the future. The colored varieties of beryl (emerald, aquamarine, morganite) are valued gemstones; chrysoberyl, phenakite, and a few of the other minerals are less common gemstones. *See* CHRYSOBERYL; EMERALD; GEM.   [M.D.B.]

Biodegradation   The destruction of organic compounds by microorganisms. Microorganisms, particularly bacteria, are responsible for the decomposition of both natural and synthetic organic compounds in nature. Mineralization results in complete conversion of a compound to its inorganic mineral constituents (for example, carbon dioxide from carbon, sulfate or sulfide from organic sulfur, nitrate or ammonium from organic nitrogen, phosphate from organophosphates, or chloride from organochlorine). Since carbon comprises the greatest mass of organic compounds, mineralization can be considered in terms of CO_2 evolution. Radioactive carbon-14 (^{14}C) isotopes enable scientists to distinguish between mineralization arising from contaminants and soil organic matter. However, mineralization of any compound is never 100% because some of it (10–40% of the total amount degraded) is incorporated into the cell mass or products that become part of the amorphous soil organic matter, commonly referred to as humus. Thus, biodegradation comprises mineralization and conversion to innocuous products, namely biomass and humus. Primary biodegradation is more limited in scope and refers to the disappearance of the compound as a result of its biotransformation to another product. *See* HUMUS.

Compounds that are readily biodegradable are generally utilized as growth substrates by single microorganisms. Many of the components of petroleum products (and frequent ground-water contaminants), such as benzene, toluene, ethylbenzene, and xylene, are utilized by many genera of bacteria as sole carbon sources for growth and energy.

The process whereby compounds not utilized for growth or energy are nevertheless transformed to other products by microorganisms is referred to as cometabolism. Chlorinated aromatic hydrocarbons, such as diphenyldichloroethane (DDT) and polychlorinated biphenyls (PCBs), are among the most persistent environmental contaminants; yet they are cometabolized by several genera of bacteria, notably *Pseudomonas*, *Alcaligenes*, *Rhodococcus*, *Acinetobacter*, *Arthrobacter*, and *Corynebacterium*. Cometabolism is caused by enzymes that have very broad substrate specificity.

The use of microorganisms to remediate the environment of contaminants is referred to as bioremediation. This process is most successful in contained systems such as surface soil or ground water where nutrients, mainly inorganic nitrogen and phosphorus, are added to enhance growth of microorganisms and thereby increase the rate of biodegradation. The process has little, if any, applicability to a large open system such as a bay or lake because the nutrient level (that is, the microbial density) is too low to effect substantive biodegradation and the system's size and distribution preclude addition of nutrients.

Remediation of petroleum products from ground waters is harder to achieve than surface soil because of the greater difficulty in distributing the nutrients throughout the zone of contamination, and because of oxygen (O_2) limitations.   [D.D.Fo.]

Biodiversity   The variety of all living things; a contraction of biological diversity. Biodiversity can be measured on many biological levels ranging from genetic diversity within a species to the variety of ecosystems on Earth, but the term most commonly refers to the number of different species in a defined area.

Numbers of extant species for selected taxonomic groups

Kingdom	Phylum	Number of species described	Estimated number of species	Percent described
Protista		100,000	250,000	40.0
Fungi	Eumycota	80,000	1,500,000	5.3
Plantae	Bryophyta	14,000	30,000	46.7
	Tracheophyta	250,000	500,000	50.0
Animalia	Nematoda	20,000	1,000,000	2.0
	Arthropoda	1,250,000	20,000,000	5.0
	Mollusca	100,000	200,000	50.0
	Chordata	40,000	50,000	80.0

*With permission, modified from G. K. Meffe and C. R. Carroll, *Principles of Conservation Biology*, 1997.

Recent estimates of the total number of species range from 7 to 20 million, of which only about 1.75 million species have been scientifically described. The best-studied groups include plants and vertebrates (phylum Chordata), whereas poorly described groups include fungi, nematodes, and arthropods (see table). Species that live in the ocean and in soils remain poorly known. For most groups of species, there is a gradient of increasing diversity from the Poles to the Equator, and the vast majority of species are concentrated in the tropical and subtropical regions.

Human activities, such as direct harvesting of species, introduction of alien species, habitat destruction, and various forms of habitat degradation (including environmental pollution), have caused dramatic losses of biodiversity; current extinction rates are estimated to be 100–1000 times higher than prehuman extinction rates.

Some measure of biodiversity is responsible for providing essential functions and services that directly improve human life. For example, many medicines, clothing fibers, and industrial products and the vast majority of foods are derived from naturally occurring species. In addition, species are the key working parts of natural ecosystems. They are responsible for maintenance of the gaseous composition of the atmosphere, regulation of the global climate, generation and maintenance of soils, recycling of nutrients and waste products, and biological control of pest species. Ecosystems surely would not function if all species were lost, although it is unclear just how many species are necessary for an ecosystem to function properly.   [M.A.Ma.]

Biogeochemistry   The study of the cycling of chemicals between organisms and the surface environment of the Earth. The chemicals either can be taken up by organisms and used for growth and synthesis of living matter or can be processed to obtain energy. The chemical composition of plants and animals indicates which elements, known as nutrient elements, are necessary for life. The most abundant nutrient elements, carbon (C), hydrogen (H), and oxygen (O), supplied by the environment in the form of carbon dioxide (CO_2) and water (H_2O), are usually present in excess. The other nutrient elements, which are also needed for growth, may sometimes be in short supply; in this case they are referred to as limiting nutrients. The two most commonly recognized limiting nutrients are nitrogen (N) and phosphorus (P).

Biogeochemistry is concerned with both the biological uptake and release of nutrients, and the transformation of the chemical state of these biologically active substances, usually by means of energy-supplying oxidation-reduction reactions, at the Earth's surface. Emphasis is on how the activities of organisms affect the chemical composition of natural waters, the atmosphere, rocks, soils, and sediments. Thus, biogeochemistry is complementary to the science of ecology, which includes a concern with how the chemical composition of the atmosphere, waters, and so forth affects life. *See* ECOLOGY.

The two major processes of biogeochemistry are photosynthesis and respiration. Photosynthesis involves the uptake, under the influence of sunlight, of carbon dioxide, water, and other nutrients by plants to form organic matter and oxygen. Respiration is the reverse of photosynthesis and involves the oxidation and breakdown of organic matter and the return of nitrogen, phosphorus, and other elements, as well as carbon dioxide and water, to the environment.

Biogeochemistry is usually studied in terms of biogeochemical cycles of individual elements. There are short-term cycles ranging from days to centuries and long-term (geological) cycles ranging from thousands to millions of years.

There has been increasing interest in biogeochemistry because the human influence on short-term biogeochemical cycling has become evident. Perhaps the best-known example is the changes in the biogeochemical cycling of carbon due to the burning of fossil fuels and the cutting and burning of tropical rainforests. The cycles of nitrogen and phosphorus have been altered because of the use of fertilizer and the addition of wastes to lakes, rivers, estuaries, and the oceans. Acid rain, which results from the addition of sulfur and nitrogen compounds to the atmosphere by humans, affects biological systems in certain areas.

Carbon cycle. Carbon is the basic biogeochemical element. The atmosphere contains carbon in the form of carbon dioxide gas. There is a large annual flux of atmospheric carbon dioxide to and from forests and terrestrial biota, amounting to nearly 7% of total atmospheric carbon dioxide. This is because carbon dioxide is used by plants to produce organic matter through photosynthesis, and when the organic matter is broken down through respiration, carbon dioxide is released to the atmosphere. The concentration of atmospheric carbon dioxide shows a yearly oscillation because there is a strong seasonal annual cycle of photosynthesis and respiration in the Northern Hemisphere.

Photosynthesis and respiration in the carbon cycle can be represented by the reaction below. Breakdown of organic matter via respiration is accomplished mainly by

$$CO_2 + H_2O \underset{\text{respiration}}{\overset{\text{photosynthesis}}{\rightleftharpoons}} CH_2O + O_2$$

bacteria that live in soils, sediments, and natural waters. There is a very large reservoir of terrestrial carbon in carbonate rocks, which contain calcium carbonate ($CaCO_3$), and in rocks such as shales which contain organic carbon. Major exchange of carbon between rocks and the atmosphere is very slow, on the scale of thousands to millions of years, compared to exchange between plants and the atmosphere, which can even be seasonal.

The oceans taken as a whole represent a major reservoir of carbon. Carbon in the oceans occurs primarily as dissolved $(HCO_3)^-$ and to a lesser extent as dissolved carbon dioxide gas and carbonate ion $[(CO_3)^{2-}]$. The well-mixed surface ocean (the top 250 ft or 75 m) rapidly exchanges carbon dioxide with the atmosphere. However, the deep oceans are cut off from the atmosphere and mix with it on a long-term time scale of about 1000–2000 years. Most of the biological activity in the oceans occurs in the surface (or shallow) water where there is light and photosynthesis can occur. *See* MARITIME METEOROLOGY.

The main biological process in seawater is photosynthetic production of organic matter by phytoplankton. Some of this organic matter is eaten by animals, which are in turn eaten by larger animals farther up in the food chain. Almost all of the organic matter along the food chain is ultimately broken down by bacterial respiration, which occurs primarily in shallow water, and the carbon dioxide is quickly recycled to the atmosphere. *See* FOOD WEB; NEARSHORE PROCESSES; SEAWATER.

Another major biological process is the secretion of shells and other hard structures by marine organisms. A biogeochemical cycle of calcium and bicarbonate exists within the oceans, linking the deep and shallow water areas. Bottom dwellers in shallow water, such as corals, mollusks, and algae, provide calcium carbonate skeletal debris. Since the shallow waters are saturated with respect to calcium carbonate, this debris accumulates on the bottom and is buried, providing the minerals that form carbonate rocks such as limestone and dolomite. Calcium carbonate is also derived from the shells of organisms inhabiting surface waters of the deep ocean; these are tiny, floating plankton such as foraminiferans, pteropods, and coccoliths. Much of the calcium carbonate from this source dissolves as it sinks into the deeper ocean waters, which are undersaturated with respect to calcium carbonate. The undissolved calcium carbonate accumulates on the bottom to form deep-sea limestone. The calcium and the bicarbonate ions [Ca^{2+} and $(HCO_3)^-$] dissolved in the deep ocean water eventually are carried to surface and shallow water, where they are removed by planktonic and bottom-dwelling organisms to form their skeletons. *See* CARBONATE MINERALS; LIMESTONE.

The long-term biogeochemical carbon cycle occurs over millions of years when the calcium carbonate and organic matter that are buried in sediments are returned to the Earth's surface. There, weathering occurs which involves the reaction of oxygen with sedimentary organic matter with the release of carbon dioxide and water (analogous to respiration), and the reaction of water and carbon dioxide with carbonate rocks with the release of calcium and bicarbonate ions. *See* WEATHERING PROCESSES.

Fossil fuels (coal and oil) represent a large reservoir of carbon. Burning of fossil fuels releases carbon dioxide to the atmosphere, and an increase in the atmospheric concentration of carbon dioxide has been observed since the mid-1950s. While much of the increase is attributed to fossil fuels, deforestation by humans accompanied by the decay or burning of trees is another possible contributor to the problem.

When estimates are made of the amount of fossil fuels burned from 1959 to 1980, only about 60% of the carbon dioxide released can be accounted for in the atmospheric increase in carbon dioxide. The remaining 40% is known as excess carbon dioxide. The surface oceans are an obvious candidate for storage of most of the excess carbon dioxide by the reaction of carbon dioxide with dissolved carbonate to form bicarbonate. Because the increase in bicarbonate concentration in surface waters due to excess carbon dioxide uptake would be small, it is difficult to detect whether such a change has occurred. Greater quantities of excess carbon dioxide could be stored as bicarbonate in the deeper oceans, but this process takes a long time because of the slow rate of mixing between surface and deep oceans.

An increase in atmospheric carbon dioxide is of concern because of the greenhouse effect. The carbon dioxide traps heat in the atmosphere; notable increases in atmospheric carbon dioxide should cause an increase in the Earth's surface temperature by as much as several degrees. This temperature increase would be greater at the poles, and the effects could include melting of polar ice, a rise in sea level, and changes in rainfall distribution, with droughts in interior continental areas such as the Great Plains of the United States. *See* DROUGHT; GREENHOUSE EFFECT.

Nitrogen cycle. Nitrogen is dominantly a biogenic element and has no important mineral forms. It is a major atmospheric constituent with a number of gaseous forms, including molecular nitrogen gas (N_2), nitrogen dioxide (NO_2), nitric oxide (NO), ammonia (NH_3), and nitrous oxide (N_2O). As an essential component of plant and animal matter, it is extensively involved in biogeochemical cycling. On a global basis, the nitrogen cycle is greatly affected by human activities.

Nitrogen gas (N_2) makes up 80% of the atmosphere by volume; however, nitrogen is unreactive in this form. In order to be available for biogeochemical cycling by organisms, nitrogen gas must be fixed, that is, combined with oxygen, carbon, or hydrogen. There are three major sources of terrestrial fixed nitrogen: biological nitrogen fixation by plants, nitrogen fertilizer application, and rain and particulate dry deposition of previously fixed nitrogen. Biological fixation occurs in plants such as legumes (peas and beans) and lichens in trees, which incorporate nitrogen from the atmosphere into their living matter; about 30% of worldwide biological fixation is due to human cultivation of these plants. Nitrogen fertilizers contain industrially fixed nitrogen as both nitrate and ammonium.

Fixed nitrogen in rain is in the forms of nitrate [$(NO_3)^-$] and ammonium [$(NH_4)^+$] ions. Major sources of nitrate, which is derived from gaseous atmospheric nitrogen dioxide (and nitric oxide), include (in order of importance) combustion of fossil fuel, especially by automobiles; forest fires (mostly caused by humans); and lightning. Nitrate in rain, in addition to providing soluble fixed nitrogen for photosynthesis, contributes nitric acid (HNO_3), a major component of acid rain. Sources of ammonium, which is derived from atmospheric ammonia gas (NH_3), include animal and human wastes, soil loss from decomposition of organic matter, and fertilizer release.

The basic land nitrogen cycle involves the photosynthetic conversion of the nitrate and ammonium ions dissolved in soil water into plant organic material. Once formed, the organic matter may be stored or broken down. Bacterial decomposition of organic matter (ammonification) produces soluble ammonium ion which can then be either taken up again in photosynthesis, released to the atmosphere as ammonia gas, or oxidized by bacteria to nitrate ion (nitrification).

Nitrate ion is also soluble, and may be used in photosynthesis. However, part of the nitrate may undergo reduction (denitrification) by soil bacteria to nitrogen gas or to nitrous oxide which are then lost to the atmosphere. Compared to the land carbon cycle, the land nitrogen cycle is considerably more complex, and because of the large input of fixed nitrogen by humans, it is possible that nitrogen is building up on land. However, this is difficult to determine since the amount of nitrogen gas recycled to the atmosphere is not known and any changes in the atmospheric nitrogen concentration would be too small to detect. *See* Nitrogen cycle.

The oceans are another major site of nitrogen cycling: the amount of nitrogen cycled biogenically, through net primary photosynthetic production, is about 13 times that on land. The main links between the terrestrial and the oceanic nitrogen cycles are the atmosphere and rivers. Nitrogen gases carried in the atmosphere eventually fall as dissolved inorganic (mainly nitrate) and organic nitrogen and particulate organic nitrogen in rain on the oceans. The flux of river nitrogen lost from the land is only about 9% of the total nitrogen recycled biogeochemically on land each year and only about 25% of the terrestrial nitrogen flux from the biosphere to the atmosphere.

River nitrogen is an important nitrogen source to the oceans; however, the greatest amount of nitrogen going into ocean surface waters comes from the upwelling of deeper waters, which are enriched in dissolved nitrate from organic recycling at depth. Dissolved nitrate is used extensively for photosynthesis by marine organisms, mainly plankton. Bacterial decomposition of the organic matter formed in photosynthesis results in the release of dissolved ammonium, some of which is used directly in photosynthesis. However, most undergoes nitrification to form nitrate, and much of the nitrate may undergo denitrification to nitrogen gas which is released to the atmosphere. A small amount of organic-matter nitrogen is buried in ocean sediments, but this accounts for a very small amount of the nitrogen recycled each year. There are no important inorganic nitrogen minerals such as those that exist for carbon and phosphorus, and thus there is no mineral precipitation and dissolution. *See* UPWELLING.

Phosphorus cycle. Phosphorus, an important component of organic matter, is taken up and released in the form of dissolved inorganic and organic phosphate. Phosphorus differs from nitrogen and carbon in that it does not form stable atmospheric gases and therefore cannot be obtained from the atmosphere. It does form minerals, most prominently apatite (calcium phosphate), and insoluble iron (Fe) and aluminum (Al) phosphate minerals, or it is adsorbed on clay minerals. The amount of phosphorus used in photosynthesis on land is large compared to phosphorus inputs to the land. The major sources of phosphorus are weathering of rocks containing apatite and mining of phosphate rock for fertilizer and industry. A small amount comes from precipitation and dry deposition. *See* PHOSPHATE MINERALS.

Phosphorus is lost from the land principally by river transport, which amounts to only 7% of the amount of phosphorus recycled by the terrestrial biosphere; overall, the terrestrial biosphere conserves phosphorus. Humans have greatly affected terrestrial phosphorus: deforestation and agriculture have doubled the amount of phosphorus weathering; phosphorus is added to the land as fertilizers and from industrial wastes, sewage, and detergents. Thus, about 75% of the terrestrial input is anthropogenic; in fact, phosphorus may be building up on the land.

In the oceans, phosphorus occurs predominantly as dissolved orthophosphates [PO_4^{3-}, $(HPO_4)^{2-}$ and $(H_2PO_4)^-$]. Since it follows the same cycle as do carbon and nitrogen, dissolved orthophosphate is depleted in surface ocean waters where both photosynthesis and respiration occur, and the concentration builds up in deeper water where organic matter is decomposed by bacterial respiration. The major phosphorus input to the oceans is from rivers, with about 5% coming from rain. However, 75% of the river phosphorus load is due to anthropogenic pollutants; humans have changed the ocean balance of phosphorus. Most of the dissolved oceanic orthophosphate is derived from recycled organic matter. The output of phosphorus from the ocean is predominantly biogenic: organic phosphorus is buried in sediments; a smaller amount is removed by adsorption on volcanic iron oxides. In the geologic past, there was a much greater inorganic precipitation of phosphorite (apatite) from seawater than at present, and this has resulted in the formation of huge deposits which are now mined.

Nutrients in lakes. Biogeochemical cycling of phosphorus and nitrogen in lakes follows a pattern that is similar to oceanic cycling: there is nutrient depletion in surface waters and enrichment in deeper waters. Oxygen consumption by respiration in deep water sometimes leads to extensive oxygen depletion with adverse effects on fish and other biota. In lakes, phosphorus is usually the limiting nutrient.

Many lakes have experienced greatly increased nutrient (nitrogen and phosphorus) input due to human activities. This stimulates a destructive cycle of biological activity: very high organic productivity, a greater concentration of plankton, and more photosynthesis. The result is more organic matter falling into deep water with increased depletion of oxygen and greater accumulation of organic matter on the lake bottom. This process, eutrophication, can lead to adverse water quality and even to the filling up of small lakes with organic matter. *See* EUTROPHICATION; LIMNOLOGY.

Biogeochemical sulfur cycle. A dominant flux in the global sulfur cycle is the release of 65–70 teragrams of sulfur per year to the atmosphere from burning of fossil fuels. Sulfur contaminants in these fuels are released to the atmosphere as sulfur dioxide (SO_2) which is rapidly converted to aerosols of sulfuric acid (H_2SO_4), the primary contributor to acid rain. Forest burning results in an additional release of sulfur dioxide.

Overall, the broad range of human activities contribute 75% of sulfur released into the atmosphere. Natural sulfur sources over land are predominantly the release of reduced biogenic sulfur gases [mainly hydrogen sulfide (H_2S) and dimethyl sulfide] from marine tidal flats and inland waterlogged soils and, to much lesser extent, the release of volcanic sulfur. The atmosphere does not have an appreciable reservoir of sulfur because most sulfur gases are rapidly returned (within days) to the land in rain and dry deposition. There is a small net flux of sulfur from the atmosphere over land to the atmosphere over the oceans.

Ocean water constitutes a large reservoir of dissolved sulfur in the form of sulfate ions [$(SO_4)^{2-}$]. Some of this sulfate is thrown into the oceanic atmosphere as sea salt from evaporated sea spray, but most of this is rapidly returned to the oceans. Another major sulfur source in the oceanic atmosphere is the release of oceanic biogenic sulfur gases (such as dimethyl sulfide) from the metabolic activities of oceanic organisms and organic matter decay. Marine organic matter contains a small amount of sulfur, but sulfur is not a limiting element in the oceans.

Another large flux in the sulfur cycle is the transport of dissolved sulfate in rivers. However, as much as 43% of this sulfur may be due to human activities, both from burning of fossil fuels and from fertilizers and industrial wastes. The weathering of sulfur minerals, such as pyrite (FeS_2) in shales, and the evaporite minerals, gypsum and anhydrite, make an important contribution to river sulfate. The major mechanism for removing sulfate from ocean water is the formation and burial of pyrite in oceanic sediments, primarily nearshore sediments. (The sulfur fluxes of sea salt and biogenic sulfur gases do not constitute net removal from the oceans since the sulfur is recycled to the oceans.)

Biogeochemical cycles and atmospheric oxygen. The main processes affecting atmospheric oxygen are photosynthesis and respiration; however, these processes are almost perfectly balanced against one another and, thus, do not exert a simple effect on oxygen levels. Only the very small excess of photosynthesis over respiration, manifested by the burial of organic matter in sediments, is important in raising the level of oxygen. This excess is so small, and the reservoir of oxygen so large, that if the present rate of organic carbon burial were doubled and the other rates remained constant, it would take 5–10 million years for the amount of atmospheric oxygen to double. Nevertheless, this is a relatively short time from a geological perspective. *See* ATMOSPHERE; BIOSPHERE; GEOCHEMISTRY; HYDROSPHERE; MARINE SEDIMENTS.   [E.K.B.; R.A.Ber.]

Biogeography   A synthetic discipline that describes the distributions of living and fossil species of plants and animals across the Earth's surface as consequences of ecological and evolutionary processes. Biogeography overlaps and complements many biological disciplines, especially community ecology, systematics, paleontology, and evolutionary biology.

Based on relatively complete compilations of species within well-studied groups, such as birds and mammals, biogeographers identified six different realms within which species tend to be closely related and between which turnovers in major groups of species are observed (see table). The boundaries between biogeographic realms are less distinct than was initially thought, and the distribution of distinctive groups such as parrots, marsupials, and southern beeches (*Nothofagus* spp.) implies that modern-day biogeographic realms have been considerably mixed in the past. *See* PALEONTOLOGY.

Two patterns of species diversity have stimulated a great deal of progress in developing ecological explanations for geographic patterns of species richness. The first is that the number of species increases in a regular fashion with the size of the geographic area

Biogeographic realms		
Realm	Continental areas included	Examples of distinctive or endemic taxa
Palearctic	Temperate Eurasia and northern Africa	Hynobiid salamanders
Oriental	Tropical Asia	Lower apes
Ethiopian	Sub-Saharan Africa	Great apes
Australian	Australia, New Guinea, and New Zealand	Marsupials
Nearctic	Temperate North America	Pronghorn antelope, ambystomatid salamanders
Neotropic	Subtropical Central America and South America	Hummingbirds, antbirds, marmosets

being considered. The second is the nearly universal observation that there are more species of plants and animals in tropical regions than in temperate and polar regions.

In order to answer questions about why there are a certain number of species in a particular geographic region, biogeography has incorporated many insights from community ecology. Species number at any particular place depends on the amount of resources available there (ultimately derived from the amount of primary productivity), the number of ways those resources can be apportioned among species, and the different kinds of ecological requirements of the species that can colonize the region. The equilibrium theory of island biogeography arose as an application of these insights to the distribution of species within a specified taxon across an island archipelago. This theory generated specific predictions about the relationships among island size and distance from a colonization source with the number and rate of turnover of species. Large islands are predicted to have higher equilibrium numbers of species than smaller islands; hence, the species area relationship can be predicted in principle from the ecological attributes of species. Experimental and observational studies have confirmed many predictions made by this theory. *See* ECOLOGICAL COMMUNITIES; ISLAND BIOGEOGRAPHY.

The latitudinal gradient in species richness has generated a number of explanations, none of which has been totally satisfactory. One explanation is based on the observation that species with more temperate and polar distributions tend to have larger geographic ranges than species from tropical regions. It is thought that since species with large geographic ranges tend to withstand a wider range of physical and biotic conditions, this allows them to penetrate farther into regions with more variable climates at higher latitudes. If this were true, then species with smaller geographic ranges would tend to concentrate in tropical regions where conditions are less variable. While this might be generally true, there are many examples of species living in high-latitude regions that have small geographic regions. *See* ALTITUDINAL VEGETATION ZONES.

Biogeography is entering a phase where data on the spatial patterns of abundance and distribution of species of plants and animals are being analyzed with sophisticated mathematical and technological tools. Geographic information systems and remote sensing technology have provided a way to catalog and map spatial variation in biological processes with a striking degree of detail and accuracy. These newer technologies have stimulated research on appropriate methods for modeling and analyzing biogeographic patterns. Modern techniques of spatial modeling are being applied to geographic information systems data to test mechanistic explanations for biogeographic patterns that could not have been attempted without the advent of the appropriate technology. *See* GEOGRAPHIC INFORMATION SYSTEMS.   [B.A.M.]

Bioherm A lenslike to moundlike structure of strictly organic origin. This term involves two concepts: shape and organic internal composition.

The term shape denotes original topographic relief above the sea floor as well as a three-dimensional quality: crudely conical (sugar loaf–shaped) or ellipsoidal (bread loaf–shaped). Such forms are massive or unbedded, their upbuilding resulting from the very rapid rate of accretion of organic carbonate once it starts in a favorable locality. There are size limitations: bioherms a meter or so in diameter are known, and some rise 300 ft (100 m) or more above the sea floor.

The second concept, organic internal composition, not only embraces sessile, bottom-dwelling organisms forming frame-building reefy bondstone but also includes piles of organically derived debris replete with organisms which encrust it and cement it in place. Even inorganic cement precipitated from marine and meteoric water is known to play a role in a buildup of massive, moundlike structures. When the internal material is coarse and identifiable, no problem is encountered in applying the term bioherm in its original sense. In some buildups, however, an appreciable amount of lime mud is present, and relatively few organisms are identifiable which could have secreted, bound, encrusted, or trapped carbonate mud (as in some early Carboniferous mounds). In such cases, problems arise in applying that part of the definition based on internal composition. In fact, as a field term, bioherm can hardly ever be completely diagnostic because careful petrographic study is commonly necessary for details of internal composition to be ascertained.

Bioherms may occur on shelves (where they are normally lens-shaped) or in shallow basins, often at the basin margin. In the latter position, they have been called reef knolls or pinnacle reef. See BIOSTROME; REEF; STROMATOLITE.   [J.L.Wi.]

Biomass The organic materials produced by plants, such as leaves, roots, seeds, and stalks. In some cases, microbial and animal metabolic wastes are also considered biomass. The term "biomass" is intended to refer to materials that do not directly go into foods or consumer products but may have alternative industrial uses. Common sources of biomass are (1) agricultural wastes, such as corn stalks, straw, seed hulls, sugarcane leavings, bagasse, nutshells, and manure from cattle, poultry, and hogs; (2) wood materials, such as wood or bark, sawdust, timber slash, and mill scrap; (3) municipal waste, such as waste paper and yard clippings; and (4) energy crops, such as poplars, willows, switchgrass, alfalfa, prairie bluestem, corn (starch), and soybean (oil).

Biomass is a complex mixture of organic materials, such as carbohydrates, fats, and proteins, along with small amounts of minerals, such as sodium, phosphorus, calcium, and iron. The main components of plant biomass are carbohydrates (approximately 75%, dry weight) and lignin (approximately 25%), which can vary with plant type. The carbohydrates are mainly cellulose or hemicellulose fibers, which impart strength to the plant structure, and lignin, which holds the fibers together. Some plants also store starch (another carbohydrate polymer) and fats as sources of energy, mainly in seeds and roots (such as corn, soybeans, and potatoes).

A major advantage of using biomass as a source of fuels or chemicals is its renewability. Utilizing sunlight energy in photosynthesis, plants metabolize atmospheric carbon dioxide to synthesize biomass. An estimated 140 billion metric tons of biomass are produced annually.

Major limitations of solid biomass fuels are difficulty of handling and lack of portability for mobile engines. To address these issues, research is being conducted to convert solid biomass into liquid and gaseous fuels. Both biological means (fermentation) and chemical means (pyrolysis, gasification) can be used to produce fluid biomass fuels. For example, methane gas is produced in China for local energy needs by anaerobic microbial digestion of human and animal wastes. Ethanol for automotive fuels is currently produced from starch biomass in a two-step process: starch is enzymatically hydrolyzed into glucose; then yeast is used to convert the glucose into ethanol. About 1.5 billion gallons of ethanol are produced from starch each year in the United States.

[B.Y.Ta.]

Biome   A major community of plants and animals having similar life forms or morphological features and existing under similar environmental conditions. The biome, which may be used at the scale of entire continents, is the largest useful biological community unit. In Europe the equivalent term for biome is major life zone, and throughout the world, if only plants are considered, the term used is formation. *See* ECOLOGICAL COMMUNITIES.

Each biome may contain several different types of ecosystems. For example, the grassland biome may contain the dense tallgrass prairie with deep, rich soil, while the desert grassland has a sparse plant canopy and a thin soil. However, both ecosystems have grasses as the predominant plant life form, grazers as the principal animals, and a climate with at least one dry season. Additionally, each biome may contain several successional stages. A forest successional sequence may include grass dominants at an early stage, but some forest animals may require the grass stage for their habitat, and all successional stages constitute the climax forest biome. *See* DESERT; ECOLOGICAL SUCCESSION; ECOSYSTEM.

Distributions of animals are more difficult to map than those of plants. The life form of vegetation reflects major features of the climate and determines the structural nature of habitats for animals. Therefore, the life form of vegetation provides a sound basis for ecologically classifying biological communities. Terrestrial biomes are usually identified by the dominant plant component, such as the temperate deciduous forest. Marine biomes are mostly named for physical features, for example, for marine upwelling, and for relative locations, such as littoral. Many biome classifications have been proposed, but a typical one might include several terrestrial biomes such as desert, tundra, grassland, savanna, coniferous forest, deciduous forest, and tropical forest. Aquatic biome examples are fresh-water lotic (streams and rivers), fresh-water lentic (lakes and ponds), and marine littoral, neritic, upwelling, coral reef, and pelagic. *See* FRESH-WATER ECOSYSTEM; MARINE ECOLOGY.   [P.Ri.]

Biometeorology   A branch of meteorology and ecology that deals with the effects of weather and climate on plants, animals, and humans.

The principal problem for living organisms is maintaining an acceptable thermal equilibrium with their environment. Organisms have natural techniques for adapting to adverse conditions. These techniques include acclimatization, dormancy, and hibernation, or in some cases an organism can move to a more favorable environment or microenvironment. Humans often establish a favorable environment through the use of technology. *See* MICROMETEOROLOGY.

Homeotherms, that is, humans and other warm-blooded animals, maintain relatively constant body temperatures under a wide range of ambient thermal and radiative conditions through physiological and metabolic mechanisms. Poikilotherms, that is, cold-blooded animals, have a wide range in body temperature that is modified almost exclusively by behavioral responses. Plants also experience a wide range of temperatures, but because of their immobility they have less ability than animals to adapt to extreme changes in environment.

Humans are physically adapted to a narrow range of temperature, with the metabolic mechanism functioning best at air temperatures around 77°F (25°C). There is a narrow range above and below this temperature where survival is possible. To regulate heat loss, warm-blooded animals developed hair, fur, and feathers. Humans invented clothing and shelter. The amount of insulation required to maintain thermal equilibrium is governed by the conditions in the atmospheric environment. There are a limited number of physiological mechanisms, controlled by the hypothalamus, that regulate body heat.

Clothing, shelter, and heat-producing objects can largely compensate for environmental cold, but with extensive exposure to cold, vasoconstriction in the peripheral organs can lead to chilblains and frostbite on the nose, ears, cheeks, and toes. This exposure is expressed quantitatively as a wind chill equivalent temperature that is a function of air temperature and wind speed. The wind chill equivalent temperature is a measure of convective heat loss and describes a thermal sensation equivalent to a lower-than-ambient temperature under calm conditions, that is, for wind speeds below 4 mi/h (1.8 m/s). Persons exposed to extreme cold develop hypothermia, which may be irreversible when the core temperature drops below 91°F (33°C.)

The combination of high temperature with high humidity leads to a very stressful thermal environment. The combination of high temperature with low humidity leads to a relatively comfortable thermal environment, but such conditions create an environment that has a very high demand for water.

Conditions of low humidity exist principally in subtropical deserts, which have the highest daytime temperatures observed at the Earth's surface. Human and animal bodies are also exposed to strong solar radiation and radiation reflected from the surface of the sand. This combination makes extraordinary demands on the sweat mechanism. Water losses of 1.06 quarts (1 liter) per hour are common in humans and may be even greater with exertion. Unless the water is promptly replaced by fluid intake, dehydration sets in. *See* DESERT; HUMIDITY.

When humans are exposed to warm environments, the first physiological response is dilation of blood vessels, which increases the flow of blood near the skin. The next response occurs through sweating, panting, and evaporative cooling. Since individuals differ in their physiological responses to environmental stimuli, it is difficult to develop a heat stress index based solely on meteorological variables. Nevertheless, several useful indices have been developed.

Since wind moves body heat away and increases the evaporation from a person, it should be accounted for in developing comfort indices describing the outdoor environment. One such index, used for many years by heating and ventilating engineers, is the effective temperature. People will feel uncomfortable at effective temperatures above 81°F (27°C) or below 57°F (15°C); between 63°F (17°C) and 77°F (25°C) they will feel comfortable.

Both physiological and psychological responses to weather changes (meteorotropisms) are widespread and generally have their origin in some bodily impairment. Reactions to weather changes commonly occur in anomalous skin tissue such as scars and corns; changes in atmospheric moisture cause differential hygroscopic expansion and contraction between healthy and abnormal skin, leading to pain. Sufferers from rheumatoid arthritis are commonly affected by weather changes; both pain and swelling of affected joints have been noted with increased atmospheric humidity. Sudden cooling can also trigger such symptoms. Clinical tests have shown that in these individuals the heat regulatory mechanism does not function well, but the underlying cause is not understood.

Weather is a significant factor in asthma attacks. Asthma as an allergic reaction may, in rare cases, be directly provoked by sudden changes in temperature that occur after passage of a cold front. Often, however, the weather effect is indirect, and attacks are caused by airborne allergens, such as air pollutants and pollen. An even more indirect relationship exists for asthma attacks in autumn, which often seem to be related to an early outbreak of cold air. This cold air initiates home or office heating, and dormant dust or fungi from registers and radiators are convected into rooms, irritating allergic persons.

A variety of psychological effects have also been attributed to heat. They are vaguely described as lassitude, decrease in men-j tal and physical performance, and increased irritability. Similar reactions to weather have been described for domestic animals, particularly dogs; hence hot, humid days are sometimes known as dog days.

Meteorological and seasonal changes in natural illumination have a major influence on animals. Photoperiodicity is widespread. The daily cycle of illumination triggers the feeding cycle in many species, especially birds. In insectivores the feeding cycle may result from the activities of the insects, which themselves show temperature-influenced cycles of animation. Bird migration may be initiated by light changes, but temperature changes and availability of food are also involved. In the process of migration, especially over long distances, birds have learned to take advantage of prevailing wind patterns. In humans, light deprivation, as is common in the cold weather season in higher latitudes, is suspected as a cause of depression. Exposure to high-intensity light for several hours has been found to be an effective means of treating this depression.

Humans and animals often exhibit a remarkable ability to adapt to harsh or rapidly changing environmental conditions. An obvious means of adaptation is to move to areas where environmental conditions are less severe; examples are birds and certain animals that migrate seasonally, animals that burrow into the ground, and animals that move to shade or sunshine depending on weather conditions. Animals can acclimatize to heat and cold. The acclimatization process is generally complete within 2–3 weeks of exposure to the stressful conditions. For example, in hot climates heat regulation is improved by the induction of sweating at a lower internal body temperature and by the increase of sweating rates. The acclimatization to cold climates is accomplished by increase in the metabolic rate, by improvement in the insulating properties of the skin, and by the constriction of blood vessels to reduce the flow of blood to the surface.

Unlike humans and animals, plants cannot move from one location to another; therefore, they must adapt genetically to their atmospheric environment. Plants are often characteristic for their climatic zone, such as palms in the subtropics and birches or firs in regions with cold winters. Whole systems of climatic classification are based on the native floras. See ALTITUDINAL VEGETATION ZONES; ECOLOGY; METEOROLOGY.   [B.L.B.; H.E.L.]

Biopyribole   A member of a chemically diverse, structurally related group of minerals that comprise substantial fractions of both the Earth's crust and upper mantle. The term was coined by Albert Johannsen in 1911; it is a contraction of biotite (a mica), pyroxene, and amphibole.

The pyroxene minerals contain single chains of corner-sharing silicate (SiO_4) tetrahedra, and the amphiboles contain double chains. Likewise, the micas and other related biopyriboles (talc, pyrophyllite, and the brittle micas) contain two-dimensionally infinite silicate sheets, which result in their characteristic sheetlike physical properties. In the pyroxenes and amphiboles, the silicate chains are articulated to strips of octahedrally coordinated cations, such as magnesium and iron; and in the sheet biopyriboles, the silicate sheets are connected by two-dimensional sheets of such cations. In addition to the classical single-chain, double-chain, and sheet biopyriboles, several biopyriboles that contain triple silicate chains have been discovered. See BIOTITE; MICA; PYROXENE; SILICATE MINERALS.

Pyroxenes, amphiboles, and micas of various compositions can occur in igneous, metamorphic, and sedimentary rocks. Pyroxenes are the second most abundant minerals in the Earth's crust (after feldspars) and in the upper mantle (after olivine). Unlike the pyroxenes, amphiboles, and micas, the wide-chain biopyriboles do not occur as abundant minerals in a wide variety of rock types. However, some of them may be widespread in nature as components of fine-grain alteration products of pyroxenes and amphiboles and as isolated lamellae in other biopyriboles. *See* IGNEOUS ROCKS; METAMORPHIC ROCKS; SEDIMENTARY ROCKS.   [D.R.V.]

Biosphere   All living organisms and their environments at the surface of the Earth. Included in the biosphere are all environments capable of sustaining life above, on, and beneath the Earth's surface as well as in the oceans. Consequently, the biosphere overlaps virtually the entire hydrosphere and portions of the atmosphere and outer lithosphere. *See* ATMOSPHERE; HYDROSPHERE; LITHOSPHERE.

Neither the upper nor lower limits of the biosphere are sharp. Spores of microorganisms can be carried to considerable heights in the atmosphere, but these are resting stages that are not actively metabolizing. A variety of organisms inhabit the ocean depths, including the giant tubeworms and other creatures that were discovered living around hydrothermal vents. Evidence exists for the presence of bacteria in oil reservoirs at depths of about 6600 ft (2000 m) within the Earth. The bacteria are apparently metabolically active, utilizing the paraffinic hydrocarbons of the oils as an energy source. These are extreme limits to the biosphere; most of the mass of living matter and the greatest diversity of organisms are within the upper 330 ft (100 m) of the lithosphere and hydrosphere, although there are places even within this zone that are too dry or too cold to support much life. Most of the biosphere is within the zone which is reached by sunlight and where liquid water exists.

The biosphere is characterized by the interrelationship of living things and their environments. Communities are interacting systems of organisms tied to their environments by the transfer of energy and matter. Such a coupling of living organisms and the non-living matter with which they interact defines an ecosystem. An ecosystem may range in size from a small pond, to a tropical forest, to the entire biosphere. Ecologists group the terrestrial parts of the biosphere into about 12 large units called biomes. Examples of biomes include tundra, desert, grassland, and boreal forest. *See* BIOME; ECOLOGICAL COMMUNITIES; ECOSYSTEM.

Human beings are part of the biosphere, and some of their activities have an adverse impact on many ecosystems and on themselves. As a consequence of deforestation, urban sprawl, spread of pollutants, and overharvesting, both terrestrial and marine ecosystems are being destroyed or diminished, populations are shrinking, and many species are dying out. In addition to causing extinctions of some species, humans are expanding the habitats of other organisms, sometimes across oceanic barriers, through inadvertent transport and introduction into new regions. Humans also add toxic or harmful substances to the outer lithosphere, hydrosphere, and atmosphere. Many of these materials are eventually incorporated into or otherwise affect the biosphere, and water and air supplies in some regions are seriously fouled.   [R.M.M.]

Biostrome   An evenly bedded and generally horizontally layered stratum composed mostly of organic remains, normally considered to be those of sedentary organisms which lived, died, and were buried essentially in place.

The criterion of formation by in-place growth of organisms is subject to some interpretation. Crinoidal limestones obviously resulted from the accumulation of decayed pieces of millions of these stalked echinoderms, often with accompanying detritus of associated fenestrate bryozoans. Usually it is impossible to ascertain whether such debris dropped vertically a few centimeters or meters through the water column as the organisms died and collapsed (an essentially in-place deposit) or whether the layers of crinoidal grainstone were piled mechanically by currents. The same is true of coquinas of many other thin-shelled calcareous tests, such as those of brachiopods, bryozoans, and trilobites.

Biostromal layers need not have been horizontally deposited when they occur as flanking beds around organic buildups. Dips of up to 25 or 30° are possible here. Such biostromes are probably veneers of sessile organisms which lived somewhat above the realm of deposition and were buried as sediment cascaded down the flank of a mound.

[J.L.Wi.]

Biotite   An iron-magnesium-rich layer silicate; it is also known as black mica. Biotite is the most abundant of the mica group of minerals. The name is derived from that of the French chemist J. Biot. The formula for the ideal end member, phlogopite, is $KMg_3AlSi_3O_{10}(OH)_2$. The more general formula is $AX_3Y_4O_{12}(Z)_2$, where A (interlayer cation) = K, Na, Ca, Ba, or vacancies; X (octahedral cations) = Li, Mg, Fe^{2+}, Fe^{3+}, Al, Ti, or vacancies; and Y (tetrahedral cation) = Fe^{3+}, Al, Si; Z = (OH), F, Cl, O^{2-}. This formula is more indicative of the wide range of compositions known for this mineral. Biotite has no commercial value, but vermiculite, an alteration product of magnesium-rich biotite, is used as insulation, as packing material, and as an ingredient for potting soils. *See* VERMICULITE.

Biotites are found commonly in igneous and metamorphic rocks. They are the common ferromagnesian phase in most granitic rocks, and are also found in some siliceous and intermediate volcanic rocks. In basaltic rocks biotite sometimes occurs in the crystalline groundmass, and is a common late interstitial phase in gabbroic rocks. It has been recognized in samples of the Earth's mantle found as inclusions in volcanic rocks. Biotites are not stable at the surface of the Earth, as they decompose by both hydrolysis and oxidation when exposed to the Earth's atmosphere. They alter to vermiculite, chlorite, and iron oxides, and thus are uncommon in sedimentary rocks. Biotites are important constituents of metamorphic rocks such as schist and gneiss, and the first appearance of biotite is an important marker in metamorphism. Biotite persists to very high grades of metamorphism, where it reacts with quartz to form granulites made up of potassium feldspar and orthopyroxene, garnet, or cordierite, in addition to quartz and plagioclase. Under conditions of ultrametamorphism, biotite reacts with quartz, plagioclase, and alkali feldspar to form siliceous melts. Biotite is also a common gangue mineral in ore deposits. The mineral has been used as an indicator of H_2O, HF, O_2, and S_2 activities in both rock- and ore-forming processes. *See* IGNEOUS ROCKS; METAMORPHIC ROCKS; METAMORPHISM; MICA; SILICATE MINERALS.

[D.R.W.]

Bitumen   A term used to designate naturally occurring or pyrolytically obtained substances of dark to black color consisting almost entirely of carbon and hydrogen with very little oxygen, nitrogen, and sulfur. Bitumen may be of variable hardness and volatility, ranging from crude oil to asphaltites, and is largely soluble in carbon disulfide. *See* ASPHALT AND ASPHALTITE.

[I.A.B.]

Black Sea   A semienclosed marginal sea with an area of 420,000 km² (160,000 mi²) bounded by Turkey to the south, Georgia to the east, Russia and Ukraine to the north, and Romania and Bulgaria to the west. The physical and chemical structure of the Black Sea is critically dependent on its hydrological balance. As a result, it is the world's largest anoxic basin. It has recently experienced numerous types of environmental stress.

The Black Sea consists of a large basin with a depth of about 2200 m (7200 ft). The continental shelf is mostly narrow except for the broad shelf in the northwest region. Fresh-water input from rivers, especially the Danube, Dniester, and Don, and precipitation exceeds evaporation. Low-salinity surface waters are transported to the Mediterranean as a surface outflow. High-salinity seawater from the Mediterranean enters the Black Sea as a subsurface inflow through the Bosporus. This estuarine circulation (seawater inflow at depth and fresh-water outflow at the surface) results in an unusually strong vertical density gradient determined mainly by the salinity. Thus the Black Sea has a two-layered structure with a lower-salinity surface layer and a higher-salinity deep layer.

The vertical stratification has a strong effect on the chemistry of the sea. Respiration of particulate organic carbon sinking into the deep water has used up all the dissolved oxygen. Thus, conditions favor bacterial sulfate reduction and high sulfide concentrations. As a result, the Black Sea is the world's largest anoxic basin and is commonly used as a modern analog of an environment favoring the formation of organic-rich black shales observed in the geological sedimentary record.

Before the 1970s the Black Sea had a highly diverse and healthy biological population. Its species composition was similar to that of the Mediterranean but with less quantity. The phytoplankton community was characterized by a large diatom bloom in May-June followed by a smaller dinoflagellate bloom. The primary zooplankton were copepods, and there were 170 species of fish, including large commercial populations of mackerel, bonito, anchovies, herring, carp, and sturgeon.

Since about 1970 there have been dramatic changes in the food web due to anthropogenic effects and invasions of new species. It is now characterized as a nonequilibrium, low-diversity, eutrophic state. The large increase in input of nitrogen due to eutrophication and decrease in silicate due to dam construction have increased the frequency of noxious algal blooms and resulted in dramatic shifts in phytoplankton from diatoms (siliceous) to coccolithophores and flagellates (nonsiliceous). The most dramatic changes have been observed in the northwestern shelf and the western coastal regions, which have the largest anthropogenic effects. The water overlying the sediments in these shallow areas frequently go anoxic due to this eutrophication. In the early 1980s the grazer community experienced major increases of previously minor indigenous species such as the omnivorous dinoflagellate *Noctilluca scintillans* and the medusa *Aurelia aurita*. The ctenophore *Mnemopsis leidyi* was imported at the end of the 1980s from the east coast of the United States as ballast water in tankers and experienced an explosive unregulated growth. These changes plus overfishing resulted in a collapse of commercial fish stocks during the 1990s.   [J.W.M.]

Black Shale   A dark mud rock rich in organic carbon. Black shales are typically very fine-grained and contain pyrite, phosphate, and abnormally large amounts of heavy metals. They commonly display excellent fissility and well-preserved planktonic and nektonic faunas and plant debris. Benthic fossils are rare or absent. Some black shales are sources of hydrocarbons. *See* SHALE.

Black shales are enigmatic deposits. Although the large organic carbon content (3–15%) must have required reducing conditions of deposition, there are few unambiguous indicators of the specific environment, most especially of the depth of water. *See* MARINE SEDIMENTS; SEDIMENTOLOGY.

Black shales are typically well laminated on a scale of millimeters. Laminae are produced by variations in the supply of sediment, such as seasonal alternations of clay and planktonic algae. Delicate laminae can be preserved only in the total absence of benthic life, for burrowing animals disrupt lamination, producing bioturbated texture. Hence it is possible to differentiate between totally anaerobic conditions of deposition and marginally oxygenated (dysaerobic) conditions by recognizing laminated or bioturbated fabrics in a shale. [C.W.By.]

Blueschist   Metamorphic rock formed at high pressure and low temperature, commonly above 5 kilobars (500 megapascals) and below 750°F (400°C). Metamorphic rocks of the relatively uncommon blueschist facies contain assemblages of minerals that record these high pressures and low temperatures. The name "blueschist" derives from the fact that at this metamorphic grade, rocks of ordinary basaltic composition are often bluish because they contain the sodium-bearing blue amphiboles glaucophane or crossite rather than the calcium-bearing green or black amphiboles actinolite or hornblende, which are developed in the more common greenschist- or amphibolite-facies metamorphism.

Blueschist metamorphic rocks are found almost exclusively in the young mountain belts of the circum-Pacific and Alpine-Himalayan chains. The rocks are usually metamorphosed oceanic sediments and basaltic oceanic crust. Previously continental rocks rarely exhibit blueschist metamorphism. The tectonic mechanism for blueschist metamorphism must move the rocks to depths of more than 6 to 12 mi (10 to 20 km) while maintaining relatively cool temperatures (390–750°F or 200–400°C). These temperatures are much cooler than for continental crust at those depths. For example, surface geothermal gradients of the order of 30°C per kilometer are common in continental crust and in thick sedimentary basins. In contrast, a steady-state surface gradient of about 44 to 58°F per mile (15 to 20°C per kilometer) would be required for typical blueschist metamorphism. Heat flow measurements above long-lived subduction zones, together with thermal models, suggest that the conditions of blueschist metamorphism exist today above subduction zones just landward of deep-sea trenches. This tectonic setting at the time of blueschist metamorphism is independently inferred for a number of metamorphic terranes.

What is not well understood is how the blueschist metamorphic rocks return to the surface; clearly the mechanism is not simple uplift and erosion of 12–18 mi (20–30 km) of the Earth's crust. Blueschist metamorphic rocks are usually in immediate fault contact with much less metamorphosed or unmetamorphosed sediments, indicating they have been tectonically displaced relative to their surroundings since metamorphism. *See* METAMORPHIC ROCKS; METAMORPHISM. [J.Sup.]

Bog   Nutrient-poor, acid peatlands with a vegetation in which peat mosses (*Sphagnum* spp.), ericaceous dwarf shrubs, and to a lesser extent, various sedges (Cyperaceae) play a prominent role. The terms muskeg, moor, heath, and moss are used locally to indicate these sites.

Bogs are most abundant in the Northern Hemisphere, especially in a broad belt including the northern part of the deciduous forest zone and the central and southern parts of the boreal forest zone. Farther south, and in drier climates farther inland, they become sporadic and restricted to specialized habitats. To the north, peatlands controlled by mineral soil water (aapa mires) replace them as the dominant wetlands.

Bogs are much less extensive in the Southern Hemisphere because there is little land in cold temperate latitudes. In these Southern Hemisphere peatlands, *Sphagnum* is much less important, and Epacridaceae and Restionaceae replace the Ericaceae and Cyperaceae of the Northern Hemisphere.

Bogs have a fibric, poorly decomposed peat consisting primarily of the remains of *Sphagnum*. Peat accumulation is the result of an excess of production over decomposition. Obviously, the very presence of bogs shows that production exceeded decay over the entire period of bog formation. However, in any given bog present production can exceed, equal, or be less than decomposition, depending on whether it is actively developing, in equilibrium, or eroding. In most bogs, production and decomposition appear to be in equilibrium at present.

Slow decay rather than high productivity causes the accumulation of peat. Decomposition of organic matter in peat bogs is slow due to the high water table, which causes the absence of oxygen in most of the peat mass, and to the low fertility of the peat. Bogs, in contrast to other peatlands, can accumulate organic matter far above the groundwater table.

Bogs show large geographic differences in floristic composition, surface morphology, and development. Blanket bogs, plateau bogs, domed bogs, and flat bogs represent a series of bog types with decreasing climatic humidity. Concentric patterns of pools and strings (peat dams) become more common and better developed northward. Continental bogs are often forest-covered, whereas oceanic bogs are dominated by dwarf shrub heaths and sedge lawns, with forests restricted to the bog slope if the climate is not too severe.

Bogs have long been used as a source of fuel. In Ireland and other parts of western Europe, the harvesting of peat for domestic fuel and reclamation for agriculture and forestry have affected most of the peatlands, and few undisturbed bogs are left. Other uses are for horticultural peat, air layering in greenhouses, litter for poultry and livestock, and various chemical and pharmaceutical purposes. Mechanical extraction of peat for horticultural purposes has affected large bog areas worldwide. *See* BIOMASS. [A.W.H.D.]

Boracite A borate mineral with chemical composition $Mg_3B_7O_{13}Cl$. It occurs in Germany, England, and the United States, usually in bedded sedimentary deposits of anhydrite, gypsum, and halite, and in potash deposits of oceanic type. The chemical composition of natural boracites varies, with Fe^{2+} or Mn^{2+} replacing part of the Mg^{2+} to yield ferroan boracite or manganoan boracite. *See* BORATE MINERALS.

The hardness is $7-7^1/_2$ on Mohs scale, and specific gravity is 2.91–2.97 for colorless crystals and 2.97–3.10 for green and ferroan types. Luster is vitreous, inclining toward adamantine. Boracite is colorless to white, inclining to gray, yellow, and green, and rarely pink (manganoan); its streak is white; and it is transparent to translucent. It is strongly piezoelectric and pyroelectric and does not cleave. [C.L.Ch.]

Borate minerals A large group of minerals in which boron is chemically bonded to oxygen. Boron is a fairly rare element. However, because of its chemical character, it is very susceptible to fractionation in earth processes and can become concentrated to a degree not found in other elements of similar abundance. Boron is symbolized B, carries atomic number 5, and has the ground-state electronic structure $[He]2s^22p^1$. The very high ionization potentials for boron mean that the total energy required to produce the B^{3+} ion is greater than the compensating structure energy of the resulting ionic solid, and hence bond formation involves covalent (rather than ionic) mechanisms. However, boron has only three electrons to contribute to covalent bonding involving four orbitals, (s, p_x, p_y, p_z). This results in boron being a strong electron-pair acceptor (that is, a strong Lewis acid) with a very high affinity for oxygen. The structural chemistry of boron and silicon (Si), when associated with oxygen (O), is quite similar. The BO_3, BO_4, and SiO_4

groups have a marked tendency to polymerize in the solid state, and this aspect of their behavior gives rise to the structural complexity of both groups. However, subtle differences in chemical bonding do give rise to differences in the character of this polymerization, particularly when water is also involved. These differences result in the very different properties of the resultant minerals and their very different behavior in earth processes.

Boron has an estimated primitive-mantle abundance of 0.6 part per million and a crustal abundance of 15 ppm. Despite this low abundance, fractionation in crustal processes results in concentration of boron to the extent that it forms an extensive array of minerals in which it is an essential constituent, and very complex deposits of borate minerals. Major concentrations of borate minerals occur in continental evaporite deposits (common in the desert regions of California and Nevada). Borate minerals are often very soluble in aqueous environments. In areas of internal drainage, saline lakes are formed, and continued evaporation leads to accumulation of large deposits of borate minerals. Borates may also occur in marine evaporites. Isolated-cluster borates are characteristic of metamorphosed boron-rich sediments and skarns. Most borosilicate minerals are characteristic of granitic-pegmatite environments, either as a pegmatite phase or as a constituent of their exocontact zone. In particular, tourmaline is the most widespread of the mixed-anion borate minerals, occurring in a wide variety of igneous, hydrothermal, and metamorphic rocks. *See* SALINE EVAPORITES; ORE AND MINERAL DEPOSITS; PEGMATITE.

Despite its low crustal abundance, fractionation of boron in crustal processes leads to formation of deposits of borate minerals from which boron and (borates) can be easily extracted in large quantities. The easy availability and unique chemical properties result in boron being a major industrial material. It is widely used in soaps and washing powders. Boron combines well with silicon and other elements to form a wide variety of special-property glasses and ceramics; it also alloys with a variety of metals, producing lightweight alloys for specialty uses. Boron compounds usually have very low density; hence borates in particular are used as lightweight fillers in medicines, and also are used as insulation. The fibrous nature of some borate minerals results in their use in textile-grade fibers and lightweight fiber-strengthened materials. The mineral borax is used as a water softener and as a cleaning flux in welding and soldering. Boric acid has long been used as an antiseptic and a drying agent. Boron is also important as a constituent of inflammatory materials in fireworks and rocket fuel. Some mixed-anion borate minerals are used as gemstones. Tourmaline is of particular importance in this respect, forming pink (rubellite), blue-green (Paraiba tourmaline), green (chrome tourmaline), and pink + green (watermelon tourmaline) from a wide variety of localities. Kornerupine and sinhalite are also used as gemstones but are far less common. *See* TOURMALINE.   [F.C.Ha.]

Bornite   A sulfide of composition Cu_5FeS_4, specific gravity 5.07, and hardness 3 (Mohs scale), commonly occurring as a primary mineral in many copper ore deposits. Crystals are rare; bornite is usually massive or granular. The metallic and brassy color of a fresh surface rapidly tarnishes upon exposure to air to a characteristic iridescent purple, giving rise to the name "peacock ore." Though of lesser importance as an ore than chalcocite or chalcopyrite, masses of bornite have been mined in Chile, Peru, Bolivia, and Mexico and in the United States in Arizona and Montana. *See* CHALCOCITE; CHALCOPYRITE.   [B.J.Wu.]

Breccia   A clastic rock composed of angular gravel-size fragments; the consolidated equivalent of rubble. The designation gravel-size refers to a mean particle diameter greater than 0.08 in. (2 mm), which means that 50% or more of the particles (by volume) are this size or larger. Various classifications specify different values for the degree of angularity. One system specifies angular or subangular fragments (roundness ≤ 0.25), whereas another restricts the term breccia to aggregates with angular fragments (roundness ≤ 0.10). *See* GRAVEL.

Sedimentary breccias, also known as sharpstone conglomerates, are significant because the angularity of their fragments indicates either proximity to the source or transportation by a mechanism that does not cause significant rounding of the fragments. Examples of the first condition are talus breccia formed at the base of a scarp, and reef breccia deposited adjacent to a reef margin. Transport mechanisms that can preserve the angularity of clasts over significant distances include debris flows, slumps, and glacial transport, although rounded fragments may also be carried. All of these mechanisms incorporate a large proportion of fine sediment in the transporting medium, which effectively cushions interparticle collisions and inhibits rounding. *See* CONGLOMERATE; REEF; SEDIMENTOLOGY.

Intraformational or intraclastic breccias are an important class of sedimentary breccias. They are formed by the breakup and incorporation of sediment aggregates from within the same formation, which requires either early cementation (for example, the formation of nodules or duricrusts) or uncemented aggregates sufficiently cohesive to be transported a short distance without disaggregation. Thus, uncemented aggregates are basically limited to sediments that are rich in clay or clay-size carbonates (calcilutites). The mechanisms for formation of intraformational breccias include bank slumping or desiccation fracturing of mud in river or tidal channels, and erosion and incorporation of mud blocks in mass flows such as slumps or turbidity currents. *See* SEDIMENTARY ROCKS.

Igneous breccias are mainly of pyroclastic origin but may also form as intrusive breccias by forceful intrusion of magma. In the latter case the operative agent is fluid pressure; in the former it is the explosive escape of gas from solidifying viscous lava. These rocks, termed pyroclastic or volcanic breccias, are distinct from agglomerates, which accumulate mainly as lava bombs solidified during flight and which are commonly rounded. *See* PYROCLASTIC ROCKS.

Cataclastic breccias result from the fracture of rocks by tectonic or gravitational stresses. However, since many tectonic processes are at least partly gravitational, the two processes can be considered together. Tectonic breccias include fault and fold breccias, the latter formed by fracturing of brittle layers within incompetent plastic strata during folding. In one classification, landslide and slump breccias are included in the gravitational category, but here they are considered to be sedimentary, commonly intraformational. Solution or collapse breccias are a type of nontectonic gravitational breccia. They result from the creation by groundwater solution of unsupported rock masses which collapse under their own weight to form breccia.   [B.Rus.]

Brucite A magnesium hydroxide mineral, $Mg(OH)_2$, crystallizing in the trigonal system. It is a member of the important $Cd(OH)_2$ structure type, consisting of hexagonal close-packed oxygen atoms with alternate octahedral layers occupied by Mg. The "brucite layer'" is an important structural component in the clay, mica, and chlorite mineral groups. Brucite occurs as tabular crystals and as elongated fibers (as the variety nemalite), hardness $2^1/_2$ (Mohs scale), color white to greenish, and specific gravity 2.4. Fe^{2+} and Mn^{2+} commonly substitute for Mg^{2+}.

Brucite often occurs in a low-temperature vein paragenesis, usually with serpentine and accessory magnesite. It is also derived by the action of water on periclase, MgO, which results from the thermal metamorphism of dolomites and limestones. Carbonate rocks rich in periclase and brucite are called predazzites. *See* DOLOMITE; MAGNESITE; SERPENTINE.   [P.B.M.]

Burgess Shale   Part of a clay and silt sequence that accumulated at the foot of a colossal "reef" during the Cambrian explosion, a dramatic evolutionary radiation of animals beginning about 545 million years ago. Although this explosion is most obvious from the geologically abrupt appearance of skeletons, the bulk of the radiation consisted of soft-bodied animals (see illustration). The Burgess Shale fauna, located near Field in southern British Columbia, is Middle Cambrian, approximately 520 million years old.

The Burgess Shale fauna is remarkably diverse, with about 120 genera. Its approximate composition is arthropods 37%, sponges 15%, brachiopods 4%, priapulids 5%, annelids 5%, chordates and hemichordates 5%, echinoderms 5%, cnidarians and ctenophores 2%, mollusks 3%, and "other fauna" 19%. Although arthropods are the most important group, the trilobites, normally dominant among Cambrian arthropods, are entirely overshadowed both in number of species and in absolute number of specimens by a remarkable variety of other arthropods with delicate exoskeletons. The priapulids, which today are a more or less relict group of marine worms, also show a wide diversity of anatomical form, as do the polychaete annelids. Only one species of polychaete annelids has a close parallel among the Recent assemblages.

The Burgess Shale has revealed many other aspects of the Cambrian explosion. First, a census of the collections reveals a marine ecology that is fundamentally unchanged to the present day. Predators, long thought to be insignificant in the Cambrian, are an important component. Second, groups with a minimal fossilization potential are preserved. One example is the gelatinous and delicate ctenophores, an important pelagic group in today's oceans but practically unknown as fossils. Third, although many of

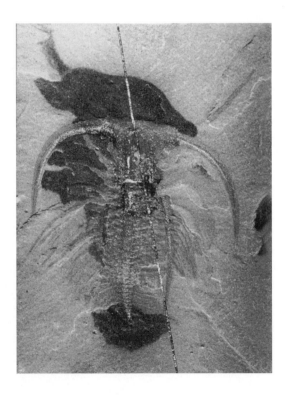

Marrella splendens, a characteristic Burgess Shale arthropod. The head shield bears two prominent pairs of spinose extensions; also visible are various appendages that include walking legs and gills. The prominent dark areas appear to represent body contents that oozed into the newly deposited sediment. This indicates that some decay occurred before an unknown factor intervened. (*Copyright by Simon Conway Morris*)

the species are a product of the Cambrian explosion, rare species are clear holdovers from the primitive Ediacaran faunas of late Precambrian age. Finally, some species are of particular evolutionary importance. Most significant is the worm *Pikaia*, which is interpreted as an early chordate, and as a predecessor of fish it lies near the beginning of the evolutionary path that ultimately leads to humans. *See* CAMBRIAN; FOSSIL; PALEONTOLOGY.   [S.C.M.]

Bytownite   A member of the plagioclase feldspar solid-solution series with a composition ranging from $Ab_{30}An_{70}$ to $Ab_{10}An_{90}$ (Ab = $NaAlSi_3O_8$ and An = $CaAl_2Si_2O_8$). Bytownite is very abundant in basic igneous rocks where it is the first plagioclase to crystallize under plutonic conditions; it forms the cores of zoned plagioclase phenocrysts in basaltic volcanics, and it sometimes occurs in anorthosites. *See* FELDSPAR.   [L.Gr.]

C

Calamine A term that may refer to either a zinc mineral, $Zn_4Si_2O_7(OH)_2 \cdot H_2O$, which is also known as hemimorphite, or to zinc oxide, ZnO, which is used in medicinal or pharmaceutical products and in cosmetics. *See* HEMIMORPHITE.   [E.E.W.]

Calcite A mineral composed of calcium carbonate ($CaCO_3$); one of the most common and widespread minerals in the Earth's crust. Calcite may be found in a great variety of sedimentary, metamorphic, and igneous rocks. It is also an important rock-forming mineral and is the sole major constituent in limestones, marbles, and many carbonatites. Calcite in such rocks is the main source of the world's quicklime and hydrated, or slaked, lime. It is also widely used as a metallurgical flux to a scavenge siliceous impurities by forming a slag in smelting furnaces. It provides the essential calcium oxide component in common glasses and cement. Limestones and marbles of lower purity may find uses as dimension stone, soil conditioners, industrial acid neutralizers, and aggregate in concrete and road building. Calcite in transparent well-formed crystals is used in certain optical instruments. *See* LIMESTONE; STONE AND STONE PRODUCTS.

When pure, calcite is either colorless or white, but impurities can introduce a wide variety of colors: blues, pinks, yellow-browns, greens, and grays have all been reported. Hardness is 3 on Mohs scale. The specific gravity of pure calcite is 2.7102 ± 0.0002 at $68°F$ ($20°C$). Calcite has a very low solubility in pure water (less than 0.001% at $77°F$ or $25°C$), but the solubility increases considerably with CO_2 added, as in natural systems from the atmosphere, when more bicarbonate ions and carbonic acid are formed. The solubility is also increased by falling temperature and rising total pressure. Shallow warm seas are supersaturated with calcite, while enormous quantities of calcite are dissolved in the unsaturated deep oceans. *See* CARBONATE MINERALS.   [R.I.Ha.]

Caldera A large volcanic collapse depression, typically circular to slightly elongate in shape, the dimensions of which are many times greater than any included vent. Calderas range from a few miles to 37 mi (60 km) in diameter. A caldera may resemble a volcanic crater in form, but differs genetically in that it is a collapse rather than a constructional feature. The topographic depression resulting from collapse is commonly widened by slumping of the sides along concentric faults, so that the topographic crater wall lies outside the caldera wall. As originally defined, the term caldron referred to volcanic subsidence structures, and caldera referred only to the topographic depression formed at the surface by collapse. However, the term caldera is now common as a synonym for caldron, denoting all features of collapse, both topographic and structural. *See* PETROLOGY.

Calderas occur primarily in three different volcanic settings, each of which affects their shape and evolution: basaltic shield cones, stratovolcanoes, and volcanic centers consisting of preexisting clusters of volcanoes. These last calderas, associated with broad, large-volume andesitic to rhyolitic ignimbrite sheets, are generally the largest and most impressive, and are those generally denoted by the term. Calderas have been formed throughout much of the Earth's history, ranging in age from Precambrian (greater than 1.4 billion years old) to Holocene (for example, Krakatau in Indonesia, which erupted in 1883). *See* RHYOLITE; VOLCANO.

In addition to Earth, large calderas occur on Mars, Venus, and Jupiter's moon Io. The presence of calderas on four solar system bodies indicates that the underlying mechanisms of shallow intrusion and caldera collapse are basic processes in planetary geology. *See* MAGMA.

Collapse occurs because of withdrawal of magma from an underlying chamber some 2.4–3.6 mi (4–6 km) beneath the surface, resulting in foundering of the roof into the chamber. Withdrawal of magma may occur either by relatively passive eruption of lavas, as in the case of calderas formed on basaltic shield cones, or by catastrophic eruption of pyroclastic material, as accompanies formation of the largest calderas.

Caldera-forming eruptions probably last only a few hours or days. Eruption of pyroclastic material begins as gases (predominantly water) that are dissolved in the magma come out of solution at shallow depths. Magma is explosively fragmented into particles ranging in size from micrometers to meters. An eruption column develops, rising several miles into the atmosphere. This first and most explosive phase of the eruption, known as the Plinian phase, covers the area around the vent with pumice. Caldera subsidence occurs during eruption. As caldera subsidence proceeds and eruption becomes less explosive, the Plinian eruption column collapses. This collapse produces hot, ground-hugging pyroclastic flows that can travel as far as 93 mi (150 km) outward from the vent at speeds of 330 ft/s (100 m/s). Successive collapses of the column produce multiple flow units with an aggregate thickness that may be several hundreds of feet thick near the caldera.

The floors of many of the largest calderas (typically those with diameters exceeding 6 mi or 10 km) have been domed upward, resulting in a central massif or resurgent dome. Resurgence results from the continued or renewed buoyant rise of magma after collapse.

Calderas typically contain or are associated with extensive hydrothermal systems, because of two factors: (1) the shallow magma chambers that underlie them provide a readily available source of heat: and (2) the floors of calderas may be extensively fractured, which, along with the main ring faults, allows meteoric water to penetrate deeply into the crust beneath calderas. Hydrothermal activity related to a caldera system can occur any time after magmas rise to shallow crustal levels, but it is dominant late in caldera evolution.

Many metals, including such base and precious metals as molybdenum, copper, lead, zinc, silver, gold, mercury, uranium, tungsten, and antimony, are mobile in hydrothermal circulation systems driven by the shallow intrusions which underlie and give rise to large calderas. Many economically important ore deposits in the western United States lie within calderas. *See* ORE AND MINERAL DEPOSITS.   [W.S.Ba.]

Caliche   A soil that is mineralogically an impure limestone. Such soils are also known as duricrust, kunkar, nari, kafkalla, Omdurman lime, croute, and race. Many soil profiles in semiarid climates (that is, those characterized by a rainfall of 4–20 in. or 10–50 cm per year) contain concentrations of calcium carbonate ($CaCO_3$). This calcium carbonate is not an original feature of the soils but has been added during soil formation either by direct precipitation in soil pores or by replacement of preexisting material. Fossil analogs of caliche, which are widely reported in ancient sedimentary sequences, are referred to as calcrete or cornstone. *See* LIMESTONE; SOIL.

The principal control on the formation of caliche is a hydrologic regime in which there is sufficient moisture to introduce calcium carbonate in solution to the soil but not enough to leach it through the system. As a result, calcium carbonate precipitates in the soil during periods of evaporation, and it will slowly increase in amount as long as the hydrologic setting remains stable. The source of the carbonate may be from the dissolution of adjacent limestones, from the hydrolysis of plagioclase and other silicates, or from carbonate loess.

Within the climatic constraints noted above, most caliche forms in river floodplains and near the surface of alluvial fans. In addition, caliche deposits may form within exposed marine and lacustrine limestones during periods of sea-level fall or lake desiccation. Caliche may also form at inert pediment (eroded rock) surfaces; in the geological record such surfaces will be seen as unconformities. In this context it is interesting that the first unconformity ever recognized as such, by James Hutton in 1787 on the Isle of Arran, western Scotland, is characterized by a development of caliche. *See* UNCONFORMITY.

The mineralogy of the host soil or rock in which a caliche develops may vary considerably; it is not essential for there to be any preexisting carbonate grains within the regolith. The most favorable medium is a clay-rich soil of limited permeability. Low permeability provides the residence time in the soil pores necessary for calcite to precipitate. *See* CALCITE; REGOLITH.   [N.Do.]

Cambrian   An interval of time in Earth history (Cambrian Period) and its rock record (Cambrian System). The Cambrian Period spanned about 60 million years and began with the first appearance of marine animals with mineralized (calcium carbonate, calcium phosphate) shells. The Cambrian System includes many different kinds of marine sandstones, shales, limestones, dolomites, and volcanics. Apart from the occurrence of an alkaline playa containing deposits of trona (hydrated basic sodium carbonate) in the Officer Basin of South Australia, there is very little provable record of nonmarine Cambrian environments. The best present estimates suggest that Cambrian time began about 545 million years ago (Ma) and ended at about 485 Ma. It is the longest of the Paleozoic periods and the fourth longest of the Phanerozoic periods.

The Cambrian world can be resolved into at least four major continents that were quite different from those of today (see illustration). These were (1) Laurentia, which is essentially North America, minus a narrow belt along the eastern coast from eastern Newfoundland to southern New England that belonged to a separate microcontinent, Avalonia. This microcontinent, which also included present-day England, and another microcontinent now incorporated in South Carolina were originally marginal to Gondwana; (2) Baltica, consisting of present-day northern Europe north of France and west of the Ural Mountains but excluding most of Scotland and northern Ireland, which are fragments of Laurentia; (3) Gondwana, a giant continent whose present-day fragments are Africa, South America, India, Australia, Antarctica, parts of southern Europe, the Middle East, and Southeast Asia; and (4) Siberia, including much of the northeastern quarter of Asia. *See* CONTINENTAL MARGIN; CONTINENTS, EVOLUTION OF; PLATE TECTONICS.

For most practical purposes, rocks of Cambrian age are recognized by their content of distinctive fossils. On the basis of the successive changes in the evolutionary record of Cambrian life that have been worked out during the past century, the Cambrian System has been divided globally into three or four series, each of which has been further divided on each continent into stages, each stage consisting of several zones. Despite the amount of work already done, precise intercontinental correlation of series

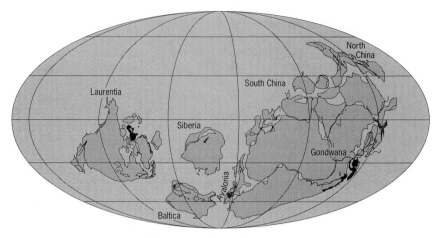

Reconstruction of the Lower Cambrian world. (*After W. S. McKerrow, C. R. Scotese, and M. D. Brasier, Early Cambrian continental reconstructions. J. Geol. Soc., 149:599–606, 1992*)

and stage boundaries, and of zones, is still difficult, especially in the Early Cambrian due to marked faunal provinciality. Refinement of intercontinental correlation of these ancient rocks is a topic of research.

The record preserved in rocks indicates that essentially all Cambrian plants and animals lived in the sea. The few places where terrestrial sediments have been preserved suggest that the land was barren of major plant life, and there are no known records of Cambrian insects or of terrestrial vertebrate animals of any kind.

The plant record consists entirely of algae, preserved either as carbonized impressions in marine black shales or as filamentous or blotchy microstructures within marine buildups of calcium carbonate, called stromatolites, produced by the actions of these organisms. Cambrian algal stromatolites were generally low domal structures, rarely more than a few meters high or wide, which were built up by the trapping or precipitation of calcium carbonate by one or more species of algae. Such structures, often composed of upwardly arched laminae, were common in regions of carbonate sedimentation in the shallow Cambrian seas. *See* STROMATOLITE.

The most abundant remains of organisms in Cambrian rocks are of trilobites. They are present in almost every fossiliferous Cambrian deposit and are the principal tools used to describe divisions of Cambrian time and to correlate Cambrian rocks. These marine arthropods ranged from a few millimeters to 20 in. (50 cm) in length, but most were less than 4 in. (10 cm) long. The next most abundant Cambrian fossils are brachiopods. These bivalved animals were often gregarious and lived on the sediment surface or on the surfaces of other organisms. Limestones of Early Cambrian age may contain large reeflike structures formed by an association of algae and an extinct phylum of invertebrates called Archaeocyatha. The Cambrian record of mollusks and echinoderms is characterized by many strange-looking forms. Some lived for only short periods of time and left no clear descendants. Except for rare jellyfish impressions, the Coelenterata were thought to be unrepresented in Cambrian rocks. Corals have now been discovered in early Middle Cambrian rocks in Australia. However, like clams, they are not seen again as fossils until Middle Ordovician time, many tens of millions of years later.

Throughout Cambrian time, terrestrial landscapes were stark and barren. Life in the sea was primitive and struggling for existence. Only in post-Cambrian time did the shallow marine environment stabilize and marine life really flourish. Only then did vertebrates evolve and plants and animals invade the land.   [A.R.P.; J.H.Sh.]

Cameo   A type of carved gemstone in which the background is cut away to leave the subject in relief. Often cameos are cut from stones in which the coloring is layered, resulting in a figure of one color and a background of another. The term cameo, when used without qualification, is usually reserved for those cut from a gem mineral, although they are known also as stone cameos. The commonly encountered cameo cut from shell is properly called a shell cameo.

Most cameos are cut from onyx or agate, but many other varieties of quartz, such as tiger's-eye, bloodstone, sard, carnelian, and amethyst, are used; other materials used include beryl, malachite, hematite, labradorite, and moonstone. *See* GEM; INTAGLIO (GEMOLOGY).   [R.T.L.]

Cancrinite   A family of minerals, related to the scapolite family, characteristically occurring in basic rocks such as nepheline syenites and sodalite syenites. Cancrinite is hexagonal. Four-, six-, and twelve-membered aluminosilicate rings can be discerned in the structure. Large anions such as $[SO_4]^{2-}$ and $[CO_3]^{2-}$ occur in the hexagonal channels of the structure. Four members of the cancrinite family are cancrinite, vishnevite, hauyne, and afghanite. The cancrinite member is white, yellow, greenish, or reddish. It has perfect prismatic cleavage, hardness is 5–6 on Mohs scale, and the specific gravity is 2.45. Localities include the Fen area, southern Norway; the Kola Peninsula, Russia; Bancroft, Ontario, Canada; and Litchfield, Maine, U.S.A. *See* SILICATE MINERALS.
                                                                                         [P.B.M.]

Carat   The unit of weight now used for all gemstones except pearls. It is also called the metric carat (m.c.). By international agreement, the carat weight is set at 200 milligrams. Pearls are weighed in grains, a unit of weight equal to 50 mg, or $1/4$ carat. The application of the term carat as a unit of weight must not be confused with the term karat used to indicate fineness or purity of the gold in which gems are mounted. *See* GEM.   [R.T.L.]

Carbonate minerals   Mineral species containing the carbonate ion as the fundamental anionic unit. The carbonate minerals can be classified as (1) anhydrous normal carbonates, (2) hydrated normal carbonates, (3) acid carbonates (bicarbonates), and (4) compound carbonates containing hydroxide, halide, or other anions in addition to the carbonate.

Most of the common carbonate minerals belong to group (1), and can be further classified according to their structures. The rhombohedral carbonates are typified by calcite, $CaCO_3$, and by dolomite, $CaMg(CO_3)_2$, The other structural type within this group is that of aragonite, which has orthorhombic symmetry. *See* ANKERITE; ARAGONITE; CALCITE; CERUSSITE; DOLOMITE; MAGNESITE; RHODOCHROSITE; SIDERITE; SMITHSONITE; STRONTIANITE; WITHERITE.

The minerals in groups (2) and (3) all decompose at relatively low temperatures and therefore occur only in sedimentary deposits (typically evaporites) and as low-temperature hydrothermal alteration products. The only common mineral in these groups is trona, $Na_3H(CO_3)_2 \cdot 2H_2O$.

Similarly, the group (4) minerals are relatively rare and are characteristically low-temperature hydrothermal alteration products. The commonest members of this group are malachite, $Cu_2CO_3(OH)_2$, and azurite, $Cu_3(CO_3)_2(OH)_2$, which are often found in copper ore deposits. *See* AZURITE; MALACHITE.

Important occurrences of carbonates include ultrabasic igneous rocks such as carbonatites and serpentinites, and metamorphosed carbonate sediments, which may recrystallize to form marble. The major occurrences of carbonates, however, are in sedimentary deposits as limestone and dolomite rock. *See* DOLOMITE ROCK; LIMESTONE; MARBLE; SEDIMENTARY ROCKS.   [A.M.G.]

Carbonatite   An igneous rock in which carbonate minerals make up at least half the volume. Individual occurrences of carbonatite are not numerous (about 330 have been recognized) and generally are small, but they are widely distributed. Carbonatites are scientifically important because they reveal clues concerning the composition and thermal history of the Earth's mantle. *See* CARBONATE MINERALS.

The carbonate minerals that dominate the carbonatites are, in order of decreasing abundance, calcite, dolomite, ankerite, and rarely siderite and magnesite. Sodium- and potassium-rich carbonate minerals have been confirmed in igneous rocks at only one locality, the active volcano Oldoinyo Lengai in Tanzania. Noncarbonate minerals that typify carbonatites are apatite, magnetite, phlogopite or biotite, clinopyroxene, amphibole, monticellite, perovskite, and rarely olivine or melilite. Secondary minerals, produced by alteration of primary magmatic minerals, include barite, alkali feldspar, quartz, fluorite, hematite, rutile, pyrite, and chlorite. Minerals that are important in some carbonatites because they carry niobium, rare-earth elements, and other metals in concentrations high enough for profitable extraction are pyrochlore, bastnaesite, monazite, baddeleyite, and bornite.

Carbonatites, compared to the inferred composition of the Earth's mantle and to other igneous rocks, are greatly enriched in niobium, rare-earth elements, barium, strontium, phosphorus, and fluorine, and they are relatively depleted in silicon, aluminum, iron, magnesium, nickel, titanium, sodium, potassium, and chlorine. These extreme differences are attributed to strong fractionation between carbonate liquid on the one hand and silicate and oxide solid phases on the other during separation of the carbonate liquid from its source. Strontium and neodymium isotope ratios indicate that the sources of carbonatites are geologically old, inhomogeneous, and variably depleted in the radioactive parent elements rubidium and samarium.

Ultramafic xenoliths from lithospheric mantle commonly show textures and mineral assemblages that indicate modification of the original rock. This alteration typically results in strong enrichment in light rare-earth elements, uranium, thorium, and lead, but much less enrichment in titanium, zirconium, niobium, and strontium. These changes are commonly attributed to interaction of lithospheric mantle with an invading carbonate-rich magma. The wide geographic dispersal of these altered xenoliths suggests that carbonate-rich liquid has been more common in the upper mantle than the low abundance of carbonatites in the upper crust would suggest. According to the testimony of these samples, carbonatite magma, ascending through lithospheric mantle, commonly is trapped before it can invade the crust. In addition to the factors that can stop the rise of any magma (heat loss, increase of solidus temperature with decrease in pressure, decrease in density and increase in strength of wall rock), carbonatite magma can be halted by reaction with wall rock to form calcium and magnesium silicates plus carbon dioxide (CO_2), and by less oxidizing conditions to reduce carbonate to elemental carbon (graphite or diamond) or to methane. Both of these changes subtract dissolved CO_2 from the magma, causing crystallization. *See* EARTH INTERIOR.

Carbonatites are not restricted to a single tectonic regime. They occur in oceanic and continental crust and have formed in compressional fold belts and stable cratons as well as regions of crustal extension. Rather than indicating the stress field in the shallow crust in which they were emplaced, carbonatites are useful in modeling the long-term thermal and chemical development of the mantle. *See* IGNEOUS ROCKS.

Carbonatites yield a variety of mineral commodities, including phosphate, lime, niobium, rare-earth elements, anatase, fluorite, and copper. Agricultural phosphate for fertilizer is the most valuable single product from carbonatites; most is obtained from apatite in lateritic soils that have developed by tropical weathering of carbonatites, dissolving the carbonates and thereby concentrating the less soluble apatite. Lime for agriculture and for cement manufacture is obtained from carbonatites in regions where limestones are lacking. The carbonatites at Bayan Obo, China, and Mountain Pass, California, dominate the world suppliers of rare-earth elements, but many other carbonatites contain unexploited reserves. Tropical weathering at several carbonatites in Brazil has produced economically important concentrations of anatase (TiO_2) from decomposition of perovskite ($CaTiO_3$). [D.S.Ba.]

Carboniferous The fifth period of the Paleozoic Era. The Carboniferous Period spanned from about 355 million years to about 295 million years ago. The rocks that formed during this time interval are known as the Carboniferous System; they include a wide variety of sedimentary, igneous, and metamorphic rocks. Sedimentary rocks in the lower portion of the Carboniferous are typically carbonates, such as limestones and dolostones, and locally some evaporites. The upper portions of the system are usually composed of cyclically repeated successions of sandstones, coals, shales, and thin limestones. *See* SEDIMENTARY ROCKS.

The economic importance of the Carboniferous is evident in its name, which refers to coal, the important energy source that fueled the industrialization of northwestern Europe in the early 1800s and led to the Carboniferous being one of the first geologic systems to be studied in detail. Carboniferous coals formed in coastal and fluvial environments in many parts of the world. Petroleum, another important energy resource, accumulated in many Carboniferous marine carbonate sediments, particularly near shelf margins adjacent to basinal black shale source rocks. In many regions the cyclical history of deposition and exposure has enhanced the permeability and porosity of the shelfal rocks to make them excellent petroleum reservoirs. The limestones of the Lower Carboniferous are extensively quarried and used for building stone, especially in northwestern Europe and the central and eastern United States. *See* COAL; PETROLEUM.

The base of the Carboniferous is placed at the first appearance of the conodont *Siphonodella sulcata*, a fossil that marks a widely recognized biozone in most marine sedimentary rocks. The reference locality for this base is an outcrop in Belgium. The top of the Carboniferous is placed at the first appearance of the conodont *Streptognathus isolatus* a few meters below the first appearance of the Permian fusulinacean foraminiferal zone of *Sphaeroschwagerina fusiformis*. The reference locality is in the southern Ural region in Kazakhstan. The equivalent biozone is at the base of *Pseudoschwagerina* in North America.

The International Subcommission on Carboniferous Stratigraphy reached general agreement in the 1970s and 1980s that the Carboniferous would be divided into two parts: a Lower Carboniferous Mississippian Subsystem and an Upper Carboniferous Pennsylvanian Subsystem. The two Carboniferous subsystems are subdivided into a number of series and stages that are variously identified in different parts of the world, based on biostratigraphic evidence using evolutionary successions in fossils or overlapping assemblage zones.

Perhaps the strongest of the many ecological factors that controlled biotic distributions were the paleogeographic changes within the Carboniferous that were brought about by the initial assembling of the supercontinent Pangaea and the associated mountain-building activities, which greatly modified climate, ocean currents, and seaways. In the Early Carboniferous, a nearly continuous equatorial seaway permitted extensive tropical and subtropical carbonate sedimentation on the shelves and platforms in North America, northern and southern Europe, Kazakhstan, North and South China, and the northern shores of the protocontinent Gondwana (such as northern Africa).

The gradual collision of northern Gondwana against northern Europe-North America (also called Euramerica or Laurussia) started the formation of the supercontinent of Pangaea. See CONTINENTAL DRIFT; CONTINENTS, EVOLUTION OF; PALEOGEOGRAPHY; PLATE TECTONICS.

An additional ramification of the formation of Pangaea was the beginning of very extensive glaciation in the Southern Hemisphere polar and high-latitude regions of the supercontinent. Glacial deposits are also known from smaller continental fragments that were at high paleolatitudes in the Northern Hemisphere. The Earth's climate cooled, tropical carbonate-producing areas became restricted toward the Equator, and eustatic sea-level fluctuations became prominent in the sedimentary record. See GLACIAL EPOCH.

During the Carboniferous, life evolved to exploit fully the numerous marine and nonmarine aquatic environments and terrestrial and aerial habitats. Single-cell protozoan foraminifers evolved new abilities to construct layered, calcareous walls. Insects have remarkable evolutionary histories during the Carboniferous. They adapted to flight and dispersed into many terrestrial and fresh-water habitats. Vertebrates also evolved rapidly. Although acanthodian fish declined from their Devonian peak, sharklike fishes and primitive bony fishes adapted well to the expanded environments and the new ecological food chains of the Carboniferous. Some sharklike groups invaded freshwater habitats, where they were associated with coal swamp deposits. Carboniferous amphibians evolved rapidly in several directions. The earliest were the labyrinthodont embolomeres, which had labyrinthodont teeth and were mainly aquatic. Another significant labyrinthodont group was the rhachitomes, which originated in the Early Carboniferous and became abundant, commonly reaching about 1 m (3 ft) or more; they were widespread in terrestrial habitats during the Late Carboniferous and Permian. Primitive reptiles evolved from one of the embolomere amphibian lineages during the Late Carboniferous. They formed the basal stock from which all other reptiles have evolved including the earliest mammallike reptiles in the Late Carboniferous. During the Late Carboniferous, early reptiles coexisted with several advanced amphibian groups which shared at least some, but probably not all, of their reptilelike characters.

Terrestrial plants also showed major diversification of habitats and the evolution of important new lineages during the Carboniferous. Initially, Early Carboniferous plants were predominantly a continuation of latest Devonian groups; however, they were distinguished in part by their large sizes with many arborescent lycopods and large articulates, and pteridosperms (seed ferns) and ferns became increasingly abundant and varied. By the Late Carboniferous, extensive swamps formed along the broad, nearly flat coastal areas; and these coal-forming environments tended to move laterally across the coastal plain areas as the sea level repeatedly rose. Other coal-forming marshes were common in the floodplains and channel fills of the broad rivers of upper delta distributary systems. During the Late Carboniferous, primitive conifers appeared and included araucarias, which became common in some, probably drier ecological habitats. See PALEOZOIC.   [C.A.R.; J.R.P.R.]

Carnotite   A mineral that is a hydrous vanadate of potassium and uranium, $K_2(UO_2)_2(VO_4)_2 \cdot nH_2O$. The water content varies at ordinary temperatures from one to three molecules. Carnotite generally occurs as a powder or as a slightly coherent microcrystalline aggregate. Color ranges from bright yellow to lemon- and greenish-yellow.

In the United States the principal region of carnotite mineralization is the Colorado Plateau and adjoining districts of Utah, New Mexico, and Arizona. Carnotite is found also in Wyoming and in Carbon County, Pennsylvania. Deposits are located at Radium Hill near Olary, Australia, and in Katanga (Zaire). Carnotite is the chief source of uranium in the United States. It is also a source of radium and vanadium. *See* RADIOACTIVE MINERALS. [W.R.Lo.]

Cartography The techniques concerned with constructing maps from geographic information. Maps are spatial representations of the environment. Typically, maps take graphic form, appearing on computer screens or printed on paper, but they may also take tactile or auditory forms for the visually impaired. Other representations such as digital files of locational coordinates or even mental images of the environment are also sometimes considered to be maps, or virtual maps.

Maps are composed of two kinds of geographic information: attribute data and locational data. Attribute data are quantitative or qualitative measures of characteristics of the landscape, such as terrain elevation, land use, or population density. Locations of features on the Earth's surface are specified by use of coordinate systems; among these, the most common is the geographical coordinate system of latitudes and longitudes.

Geographical coordinates describe positions on the spherical Earth. These must be transformed to positions on a two-dimensional plane before they can be depicted on a printed sheet or a computer screen. Hundreds of map projections—mathematical transformations between spherical and planar coordinates—have been devised, but no map projection can represent the spherical Earth in two dimensions without distorting spatial relationships among features on Earth's surface in some way. One specialized body of knowledge that cartographers bring to science is the ability to specify map projections that preserve the subset of geometric characteristics that are most important for particular mapping applications.

Although many broadly applicable map design principles have been established, the goal of specifying an optimal map for a particular task is less compelling than it once was. Instead, there is interest in the potential of providing map users with multiple, modifiable representations via dynamic media. Maps, graphs, diagrams, movies, text, and sound can be incorporated in multimedia software applications that enable users to navigate through vast electronic archives of geographic information. Interactive computer graphics are eliminating the distinction between the mapmaker and the map user. Modern cartography's challenge is to provide access to geographic information and to cartographic expertise through well-designed user interfaces. *See* MAP DESIGN; MAP PROJECTIONS. [D.DiB.]

Cassiterite A mineral having the composition SnO_2. It is the principal ore of tin. Cassiterite is usually massive granular, but may be in radiating fibrous aggregates with reniform shapes (wood tin). The hardness is 6–7 (Mohs scale), and the specific gravity is 6.8–7.1 (unusually high for a nonmetallic mineral). The luster is adamantine to submetallic. Pure tin oxide is white, but cassiterite is usually yellow, brown, or black because of the presence of iron.

Cassiterite is most abundantly found as stream tin (rolled pebbles in placer deposits). The world's supply comes mostly from placer or residual deposits in the Malay Peninsula, Indonesia, Zaire, and Nigeria. It is also mined in Bolivia. [C.S.Hu.]

Cave   A natural cavity located underground or in the side of a hill or cliff, generally of a size to admit a human. Caves occur in all types of rocks and topographic situations. They may be formed by many different erosion processes. The most important are created by ground waters that dissolve the common soluble rocks—limestone, dolomite, gypsum, and salt. Limestone caves are the most frequent, longest, and deepest. Lava-tube caves, sea caves created by wave action, and caves caused by piping in unconsolidated rocks are the other important types. The science of caves is known as speleology. *See* DOLOMITE; GYPSUM; HALITE; LIMESTONE.

Caves are important sediment traps, preserving evidences of past erosional, botanic, and other phases that may be obliterated aboveground. Chemical deposits are very important. More than 100 different minerals are known to precipitate in caves. Most abundant and significant are stalactites, stalagmites, and flowstones of calcite. These may be dated with uranium series methods, thus establishing minimum ages for the host caves. They contain paleomagnetic records. Their oxygen and carbon isotope ratios and trapped organic materials may record long-term changes of climate and vegetation aboveground that can be dated with great precision. As a consequence, cave deposits are proving to be among the most valuable paleoenvironmental records preserved on the continents. *See* STALACTITES AND STALAGMITES.   [D.C.F.]

Celestite   A mineral with the chemical composition $SrSO_4$. Celestite occurs commonly in colorless to sky-blue, orthorhombic, tabular crystals. Fracture is uneven and luster is vitreous. Hardness is 3–3.5 on Mohs scale and specific gravity is 3.97. It fuses readily to a white pearl. The strontium present in celestite imparts a characteristic crimson color to the flame.

Celestite occurs in association with gypsum, anhydrite, salt beds, limestone, and dolomite. Large crystals are found in vugs or cavities of limestone. It is deposited directly from sea water, by groundwater, or from hydrothermal solutions. Celestite is the major source of strontium: Although celestite deposits occur in Arizona and California, domestic production of celestite has been small and sporadic. Much of the strontium demand is satisfied by imported ores from England and Mexico.   [E.C.T.C.]

Cenozoic   Cenozoic (Cainozoic) is the youngest and the shortest of the three Phanerozoic geological eras. It represents the geological time (and rocks deposited during that time) extending from the end of the Mesozoic Era to the present day.

Traditional classifications subdivide the Cenozoic Era into two periods (Tertiary and Quaternary) and seven epochs (from oldest to youngest): Paleocene, Eocene, Oligocene, Miocene, Pliocene, Pleistocene, and Holocene. The older five epochs, which together constitute the Tertiary Period, span the time interval from 65 to 1.8 million years before present. The Tertiary is often separated into two subperiods, the Paleogene (Paleocene through Oligocene epochs, also collectively called the Nummulitic in older European literature) and the Neogene (Miocene and Pliocene epochs). These subperiods were introduced by M. Hornes in 1853. The Quaternary Period, which encompasses only the last 1.8 million years, includes the two youngest epochs (Pleistocene and Holocene). Holocene is also often referred to as the Recent, from the old Lyellian classification. Recent stratigraphic opinions are leaning toward abandoning the use of Tertiary and Quaternary (which are seen as the unnecessary holdovers from obsolete classifications) and in favor of retaining Paleogene and Neogene as the prime subdivisions of Cenozoic. *See* EOCENE; HOLOCENE; MIOCENE; OLIGOCENE; PALEOCENE; PLEISTOCENE; PLIOCENE.

Many of the tectonic events (mountain-building episodes or orogenies, changes in the rates of sea-floor spreading, or tectonic plate convergences) that began in the Mesozoic continued into the Cenozoic. The Laramide orogeny that uplifted the Rocky Mountains in North America, which began as early as Late Jurassic, continued into the Cretaceous and early Cenozoic time. In its post-Cretaceous phase the orogeny comprised a series of diastrophic movements that deformed the crust until some 50 million years ago, when it ended abruptly. The Alpine orogeny, which created much of the Alps, also began in the Mesozoic, but it was most intense in the Cenozoic when European and African plates converged at an increased pace. *See* CRETACEOUS; JURASSIC; MESOZOIC; OROGENY.

Another major long-term affect of the tectonic uplift of Tibetan Plateau, which is dated to have been significant by 40 million years ago, may have been the initiation of the general global cooling trend that followed this event. The uplifted plateau may have initiated a stronger deflection of the atmospheric jet stream, strengthening of the summer monsoon, and increased rainfall and weathering in the Himalayas. Increased weathering and dissolution of carbonate rock results in greater carbon dioxide drawdown from the atmosphere. The decreased partial pressure of carbon dioxide levels may have ultimately led to the Earth entering into a renewed glacial phase.

The modern circulation and vertical structure of the oceans and the predominantly glacial mode that the Earth is in at present was initiated in the mid-Cenozoic time. The early Cenozoic was a period of transition between the predominantly thermospheric circulation of the Mesozoic and the thermohaline circulation that developed in the mid-Cenozoic. By the mid-Cenozoic the higher latitudes had begun to cool down, especially in the Southern Hemisphere due to the geographic isolation of Antarctica, leading to steeper latitudinal thermal gradients and accentuation of seasonality. The refrigeration of the polar regions gave rise to the cold high-latitude water that sank to form cold bottom water. The development of the psychrosphere (cold deeper layer of the ocean) and the onset of thermohaline circulation are considered to be the most significant events of Cenozoic ocean history, which ushered the Earth into its modern glacial-interglacial cyclic mode.

The Quaternary climatic history is one of repeated alternations between glacial and interglacial periods. At least five major glacial cycles have been identified in the Quaternary of northwestern Europe. The most recent glacial event occurred between 30,000 and 18,000 years ago when much of North America and northern Europe was covered with extensive ice sheets. The late Pliocene and Pleistocene glacial cyclicity led to repeated falls in global sea level as a result of sequestration of water as ice sheets in higher latitudes during the glacial intervals. For example, the sea level is estimated to have risen some 110 m (360 ft) since the end of the last glacial maximum. As a by-product of these repeated drops in sea level and movement of the shorelines toward the basins, large deltas developed at the mouths of the world's major drainage systems during the Quaternary. These bodies of sand and silt constitute ideal reservoirs for hydrocarbon accumulation. *See* DELTA; PALEOCLIMATOLOGY.

At the end of the Cretaceous a major extinction event had decimated marine biota and only a few species survived into the Cenozoic. The recovery, however, was relatively rapid. During the Paleocene through middle Eocene interval, the overall global sea-level rise enlarged the ecospace for marine organisms, and an associated climatic optimum led to increased speciation through the Paleocene, culminating in high marine diversities during the early and middle Eocene. Limestone-building coral reefs were also widespread in the tropical-temperate climatic belt of the early Cenozoic, and the tropical Tethyan margins were typified by expansive distribution of the larger foraminifera known as *Nummulites* (giving the Paleogene its informal name of the Nummulitic period).

The late Eocene saw a rapid decline in diversities of marine phyto- and zooplankton due to a global withdrawal of the seas from the continental margins and the ensuing deterioration in climate. Marine diversities reached a new low in the mid-Oligocene, when the sea level was at its lowest, having gone through a major withdrawal of seas from the continental margins. The climates associated with low seas were extreme and much less conducive to biotic diversification. The late Oligocene and Neogene as a whole constitute an interval characterized by increasing partitioning of ecological niches into tropical, temperate, and higher-latitude climatic belts, and greater differentiation of marine fauna and flora.

Mammals evolved and spread rapidly to become dominant in the Cenozoic. The evolution of grasses in the early Eocene and the wide distribution of grasslands thereafter may have been catalytic in the diversification of browsing mammals. Marsupials and insectivores as well as rodents (which first appeared in the Eocene) diversified rapidly, as did primates, carnivores, and ungulates. The ancestral horse first appeared in the early Eocene in North America, where its lineage evolved into the modern genus *Equus*, only to disappear from the continent in the late Pleistocene. A complete evolution of the horse can be followed in North America during the Cenozoic. Increase in overall size, reduction in the number of toes, and increasing complexity of grinding surface of the molars over time are some of the obvious trends. Hominoid evolution began during the Miocene in Africa. Modern hominids are known to have branched off from the hominoids some 5 million years ago. Over the next 4.5 million years the hominids went through several evolutionary stages to finally evolve into archaic *Homo sapiens* about 1 million years ago. Truly modern *Homo sapiens* do not enter the scene until around 100 thousand years ago.   [B.U.H.]

Cerargyrite   A mineral with composition AgCl. Its structure is that of the isometric NaCl type, but well-formed cubic crystals are rare. The hardness is $2^1/_2$ on Mohs scale and specific gravity 5.5. Cerargyrite is colorless to pearl-gray but darkens to violet-brown on exposure to light. It is perfectly sectile and can be cut with a knife-like horn; hence the name horn silver. Bromyrite, AgBr, is physically indistinguishable from cerargyrite and the two minerals form a complete series. Both minerals are secondary ores of silver and occur in the oxidized zone of silver deposits.   [C.S.Hu.]

Cerussite   The mineral form of lead carbonate, $PbCO_3$. Cerussite is common as a secondary mineral associated with lead ores. In the United States it occurs mostly in the central and far western regions. Cerussite is white when pure but is sometimes darkened by impurities. Hardness is $3^1/_4$ on Mohs scale and specific gravity is 6.5. Crystals may be tabular, elongated, or arranged in clusters. See CARBONATE MINERALS.
                                                                  [R.I.Ha.]

Chabazite   A mineral belonging to the zeolite family of silicates. Hardness (on the Mohs hardness scale) is in the range of 4–5. Colors range from white to yellow, pink, and red. The ideal composition is $Ca_2Al_2Si_4O_{12} \cdot 6H_2O$ (where Ca = calcium, Al = aluminum, Si = silicon, O = oxygen, H_2O = water), but there is considerable chemical substitution of Ca by sodium (Na) and potassium (K), as well as (Na,K)Si for CaAl. The internal structure of chabazite consists of a framework linkage of (AlO_4) and (SiO_4) tetrahedra, with large cagelike openings bounded by rings of tetrahedra. The cages are connected to each other by open structural channels that allow for the diffusion of molecules through the structure of a size comparable to that of the diameter of the channels (about 0.39 nanometer in diameter). For example, argon (0.384 nm in diameter) is quickly absorbed by the chabazite structure, but isobutane (0.56 nm in diameter) cannot enter the structure. In this manner, chabazite can be used as a sieve on a molecular level. *See* ZEOLITE. [C.K.]

Chalcanthite   A mineral with the chemical composition $CuSO_4 \cdot 5H_2O$. Chalcanthite commonly occurs in blue to greenish-blue triclinic crystals or in massive fibrous veins or stalactites. Fracture is conchoidal and luster is vitreous. Hardness is 2.5 on Mohs scale and specific gravity is 2.28. It has a nauseating taste and is readily soluble in water. It dehydrates in dry air to a greenish-white powder. Although deposits of commercial size occur in arid areas, chalcanthite is generally not an important source of copper ore. Its occurrence is widespread in the western United States. [E.C.T.C.]

Chalcedony   A fine-grained fibrous variety of quartz, silicon dioxide. The individual fibers that compose the mineral aggregate usually are visible only under the microscope. Subvarieties of chalcedony recognized on the basis of color differences, some valued since ancient times as semiprecious gem materials, include carnelian (translucent, deep flesh red to clear red in color), sard (orange-brown to reddish-brown), and chrysoprase (apple green). *See* GEM; QUARTZ.

Chalcedony occurs as crusts with a rounded, mammillary, or botryoidal surface and as a major constituent of nodular and bedded cherts. The hardness is 6.5–7 on Mohs scale. The specific gavity is 2.57–2.64.

Crusts of chalcedony generally are composed of fairly distinct layers concentric to the surface. Agate is a common and important type of chalcedony in which successive layers differ markedly in color and degree of translucency. In the most common kind of agate the layers are curved and concentric to the shape of the cavity in which the material formed. *See* AGATE. [R.Si.]

Chalcocite   A mineral having composition Cu_2S and crystallizing in the orthorhombic system (below $217°F$ or $103°C$). Crystals are rare and small, usually with hexagonal outline because of twinning. Most commonly, the mineral is fine-grained and massive with a metallic luster and a lead-gray color which tarnishes to dull black on exposure. The Mohs hardness is 2.5–3, and the density 5.5–5.8. Chalcocite is an important copper ore found at Miami, Morenci, and Bisbee, Arizona; Butte, Montana; Kennecott, Alaska; and Tsumeb, South-West Africa. [L.Gr.]

Chalcopyrite   A mineral having composition $CuFeS_2$. Crystals are usually small and resemble tetrahedra. Chalcopyrite is usually massive with a metallic luster, brass-yellow color, and sometimes an iridescent tarnish. The Mohs hardness is 3.5–4.0, and the density 4.1–4.3. Chalcopyrite is a so-called fool's gold, but is brittle while gold is sectile. Pyrite, the most widespread fool's gold, is harder than chalcopyrite. *See* PYRITE.

Chalcopyrite is the most widespread primary copper ore mineral. It is commonly found in veins (Braden mine, Chile; Cornwall, England; Butte, Montana; Freiberg, Saxony; Tasmania; Rio Tinto, Spain). Chalcopyrite is also found in contact metamorphic deposits in limestone (Bisbee, Arizona) and as sedimentary deposits (Mansfeld, Germany). [L.Gr.]

Chalk   The term sometimes used in a broad sense for any soft, friable, or weathered fine-grained limestone; however the term is mostly restricted to pelagic (biogenic) limestones. Chalk is a uniformly fine-grained, typically light-colored marine limestone primarily composed of the remains of calcareous nannofossils and microfossils. These minute pelagic organisms live in surface and near-surface oceanic waters and include coccolithophores (algae) and planktic foraminifers (Protozoa). Larger fossil constituents may be present, but only in subordinate amounts. The dominant pelagic skeletal remains are composed of low-magnesium calcite and have accumulated where the sea floor lies at a depth of less than about 4 km or 13,000 ft (the carbonate is redissolved at greater depths). Typical chalk sedimentation rates are 30 m (100 ft) per million years, so chalk accumulation is also dependent on the exclusion of diluting materials such as reefal detritus or terrigenous debris (clay, silt, or sand) transported from land areas by rivers. Chalks therefore form mainly in isolated outer shelf or deeper-water settings that are far from land areas. *See* CALCITE; LIMESTONE.

The unique combination of light color, compositional purity, softness, and fine texture led to many of the early uses of chalk for writing on blackboards. Chalks are also widely used in the manufacture of portland cement, as lime for fertilizers, and in powders, abrasives, and coatings.   [P.A.Sc.]

Chaos   System behavior that depends so sensitively on the system's precise initial conditions that it is, in effect, unpredictable and cannot be distinguished from a random process, even though it is deterministic in a mathematical sense.

Throughout history, sequentially using magic, religion, and science, people have sought to perceive order and meaning in a seemingly chaotic and meaningless world. This quest for order reached its ultimate goal in the seventeenth century when newtonian dynamics provided an ordered, deterministic view of the entire universe epitomized in P. S. de Laplace's statement, "We ought then to regard the present state of the universe as the effect of its preceding state and as the cause of its succeeding state."

But if the determinism of Laplace and Newton is totally accepted, it is difficult to explain the unpredictability of a gambling game or, more generally, the unpredictably random behavior observed in many newtonian systems. Commonplace examples of such behavior include smoke that first rises in a smooth, streamlined column from a cigarette, only to abruptly burst into wildly erratic turbulent flow (see illustration); and the unpredictable phenomena of the weather.

At a more technical level, flaws in the newtonian view had become apparent by about 1900. The problem is that many newtonian systems exhibit behavior which is so exquisitely sensitive to the precise initial state or to even the slightest outside perturbation that, humanly speaking, determinism becomes a physically meaningless though mathematically valid concept. But even more is true. Many deterministic newtonian-system orbits are so erratic that they cannot be distinguished from a random process even though they are strictly determinate, mathematically speaking. Indeed, in the totality of newtonian-system orbits, erratic unpredictable randomness is overwhelmingly the most common behavior.

One example of chaos is the evolution of life on Earth. Were this evolution deterministic, the governing laws of evolution would have had built into them anticipation of every natural crisis which has occurred over the centuries plus anticipation of every possible ecological niche throughout all time. Nature, however, economizes and uses the richness of opportunity available through chaos. Random mutations provide choices sufficient to meet almost any crisis, and natural selection chooses the proper one.

Another example concerns the problem that the human body faces in defending against all possible invaders. Again, nature appears to choose chaos as the most economical solution. Loosely speaking, when a hostile bacterium or virus enters the body, defense strategies are generated at random until a feedback loop indicates that the correct strategy has been found. A great challenge is to mimic nature and to find new and useful ways to harness chaos.

Transition from order to chaos (turbulence) in a rising column of cigarette smoke. The initial smooth streamline flow represents order, while the erratic flow represents chaos.

Another matter for consideration is the problem of predicting the weather or the world economy. Both these systems are chaotic and can be predicted more or less precisely only on a very short time scale. Nonetheless, by recognizing the chaotic nature of the weather and the economy, it may eventually be possible to accurately determine the probability distribution of allowed events in the future given the present. At that point it may be asserted with mathematical precision that, for example, there is a 90% chance of rain 2 months from today. Much work in chaos theory seeks to determine the relevant probability distributions for chaotic systems. *See* WEATHER FORECASTING AND PREDICTION.

Finally, many physical systems exhibit a transition from order to chaos, as exhibited in the illustration, and much work studies the various routes to chaos. Examples include fibrillation of the heart and attacks of epilepsy, manic-depression, and schizophrenia. Physiologists are striving to understand chaos in these systems sufficiently well that these human maladies can be eliminated.

Reduced to basics, chaos and noise are essentially the same thing. Chaos is randomness in an isolated system; noise is randomness entering this previously isolated system from the outside. If the noise source is included to form a composite isolated system, there is again only chaos.   [J.Fo.]

Chemical ecology   The study of ecological interactions mediated by the chemicals that organisms produce. These substances, known as allelochemicals, serve a variety of functions. They influence or regulate interspecific and intraspecific interactions of microorganisms, plants, and animals, and operate within and between all trophic levels—producers, consumers, and decomposers—and in terrestrial, fresh-water, and marine ecosystems.

Function is an important criterion for the classification of allelochemicals. Allelochemicals beneficial to the emitter are called allomones; those beneficial to the recipient are called kairomones. An allomone to one organism can be a kairomone to another. For example, floral scents benefit the plant (allomones) by encouraging pollinators, but also benefit the insect (kairomones) by providing a cue for the location of nectar.

The chemicals involved are diverse in structure and are often of low molecular weight (<10,000). They may be volatile or nonvolatile; water-soluble or fat-soluble. Proteins, polypeptides, and amino acids are also found to play an important role.

Plant allelochemicals are often called secondary compounds or metabolites to distinguish them from those chemicals involved in primary metabolism, although this distinction is not always clear.

Chemical defense in plants. Perhaps to compensate for their immobility, plants have made wide use of chemicals for protection against competitors, pathogens, herbivores, and abiotic stresses. A chemically mediated competitive interaction between higher plants is referred to as allelopathy. Allelopathy appears to occur in many plants, may involve phenolics or terpenoids that are modified in the soil by microorganisms, and is at least partly responsible for the organization of some plant communities.

Chemicals that are mobilized in response to stress or attack are referred to as active or inducible chemicals, while those that are always present in the plant are referred to as passive or constitutive. In many plants, fungus attack induces the production of defensive compounds called phytoalexins, a diverse chemical group that includes isoflavonoids, terpenoids, polyacetylenes, and furanocoumarins.

Defensive chemicals can be induced by herbivore attack. There has been increasing evidence that inducible defenses, such as phenolics, are important in plant-insect interactions.

Constitutive defenses include the chemical hydrogen cyanide. Trefoil, clover, and ferns have been found to exist in two genetically different forms, one containing cyanide (cyanogenic) and one lacking it (acyanogenic); acyanogenic forms are often preferred by several herbivores.

Chemical defenses frequently occur together with certain structures which act as physical defenses, such as spines and hairs. While many chemicals protect plants by deterring herbivore feeding or by direct toxic effects, other defenses may act more indirectly. Chemicals that mimic juvenile hormones, the antijuvenile hormone substances found in some plants, either arrest development or cause premature development in certain susceptible insect species.

Plant chemicals potentially affect not only the herbivores that feed directly on the plant, but also the microorganisms, predators, or parasites of the herbivore. For example, the tomato plant contains an alkaloid, tomatine, that is effective against certain insect herbivores. The tomato hornworm, however, is capable of detoxifying this alkaloid and can thus use the plant successfully—but a wasp parasite of the hornworm cannot detoxify tomatine, and its effectiveness in parasitizing the hornworm is reduced. Therefore, one indirect effect of the chemical in the plant may be to reduce the effectiveness of natural enemies of the plant pest, thereby actually working to the disadvantage of the plant.

Most plant chemicals can affect a wide variety of herbivores and microorganisms, because the modes of action of the chemicals they manufacture are based on a similarity of biochemical reaction in most target organisms (for example, cyanide is toxic to most organisms). In addition, many plant chemicals may serve multiple roles: resins in the creosote bush serve to defend against herbivores and pathogens, conserve water, and protect against ultraviolet radiation.

It is argued that there are two different types of defensive chemicals in plants. The first type occurs in relatively small amounts, is often toxic in small doses, and poisons the herbivore. These compounds may also change in concentration in response to plant damage; that is, they are inducible. These kinds of qualitative defensive compounds are the most common in short-lived or weedy species that are often referred to as unapparent. They are also characteristic of fast-growing species with short-lived leaves. In contrast, the second type of defensive chemicals often occurs in high concentrations, is not very toxic, but may inhibit digestion by herbivores and is not very inducible. These quantitative defenses are most common in long-lived, so-called apparent plants such as trees that have slow growth rates and long-lived leaves. Some plants may use both types of defenses.

There is accumulating evidence that marine plants may be protected against grazing by similar classes of chemicals to those found in terrestrial plants. One interesting difference in the marine environment is the large number of halogenated organic compounds that are rare in terrestrial and fresh-water systems.

Through evolution, as plants accumulate defenses, herbivores that are able to bypass the defense in some way are selected for and leave more offspring than others. This in turn selects for new defenses on the part of the plant in a continuing process called coevolution.

Animals that can exploit many plant taxa are called generalists, while those that are restricted to one or a few taxa are called specialists. Specialists often have particular detoxification mechanisms to deal with specific defenses. Some generalists possess powerful, inducible detoxification enzymes, while others exhibit morphological adaptations of the gut which prevent absorption of compounds such as tannins, or provide reservoirs for microorganisms that accomplish the detoxification. Animals may avoid eating plants, or parts of plants, with toxins.

Some herbivores that have completely surmounted the plant toxin barrier use the toxin itself as a cue to aid in locating plants. The common white butterfly, *Pieris rapae*, for example, uses mustard oil glycosides, which are a deterrent and toxic to many organisms, to find its mustard family hosts.

Chemical defense in animals. Many animals make their own defensive chemicals—such as all of the venoms produced by social insects (bees, wasps, ants), as well as snakes and mites. These venoms are usually proteins, acids or bases, alkaloids, or combinations of chemicals. They are generally injected by biting or stinging, while other defenses are produced as sprays, froths, or droplets from glands.

Animals frequently make the same types of toxins as plants, presumably because their function as protective agents is similar. Other organisms, particularly insects, use plant chemicals to defend themselves. Sequestration may be a low-cost defense mechanism and probably arises when insects specialize on particular plants.

Microbial defenses. Competitive microbial interactions are regulated by many chemical exchanges involving toxins. They include compounds such as aflatoxin, botulinus toxin, odors of rotting food, hallucinogens, and a variety of antibiotics.

Microorganisms also play a role in chemical interaction with plants and animals that range from the production of toxins that kill insects, such as those produced by the common biological pest control agent *Bacillus thuringiensis*, to cooperative biochemical detoxification of plant toxins by animal symbionts.

Information exchange. A large area of chemical ecology concerns the isolation and identification of chemicals used for communication. Pheromones, substances produced by an organism that induce a behavioral or physiological response in an individual of the same species, have been studied particularly well in insects. These signals are compounds that are mutually beneficial to the emitter and sender, such as sex attractants, trail markers, and alarm and aggregation signals. Sex pheromones are volatile substances, usually produced by the female to attract males. Each species has a characteristic compound that may differ from that of other species by as little as a few atoms.

Pheromones are typically synthesized directly by the animal and are usually derived from fatty acids. In a few cases the pheromone or its immediate precursors may be derived from plants, as in danaid butterflies.

Very little work has been done in identifying specific pheromones in vertebrates, particularly mammals. It is known, however, that they are important in marking territory, in individual recognition, and in mating and warning signals. Chemical communication may also occur among plants and microorganisms, although it is rarer and less obvious than in animals.   [C.G.J.; A.C.L.]

Chemostratigraphy   A subdiscipline of stratigraphy and geochemistry that involves correlation and dating of marine sediments and sedimentary rocks through the use of trace-element concentrations, molecular fossils, and certain isotopic ratios that can be measured on components of the rocks. The isotopes used in chemostratigraphy can be divided into three classes: radiogenic (strontium, neodymium, osmium), radioactive (radiocarbon, uranium, thorium, lead), and stable (oxygen, carbon, sulfur). Trace-element concentrations (that is, metals such as nickel, copper, molybdenum, and vanadium) and certain organic molecules (called biological markers or bio-markers) are also employed in chemostratigraphy. See DATING METHODS; ROCK AGE DETERMINATION.

Radiogenic isotopes are formed by the radioactive decay of a parent isotope to a stable daughter isotope. The application of these isotopes in stratigraphy is based on natural cycles of the isotopic composition of elements dissolved in ocean water, cycles which are recorded in the sedimentary rocks.

The elements hydrogen, carbon, nitrogen, oxygen, and sulfur owe their isotopic distributions to physical and biological processes that discriminate between the isotopes because of their different atomic mass. The use of these isotopes in stratigraphy is also facilitated by cycles of the isotopic composition of seawater, but the isotopic ratios in marine minerals are also dependent on water temperature and the mineral-forming processes. See SEAWATER.

Certain organic molecules that can be linked with a particular source (called biomarkers) have become useful in stratigraphy. The sedimentary distributions of biomarkers reflect the biological sources and inputs of organic matter (such as that from algae, bacteria, and vascular higher plants), and the depositional environment.

Certain trace metals, such as nickel, copper, vanadium, magnesium, iron, uranium, and molybdenum, are concentrated in organic-rich sediments in proportion to the amount of organic carbon. Although the processes controlling their enrichment are complex, they generally form in an oxygen-poor environment (such as the Black Sea) or at the time of global oceanic anoxic events, during which entire ocean basins become oxygen poor, resulting in the death of many organisms; hence large amounts of organic carbon are preserved in marine sediments. The trace-metal composition of individual stratigraphic units may be used as a stratigraphic marker, or "fingerprint." See GEOCHEMISTRY; MARINE SEDIMENTS; STRATIGRAPHY.   [B.L.I.; D.J.DeP.]

Chert   A hard, dense sedimentary rock composed of fine-grained silica (SiO_2). Chert is characterized by a semivitreous to dull luster and a splintery to conchoidal fracture, and is most commonly gray, black, reddish brown, or green. Chert is also used as a field term to describe silica-rich rocks which may be impure; common impurities include carbonates, iron and manganese oxides, and clay minerals. When impurities change the texture of the rock to the extent that it is less dense and hard than chert, and has the appearance of unglazed porcelain, the rock is then called porcellanite or siliceous shale. The term flint is synonymous with chert, but its use has become restricted to archeological artifacts and to nodular chert that occurs in chalk. The term chert, however, is preferred for the nodular deposits. Jasper refers to red or yellow quartz chert associated with iron ore or containing iron oxide. Novaculite is a white chert of great purity and uniform grain size, and is composed chiefly of quartz; the term is mostly restricted to descriptions of Paleozoic cherts in Oklahoma and Arkansas. Chert synonyms that have become obsolete include silexite, petrosilex, phthanite, and hornstone. *See* JASPER.

Chert occurs mainly in three forms: bedded sequences, nodular, and massive. Bedded chert (called ribbon chert if beds show pinch-and-swell structure) consists of rhythmically interlayered beds of chert and shale; chert and carbonates; or in some pre-Phanerozoic formations, alternations of chert and siderite or hematite. Bedded sequences can be hundreds of feet thick stratigraphically and cover areas of hundreds of square miles. Individual beds are commonly $\frac{1}{2}$–8 in. (1–20 cm) thick. Chert nodules and lenses occur primarily in chalk, limstone, and dolomite. Nodules and lenses vary in size from $\frac{1}{2}$ in. to 30 ft (1 cm to 9 m). Fossils and sedimentary structures characteristic of the host rock are preserved within the nodules. Massive cherts occur in the interstices between basalt pillows, and as the basal member of bedded chert that overlies pillow basalts in ophiolites. *See* BASALT; CHALK; DOLOMITE ROCK; LIMESTONE.

When a supply of silica is available, chert forms in four ways: by replacement of mainly carbonate rock; by deposition from turbidity currents composed primarily of biogenic silica; by increasing the deposition of silica relative to terrigenous input, commonly by increased productivity of biogenic silica; and by precipitation of silica from water under either hydrothermal or low-temperature hypersaline conditions. *See* TUR-BIDITY CURRENT.   [J.R.He.]

Chinook   A mild, dry, extremely turbulent westerly wind on the eastern slopes of the Rocky Mountains and closely adjoining plains. The term is an Indian word which means "snow-eater," appropriately applied because of the great effectiveness with which this wind reduces a snow cover by melting or by sublimation. The chinook is a particular instance of a type of wind known as a foehn wind. Foehn winds, initially studied in the Alps, refer to relatively warm, rather dry currents descending the lee slope of any substantial mountain barrier. The dryness is an indirect result of the condensation and precipitation of water from the air during its previous ascent of the windward slope of the mountain range. The warmth is attributable to adiabatic compression, turbulent mixing with potentially warmer air, and the previous release of latent heat of condensation in the air mass and to the turbulent mixing of the surface air with the air of greater heat content aloft. In winter the chinook wind sometimes impinges upon much colder stagnant polar air along a sharp front located in the foothills of the Rocky Mountains or on the adjacent plain. Small horizontal oscillations of this front have been known to produce several abrupt temperature rises and falls of as much as 45–54°F (25–30°C) at a given location over a period of a few hours. Damaging winds sometimes occur as gravity waves, which are triggered along the interface between the two air masses. *See* FRONT; ISENTROPIC SURFACES; PRECIPITATION (METEOROLOGY); WIND; WIND STRESS.   [F.S.; H.B.B.]

Chlorite   One of a group of layer silicate minerals, usually green in color, characterized by a perfect cleavage parallel to (001). The cleavage flakes are flexible but inelastic, with a luster varying from pearly or vitreous to dull and earthy. The hardness on the cleavage is about 2.5. The specific gravity of chlorate varies between 2.6 and 3.3 as a function of composition.

Chlorite is a common accessory mineral in low- to medium-grade regional metamorphic rocks and is the dominant mineral in chlorite schist. It can form by alteration of ferromagnesian minerals in igneous rocks and is found occasionally in pegmatites and vein deposits. It is a common constituent of altered basic rocks and of alteration zones around metallic ore bodies. Chlorite also can form by diagenetic processes in sedimentary rocks. *See* AUTHIGENIC MINERALS; CLAY MINERALS; DIAGENESIS; SILICATE MINERALS. [S.W.Ba.]

Chloritoid   A hydrous iron aluminum silicate mineral with an ideal formula of $Fe_2^{2+}Al_4O_2(SiO_4)_2(OH)_4$. Chloritoid occurs as platy, black or dark green crystals, rarely more than a few millimeters in size. Its density ranges from 3.46 to 3.80 g/cm^3, and its hardness on the Mohs scale is 6.5.

Chloritoid is increasingly being recognized as a constituent in rocks that formed under high-pressure conditions. It is found in association with glaucophane in blueschist-facies metamorphic rocks, and with amphibole and pyroxene in eclogite-facies metamorphic rocks. These are rocks that formed under conditions thought to prevail at the base of the Earth's crust or in the mantle. Experimental studies of metamorphism of basalt under these conditions indicate the formation of chloritoid at pressures exceeding 2300 MPa (23 kbar) and at a temperature of 650°C (1200°F). In these high-pressure occurrences of chloritoid, the chloritoid is rich in magnesium, having a value of Mg/(Mg + Fe) ranging from 0.38 to 0.40. *See* METAMORPHIC ROCKS. [T.C.L.]

Chromite   The only ore mineral of chromium. Chromite is jet black to brownish black, has a submetallic luster. It belongs to the spinel group of minerals and has cubic symmetry. Naturally occurring chromite has the general formula $(Mg,Fe^{2+})(Cr,Al,Fe^{3+})_2O_4$ and ranges from 15 to 64 wt % Cr_2O_3, with minor amounts of nickel, titanium, zinc, cobalt, and manganese. The specific gravity of chromite ranges from 4 to 5, depending on its composition. Pure chromite, $Fe^{2+}Cr_2O_4$, is extremely rare in nature and has been found only in meteorites.

Chromite has a variety of uses. Chromium is extracted from it to make stainless steel and other alloys for which resistance to oxidation and corrosion is important. Chromium is also used as a plating and tanning agent. The mineral chromite is made into refractory lining for steel-making furnaces. *See* SPINEL. [B.R.L.]

Chrysoberyl   A mineral having composition $BeAl_2O_4$ and crystallizing in the orthorhombic system. The hardness is 8.5 (Mohs scale) and the specific gravity is 3.7–3.8. The luster is vitreous and the color various shades of green, yellow, and brown. There are two gem varieties of chrysoberyl. Alexandrite, one of the most prized of gemstones, is an emerald green but in transmitted or in artificial light is red. Cat's eye, or cymophane, is a green chatoyant variety with an opalescent luster. When cut en cabochon, it is crossed by a narrow beam of light. This property results from minute tabular cavities that are arranged in parallel position.

Chrysoberyl is a rare mineral found most commonly in pegmatite dikes and occasionally in granitic rocks and mica schists. Gem material is found in stream gravels in Ceylon and Brazil. The alexandrite variety is found in the Ural Mountains. In the United States chrysoberyl is found in pegmatites in Maine, Connecticut, and Colorado. *See* GEM. [C.S.Hu.]

Chrysocolla   A silicate mineral, composition $CuSiO_3 \cdot 2H_2O$. Small acicular crystals have been observed, but it ordinarily occurs in impure cryptocrystalline crusts and masses with conchoidal fracture. The hardness varies from 2 to 4 on Mohs scale, and the specific gravity varies from 2.0 to 2.4. The luster is vitreous and it is normally green to greenish-blue, but may be brown to black when impure. Chrysocolla is a secondary mineral occurring in the oxidized zones of copper deposits, where it is associated with malachite, azurite, native copper, and cuprite. It is a minor ore of copper.   [C.S.Hu.]

Chrysotile   Chrysotile is a fibrous mineral with a tubular morphology for each fibril. It is a member of the serpentine mineral group, as are antigorite and lizardite. Chrysotile aggregates make up serpentine asbestos, which is the most important type of commercially mined asbestos. Russia and Canada are the main producing countries. Chrysotile displays interesting properties such as being thermally and electrically insulating, sound insulating, chemically inert, fire-resistant, mechanical energy-absorbing, and flexible with enough high tensile strength to be woven. There are hundreds of applications for chrysotile including fire retarder in buildings, roofing tiles, brake pads, weavable material for refractory clothes, filters, and fibers in fibrocement and road surfaces. *See* ASBESTOS; SERPENTINE; SERPENTINITE.

Intensive inhalation of long and thin asbestos fibers over a considerable time period can induce pulmonary deseases such as asbestosis and lung cancers, as well as pleural diseases such as plaques, fibrosis, and mesothelioma. Such health hazards have drastically reduced the use of chrysotile, which is strictly regulated by law in western countries.   [A.J.Ba.]

Cinnabar   A mineral of composition HgS, crystallizing in the hexagonal system. Crystals are rare, usually of rhombohedral habit and often in penetration twins. Cinnabar most commonly occurs in fine, granular, massive form. It has perfect prismatic cleavage, a Mohs hardness of 2.0–2.5, and a density of 8.09. It has either an adamantine luster and vermilion red color or a dull luster and brownish-red color.

Cinnabar is deposited from hydrothermal solutions in veins and as impregnations near recent volcanic rocks and hot springs. It is the principal ore of mercury. Notable occurrences are Almaden, Spain; Idria, Italy; near Belgrade, former Yugoslavia; Kweichow and Hunan provinces, China; Turkistan; New Almaden and New Idria, California; Terlingua, Texas; and several localities in Utah, Nevada, New Mexico, Oregon, and Idaho.   [L.Gr.]

Cirque   A cliffed rock basin, shaped like half a bowl, at the head of a mountain valley. Cirques may be shallow if glacier erosion is slight, but may be 1600–2600 ft (500–800 m) from the top of the headwall to the cirque floor if erosion is deep. Cirques occur in glaciated mountains all over the world. Some cirques used repeatedly by glaciers over long periods of years are excavated profoundly. Many cirques contain small glaciers today that were inherited from a huge cirque cut by glaciers long ago. Well-formed cirques have steep rock walls and floors that slope down-valley or back toward the base of the headwall. Some cirques are cup-shaped rock basins holding rainwater; a lake formed in them is a tarn. If glacial deposits, such as till or a small moraine on the cirque floor, form a depression for a lake the lake is known as a moraine-dammed lake.   [S.E.Wh.]

Clay   The finest-grain particles in a sediment, soil, or rock. Clay is finer than silt, characterized by a grain size of less than approximately 4 micrometers. However, the term clay can also refer to a rock or a deposit containing a large component of clay-size material. Thus clay can be composed of any inorganic materials, such as clay minerals, allophane, quartz, feldspar, zeolites, and iron hydroxides, that possess a sufficiently fine grain size. Most clays, however, are composed primarily of clay minerals. *See* CLAY MINERALS; FELDSPAR; QUARTZ; ZEOLITE.

Although the composition of clays can vary, clays can share several properties that result from their fine particle size. These properties include plasticity when wet, the ability to form colloidal suspensions when dispersed in water, and the tendency to flocculate (clump together) and settle out in saline water.

Clays, together with organic matter, water, and air, are one of the four main components of soil. Clays can form directly in a soil by precipitation from solution (neoformed clays); they can form from the partial alteration of clays already present in the soil (transformed clays); or they can be inherited from the underlying bedrock or from sediments transported into the soil by wind, water, or ice (inherited clays). *See* SOIL.

The type of clays neoformed in a soil depends on the composition of the soil solution, which in turn is a function of climate, drainage, original rock type, vegetation, and time. Generally, neoformed clays that have undergone intense leaching, such as soils formed under wet, tropical climates, are composed of the least soluble elements, such as ferric iron, aluminum, and silicon. These soils contain clays such as gibbsite, kaolinite, goethite, and amorphous oxides and hydroxides of aluminum and iron. Clays formed in soils that are found in dry climates or in soils that are poorly drained can contain more soluble elements, such as sodium, potassium, calcium, and magnesium, in addition to the least soluble elements. These soils contain clays such as smectite, chlorite, and illite, and generally are more fertile than those formed under intense leaching conditions. *See* CHLORITE; GOETHITE; ILLITE; KAOLINITE.

Examples of clays formed by the transformation of other clays in a soil include soil chlorite and soil vermiculite, the first formed by the precipitation of aluminum hydroxide in smectite interlayers, and the second formed by the leaching of interlayer potassium from illite. Examples of inherited clays in a soil are illite and chlorite-containing soils formed on shales composed of these minerals. *See* SHALE.

Clays also occur abundantly in sediments and sedimentary rocks. For example, clays are a major component of many marine sediments. These clays generally are inherited from adjacent continents, and are carried to the ocean by rivers and wind, although some clays (such as smectite and glauconite) are neoformed abundantly in the ocean. Hydrothermal clays can form abundantly where rock has been in contact with hot water or steam. Illite and chlorite, for example, form during the deep burial of sediments, and smectite and chlorite form by the reaction of hot, circulating waters at ocean ridges. *See* MARINE SEDIMENTS; SEDIMENTARY ROCKS.

Various clays possess special properties which make them important industrially. For example, bentonite, a smectite formed primarily from the alteration of volcanic ash, swells; is readily dispersible in water; and possesses strong absorptive powers, including a high cation exchange capacity. These properties lead to uses in drilling muds, as catalysts and ion exchangers, as fillers and absorbents in food and cosmetics, and as binders for taxonite and fertilizers. Other important uses for clays include the manufacture of brick, ceramics, molding sands, decolorizers, detergents and soaps, medicines, adhesives, liners for ponds and landfills, lightweight aggregate, desiccants, molecular sieves, pigments, greases, paints, plasticizing agents, emulsifying, suspending, and stabilizing agents, and many other products. *See* BENTONITE; CLAY, COMMERCIAL.   [D.D.E.]

Clay, commercial   Clays utilized as raw material in manufacturing, which are among the most important nonmetallic mineral resources. The value of clays is related to their mineralogical and chemical composition, particularly the clay mineral constituents kaolinite, montmorillonite, illite, chlorite, and attapulgite. The presence of minor amounts of mineral or soluble salt impurities in clays can restrict their use. The more common mineral impurities are quartz, mica, carbonates, iron oxides and sulfides, and feldspar. In addition, many clays contain some organic material. See CLAY MINERALS.

Kaolinitic clays. Clays containing a preponderance of the clay mineral kaolinite are known as kaolinitic clays. Several commercial clays are composed predominantly of kaolinite. These are china clays, kaolines, ball clays, fireclays, and flint clays. The terms china clay and kaolin are used interchangeably in industry. See KAOLINITE.

China clays are high-grade white kaolins found in the south-eastern United States, England, and many other countries. Many grades of kaolin are used in the manufacture of ceramics, paper, rubber, paint, plastics, insecticides, adhesives, catalysts, and ink. By far the largest consumer of white kaolins is the paper industry, which uses them to make paper products smoother, whiter, and more printable. The kaolin is used both as a filler in the sheet to enhance opacity and receptivity to ink and as a thin coating on the surface of the sheet to make it smoother and whiter for printing.

Ball clays are composed mainly of the mineral kaolinite but usually are much darker in color than kaolin. The term ball clay is used for a fine-grained, very plastic, refractory bond clay. Most ball clays contain minor amounts of organic material and the clay mineral montmorillonite, and are finer grained than china clays. This fineness, together with the montmorillonite and organic material, gives ball clays excellent plasticity and strength. For these reasons and because they fire to a light-cream color, ball clays are commonly used in whitewares and sanitary ware.

The term fireclay is used for clays that will withstand temperatures of 2730°F (1500°C) or higher. Such clays are composed primarily of the mineral kaolinite. Fireclays are generally light to dark gray in color, contain minor amounts of mineral impurities such as illite and quartz, and fire to a cream or buff color. Most fireclays are plastic, but some are nonplastic and very hard; these are known as flint clays. Fireclays are used primarily by the refractories industry. The foundry industry uses fireclay to bind sands into shapes in which metals can be cast.

Diaspore clay. This clay is composed of the minerals diaspore and kaolinite. Diaspore is a hydrated aluminum oxide with an Al_2O_3 content of 85% and a water content of 15%. Diaspore clay is used almost exclusively by the refractories industry in making refractory brick. However, after calcination, it is sometimes used as an abrasive material.

Mullite. Mullite is a high-temperature conversion product of many aluminum silicate minerals, including kaolinite, pinite, topaz, dumortierite, pyrophyllite, sericite, andalusite, kyanite, and sillimanite. Mullite is used in refractories to produce materials of high strength and great refractoriness. Mullite does not spall, withstands the shock of heating and cooling exceptionally well, and is resistant to slag erosion. It is used in making spark plugs, laboratory crucibles, kiln furniture, saggars, and other special refractories.

Bentonites. Those clays that are composed mainly of the clay mineral montmorillonite and are formed by the alteration of volcanic ash are known as bentonites. The term bentonite is a rock term, but in industrial usage it has become almost synonymous with swelling clay. Bentonites are used in many industries; the most important uses are as drilling muds and catalysts in the petroleum industry, as bonding clays in foundries, as bonding agents for taconite pellets, and as adsorbents in many industries. See BENTONITE .

Attapulgite clays. Attapulgite is a hydrated magnesium aluminum silicate with a needlelike shape. Each individual needle is exceedingly small, about 1 micrometer in length and approximately 0.01 micrometer across. Attapulgite is used as a suspending agent, and gives high viscosity because of the interaction of the needles. Some commercial uses are as an oil well drilling fluid, in adhesives as a viscosity control, in oil base foundry sand binders, as thickeners in latex paints, in liquid suspension fertilizers, and as a suspending agent and thickener in pharmaceuticals.

Properties. Most clays become plastic when mixed with varying proportions of water. Plasticity of a material can be defined as the ability of the material to undergo permanent deformation in any direction without rupture under a stress beyond that of elastic yielding. Clays range from those which are very plastic, called fat clay, to those which are barely plastic, called lean clay. The type of clay mineral, particle size and shape, organic matter, soluble salts, adsorbed ions, and the amount and type of nonclay minerals are all known to affect the plastic properties of a clay.

Green strength and dry strength properties are very important because most structural clay products are handled at least once and must be strong enough to maintain shape. Green strength is the strength of the clay material in the wet, plastic state. Dry strength is the strength of the clay after it has been dried.

Both drying and firing shrinkages are important properties of clay used for structural clay products. Shrinkage is the loss in volume of a clay when it dries or when it is fired. Drying shrinkage is high in most very plastic clays and tends to produce cracking and warping. It is low in sandy clays or clays of low plasticity and tends to produce a weak, porous body. Firing shrinkage depends on the volatile materials present, the types of crystalline phase changes that take place during firing, and the dehydration characteristics of the clay minerals.

The temperature range of vitrification, or glass formation, is a very important property in structural products. Vitrification is due to a process of gradual fusion in which some of the more easily melted constituents begin to produce an increasing amount of liquid which makes up the glassy bonding material in the final fired product. The degree of vitrification depends on the duration of firing as well as on the temperature attained.

Color is important in most structural clay products, particularly the maintenance of uniform color. The color of a product is influenced by the state of oxidation of iron, the state of division of the iron minerals, the firing temperature and degree of vitrification, the proportion of alumina, lime, and magnesia in the clay material, and the composition of the fire gases during the burning operation.   [H.H.Mu.]

Structural uses. All types of clay and shale are used in the structural products industry but, in general, the clays that are used are considered to be relatively low grade. Clays that are used for conduit tile, glazed tile, and sewer pipe are underclays and shales that contain large proportions of kaolinite and illite. The semirefractory plastic clays found directly beneath the coal seams make the best raw material for the above mentioned uses. Brick and drain tile can be made from a wide variety of clays depending on their location and the quality of product desired. Clays used for brick and drain tile must be plastic enough to be shaped. In addition, color and vitrification range are very important. For common brick, drain tile, and terra-cotta, shales and surface clays are usually suitable, but for high-quality face bricks, shales and underclays are used. Geographic location is a prime factor in the type of clay used for structural clay products because, in general, these products cannot be shipped great distances without excessive transportation costs. Many raw materials of questionable quality are utilized in certain areas because no better raw material is available nearby.

The cement industry uses large quantities of impure clays and shales. Clays are used to provide alumina and silica to the charge for the cement kiln. Generally, a suitable clay can be found in the area in which the cement is being manufactured.

Clay minerals   Fine-grained, hydrous, layer silicates that belong to the larger class of sheet silicates known as phyllosilicates. Their structure is composed of two basic units. (1) The tetrahedral sheet is composed of silicon-oxygen tetrahedra linked to neighboring tetrahedra by sharing three corners to form a hexagonal network. The fourth corner of each tetrahedron (the apical oxygen) points into and forms a part of the adjacent octahedral sheet. (2) The octahedral sheet is usually composed of aluminum or magnesium in sixfold coordination with oxygen from the tetrahedral sheet and with hydroxyl. Individual octahedra are linked laterally by sharing edges. Tetrahedral and octahedral sheets taken together form a layer, and individual layers may be joined to each other in a clay crystallite by interlayer cations, by van der Waals and electrostatic forces, or by hydrogen bonding.

Clay minerals are classified by their arrangement of tetrahedral and octahedral sheets. Thus, 1:1 clay minerals contain one tetrahedral and one octahedral sheet per clay layer; 2:1 clay minerals contain two tetrahedral sheets with an octahedral sheet between them; and 2:1:1 clay minerals contain an octahedral sheet that is adjacent to a 2:1 layer.

Ionic substitutions may occur in any of these sheets, thereby giving rise to a complex chemistry for many clay minerals. For example, cations small enough to enter into tetrahedral coordination with oxygen, cations such as Fe^{3+} and Al^{3+}, can substitute for Si^{4+} in the tetrahedral sheet. Cations such as Mg^{2+}, Fe^{2+}, Fe^{3+}, Li^+, Ni^{2+}, Cu^{2+}, and other medium-sized cations can substitute for Al^{3+} in the octahedral sheet. Still larger cations such as K^+, Na^+, and Cs^+ can be located between layers and are called interlayer cations. F^- may substitute for $(OH)^-$ in some clay minerals.

Clay minerals and related phyllosilicates are classified further according to whether the octahedral sheet is dioctahedral or trioctahedral. In dioctahedral clays, two out of three cation positions in the octahedral sheet are filled, every third position being vacant. This type of octahedral sheet is sometimes known as the gibbsite sheet, with the ideal composition $Al_2(OH)_6$. In trioctahedral clay minerals, all three octahedral positions are occupied, and this sheet is called a brucite sheet, composed ideally of $Mg_3(OH)_6$.

Clay minerals can be classified further according to their polytype, that is, by the way in which adjacent 1:1, 2:1, or 2:1:1 layers are stacked on top of each other in a clay crystallite. For example, kaolinite shows at least four polytypes: b-axis ordered kaolinite, b-axis disordered kaolinite, nacrite, and dickite. Serpentine shows many polytypes, the best-known of which is chrysotile, a mineral that is used to manufacture asbestos products. *See* KAOLINITE; SERPENTINE.

Finally, clays are named on the basis of chemical composition. For example, two types of swelling clay minerals are the 2:1, dioctahedral smectites termed beidellite and montmorillonite. The important difference between them is in the location of ionic substitutions. In beidellite, charge-building substitutions are located in the tetrahedral sheet; in montmorillonite, the majority of these substitutions are located in the octahedral sheet.

Because clay minerals are composed of only two types of structural units (octahedral and tetrahedral sheets), different types of clay minerals can articulate with each other, thereby giving rise to mixed-layer clays. The most common type of mixed-layer clay is mixed-layer illite/smectite, which is composed of an interstratification of various proportions of illite and smectite layers. The interstratification may be random or ordered. The ordered mixed-layer clays may be given separate names. For example, a dioctahedral mixed-layer clay containing equal proportions of illite and smectite layers that are regularly interstratified is termed potassium rectorite. A regularly interstratified trioctahedral mixed-layer clay mineral containing approximately equal proportions of chlorite and smectite layers is termed corrensite.

A primary requirement for the formation of clay minerals is the presence of water. Clay minerals form in many different environments, including the weathering environment, the sedimentary environment, and the diagenetic-hydrothermal environment. Clay minerals composed of the more soluble elements (for example, smectite and sepiolite) are formed in environments in which these ions can accumulate (for example, in a dry climate, in a poorly drained soil, in the ocean, or in saline lakes), whereas clay minerals composed of less soluble elements (for example, kaolinite and halloysite) form in more dilute water such as that found in environments that undergo severe leaching (for example, a hilltop in the wet tropics), where only sparingly soluble elements such as aluminum and silicon can remain. Illite and chlorite are known to form abundantly in the diagenetic-hydrothermal environment by reaction from smectite. *See* CHLORITE; CLAY; CLAY, COMMERCIAL; HALLOYSITE; ILLITE; LITHOSPHERE; SEPIOLITE; SILICATE MINERALS.   [D.D.E.]

Clear-air turbulence   Turbulence above the boundary layer but not associated with cumulus convection. The atmosphere is a fluid in turbulent motion. That turbulence of a scale sensed by humans in aircraft is primarily associated with the boundary layer within a kilometer or so of the Earth, where it is induced by the surface roughness, or in regions of deep convection such as cumulus cloud development or thunderstorms. However, aircraft occasionally encounter turbulence when flying at altitudes well above the surface and far from convective clouds. This phenomenon has been given the rather unsatisfactory name of clear-air turbulence (CAT).

What is primarily sensed in CAT by the human is vertical acceleration. This acceleration will depend on the person's location in the plane, the speed of flight relative to the air, and the response characteristics of the airframe. A plane with a wing that generates aerodynamic lift more efficiently or an air-frame with less weight per unit wing area will respond more strongly to a given gust magnitude.

CAT is encountered in the atmosphere with a probability depending on flight altitude, geographical location, season of the year, and meteorological conditions. Given this variability and the small scale of the phenomenon, it is difficult to establish reliable statistics on the frequency of its occurrence. Although CAT may be encountered in unexpected meteorological contexts, there are highly favored locations for its occurrence. One is in the vicinity of the jet stream, particularly in ridges and troughs where the wind direction is turning sharply. A second and even more common location of occurrence is in the lee of a mountain range when a strong air flow is distorted by being forced over the range. In this situation a gravity lee wave is generated, which propagates to stratospheric heights. At various altitudes and distances from the mountain, this wave may break, and as many as a dozen or more CAT patches, light to severe, may be formed. Despite knowledge of these favored meteorological areas, it is not possible to forecast with confidence the precise location of a CAT patch. Warning forecasts for substantial portions of routes are typically given to pilots when CAT conditions prevail. *See* JET STREAM.

Aside from the practical implications of CAT for air transport, this phenomenon plays a role of undetermined magnitude in the dissipation of the kinetic energy of the atmosphere.   [M.G.W.; L.J.E.]

Climate history   The long-term records of precipitation, temperature, wind, and all other aspects of the Earth's climate. The climate, like the Earth itself, has a history extending over several billion years. Climatic changes have occurred at time scales ranging from hundreds of millions of years to centuries and decades. Processes in the atmosphere, oceans, cryosphere (snow cover, sea ice, continental ice sheets), biosphere, and lithosphere (such as plate tectonics and volcanic activity) and certain extraterrestrial factors (such as the Sun) have caused these changes of climate.

The present climate can be described as an ice age climate, since large land surfaces are covered with ice sheets (for example, Antartica and Greenland). The origins of the present ice age may be traced, at least in part, to the movement of the continental plates. With the gradual movement of Antarctica toward its present isolated polar position, ice sheets began to develop there about 30 million years ago. For the past several million years, the Antarctic ice sheet reached approximately its present size, and ice sheets appeared on the lands bordering the northern Atlantic Ocean. During the past million years of the current ice age, about 10 glacial-interglacial cycles have been documented. Changes in the Earth's orbital parameters, eccentricity, obliquity, and longitude of perihelion are thought to have initiated, or paced, these cycles through the associated small changes in the seasonal and latitudinal distribution of solar radiation. The most recent glacial period ended between about 15,000 and 6000 years ago with the rapid melting of the North American and European ice sheets and an associated rise in sea level, and the atmospheric concentration of carbon dioxide.

The climates of the distant geologic past were strongly influenced by the size and location of continents and by large changes in the composition of the atmosphere. For example, around 250 million years ago the continents were assembled into one supercontinent, Pangaea, producing significantly different climatic patterns than are seen today with widely distributed continents. In addition, based upon models of stellar evolution, it is hypothesized that the Sun's radiation has gradually increased by 10–20% over the past several billion years and, if so, this has contributed to a significant warming of the Earth. *See* CONTINENTS, EVOLUTION OF; GLACIAL EPOCH; PALEOCLIMATOLOGY.

Instrumental records of climatic variables such as temperature and precipitation exist for the past 100 years in many locations and for as long as 200 years in a few locations. These records provide evidence of year-to-year and decade-to-decade variability, but they are completely inadequate for the study of century-to-century and longer-term variability. Even for the study of short-term climatic fluctuations, instrumental records are incomplete, because most observations are made from the continents (covering only 29% of the Earth's surface area). Aerological observations, which permit the study of atmospheric mass, momentum and energy budgets, and the statistical structure of the large-scale circulation, are available only since about the mid-1960s. Again there is a bias toward observations over the continents. It is only with the advent of satellites that global monitoring of the components of the Earth's radiation budget (clouds; planetary albedo, from which the net incoming solar radiation can be estimated; and the outgoing terrestrial radiation) became possible. *See* METEOROLOGICAL SATELLITES.

Evidence of climatic changes prior to instrumental records comes from a wide variety of sources. Tree rings, banded corals, and pollen and trace minerals retrieved from laminated lake sediments and ice sheets yield environmental records for past centuries and millennia. Advanced drilling techniques have made it possible to obtain long cores from ocean sediments that provide geologic records of climatic conditions going back hundreds of millions of years. *See* DENDROCHRONOLOGY.

Many extraterrestrial and terrestrial processes have been hypothesized to be possible causes of climatic fluctuations. These include solar irradiance, variations in orbital parameters, motions of the lithosphere, volcanic activity, internal variations of the climate system, and human activities. It is likely that all of the natural processes have played a role in past climatic changes. Also, the climatic response to some particular causal process may depend on the initial climatic state, which in turn depends upon previous climatic states because of the long time constants of lithosphere, oceans, and cryosphere. True equilibrium climates may not exist, and the climate system may be in a continual state of adjustment.

Because of the complexity of the real climate system, simplified numerical models of climate are being used to study particular processes and interactions. Some models treat only the global-average conditions, whereas others, particularly the dynamical atmosphere and ocean models, simulate detailed patterns of climate. These models will undoubtedly be of great importance in attempts to understand climatic processes and to assess the possible effects of human activities on climate. *See* CLIMATE MODELING; CLIMATOLOGY.   [J.E.K.]

Climate modeling   Construction of a mathematical model of the climate system of the Earth capable of simulating its behavior under present and altered conditions. The Earth's climate is continually changing over time scales ranging from millions of years to a few years. Since the climate is determined by the laws of classical physics, it should be possible in principle to construct such a model. The advent of a worldwide weather observing system capable of gathering data for validation and the development and widespread routine use of digital computers have made this undertaking possible.

The Earth's average temperature is determined mainly by the balance of radiant energy absorbed from sunlight and the radiant energy emitted by the Earth system. About 30% of the incoming radiation is reflected directly to space, and 72% of the remainder is absorbed at the surface. The radiation is absorbed unevenly over the Earth, which sets up thermal contrasts that in turn induce convective circulations in the atmosphere and oceans. Climate models attempt to calculate from mathematical algorithms the effects of these contrasts and the resulting motions in order to understand better and perhaps predict future climates in some probabilistic sense. *See* TERRESTRIAL RADIATION.

Climate models differ in complexity, depending upon the application. The simplest models are intended for describing only the surface thermal field at a fairly coarse resolution. These mainly thermodynamical formulations are successful at describing the seasonal cycle of the present climate, and have been used in some simulations of past climates, for example, for different continental arrangements millions of years ago. At the other end of the spectrum are the most complex climate models, which are extensions of the models in weather forecasts. These models aim at simulating seasonal and even monthly averages just shortly into the future, based upon conditions such as the temperatures of the tropical-sea surfaces. Intermediate to these extremes are models that attempt to model climate on a decadal basis, and these are used mainly in studies of the impact of hypothesized anthropogenically induced climate change. *See* WEATHER FORECASTING AND PREDICTION.

Attempts at modeling climate have demonstrated the extreme complexity and subtlety of the problem. This is due largely to the many feedbacks in the system. One of the simplest and yet most important feedbacks is that due to water vapor. If the Earth is perturbed by an increase in the solar radiation, for example, the first-order response of the system is to increase its temperature. But an increase in air temperature leads to more water vapor evaporating into the air; this in turn leads to increased absorption of space-bound long-wave radiation from the ground (greenhouse effect), which leads to an increased equilibrium temperature. Water vapor feedback is not the only amplifier in the system. Another important one is snowcover: a cooler planet leads to more snow and hence more solar radiation reflected to space, since snow is more reflecting of sunlight than soil or vegetation. Other, more subtle mechanisms that are not yet well understood include those involving clouds, oceans, and the biosphere.

While water vapor and snowcover feedback are fairly straightforward to model, the less understood feedbacks differ in their implementations from one climate model to another. These differences as well as the details of their different numerical formulations have led to slight differences in the sensitivity of the various models to such standard experimental perturbations as doubling carbon dioxide in the atmosphere. All models agree that the planetary average temperature should increase if carbon dioxide concentrations are doubled. However, the predicted response in planetary temperatures ranges from 4.5 to 9°F (2.5 to 5.0°C). Regional predictions of temperature or precipitation are not reliable enough for detailed response policy formulation. Many of the discrepancies are expected to decrease as model resolution increases (more grid points), since it is easier to include such complicated phenomena as clouds in finer-scale formulations and coupling with dynamic models of the ocean. Similarly, it is anticipated that some observational data (such as rainfall over the oceans) that are needed for validation of the models will soon be available from satellite sensors. *See* METEOROLOGICAL SATELLITES.   [R.E.D.; E.S.S.]

Climate modification   Alteration of the Earth's climate by human activities. This can occur on various scales. For example, conventional agriculture alters the microclimate in the lowest few meters of air, causing changes in the evapotranspiration and local heating characteristics of the air-surface interface. These changes lead to different degrees of air turbulence over the plants and to different moisture and temperature distributions in the local air. An example at a larger scale is that the innermost parts of cities are several degrees warmer than the surrounding countryside, and have slightly more rainfall. These changes are brought about by the differing surface features of urban land versus natural countryside and the ways that cities dispose of water (for example, storm sewers). The urban environment prevents evaporation cooling of surfaces in the city. The modified surface texture of cities (horizontal and vertical planes of buildings and streets versus gently rolling surfaces over natural forest or grassland) leads to a more efficient trapping of solar heating of the near-surface air. The scales of buildings and other structures also lead to a different pattern of atmospheric boundary-layer turbulence, modifying the stirring efficiency of the atmosphere. *See* MICROMETEOROLOGY; URBAN CLIMATOLOGY.

At the next larger scale, human alteration of regional climates is caused by changes in the Earth's average reflectivity to sunlight. For example, the activities of building roads and highways and deforestation alter the amount of sunshine that is reflected to space as opposed to being absorbed by the surface and thereby heating the air through contact. Such contact heating leads to temperature increases and evaporation of liquid water at the surface. Vapor wakes from jet airplanes are known to block direct solar radiation near busy airports by up to 20%. Human activities also inject dust, smoke, and other aerosols into the air, causing sunlight to be scattered back to space. Dust particles screen out sunlight before it can enter the lower atmosphere and warm the near-surface air. *See* AIR POLLUTION; SMOG.   [G.R.N.]

One of the most important and best understood features of the atmosphere is the process that keeps the Earth's surface much warmer than it would be with no atmosphere. This process involves several gases in the air that trap infrared radiation, or heat, emitted by the surface and reradiate it in all directions, including back to the surface. The heat-trapping gases include water vapor, carbon dioxide (CO_2), methane (CH_4), and nitrous oxide (N_2O). These gases constitute only a small fraction of the atmosphere, but their heat-trapping properties raise the surface temperature of the

Earth by a large amount, estimated to be more than 55°F (30°C). Human activities, however, are increasing the concentrations of CO_2, CH_4, and N_2O in the atmosphere, and in addition industrially synthesized chemicals—chlorofluorocarbons (CFCs) and related compounds—are being released to the air, where they add to the trapping of infrared radiation. Carbon dioxide is released to the air mainly from fossil fuel use, which also contributes to the emissions of CH_4 and N_2O. Agricultural and industrial processes add to the emissions of these gases. These concentration increases add to the already powerful heat-trapping capability of the atmosphere, raising the possibility that the surface will warm above its past temperatures, which have remained roughly constant, within about 3°F (1.7°C) for the past 10,000 years. Treaties are in existence that control internationally the production and use of many of the CFCs and related compounds, so their rate of growth has slowed, and for some a small decrease in atmospheric concentration has been observed. Large emissions of sulfur dioxide in industrial regions are thought to result in airborne sulfate particles that reflect sunlight and decrease the amount of heating in the Northern Hemisphere. *See* ATMOSPHERE; GREENHOUSE EFFECT. [J.Fi.]

Climatology   The scientific study of climate. Climate is the expected mean and variability of the weather conditions for a particular location, season, and time of day. The climate is often described in terms of the mean values of meteorological variables such as temperature, precipitation, wind, humidity, and cloud cover. A complete description also includes the variability of these quantities, and their extreme values. The climate of a region often has regular seasonal and diurnal variations, with the climate for January being very different from that for July at most locations. Climate also exhibits significant year-to-year variability and longer-term changes on both a regional and global basis.

The goals of climatology are to provide a comprehensive description of the Earth's climate over the range of geographic scales, to understand its features in terms of fundamental physical principles, and to develop models of the Earth's climate for sensitivity studies and for the prediction of future changes that may result from natural and human causes. *See* CLIMATE HISTORY; CLIMATE MODELING; CLIMATE MODIFICATION; WEATHER. [D.L.Ha.]

Cloud   Suspensions of minute droplets or ice crystals produced by the condensation of water vapor (the ordinary atmospheric cloud). Other clouds, less commonly seen, are composed of smokes or dusts. *See* AIR POLLUTION; DUST STORM.

If water vapor is cooled sufficiently, it becomes saturated, that is, in equilibrium with a plane surface of liquid water (or ice) at the same temperature. Further cooling in the presence of such a surface causes condensation upon it. In the atmosphere, even in the apparent absence of any surfaces, there are invisible motes upon which the condensation proceeds at barely appreciable cooling beyond the state of saturation. Consequently, when atmospheric water vapor is chilled sufficiently, such motes, or condensation nuclei, swell into minute waterdroplets and form a visible cloud.

The World Meteorological Organization (WMO) uses a classification which divides clouds into low-level (base below about 1.2 mi or 2 km), middle-level (about 1.2–4 mi or 2–7 km), and high-level (4–8 mi or 7–14 km) forms within the middle latitudes. The names of the three basic forms of clouds are used in combination to define 10 main characteristic forms, or "genera."

1. Cirrus are high white clouds with a silken or fibrous appearance.

2. Cumulus are detached dense clouds which rise in domes or towers from a level low base.

3. Stratus are extensive layers or flat patches of low clouds without detail.

4. Cirrostratus is cirrus so abundant as to fuse into a layer.

5. Cirrocumulus is formed of high clouds broken into a delicate wavy or dappled pattern.

6. Stratocumulus is a low-level layer cloud having a dappled, lumpy, or wavy structure.

7. Altocumulus is similar to stratocumulus but lies at intermediate levels.

8. Altostratus is a thick, extensive, layer cloud at intermediate levels.

9. Nimbostratus is a dark, widespread cloud with a low base from which prolonge drain or snow falls.

10. Cumulonimbus is a large cumulus which produces a rain or snow shower.

See CLOUD PHYSICS.   [F.H.Lu.]

Cloud physics   The study of the physical and dynamical processes governing the structure and development of clouds and the release from them of snow, rain, and hail (collectively known as precipitation).

The factors of prime importance are the motion of the air, its water-vapor content, and the numbers and properties of the particles in the air which act as centers of condensation and freezing. Because of the complexity of atmospheric motions and the enormous variability in vapor and particle content of the air, it seems impossible to construct a detailed, general theory of the manner in which clouds and precipitation develop. However, calculations based on the present conception of laws governing the growth and aggregation of cloud particles and on simple models of air motion provide reasonable explanations for the observed formation of precipitation in different kinds of clouds.

Clouds are formed by the lifting of damp air which cools by expansion under continuously falling pressure. The relative humidity increases until the air approaches saturation. Then condensation occurs on some of the wide variety of aerosol particles present; these exist in concentrations ranging from less than 2000 particles/in.^{3} (100/cm^3) in clean, maritime air to perhaps 10^7/in.^{3} (10^6/cm^3) in the highly polluted air of an industrial city. A portion of these particles are hygroscopic and promote condensation at relative humidities below 100%; but for continued condensation leading to the formation of cloud droplets, the air must be slightly supersaturated. Among the highly efficient condensation nuclei are the salt particles produced by the evaporation of sea spray, but it appears that particles produced by human-made fires and by natural combustion (for example, forest fires) also make a major contribution. Condensation onto the nuclei continues as rapidly as the water vapor is made available by cooling of the air and gives rise to droplets of the order of 0.0004 in. (0.01 mm) in diameter. These droplets, usually present in concentrations of several thousand per cubic inch, constitute a nonprecipitating water cloud.

Cloud droplets are seldom of uniform size. Droplets arise on nuclei of various sizes and grow under slightly different conditions of temperature and supersaturation in different parts of the cloud. A droplet appreciably larger than average will fall faster than the smaller ones, and so will collide and fuse (coalesce) with some of those which it overtakes.

The second method of releasing precipitation can operate only if the cloud top reaches elevations where temperatures are below 32°F (0°C) and the droplets in the upper cloud regions become supercooled. At temperatures below $-40°F$ ($-40°C$) the droplets freeze automatically or spontaneously; at higher temperatures they can freeze only if they are infected with special, minute particles called ice nuclei. As the temperature falls below $32°F$ ($0°C$), more and more ice nuclei become active, and ice crystals appear in increasing numbers among the supercooled droplets. Such a mixture of supercooled droplets and ice crystals is unstable. After several minutes the growing crystals will acquire definite falling speeds, and several of them may become joined together to form a snowflake. In falling into the warmer regions of the cloud, however, the snowflake may melt and reach the ground as a raindrop.

The deep, extensive, multilayer-cloud systems, from which precipitation of a usually widespread, persistent character falls, are generally formed in cyclonic depressions (lows) and near fronts. Although the structure of these great raincloud systems, which are being explored by aircraft and radar, is not yet well understood, radar signals from these clouds usually take a characteristic form which has been clearly identified with the melting of snowflakes.

Precipitation from shower clouds and thunderstorms, whether in the form of raindrops, pellets of soft hail, or true hailstones, is generally of greater intensity and shorter duration than that from layer clouds and is usually composed of larger particles. The clouds themselves are characterized by their large vertical depth, strong vertical air currents, and high concentrations of liquid water, all these factors favoring the rapid growth of precipitation elements by accretion.

The development of precipitation in convective clouds is accompanied by electrical effects culminating in lightning. The mechanism by which the electric charge dissipated in lightning flashes is generated and separated within the thunderstorm has been debated for more than 200 years, but there is still no universally accepted theory. However, the majority opinion holds that lightning is closely associated with the appearance of the ice phase, and the most promising theory suggests that the charge is produced by the rebound of ice crystals or a small fraction of the cloud droplets that collide with the falling hail pellets. *See* LIGHTNING.

The various stages of the precipitation mechanisms raise a number of interesting and fundamental problems in classical physics. Worthy of mention are the supercooling and freezing of water; the nature, origin, and mode of action of the ice nuclei; and the mechanism of ice-crystal growth which produces the various snow crystal forms.

The maximum degree to which a sample of water may be supercooled depends on its purity, volume, and rate of cooling. The freezing temperatures of waterdrops containing foreign particles vary linearly as the logarithm of the droplet volumes for a constant rate of cooling. This relationship, which has been established for drops varying between 10 micrometers and 1 centimeter in diameter, characterizes the heterogeneous nucleation of waterdrops and is probably a consequence of the fact that the ice-nucleating ability of atmospheric aerosol increases logarithmically with decreasing temperature.

Measurements made with large cloud chambers on aircraft indicate that the most efficient nuclei, active at temperatures above $14°F$ ($-10°C$), are present in concentrations of only about 10 in a cubic meter of air, but as the temperature is lowered, the numbers of ice crystals increase logarithmically to reach concentrations of about 1 per liter at $-4°F$ ($-20°C$) and 100 per liter at $-22°F$ ($-30°C$). Since these measured concentrations of nuclei are less than one-hundredth of the numbers that apparently are consumed in the production of snow, it seems that there must exist processes by which the original number of ice crystals are rapidly multiplied, Laboratory experiments suggest the fragmentation of the delicate snow crystals and the ejection of ice splinters from freezing droplets as probable mechanisms.

The most likely source of atmospheric ice nuclei is provided by the soil and mineral-dust particles carried aloft by the wind. Laboratory tests have shown that, although most common minerals are relatively inactive, a number of silicate minerals of the clay family produce ice crystals in a supercooled cloud at temperatures above $-4°F$ ($-18°C$). A major constituent of some clays, kaolinite, which is active below $16°F$ ($-9°C$), is probably the main source of highly efficient nuclei.

The fact that there may often be a deficiency of efficient ice nuclei in the atmosphere has led to a search for artificial nuclei which might be introduced into supercooled clouds in large numbers. In general, the most effective ice-nucleating substances, both natural and artificial, are hexagonal crystals in which spacings between adjacent rows of atoms differ from those of ice by less than 16%. The detailed surface structure of the nucleus, which is determined only in part by the crystal geometry, is of even greater importance.

Collection of snow crystals from clouds at different temperatures has revealed their great variety of shape and form. This multiple change of habit over such a small temperature range is remarkable and is thought to be associated with the fact that water molecules apparently migrate between neighboring faces on an ice crystal in a manner which is very sensitive to the temperature. The temperature rather than the supersaturation of the environment is primarily responsible for determining the basic shape of the crystal, though the supersaturation governs the growth rates of the crystals, the ratio of their linear dimensions, and the development of dendritic forms.

The presence of either ice crystals or some comparatively large waterdroplets (to initiate the coalescence mechanism) appears essential to the natural release of precipitation. Rainmaking experiments are conducted on the assumption that some clouds precipitate inefficiently, or not at all, because they are deficient in natural nuclei; and that this deficiency can be remedied by "seeding" the clouds artificially with dry ice or silver iodide to produce ice crystals, or by introducing waterdroplets or large hygroscopic nuclei. See PRECIPITATION (METEOROLOGY); WEATHER MODIFICATION.   [B.J.M.]

Coal   A brown to black combustible rock that originated by accumulation and subsequent physical and chemical alteration of plant material over long periods of time, and that on a moisture-free basis contains no more than 50% mineral matter. The plant debris accumulated in various wet environments, commonly called peat swamps, where dead plants were largely protected from decay by a high water table and oxygen-deficient water. The accumulating spongy, water-saturated, plant-derived organic material known as peat is the precursor of coal. Over time, many changes of the original vegetable matter are brought about by bacteria, fungi, and chemical agents. The process progressively transforms peat into lignite or brown coal, subbituminous coal, bituminous coal, and anthracite. This progression is known as the coalification series. The pressure exerted by the weight of the overlying sediment and the heat that increases with depth, as well as the length of exposure to them, determine the degree of coalification reached. See KEROGEN; LIGNITE; PEAT.

Minable coal seams occur in many different shapes and compositions. Some coal seams can be traced over tens, even hundreds, of miles in relatively uniform thickness and structure. The extensively mined Herrin coal bed of the Illinois Basin and the Pittsburgh coal bed of the northern Appalachian Basin are examples. They are 6–8 ft (2–2.5 m) thick over thousands of square miles. These coals originated in peat swamps that developed on vast coastal plains during the Pennsylvanian Period. The German brown coal deposits near Cologne are characterized by very thick coal deposits (300 ft or 100 m). However, their lateral extent is much more limited than are the two examples from the United States. These peat deposits formed in a gradually subsiding structural

American Society for Testing and Materials classification of coals by rank (in box, ASTM standard D 388) and other related coal properties

ASTM class	ASTM group	1000 Btu/lb[1]	Agglomerating[2]	Volatile matter,[3] %	MJ/kg[1]	Maximum reflectance,[4] %	Moisture,[1] %	C,[3] %	O,[3] %	H,[3] %
	Peat	1.0-6.0	No	72-62	2.3-14.0	0.2-0.4	95-50	50-65	42-30	7-5
Lignite	Lignite B	>6.3[5]	No	65-40	<14.7[5]	0.2-0.4	60-40[5]	55-73	35-23	7-5
	Lignite A	6.3-8.3[5]	No	65-40	14.7-19.3[5]	0.2-0.4	50-31[5]	55-73	35-23	7-5
Subbituminous	Subbituminous C	8.3-9.5[5]	No	55-35	19.3-22.1[5]	0.3-0.6	38-25[5]	60-80[6]	28-15[6]	6.0-4.5
	Subbituminous B	9.5-10.5[5]	No	55-35	22.1-24.4[5]	0.3-0.6	30-20[5]	60-80[6]	28-15[6]	6.0-4.5
	Subbituminous A	10.5-11.5[5]	No	55-35	24.4-26.7[5]	0.3-0.7	25-18[5]	60-80[6]	28-15[6]	6.0-4.5
Bituminous	High volatile C	10.5-13.0[5]	Yes	55-35	24.4-30.2[5]	0.4-0.7	25-10[5]	76-83[6]	18-8[6]	6.0-4.5
	High volatile B	13.0-14.0[5]	Yes	50-35	30.2-32.6[5]	0.5-0.8[6]	12-5[6]	77-84[6]	12-7[6]	6.0-4.5
	High volatile A	≥14.0	Yes	45-31	≥32.6	0.6-1.2[6]	7-1[5]	78-88[6]	10-6[6]	6.0-4.5
	Medium volatile	>14.0	Yes	31-22[5]	>32.6	1.0-1.7[5]	<1.5	84-91	9-4	6.0-4.5
	Low volatile	>14.0	Yes	22-14[5]	>32.6	1.4-2.2[5]	<1.5	87-92	5-3	6.0-4.5
Anthracite	Semianthracite	>14.0	No	14-8[5]	>32.6	2.0-3.0[5]	<1.5	89-93[6]	5-3[6]	5-3[5]
	Anthracite	>14.0	No	8-2[5]	>32.6	2.6-6.0[5]	0.5-2	90-97[6]	4-2[6]	4-2[5]
	Meta-anthracite	>14.0	No	≤2	>32.6	>5.5[5]	1-3	>94[6]	2-1[6]	2-1[5]

[1] Moist, mineral-matter-free.
[2] Agglomerating coals form a button of cokelike residue in the standard volatile-matter determination that shows swelling or cell structure or supports a 500-g weight without pulverizing. The residue of nonagglomerating coal lacks these characteristics.
[3] Dry, mineral-matter-free.
[4] Reflectance of virinite under oil immersion.
[5] Well suited for rank discrimination in range indicated.
[6] Moderately well suited for rank discrimination.
SOURCE: Modified from Damberger et al., in B. R. Cooper and W. A. Ellingson (eds.), *The Science and Technology of Coal and Coal Utilization*, 1984.

graben bounded by major faults. Land lay to the south and the sea to the north. Only a relatively small portion of the subsiding graben block provided optimal conditions for peat accumulation over a long period of time. Thus each coal bed has its own depositional history that determined many of its characteristics. *See* CARBONIFEROUS; CRETACEOUS; PENNSYLVANIAN; TERTIARY.

Coal seams are commonly composed of a number of benches of alternating coal and more or less carbonaceous shale. The shale represents periods when the peat accumulation was interrupted by flooding from a river or the sea, or, more rarely, interrupted by volcanic ash deposition (tonsteins). The individual benches of a coal bed vary laterally in thickness and composition, sometimes quite rapidly. The degree of variability is related to the stability of conditions during accumulation. Fluvial and lacustrine depositional environments produce greater lateral variability than deltaic or coastal plain environments. *See* SHALE.

Both physical (pressure, heat) and chemical (biochemical, thermochemical) factors are influential in the transformation of peat into the other members of the coalification series. The boundaries between the members of the series are transitional and must be chosen somewhat arbitrarily. The term rank is used to identify the stage of coalification reached in the course of coal metamorphism. Rank is a fundamental property of coal, and its determination is essential in the characterization of a coal. Classification of coal by rank generally is based upon the chemical composition of the coal's ash-free or mineral-matter-free organic substance (see table), but parameters derived from empirical tests indicative of technological properties, such as agglomerating characteristics, are commonly used in several countries outside the United States in addition to coal rank parameters. *See* METAMORPHISM.

Coal rank increases with depth at differing rates from place to place, depending primarily on the rate of temperature increase with depth (geothermal gradient) at the time of coalification. Coal rank also changes laterally, even in the same coal seam, as former depth of burial and thus exposure to different pressure and temperature vary. Originally established vertical and regional coalification patterns can be significantly altered by various kinds of geologic events, such as the intrusion of large magma bodies at depth (plutonism), or renewed subsidence of a region. Volcanic activity may cause significant local anomalies in coal rank, but rarely leads to regional changes in coalification pattern. *See* MAGMA.

Coal is used primarily for producing steam in electric power plants. Other important uses are by industry for producing steam and heat, and by the steel industry for coke making. Conversion of coal to synthetic liquid or gaseous fuels does not constitute a major use of coal worldwide or in most countries.   [H.H.D.]

Coal balls   Variously shaped nodules consisting of fossilized peat in which the individual cells and tissue systems of the plant parts are infiltrated by minerals, principally calcium carbonate, along with pyrite, dolomite, and occasionally silica. This type of fossilization, in which the cell walls are filled with minerals, is termed permineralization. Coal balls occur principally in Pennsylvanian (upper Carboniferous) bituminous and anthracite coals, but permineralized plants have also been reported in coals from as early as the Devonian and extending well into the Paleocene.

How coal balls formed is not well understood, but some are believed to represent accumulations of peat in which the plants were growing in low-lying swampy areas close to the sea. According to this model, seawater provided the high source of calcium carbonate in the permineralization processes.

Because the individual plant cells in coal balls have not been crushed, they offer a wealth of information to paleobiologists about the structure, morphology, and biology of the plants. *See* COAL PALEOBOTANY.   [T.N.T.]

Coal paleobotany   A special branch of the paleobotanical sciences concerned with the origin, composition, mode of occurrence, and significance of the fossil plant materials that occur in, or are associated with, coal seams. Information developed in this field of science provides knowledge useful to the biologist in attempting to describe the development of the plant world, aids the geologist in unraveling the complexities of coal measure stratigraphy in order to reconstruct the geography of past ages and to describe ancient climates, and has practical application in the coal, coke, and coal chemical industries.

All coal seams consist of countless fragments of fossilized plant material admixed with varying percentages of mineral matter. The organic and inorganic materials initially accumulate in some type of swamp environment. In some instances, the fossilized plant fragments can be recognized as remnants of a plant of some particular family, genus, or species. When this is possible, information can be obtained on the vegetation extant at the time the source peat was formed, and such data aid greatly in reconstructing paleogeographies and paleoclimatic patterns.

Pollen grains and spores are more adequately preserved in coals than are most other plant parts. Recognition of this fact, coupled with an appreciation of the high degree to which these fossils are diagnostic of floral composition, has led to the rapid development of the paleobotanical subscience of palynology. Fruits, seeds, and identifiable woods also occur as coalified fossils. Occasionally, coal seams contain fossil-rich coal balls. Essentially all types of plant fossil, including entire leaves, cones, and seeds, are encountered in these discrete nodular masses. *See* COAL BALLS.

Coal seams generally are composed of several superposed sedimentary layers, each having formed under somewhat different environmental conditions. The coal petrologist and paleobotanist recognize these layers because of their distinctive textural appearance and because each consists of a particular association of organic and inorganic materials. Accordingly, each coal seam usually contains several types of coal. These coal types, or lithotypes, possess characteristic suites of physical properties, and knowledge of these properties is very profitably employed in manipulating coal composition in coal preparation and beneficiation plants.   [W.Sp.]

Coastal landforms   The characteristic features and morphology of the land in the coastal zone. They are subject to processes of erosion and deposition as produced by winds, waves, tides, and river discharge. The interactions of these processes and the coastal environments produce a wide variety of landforms. Processes directed seaward from the land are dominated by the transport of sediment by rivers, but also include gravity processes such as landslides, rockfalls, and slumping. The dominant processes on the seaward side are wind, waves, and wave-generated currents. Mixed among these locations are tidal currents which also carry large volumes of sediment.

Subcontinental- to continental-scale coastal landform patterns are related to plate tectonics. The three major tectonic coastal types are leading-edge, trailing-edge, and marginal sea coasts. Leading-edge coasts are associated with colliding plate boundaries where there is considerable tectonic activity. Trailing-edge coasts are on stable continental margins, and marginal seas have fairly stable coasts with plate margins, commonly characterized by island arcs and volcanoes, that form their seaward boundaries. *See* CONTINENTAL MARGIN; PLATE TECTONICS.

The primary characteristics of leading-edge coasts is the rugged and irregular topography, commonly displaying cliffs or bluffs right up to the shoreline. Seaward, the topography reflects this with an irregular bottom and deep water near the shoreline. The geology is generally complex with numerous faults and folds in the strata of the coastal zone. These coasts tend to be dominated by erosion with only local areas of deposition, typically in the form of small beaches or spits between headlands. Waves tend to be large because of the deep nearshore water, and form wave-cut platforms, terraces, notches, sea stacks, and caves.

The most diverse suite of coastal landforms develops along trailing-edge coasts. These coasts are generally developed along the margin of a coastal plain, they are fed by well-developed and large river systems, and they are subjected to a low-to-modest wave energy because of the gently sloping adjacent continental shelf. The overall appearance is little topographic relief dominated by deposition of mud and sand. The spectrum of environments and their associated landforms includes deltas, estuaries, barrier islands, tidal inlets, tidal flats, and salt marshes. *See* BARRIER ISLANDS; COASTAL PLAIN; DELTA; FLOODPLAIN.

Marginal coastal settings are along a stable continental mass and are protected from open ocean processes, commonly an island arc system or other form of a plate boundary. The consequence is a coastal zone that tends to be subjected to small waves and where considerable mud is allowed to accumulate in the coastal zone. Many coastal landforms are present in much the same fashion as on the trailing-edge coasts. Marginal sea coasts tend to have large river deltas. Examples are the eastern margin of Asia along the Gulf of Korea and the China Sea, the entire Gulf of Mexico, and the Mediterranean Sea. *See* DEPOSITIONAL SYSTEMS AND ENVIRONMENTS; NEARSHORE PROCESSES.   [R.A.D.]

Coastal plain   An extensive, low-relief area that is bounded by the sea on one side and by some type of relatively high-relief province on the landward side. The geologic province of the coastal plain actually extends beyond the shoreline across the continental shelf. It is only during times of glacial melting and high sea level that much of the coastal plain is drowned. *See* CONTINENTAL MARGIN.

The coastal plain is a geologic province that is linked to the stable part of a continent on the trailing edge of a plate. The extent and nature of the coastal plains of the world range widely. Some are very large and old, whereas others are small and geologically young. For example, the Atlantic and Gulf coasts of the United States are among the

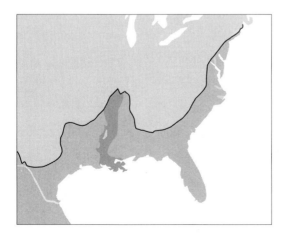

Coastal plains of the United States.

largest in the world. (see illustration). In some areas the coastal plain is hundreds of kilometers wide and extends back about 100 million years. By contrast, local coastal plains in places like the east coast of Australia and New Zealand are only 1 or 2 million years old and extend only tens of kilometers landward from the shoreline.

The typical character of a coastal plain is one of strata that dip gently and uniformly toward the sea. There may be low ridges that are essentially parallel to the coast and that have developed from erosion of alternating resistant and nonresistant strata. These strata are commonly a combination of mudstone, sandstone, and limestone, although the latter is typically a subordinate amount of the total. These strata resulted from deposition in fluvial, deltaic, and shelf environments as sea level advanced and retreated over this area. Coastal plain strata have been a source of considerable oil and gas as well as various economic minerals. Although the coastal plain province is typically stable tectonically, there may be numerous normal faults and salt dome intrusions. *See* COASTAL LANDFORMS; DELTA; GEOMORPHOLOGY; PLAINS.   [R.A.D.]

Cobaltite   A mineral having composition (Co,Fe)AsS. Cobaltite is one of the chief ores of cobalt. It crystallizes in the isometric system, commonly in cubes or pyritohedrons, resembling crystals of pyrite. There is perfect cubic cleavage. The luster is metallic and the color silver-white but with a reddish tinge. The hardness is 5.5 (Mohs scale) and the specific gravity is 6.33. Notable occurrences of cobaltite are at Skutterud, Norway; Lunaberg, Sweden; Ravensthorpe, Australia; and Cobalt, Ontario, Canada.
[C.S.Hu.]

Coesite   Naturally occurring coesite, a mineral of wide interest and the high-pressure polymorph of SiO_2, was first discovered and identified from shocked Coconino sandstone of the Meteor Crater in Arizona in 1960. Since then coesite has been identified from the Wabar (meteorite) Crater in Saudi Arabia, from the Ries Crater in Bavaria in southern Germany, from the Lake Bosumtwi Crater in Ashanti, Ghana, Africa, and from Lake Mien in Sweden. Coesite has also been identified from some Thailand tektites which are considered to have been formed also by an impact cratering process. The finding of natural coesite elevates it as a true mineral species. Because it requires a unique physical condition, extremely high pressure, for its formation, its occurrence is diagnostic of a special natural phenomenon, in this case, the hypervelocity impact of a meteorite.

Synthetic coesite was first produced in the laboratory by L. Coes, Jr., as a chemical compound at pressures of about 35 kilobars (3.5×10^9 pascals) in the temperature range of 500–800°F (246–404°C). Coesite has also been found in synthetic diamonds and has been formed by transformation from alpha quartz by the application of shearing stress. *See* DIAMOND.

Coesite occurs in grains that are usually less than 0.0002 in. (5 micrometers) in size and are generally present in small amounts. The properties of the mineral are known mainly from studies of synthesized crystals. It is colorless with vitreous luster and has no cleavage. It has a specific gravity of 2.915 ± 0.015 and a hardness of about 8 on Mohs scale.

Coesite has as yet no evident commercial use and therefore has no obvious economic value. As a stepping-stone in scientific research, it serves in at least two ways: where it occurs naturally, coesite is diagnostic of a past history of high pressure; the occurrence of coesite from Meteor Crater clearly suggests that shock as a process can transform a low-density ordinary substance to one of high density and unique properties. Coesite has been found in materials ejected from craters formed by the explosion of 500,000 tons (450,000 metric tons) of TNT. Research on the occurrence of coesite from rocks deformed by other energy sources, such as volcanic explosions and deep-seated tectonic movement, is continuing. The study of coesite and craters may be useful in understanding the impact craters on the Earth as well as those on the Moon.   [E.C.T.C.]

Colemanite   One of the more common minerals of the borate group (in which boron is chemically bonded to oxygen), colemanite is a hydroxy-hydrated calcium borate with the chemical formula $Ca[B_3O_4(OH)_3](H_2O)$ [Ca = calcium, B = boron, O = oxygen, OH = hydroxyl, H_2O = water].

Colemanite is white to gray with a white streak, is transparent to translucent, and has a vitreous luster. It occurs either as short highly modified prismatic crystals or as massive to granular aggregates. It has perfect cleavage in one direction, parallel to the sheet nature of its crystal structure. Colemanite has hardness of 4–$4^{1}/_2$ on Mohs scale and a specific gravity of 2.42.

Colemanite typically occurs in lacustrine (lake) evaporite deposits of Tertiary age. It forms by thermal diagenesis of primary borate minerals, such as ulexite, $NaCa[B_5O_6(OH)_6](H_2O)_6$, and borax, $Na_2[B_4O_5(OH)_4](H_2O)_8$ (Na = sodium). *See* BORATE MINERALS; DIAGENESIS; SALINE EVAPORITES.

Colemanite is a principal source of boron, together with borax and ulexite, and is mined extensively in California and Nevada, Turkey, and Argentina. Boron is used in a wide variety of industrial commodities: soaps and washing powders, glasses and ceramics, specialty alloys, and fillers in many products.   [F.C.Ha.]

Columbite   A mineral with the composition $(Fe,Mn)Nb_2O_6$. Tantalum may substitute in all proportions for the niobium; and a complete series extends to the pure end-member tantalite [$(Fe,Mn)Ta_2O_6$], a relatively rare mineral. With a complete solid solution of iron and manganese, four end-member compositions are possible: $FeNb_2O_6$ (ferrocolumbite), $MnNb_2O_6$ (mangancolumbite), $FeTa_2O_6$ (ferrotantalite), and $MnTa_2O_6$ (mangantantalite).

Physical properties vary with composition. The specific gravity may range from 5.0 for $MnNb_2O_6$ to 5.4 for $FeNb_2O_6$; $FeTa_2O_6$ has a specific gravity of 7.9. The hardness of columbite on Mohs scale is 6, while that of tantalite is 6.5. When the mineral is studied as a hand specimen, the color is black; however, the streak may be dark red to black. Manganoan varieties, end-member compositions rich in manganese, are often reddish brown. The luster is submetallic to weakly vitreous.

Columbite is a common accessory mineral in granitic pegmatites, and it may occur as a heavy mineral in placer deposits in streams. Minerals of the columbite-tantalite series are the most abundant and widespread of the natural columbates and tantalates, and columbite is the chief ore mineral of niobium. Niobium is used chiefly as an alloying element in the manufacture of specialty steels and alloys with nuclear and aerospace applications. *See* MINERALOGY; PEGMATITE; TANTALITE.   [R.C.Ew.]

Concretion   A loosely defined term used for a sedimentary mineral segregation that may range in size from inches to many feet. Concretions are usually distinguished from the sedimentary matrix enclosing them by a difference in mineralogy, color, hardness, and weathering characteristics. Some concretions show definite sharp boundaries with the matrix, while others have gradational boundaries. Most concretions are composed dominantly of calcium carbonate, with or without an admixture of various amounts of silt, clay, or organic material. Coal balls are calcareous concretions found in or immediately above coal beds. Concretions are normally spherical or ellipsoidal; some are flattened to disklike shapes. Frequently a concretion is dumbbell-shaped, indicating that two separate concretionary centers have grown together. *See* COAL BALLS; SEDIMENTARY ROCKS. [R.Si.]

Conglomerate The consolidated equivalent of gravel. Conglomerates are aggregates of more or less rounded particles greater than 0.08 in. (2 mm) in diameter. Frequently they are subdivided on the basis of size of particles into pebble (fine), cobble (medium), and boulder (coarse) conglomerates. The common admixture of sand-sized and gravel-sized particles in the same deposit leads to further subdivisions, into conglomerates (50% or more pebbles), sandy conglomerates (25–50% pebbles), and pebbly or conglomeratic sandstones (less than 25% pebbles). The pebbles of conglomerates are always somewhat rounded, giving evidence of abrasion during transportation; this distinguishes them from some tillites and from breccias, whose particles are sharp and angular (see illustration).

Conglomerates fall into two general classes: the well-sorted, matrix-poor conglomerates with homogeneous pebble lithology, and the poorly sorted, matrix-rich conglomerates with heterogeneous pebble lithology. The well-sorted class includes quartz-pebble, chert-pebble, and limestone-pebble conglomerates which tend to be distributed in thin, widespread sheets, normally interbedded with well-sorted, quartzose sandstones. The poorly sorted conglomerates include many different types, all related in having very large amounts of sandy or clayey matrix and pebbles of many different rock classes. The graywacke conglomerates are the outstanding representatives. All poorly sorted conglomerates tend to occur in fairly thick sequences, and some of them, typically the

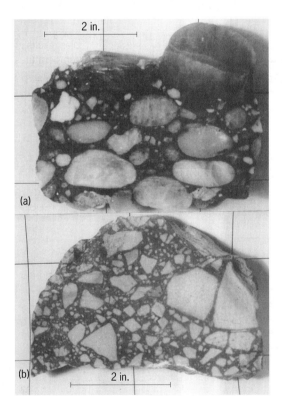

Lithified gravels.
(*a*) Conglomerate, composed of rounded pebbles. (*b*) Brecica, containing many angular fragments. 2 in. = 5.2 cm. (*Specimens from Princeton University Museum of Natural History; photo by Willard Starks*)

fanglomerates (conglomerates formed on alluvial fans) are wedge-shaped accumulations. See Sedimentary rocks.

Special types of conglomerates, such as volcanic conglomerates and agglomerates and some intraformational conglomerates composed of shale pebbles or deformed limestone pebbles, do not seem to fall easily into either class. See Breccia; Gravel; Graywacke; Till.   [R.Si.]

Contact aureole   The zone of alteration surrounding a body of igneous rock caused by heat and volatiles given off as the magma crystallized. Changes can be in mineralogy, texture, or elemental and isotopic composition of the original enclosing (country or wall) rocks, and progressively increase closer to the igneous contact. The contact aureole is the shell of metamorphosed or metasomatized rock enveloping the igneous body (see illustration). The ideal contact aureole forms locally around a single magma after it is emplaced. Metamorphism over a much larger area can result from coalescing of several contact aureoles. This is termed a contact-regional metamorphic aureole and is thought responsible for the regional metamorphism of several mountain areas. Other contact aureoles develop at greater depths and may be physically emplaced to shallower levels along with the igneous body. These are termed dynamothermal aureoles. See Igneous rocks; Magma; Metamorphic rocks; Metamorphism; Metasomatism; Pluton.

The aureole extends from the igneous contact, where the metamorphic effects are the greatest, out into the country rocks to where the temperature or heat energy is insufficient to effect any changes. This temperature lies between 400 and 750°F (200 and 400°C), and actual widths of contact aureoles range from several inches to miles.

Contact metamorphism can occur over a wide range of temperatures, pressures, or chemical gradients in rocks of any composition. Thus any mineral assemblage or facies of metamorphic rocks an be found. However, the nature of contact aureoles results in minerals characteristic of low to moderate pressures and moderate to high temperatures usually in common rock types: shales, basalt, limestone, and sandstone. Characteristic minerals developed in shales are andalusite, sillimanite, cordierite, biotite, orthopyroxene, and garnet. At the highest temperatures, tridymite, sanidine, mullite, and pigeonite form; whereas in limestone unusual calcium silicates form, including tilleyite, spurrite, rankinite, larnite, merwinite, akermanite, monticellite, and melilite. See Facies (Geology).

Compositional changes in a contact aureole range from none to great, but as a rule, contact metamorphism entails relatively little change in bulk rock composition. Because metamorphic changes are largely brought about by heat, contact aureoles are often termed thermal aureoles. However, there is a tendency for volatiles (water, carbon dioxide, oxygen) and alkalies (sodium, potassium) to be lost from rocks in the

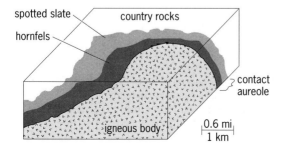

spotted slate   country rocks
hornfels
contact aureole
igneous body
0.6 mi
1 km

Zoned contact aureole developed around an igneous body.

aureole. Stable isotope compositions (oxygen, sulfur) change in response to the thermal gradient and flow of fluids through the rocks. In some cases, volatiles (boron, fluorine, and chlorine) and other elements from the crystallizing magma are gained.

Some wall rock compositions, such as limestone, can be greatly changed and form rocks termed skarn. These contact aureoles are economically important because they often contain ore deposits of iron, copper, tungsten, graphite, zinc, lead, molybdenum, and tin. Conversely, the magma can incorporate material from the wall rocks by assimilation or mixing with any partial melts formed. Mixing results in elemental and isotopic contamination of the magma, crystallization of different minerals from the melt, and hybrid rock types at the margin of the igneous body. *See* ORE AND MINERAL DEPOSITS; PNEUMATOLYSIS; SKARN.   [J.A.Sp.]

Continent   A protuberance of the Earth's crustal shell with an area of several million square miles and with sufficient elevation above neighboring depressions (the ocean basins) so that much of it is above sea level.

The great majority of maps now in use imply that the boundaries of continents are their shorelines. From the geological point of view, however, the line of demarcation between a continent and an adjacent ocean basin lies offshore, at distances ranging from a few to several hundred miles, where the gentle slope of the continental shelf changes somewhat abruptly to a steeper declivity. This change occurs at depths ranging from a few to several hundred fathoms (1 fathom = 6 feet) at different places around the periphery of various continents. *See* CONTINENTAL MARGIN.

On such a basis, numerous offshore islands, including the British Isles, Greenland, Borneo, Sumatra, Java, New Guinea, Tasmania, Taiwan, Japan, and Sri Lanka, are parts of the nearby continent. Thus, there are six continents: Eurasia (Europe, China, and India are parts of this largest continent), Africa, North America, South America, Australasia (including Australia, Tasmania, and New Guinea), and Antarctica.

All continents have similar structural features but display great variety in detail. Each includes a basement complex (shield) of metamorphosed sedimentary and volcanic rocks of Precambrian age, with associated igneous rocks, mainly granite. Originally formed at considerable depths below the surface, this shield was later exposed by extensive erosion, then largely covered by sediments of Paleozoic, Mesozoic, and Cenozoic age, chiefly marine limestones, shales, and sandstones. In at least one area on each continent, these basement rocks are now at the surface (an example is the Canadian Shield of North America). In some places they have disappeared beneath a sedimentary platform occupying a large fraction of the area of each continent, such as the area in the broad lowland drained by the Mississippi River in the United States. In each continent there are long belts of mountains in which thick masses of sedimentary rocks have been compressed into folds and broken by faults. *See* CONTINENTS, EVOLUTION OF.

Continents are the less dense, subaerially exposed portion of the plates that make up the Earth's lithosphere, or outer shell of rigid rock material. As such, continents together with part of the ocean's floor are intimately joined portions of the lithospheric plates. As plates rip apart and migrate horizontally over the Earth's surface, so too do continents rip apart and migrate, sometimes colliding with other continental segments scores of millions of years later. Mountain systems, such as the Appalachian-Ouachita, the Arbuckle-Wichita, and the Urals systems, are now believed to represent the sutures of former continents attached to their respective plates which collided long ago. The Red Sea and the linear rift-volcano-lakes district of Africa are also believed by many to manifest continental ripping and early continental drifting. Such continental collision and accretion are believed to have occurred throughout most of the Earth's history.

See AFRICA; ANTARCTICA; ASIA; EUROPE; ISOSTASY; LITHOSPHERE; NORTH AMERICA; PLATE TECTONICS; SEISMOLOGY; SOUTH AMERICA.   [D.L.J.]

Continental drift   The concept that the world's continents once formed part of a single mass and have since drifted into their present positions. Although it was outlined by Alfred Wegener in 1912, the idea was not particularly new. Paleontological studies had already demonstrated such strong similarities between the flora and fauna of the southern continents between 300,000,000 and 150,000,000 years ago that a huge supercontinent, Gondwana, containing South America, Africa, India, Australia, and Antarctica, had been proposed. However, Gondwana was thought to be the southern continents linked by land bridges, rather than contiguous units.

Wegener's ideas were almost universally rejected in 1928; the fundamental objection was the lack of a suitable mechanism. Almost simultaneously with the temporary eclipse of Wegener's theory, Arthur Holmes was considering a mechanism that is still widely accepted. Holmes conceived the idea of convective currents within the Earth's mantle which were driven by the radiogenic heat produced by radioactive minerals within the mantle. At that time, Holmes's ideas, like those of Wegener, were largely ignored. Nonetheless, several geologists, particularly those living in the Southern Hemisphere, continued to believe the theory and accumulate more data in its support.

By the 1950s, convincing evidence had accumulated, with studies of the magnetization of rocks, paleomagnetism, beginning to provide numerical parameters on the past latitude and orientation of the continental blocks. Early work in North America and Europe clearly indicated how these continents had once been contiguous and had since separated. The discovery of the midoceanic ridge system also provided further evidence for the geometric matching of continental edges, but the discovery of magnetic anomalies parallel to these ridges and their interpretation in terms of sea-floor spreading finally led to almost universal acceptance of continental drift as a reality.

In the 1970s and 1980s, the interest changed from proving the reality of the concept to applying it to the geologic record, leading to a greater understanding of how the Earth has evolved through time. The fundamental change in concept was that not only have the continents drifted, but the continents are merely parts of thicker tectonic plates comprising both oceanic and continental crust, with 50–300 km (31–186 mi) of the Earth's mantle moving along with them. *See* CONTINENTS, EVOLUTION OF; GEODYNAMICS; PALEOMAGNETISM; PLATE TECTONICS.   [D.H.T.]

Continental margin   The submerged portions of the continental masses on crustal plates, including the continental shelf, the continental slope, and the continental rise. All continental masses have some continental margin, but there is great variety in the size, shape, and geology depending upon the tectonic setting.

The most common settings are the trailing-edge margin and the leading-edge margin. The former is associated with tectonic stability, as exemplified by the Atlantic side of the North American landmass (see illustration). Here the margin is wide and geologically relatively uncomplicated, with thick sequences of coastal plain to shallow marine strata dipping slightly toward the ocean basin. By contrast, the leading-edge margin (for example, the Pacific side of the United States) is narrow, rugged, and geologically complicated. The global distribution of these widespread continental margin types is controlled by the plate tectonic setting in which the landmass resides. Some major landmasses, such as Australia, are surrounded by wide margins, but most, such as North and South America, have some of both types. *See* PLATE TECTONICS.

Any consideration of the continental margin must include a general understanding of global seal-level history over the past few million years. As glaciers expanded greatly

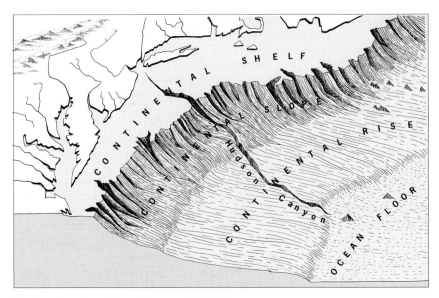

Continental margin off the northeastern United States.

just over 2 million years ago, sea level was lowered more than 300 ft (100 m). The cyclic growth and decay of glaciers during this period caused the shoreline to move from near its present position to near the edge of the present continental shelf on multiple occasions. These sea-level changes had a profound effect on the entire continental margin, particularly the shelf and the rise. During times of glacial advance, the coast was near the shelf edge, causing large volumes of river-borne sediment to flow down the continental slope and pile up on the rise; deltas were poorly developed for lack of place for sediment to accumulate. During times of high sea-level stand similar to the present time, little sediment crossed the shelf and large volumes of riverine sediment accumulated in large fluvial deltas. *See* DELTA.

The continental shelf is simply an extension of the adjacent landmass. It is characterized by a gentle slope and little relief except for shelf valleys (see illustration), which are old rivers that were active during times of low sea level. The outer limit of the shelf shows a distinct change in gradient to the much steeper slope.

The continental slope and rise of the outer continental margin includes the relatively steep slope and the rise that accumulates at the base of the slope. This continental material has the same general composition as the landmass.

The leading-edge continental margin that is commonly associated with a crustal plate boundary displays a very different geology, geomorphology, and bathymetry than the outer continental margin. In this type there is no distinct shelf, slope, and rise. Like the trailing-edge margin, the leading-edge margin exhibits the same characteristics as the adjacent landmass, in this case a structurally complex geology with numerous fault basins and high relief. The borderland is narrow and overall steep. Its geomorphology consists of numerous local basins that receive sediment through numerous submarine canyons. The canyons commonly extend nearly to the beach; there is no shelf as such. *See* SUBMARINE CANYON.

The continental margin contains a vast amount and array of natural resources, most of which are being harvested. The primary fishing grounds around the globe are in shelf waters. The Grand Banks off northeastern North America and the North Sea adjacent to Europe are among the most heavily fished. There are also many mineral resources that are taken from shelf sediments, including heavy minerals that are sources of titanium, phosphate, and even placer gold. Important commodities such as sand, gravel, and shell are also taken in large quantities from the inner shelf. Salt domes that underlie the shelf, especially in the northern Gulf of Mexico, provide salt and sulfur.

Probably the most important resource obtained from the continental margin is petroleum, in the form of both oil and gas. Production is extremely high is some places, ranging from the deltas at the coast across the entire shelf and onto the outer margin, and reserves are high. *See* MARINE GEOLOGY.   [R.A.D.]

Shelf circulation is the pattern of flow over continental shelves. An important part of this pattern is any exchange of water with the deep ocean across the shelf-break and with estuaries or marginal seas at the coast. The circulation transports and distributes materials dissolved or suspended in the water, such as nutrients for marine life, fresh-water and fine sediments originating in rivers, and domestic and industrial waste. Water movements over continental shelves include tidal motions, wind-driven currents, and long-term mean circulation. The inflow of fresh water from land also contributes to shelf circulation, because such water would tend to spread out on the surface on account of its low density. Rapid nearshore mixing reduces the density contrast, and the Earth's rotation deflects the offshore flow into a shore-parallel direction, leaving the coast to the right. A compensating shoreward flow at depth is deflected in the opposite direction, adding to the complexity of shelf circulation. *See* NEARSHORE PROCESSES; OCEAN CIRCULATION.   [G.T.C.]

Continents, evolution of   The process that led to the formation of the continents. The Earth's crust is distinctively bimodal in thickness. Oceanic crust is normally about 4 mi (7 km) thick, varying mainly with the temperature of the mantle beneath the sea-floor spreading ridges when the crust was formed. In contrast, the typical 22–25-mi (35–40-km) thickness of continental crust is controlled ultimately—through the agents of erosion, sedimentation, and isostatic adjustment—by sea level. Oceanic crust is formed at spreading ridges, continental crust at subduction zones. Both are recycled to the mantle, but oceanic crust, being less buoyant, is recycled about 30 times faster than continental crust. Consequently, continents, having a mean age of almost 2 billion years and a maximum age of 4 billion years, provide the only directly accessible record

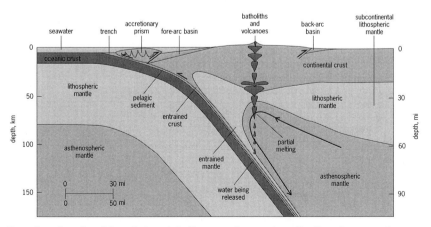

Recycling of continental crust at a subduction zone. Arrows show direction of movement.

spanning most of Earth history. They are, however, structurally more complex than ocean basins because of their great antiquity and weak rheology. *See* EARTH CRUST; MID-OCEANIC RIDGE.

Constructive processes. Subduction zones are the main factories for making continental crust (see illustration). The primary constructive processes are trench accretion, arc magmatism, and arc-continent collision. Mantle plumes and lithospheric stretching cause secondary magmatic additions to continental crust. *See* LITHOSPHERE; MAGMA; SUBDUCTION ZONES.

Trench accretion occurs where an oceanic plate sinks beneath a continental plate or another oceanic plate; sediment scraped off the top of the descending plate accumulates as an accretionary prism at the leading edge of the overriding plate; Trench accretion is dominantly a process of crustal reworking, not crustal growth, because the bulk of the sediment is derived from the erosion of preexisting crust.

A volcanic arc is a surface manifestation of partial melting near the tip of the mantle wedge above the subducting slab. Melting in the wedge is induced by the infiltration of aqueous fluids, which lower the temperature required for melting to begin (hydration melting). Although the main source of arc magmas is the mantle wedge, sediment and ablated crust entrained in the slab make subordinate contributions to arc magmas. The mantle-derived fraction represents new crustal addition.

Where subduction occurs beneath continental lithosphere, trench accretion and arc magmatism add crust to the continent directly. Continental-type protocrust is also formed at subduction zones (often having complex developmental histories) situated entirely within oceanic lithosphere. Incorporation of such protocrust in a continent is accomplished by arc-continent collision.

Plumes are jets of anomalously hot mantle that partially melts as it reaches the lithosphere, causing the volcanism of Hawaiian-type islands and related seamount chains. Plumes also cause volcanism on continents, for example, Yellowstone in North America. In addition to surface volcanism, the melts may pond at or near the base of the crust, causing magmatic underplating. Seamounts and oceanic plateaus are fragmentarily accreted to continents at trenches. *See* BASALT; OCEANIC ISLANDS; SEAMOUNT AND GUYOT; VOLCANOLOGY.

Destructive processes. Subduction zones may also cause destruction of continental crust (see illustration). The destructive processes are sediment subduction and subduction ablation. Constructive and destructive processes may be at work simultaneously, and the net balance may swing one way or the other with time. Selective destruction of lower crust may occur in continent-continent collision zones as a result of convective dripping of tectonically thickened lithosphere.

Even at subduction zones with well-developed accretionary prisms, some of the sediments disappear beneath the deformation front of the prism. Some is accreted structurally to the base of the prism, and some is transferred to the magmatic arc by melting. Some escapes melting and sinks deeply into the mantle, where it constitutes an isotopically recognizable component in the source of plume basalts. Sediment subduction, being fed by surface erosion, preferentially destroys the upper crust.

It is postulated that crust and lithospheric mantle from the overriding plate may become entrained in the subducting slab, causing tectonic ablation. Like subducted sediment, ablated crust is potentially capable of melting as it passes beneath the convecting mantle wedge. Unlike subducted sediment, ablated material may include lower crust and lithospheric mantle as well as upper crust.

Where continental lithosphere is tectonically thickened, cold lithospheric mantle is forced downward into hotter convecting mantle. Lithospheric thickening generates lateral thermal gradients, which drive mantle convection. The thickened lithospheric mantle is therefore unstable and may drip away. Unlike sediment subduction, upper crust is not lost by dripping; and unlike subduction ablation, continental drips are not susceptible to hydration melting.

Continents and Earth history. The mean age of extant continental crust is about 2 billion years. The conventional interpretation is that there was little continental crust following a period of high impact flux and that continental growth was slow at first, then rose to a peak 2–3 billion years ago, after which it slowly tapered off. An alternative interpretation, however, holds that the volume of continental crust has been in a near steady state since the impact flux waned. The present age distribution is explained by assuming a secular decline in the rate of crustal recycling, presumably modulated by the decreasing vigor of mantle convection as the Earth cooled. The difference in interpretation hinges on the importance assigned to recycling of continental crust. In crustal growth models, there is little crust older than 3 billion years, because little was formed. In steady-state models, much crust was formed early on but little of it survives. The steady-state interpretation is consistent with isotopic data showing that the oldest crustal relics contain highly evolved as well as juvenile components, and that the contemporaneous mantle was also heterogeneous and included strongly depleted regions. Furthermore, the near steady-state alternative is supported by comparative planetology, which indicates that for the hot young mantle not to be thoroughly differentiated by near-surface melting is physically implausible. *See* CONTINENTAL DRIFT; EARTH; FAULT AND FAULT STRUCTURES; GEOPHYSICS; MARINE GEOLOGY; PLATE TECTONICS; SEISMOLOGY.

[P.Ho.]

Contour The locus of points of equal elevation used in topographic mapping. Contour lines represent a uniform series of elevations, the difference in elevation between adjacent lines being the contour interval of the given map. Thus, contours represent the shape of terrain on the flat map surface (*see* illustration). Closely spaced

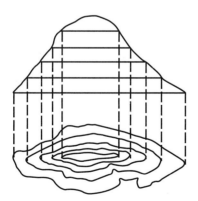

Contour representation.

contours indicate steep ground; sparse-ness or absence of contours indicates gentle slope or flat ground. Contours do not cross each other unless there is an overhang. *See* TOPOGRAPHIC SURVEYING AND MAPPING.

[R.H.Do.]

Convective instability A state of fluid flow in which the distribution of body forces along the direction of the net body force is unstable and will thus break down. Fluid flows are subject to a variety of instabilities, which may be broadly viewed as the means by which relatively simple flows become more complex. Instabilities are an important step in the transition between smooth and turbulent flow, and in the atmosphere they are responsible for phenomena ranging from thunderstorms to low- and high-pressure systems. Meteorologists and oceanographers divide instabilities into two broad classes: convective and dynamic. *See* DYNAMIC INSTABILITY.

In the broadest terms, convective instabilities arise when the displacement of a small parcel of fluid causes a force on that parcel which is in the same direction as the displacement. The parcel of fluid will then continue to accelerate away from its initial position, and the fluid is said to be unstable. In most geophysical flows, the convective motions that result from convective instabilities operate very quickly compared with the processes acting to destabilize the fluid; the result is that such fluids *seem* to be nearly neutrally stable to convection.

The simplest type of convective instability arises when a fluid is heated from below or cooled from above. Warm air rises and cold air sinks; thus a fluid whose temperature decreases with altitude is convectively unstable, while one in which the temperature increases with height is convectively stable. A fluid at constant temperature is said to be convectively neutral.

The above description assumes that density depends on temperature alone and that density is conserved in parcels of fluid, so that when the parcels are displaced their density does not change. However, in the Earth's atmosphere, neither of these assumptions is true. In the first place, the density of air depends on pressure and on the amount of water vapor in the air, in addition to temperature. Second, the density will change when the parcel is displaced because both its pressure and its temperature will change. Because of these conditions it is convenient to define a quantity known as virtual potential temperature (θ_v) that both is conserved and reflects the actual density of air. This quantity is given by the equation below, where T is the temperature in kelvins,

$$\theta_v \equiv T \left(\frac{1 + (r/0.622)}{1 + r} \right) \left(\frac{1000}{p} \right)^{0.287}$$

p is the pressure in millibars, and r is the number of grams of water vapor in each gram of dry air. When θ_v decreases with height in the atmosphere, it is convectively unstable, while θ_v increasing with height denotes stability.

In the oceans, density is a function of pressure, temperature, and salinity; convection there is driven by cooling of the ocean surface by evaporation of water into the atmosphere and by direct loss of heat when the air is colder than the water. It is also driven by salinity changes resulting from precipitation and evaporation. In many regions of the ocean, a convectively driven layer exists near the surface in analogy with the atmospheric convective layer. This oceanic mixed layer is also nearly neutral to convection.

Convective instabilities are also responsible for convection in the Earth's mantle, which among other things drives the motion of the plates, and for many of the motions of gases within other planets and in stars.   [K.A.Em.]

Coquina   A calcarenite or clastic limestone whose detrital particles are chiefly fossils, whole or fragmented. The term is most frequently used for an aggregate of large shells more or less cemented by calcite. If the rock consists of fine-sized shell debris, it is called a microcoquina. Some coquinas show little evidence of any transportation by currents. *See* LIMESTONE.   [R.S.]

Cordilleran belt   A mountain belt or chain which is an assemblage of individual mountain ranges and associated plateaus and intermontane lowlands. A cordillera is usually of continental extent and linear trend; component elements may trend at angles to its length or be nonlinear.

The term cordillera is most frequently used in reference to the mountainous regions of western South and North America, which lie between the Pacific Ocean and interior lowlands to the east. Farther north, the extensive and geologically diverse mountain terrane of western North America is formally known as the Cordilleran belt or orogen. This belt includes such contrasting elements within the United States as the Sierra Nevada, Central Valley of California, Cascade Range, Basin and Range Province, Colorado Plateau, and Rocky Mountains. *See* MOUNTAIN SYSTEMS.

Cordilleras represent zones of intense deformation of the Earth's crust produced by the convergence and interaction of large, relatively stable areas known as plates. Mountain belts have been analyzed in terms of different modes of plate convergence. Cordilleran-type mountain belts, such as the North American Cordillera, are contrasted with collision-type belts, such as the Himalayas. The former develop during long-term convergence of an oceanic plate toward and beneath a continental plate, whereas the latter are produced by the convergence and collision of one continental plate with another or with an island arc. Characteristics of cordilleran-type mountain belts include their position along a continental margin, their widespread volcanic and plutonic igneous activity, and their tendency to be bordered on both sides by zones of low-angle thrust faulting directed away from the axis of the belt. *See* OROGENY.   [G.A.D.]

Coriolis acceleration   An acceleration which arises as a result of motion of a particle relative to a rotating system. Only the components of motion in a plane parallel to the equatorial plane are influenced. Coriolis accelerations are important to the circulation of planetary atmospheres, and also in ballistics.

Newton's second law of motion is valid only when the motions and accelerations are those observed in a coordinate system that is not itself accelerating, that is, an inertial reference frame. In order to utilize familiar concepts in mathematical treatment, the Earth is commonly treated as if it were fixed, as it appears to one observing from a point on the surface, and the Coriolis force is introduced to balance the acceleration observed by virtue of the observer's motion in the rotating frame. As with the influenced components of motion, the Coriolis force is directed perpendicularly to the Earth's axis, that is, in a plane parallel to the equatorial plane. Since the direction of its action is also perpendicular to the particle velocity itself, the Coriolis force affects only the direction of motion, not the speed. This is the basis for referring to it as the deflecting force of the Earth's rotation.

A simple illustration of a Coriolis effect in the Northern Hemisphere is afforded by a turntable in counterclockwise rotation, and an external observer who moves a marker steadily in a straight line from the axis to the rim of the turntable. The trace on the turntable is a right-turning curve, and obviously this is also the nature of the path apparent to an observer who rotates with the table. Now consider the contrasting case of an air parcel near the North Pole that moves directly south (away from the Earth's axis of rotation) so that its motion to an observer on the Earth is in a straight line. To a nonrotating observer in space, this same motion appears curved toward the east because of the increased linear velocity of the meridian at lower latitudes. The force necessary to produce the eastward acceleration in the inertial frame is equal to the Coriolis force and would be produced by a gradient of air pressure from west to east, and shown by north-south-oriented isobars. In the absence of the pressure gradient force, the Coriolis force would cause the trajectory of the southward-moving air to curve westward on the Earth's surface. Then an air parcel moving uniformly away from the North Pole along a line which appears straight to an observer in space would appear to earthbound observers to curve westward. *See* GEOSTROPHIC WIND; ISOBAR (METEOROLOGY).   [E.Ke.]

Corundum   A mineral with the ideal composition Al_2O_3. It is one of a large group of isostructural compounds including hematite (Fe_2O_3) and ilmenite ($FeTiO_3$), all of which crystallize in the hexagonal crystal system, trigonal subsystem. Corundum has the high hardness of 9 on Mohs scale and is therefore commonly used as an abrasive, either alone or in the form of the rock called emery, which consists principally of the minerals corundum and magnetite. Crystals occurring in igneous rocks usually have an elongated barrellike shape, while crystals from metamorphic rocks are generally tabular. The specific gravity is approximately 3.98. *See* HEMATITE; ILMENITE.

Pure corundum is transparent and colorless, but most specimens contain some transition elements substituting for aluminum, resulting in the presence of color. Substitution of chromium results in a deep red color; such red corundum is known as ruby. The term "sapphire" is used in both a restricted sense for the "cornflower blue" variety containing iron and titanium, and in a general sense for gem-quality corundums of any color other than red. Star ruby and star sapphire contain tiny needles of the mineral rutile. *See* RUBY; SAPPHIRE.

Corundum occurs as a rock-forming mineral in both metamorphic and igneous rocks, but only in those which are relatively poor in silica, and never in association with free silica. Igneous rocks which most commonly contain corundum include syenites, nepheline syenites, and syenite pegmatites. Both contact and regionally metamorphosed silica-poor rocks may contain corundum. *See* IGNEOUS ROCKS; METAMORPHIC ROCKS.   [D.R.P.]

Cosmic spherules   Solidified droplets of extraterrestrial materials that melted either during high-velocity entry into the atmosphere or during hypervelocity impact of large meteoroids onto the Earth's surface. Cosmic spherules are rounded particles that are millimeter to microscopic in size and that can be identified by unique physical properties. Although great quantities of the spheres exist on the Earth, they are ordinarily found only in special environments where they have concentrated and are least diluted by terrestrial particulates.

The most common spherules are ablation spheres produced by aerodynamic melting of meteoroids as they enter the atmosphere. Typical ablation spheres are produced by melting of submillimeter asteroidal and cometary fragments that enter the atmosphere at velocities ranging from 6.5 to 43 mi (11 to 72 km) per second. The spheres are formed near 48 mi (80 km) altitude, where deceleration, intense frictional heating, melting, partial vaporization, and solidification all occur within a few seconds. During formation, the larger particles can be seen as luminous meteors or shooting stars.

Ablation spheres fall to Earth at a rate of one 0.1-mm-diameter sphere per square meter per year, and every rooftop contains these particles. Unfortunately they are usually mixed in with vast quantities of terrestrial particulates, and they are very difficult to locate. They can, however, be easily found in special environments, such as the ocean floor, that do not contain high concentrations of terrestrial particles that could be confused with cosmic spheres larger than 0.004 in. (0.1 mm) in diameter. *See* MARINE SEDIMENTS.

Impact spheres constitute a second and rarer class of particles that are produced when giant meteoroids impact the Earth's surface with sufficient velocity to produce explosion craters that eject molten droplets of both meteoroid and target materials. They are very abundant on the Moon, but they are rare on the Earth, and they have been found in only a few locations. Impacts large enough to produce explosion craters occur on the Earth every few tens of thousands of years, but the spheres and the craters themselves are rapidly degraded by weathering and geological processes. Meteoritic spherules have been found around a number of craters. Silica-rich glass spheroids (microtektites) are found in thin layers that are contemporaneous with the conventional tektites. Microtektites are believed to be shock-melted sedimentary materials that were ejected from large impact craters. They were ejected as plumes that covered substantial fractions of the surface of the Earth. Microspherules of a different composition have been found in the thin iridium-rich layer associated with the global mass extinctions at the Cretaceous–Tertiary boundary. *See* TEKTITE.

Cosmic spherules are of particular scientific interest because they provide information about the composition of comets and asteroids and also because they can be used as tracers to identify debris resulting from the impact of large extraterrestrial objects. In general, cosmic spherules can be confidently identified on the basis of their elemental and mineralogical compositions, which are radically different from nearly all spherical particles of terrestrial origin.   [D.E.Br.]

Cosmogenic nuclide A rare nuclide produced by nuclear reactions between high-energy cosmic radiation and terrestrial or extraterrestrial material. Cosmogenic nuclides may be used to examine the history of exposure to cosmic rays, and have numerous applications in earth science and archeology.

Primary cosmic radiation is present in space and consists of nuclear particles (mostly protons and alpha particles) with energies that are greater than typical nuclear binding energies. Collisions between cosmic-ray particles and atomic nuclei produce fragments, including cosmogenic nuclides, and also alter the characteristics of the cosmic radiation. As cosmic radiation enters the atmosphere, its flux decreases and its composition becomes dominated by neutrons rather than protons and alpha particles (which react with atmospheric gases). This reduction in flux continues through the atmosphere; at ground level, small but measurable quantities of cosmogenic nuclides are produced in solid material within about 1 m (3.3 ft) of the Earth's surface. The cosmogenic nuclides produced in these two reservoirs (atmosphere and lithosphere) may be employed for examination of different geological processes. Because of their wide range of half-lives and chemical properties, cosmogenic nuclides have a wide range of applications in geological, geomorphological, and biogeochemical studies.

Numerous nuclides are produced in measurable quantities in the atmosphere. Their half-lives and their chemical reactivities determine their applications in earth science.

Short-lived nuclides, including beryllium-7 (^{7}Be), sodium-22 (^{22}Na), phosphorus-32 (^{32}P), and phosphorus-33 (^{33}P), have been employed in studies of atmospheric circulation, particularly stratospheric-tropospheric exchange. Because of its highly successful applications in dating, carbon-14 (^{14}C; radiocarbon) is the best-known cosmogenic nuclide. Essentially all atmospheric ^{14}C is in the form of gaseous carbon dioxide (^{14}CO$_2$); it thus has an atmospheric residence time long enough for the ratio of ^{14}C to ^{12}C (the stable isotope) to become homogeneous throughout the atmosphere. Living organisms and inorganic carbonates incorporate carbon with an isotope ratio reflecting that of the atmosphere. Upon removal from contact with the atmosphere, the ratio ^{14}C/^{12}C decreases through radioactive decay of ^{14}C. With knowledge of the initial ratio and the half-life of ^{14}C, measurement of ^{14}C/^{12}C allows calculation of the age of a sample. This technique has been extensively used in archeology and geology for samples as old as 4×10^4 years. High-precision ^{14}C dates of inorganic carbon dissolved in seawater are also used to deduce global oceanic circulation rates and patterns by determining the time since a particular water mass had contact with the atmosphere. *See* OCEAN CIRCULATION; RADIOCARBON DATING; SEAWATER.

A small but significant flux of cosmic rays is present at ground level. This radiation produces cosmogenic nuclides within mineral lattices of exposed rock. In-situ-produced cosmogenic nuclides are used to examine geological problems. In-situ-produced cosmogenic nuclides are useful for dating periods of surface exposure. Their production rates decrease by a factor of 2 for every ~40 cm (16 in.) depth below an exposed rock surface, so accumulation of cosmogenic nuclides can date geological events that bring material to the Earth's surface. For example, the theoretical evolution of the concentration of ^{10}Be as a function of time for various erosion rates can be calculated. The method has been used to date exposure of rocks brought to the surface by processes including glaciation, volcanic activity, and meteor impact. It has also been used to date the formation of geomorphological features (such as alluvial fans or glacial moraines) deposited over active faults and subsequently offset by fault movement. This provides a quantitative approach to determine slip rates on faults and earthquake recurrence intervals. Such results could not have been obtained by conventional methods. *See* GEOMORPHOLOGY.

Because cosmic rays penetrate just a few meters of solid matter, cosmogenic nuclides cannot form deep in the interiors of large bodies such as asteroids. The accumulation of cosmogenic nuclides begins when a collision lifts deep-seated material to the surface or ejects it into space as part of a small body. When a meteorite hits the Earth, the accumulation of cosmogenic nuclides effectively stops. The reason is that the Earth's atmosphere and magnetic field screen out most cosmic rays. The total amounts of the cosmogenic nuclides in a meteorite are related to its exposure age. They also indicate the size of the meteoroid. To get this information, it is necessary to be able to identify cosmogenic nuclides as such and measure their concentrations. [G.F.H.]

Covellite A mineral having composition CuS and crystallizing in the hexagonal system. It is usually massive or occurs in disseminations through other copper minerals. The luster is metallic and the color indigo blue. The hardness is 1.5 (Mohs scale), the specific gravity 4.7. Covellite is a common though not abundant mineral in most copper deposits. It is a supergene mineral and is thus found in the zone of sulfide enrichment associated with other copper minerals, principally chal-cocite, chalcopyrite, bornite, and enargite, and is derived by their alteration. *See* BORNITE; CHALCOCITE; CHALCOPYRITE; ENARGITE. [C.S.Hu.]

Craton A large, relatively stable portion of the Earth's crust. Although ocean basins originally were considered low cratons, today the term applies only to continents. Continental (high) cratons are the broad heartlands of continents with subdued topography, encompassing the largest areas of most continents. Cratons experience only broad (epeirogenic) warping and occasional faulting in contrast to the much more structurally mobile or unstable zones of continents, which include mountain ranges, such as the Himalaya, and rift zones, such as those of East Africa. The terminology used today to express these contrasts was first proposed by the German geologist L. Kober in 1921. *Kratogen* referred to stable continental platforms, and *orogen* to mountain or orogenic belts. Later authors shortened the former term to kraton or craton. A complementary term, taphrogen, coined in the 1940s, encompasses the rift structures.

The present North American craton comprises the low continental interior extending from the young Rocky Mountains east to the older Paleozoic Appalachian Mountains, and north to the Paleozoic Franklin mountain belt along the Arctic margin of Canada and Greenland. The Canadian shield is that part of the craton where pre-Paleozoic rocks are widely exposed today. In contrast, the remainder of the craton in the plains of western Canada and most of the central United States has little-deformed younger strata at the surface. Most continents have similar large Phanerozoic cratons. *See* CONTINENTS, EVOLUTION OF; EARTH CRUST.

Cratons are believed to comprise bits of old continental crust, which have had long and complex histories involving the overprinting of many tectonic events. Continents can be viewed as great collages of tectonic elements amalgamated together at different times and in different ways. Therefore, the delineation of both cratons and orogens has an important temporal element: What is now a craton may be made up of interlaced orogens of differing ages, each of which may include bits of mangled older cratons within them. Refined studies of seismic waves that have penetrated the Earth's deep interior indicate that beneath cratons there is anomalous mantle with distinctive compositional or thermal characteristics. Such roots imply a long-term linkage between the evolution of early continental crust and underlying mantle, which somehow contributed to the survival of fragments of cratons older than 2.5–3 billion years. *See* OROGENY; PLATE TECTONICS. [R.H.Dot.]

Cretaceous In geological time, the last period of the Mesozoic Era, preceded by the Jurassic Period and followed by the Tertiary Period. The rocks formed during Cretaceous time constitute the Cretaceous System. Omalius d'Halloys first recognized the widespread chalks of Europe as a stratigraphic unit. W. O. Conybeare and W. Phillips (1822) formally established the period, noting that whereas chalks were remarkably widespread deposits at this time, the Cretaceous System includes rocks of all sorts and its ultimate basis for recognition must lie in its fossil remains. *See* CHALK; FOSSIL; JURASSIC; ROCK AGE DETERMINATION; STRATIGRAPHY; TERTIARY.

The parts of the Earth's crust that date from Cretaceous time include three components: a large part of the ocean floor, formed by lateral accretion; sediments and extrusive volcanic rocks that accumulated in vertical succession on the ocean floor and on the continents; and intrusive igneous rocks such as the granitic batholiths that invaded the crust of the continents from below or melted it in situ. The sedimentary accumulations contain, in fossils, the record of Cretaceous life. The plutonic and volcanic rocks are the chief source of radiometric data from which actual ages can be estimated, and suggest that the Cretaceous Period extended from 144 million years to 65 ± 0.5 million years before present (see illustration).

There are 12 globally recognized subdivisions, or stages, in the Cretaceous, based on species development (see illustration). In marine sediments the appearance and disappearance of individual, widely distributed species allows further time resolution by so-called zones. Initially these zones were largely based on ammonites, a now extinct group of cephalopods, closest to the squids and octopuses but resembling the pearly nautilus. Due to the provinciality of some ammonite species and the rarity of many, the dating of Cretaceous marine sediments now mainly devolves on microscopic fossils of calcareous plankton. The most important of these are protozoans. Their shells, in the range of 0.1–1 mm, occurring by hundreds if not thousands in a handful of chalk, have furnished about 38 pantropical zones. Next in importance are the even tinier (0.01-mm) armor plates of "nanoplanktonic" coccolithophores. Thousands may be present in a pinch of chalk, and 24 zones have been recognized. *See* MARINE SEDIMENTS; MICROPALEONTOLOGY; PALEONTOLOGY.

The Earth's magnetic field reversed about 60 times during Cretaceous time, and the resulting polarity chrons have been recorded in the remanent magnetism of many rock types (illustration). The actual process of reversal occurs in a few thousand years and affects the entire Earth simultaneously, providing geologically instantaneous time signals by which the continental and volcanic records can be linked to marine sequences

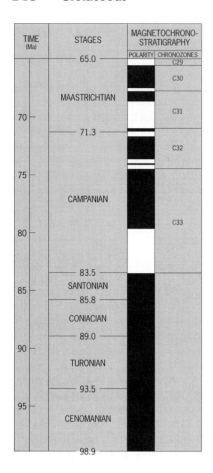
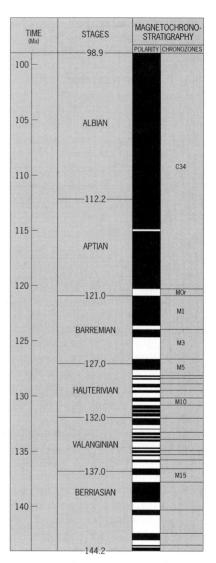

Stages of the Cretaceous Period, their estimated ages in years before present, and polarity chrons representing the alternation between episodes of normal (black) and reversed (white) orientations of the Earth's magnetic field. (*After F. M. Gradstein et al., in W. A. Berggren et al., eds., Geochronology, Time Scales and Global Stratigraphic Correlations, SEPM Spec. Publ., no. 54, 1995*)

and their fossil zonation. Noteworthy here is the occurrence of a very long (32-million-year) interval during which the field remained in normal polarity. *See* PALEOMAGNETISM.

During Cretaceous time the breakup of Gondwana, the great late Paleozoic-Triassic supercontinent, became complete. Laurasia had already separated from Africa by the development of Tethys and became split into North America and Eurasia by the opening of the North Atlantic. These new, deep oceanic areas continued to grow in Cretaceous time. India broke away from Australia and Australia from Antarctica. South America tore away from Africa by the development of the South Atlantic Ocean, while India brushed past Madagascar on its way north to collide with southeast Asia. As these new oceanic areas grew, comparable areas of old ocean floor plunged into the mantle in subduction zones such as those that still ring the Pacific Ocean, marked by deep oceanic trenches and by the development of mountain belts and volcanism on adjacent continental margins. *See* CONTINENTS, EVOLUTION OF; PLATE TECTONICS; SUBDUCTION ZONES.

The face of the globe was also affected by changes in sea level. Sea level at times in the early Cretaceous stood at levels comparable to the present, but subsequently the continents were flooded with relatively shallow seas to an extent probably not attained since Ordovician-Silurian times. Maximal flooding, in the Turonian Stage, inundated at least 40% of present land area. Cretaceous seas covered most of western Europe, though old mountain belts such as the Caledonides of Scandinavia and Scotland remained dry and archipelagos began to emerge in the Alpine belt. In America, seas flooded the southeastern flank of the Appalachian Mountains, extended deep into what is now the Mississippi Valley, and advanced along the foredeep east of the rising Western Cordillera to link at times the Gulf of Mexico with the Arctic Ocean.

Large seas extended over parts of Asia, Africa, South America, and Australia. The wide spread of the chalk facies is essentially due to this deep inundation of continents, combined with the trapping of detrital sediments near their mountain-belt sources, in deltas or in turbidite-fed deep-water fans. At the same time, carbonate platforms were still widespread, and the paratropical dry belts were commonly associated with evaporite deposits. *See* PALEOGEOGRAPHY; SALINE EVAPORITES.

In parts of early Cretaceous time, ice extended to sea level in the polar regions. But during most of Cretaceous time, climates were in the hothouse or greenhouse mode, showing lower latitudinal temperature gradients. Tropical climates may have been much like present ones, and paratropical deserts existed as they do now, but terrestrial floras and faunas suggest that nearly frost-free climates extended to the polar circles as did abundant rainfall, and no ice sheets appear to have reached sea level.

On land, flowering plants (angiosperms) first appeared in early Cretaceous time, as opportunistic plants in marginal settings, and then spread to the understory of woodlands, replacing cycads and ferns. In late Cretaceous time, evergreen angiosperms, including palms, thus came to dominate the tropical rainforests. Evergreen conifers maintained dominance in the drier midlatitude settings, while in the moist higher latitudes forests of broad-leaved deciduous trees dominated. Insects became highly diverse, and many modern families have their roots in the Cretaceous. Amphibians and small reptiles were present. Larger land animals included crocodiles and crocodilelike reptiles, turtles, and dinosaurs. The mammals remained comparatively minor elements in the Cretaceous faunas. In early Cretaceous time, egg-laying and marsupial mammals were joined by placentals, but Cretaceous mammals were in general small, and lack of color vision in most modern mammalians suggests a nocturnal ancestry and a furtive existence in a dinosaurian world. Birds had arisen, from dinosaurs in Jurassic time, but their fossil record from the Cretaceous is poor and largely one of water birds. More common are the remains of flying reptiles, the pterosaurs.

About 65 million years ago, during the reversed magnetic interval known as chron 29R, a collision with an asteroid or comet showered the entire Earth with impact debris, preserved in many places as a thin "boundary clay" enriched in the trace element iridium. This event coincided with the great wave of extinctions—the K/T crisis—which serve to bound the Cretaceous (Kreide) Period against the Tertiary.

Global effects of the impact must have included earthquake shock many orders of magnitude greater than any found in human history; associated land slips and tidal waves; a dust blackout of sunlight that must have taken many months to clear; a sharp drop in temperatures that would have brought frost to the tropics; changes in atmospheric and water chemistry; and disturbance of existing patterns of atmospheric and oceanic circulation. It is possible that earthquakes influenced volcanic eruptions. Different biotic communities were affected to different degrees. The pelagic community, sensitive to photosynthetic productivity, was severely struck, with coccolithophores and planktonic foraminiferans reduced to a few species, while ammonites, belemnites, plesiosaurs, and mosasaurs were eliminated. Dinoflagellates, endowed with the capacity to encyst under stress, suffered no great loss. Benthic life was only moderately damaged, excepting destruction of the reef community. While North American trees underwent far more extinction at the specific level than formerly believed, land floras escaped with little damage, presumably because they were generally equipped to handle stress by dormancy and seed survival. The plant-fodder-dependent dinosaurs perished, as did their predators and scavengers. The fresh-water community, buffered by ground water against temperature change and food-dependent mainly on terrestrial detritus, was little affected. While a great many individual organisms must have been killed by the immediate effects of the impact, the loss of species and higher taxa must have occurred on land and in shallowest waters mainly in the aftermath of darkness, chill, and starvation, and in the deeper waters in response to changed regimes in currents, temperatures, and nutrition. The Cretaceous crash led above all to an evolutionary outburst of the mammals, which in the succeeding tens of millions of years not only filled and multiplied the niches left by dinosaurs but also invaded the seas.   [A.G.F.]

Crocoite   A mineral with the chemical composition $PbCrO_4$. Crocoite occurs in yellow to orange or hyacinth red, monoclinic, prismatic crystals with adamantine to vitreous luster; it is also massive granular. Hardness is 2.5–3 on Mohs scale and specific gravity is 6.0. Streak, or color of the mineral powder, is orangish-yellow. It fuses easily.

Crocoite is a secondary mineral associated with other secondary minerals of lead such as pyromorphite and of zinc such as cerussite. It has been found in mines in California and Colorado.   [E.C.T.C.]

Cryolite   A mineral with chemical composition Na_3AlF_6. It crystallizes in the monoclinic system. Hardness is $2^1/_2$ on Mohs scale and the specific gravity is 2.95. Crystals are usually snow-white but may be colorless and more rarely brownish, reddish, or even black. The mean refraction index is 1.338, approximately that of water, and thus fragments become invisible when immersed in water.

Cryolite was once used as a source of metallic sodium and aluminum, but now is used chiefly as a flux in the electrolytic process in the production of aluminum from bauxite.   [C.S.Hu.]

Cummingtonite   An amphibole (a double-chain silicate mineral) with the idealized chemical formula $Mg_7Si_8O_{22}(OH)_2$ that crystallizes with monoclinic symmetry. Naturally occurring samples generally are solid solutions between $Mg_7Si_8O_{22}(OH)_2$ and the corresponding iron (Fe)-end member with the general formula $(Mg,Fe)_7Si_8O_{22}(OH)_2$. The name cummingtonite (derived from the location Cummington, Massachusetts) is applied to all solid solutions with $Mg/(Mg + Fe^{2+}) \geq 0.5$, whereas those with $Mg/(Mg + Fe^{2+}) < 0.5$ are termed grunerite. Up to a total of 1.0 (Ca + Na) atom per formula unit may also be present in cummingtonite.

Cummingtonite commonly occurs as aggregates of fibrous crystals, often in radiating clusters. It is transparent to translucent, varies in color from white to green to brown, and may be pale to dark depending primarily on the iron content. Hardness is 5–6 on

Mohs scale; density is 3.1–3.6 g/cm^3 (1.8–2.1 oz/in.^{3}), increasing with increasing iron content.

Cummingtonite is generally considered to be a metamorphic mineral, but it has been found in silicic volcanic rocks and, rarely, plutonic igneous rocks. It occurs in a variety of metamorphic rock types (amphibolite, schist, gneiss, granulite) that have undergone medium- to high-grade metamorphism. It commonly occurs in the calcium- and aluminum-poor environment of metamorphosed iron formation. It can also be a constituent of metamorphosed mafic and ultramafic igneous rocks, where it may coexist with other amphibole minerals such as hornblende, tremolite, and anthophyllite. With increasing intensity of metamorphism, cummingtonite is commonly replaced by pyroxene-bearing mineral assemblages. *See* AMPHIBOLE; HORNBLENDE; IGNEOUS ROCKS; METAMORPHIC ROCKS; TREMOLITE.   [R.K.P.]

Cuprite   A mineral having composition Cu_2O and crystallizing in the isometric system. Cuprite is commonly in crystals showing the cube, octahedron, and dodecahedron. It is various shades of red and a fine ruby red in transparent crystals which have a metallic to adamantine luster. The hardness is 3.5–4 (Mohs scale) and the specific gravity is 6.1.

Cuprite is a widespread supergene copper ore. Fine crystals have been found at Cornwall, England, and Chessy, France. It has served as an ore in the Congo, Chile, Bolivia, and Australia. In the United States it has been found at Clifton, Morenci, Globe, and Bisbee, all in Arizona.   [C.S.Hu.]

Cyclone   An atmospheric circulation system in which the sense of rotation of the wind about the local vertical is the same as that of the Earth's rotation. Thus, a cyclone rotates clockwise in the Southern Hemisphere and counterclockwise in the Northern Hemisphere. In meteorology the term cyclone is reserved for circulation systems with horizontal dimensions of hundreds (tropical cyclones) or thousands (extratropical cyclones) of kilometers. For such systems the Coriolis force due to the Earth's rotation, which is directed to the right of the flow in the Northern Hemisphere, and the pressure gradient force, which is directed toward low pressure, are in opposite directions. Thus, there must be a pressure minimum at the center of the cyclone, and cyclones are sometimes simply called lows. *See* AIR PRESSURE.

Extratropical cyclones are the common weather disturbances which travel around the world from west to east in mid-latitudes. They are generally associated with fronts, which are zones of rapid transition in temperature. Extratropical cyclones arise due to the hydrodynamic instability of the upper-level jet stream flow. *See* FRONT; JET STREAM.

Tropical cyclones, by contrast, derive their energy from the release of latent heat of condensation in precipitating cumulus clouds. Over the tropical oceans, where moisture is plentiful, tropical cyclones can develop into intense vortical storms (hurricanes and typhoons), which can have wind speeds in excess of 200 mi/h (100 m · s^{-1}). *See* HURRICANE; STORM; WIND.   [J.R.H.]

Cyclothem   A vertical sequence of several different kinds of distinctive sedimentary rock units that is repeated upward through the stratigraphic succession. Originally defined in the rock succession of Pennsylvanian age in the Illinois Basin in the 1930s, the rock types include coal, limestone, sandstone, and several types of shale and mudstone. Cyclothems were soon recognized elsewhere in rocks of this age in the central and eastern United States. Those in the Midcontinent (Kansas and states to the northeast) are dominated by several types of limestone and shale, with less coal and sandstone.

Those in the Appalachian region are dominated by coal, mudstone, shale, and sandstone, with less limestone. One proposed reason for the Pennsylvanian cyclothems is the periodic rise and fall of sea level (eustacy) that was driven by repeated episodes of continental glaciation in the southern continents of that time. *See* PENNSYLVANIAN; SEDIMENTARY ROCKS; STRATIGRAPHY.

Cyclothems were first noticed in the United States because of the many distinctive rock types involved. These resulted from the interplay between the then tropical to equatorial humid climate with ready access to detrital sand and mud and the marine circulation changes caused by the changing water depth of the sea. Coal-rich cyclothems were soon recognized in strata of the same age in western Europe. Currently, cyclothems without coal are being identified in late Paleozoic strata elsewhere in the world. *See* PALEOZOIC.   [P.H.H.]

D

Dacite Aphanitic (very finely crystalline or glassy) rock of volcanic origin, composed chiefly of sodic plagioclase (oligoclase or andesine) and free silica (quartz or tridymite) with subordinate dark-colored (mafic) minerals (biotite, amphibole, or pyroxene). If alkali feldspar exceeds 5% of the total feldspar, the rock is a quartz latite. As quartz decreases in abundance, dacite passes into andesite. Thus, dacite is roughly intermediate between andesite and quartz latite. *See* ANDESITE. [C.A.C.]

Dating methods Relative and quantitative techniques used to arrange events in time and to determine the numerical age of events in history, geology, paleontology, archeology, paleoanthropology, and astronomy. Relative techniques allow the order of events to be determined, whereas quantitative techniques allow numerical estimates of the ages of the events. Most numerical techniques are based on decay of naturally occurring radioactive nuclides, but a few are based on chemical changes through time, and others are based on variations in the Earth's orbit. Once calibrated, some relative techniques also allow numerical estimates of age. *See* ARCHEOLOGY; GEOLOGY.

Relative dating methods rely on understanding the way in which physical processes in nature leave a record that can be ordered. Once the record of events is ordered, each event is known to be older or younger than each other event. In most cases the record is contained within a geological context, such as a stratigraphic sequence; in other cases the record may be contained within a single fossil or in the arrangement of astronomical bodies in space and time. The most important relative dating methods are stratigraphic dating and paleontologic dating. Other relative dating methods include paleomagnetic dating, dendrochronology, and tephrostratigraphy. *See* DENDROCHRONOLOGY; PALEOMAGNETISM; SEQUENCE STRATIGRAPHY; STRATIGRAPHY.

Several chemical processes occur slowly, producing changes over times of geological interest; among these are the hydration of obsidian, and the conversion of L- to D-amino acids (racemization or epimerization). Determination of age requires measurement of a rate constant for the process, knowledge of the temperature history of the material under study, and (particularly for amino acid racemization) knowledge of the chemical environment of the materials. *See* AMINO ACID DATING; OBSIDIAN.

Unlike chemical methods, in which changes depend both on time and on environmental conditions, isotopic methods which are based on radioactive decay depend only on time. A parent nuclide may decay to one stable daughter in a single step by simple decay [for example, rubidium decays to strontium plus a beta particle $(^{87}\text{Rb} \rightarrow \,^{87}\text{Sr} + \beta)$]; to two daughters by branched decay through different processes [for example, potassium captures an electron to form argon, or loses a beta particle to form calcium $(^{40}\text{K} + e^- \rightarrow \,^{40}\text{Ar}; \,^{40}\text{K} \rightarrow \,^{40}\text{Ca} + \beta)$]; to one stable daughter through a series of steps (chain decay); or into two unequal-sized fragments by fission. In all cases, the number of parent atoms decreases as the number of daughter atoms increases, so

Principal parent and daughter isotopes used in radiometric dating		
Radioactive parent isotope	Stable daughter isotope	Half-life, years
Carbon-14	Nitrogen-14	5730
Potassium-40	Argon-40	1.25×10^9
Rubidium-87	Strontium-87	4.88×10^{10}
Samarium-147	Neodymium-143	1.06×10^{11}
Lutetium-176	Hafnium-176	3.5×10^{10}
Rhenium-187	Osmium-187	4.3×10^{10}
Thorium-232	Lead-208	1.4×10^{10}
Uranium-235	Lead-207	7.04×10^8
Uranium-238	Lead-206	4.47×10^9

that for each method there is an age-sensitive isotopic ratio of daughter to parent that increases with time. Many different isotopes have been exploited for measuring the age of geological and archeological materials. For example, the table shows parent isotopes, their half-lives, and the resulting daughter products.

Astronomers have estimated the age of the universe, and of the Milky Way Galaxy, by various methods. It is well known that the universe is expanding equally from all points, and that the velocity of recession of galaxies observed from Earth increases with distance. The rate of increase of recession velocity with distance is called the Hubble constant; and knowing the recession rate and distance of galaxies at some distance, it is simple to find how long it took them to get there. Initial estimates for the age of the universe were approximately 20 billion years; but as the rate of expansion decreases with time, revised estimates are nearer 13 billion years. By contrast, the Earth and other bodies in the solar system are only about 4.5 billion years old. Comparison of present-day osmium isotope ratios with theoretically estimated initial ratios yields estimates of 8.6–15.7 billion years for the age of the Galaxy.   [F.S.B.]

Datolite   A mineral nesosilicate, composition $CaBSiO_4(OH)$, crystallizing in the monoclinic system. It usually occurs in crystals showing many faces and having an equidi-mensional habit. It may also be fine granular or compact and massive. Hardness is 5–5$\frac{1}{2}$ on Mohs scale; specific gravity is 2.8–3.0. The luster is vitreous, the crystals colorless or white with a greenish tinge. Datolite is found in the Harz Mountains, Germany; Bologna, Italy; and Arendal, Norway. In the United States fine crystals have come from Westfield, Massachusetts; Bergen Hill, New Jersey; and various places in Connecticut. In Michigan, in the Lake Superior copper district, datolite occurs in fine-grained porcelainlike masses which may be coppery red because of inclusions of native copper. *See* SILICATE MINERALS.   [C.S.Hu.]

Deep-marine sediments   The term "deep marine" refers to bathyal sedimentary environments occurring in water deeper than 200 m (656 ft), seaward of the continental shelf break, on the continental slope and the basin (see illustration). The continental rise, which represents that part of the continental margin between continental slope and abyssal plain, is included under the broad term "basin." On the slope and basin environments, sediment-gravity processes (slides, slumps, debris flows, and turbidity currents) and bottom currents are the dominant depositional mechanisms, although pelagic and hemipelagic deposition is also important. *See* BASIN; CONTINENTAL MARGIN; GULF OF MEXICO; MARINE SEDIMENTS.

Types of processes. The mechanics of deep-marine processes is critical in understanding the nature of transport and deposition of sand and mud in the deep sea. In

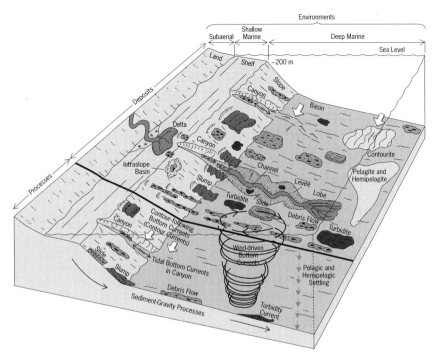

Slope and basinal deep-marine sedimentary environments occurring at water depths greater than 200 m (656 ft). Slides, slumps, debris flows, turbidity currents, and various bottom currents are important processes in transporting and depositing sediment in the deep sea. Note the complex distribution of deep-marine deposits.

deep-marine environments, gravity plays the most important role in transporting and depositing sediments. Sediment failure under gravity near the shelf edge commonly initiates gravity-driven deep-marine processes, such as slides, slumps, debris flows, and turbidity currents (see illustration). Sedimentary deposits reflect only depositional mechanisms, not transportational mechanisms. *See* SEDIMENTOLOGY.

A slide is a coherent mass of sediment that moves along a planar glide plane and shows no internal deformation. Slides represent translational movement. Submarine slides can travel hundreds of kilometers. For example, the runout distance of Nuuanu Slide in offshore Hawaii is 230 km (143 mi). Long runout distances of 50–100 km (31–62 mi) of slides are common.

A slump is a coherent mass of sediment that moves on a concave-up glide plane and undergoes rotational movements causing internal deformation.

A downslope increase in mass disaggregation results in the transformation of slumps into debris flows. Sediment is now transported as an incoherent viscous mass, as opposed to a coherent mass in slides and slumps. A debris flow is a sediment-gravity flow with plastic rheology (that is, fluids with yield strength) and laminar state. Deposition from debris flows occurs through freezing. The term "debris flow" is used here for both the process and the deposit of that process. The terms "debris flow" and "mass flow" are used interchangeably because each exhibits plastic flow behavior with shear stress distributed throughout the mass. Although only muddy debris flows (debris flows with mud matrix) received attention in the past, recent experimental and field studies show that sandy debris (debris flows with sand matrix) flows are equally important.

Rheology is more important than grain-size distribution in controlling sandy debris flows, and the flows can develop in slurries of any grain size (very fine sand to gravel), any sorting (poor to well), any clay content (low to high), and any modality (unimodal and bimodal).

With increasing fluid content, plastic debris flows tend to become turbidity currents. Turbidity currents can occur in any part of the system (proximal and distal), and can also occur above debris flows due to flow transformation in density-stratified flows. A turbidity current is a sediment-gravity flow with newtonian rheology (that is, fluids without yield strength) and turbulent state. Deposition from turbidity currents occurs through suspension settling. Deposits of turbidity currents are called turbidites. Although turbidity currents have received a lot of emphasis in the past, other processes are equally important in the deep sea (see illustration). In terms of transporting coarse-grained sediment into the deep sea, sandy debris flows and other mass flows appear to play a greater role than turbidity currents.

Bottom currents. In large modern ocean basins, such as the Atlantic, thermohaline-induced geostrophic bottom currents within the deep and bottom water masses commonly flow approximately parallel to bathymetric contours (that is, along the slope (see illustration). They are generally referred to as contour currents. However, because not all bottom currents follow regional bathymetric contours, it is preferred that the term "contour current" be applied only to currents flowing parallel to bathymetric contours, and other currents be termed bottom currents. For example, wind-driven surface currents may flow in a circular motion (see illustration) and form eddies that reach the deep-sea floor, such as the Loop Current in the Gulf of Mexico, and the Gulf Stream in the North Atlantic. Local bottom currents that move up- and downslope can be generated by tides and internal waves, especially in submarine canyons. These currents are quite capable of erosion, transportation, and redeposition of fine-to-coarse sand in the deep sea. See also GULF STREAM; OCEAN CIRCULATION.

Pelagic and hemipelagic settling. Pelagic and hemipelagic processes generally refer to settling of mud fractions derived from the continents and shells of microfauna down through the water column throughout the entire deep-ocean floor (see illustration). Hemipelagites are deposits of hemipelagic settling of deep-sea mud in which more than 25% of the fraction coarser than 5 micrometers is of terrigenous, volcanogenic, or neritic origin. Although pelagic mud and hemipelagic mud accumulate throughout the entire deep-ocean floor, they are better preserved in parts of abyssal plains (see illustration). Rates of sedmentation vary from millimeters to greater than 50 cm (20 in.) per 1000 years, with the highest rates on the upper continental margin.

Submarine slope environments. Submarine slopes are considered to be of the sea floor between the shelf-slope break and the basin floor (see illustration). Modern continental slopes around the world average $4°$, but slopes range from less than $1°$ to greater than $40°$. Slopes of active margins (for example, California and Oregon, about $2°$) are relatively steeper than those of passive margins (for example, Louisiana, about $0.5°$). On constructive continental margins with high sediment input, gravity tectonics involving salt and shale mobility and diapirism forms intraslope basins of various sizes and shapes (for example, Gulf of Mexico). Erosional features, such as canyons and gullies, characterize intraslope basins. Deposition of sand and mud occurs in intraslope basins. Slope morphology plays a major role in controlling deep-marine deposition through (1) steep versus gentle gradients, (2) presence or absence of canyons and gullies, (3) presence or absence of intraslope basins, and (4) influence of salt tectonics. See DIAPIR; EROSION.

Submarine canyon and gully environments. Submarine canyons and gullies are erosional features that tend to occur on the slope. Although canyons are larger than gullies, there are no standardized size criteria to distinguish between them. Submarine canyons are steep-sided valleys that incise the continental slope and shelf. They serve as major conduits for transporting sediment from land and the continental shelf to the basin floor. Modern canyons are relatively narrow, deeply incised, steeply walled, often sinuous valleys with predominantly V-shaped cross sections. Most canyons originate near the continental shelf break and generally extend to the base of the continental slope. Canyons commonly occur off the mouths of large rivers such as the Hudson and Mississippi, although many others, such as the Bering Canyon in the southern Bering Sea, have developed along structural trends. *See* BERING SEA.

Modern submarine canyons vary considerably in their dimensions. Their average length of canyons has been estimated to be about 55 km (34 mi), although the Bering Canyon, the world's longest, is nearly 1100 km (684 mi). The shortest canyons are those off the Hawaiian Islands, with average lengths of about 10 km (6.2 mi).   [G.Sh.]

Deep-sea fauna   The deep sea may be regarded as that part of the ocean below the upper limit of the continental slopes (see illustration). Its waters fill the deep ocean basins, cover about two-thirds of the Earth's surface, have an average depth of about 12,000 ft (4000 m), and provide living space for communities of animals that are quite different from those inhabiting the land-fringing waters which overlie the continental shelves (neritic zone). *See* ECOLOGICAL COMMUNITIES.

The deep-sea fauna consists of pelagic animals (swimming and floating forms between the surface and deep-sea floor) and below these the benthos, or bottom dwellers, which live on or near the ocean bottom. Pelagic animals can be divided into the usually smaller forms that tend to drift with the currents (zooplankton) and the larger and more active nekton, such as squids, fishes, and cetaceans. Pelagic, deep-sea animals are frequently termed bathypelagic in contrast to the epipelagic organisms of the surface waters (see illustration).

All animal life in the sea, pelagic and benthic, depends on the growth of microscopic plants (phytoplankton). From the surface down to a maximum depth of about 300 ft (100 m) there is sufficient light for photosynthesis and vigorous phytoplanktonic growth. This layer is known as the photic zone.

Bathypelagic fauna. The typical bathypelagic animals begin to appear below depths of about 600 ft (200 m). The bathypelagic fauna is most diverse in the tropical and temperate parts of the ocean. Numerous species are found in all three temperature zones, but many appear to have a more limited distribution. Each species also has a definite vertical occurrence. Findings suggest that there are three main vertical zones, each with a characteristic community. Here the term bathypelagic is used for the fauna

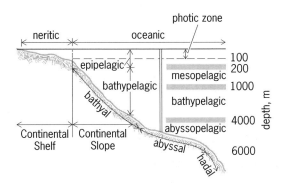

Classification of marine environments. Right side of diagram illustrates the proposal to divide the bathypelagic zone into mesopelagic, bathypelagic, and abyssopelagic zones. Division of benthic region into bathyal, abyssal, and hadal zones also is shown. 1 m = 3.3 ft.

between about 3000 and 6000 ft (1000 and 2000 m), that above (between 600 and 3000 ft or 200 and 1000 m) being called mesopelagic and that below 6000 ft (2000 m) abyssopelagic (see illustration). The typical forms of the mesopelagic fauna (stomiatoids and lantern fishes) live in the twilight zone of the deep sea (between the 68 and 50°F or 20 and 10°C isotherms), while the bathypelagic species (ceratioid angler fishes and *Vampyroteuthis*) occur in the dark, cooler parts below the 50°F (10°C) isotherm.

Perhaps the most conspicuous feature of pelagic deep-sea life is the widespread occurrence of luminescent species bearing definite light organs (photophores). Many of the squids and fishes have definite patterns of such lights, as do some of the larger crustaceans (hoplophorid and sergestid prawns and euphausiids). Investigations have shown that flashes from luminescent organisms could be detected down to depths of 12,300 ft (3750 m).

Benthic fauna. There are two main ecological groups of bottom-living animals in the ocean: organisms that attach to the bottom and those that freely move over the bottom. The benthic fauna is most diverse in the temperate and tropical ocean, although the arctic and antarctic areas have their characteristic species. As in the pelagic fauna, certain species occur in all three oceanic zones, while others appear to have a more restricted occurrence. While a number of species—particularly among the polychaete worms, gastropod mollusks, and the brittle stars (Ophiuroidea)—range from littoral to abyssal regions, most forms tend to live within smaller ranges of depth. Data suggest that there are typical communities of animals over the continental slopes (see illustration) extending down to about 9000 ft (3000 m; bathyal zone); others occur below this in the abyssal zone. *See* MARINE ECOLOGY.   [N.B.M.]

Deep-sea trench   A long, narrow, very deep, and asymmetrical depression of the sea floor, with relatively steep sides. Oceanic trenches characterize active margins at the ocean-basin–continent or ocean-basin–island-arc boundaries. They contain the greatest oceanic depths and are associated with the most active volcanism, largest negative gravity anomalies, most frequent shallow seismicity, and almost all of the intermediate and deep-focus earthquake activity. As the surface expression of the widely accepted process of subduction by which oceanic crustal material is returned to the upper mantle, they are key elements in current models of plate tectonic evolution on Earth and possibly on Venus. *See* PLATE TECTONICS; VOLCANOLOGY.

Deep-sea trenches are the signature relief form of the Pacific; in a counterclockwise direction, they occur from southern Chile to just northeast of North Island, New Zealand. A secondary or outer branch trends southward from near Tokyo Bay in a festoon of arcs to south of Palau. The principal gaps in the circum-Pacific chain are from Baja California to south-eastern Alaska, and off the northern coast of New Guinea. From eastern New Guinea to southern Vanuatu the trenches lie southwest, or "inside," the island chains; otherwise their characteristics are like those facing the Pacific. The Indian Ocean contains only the very long contorted Sunda Trench that appears near the northwestern end of Sumatra and extends southeast and east past Timor, to curve north and west near Aru and end adjacent to Buru. In the Atlantic, the Puerto Rico–Antillean trench system extends outside the island arc from eastern Hispaniola around to Trinidad; but south of 14°N, off Barbados, the trench is filled with sediment. In the far South Atlantic a typical island-arc–trench complex extends from near South Georgia through the South Sandwich Archipelago.

A series of pioneering gravity observations with pendulum instruments on Dutch submarines during the 1920s and 1930s established that the East Indian trenches, and several others, were characterized by a belt of negative gravity anomalies of 150–200+

milligals, that is, values 150–200 parts per million less than normal, interpretable as deficiency of mass near and at their axes.

It was established that oceanic crust is thin, that crust under island arcs is thicker, and its layers display different sound transmission velocities, indicating different composition. Shipboard studies in the Middle America, Tonga, Cedros, Aleutian, Peru-Chile, and Sunda trenches established that the characteristic oceanic crustal layer [that is, 6.8–7.0 km/s (4.1–4.2 mi/s) compressional wave velocity] does not end or thin under the trench; rather, it may thicken slightly but does deepen steeply as it passes beneath the island arc or continental slope by the process of subduction. *See* EARTH, GRAVITY FIELD OF; EARTH, HEAT FLOW IN; EARTH CRUST; FAULT AND FAULT STRUCTURES; OCEANIC ISLANDS; SEISMOLOGY; SUBDUCTION ZONES. [R.L.Fl.]

Delta   A deposit of sediment at the mouth of a river or tidal inlet. It is also used for storm washovers of barrier islands and for sediment accumulations at the mouths of submarine canyons. *See* FLOODPLAIN.

The shape and internal structure of a delta depend on the nature and interaction of two forces: the sediment-carrying stream from a river, tidal inlet, or submarine canyon, and the current and wave action of the water body in which the delta is building. This interaction ranges from complete dominance of the sediment-carrying stream (still-water deltas) to complete dominance of currents and waves, resulting in redistribution of the sediment over a wide area (no deltas). This interaction has a large effect on the shape and structure of the delta body.

Most of the sediment carried into the basin is deposited when the inflowing stream decelerates. If there is little density contrast, this deceleration is sudden and most sediment is deposited near the mouth of the river. If the inflowing water is much lighter than the basin water, for example, fresh water flowing into a colder sea, the outflow spreads at the surface over a large distance away from the outlet. If the inflow is very dense, for instance, cold muddy water in a warm lake, it may form a density flow on or near the bottom, and the principal deposition may occur at great distance from the outlet.

Three principal components make up the bodies of most deltas in varying proportions: topset, foreset, and bottomset beds (see illustration). As defined for most deltas, the topset beds comprise the sediments formed on the subaerial delta: channel deposits, natural levees, floodplains, marshes, and swamp and bay sediments. The foreset beds are those formed in shallow water, mostly as a broad platform fronting the delta shore, and the bottomset beds are the deep-water deposits beyond the deltaic bulge. In

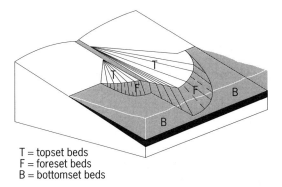

T = topset beds
F = foreset beds
B = bottomset beds

Schematic diagram showing two stages of growth and a Gilbert-type delta. (*After P. H. Kuenen, Marine Geology, John Wiley, 1950*).

marine deltas the fluviatile influence decreases and the marine influence increases from the topset to the bottomset beds.

In a different way, deltas can be viewed as being composed of three structural elements: (1) a framework of elongate coarse bodies (channels, river-mouth bars, levee deposits), which radiate from the apex to the distributary mouths (sand fingers); (2) a matrix of fine-grained floodplain, marsh, and bay sediments; and (3) a littoral zone, usually of beach and dune sands which result from sorting and longshore transport of river-mouth deposits by waves, currents, tides, and wind. The relative proportions of these components vary widely. The Mississippi delta consists almost entirely of framework and matrix; its rapid seaward growth is the result of deposition of river-mouth bars and extension of levees, and the areas in between are filled later with matrix. This gives the delta its characteristic bird-foot outline. A different makeup is presented by the Rhone delta, where the supply of coarse material at the distributary mouths is slow, and dispersal by wave action and longshore drift fairly efficient, so that nearly all material is evenly redistributed as a series of coastal bars and dunes across a large part of the delta front. This delta advances as a broad lobate front, while the present Mississippi delta grows at several localized and sharply defined points.

Despite difficult engineering problems, many cities, such as Calcutta, Shanghai, Venice, Alexandria (Egypt), and New Orleans, were constructed on deltas. These problems include shifting and extending shipping channels; lack of firm footing for construction except on levees; steady subsidence; poor drainage; and extensive flood danger. Moreover, in certain deltas the tendency of the main flow to shift away to entirely different areas, with resulting disappearance of the main channels for water traffic, is a constant problem that is difficult and costly to counter. *See* ESTUARINE OCEANOGRAPHY.

[T.H.V.A.]

Dendrochronology   The science that uses annual tree rings dated to their exact year of formation for dating historical and environmental events and processes. Trees, like most plants, are sensitive to both natural (precipitation and temperature patterns) and human-related (air and water pollution) events that trigger certain responses in the vigor of the tree as seen in its growth rate. In most geographic regions, climate patterns in any year cause a response by trees in the volume of wood the tree produces, and often leave indelible "fingerprints" in certain physical and chemical properties of the wood. These fingerprints can be seen in the varying widths of tree rings. In some years, environmental conditions may be favorable for tree growth, allowing trees to produce greater volumes of wood. In other years, climate conditions may be generally unfavorable for tree growth, causing a reduction in the volume of wood produced. *See* TREE.

Crossdating is the primary guiding principle in dendrochronology, and concerns the matching of patterns of ring widths from one tree with corresponding patterns for the same years from another tree. Crossdating allows scientists to accurately assign calendar dates to tree rings by matching the sequence of tree-ring widths against a known reference chronology. Crossdating is possible because climate is largely a regional phenomenon, affecting trees in a like manner, so that similar patterns of ring widths are produced among many trees. Furthermore, crossdating helps identify false and locally absent rings that may otherwise be recorded as true rings.

Dendrochronology has become a useful tool in many areas of research. Dendroarcheology uses tree rings to date wood material from archeological sites or artifacts, and is most often applied in the southwestern United States and Europe. In dendroclimatology climatic information is mathematically extracted from the tree-ring record and reconstructed back in time for the length of the tree-ring record. Dendroclimatologists also use tree-ring records to quantify the rising levels of atmospheric carbon dioxide to better understand global warming. Dendroecology analyzes changes in ecological processes over time using tree-ring information. Dendropyrochronology reconstructs the history of wildfires from tree rings. Dendrogeomorphology studies earth surface processes using tree-ring data. Dendrohydrology uses tree-ring data to investigate and reconstruct hydrologic properties, such as streamflow and riverflow, runoff, and past lake levels. Dendrochemistry is an important, emerging field of dendrochronology that analyzes the chemical composition of tree rings, especially the mineral elements. *See* ARCHEOLOGY; ECOLOGY; GEOMORPHOLOGY; HYDROLOGY.   [H.D.G.M.]

Denudation   All the weathering and erosional processes that contribute to the lowering of the land surface. Denudation is, thus, the complement of deposition, which is the accumulation of the products of denudation in sedimentation basins. *See* BASIN; DEPOSITIONAL SYSTEMS AND ENVIRONMENTS; EROSION; WEATHERING PROCESSES.

Contemporary records of sediment and solute flux through river systems, representing the mass removed from the continental surfaces, have been used to estimate rates of denudation since the mid-nineteenth century. These are usually expressed as an average rate of lowering of the land surface, in units of millimeters per thousand years (mm/ka).

In recent decades methods which allow the definition of denudation rates over geologic time scales have been developed. Attempts to invert sedimentary and stratigraphic records, and so define rates of erosion on the contributing continental surface, have greatly benefitted from improved seismic surveys, core extraction, and modeling of sedimentation basin processes. On the land surface itself, estimates of denudation rates based on the dissection of surfaces of known age have improved with the use of digital terrain models and absolute dating techniques. Equivalent estimates show that very rapid rates of bedrock denudation (up to 15 mm/yr) have been maintained in the western Himalaya during the past few million years. *See* DATING METHODS.

Developments in the dating of bedrock surfaces and surficial material in the 1990s represent the greatest improvement in defining rates of denudation during the late Cenozoic era. Potassium-argon (K-Ar) dating of lavas and the magnitude of dissection on volcanic cones had been used earlier to define erosion rates. More recently, exposure dating of bedrock through the accumulation of cosmogenic nuclides produced within them is a development that allows the estimation of average surface lowering rates over long time intervals. Cosmogenic nuclides have also been used to define rates of regolith production and erosion, contributing to denudation, over more recent time periods. Fission track analysis of apatite is another recent development that has had success in defining the long-term tectonic and denudational history of continental margins and mountain systems. *See* APATITE; CENOZOIC; COSMOGENIC NUCLIDE; FISSION TRACK DATING; REGOLITH.   [N.C.]

Depositional systems and environments   Depositional systems are descriptions of the interrelationships of form and the physical, chemical, or biological processes involved in the development of stratigraphic sequences. Depositional environments are the locations where accumulations of sediment have been deposited by either mechanical or chemical processes.

Depositional systems. Traditional stratigraphic analysis, which emphasized the physically descriptive aspects of strata, has changed; a critical genetic dimension has been added. Where once, for example, a formation may have been described in physical terms as a fine- to medium-grained quartzose sandstone, overlying a thick dark-gray shale sequence and underlying a coal-bearing sequence with discontinuous sands, the same sandstone may now be recognized as a delta-front sandstone, the underlying shale as a prodelta facies, and the overlying coal-bearing formation as the product of deposition on a delta plain. In this interpretation, the three distinct stratigraphic units become part of a genetically related sequence, each a component facies of a prograding delta system. *See* STRATIGRAPHY.

In a modern setting, a particular system of deposition is directly observable and known, whether it is a major delta, an alluvial fan, a meandering river, a barrier bar, a carbonate platform, or the like. Through specific observation, description, and delineation, it may be determined that a variety of depositional processes are active. At the terminus of rivers, sands may be deposited as bars in river mouths, creating the delta front; muds may be carried by suspension into the oceanic waters and deposited through flocculation, creating the prodelta; on the delta plain, a distributary channel may be depositing bed-load sands and, during flooding, may be carrying suspended muds to flanking flood basins or breaching the natural levee to form crevasse splays. Each process or combination of processes gives rise to distinct, specific environments of deposition, with each resulting in a deposit which can be characterized by such features as lithologic composition, texture, sedimentary structures, geometry, size, and relationship to other deposits. *See* DELTA; FLOODPLAIN; RIVER.

Distinct physical, or in some cases biologic or chemical, products of deposition can be related directly to definable, operative processes. Such data permit the development of models of modern deposition in which processes and resulting deposits or facies are linked. By recognizing comparable physical, chemical, and biologic attributes of ancient strata, modern depositional analogs can be applied, and the original processes forming the ancient deposit can be inferred. Such an ancient deposit is called a genetic facies. It contains the sedimentary record and constitutes a three-dimensional stratigraphic (ancient) depositional system. *See* FACIES (GEOLOGY).

An ancient depositional system is a three-dimensional, genetically defined, physical stratigraphic unit that consists of a contiguous set of process-related sedimentary facies. Several corollaries have evolved from the application of this concept. Depositional systems, such as delta, fluvial, and shelf systems, (1) are the stratigraphic equivalents of major physical geographic units; (2) form the principal building blocks of the sedimentary basin fill; and (3) can be applied where principal boundaries of the systems are preserved and where the geometry of the framework facies can be mapped.

The major realms of deposition may be classed broadly as terrigenous clastic depositional systems and biogenic-chemical depositional systems. Each of these major systems is subdivided according to particular systems of deposition, and within each of the subdivisions is an assemblage of genetic facies, which are the fundamental units of depositional systems.

Terrigenous clastic systems, composed chiefly of sands and shales, embrace eight major systems: (1) fluvial or river systems; (2) delta systems; (3) strike coastal systems; (4) fan or clastic wedge systems; (5) lacustrine systems; (6) continental eolian systems; (7) shelf systems; and (8) slope and abyssal systems.

The biogenic-chemical systems consist of three major systems: (1) carbonate systems; (2) glauconitic and authigenic shelf systems; and (3) evaporite systems.   [W.L.Fi.]

Depositional environments. Depositional environments may be distinguished from erosional environments, in which erosion of the Earth's surface is taking place. Both depositional and erosional environments are of interest to geomorphologists. However, most attention to depositional environments has come from sedimentologists, particularly in order to understand the origin of sedimentary rocks. *See* EROSION.

Sediment is derived mainly from source areas that are actively undergoing uplift and erosion, and is deposited mainly in areas that are undergoing subsidence. Location of the source and basin of deposition is mainly controlled by large-scale geophysical processes acting within the Earth's mantle, so a major factor affecting the nature and distribution of sedimentary environments is the overall structural development, or tectonics, of the area. *See* PLATE TECTONICS.

Tectonics determines the major geological structure or setting of an environment of deposition, including the location and nature of the main areas undergoing uplift or subsidence. Areas with high relief, such as mountains and volcanoes, suffer rapid erosion and supply much more sediment to basins of deposition than larger areas of low relief. One investigation, for example, found that 82% of the suspended solids (mud) discharged by the Amazon River were supplied by the 12% of the drainage basin located within the Andes Mountains.

A second important major control is climate. This includes the average temperature, the range of temperature variation, the aridity or humidity (ratio of evaporation to precipitation), and the magnitude and frequency of floods and storms. Climate in turn has an important influence on such physical factors as the salinity and energy of the environment (wind and water speeds and degree of turbulence, for example), as well as on the abundance and types of plants and animals.

In areas of subsidence and sedimentation, topography results from and controls sedimentary environments. Along a coastline of low relief, for example, spits and barrier islands are produced by waves generated in the open sea. Shallow lagoons on the landward side of barrier islands are protected from wave action by the islands themselves. The distinctive features of the lagoon environment are a result of a topography which has been produced by the accumulation of sediment in another sedimentary environment (the barrier island). *See* BARRIER ISLANDS.

Sedimentary environments can be classified into three categories: terrestrial, including alluvial fans, fluvial plains, sandy deserts, lakes, and glacial regions; mixed (shore-related), including deltas, estuaries, barrier island complexes, and glacial marine environments; and marine, including terrigenous shelves or shallow seas, carbonate shelves or platforms, continental slopes, continental rises, basin plains, ocean ridges, and ocean trenches.

Although the importance of tectonics and climate in controlling sedimentary environments is widely recognized, most classifications are based mainly on topography. Almost all distinguish terrestrial (sub-aerial or fresh-water) from marine environments, and also recognize an important group of mixed or shore-related environments.

A number of major processes operate within environments and determine the types of sediment deposited in the environment, including water depth, energy (waves and current), temperature, and salinity. Biological factors also exert a very strong influence. There would be little or no oxygen in the atmosphere if it were not for the photosynthetic activity of plants. Deposition of calcium carbonate and silica in lakes and the oceans takes place largely through the action of plants and animals, and organic matter deposited along with mineral particles is largely responsible for the development of reducing conditions within sediments after deposition. Vegetable material accumulates in swamps to form peat and coal, and fine organic detritus settles with marine muds and is the ultimate source of oil and gas. Both terrestrial and aquatic plants exert a trapping and binding action that tends to immobilize sedimentary particles, as, for example, when coastal sand dunes or tidal flats become stabilized by the growth of salt-tolerant grasses. Terrestrial vegetation plays an important role in rock weathering.

[G.V.M]

Desert No precise definition of a desert exists. From an ecological viewpoint the scarcity of rainfall is all important, as it directly affects plant productivity which in turn affects the abundance, diversity, and activity of animals. It has become customary to describe deserts as extremely arid where the mean precipitation is less than 2.5–4 in. (60–100 mm), arid where it is 2.5–4 to 6–10 in. (60–100 to 150–250 mm), and semiarid where it is 6–10 to 10–20 in. (150–250 to 250–500 mm). However, mean figures tend to distort the true state of affairs because precipitation in deserts is unreliable and variable. In some areas, such as the Atacama in Chile and the Arabian Desert, there may be no rainfall for several years. It is the biological effectiveness of rainfall that matters and this may vary with wind and temperature, which affect evaporation rates. The vegetation cover also alters the evaporation rate and increases the effectiveness of rainfall. Rainfall, then, is the chief limiting factor to biological processes, but intense solar radiation, high temperatures, and a paucity of nutrients (especially of nitrogen) may also limit plant productivity, and hence animal abundance. Of the main desert regions of the world, most lie within the tropics and hence are hot as well as arid. The Namib and Atacama coastal deserts are kept cool by the Benguela and Humboldt ocean currents, and many desert areas of central Asia are cool because of high latitude and altitude.

The diversity of species of animals in a desert is generally correlated with the diversity of plant species, which to a considerable degree is correlated with the predictability and amount of rainfall. There is a rather weak latitudinal gradient of diversity with relatively more species nearer the Equator than at higher latitudes. This gradient is much more conspicuous in wetter ecosystems, such as forests, and in deserts appears to be overridden by the manifold effects of rainfall. Animals, too, may affect plant diversity: the burrowing activities of rodents create niches for plants which could not otherwise survive, and mound-building termites tend to concentrate decomposition and hence nutrients, which provide opportunities for plants to colonize.

Each desert has its own community of species, and these communities are repeated in different parts of the world. Very often the organisms that occupy similar niches in different deserts belong to unrelated taxa. The overall structural similarity between American cactus species and African euphorbias is an example of convergent evolution, in which separate and unrelated groups have evolved almost identical adaptations under similar environmental conditions in widely separated parts of the world. Convergent structural modification occurs in many organisms in all environments, but is especially noticeable in deserts where possibly the small number of ecological niches has necessitated greater specialization and restriction of way of life. The face and especially the large ears of desert foxes of the Sahara and of North America are remarkably similar, and there is an extraordinary resemblance between North American sidewinding rattlesnakes and Namib sidewinding adders. *See* ECOLOGY; PRECIPITATION (METEOROLOGY).   [D.F.Ow.]

Desert erosion features   A distinctive topography carved by erosion in regions of low rainfall and high evaporation where vegetation is scanty or absent. Although rainfall is low, it is the most important climatic factor in the formation of desert erosion features. Desert rains commonly occur as torrential downpours of short duration with a consequent high percentage of runoff. As a result of the dryness, wind and mechanical weathering also play an important part in desert erosion. *See* SEDIMENTOLOGY; WEATHERING PROCESSES.

When storms of the so-called cloudburst type occur in the desert, sudden rushes of water, or flash floods, sweep down the normally dry washes or the narrow canyons in the mountains bordering the basins. The comparatively large volume of water combined with a high velocity due to the steepness of the slopes give the short-lived streams power to carry large amounts of fine and coarse rock fragments. As a result, the streams have great erosive power.

When intermittent streams leave the canyons and spread out at the foot of a desert mountain, they lose velocity and quickly drop the coarsest of the transported material to build an alluvial fan. Some of the water sinks into the fan, and some evaporates, but whatever remains may follow one of the channels on the fan or spread out in the form of a sheetflood, in either case carrying coarse sand, silt, and clay, and perhaps rolling some larger rock fragments along.

When the water reaches the toe of the fan, it spreads still more, dropping all but the finest silt and clay. Any excess water follows shallow washes to the lowest part of the basin, where it may form a playa lake. This evaporates in a few hours or a few days, depositing the silt and clay, mixed perhaps with soluble salts. The flat-surfaced area resulting from the silt and clay deposition is a playa. *See* PLAYA.

The lack of moisture during most of the year and the scanty vegetation make the wind a more potent agent of erosion in deserts than in humid lands. The finest material is blown high in the air and may be carried entirely out of the area, a process known as deflation. The larger sand grains are rolled along the surface, bouncing into the air when they strike an obstacle, knocking more grains into the air as they hit the ground again, until eventually a sheet of sand is moving along in the 3 or 4 ft (1 or 1.3 m) above the surface. This moving sand abrades rocks and other objects with which it comes in contact; at the same time the grains themselves become rounded and frosted. If movement is impeded by vegetation or other obstacles, sand accumulates to form dunes. *See* DUNE.

Desert landscapes evolve in three stages. In the early, or youthful, stage, alluvial fans are built, washes develop, playas form, and the basins slowly fill with detritus. As this stage progresses, some alluvial fans coalesce to form bajadas or piedmont alluvial plains along the mountain fronts, and individual basins may become deeply filled with waste to form bolsons. Desert flats develop between alluvial fans (or bajadas) and playas, and isolated dunes accumulate on the lee sides of the latter. If the original highlands are flat-topped rather than tilted mountain blocks, mesas develop. As the mountain fronts slowly retreat under the attack of the atmosphere and running water, small bare rock surfaces or pediments form at the canyon mouths, the result of lateral cutting by the intermittent streams. The general tendency during youth is for relief to decrease.

The middle, or mature, stage is initiated by the development of exterior drainage or the capture of higher basins by lower ones as drainage channels erode headward through low divides. The fill deposited during youth undergoes erosion, and pediments become more widely developed. The mountains are worn still lower, and more and more channels extend completely through them, cut by the streams engaged in draining and dissecting the higher basins. Playa deposits or other easily eroded sediments are cut into badlands before being entirely removed, and mesas are reduced to buttes. Undissected remnants of older deposits become covered with desert pavement (flat-lying, interlocking, angular stones left after finer particles are removed by deflation). Where winds are turbulent and large supplies of sand are available, complex dune areas develop. Relief shows some net increase during maturity.

At the late, or old-age, stage of desert evolution, the original mountains are so reduced in elevation that the winds sweep over them with little or no condensation of moisture, and rains become still more infrequent. Great expanses of wind-scoured bare rock, or hammada, are exposed, with here and there a more resistant remnant standing above the general level as an inselberg. Buttes are reduced to smaller bornhardts and finally disappear. Those parts of the flat surface floored by earlier deposits are covered and protected by extensive areas of desert pavement. The rock fragments may be colored brown to black by desert varnish, a coating of manganese and iron oxides. Sand blown from the bare rock surfaces and from the sediments may form large dune areas. If there are no obstacles to obstruct movement or cause wind turbulence, the sand may move as a sheet, forming large expanses of flat or gently undulating sand surfaces. Relief slowly decreases in old age. *See* DESERT.   [T.C.]

Desertification   Land degradation in low-rainfall and seasonally dry areas of the Earth. It can be viewed as both a process and the resulting condition. Desertification involves the impoverishment of vegetation and soil resources. Key characteristics include the degradation of natural vegetation cover and undesirable changes in the composition of forage species, deterioration in soil quality, decreasing water availability, and increased soil erosion from wind and water. Various stages of desertification can be seen in most of the world's drylands. In rare cases, desertification leads to abandoned, desertlike landscapes.

It is generally agreed that human activities, particularly excessive resource use and abusive land-use practices, are the primary cause of desertification. Specific activities leading to desertification include clearing and cultivation of low-rainfall areas where such cultivation is not sustainable, overgrazing of rangelands, clearing of woody plant species for fuelwood and building materials, and mismanagement of irrigated cropland leading to the buildup of mineral salts in the soil (salinization). Drought is often cited as a basic cause of desertification; however, it merely accelerates or accentuates land degradation processes already under way. *See* DROUGHT.

Consequences of desertification include reduced biological productivity, reduction of biodiversity, a gradual loss of agricultural potential and resource value, loss of food security, reduced carrying capacity for humans and livestock, increased risks from drought and flooding, and in extreme cases, barren lands that are effectively beyond restoration. Paleostudies, supported by model simulations, have shown that the intensity of Northern Hemisphere desert conditions has waxed and waned over the past 9000 years in response to the precession of the Earth's orbit about the Sun. Thus, it may be that the causal factors of desertification, whether climate change or human activities, depend on the time scale being addressed. *See* CLIMATE MODIFICATION; DESERT.   [W.Swe.]

Devonian   The fourth period of the Paleozoic Era, encompassing an interval of geologic time between 418 and 362 million years before present based on radiometric data. The Devonian System encompasses all rocks deposited or formed during the Devonian Period. *See* PALEOZOIC.

The base of the Devonian System has been fixed, by international agreement, at an outcrop of sedimentary rocks at Klonk in the Czech Republic, where it corresponds to the base of the *Monograptus uniformis* graptolite zone. The top of the Devonian System, corresponding to the base of the Carboniferous System, was fixed at LaSerre in southern France, recognized by the base of the *Siphonodella sulcata* conodont zone. The Devonian is customarily divided into Lower, Middle, and Upper series and their corresponding epochs.

Devonian fossils are found to be distributed in three realms. Within each realm there is taxonomic similarity, which indicates that there was reproductive interchange among members of the same phyletic groups, but between each two realms there are various degrees of taxonomic dissimilarity, which indicates that there were various degrees of reproductive isolation among members of the same phylogenetic groups. The largest realm covered Australia, Asia, Europe, western North America, and the Morocco-India fringe of Gondwana, and is termed the Old World Realm. It was unified by relatively free flowage of the warm equatorial currents and their immediate branches among the continental masses throughout this tropical to subtropical region. The Appalachian

Realm covered most of eastern North America and the Colombia-Venezuela-Amazon part of northern South American Gondwana, which was adjacent to Appalachian North America during the Devonian. This region was bathed by the temperate southern west-wind current, which crossed a sufficiently broad stretch of ocean so that many Old World larvae could not make the journey, allowing endemic Appalachian forms to develop locally. The Malvinokaffric Realm covered central Gondwana, including southern South America, southern Africa, and Antarctica.

Devonian mountain building was particularly noticeable along the margins of Euramerica. The Acadian orogeny formed mountainous highlands accompanied by a chain of granitic intrusions from Nova Scotia to Pennsylvania during much of Devonian time. These mountains formed a barrier that prevented mixing between organisms of the Appalachian Realm and those of the Old World Realm at the same latitude in central Europe. Erosion from the Acadian mountains produced the thick Catskill deltaic complex of New York and Pennsylvania, which spread its fine-grained sediments far into the interior of eastern North America during later Devonian time. Simultaneously, along the Arctic margin of Canada, the Ellesmerian orogeny was producing folded mountains whose erosional products formed a clastic wedge that was a mirror image of the Catskill deposits. During latest Devonian time, the Roberts Mountains thrust, of the Antler orogeny in Nevada and Idaho, formed at the top of a subduction zone along which continental crust and overlying sediments were descending beneath oceanic sediments. A Late Devonian orogeny also affected eastern Australia. *See* OROGENY; PLATE TECTONICS.

Among the marine invertebrates, trilobites (Arthropoda) were much less abundant than during the Cambrian. The planktic members of the extinct graptolites died out during the Early Devonian, at about the same time as the pelagic ammonoid cephalopods first evolved. The externally two-shelled brachiopods were at their greatest diversity. Lime-secreting corals and stromatoporoids were important and widespread in warm-water environments, and formed reefs during the Middle and Late Devonian. The extinct microfossil group known as conodonts was abundant, widespread, and rapidly evolving during the Devonian, so that conodont fossils are now regarded as the principal tools to be used for international correlation and relative age determination. *See* MICROPALEONTOLOGY.

The great diversification and radiation of fish in the Devonian has led to the term "Age of Fishes" for the period. Placoderm fish, among the most primitive of the jawed vertebrates, were successful predators in Devonian waters, and some grew to lengths up to 8 m (25 ft) just before their extinction at the end of the Devonian. The sharks, with a cartilaginous skeleton but lacking a swimbladder, may have evolved from an early placoderm. Bony fishes or Osteichthyes, a class that includes all modern fish other than sharks and agnathans, were represented in the Devonian by the primitive acanthodians, but more modern groups of bony fishes appeared in the Early Devonian. Lobe-finned bony fishes, or sarcopterygians, include both the lungfish (Dipnoi) and crossopterygians in the Devonian. The oldest known amphibians, including *Acanthostega* and *Ichthyostega*, occur in strata thought to be high Upper Devonian.

Land plants began to flourish near the beginning of Devonian time, and were exemplified by the vascular genus *Psilophyton* of the phylum Psilopsida. The latter gave rise in the Devonian to the Lycopsida (scale trees) and Pteropsida (true ferns).

[J.G.J.; P.H.H.; D.J.Ov.]

Dew   The deposit of liquid water resulting from condensation of atmospheric water vapor to exposed surfaces that cool during the night. Dewfall is noticeable in the early morning after a calm, cool, clear night, usually as beads of liquid water on the outside and upward-facing surfaces of trees, buildings, and so forth. If the ground is moist, some of the condensed water can be evaporated surface moisture. Dew forms when the surface temperature drops sufficiently to saturate air in contact with the surface (that is, when the surface drops to below atmospheric dew-point temperature); when the surface cools to below freezing temperature, frost occurs. *See* DEW POINT; FROST.

Hygroscopic particles on surfaces can act as sites for condensation at temperatures higher than the atmospheric dew-point temperature. Thus, if a surface is not clean, the first deposit of moisture from the air can occur well before the surface cools to dew-point temperature. For some chemicals, such as common salt, condensation can start to occur when the local relative humidity reaches 80%; the humidity must be 100% in the case of a clean surface. Some desert plants exude hygroscopic salts from the interior of leaves which provide preferred sites for condensation and thereby create a supply of water for the plant. *See* HUMIDITY.   [B.Hi.]

Dew point   The temperature at which air becomes saturated when cooled without addition of moisture or change of pressure. Any further cooling causes condensation; fog and dew are formed in this way.

Frost point is the corresponding temperature of saturation with respect to ice. At temperatures below freezing, both frost point and dew point may be defined because water is often liquid (especially in clouds) at temperatures well below freezing; at freezing (more exactly, at the triple point, $+.01°C$) they are the same, but below freezing the frost point is higher. For example, if the dew point is $-9°C$, the frost point is $-8°C$. Both dew point and frost point are single-valued functions of vapor pressure. *See* DEW; FOG; HUMIDITY.   [J.R.F.]

Diagenesis   All the chemical, biochemical, and physical changes that sediments undergo from the time of deposition until the stage of metamorphism is reached. Diagenetic changes are gradational, with metamorphism at elevated temperatures or pressures, and with atmospheric weathering effects when sedimentary rocks become exposed at the surface. Sandstones, shales, and carbonate sediments are particularly susceptible to diagenetic modifications. *See* CARBONATE MINERALS; SANDSTONE; SHALE.

Diagenetic processes include purely physical ones that involve rearrangement of the sediments such as compaction, slumping, bioturbation by organisms, infiltration, and soft sediment deformation; biochemical or organic processes such as particle accretion, flocculation, boring, and decomposition; and physiochemical processes such as cementation, authigenesis (formation of new minerals), inversion, recrystallization, grain growth, replacement, and interstratal solution. Most of these processes involve a reduction in porosity and permeability, which are two important sediment properties in considering the migration of subsurface fluids and the accumulation of oil and gas and certain types of mineral deposits in subsurface rock units. However, one diagenetic process, interstratal solution, is of major importance in creating secondary porosity. *See* AUTHIGENIC MINERALS.

There have been attempts to group diagenetic processes into phases (or stages) in order to develop a comprehensive model for diagenetic evolution. The boundary limits of phases are a function of chemical as well as physical conditions and include pH, oxidation potential (Eh), ionic adsorption phenomena, temperature, pressure, depth of burial, and geologic time. In one model, three diagenetic phases are recognized: syndia-genesis, anadiagenesis, and epidiagenesis. Syndiagenesis includes sediment modifications that take place during and immediately following deposition. Ana-diagenesis refers to the diagenesis processes that are characterized by expulsion and upward migration of connate water and other fluids, such as petroleum. Epidiagenesis includes those sediment-modifying processes that take place during and after uplift and emergence. *See* SEDIMENTARY ROCKS.   [F.G.Et.]

Diamond   A mineral composed entirely of carbon; the hardest substance known. Diamond is a polymorph of carbon; lonsdaleite, another polymorph, is sometimes referred to as hexagonal diamond. Diamond is found on all continents except Antarctica, which has not yet been explored for it. It occurs in nature as single crystals of gem or industrial quality, and as polycrystalline masses referred to as boart, framesite, or carbonado. It has also been found as minute black grains in some meteorites. Diamond can be synthesized in the laboratory and is produced commercially in large amounts for industrial uses.

Diamond has a cubic (isometric) crystal structure in which all carbon atoms have covalent (sp^3) bonds. It is this strong bonding that makes diamond hard. Nevertheless, if diamond is struck in specific directions it will readily cleave—a property utilized in the preparation of polished gem diamonds. The combination of refractive index and dispersion gives diamond its brilliance and so-called fire when cut and polished. The thermal conductivity of diamond is the highest of any material (five times that of copper). This property, plus hardness, makes diamond an ideal material for use as a cutting tool in industry and also as a heat sink in electronics.

Although diamond consists of carbon, at least 58 other elements have been found (for example, aluminum, 10 parts per million; hydrogen, 1000 ppm; silicon, 80 ppm) as impurities in natural diamond. However, only two, nitrogen and boron, replace carbon atoms in the diamond lattice. Nitrogen is the major impurity and may substitute for carbon in a number of ways, commonly as either isolated or paired nitrogen atoms, and as discrete platelets of nitrogen within the diamond structure. The presence or absence of nitrogen and the manner of its substitution leads to different physical properties, such as thermal conductivity, electrical restivity, and infrared spectra.

Diamond is resistant to chemical attack, other than by strong oxidizing agents. In vacuum or an inert atmosphere, a clear, colorless gem diamond transforms to a gray-black mass of graphite at about 1500°C (2700°F). In air, diamond oxidizes (burns) to carbon dioxide at and above 800°C (1500°F). At high temperature, some metals (for example, tungsten, titanium, and tantalum) react with diamond to form metal carbides. Metals, such as iron, nickel, cobalt, and platinum, in the molten state are solvents for carbon and dissolve diamond; this phenomenon is used as a basis for the synthesis of diamond.

Most natural diamond, apart from that in meteorites, crystallizes at depths of approximately 110 mi (180 km) in the Earth's upper mantle at temperatures in the range 900–1200°C (1650–2200°F). The host rock in which diamond forms is either a magnesian-rich silica-deficient ultramafic (peridotitic) rock or an ultrabasic eclogitic rock. Minerals that constitute the ultramafic type of diamond-host rock are magnesian rich and include varieties of olivine, pyroxene, and pyrope garnet. The eclogitic rock consists of sodium-bearing pyroxene and an almandine garnet. These various constituent minerals may also occur as inclusions in diamond and result in the host being identified as either an ultramafic (peridotitic) or eclogitic diamond. Diamonds in each group have formed in a distinct and different geochemical environment in the upper mantle. *See* ECLOGITE; LITHOSPHERE; PERIDOTITE.

Diamond is eventually transported to the Earth's surface by unique types of volcanic eruption in which gases play a major role. The eruptions drill narrow (much less than 3000 ft or 1000 m) explosive vents or pipes through the crust of the Earth. Two different rock types, each containing diamond, may result and infill the volcanic neck or pipe. The first type is known as kimberlite, and the second as lamproite. The most productive mine in the world based on the number of diamonds produced per unit of host rock is based on lamproite—the Argyle mine in Western Australia. In general, diamonds are considerably older than the volcanic eruption that transported them to the surface. Thus diamonds that are 3.2 billion years old reached the surface only 85 million years ago. Diamond-dearing kimberlites occur in South Africa, Botswana, Angola, Sierra Leone, Guinea, Tanzania, Brazil, Venezuela, the United States, Canada, Russia, Siberia, China, India, and Australia. Only lamproites in Western Australia and one in Arkansas in the United States are diamond bearing.

The major production of diamond is from the primary sources in South Africa, Botswana, Zaire, Australia, and Siberia, with minor amounts coming from Tanzania and China. The diamond mines in all these countries are based on kimberlite, except for the Argyle mine in Australia, where the source rock is lamproite. Although Botswana and South Africa produce the most gem diamonds, as well as Russia whose production is difficult to assess, the major worldwide production is from the Argyle mine in Australia, albeit mostly industrial diamonds. Diamond production from secondary (alluvial or placer) deposits, apart from the extensive mining of the marine gravels off the west coast of southern Africa, is relatively small compared to the output from mines based on kimberlite and lamproite. Alluvial deposits in Guinea, Ghana, Russia, and Australia are mined by large companies or government agencies. All other alluvial diamond deposits are worked by small local groups or individual miners.

Rough diamonds occur in a variety of shapes, including octahedra, dodecahedra, twinned octahedra (macle), and broken or cleavage fragments. The largest diamond found, the Cullinan, was a cleavage fragment. After the rough is sorted into cuttable (gem and near gem) and industrial stones, the decision is made as to how a specific diamond will be shaped and made into a polished gem. The cutting and polishing process can result in the loss of as much as 60% of the original diamond.

Polished diamonds are graded on the basis of the 4 C's—carat, cut, clarity, and color. The carat is the unit of weight in the diamond industry and is standardized as 0.2 gram (0.0071 oz or 200 milligrams) and is divided into 100 points. Thus a 10-point diamond weighs 0.1 ct (0.00071 oz or 20 mg). The largest diamond, the Cullinan, weighed about 3000 ct (4.27 oz or 600 g) and was the size of an average human fist. The grading cut is based on how well the facets and the shape of a polished diamond compare to a standard model. *See* GEM.

Diamonds, although commonly considered to be mostly colorless, actually exist in all the colors of the rainbow. Colored stones are known as fancies, and if of excellent uniform color they are most desirable.

Diamonds were first synthesized in Sweden in 1953. These early experiments used the principle that carbon dissolves in the transition elements of groups 8–10 (such as iron or nickel). At high pressures (50–60 kilobars or 5–6 gigapascals) and temperatures (1500°C or 2700°F), the dissolved carbon nucleates and crystallizes as diamond. Direct conversion of graphite to diamond was achieved in 1961 in shock-wave experiments in which transient high pressures in excess of 300 kb (30 GPa) and temperatures of about 1100°C (2000°F) existed.

Synthetic diamonds generally are not large; most are produced in sizes below 0.004 in. (0.1 mm). These are used extensively as grit for industrial grinding purposes. Colorless gem-quality diamond can also be synthesized, but the cost of synthesis has proved to be greater than the cost of the natural product.

Diamond films can be grown in several ways. For example, diamond crystals up to 0.02 in. (0.5 mm) in size can be formed from a mixture of methane and hydrogen at about 50 torr (6.7 kPa) pressure and 1000°C (1900°F) on a silicon substrate. This method is known as thermally induced chemical vapor deposition, but other techniques may be used, including plasma chemical vapor ion depostion and electron-beam deposition. The diamond films display properties similar to those of natural diamond and have similar hardness and thermal conductivity, both significant properties for the uses of diamond films.

Diamond, apart from its use as a gem, has numerous applications in industry, and it is designated a strategic mineral. Many of the uses of natural and synthetic diamond are equivalent. Originally, natural diamond, including boart, carbonado, and framesite, was crushed to various sizes of powder and used as grinding and polishing agents for glasses, ceramics, and nonferrous metals. Diamonds, as single crystals or powders, are also bonded in metal drills and bits. Small drills are used in applications such as dental work; large drills are used in drilling for oil and other minerals. Diamond-impregnated wheels are used for cutting many hard materials, including concrete and dimension stone for architectural purposes. Synthetic diamond is sometimes preferred for various uses, as it is grown to the specific grain size rather than crushed as in the case of natural diamond. Diamonds are used in eye surgery, and also as heat sinks and semiconductors in the electronics industry. Diamond films have potential uses as scratchproof coatings on optical lenses, compact discs, and even on nondiamond jewelry; bearings in machines; heat sinks and semiconductors in electronics; and general inert coatings or surfaces in areas of high chemical corrosion. Natural diamond has also been used as optical windows in spacecraft. [H.O.A.M.]

Diapir A buoyant mass of ductile rock or sediment that has pierced, or appears to have pierced, overlying rock, known as overburden. The overburden can yield by ductile processes or by brittle faulting. Diapirs form by lateral and vertical intrusion of buoyant or nonbuoyant rock.

Diapirs are composed of salt, gypsum, shale, mud, sand, peat, coal, limestone, ice, serpentinite, granite, gneiss, and migmatite. Salt diapirs are typically several miles wide and high, but sheetlike varieties that have spread or coalesced laterally can be as much as 200 mi (300 km) wide. This type of diapir is economically important. Gneiss and migmatite diapirs are typically 6–13 mi (10–20 km) wide but locally approach 60 mi (100 km) in width. Diapirs doming the sea floor form large islands of salt in the Persian Gulf and islands of shale in the Caspian and Banda seas. See GNEISS; GRANITE; MIGMATITE. [M.P.A.J.]

Diastem A break in the stratigraphic record produced by local erosion or nondeposition and representing a short interval of geologic time. The breaks in deposition are those that would occur within a particular sedimentary environment, rather than those associated with a major change in environment. See STRATIGRAPHY; UNCONFORMITY.

Nondeposition can result either from excessive turbulence in the environment or from a lack in sediment supply. Discontinuity in sedimentation occurs on many scales. Short breaks, from seconds to days, are associated with migration of bedforms, variation in wave or current energy, and tidal cyclicity. Seasonal deposition occurs due to floods, storms, and plankton blooms. Sedimentation by major floods or hurricanes may occur on a scale of decades to centuries. In the deep ocean, the interval between successive turbidity current flows can be thousands of years. See MARINE SEDIMENTS; SEDIMENTOLOGY; TURBIDITY CURRENT. [C.W.By.]

Diopside The monoclinic pyroxene mineral which in pure form has the formula $CaMgSi_2O_6$. Pure diopside melts congruently at 1391°C (2536°F) at atmospheric pressure. Diopside has no known polymorphs. Its structure consist of chains of SiO_4 tetrahedrons in which each silicon ion shares an oxygen with each of its two nearest silicon neighbors. These chains are linked together by Ca and Mg ions in octahedral coordination.

Diopside forms gray to white, short, stubby, prismatic, often equidimensional, crystals. Small amounts of iron impart a greenish color to the mineral. Pure diopside is common and occurs as a metamorphic alteration of impure dolomites in medium and high grades of metamorphism.

Natural diopsidic pyroxenes which show extensive solid solution with jadeite and to a lesser extent with acmite are called omphacite. Omphacite is a principal constituent of eclogites, rocks of basaltic composition which have formed at high pressure. *See* DOLOMITE; ECLOGITE; PYROXENE.   [F.R.B.]

Diorite   A phaneritic (visibly crystallized) plutonic rock having intermediate SiO_2 content (53–66%), composed mainly of plagioclase (oligoclase or andesine) and one or more ferromagnesian minerals (hornblende, biotite, or pyroxene), and having a granular texture. Diorite is the plutonic equivalent of andesite (a volcanic rock). This dark gray rock is used occasionally as a building stone and is known commercially as black granite. *See* IGNEOUS ROCKS.

Gray or white plagioclase feldspar is the dominant mineral. Rocks with more calcic plagioclases and more abundant ferromagnesian minerals are gabbros. Rocks with greater proportions of alkali feldspar are called monzonite. Those with more quartz are called quartz-diorite or tonalite.

The texture of diorites is notably variable. Most often diorites are equigranular, with coarse, partly or mostly anhedral plagioclase and hornblende crystals, subordinate biotite, and interstitial quartz and orthoclase.

Diorite is found as isolated small bodies such as dikes, sills, and stocks, but it is also found in association with other plutonic rocks in batholithic bodies. It is closely associated with convergent plate boundaries where calc-alkalic magmatism and mountain building are taking place. *See* MAGMA.   [W.I.R.]

Diving   Skin diving, scuba diving, saturation diving, and "hard hat" diving are techniques used by scientists to investigate the underwater environment. Skin diving is usually without breathing apparatus and is done with fins and faceplate. The diver's underwater observation is limited to the time that breath can be held (1–2 min). Diving with scuba (self-contained underwater breathing apparatus) and "hard hat" provide the diver with a breathable gas, thus expanding the submerged time and the depth range of underwater observations. This type of diving is limited by human physiology and the diver's reaction to the pressure and nature of the breathing gas. Saturation diving permits almost unlimited time down to depths of 100 ft (30 m).

Scuba diving. Scuba is used by trained personnel as a tool for direct observation in marine research and underwater engineering. This equipment is designed to deliver through a demand-type regulator a breathable gas mixture at the same pressure as that exerted on the diver by the overlying water column. The gas which is breathed is carried in high-pressure cylinders (at starting pressures of 2000–3000 psi or 14–20 megapascals) worn on the back.

Scuba can be divided into three types: closed-circuit, semiclosed-circuit, and open-circuit. In the first two, which use pure oxygen or various combinations of oxygen, helium, and nitrogen, exhaled gas is retained and passed through a canister containing a carbon dioxide absorbent for purification and then recirculated to a bag worn by the diver. During inhalation additional gas is supplied to the bag by various automatic devices from the high-pressure cylinders. These two types of equipment are much more efficient than the open-circuit system, in which the exhaled gas is discharged directly into the water after breathing. Most open-circuit systems use compressed air because it is relatively inexpensive and easy to obtain. Although open-circuit scuba is not as efficient as the other types, it is preferred because of its safety, the ease in learning its use, and its relatively low cost.

For physiological reasons scuba diving is limited to about 165 ft (50 m) of water depth. Below this depth when using compressed air as a breathing gas, the diver is limited, not by equipment, but by the complex temporary changes which take place in the body chemistry while breathing gas (air) under high pressure.

Saturation diving. This type of diving permits long periods of submergence (1–2 weeks). It allows the diver to take advantage of the fact that at a given depth the body will become fully saturated with the breathing gas and then, no matter how long the submergence period, the decompression time needed to return to the surface will not be increased. Using this method, the diver can live on the bottom and make detailed measurements and observations, and work with no ill effects. This type of diving requires longer periods of decompression in specially designed chambers to free the diver's body of the high concentration of breathing gas. Decompression times of days or weeks (the time increases with depth) are common on deep dives of over 200 ft (60 m).   [R.F.Di.]

Physiology. Environmental effects on the submerged diver are quite different from those experienced at sea level. Two elements are very evident during the dive. As the depth of surrounding water increases, pressure on the air the diver breathes also increases. In addition, as the pressure increases, the solubility of the gases in the diver's tissues increases. The tissues, therefore, accumulate certain gases which are not metabolized. The increased presence of certain gases causes specific and often dangerous physiological effects.

The total pressure of the atmosphere at sea level is approximately 760 mmHg or 30.4 in. Hg. Pressure increases underwater at the rate of 1 atm (10 kilopascals) for each 33 ft (10 m) that the diver descends. The total pressure applied to the body and to the breathing gas increases proportionately with depth. As pressure of the gas increases, the amount of gas that is absorbed by the body increases. This is particularly evident if the diver is breathing air within a caisson since the percentage of nitrogen in the breathing gas increases proportionately to the amount of oxygen that is removed. Likewise, the amount of carbon dioxide in the body increases during the dive, particularly if the exhaled air is not separated from the inhaled air.

One of the more obvious effects of gases on divers is caused by nitrogen. This gas makes up about 78% of the air that is normally breathed, and its solubility in the tissues increases as atmospheric pressure increases. When nitrogen is dissolved in the body, more than 50% is contained in the fatty tissues; this includes the myelin sheaths which surround many nerve cells. When divers undergo increased pressure, amounts of nitrogen in nerve tissue increase and lead to nitrogen narcosis or "rapture of the deep." Nitrogen narcosis can occur at 415 ft (130 m) or 5 atm (500 kPa), and increases in severity as the diver descends below this depth. The irrational behavior and euphoria often seen in nitrogen narcosis can result in serious, even deadly mistakes during a dive. The maximum time that divers can remain underwater without showing symptoms decreases with increasing depth.

One of the earlier recognized problems associated with human diving is known as decompression sickness, the bends, or caisson disease. If a diver is allowed to stay beneath the surface for long periods of time, the volume of dissolved gases in the tissues will increase. This is particularly true of nitrogen in the case of air breathing. When nitrogen accumulates in the tissues, it remains in solution as long as the pressure remains constant. However, when pressure decreases during the ascent, bubbles form in the tissues. Nitrogen bubbles can occur in nerves or muscles and cause pain, or they can occur in the spinal cord or brain and result in paralysis, dizziness, blindness, or even unconsciousness. Bubbles forming in the circulatory system result in air embolism. If the embolism occurs in the circulation of the lungs, a condition known as the chokes occurs.

A method of prevention of decompression sickness was suggested by J. S. Haldane in 1907. Haldane introduced the method of stage decompression, in which the diver is allowed to ascend a few feet and then remain at this level until the gases in the tissues have been allowed to reequilibrate at the new pressure. This stepwise ascent is continued until the diver finally reaches the surface. A modern variation of this method consists of placing the diver in a decompression chamber after the surface is reached, to allow for periods of decompression which simulate ocean depths.   [J.H.F.]

Dolerite   A fine-textured, dark-gray to black igneous rock composed mostly of pla-gioclase feldspar (labradorite) and pyroxene and exhibiting ophitic texture. It is com-monly used for crushed stone. Its resistance to weathering and its general appearance make it a first-class material for monuments. *See* STONE AND STONE PRODUCTS.

The most diagnostic feature is the ophitic texture, in which small rectangular plagio-clase crystals are enclosed or partially wrapped by large crystals of pyroxene. As the quantity of pyroxene decreases, the mineral becomes more interstitial to feldspar. The rock is closely allied chemically and mineralogically with basalt and gabbro. As grain size increases, the rock passes into gabbro; as it decreases, diabase passes into basalt. *See* BASALT; GABBRO.

Diabase forms by relatively rapid crystallization of basaltic magma (rock melt). It is a common and extremely widespread rock type. It forms dikes, sills, sheets, and other small intrusive bodies. The Palisades of the Hudson, near New York City, are formed of a thick horizontal sheet of diabase. In the lower part of this sheet is a layer rich in the mineral olivine. This concentration is attributed by some investigators to settling of heavy olivine crystals through the molten diabase and by others to movement of early crystals away from the walls of the passageway along which the melt flowed upward from depth, before it spread horizontally to form the sill. *See* MAGMA.

As defined, diabase is equivalent to the British term dolerite. The British term diabase is an altered diabase in the sense defined here. *See* BASALT; GABBRO; IGNEOUS ROCKS.
                                                                                            [C.A.C.]

Dolomite   The carbonate mineral $CaMg(CO_3)_2$. Often small amounts of iron, man-ganese, or excess calcium replace some of the magnesium; cobalt, zinc, lead, and barium are more rarely found. Dolomite is normally white or colorless with a specific gravity of 2.9 and a hardness of 3.5–4 on Mohs scale. It can be distinguished from calcite by its extremely slow reaction with cold dilute acid. Dolomite is a very common mineral, occurring in a variety of geologic settings. It is often found in ultrabasic ig-neous rocks, notably in carbonatites and serpentinites, in metamorphosed carbonate sediments, where it may recrystallize to form dolomite marbles, and in hydrothermal veins. The primary occurrence of dolomite is in sedimentary deposits, where it consti-tutes the major component of dolomite rock and is often present in limestones. *See* DOLOMITE ROCK; LIMESTONE; SEDIMENTARY ROCKS.   [A.M.G.]

Dolomite rock   Sedimentary rock containing more than 50% by weight of the mineral dolomite [$CaMg(CO_3)_2$]. The term dolostone is used synonymously. Since dolomites usually form by replacement of preexisting limestones, it is often possible to see relict grains, fossils, and sedimentary structures preserved in dolomite rocks. More often, however, original textures are obliterated. *See* DOLOMITE; LIMESTONE.

Field differentiation of dolomite from calcite ($CaCO_3$) is most easily accomplished by application of dilute hydrochloric acid. Calcite is strongly effervescent in acid, whereas dolomite is usually very weakly effervescent unless scratched or ground into a fine powder. Dolomite also often weathers to a brown color on outcrop because of the common substitution of iron for magnesium in the dolomite structure. X-ray diffraction analysis is the most reliable method for the differentiation of dolomite from calcite, and also the best method for characterization of the degree of ordering of a particular dolomite sample.

Dolomite rocks are predominantly monomineralic. The most common noncarbonate minerals are quartz (either authigenic or detrital), clay minerals, pyrite, and glauconite. Evaporite minerals or their replacements are also common.

Geologists have been unable to decipher the exact conditions of dolomite formation. This so-called dolomite problem revolves around several questions relating to the stoichiometry of the reaction in which dolomite is formed, the fact that dolomite is not common in young marine sediments, and the type of geological setting in which ancient dolomites formed.

The relative proportions of dolomite and calcite have changed through geologic time. Dolomitic rocks are dominant in the Precambrian and early Paleozoic, whereas calcitic rocks become dominant in the late Mesozoic and continue to dominate through the present. Several alternative hypotheses have been offered to explain this.

In a few geological settings, dolomite is forming within the sediments at the present time. One general setting includes a wide variety of supratidal ponds, lagoons, tidal flats, and sabkhas such as the Solar Lake in Israel, the Coorong Lagoon in Australia, the bank tops of Andros Island in the Bahamas, and the Sabkha of Abu Dhabi. Each of these environments is hot, more saline than seawater, and rich in organic matter. A second general setting of modern dolomite formation is in continental margin sediments underlying productive coastal oceanic upwelling zones. Examples include the Peru-Chile margin, the southern California borderlands, the Gulf of California, and the Walvis Ridge west of Namibia (southwestern Africa). This second set of geological environments is characterized by lower water temperature and normal marine salinity. Like the first set, these environments are rich in organic matter. The association of dolomite formation with abundant organic carbon seems to be the one common thread linking each dolomite occurrence and possibly controlling dolomite precipitation.

Dolomites are extremely important oil reservoir rocks. This is partly a result of the high porosity of many dolomite rocks and partly the result of the association between dolomite formation and the presence of sedimentary organic matter.

Dolomites are also the main host rocks for lead and zinc ore deposits. Rocks of this type are known as Mississippi Valley ore deposits. Neither the origin of the ore-forming solutions nor the association between the lead-zinc ores and the dolomite rocks has been satisfactorily explained. *See* PETROLEUM GEOLOGY; SEDIMENTARY ROCKS.   [P.A.B.]

Double diffusion   A type of convective transport in fluids that depends on the difference in diffusion rates of at least two density-affecting components. This phenomenon was discovered in 1960 in an oceanographic context, where the two components are heat and dissolved salts. Besides different diffusivities, it is necessary to have an unstable or top-heavy distribution of one component.

In the oceanographic context, if the unstable component is the slower-diffusing one (salt), with the overall gravitational stability maintained by the faster-diffusing component (heat), then "salt fingers" will form. Since warm, salty tropical waters generally overlie colder, fresher waters from polar regions, this is a very common stratification in the mid- to low-latitude ocean. Salt fingers arise spontaneously when small parcels of warm, salty water are displaced into the underlying cold, fresh water. Thermal conduction then removes the temperature difference much quicker than salt diffusion can take effect. The resulting cold, salty water parcel continues to sink because of its greater density. Conversely, a parcel of cold, fresh water displaced upward gains heat but not salt, becoming buoyant and continuing to rise. The fully developed flow has intermingled columns of up- and downgoing fluid, with lateral exchange of heat but not salt, carrying advective vertical fluxes of salt and to a lesser extent heat.

Another form of double-diffusive convection occurs when the faster-diffusing component has an unstable distribution. In the ocean, this happens when cold, fresh water sits above warmer, saltier and denser water. Such stratifications are common in polar regions and in local areas above hot springs at the bottom of the deep sea. *See* OCEAN CIRCULATION; OCEANOGRAPHY; SEAWATER.

The importance of double diffusion lies in its ability to affect water mass structure with its differential transport rates for heat and salt. This is believed to play a significant role in producing certain oceanic water types with well-defined relationships between temperature and salinity.   [R.W.Sc.]

Drought   A general term implying a deficiency of precipitation of sufficient magnitude to interfere with some phase of the economy. Agricultural drought, occurring when crops are threatened by lack of rain, is the most common. Hydrologic drought, when reservoirs are depleted, is another common form. The Palmer index is used by agriculturalists to express the intensity of drought as a function of rainfall and hydrologic variables.

The meteorological causes of drought are usually associated with slow, prevailing, subsiding motions of air masses from continental source regions. These descending air motions, of the order of 660–1000 ft (200 or 300 m) per day, result in compressional warming of the air and therefore reduction in the relative humidity. Since the air usually starts out dry, and the relative humidity declines as the air descends, cloud formation is inhibited—or if clouds are formed, they are soon dissipated.   [J.N.]

Drumlin   A streamlined, oval-shaped hill which has been shaped by flowing glacial ice. The long axis is parallel to the direction of ice flow, the up-glacier slope is usually steeper than the lee slope, and composition includes a variety or combination of materials—till, outwash, or bedrock. Drumlins are highly localized, but where present, they occur in large numbers. Some drumlins are clearly erosional in origin, but in others till deposition appears to have been synchronous with drumlin formation. Thus, one or both processes must be operative at some time in the subglacial environment where drumlins form.   [W.H.J.]

Dumortierite   A nesosilicate mineral with composition $Al_7(BO)_3SiO_4)_3O_3$. Dumortierite crystallizes in the orthorhombic system but well-formed crystals are rare; the mineral usually occurs in parallel or radiating fibrous aggregates. There is one direction of poor cleavage. The hardness is 7 on Mohs scale, and the specific gravity is 3.26–3.36. The mineral has a vitreous luster and a color that varies not only from one locality to another but in a single specimen. It may be pink, green, blue, or violet. Dumortierite is found in schists and gneisses and more rarely in pegmatites. In the United States it occurs at Dehesa, California, and at Rochester, Nevada, where it has been mined for the manufacture of high-grade porcelain. *See* SILICATE MINERALS.   [C.S.Hu.]

Dune Mobile accumulation of sand-sized material that occurs along shorelines and in deserts because of wind action. Dunes are typically located in areas where winds decelerate and undergo decreases in sand-carrying capacity. Dunefields are composed of rhythmically spaced mounds of sand that range from about 3 ft (1 m) to more than 650 ft (200 m) in height and may be spaced as much as 5000 ft (1.5 km) apart. Smaller accumulations of windblown sand, typically ranging in height from 0.25 to 0.6; in. (5 to 15 mm) and in wavelength from 3 to 5 in. (7 to 12 cm), are known as wind ripples. Dunes and ripples are two distinctly different features. The lack of intermediate forms shows that ripples do not grow into dunes. Ripples commonly are superimposed upon dunes, typically covering the entire upwind (stoss) surface and much of the downwind (leeward) surface as well.

Virtually any kind of sand-sized material can accumulate as dunes. The majority of dunes are composed of quartz, an abundant and durable mineral released during weathering of granite or sandstone. Dunes along subtropical shorelines, however, are commonly composed of grains of calcium carbonate derived in part from the break-down of shells and coral. Along the margins of seasonally dry lakes, dunes may be composed of gypsum (White Sands, New Mexico) or sand-sized aggregates of clay minerals (Laguna Madre, Texas). See CLAY MINERALS; GYPSUM; QUARTZ.

The leeward side of most dunes is partly composed of a slip face, that is, a slope at the angle of repose. For dry sand, this angle is approximately 33°. When additional sand is deposited at the top of such a slope, tonguelike masses of sand avalanche to the base of the slope. The dune migrates downwind as material is removed from the gently sloping stoss side of the dune and deposited by avalanches along the slip face. Much of the sand on the leeward side of the dune is later reworked by side winds into wind ripple deposits. Because the coarsest grains preferentially accumulate at the crests of wind ripples, the layering in wind ripple deposits is distinctive and relatively easily recognized—each thin layer is coarser at its top than at its base.

Dunes can be classified on the basis of their overall shape and number of slip faces. Three kinds of dunes exist with a single slip face; each forms in areas with a single dominant wind direction. Barchans are crescent-shaped dunes; their arms point down-wind. They develop in areas in which sand is in small supply. If more sand is available, barchans coalesce to form sinuous-crested dunes called barchanoid ridges. Transverse dunes with straight crests develop in areas of abundant sand supply. The axis of each of these dune types is oriented at right angles to the dominant wind, and the dunes migrate rapidly relative to other dune types. The migration rate of individual dunes is quite variable, but in general, the larger the dune, the slower the migration rate. In the Mojave Desert of southern California, barchans having slip faces 30 ft (10 m) long migrate about 50 ft (15 m) per year.

Dunes having more than one slip face develop in areas with more complex wind regimes. Linear dunes, sometimes called longitudinal or self dunes, possess two slip faces which meet along a greatly elongated, sharp crest (see illustration). Some linear dunes in Saudi Arabia reach lengths of 120 mi (190 km). Experimental evidence has shown that linear dunes are the result of bidirectional winds that differ in direction by more than 90°. The trend of these dunes is controlled by wind direction, strength, and duration, but the nature of the wind regime cannot be deduced from a knowledge of dune trends. Star dunes bear many slip faces and consist of a central, peaked mound from which several ridges radiate. Because they do not migrate appreciably, they grow in height as sand is delivered to them, some reaching 1000 ft (300 m).

Plant growth appears to be important to the growth and maintenance of two types of dunes. Coppice dunes are small mounds of sand that are formed by the wind-baffling and sand-trapping action of desert plants. The crescentic shape of parabolic dunes

Linear dune, Imperial County, California.

gives them a superficial resemblance to barchan dunes, but their arms point upward. Plants commonly colonize and anchor only the *edges* of a dune, leaving the body of the dune free to migrate. The retarded migration rate of the dune margin leads to the formation of the trailing arms of a parabolic dune.

Other dune types are dependent on special topographic situations for their formation. Climbing dunes develop on the upwind side of mountains or cliffs; falling dunes are formed at the sheltered, downward margin of similar features.   [D.B.Lo.]

Dunite   An ultramafic igneous rock composed of at least 90% olivine. Important accessory minerals (abundances usually less than 1%) found in different occurrences of dunite include chromian spinel, low- or high-calcium pyroxenes (enstatite and diopside), and plagioclase. If these minerals constitute greater than 10% of the rock, it is called chromitite, peridotite, or troctolite, respectively. Low-temperature alteration (less than 750°F or 400°C) causes hydration of olivine to the mineral serpentine and, where extensive, may transform dunite to the metamorphic rock serpentinite. The dun color is a characteristic feature of the weathered rock. *See* CHROMITE; OLIVINE; PERIDOTITE; PYROXENE; SERPENTINITE.

Dunite is ultrabasic in composition, meaning that it is low in silica compared to most crustal rocks (<45% silica) and is generally very high in magnesium (up to 54%). In some occurrences, notably in continental layered intrusions, however, the olivine may be very iron rich (variety hortonolite), and the rock can contain as much as 40% iron. Dunite is notably poor in alumina, soda, and lime; and while it may be relatively nickel rich, many other critical trace elements are nearly missing. Thus, weathering of dunite forms soils to which most terrestrial plants are poorly adapted, and such soils often host unusual plant species such as the carnivorous cobra lily and the miniature rhododendron *Kalmiopsis leachiana* found in the coast ranges of Oregon and California.

The most important occurrence of dunite is in its association with rocks believed to have come from the Earth's mantle (principally peridotite) where it marks the location of former conduits through which magmas have been transported out of the Earth to the crust. In the oceans, it is infrequently dredged from the great oceanic transform faults cutting the ocean ridges, exposed in tectonic windows in disrupted and uplifted ocean crust in association with serpentinized mantle peridotite. Important dunite bodies crosscutting tectonically exposed mantle sections on the walls and floors of rift valleys have been found in the central Atlantic and eastern Pacific oceans far from transform faults. The suggestion is that the flow of magma out of the mantle upwelling beneath ocean ridges is not uniform, and is therefore important in controlling the formation and structure of the two-thirds of the Earth's crust that is formed in the oceans. *See* EARTH; MID-OCEANIC RIDGE; RIFT VALLEY; TRANSFORM FAULT.

Dunite has been extensively quarried as a building stone because of its unusual pale water-green color, but is little used now because of its high susceptibility to acid rain. Dunite is also used for refractory materials for furnaces. Olivine in dunite is occasionally of gem quality. Dunite is also the principal host of deposits of the mineral chromite. The hortonolite-dunite pipes found in the Bushveld Complex of South Africa have also been mined for platinum-group minerals. *See* GEM; IGNEOUS ROCKS; PETROLOGY.

[H.J.B.D.]

Dust storm   A strong, turbulent wind carrying large clouds of dust. In a large storm, clouds of fine dust may be raised to heights well over 10,000 ft (3000 m) and carried for hundreds or thousands of miles (1 mi = 1.6 km).

Sandstorms differ by the larger mass, more rapid setting speeds of the particles involved, and the stronger transporting winds required. The sand cloud seldom rises above 3.3–6.6 ft (1–2 m) and is not carried far from the place where it was raised.

Dust storms cause enormous erosion of the soil, as in the dust bowl disasters of 1933–1937 in the Great Plains of the United States. Besides causing acute physical discomfort, they present a severe hazard to transportation by reducing the visibility to very low ranges. Conditions required are an ample supply of fine dust or loose soil, surface winds strong enough to stir up the dust, and sufficient atmospheric instability for marked vertical turbulence to occur.

Small dust particles increase scattering of light, mainly in short (blue) wavelengths. The Sun often appears a deep orange or red when seen through a dust cloud; however, optical effects are variable. Large particles are effective reflectors, and an observer in an aircraft above a dust storm may see a solid sheet with an apparent dust horizon.

[C.W.N.]

Dynamic instability   A state of fluid flow in which the distribution of mass and momentum is unstable. Fluid flows are subject to a variety of instabilities which generally cause the flow to become more complex and which often lead to turbulent, chaotic flow. Instabilities are responsible for a variety of phenomena in natural flows, including cyclones, hurricanes, and thunderstorms in the atmosphere; mantle convection in the Earth's interior; and granules and supergranules in stellar atmospheres. A great deal of research in the geophysical and astrophysical sciences has focused on flow instabilities; and instability plays an important role in engineering problems ranging from naval engineering and aeronautics to the design of efficient heating and cooling systems.

Fluid instabilities may be broadly divided into two classes: convective and dynamic. While convective instabilities can be understood rather easily in terms of forces acting on displaced parcels of fluid, dynamic instabilities are more varied and more challenging to understand. This broad class of fluid instabilities is responsible for the cyclones and anticyclones that dominate the weather in middle and high latitudes, as well as meanders and rings in ocean currents such as the Gulf Stream. These instabilities are ultimately responsible for the lack of predictability of complex flows such as are found in the atmosphere. *See* CONVECTIVE INSTABILITY; CYCLONE; WIND.

Many, but not all, dynamic instabilities can be understood with the aid of a vorticity principle and an invertibility principle. The vorticity of a fluid can be thought of as the rate of rotation of a rigid paddle wheel embedded in the flow (see illustration). [Mathematically, it is the curl of the vector velocity field.] The rotation can be brought about by shear (a change in the speed of the flow in the direction across the flow) and by the curvature of the flow. Vorticity is important for two reasons. First, in two-dimensional

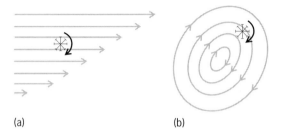

The vorticity of a fluid is related to the rate at which a rigid paddle wheel would rotate if placed in the flow. (*a*) The paddle wheel rotates because the velocity of the fluid varies in a direction perpendicular to the flow. (*b*) Rotation is generated by the curvature of the flow.

flows, vorticity is conserved; that is, if the paddle wheel is followed around it is observed that its rate of rotation remains constant. Second, in such flows, it is possible to work backward from knowledge of the vorticity at every point in the fluid to find the flow field at every point. This is called the invertibility principle. Basically, a clockwise vorticity at a point in the fluid will induce a clockwise rotation of the fluid nearby; the total velocity field in the fluid can then be found by adding the velocities induced by all the vorticities at each point.

The flow of the Earth's atmosphere is three-dimensional. Under this circumstance, vorticity is not conserved because fluid converging on or diverging from a point in the fluid causes it to spin faster or slower. (The principle at work here is the conservation of angular momentum.) Fortunately, the Earth's atmosphere is usually stably stratified; that is, a parcel displaced upward will be colder than its environment and will accelerate downward toward its original position. In this sense the atmosphere is convectively stable. It is possible to identify a quantity that reflects the air's density and at the same time is conserved as long as no heat is added to or subtracted from the air. This quantity is called the potential temperature (θ) and is related to temperature and pressure by the expression below, where T is the temperature in kelvins and p is the pressure in

$$\theta = T \left(\frac{1000}{p} \right)^{0.287}$$

millibars. Generally, θ increases upward in the atmosphere: slowly in the troposphere (a layer of air that extends upward to roughly 6 mi or 10 km) and more rapidly in the stratosphere (a layer of air between about 6 and 18 mi, or 10 and 30 km). *See* STRATOSPHERE.

The criterion for the dynamic instability of large-scale three-dimensional flows is analogous to barotropic instability of two-dimensional flows. The necessary condition for this kind of instability is that potential vorticity must be a maximum or minimum somewhere on a surface of constant θ. The type of instability that results from maxima or minima of potential vorticity in a fluid is called internal baroclinic instability. This type of instability can draw on both the kinetic energy of the original flow and the potential energy that occurs when there are horizontal variations of density.

Observations of the Earth's atmosphere show that the potential vorticity of the troposphere in middle and high latitudes is nearly uniform. But when a gently sloping θ surface is followed into the stratosphere, a large increase of potential vorticity occurs. Even so, there is no localized maximum or minimum of potential vorticity, but just a large gradient of the quantity, concentrated near the boundary between the troposphere and stratosphere. (Many meteorologists define the stratosphere as a region of large potential vorticity.) For this reason, internal baroclinic instability seldom occurs in the atmosphere. *See* ATMOSPHERE. [K.A.Em.]

Dynamic meteorology   The study of those motions of the atmosphere that are associated with weather and climate. Atmospheric motions span an enormous range of spatial and temporal scales; dynamic meteorology concentrates mainly on large-scale and mesoscale motions. Large-scale motions are those with horizontal scales in excess of a few hundred kilometers and time scales longer than a day. Such motions are strongly influenced by the rotation of the Earth and by the vertical thermal stratification of the atmosphere. Mesoscale motions have horizontal scales in the range of a few kilometers to a few hundred kilometers; they are often associated with convective clouds and precipitation.

The mechanics and thermodynamics of the unsaturated atmosphere are governed by three fundamental conservation laws: (1) conservation of mass, expressed by the mass continuity equation; (2) conservation of momentum, expressed by Newton's second law of motion; and (3) conservation of thermodynamic energy, expressed by the first law of thermodynamics. For saturated conditions, conservation of water substance must also be considered. *See* CORIOLIS ACCELERATION.

Statics. The vertical structure of the atmosphere is determined by the equation of state for an ideal gas and by the hydrostatic relationship. The former expresses the relationship among pressure, density, and temperature at any point; the latter expresses the balance between the upward-directed component of the pressure-gradient force (associated with the approximate exponential decrease of pressure with height) and the downward-directed gravity force. The equation of state and hydrostatic equation may be combined to form the hypsometric equation (1), which relates the geopotential

$$Z_2 - Z_1 = \frac{R}{g} \int_{p_2}^{p_1} T \, d\ln p \tag{1}$$

height differences $(Z_2 - Z_1)$ between two pressure surfaces p_2 and p_1 to the mean temperature T in the layer between the two surfaces; and where R (the gas constant for dry air) $= 287$ J $\cdot$ kg^{-1} $\cdot$ K^{-1} and g (the acceleration of gravity) $= 9.81$ m $\cdot$ s^{-2}. Thus, pressure decreases more rapidly with height in cold air than in warm air.

Except in regions of active precipitation, where the heating rate due to latent heat of condensation is large, temperature changes following the motion of individual parcels of air are controlled primarily by adiabatic expansion and compression as the air parcels move to lower or higher pressure. The thermodynamic state of such parcels can be characterized by the potential temperature θ, as in Eq. (2), where p_0 [$= 10^5$ pascals

$$\theta = t \left(\frac{p_0}{p} \right)^{R/C_p} \tag{2}$$

(1000 millibars)] is a reference pressure and cp ($= 1004$ J $\cdot$ kg^{-1} $\cdot$ K^{-1}) is the specific heat at constant pressure. The potential temperature is the temperature that a parcel of air at pressure p and temperature T would acquire if it were moved adiabatically to pressure p_0. When all diabatic heat sources can be neglected, θ remains constant in time for each air parcel. Normally, θ increases with altitude in the atmosphere, so that an air parcel displaced upward (downward) has a value of θ less (greater) than that of its environment, and hence experiences a net buoyancy force that tends to return it to its equilibrium level. The atmosphere is then said to be statically stable. When θ is constant with height, a condition that occurs when the temperature decreases with height at a rate $10°$C $\cdot$ km^{-1} ($30°$F $\cdot$ mi^{-1}), the atmosphere is said to be neutrally stable; if θ decreases with height, the atmosphere is absolutely unstable and convective motion develops spontaneously. If an air parcel is saturated, upward displacement causes water vapor to condense and release its latent heat of condensation; the potential temperature is then no longer conserved.

The hydrostatic relationship implies that pressure decreases monotonically with altitude. Pressure may thus be substituted for height as the independent vertical coordinate. If the atmosphere is everywhere statically stable so that θ increases monotonically with height, potential temperature can also be used as a vertical coordinate. Potential temperature coordinates are useful, for analysis of adiabatic motions, since in that reference frame prediction of adiabatic flow is reduced to a two-dimensional problem of following the motion on θ surfaces. Isobaric coordinates, in which pressure is the vertical coordinate, have the advantage of eliminating any explicit reference to the density field. These are the most commonly used vertical coordinates in dynamic meteorology. *See* CONVECTIVE INSTABILITY; DYNAMIC INSTABILITY; ISOBAR (METEOROLOGY).

Baroclinic instability. Baroclinic energy conversion processes are responsible for the growth and maintenance of most large-scale weather disturbances.

In midlatitudes in the troposphere, potential temperature normally decreases from Equator to pole on isobaric surfaces. This decrease does not occur uniformly, but tends to be concentrated in the jet stream, a narrow band of strong westerly winds in the upper troposphere that encircles the globe in midlatitudes. *See* JET STREAM.

When the shear of the zonal wind is sufficiently strong so that the meridional gradient of potential vorticity on a constant potential temperature surface is locally reversed, or when there is a nonvanishing gradient of potential temperature at the surface of the Earth, the linearized equations have solutions in the form of exponentially growing disturbances. These baroclinically unstable modes have growth rates, structures, and scales similar to those observed in developing extratropical cyclones. Baroclinic instability provides a mode whereby infinitesimal disturbances may amplify into large-amplitude storms. However, in many cases it appears that storms in the atmosphere grow through nonlinear interactions involving preexisting disturbances. Such processes must generally be studied by numerical simulations on high-speed computers.

Mesoscale convective systems. For horizontal scales less than several hundred kilometers, the major energy source is not baroclinic instability; it is latent heat release by cumulonimbus clouds. The convective storms associated with such clouds can occur only when the atmosphere is conditionally unstable [see Eq. (2)], sufficient moisture is present, and there is an initial disturbance strong enough to lift air parcels high enough to release the conditional instability. Mesoscale convective systems take a variety of forms. Among these are hurricanes, squall lines, and mesoscale convective complexes. *See* HURRICANE; MESOMETEOROLOGY; SQUALL LINE.

Numerical weather prediction. In current global weather prediction models, a mixture of techniques is often employed. Nearly all models use finite difference representations for the vertical coordinate and time; the horizontal variation is generally represented either by a network of grid points at uniform intervals of latitude and longitude or by a finite set of spherical harmonics.

To predict the weather dynamically, it is necessary to solve the dynamics equations by integrating in time, starting from an initial state of the atmospheric variables determined from observations. In practice, observational errors and poor data coverage, particularly over the oceans, make it impossible to exactly determine the initial state of the atmosphere. Moreover, the state determined from observations generally does not properly represent the true dynamical balance among the pressure and wind fields. This imbalance introduces noise that is interpreted as high-frequency inertia-gravity waves. If a prediction is attempted from such an initial state, the noise rapidly dominates the true, slowly evolving weather disturbances. In order to prevent the growth of such spurious noise, the analysis must be initialized by processing it in a manner that assures a dynamical balance between the pressure and wind fields while preserving the actual observations as closely as possible. However, even if the initial state were known perfectly, there still would be a limit beyond which errors would dominate the forecast. *See* WEATHER FORECASTING AND PREDICTION.

Global climate modeling. In addition to their role in weather prediction, dynamical models can also be used to simulate global climate. Climate is the study of the average state of the atmosphere and its seasonal and interannual variability. Climate is determined by the joint influence of energy sources and sinks at the Earth's surface, and the transformation and transport of energy in the atmosphere and the oceans. Models that simulate these processes are usually referred to as general circulation models. Such models must contain accurate representations of all the important physical processes that influence the circulation. *See* ATMOSPHERE; CLIMATE MODELING; METEOROLOGY; WIND.   [J.R.H.]

E

Earth The third planet from the Sun and the largest of the four inner, or terrestrial, planets. The Sun is an average-sized, middle-aged star situated toward the outer edge of one of the spiral arms of the Milky Way Galaxy. So far as is known, Earth is unique in the solar system in having life. Whether life exists in the universe beyond the solar system is unknown.

Earth has one natural satellite, the Moon. Otherwise, Earth's nearest neighbors in space are Venus, which is about 108×10^6 km (67×10^6 mi) from the Sun, and Mars, about 228×10^6 km (141×10^6 mi) from the Sun. Earth is about 150×10^6 km (93×10^6 mi) from the Sun.

Earth completes an orbit around the Sun in 365 days, 5 h, 48 min, 46 s; the orbit defines the length of the year. The length of the day is determined by the period of Earth's rotation about its axis. The fact that the year is not a whole number of days has affected the development of the calendar.

Earth rotates on its axis once each day. The axis of rotation is perpendicular to the Equator, and the Equator is inclined at about 23.5° to the plane of Earth's orbit around the Sun. As Earth moves in its orbit, the north spin axis, or north geographic pole, points in the direction of the star Polaris, making it the North Star or polestar. One result of the tilt of the Equator relative to the orbital plane is that different parts of Earth receive differing amounts of sunlight through the year; this is the primary cause of seasons. *See* EQUATOR.

Earth is an oblate spheroid. The mean equatorial radius is 6378.139 km (3963.37 mi), and the polar radius is 6356.779 km (3950.10 mi), the difference being 21.360 km (13.27 mi).

Earth's mass is 5.976×10^{27} g (0.2108×10^{27} oz), being the sum of 5.974×10^{27} g (0.2107×10^{27} oz) for solid Earth, 1.4×10^{24} g (0.049×10^{24} oz) for the ocean, and 5.1×10^{21} g (0.18×10^{21} oz) for the atmosphere. Earth's average density is 5.518 g/cm^3, which is just about double the density of the common rocks that form at Earth's surface, indicating that Earth's interior is more dense than the surface. Seismic studies have confirmed that Earth is layered both compositionally and mechanically (see illustration). *See* ATMOSPHERE; EARTH INTERIOR; OCEANOGRAPHY; SEISMOLOGY.

The deepest compositional layer is the core, which is divided into a solid inner core and a liquid outer core. Both the inner and outer core have the same composition, believed to be nickel-iron plus a small amount of lighter elements such as sulfur and silicon. Electric currents moving in the molten metal outer core are believed to be the origin of Earth's magnetic field. Above the core is the mantle which, on the basis of density of rare rock samples brought up from deep in the mantle in kimberlite pipes, and other evidence, is believed to be composed of silicate minerals, and in particular olivine and pyroxene. A rock composed largely of olivine and pyroxene is called a peridotite. *See* OLIVINE; PERIDOTITE; PYROXENE; SILICATE MINERALS.

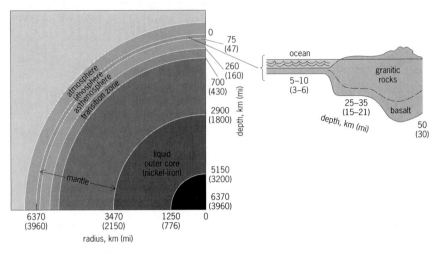

Principal layers of Earth.

Above the mantle is Earth's crust, and between the crust and the mantle there is a pronounced seismic discontinuity known as the Mohorovičić discontinuity, or Moho. The crust is of two kinds, both of which are less dense and compositionally different from the peridotitic mantle below. Beneath the ocean the crust is basaltic in composition and about 8 km (5 mi) thick. The crust beneath the continents is granitic in composition and averages 35 km (21.7 mi) in thickness but ranges up to 80 km (49.7 mi), as beneath Tibet. The oceanic crust is geologically young because it is continually created and destroyed through the process of plate tectonics. No part of the oceanic crust that is older than about 180×10^6 years has yet been discovered. The continental crust is much older than the oceanic crust. Continental rocks as old as 4×10^9 years have been discovered in Canada, and the fact that they are highly deformed indicates a long and eventful history. *See* EARTH CRUST; GRANITE; MOHO (MOHOROVIČIĆ DISCONTINUITY); PLATE TECTONICS.

The surface of solid Earth has a bimodal distribution of elevations. If the water from the ocean could be removed, it would be apparent that continents stand high (average elevation is 840 m or 2755 ft above sea level), while the ocean floor sits low (average elevation is 3800 m or 12,464 ft below sea level). This difference in elevation arises because rigid lithosphere floats on the weak asthenosphere, and because the density of oceanic lithosphere (that is, lithosphere capped by oceanic crust) is greater than the density of continental lithosphere.

On the continents, mountain belts are the most dramatic features. They range in elevation from Mount Everest, 8848 m (29,030 ft), in the Himalaya Mountains to older, deeply eroded ranges that are now barely above sea level. Granitic and metamorphic rocks are generally exposed in the cores of mountain ranges. The overlying rocks that cover most of the Earth's surface are sedimentary, mainly of shallow marine origin, that may or may not have been deformed. The deformation is the result of compression and tension that causes folding and faulting, and may be accompanied by intrusion and metamorphism. Movements and collisions of tectonic plates are the principal cause of mountain building. Mountains generally are formed over several tens of millions of years. The rocks deformed in the process are generally marine sedimentary rocks formed along the margins of continents. *See* DEEP-MARINE SEDIMENTS; MARINE SEDIMENTS; METAMORPHIC ROCKS; OROGENY; SEDIMENTARY ROCKS.

The topographic features underlying the oceans are similarly diverse and reveal more evidence of a dynamic Earth. The continental shelf, an area covered by shallow water, generally less than 150 m (500 ft) deep, surrounds the continents at most places. Such areas are generally underlain by continental, granitic rocks, and are submerged parts of the continents. Continental slopes are the transition between the continental shelf and the ocean floors. Their tops are generally less than 150 m below sea level, and they slope down to about 4400 m (14,000 ft). They are narrow, steep features, with slopes generally between 2 and 6°, but some are up to 45°. They are generally underlain by thick accumulations of sedimentary rocks. *See* SEA-FLOOR IMAGING.

Submarine trenches and their associated volcanic island arcs are formed as a result of a tectonic plate of lithosphere sinking into the mantle beneath the edge of an overriding plate. The deepest place on Earth is in the Mariana Trench, 11,022 m (36,152 ft) below sea level.

The ocean floor is the most widespread surface feature of Earth. Beneath an average of 4.4 km (2.75 mi) of seawater are about 2.3 km (1.4 mi) of sedimentary rocks with some intercalated basalt, and below that is the oceanic crust, consisting of 4–6 km (2.5–3.7 mi) of basaltic rocks. Interrupting the ocean floor at many places are submarine mountains formed by basaltic volcanoes. Some of these volcanoes are very large and form oceanic islands such as the Hawaiian Islands.

New oceanic lithosphere capped by basaltic crust is created at the mid-ocean ridges, and this newly formed plate moves away from the ridges. The tectonic plates formed in this way may carry continents on them, and are the mechanism of continental drift. Paleomagnetic data from the continents indicate that the continents have moved relative to each other. The tectonic plates capped by basaltic crust plates are consumed at the trench-volcanic island arc areas. *See* PALEOMAGNETISM; SUBDUCTION ZONES.

As well as the ridge and the trench, a third type of plate boundary occurs where two plates slide past each other at a transform fault. Such collisions account for the deformed rocks found in the crust. *See* TRANSFORM FAULT.

The evidence for continental drift in the geological past includes matching of rock types, ages, fossils, climates, and structures (mountain ranges), as well as the paleomagnetic data. Evidence showing or suggesting present movements consists of shallow earthquakes along mid-ocean ridges and transform faults that offset them; deep earthquakes associated with deep-sea trench-volcanic island arc areas; direct measurement of movement; volcanic activity at mid-ocean ridges; and volcanic activity at trench-island arc areas. *See* GEODESY.

Earth's temperature and gravitation are such that an atmosphere is present. The major constituents are nitrogen and oxygen. A thin ozone layer in the atmosphere shields the Earth from lethal ultraviolet radiation from the Sun. The atmosphere, especially oxygen, and the presence of water, both at the surface and in the atmosphere, make life possible. Precipitation, mainly rain, results in running water such as streams and rivers on the continents. Running water is the main cause of erosion of the continents, and most of the landscapes have been eroded by water, although some are eroded by wind or ice (glaciers). *See* EROSION; GLACIOLOGY.

Earth, along with the rest of the solar system, is believed to have formed about 4.55×10^9 years ago. This age is determined by dating radioactive isotopes in meteorities. Meteorites are believed to be fragments produced by collisions among small bodies formed by the same process that created the solar system. Theoretical studies of the Sun and other studies of radioactive isotopes also suggest a similar age.   [B.J.S.]

Earth, gravity field of   The field of gravitational attraction of the Earth. Since, at the Earth's surface, the small centrifugal force due to the Earth's rotation is inseparably superimposed on the attraction, the gravity field is usually understood to include also the effect of the centrifugal force.

The resultant of gravitation (pure attraction) and centrifugal force is called gravity. Gravity is the force that acts on the body at rest with respect to the Earth since the effects of attraction and of centrifugal force cannot be separated because of the equivalence of gravitational and inertial mass; thus gravity determines the weight of a body. The gravity potential W is the sum of the gravitational potential V and the potential of centrifugal force, which is given by a simple analytical expression and may be considered as known.

A body moving with respect to the Earth is also affected by the Coriolis force. Like centrifugal force, the Coriolis force is an inertial force due to the Earth's rotation, but unlike centrifugal force, it does not possess a potential and hence cannot be easily incorporated into the gravity field. Therefore Coriolis force is not considered in the context of terrestrial gravitation. This is perfectly adequate since this force is zero for bodies at rest with respect to the Earth, and almost all measuring systems are at rest. *See* CORIOLIS ACCELERATION.

The gravity vector **g** represents the force of gravity on a unit mass. It is the gradient vector of the gravity potential, $\mathbf{g} = \text{grad } W$. The magnitude of the gravity vector is the intensity of gravity, or briefly, gravity g. The dimension of g is force per unit mass, or acceleration. The SI unit is $m \cdot s^{-2}$. The cgs unit gal (1 gal = $1 \text{ cm} \cdot s^{-2}$), named after Galileo, is still frequently used, especially the milligal (1 mgal = 10^{-3} gal = $10^{-5} m \cdot s^{-2}$). Gravity g on the Earth's surface varies from about 978 gals at the Equator to about 983 gals at the poles. The direction of the gravity vector defines the vertical, or plumb line.

The surfaces of constant gravity potential, $W = $ const, are called equipotential surfaces or level surfaces. The surface of a quiet lake is part of a level surface. So is the surface of the oceans, after some obvious idealization; the whole level surface so defined is called the geoid. After C. F. Gauss, the geoid is considered as the mathematical surface of the Earth, as opposed to the visible topographical surface. The plumb lines intersect the level surfaces orthogonally; they are not quite straight but very slightly curved.

The quantity that is measured most commonly is the gravity g. The determination of g as such is called an absolute gravity measurement. Usually only relative gravity measurements are performed, determining the difference between, or the quotient of, the gravity values at two different points. The direction of the gravity vector, which gives the plumb line in space, is measured by astronomical methods. Differences in the gravity potential W are obtained by geodetic leveling. Finally, certain derivatives of g and similar quantities are measured by instruments such as the torsion balance.

Satellite methods have enormously improved knowledge of the gravity field. In the future, new satellite technologies will advance gravity-related measurements, to be studied along with classical theoretical measurements, which will fully retain their importance.

For most purposes, the Earth's gravitational field may be considered invariable in time. However, it is subject to extremely small periodic variations due to tidal effects. These are caused by the attraction of the Sun and Moon. The attraction acts directly by superimposing itself onto the Earth's gravitational attraction; and it acts also indirectly by slightly deforming the Earth and shifting the waters of the oceans, so that the attracting terrestrial masses themselves are modified.

The lunar effect on gravity attains a maximum of 0.20 mgal, and the solar effect, a maximum of 0.09 mgal; both are well within the measuring accuracy of modern gravimeters. The results of stationary gravimeters recording variations of gravity may be used to draw conclusions as to the elastic behavior of the Earth under the influence of tidal stresses. *See* EARTH TIDES.

Anomalies of the terrestrial gravitational field are caused by mass irregularities. These may be the visible irregularities of topography such as mountains; or they may be invisible subsurface density anomalies. This is the reason why it is possible to use gravity measurements for investigating the underground structure of the Earth's crust. Thus analysis of gravity is applied by geophysicists and geologists for studying general features of the crust, and by exploration geophysicists for searching for shallow density irregularities that might indicate the presence of mineral deposits.

Geodetic instruments employ spirit levels and other devices to orient them with respect to the horizontal or, what amounts to the same thing, to the plumb line. Since the plumb line is defined by the gravitational field, it can be understood why this field enters essentially into almost all geodetic measurements, even into apparently purely geometric ones. In return, geodetic techniques are among the most efficient means for determining the gravitational field. The "mathematical figure of the Earth" for the purpose of geodesists, the geoid, is defined as a surface of constant potential W. "Heights above the sea level" are heights above the geoid; their determination is therefore a physical as well as a geometric problem. (Geodetic theories have been developed which employ only quantities referred to the Earth's topographical surface; but here the gravitational field enters in an even more complicated way.) Thus geodesy is essentially concerned with the Earth's gravitational field and its determination; the theory of the figure of the Earth is to a large extent equivalent to the theory of terrestrial gravitation. *See* GEODESY.   [H.Mor.]

Earth, heat flow in   The Earth's heat flow is defined as the amount of heat escaping per unit time from the interior across each unit area of the Earth's solid surface. The movement of heat within the Earth and its eventual loss through the surface are central elements in the modern theory of plate tectonics. The plate movements over the surface of the Earth are seen as one manifestation of a heat engine at work in the interior, and the heat flow at the surface as the exhaust from the engine. *See* EARTH INTERIOR; PLATE TECTONICS.

Because heat flows from the warmer to the cooler parts of a body, the observation that the Earth's temperature increases with depth clearly implies that heat is escaping from the interior by conduction through the rocky crust. William Thomson (Lord Kelvin) used the heat flow in estimating the age of the Earth. He argued that as the Earth cooled, following its formation and solidification from molten rock, the rate at which the temperature increased with depth (the geothermal gradient) lessened over time, and thus by measuring the geothermal gradient the length of time that the Earth had been cooling could be estimated. Such an estimate is based on the assumption that all the heat being lost is drawn from the endowment of heat that the Earth possessed at the time of its initially molten condition. This assumption was later shown to be untenable because a significant fraction of the Earth's heat flow is derived not from the initial heat but from the decay of radioactive elements present in trace amounts within the Earth. Thus the Earth's heat should be thought of not as a finite quantity acquired at the time of formation and gradually dissipated over time, but as a quantity that in large part has been continually produced, albeit in diminishing amounts throughout the history of the Earth.

The most significant empirical relationship to emerge from heat flow studies is that heat flow generally decreases with increasing crustal age at the site of the measurement. In the oceans the significant age is the time of magmatic emplacement of the basaltic crust at a mid-ocean ridge. On the continents it is usually the age of the last major tectonic, magmatic, or thermal metamorphic event to have affected the measurement site. The pattern of the decrease in heat flow with age differs from ocean to continent. While both settings show a heat flow in the older segments of about 40–45 milliwatts per square meter, the oceanic heat flow shows a greater range over a shorter time interval than does the continental heat flow.

The heat production from radioactive decay contributes significantly to the heat flow from the Earth's interior. In fact, the entire surface heat flow on continents could arise within the crust if the surface concentrations of the important heat-producing isotopes such as are found in granites and gneisses persisted throughout the full thickness of the crust. However, probable lower crustal rocks such as granulites and migmatites have lesser concentrations of these isotopes, and so radiogenic heat production apparently decreases with depth in some manner. The principal heat-producing isotopes are thorium-232 (^{232}Th), uranium-238 (^{238}U), potassium-40 (^{40}K), and uranium-235 (^{235}U) with respective half-lives of 14.0, 4.47, 1.25, and 0.70 × 10^9 years. Other radioactive isotopes do not presently contribute significant heat because their decay chains are not sufficiently energetic, or their abundances are insignificant, or their half-lives are too short. The concentrations of uranium and thorium in rocks are generally in trace amounts measured in parts per million, while potassium is much more abundant, in the range of a few percent, of which a small but well-known fraction is ^{40}K. A range of values for the upper mantle is characteristic of various possible mantle rocks as inferred from xenoliths brought to the surface by igneous and tectonic processes. If the entire mantle had the lesser concentrations, then crustal and mantle radiogenic heat production would be less than half the present-day heat loss; whereas if the greater concentrations were representative of the mantle as a whole, radiogenic heat production would be about equal to the present-day heat loss.

The average heat flow over the entire Earth is 87 mW · m^{-2} and is comparatively a trickle; it is sufficient to bring a thimbleful of water to a boil in about 2 years, or if collected over the area of a football field, would be adequate to light four 100-watt incandescent bulbs. The radiant energy from the Sun intercepted by the Earth is some 4000 times greater than the geothermal flux. The absorption and reradiation of the incident solar energy are the principal processes that determine the surface temperature of the Earth. The geothermal energy is of little significance to the surface temperature but is of paramount importance in considering the Earth's internal thermal condition. See GEOLOGIC THERMOMETRY.   [H.N.P.]

Earth crust   The low-density outermost layer of the Earth above the Mohorovičić discontinuity (the Moho), a global boundary that is defined as the depth in the Earth where the compressional-wave seismic velocity increases rapidly or discontinuously to a value in excess of 4.7 mi/s (7.6 km/s; the upper mantle). The crust is also the cold, upper portion of the Earth's lithosphere, which in terms of plate tectonics is the mobile, outer layer that is underlain by the hot, convecting asthenosphere. See ASTHENOSPHERE; LITHOSPHERE; PLATE TECTONICS.

Continental crust. The Earth's continental crust has evolved over the past 4 billion years, and is highly variable in geologic composition and internal structure. The worldwide mean thickness of continental crust is 24 mi (40 km), with a standard deviation of 5.4 mi (9 km). The thinnest continental crust (found in the Afar Triangle, northeast Africa) is about 9 mi (15 km) thick, and the thickest crust (the Himalayan Mountains in China) is about 47 mi (75 km) thick. Ninety-five percent of all continental crust has a thickness within two standard deviations of the mean thickness, between 13 mi (22 km) and 37 mi (58 km). The Antarctic continent has a crustal thickness of 24 mi (40 km) in the ancient, stable (cratonic) region of East Antarctica, and about 12 mi (20 km) in the recently stretched (extended) crust of West Antarctica. Continental margins, which mark the transition from oceanic to continental crust, range in thickness from about 9 mi (15 km) to 18 mi (30 km). *See* CONTINENTAL MARGIN.

Despite its geologic complexity, the continental crust may generally be divided into four layers: an uppermost sedimentary layer, and an upper, middle, and lower crust composed of crystalline rocks. The sedimentary cover of the continental crust is an important source of natural resources. This cover averages 0.6 mi (1 km) in thickness, and varies in thickness from zero (for example, on shields) to more than 9 mi (15 km) in deep basins. In stable continental crust of average thickness (25 mi or 40 km), the crystalline upper crust is commonly 6–9 mi (10–15 km) thick and has an average composition equivalent to a granite. The middle crust is 3–9 mi (5–15 km) thick and has a composition equivalent to a diorite; and the lower crust is 3–12 mi (5–20 km) thick and has a composition equivalent to a gabbro. Due to increasing temperature and pressure with depth, the metamorphic grade of rocks increases with depth, and the rocks within the deep continental crust generally are metamorphic rocks, even if they originated as sedimentary or igneous rocks. *See* DIORITE; GABBRO; GRANITE; METAMORPHIC ROCKS.

Crustal properties vary systematically with geologic setting, which may be divided into six groups: orogens (mountain belts), shields and platforms, island arcs (volcanic arcs), continental magmatic arcs, rifts, extended (stretched) crust, and forearcs. Orogens are typified by thick crust [average thickness is 29 mi (46 km), but the maximum thickness is as much as 47 mi (75 km) in the Himalayas]. Shields and platforms, such as the Canadian Shield and the Russian Platform, commonly have an approximately 26-mi-thick (42-km) crust, including a 3–6 mi-thick (5–10 km) lower crust. In comparison with shields, island arcs (such as Japan) have thinner crusts and significantly shallower middle and lower crustal layers due to the intrusion of mafic (that is, low silica content) plutons. Continental magmatic arcs, such as the Cascades volcanoes of the northwestern United States, intrude preexisting continental crust, and therefore they are generally 3–9 mi (10–15 km) thicker than island arcs. Continental rifts, such as the East African and Rio Grande rifts, have an average crustal thickness of about 22 mi (36 km). Extended continental crust, such as the Basin and Range Province of the western United States, averages 18 mi (30 km) in thickness. Forearcs are regions that were formed oceanward of volcanic arcs, such as much of the west coast of North America. They typically have thin crust, about 15 mi (25 km), and have a thick (9 mi or 15 km) upper crustal section that consists of relatively low-density metasedimentary rocks. *See* NORTH AMERICA; OCEANIC ISLANDS; PLUTON; RIFT VALLEY; SEDIMENTARY ROCKS; VOLCANO.

At least three processes provide new continental crust. The first is the accretion and consolidation of island arcs, such as Japan or the Aleutian Islands, onto a continental margin. The second process is the tectonic underplating of oceanic crust at active subduction zones. In this process, the continental crust grows from below as oceanic crust is welded to the base of the continental margin, either when subduction stops or when subduction steps oceanward and a new trench is formed. This process has been identified in western Canada and southern Alaska. The third process is the magmatic inflation of the crust at continental arcs, rifts, and regions of crustal extension. This process has been identified in many regions. *See* GEODYNAMICS.   [W.D.Mo.]

Oceanic crust. The surface of the ocean crust, except for some locally high volcanoes and plateaus, resides some 1–3 mi (2–5 km) below sea level, and about another kilometer below the average level of the continents. The ocean crust represents the youngest and geologically most dynamic portion of the Earth's surface. Most of it was produced at mid-ocean ridges during the process of sea-floor spreading. The ridges define the trailing edges, or accreting boundaries, of the major lithospheric plates that are moving about the surface of the Earth at present. Thus, the oldest rocks of the ocean crust date back no earlier than the rifting episodes that created most of these plates and initiated the most recent phase of continental drift, the Pangaean breakup, in Late Jurassic times. *See* JURASSIC; MID-OCEANIC RIDGE.

There are fault slices of types of ocean crust on land, known as ophiolites, where nearly or entirely complete cross sections through the crust can be mapped and sampled. These strongly indicate that the ocean crust consists in downward sequence of submarine extrusives (usually pillow basalts), feeder dikes (often vertically sheeted), or sills, gabbros, and peridotites. There is much uncertainty, however, about the extent to which typical ophiolites, most of which formed in island-arc or backarc environments, can represent abyssal ocean crust, which is produced at the major accreting plate boundaries. Moreover, the physical correspondence of the rocks in ophiolites to ocean crust is often complicated by their complex structure and extent of alteration and metamorphism, particularly in the ultramafic sections.   [J.H.Na.]

Earth interior   All of the Earth beneath the land surface and the ocean bottom, including the crust, the mantle, and the core. The interior is not accessible to direct observation. Nevertheless, a rather detailed model has been constructed on the basis of measurements made at or above the surface. Measurements of gravity, the geomagnetic field, surface heat flow, and surface deformation can all be used to put constraints on the Earth model, but the most detailed information about the interior is provided by seismic measurements. In the exploration of the Earth's interior, the seismic waves being analyzed are usually generated by earthquakes, and measurements are made of waves propagating through the interior of the body, waves propagating along the surface, and standing waves bringing the whole Earth into a state of oscillation. Such measurements, when properly interpreted, provide information about seismic-wave velocities in the Earth. Seismic-wave velocities can also be measured in laboratory experiments where rock samples are subjected to the high pressures and temperatures typical of conditions in the deep interior.

Several standard Earth models have been constructed in the past. One of the more comprehensive models was constructed in 1981 by A. Dziewonski and D. Anderson, the Preliminary Reference Earth Model (PREM). The principal regions of this model are listed in the table. *See* EARTH.

Principal regions of a standard Earth model

Layer	Approximate depth range,* mi (km)
(1) Ocean layer	0–1.8 (0–3)
(2) Upper and lower crust	1.8–15 (3–24)
(3) Lithosphere below the crust	15–50 (24–80)
(4) Asthenosphere (low-velocity zone)	50–140 (80–220)
(5) Upper mantle above phase or compositional changes near 240 mi (400 km)	140–240 (220–400)
(6) Transition region between phase or compositional changes near 240 and 416 mi (400 and 670 km)	240–416 (400–670)
(7) Lower mantle above core-mantle boundary layer	416–1703 (670–2741)
(8) Core-mantle boundary layer (D″)	1703–1796 (2741–2891)
(9) Outer core	1796–3200 (2891–5150)
(10) Inner core	3200–3959 (5150–6371)

*Depth ranges are uncertain, especially in the crust and upper mantle.

Seismic structure. The seismic structure of the Earth is studied by analyzing body waves, surface waves and free oscillations, and anisotropy. Body waves are generated by earthquakes or large (nuclear) explosions and are recorded by many seismic stations throughout the world. Body waves are characterized as P (primary) and S (secondary) waves. Both wave types are supported by a solid, but P waves have a higher velocity and arrive first. It is routine practice that the stations report the arrival time of the first P wave to an earthquake information center, where these data are used to locate the earthquake and determine its origin time. By subtracting the origin time from the arrival times, the travel times of waves from the earthquake to the stations are determined, and these travel times can be used to determine the seismic velocity structure of the Earth's interior. *See* EARTHQUAKE; SEISMOLOGY.

Surface waves and free oscillations are also generated by earthquakes. The velocity with which surface waves propagate is a function of the frequency of the waves, that is, the surface waves are dispersive. The dispersion depends on the seismic velocity structure of the surface layers. The free oscillations of the Earth resonate only at certain discrete frequencies (the eigenfrequencies), that is, the free oscillation spectrum would be a line spectrum (in practice the lines are somewhat smeared because of dissipation of the oscillations). The eigenfrequencies depend on the velocity structure of the whole Earth. Thus surface waves and free oscillations provide an alternative data set for constructing velocity models. PREM is based on both travel times and free-oscillation data.

Anisotropy refers to the directional dependence of the wave velocity. Anisotropy in the Earth's interior may be induced by flow of material containing noncubic crystals or elongated grains. Anisotropy has been observed in the lithosphere, especially under the oceans, and has been postulated for the asthenosphere. This anisotropy has been explained in terms of alignment of olivine crystals, which are thought to be the dominant material in these layers. Anisotropy may occur in deeper regions of the Earth as well, but is difficult to resolve. However, to account for the differences in travel time between P waves passing through the inner core in the equatorial and polar directions, respectively, it has been proposed that the inner core is anisotropic, although the physical mechanism is not understood. *See* LITHOSPHERE; OLIVINE.

Composition and state. The Earth comprises a crust, a mantle, and a core, so there is a compositional differentiation into at least three regions. Each of these regions is differentiated again, both vertically and, at least for the crust and upper part of the mantle, laterally. *See* ELEMENTS, GEOCHEMICAL DISTRIBUTION OF.

The Earth's crust is on average about 12 mi (20 km) thick but is thinner under oceans and thicker under continents. Thickness under some high mountain ranges may be up to 36 mi (60 km; the so-called mountain roots). The boundary between the crust and the underlying mantle is called the Mohorovičić discontinuity or, more usually, the Moho. The continental and oceanic portions of the crust have different compositions. Continental crust can be divided into a lighter, granitic upper crust and a lower crust that is more basaltic, like the oceanic crust. Despite the striking heterogeneity of the crust, some generalizations can be made. Both the crust and the mantle are rich in silicates, but the mantle has much more magnesium, and the crust has proportionally more of the relatively light elements such as calcium, aluminum, sodium, and potassium in the form of oxides. *See* EARTH CRUST.

The mantle is the region between the crust and the core. It is common to make a distinction between the upper mantle and lower mantle, which are separated by the so-called 670-km discontinuity at a depth of 416 mi. The upper mantle includes the lithosphere below the crust, the asthenosphere, and the region of phase transformations (the transition zone). It contains also the subducting slabs of oceanic lithosphere. The difference between continental and oceanic crust extends to at least the bottom of the lithosphere and probably to the bottom of the asthenosphere. It is believed that the stiff lithosphere and the soft asthenosphere differ mainly in mechanical strength, not in composition. The shallow part of the upper mantle is likely to be relatively depleted in the light elements typical of the crust. On the other hand, the light elements do not seem to be underrepresented in estimates of the composition of the deeper parts of the upper mantle. It has been suggested that this region contains more eclogite, the high-pressure form of basalt, which is subducted with the oceanic lithosphere. Olivine is probably the most abundant upper-mantle mineral. At pressures prevailing near 240 mi (400 km) depth, it transforms to a denser structure (spinel). *See* BASALT; ECLOGITE; SPINEL.

The lower mantle, the mantle below 416 mi (670 km), appears to have a homogeneous composition down to the boundary layer above the core. Experimental results seem to indicate that mineralogy is dominated by ferromagnesium oxides (wüstite) and silicates (perovskite). The high-pressure perovskite phase of ferromagnesium silicate is thought to be the most abundant mineral of the Earth. *See* PEROVSKITE.

The core is the central region of the Earth. Its seismic and density structure are well matched by a solid inner core of iron or nickel iron and a liquid outer core of iron mixed with about 10% lighter material. Candidate light elements are sulfur and oxygen, with some preference for the former. The outer core can be taken as homogeneous for all practical purposes.

Core-mantle coupling is believed to be responsible for decade-long variations in the length of the day of up to about 5 milliseconds. The idea is that fluid flow in the outermost regions of the core is coupled to the mantle, so that the rotation rate of the mantle can fluctuate in response to time variations of the flow in the core. This view is supported by an observed correlation between the decade variations in the length of the day and certain variations of the geomagnetic field (the latter being associated with core flow). The coupling may be electromagnetic or topographic. It has been suggested that topographic coupling can be quite effective, provided that there is indeed topography on the core-mantle boundary (of the order of a few hundred meters). *See* GEOMAGNETISM.   [D.J.D.]

Earth sciences   Sciences that involve attempts to understand the nature, origin, evolution, and behavior of the Earth or of its parts and to comprehend its place in the universe, especially in the solar system. Understanding has advanced primarily through improved appreciation of the complex, usually cyclical interactions that take place among distinct parts of the Earth such as the lithosphere, atmosphere, hydrosphere, and biosphere. Geophysics is the study of the physics of the Earth, emphasizing its physical structure and dynamics. Geochemistry is the study of the chemistry of the Earth, dealing with its composition and chemical change. Geology is the study of the solid Earth and of the processes that have formed and modified it throughout its 4.5-billion-year history. *See* GEOCHEMISTRY; GEODESY; GEOLOGY; GEOPHYSICS.

Many branches of geology are considered separate sciences. Mineralogy is the study of the composition, structure, and properties of minerals. Petrology involves understanding how rocks originate and evolve, as well as rock description and classification. Specialties related to petrology include sedimentology and volcanology. Stratigraphy is the study of the origin, age, and development of layered, generally sedimentary rocks. Paleontology is the study of ancient (fossil) life. Historical geology is the study of the evolution of the Earth and its life. Geomorphology is the study of landscapes and their evolution. Seismology is the study of earthquakes and their effects. Structural geology is the study of deformed rocks. Engineering geology relates to the support of human constructions by underlying rock. *See* ENGINEERING GEOLOGY; GEOLOGY; GEOMORPHOL-OGY; HYDROLOGY; MINERALOGY; PALEONTOLOGY; PETROGRAPHY; PETROLOGY; SEISMOLOGY; STRATIGRAPHY; STRUCTURAL GEOLOGY; VOLCANOLOGY.

Oceanography is the study of the oceans; limnology, the study of lakes; hydrology, the study of underground and surface water; and glaciology, the study of glaciers, ice caps, and ice sheets. These disciplines address the study of water in and on the Earth. The gaseous outer parts of the planet are the province of the atmospheric sciences, including meteorology, which is concerned with the weather and weather forecasting; climatology, which deals with longer-term and regional variations; and aeronomy which, because it deals with the outermost ionized region of the atmosphere, is much concerned with solar terrestrial interactions, including the aurora borealis and aurora australis. The biosphere embodies all life on Earth, and its study includes molecular biology, zoology, botany, and ecology. Geography, the study of all that happens at the Earth's surface, has been distinct insofar as it has encompassed not only physical and biological sciences but also the social sciences, including aspects of political science and economics. This distinction is fading rapidly as other earth sciences become more involved with social considerations.   [K.Bu.]

Earth tides   Cyclic motions of the Earth, sometimes over a foot or so in height, depending on latitude, caused by the same lunar and solar forces which produce tides in the sea. These forces also react on the Moon and Sun, and thus are significant in astronomy in evaluations of the dynamics of the three bodies. For example, the secular spin-down of the Earth due to lunar tidal torques is best computed from the observed acceleration of the Moon's orbital velocity. In oceanography, earth tides and ocean tides are very closely related. *See* GEODESY; TIDE.

By far the most widely used earth tide instruments are the tiltmeter and the gravimeter. Both instruments have the merits of portability, high potential precision, and low cost. Thus they are able to advance economically an important mission—the global mapping of earth tides and ocean tides.   [L.B.S./J.T.K.]

Earthquake   The sudden movement of the Earth caused by the abrupt release of accumulated strain along a fault in the interior. The released energy passes through the Earth as seismic waves (low-frequency sound waves), which cause the shaking. Seismic waves continue to travel through the Earth after the fault motion has stopped. Recordings of earthquakes, called seismograms, illustrate that such motion is recorded all over the Earth for hours, and even days, after an earthquake.

Earthquakes are not distributed randomly over the globe but tend to occur in narrow, continuous belts of activity. Approximately 90% of all earthquakes occur in these belts, which define the boundaries of the Earth's plates. The plates are in continuous motion with respect to one another at rates on the order of centimeters per year; this plate motion is responsible for most geological activity.

Plate motion occurs because the outer cold, hard skin of the Earth, the lithosphere, overlies a hotter, soft layer known as the asthenosphere. Heat from decay of radioactive minerals in the Earth's interior sets the asthenosphere into thermal convection. This convection has broken the lithosphere into plates which move about in response to the convective motion. As the plates move past each other, little of the motion at their boundaries occurs by continuous slippage; most of the motion occurs in a series of rapid jerks. Each jerk is an earthquake. This happens because, under the pressure and temperature conditions of the shallow part of the Earth's lithosphere, the frictional sliding of rock exhibits a property known as stick-slip, in which frictional sliding occurs in a series of jerky movements, interspersed with periods of no motion—or sticking.

In the geologic time frame, then, the lithospheric plates chatter at their boundaries, and at any one place the time between chatters may be hundreds of years. *See* PLATE TECTONICS.

The periods between major earthquakes is thus one during which strain slowly builds up near the plate boundary in response to the continuous movement of the plates. The strain is ultimately released by an earthquake when the frictional strength of the plate boundary is exceeded. *See* FAULT AND FAULT STRUCTURES.

Most great earthquakes occur on the boundaries between lithospheric plates and arise directly from the motions between the plates. These may be called plate boundary earthquakes. There are many earthquakes, sometimes of substantial size, that cannot be related so simply to the movements of the plates. At many plate boundaries, earthquakes occur over a broad zone—often several hundred miles wide—adjacent to the plate boundary. These earthquakes, which may be called plate boundary-related earthquakes, are secondarily caused by the stresses set up at the plate boundary. Some earthquakes also occur, although infrequently, within plates. These earthquakes, which are not related to plate boundaries, are called intraplate earthquakes. The immediate cause of intraplate earthquakes is not understood.

In addition to the tectonic types of earthquakes described above, some earthquakes are directly associated with volcanic activity. These volcanic earthquakes result from the motion of undergound magma that leads to volcanic eruptions.

Earthquakes often occur in well-defined sequences in time. Tectonic earthquakes are often preceded, by a few days to weeks, by several smaller shocks (foreshocks), and are nearly always followed by large numbers of aftershocks. Foreshocks and aftershocks are usually much smaller than the main shock. Volcanic earthquakes often occur in flurries of activity, with no discernible main shock. This type of sequence is called a swarm.

Earthquakes range enormously in size, from tremors in which slippage of a few tenths of an inch occurs on a few feet of fault, to the greatest events, which may involve a rupture many hundreds of miles long, with tens of feet of slip.

The size of an earthquake is given by its moment: average slip times the fault area that slipped times the elastic constant of the Earth. The units of seismic moment are dyne-centimeters. An older measure of earthquake size is magnitude, which is proportional to the logarithm of moment. Magnitude 2.0 is about the smallest tremor that can be felt. Most destructive earthquakes are greater than magnitude 6; the largest shock known was the 1960 Chile earthquake, with a moment of 10^{30} dyne-centimeters (10^{23} newton-meters) or magnitude 9.5. It involved a fault 600 mi (1000 km) long slipping 30 ft (10 m).

The intensity of an earthquake is a measure of the severity of shaking and its attendant damage at a point on the surface of the Earth. The same earthquake may therefore have different intensities at different places. The intensity usually decreases away from the epicenter (the point on the surface directly above the onset of the earthquake), but its value depends on many factors and generally increases with moment. Intensity is usually higher in areas with thick alluvial cover or landfill than in areas of shallow soil or bare rock. Poor building construction leads to high intensity ratings because the damage to structures is high. Intensity is therefore more a measure of the earthquake's effect on humans than an innate property of the earthquake.

Many additional effects may be produced by earthquake shaking, including landslides and tsunamis. *See* LANDSLIDE; TSUNAMI.

Earthquake prediction research has been going on for nearly a century. Unfortunately, successful earthquake predictions are extremely rare. There are two basic categories of earthquake predictions: forecasts (months to years in advance) and short-term predictions (hours or days in advance). Forecasts are based a variety of research, including the history of earthquakes in a specific region, the identification of fault characteristics (including length, depth, and segmentation), and the identification of strain accumulation. Data from these studies are used to provide rough estimates of earthquake sizes and recurrence intervals.   [C.H.S.; K.M.S.]

East Indies   A loosely defined region in southeast Asia comprising the countries of Malaysia, Brunei, and Indonesia. The islands of the East Indies extend for about 2800 mi (4500 km) from western Sumatra to New Guinea. They form part of a region of great geological and biological diversity.

The three broad geographic subdivisions can be made: (1) Sundaland, which comprises the shallow marine Sunda Shelf, peninsular southeast Asia, and the islands west of the Makassar and Lombok straits; (2) northern Australia, which includes the shallow marine Sahul and Arafura shelf areas and islands surrounding northern Australia and New Guinea; and (3) Wallacea, which comprises the islands south of the Philippines between Sundaland and northern Australia. Each division is an important geomorphological and biological region. The boundaries of Sundaland and northern Australia can be drawn at the 660-ft (200-m) bathymetric contour. Deep trenches with depths averaging 5 mi (8 km) border the margins of Sundaland with the Indian Ocean, and northern Australia with the Pacific Ocean. In contrast, within the area of Wallacea there are a number of deep troughs, and there are large variations in relief with volcanic and nonvolcanic mountains, typically up to 6600–9900 ft (2000–3000 m) in height, separated by deep marine basins underlain by extended continental and oceanic crust with depths of several kilometers. *See* ASIA; AUSTRALIA.

At present, almost all the islands lie in a belt close to the Equator within the Intertropical Convergence Zone (ITCZ). An equatorial climate prevails, with high rainfall and, except at higher elevations, high temperatures throughout the year. Diurnal variations are greater than the difference of mean temperatures of the hottest and coldest months. High relative humidity is normal in most lowland regions. Regionally, there are significant variations in rainfall, reflecting topography and position with respect to major landmasses and oceans, and each island's climate can be different. Borneo is the only large island within which there is broadly an ever-wet tropical climate. *See* CLIMATOLOGY; TROPICAL METEOROLOGY.

The shallow seas with narrow deep-water passages of Wallacea mean that the region is particularly important for oceanic circulation. The Indonesian throughflow is a current to the Indian Ocean which transports large amounts of warm water from the Pacific, influencing sea surface temperatures, salinity, and rainfall. The magnitude and variations of this current are important controls on the thermohaline balance of the Pacific and Indian oceans, and perhaps on global thermohaline circulation. Most water passes from the North Pacific via the Celebes Sea, Makassar Strait, Flores Sea, and Banda Sea. *See* EQUATORIAL CURRENTS; OCEAN CIRCULATION.

The East Indies are characterized by intensely active seismicity and volcanic activity. The correlation of seismicity, volcanicity, deep trenches, and strong negative gravity anomalies along the Sunda and Banda arcs was noted long before the formulation of the theory of plate tectonics; these are now known to be features characteristic of the subduction of oceanic lithosphere. The history of convergence of the Pacific, Indian-Australian, and Asian plates offers a broad explanation of the geological development and complexity of the region, but many small plates also need to be considered. However, the full details of the development of the region are still far from understood because of its size, relative inaccessibility, and the nature of the terrain. *See* CONTINENTAL DRIFT; GEODYNAMICS; PLATE TECTONICS.

The entire region is immensely rich in natural resources, in particular petroleum, minerals, and timber. Petroleum has been produced in large quantities on land in Sumatra, Borneo, and the Bird's Head of New Guinea since the mid-twentieth century. Most oil and gas provinces are in Cenozoic basins, and in the late 1990s the East Indies provided about 5% of annual world oil production. In recent years, exploration and production has moved offshore and is increasingly moving into deeper waters. Many parts of the region have considerable potential for geothermal energy production. Mineral production has been historically important, with major discoveries of tin, gold, and nickel. The late-twentieth-century discovery of major mineral deposits in the northern New Guinea margin suggests that the young island arcs of the region will continue to be targets for exploration, and large deposits are likely to be found both on and off shore.

[R.Hai.]

Echo sounder   A marine instrument used primarily for determining the depth of water by means of an acoustic echo. A pulse of sound sent from the ship is reflected from the sea bottom back to the ship, the interval of time between transmission and reception being proportional to the depth of the water. An echo sounder is really a type of active sonar. It consists of a transducer located near the keel of the ship which serves (in most models) as both the transmitter and receiver of the acoustic signal; the necessary oscillator, receiver, and amplifier which generate and receive the electrical impulses to and from the transducer; and a recorder or other indicator which is calibrated in terms of the depth of water.

Echo sounders, sometimes called fathometers, are used by vessels for navigational purposes, not only to avoid shoal water but as an aid in fixing position when a good bathymetric chart of the area is available. Some sensitive instruments are used by commercial fishers or marine biologists to detect schools of fish or scattering layers of minute marine life. Oceanographic survey ships use echo sounders for charting the ocean bottom. *See* SCATTERING LAYER; UNDERWATER SOUND.   [R.W.Mo.]

Eclogite   A very dense rock composed of red-brown garnet and the grape-green pyroxene omphacite. Eclogites possess basaltic bulk chemistry, and their garnets are rich in the components pyrope, almandine, and grossular, while the pyroxenes are rich in jadeite and diopside.

Eclogite occurrences may be subdivided into three broad categories: Group a eclogites are found as layers, lenses, or boudins in schists and gneisses seemingly of the amphibolite facies. Quartz, together with zoisite or kyanite, commonly occurs in these rocks. Amphibole of barroisitic composition may also be present. Group b eclogites are found as inclusions in kimberlites and basalts. They are frequently accompanied by xenoliths of garnet peridotite. Group c eclogites are found as blocks and lenses in schists of the glaucophane-schist facies. Such eclogites do not contain kyanite, rarely contain quartz, but bear amphibole, epidote, rutile, or sphene.

Eclogite is the name given to the highest-pressure facies of metamorphism; the critical mineral assemblage defining this facies is garnet + omphacite, together with kyanite or quartz in rocks of basaltic composition. Where sedimentary and granitic rocks have been metamorphosed under eclogite facies conditions, they result in spectacular omphacite + garnet + quartz-bearing mica schists and metagranitic gneisses such as are found in the Sezia-Lanzo zone of the western Alps. *See* METAMORPHIC ROCKS.

The high density of eclogite, together with its elastic properties, makes it a candidate for upper mantle material. Large quantifies of basaltic oceanic crust are returned to the mantle through the process of subduction, where prevailing high pressures convert it to eclogite. The quantity and distribution of eclogite within the mantle is not known;

that it occurs is known from the nodules brought up in kimberlite pipes and in basalts.

[T.J.B.H.]

Ecological communities   Assemblages of living organisms that occur together in an area. The nature of the forces that knit these assemblages into organized systems and those properties of assemblages that manifest this organization have been topics of intense debate among ecologists since the beginning of the twentieth century. On the one hand, there are those who view a community as simply consisting of species with similar physical requirements, such as temperature, soil type, or light regime. The similarity of requirements dictates that these species be found together, but interactions between the species are of secondary importance and the level of organization is low. On the other hand, there are those who conceive of the community as a highly organized, holistic entity, with species inextricably and complexly linked to one another and to the physical environment, so that characteristic patterns recur, and properties arise that one can neither understand nor predict from a knowledge of the component species. In this view, the ecosystem (physical environment plus its community) is as well organized as a living organism, and constitutes a superorganism. Between these extremes are those who perceive some community organization but not nearly enough to invoke images of holistic superorganisms. *See* ECOSYSTEM.

Every community comprises a given group of species, and their number and identities are distinguishing traits. Most communities are so large that it is not possible to enumerate all species; microorganisms and small invertebrates are especially difficult to census. However, particularly in small, well-bounded sites such as lakes or islands, one can find all the most common species and estimate their relative abundances. The number of species is known as species richness, while species diversity refers to various statistics based on the relative numbers of individuals of each species in addition to the number of species. The rationale for such a diversity measure is that some communities have many species, but most species are rare and almost all the individuals (or biomass) in such a community can be attributed to just a few species. Such a community is not diverse in the usual sense of the word. Patterns of species diversity abound in the ecological literature; for example, pollution often effects a decrease in species diversity.

The main patterns of species richness that have been detected are area and isolation effects, successional gradients, and latitudinal gradients. Larger sites tend to have more species than do small ones, and isolated communities (such as those on oceanic islands) tend to have fewer species than do less isolated ones of equal size. Later communities in a temporal succession tend to have more species than do earlier ones, except that the last (climax) community often has fewer species than the immediately preceding one. Tropical communities tend to be very species-rich, while those in arctic climates tend to be species-poor. This observation conforms to a larger but less precise rule that communities in particularly stressful environments tend to have few species.

Communities are usually denoted by the presence of species, known as dominants, that contain a large fraction of the community's biomass, or account for a large fraction of a community's productivity. Dominants are usually plants. Determining whether communities at two sites are truly representatives of the "same" community requires knowledge of more than just the dominants, however. "Characteristic" species, which are always found in combination with certain other species, are useful in deciding whether two communities are of the same type, though the designation of "same" is arbitrary, just as is the designation of "dominant" or "characteristic."

Communities often do not have clear spatial boundaries. Occasionally, very sharp limits to a physical environmental condition impose similarly sharp limits on a community. For example, serpentine soils are found sharply delimited from adjacent soils in many areas, and have mineral concentrations strikingly different from those of the neighboring soils. Thus they support plant species that are very different from those found in nearby nonserpentine areas, and these different plant species support animal species partially different from those of adjacent areas.

Here two different communities are sharply bounded from each other. Usually, however, communities grade into one another more gradually, through a broad intermediate region (an ecotone) that includes elements of both of the adjacent communities, and sometimes other species as well that are not found in either adjacent community. *See* ECOTONE.

The environment created by the dominant species, by their effects on temperature, light, humidity, and other physical factors, and by their biotic effects, such as allelopathy and competition, may entrain some other species so that these other species' spatial boundaries coincide with those of the dominants. *See* POPULATION ECOLOGY.

More or less distinct communities tend to follow one another in rather stylized order. As with recognition of spatial boundaries, recognition of temporal boundaries of adjacent communities within a sere (a temporary community during a successional sequence at a site) is partly a function of the expectations that an observer brings to the endeavor. Those who view communities as superorganisms are inclined to see sharp temporal and spatial boundaries, and the perception that one community does not gradually become another community over an extended period of time confirms the impression that communities are highly organized entities, not random collections of species that happen to share physical requirements. However, this superorganismic conception of succession has been replaced by an individualistic succession. Data on which species are present at different times during a succession show that there is not abrupt wholesale extinction of most members of a community and concurrent simultaneous colonization by most species of the next community. Rather, most species within a community colonize at different times, and as the community is replaced most species drop out at different times. That succession is primarily an individualistic process does not mean that there are not characteristic changes in community properties as most successions proceed. Species richness usually increases through most of the succession, for example, and stratification becomes more highly organized and well defined. A number of patterns are manifest in aspects of energy flow and nutrient cycling. *See* ECOLOGICAL SUCCESSION.

Living organisms are characterized not only by spatial and temporal structure but by an apparent purpose or activity termed teleonomy. In the first place, the various species within a community have different trophic relationships with one another. One species may eat another, or be eaten by another. A species may be a decomposer, living on dead tissue of one or more other species. Some species are omnivores, eating many kinds of food; others are more specialized, eating only plants or only animals, or even just one other species. These trophic relationships unite the species in a community into a common endeavor, the transmission of energy through the community. This energy flow is analogous to an organism's mobilization and transmission of energy from the food it eats.

By virtue of differing rates of photosynthesis by the dominant plants, different communities have different primary productivities. Tropical forests are generally most productive, while extreme environments such as desert or alpine conditions harbor rather unproductive communities. Agricultural communities are intermediate. Algal communities in estuaries are the most productive marine communities, while open ocean communities are usually far less productive. The efficiency with which various animals ingest and assimilate the plants and the structure of the trophic web determine the secondary productivity (production of organic matter by animals) of a community. Marine secondary productivity generally exceeds that of terrestrial communities.

A final property that any organism must have is the ability to reproduce itself. Communities may be seen as possessing this property, though the sense in which they do so does not support the superorganism metaphor. A climax community reproduces itself through time simply by virtue of the reproduction of its constituent species, and may also be seen as reproducing itself in space by virtue of the propagules that its species transmit to less mature communities. For example, when a climax forest abuts a cutover field, if no disturbance ensues, the field undergoes succession and eventually becomes a replica of the adjacent forest. Both temporally and spatially, then, community reproduction is a collective rather than an emergent property, deriving directly from the reproductive activities of the component species. See ALTITUDINAL VEGETATION ZONES; BOG; DESERT; ECOLOGY.   [D.Sim.]

Ecological succession   A directional change in an ecological community. Populations of animals and plants are in a dynamic state. Through the continual turnover of individuals, a population may expand or decline depending on the success of its members in survival and reproduction. As a consequence, the species composition of communities typically does not remain static with time. Apart from the regular fluctuations in species abundance related to seasonal changes, a community may develop progressively with time through a recognizable sequence known as the sere. Pioneer populations are replaced by successive colonists along a more or less predictable path toward a relatively stable community. This process of succession results from interactions between different species, and between species and their environment, which govern the sequence and the rate with which species replace each other. The rate at which succession proceeds depends on the time scale of species' life histories as well as on the effects species may have on each other and on the environment which supports them. In some cases, seres may take hundreds of years to complete, and direct observation at a given site is not possible. Adjacent sites may be identified as successively older stages of the same sere, if it is assumed that conditions were similar when each seral stage was initiated. See ECOLOGICAL COMMUNITIES; POPULATION ECOLOGY.

The course of ecological succession depends on initial environmental conditions. Primary succession occurs on novel areas such as volcanic ash, glacial deposits, or bare rock, areas which have not previously supported a community. In such harsh, unstable environments, pioneer colonizing organisms must have wide ranges of ecological tolerance to survive. In contrast, secondary succession is initiated by disturbance such as fire, which removes a previous community from an area. Pioneer species are here constrained not by the physical environment but by their ability to enter and exploit the vacant area rapidly.

As succession proceeds, many environmental factors may change through the influence of the community. Especially in primary succession, this leads to more stable, less severe environments. At the same time interactions between species of plant tend to intensify competition for basic resources such as water, light, space, and nutrients. Successional change results from the normal complex interactions between organism and environment which lead to changes in overall species composition. Whether succession is promoted by changing environmental factors or competitive interactions, species composition alters in response to availability of niches. Populations occurring in the community at a point in succession are those able to provide propagules (such as seeds) to invade the area, being sufficiently tolerant of current environmental conditions, and able to withstand competition from members of other populations present at the same stage. Species lacking these qualities either become locally extinct or are unable to enter and survive in the community.

Early stages of succession tend to be relatively rapid, whereas the rates of species turnover and soil changes become slower as the community matures. Eventually an approximation to the steady state is established with a relatively stable community, the nature of which has aroused considerable debate. Earlier, the so-called climax vegetation was believed to be determined ultimately by regional climate and, given sufficient time, any community in a region would attain this universal condition. This unified concept of succession, the monoclimax hypothesis, implies the ability of organisms progressively to modify their environment until it can support the climatic climax community. Although plants and animals do sometimes ameliorate environmental conditions, evidence suggests overwhelmingly that succession has a variety of stable end points. This hypothesis, known as the polyclimax hypothesis, suggests that the end point of a succession depends on a complex of environmental factors that characterize the site, such as parent material, topography, local climate, and human influences.

Actions of the community on the environment, termed autogenic, provide an important driving force promoting successional change, and are typical of primary succession where initial environments are inhospitable. Alternatively, changes in species composition of a community may result from influences external to the community called allogenic.

Whereas intrinsic factors often result in progressive successional changes, that is, changes leading from simple to more complex communities, external (allogenic) forces may induce retrogressive succession, that is, toward a less mature community. For example, if a grassland is severely overgrazed by cattle, the most palatable species will disappear. As grazing continues, the grass cover is reduced, and in the open areas weeds characteristic of initial stages of succession may become established.

In some instances of succession, the food web is based on photosynthetic organisms, and there is a slow accumulation of organic matter, both living and dead. This is termed autotrophic succession. In other instances, however, addition of organic matter to an ecosystem initiates a succession of decomposer organisms which invade and degrade it. Such a succession is called heterotrophic. *See* EUTROPHICATION; FOOD WEB.

Observed changes in the structure and function of seral communities result from natural selection of individuals within their current environment. Three mechanisms by which species may replace each other have been proposed; the relative importance of each apparently depends on the nature of the sere and stage of development.

1. The facilitation hypothesis states that invasion of later species depends on conditions created by earlier colonists. Earlier species modify the environment so as to increase the competitive ability of species which are then able to displace them. Succession thus proceeds because of the effects of species on their environment.

2. The tolerance hypothesis suggests that later successional species tolerate lower levels of resources than earlier occupants and can invade and replace them by reducing resource levels below those tolerated by earlier occupants. Succession proceeds despite the resistance of earlier colonists.

3. The inhibition hypothesis is that all species resist invasion of competitors and are displaced only by death or by damage from factors other than competition. Succession proceeds toward dominance by longer-lived species.

None of these models of succession is solely applicable in all instances; indeed most examples of succession appear to show elements of all three replacement mechanisms.

Succession has traditionally been regarded as following an orderly progression of changes toward a predictable end point, the climax community, in equilibrium with the prevailing environment. This essentially deterministic view implies that succession will always follow the same course from a given starting point and will pass through a recognizable series of intermediate states. In contrast, a more recent view of succession is based on adaptations of independent species. It is argued that succession is disorderly and unpredictable, resulting from probabilistic processes such as invasion of propagules and survival of individuals which make up the community. Such a stochastic view reflects the inherent variability observed in nature and the uncertainty of environmental conditions. In particular, it allows for succession to take alternative pathways and end points dependent on the chance outcome of interactions among species and between species and their environment.

Consideration of community properties such as energy flow supports the view of succession as an orderly process. The rate of gross primary productivity typically becomes limited also by the availability of nutrients, now incorporated within the community biomass, and declines to a level sustainable by release from decomposer organisms. Species diversity tends to rise rapidly at first as successive invasions occur, but declines again with the elimination of the pioneer species by the climax community.

Stochastic aspects of succession can be represented in the form of models which allow for transitions between a series of different "states." Such models, termed Markovian models, can apply at various levels: plant-by-plant replacement, changes in tree size categories, or transitions between whole communities. A matrix of replacement probabilities defines the direction, pathway, and likelihood of change, and the model can be used to predict the future composition of the community from its initial state.   [P.Ran.]

Ecology   The subdiscipline of biology that concentrates on the relationships between organisms and their environments; it is also called environmental biology. Ecology is concerned with patterns of distribution (where organisms occur) and with patterns of abundance (how many organisms occur) in space and time. It seeks to explain the factors that determine the range of environments that organisms occupy and that determine how abundant organisms are within those ranges. It also emphasizes functional interactions between co-occurring organisms. In addition to being a unique component of the biological sciences, ecology is both a synthetic and an integrative science since it often draws upon information and concepts in other sciences, ranging from physiology to meteorology, to explain the complex organization of nature.

Environment is all of those factors external to an organism that affect its survival, growth, development, and reproduction. It can be subdivided into physical, or abiotic, factors, and biological, or biotic, factors. The physical components of the environment include all nonbiological constituents, such as temperature, wind, inorganic chemicals, and radiation. The biological components of the environment include the organisms. A somewhat more general term is habitat, which refers in a general way to where an organism occurs and the environmental factors present there. *See* ENVIRONMENT.

A recognition of the unitary coupling of an organism and its environment is fundamental to ecology; in fact, the definitions of organism and environment are not separate. Environment is organism-centered since the environmental properties of a habitat are determined by the requirements of the organisms that occupy that habitat. For example, the amount of inorganic nitrogen dissolved in lake water is of little immediate significance to zooplankton in the lake because they are incapable of utilizing inorganic nitrogen directly. However, because phytoplankton are capable of utilizing inorganic nitrogen directly, it is a component of their environment. Any effect of inorganic nitrogen upon the zooplankton, then, will occur indirectly through its effect on the abundance of the phytoplankton that the zooplankton feed upon.

Just as the environment affects the organism, so the organism affects its environment. Growth of phytoplankton may be nitrogen-limited if the number of individuals has become so great that there is no more nitrogen available in the environment. Zooplankton, not limited by inorganic nitrogen themselves, can promote the growth of additional phytoplankton by consuming some individuals, digesting them, and returning part of the nitrogen to the environment.

Ecology is concerned with the processes involved in the interactions between organisms and their environments, with the mechanisms responsible for those processes, and with the origin, through evolution, of those mechanisms. It is distinguished from such closely related biological subdisciplines as physiology and morphology because it is not intrinsically concerned with the operation of a physiological process or the function of a structure, but with how a process or structure interacts with the environment to influence survival, growth, development, and reproduction.

Major subdivisions of ecology by organism include plant ecology, animal ecology, and microbial ecology. Subdivisions by habitat include terrestrial ecology, the study of organisms on land; limnology, the study of fresh-water organisms and habitats; and oceanography, the study of marine organisms and habitats.

The levels of organization studied range from the individual organism to the whole complex of organisms in a large area. Autecology is the study of individuals, population ecology is the study of groups of individuals of a single species or a limited number of species, synecology is the study of communities of several populations, and ecosystem, or simply systems, ecology is the study of communities of organisms and their environments in a specific time and place. *See* POPULATION ECOLOGY; SYSTEMS ECOLOGY.

Higher levels of organization include biomes and the biosphere. Biomes are collections of ecosystems with similar organisms and environments and, therefore, similar ecological properties. All of Earth's coniferous forests are elements in the coniferous forest biome. Although united by similar dynamic relationships and structural properties, the biome itself is more abstract than a specific ecosystem. The biosphere is the most inclusive category possible, including all regions of Earth inhabited by living things. It extends from the lower reaches of the atmosphere to the depths of the oceans. *See* BIOME; BIOSPHERE.

The principal methodological approaches to ecology are descriptive, experimental, and theoretical. Descriptive ecology concentrates on the variety of populations, communities, and habitats throughout Earth. Experimental ecology involves manipulating organisms or their environments to discover the underlying mechanisms governing distribution and abundance. Theoretical ecology uses mathematical equations based on assumptions about the properties of organisms and environments to make predictions about patterns of distribution and abundance. *See* THEORETICAL ECOLOGY.   [S.J.McN.]

Ecology, applied   The application of ecological principles to the solution of human problems and the maintenance of a quality life. It is assumed that humans are an integral part of ecological systems and that they depend upon healthy, well-operating, and productive systems for their continued well-being. For these reasons, applied ecology is based on a knowledge of ecosystems and populations, and the principles and techniques of ecology are used to interpret and solve specific environmental problems and to plan new management systems in the biosphere. Although a variety of management fields, such as forestry, agriculture, wildlife management, environmental engineering, and environmental design, are concerned with specific parts of the environment, applied ecology is unique in taking a view of whole systems, and attempting to account for all inputs to and outputs from the systems—and all impacts. In the past, applied ecology has been considered as being synonymous with the above applied sciences.

The objective of applied ecology management is to maintain the system while altering its inputs or outputs. Often, ecology management is designed to maximize a particular output or the quantity of a specific component. Since outputs and inputs are related, maximization of an output may not be desirable; rather, the management objective may be the optimum level. Optimization of systems can be accomplished through the use of systems ecology methods which consider all parts of the system rather than a specific set of components. In this way, a series of strategies or scenarios can be evaluated, and the strategy producing the largest gain for the least cost can be chosen for implementation.

A variety of general environmental problems within the scope of applied ecology relate to the major components of the Earth: the atmosphere, water, land, and the biota. Applied ecology also is concerned with the size of the human population, since many of the impacts of human activities on the environment are a function of the number and concentration of people. *See* ECOLOGY; ECOSYSTEM; SYSTEMS ECOLOGY.

[F.B.Go.]

Ecosystem   A functional system that includes an ecological community of organisms together with the physical environment, interacting as a unit. Ecosystems are characterized by flow of energy through food webs, production and degradation of organic matter, and transformation and cycling of nutrient elements. This production of organic molecules serves as the energy base for all biological activity within ecosystems. The consumption of plants by herbivores (organisms that consume living plants or algae) and detritivores (organisms that consume dead organic matter) serves to transfer energy stored in photosynthetically produced organic molecules to other organisms. Coupled to the production of organic matter and flow of energy is the cycling of elements. *See* ECOLOGICAL COMMUNITIES; ENVIRONMENT.

All biological activity within ecosystems is supported by the production of organic matter by autotrophs (organisms that can produce organic molecules such as glucose from inorganic carbon dioxide; see illustration). More than 99% of autotrophic production on Earth is through photosynthesis by plants, algae, and certain types of bacteria. Collectively these organisms are termed photoautotrophs (autotrophs that use energy from light to produce organic molecules). In addition to photosynthesis, some production is conducted by chemoautotrophic bacteria (autotrophs that use energy stored in the chemical bonds of inorganic molecules such as hydrogen sulfide to produce organic molecules). The organic molecules produced by autotrophs are used to support the organism's metabolism and reproduction, and to build new tissue. This new tissue is consumed by herbivores or detritivores, which in turn are ultimately consumed by predators or other detritivores.

Terrestrial ecosystems, which cover 30% of the Earth's surface, contribute a little over one-half of the total global photosynthetic production of organic matter— approximately 60×10^{15} grams of carbon per year. Oceans, which cover 70% of the Earth's surface, produce approximately 51×10^{15} g C y^{-1} of organic matter. *See* BIOMASS.

Food webs.   Organisms are classified based upon the number of energy transfers through a food web (see illustration). Photoautotrophic production of organic matter represents the first energy transfer in ecosystems and is classified as primary production. Consumption of a plant by a herbivore is the second energy transfer, and thus herbivores occupy the second trophic level, also known as secondary production. Consumer organisms that are one, two, or three transfers from photoautotrophs are classified as primary, secondary, and tertiary consumers. Moving through a food web, energy is lost during each transfer as heat, as described by the second law of thermodynamics.

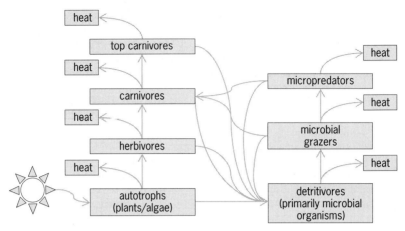

General model of energy flow through ecosystems.

Consequently, the total number of energy transfers rarely exceeds four or five; with energy loss during each transfer, little energy is available to support organisms at the highest levels of a food web. *See* FOOD WEB.

Biogeochemical cycles. In contrast to energy, which is lost from ecosystems as heat, chemical elements (or nutrients) that compose molecules within organisms are not altered and may repeatedly cycle between organisms and their environment. Approximately 40 elements compose the bodies of organisms, with carbon, oxygen, hydrogen, nitrogen, and phosphorus being the most abundant. If one of these elements is in short supply in the environment, the growth of organisms can be limited, even if sufficient energy is available. In particular, nitrogen and phosphorus are the elements most commonly limiting organism growth. This limitation is illustrated by the widespread use of fertilizers, which are applied to agricultural fields to alleviate nutrient limitation. *See* BIOGEOCHEMISTRY; NITROGEN CYCLE.

Carbon cycles between the atmosphere and terrestrial and oceanic ecosystems. This cycling results, in part, from primary production and decomposition of organic matter. Rates of primary production and decomposition, in turn, are regulated by the supply of nitrogen, phosphorus, and iron. The combustion of fossil fuels is a recent change in the global cycle that releases carbon that has long been buried within the Earth's crust to the atmosphere. Carbon dioxide in the atmosphere traps heat on the Earth's surface and is a major factor regulating the climate. This alteration of the global carbon cycle along with the resulting impact on the climate is a major issue under investigation by ecosystem ecologists. *See* AIR POLLUTION; ECOLOGY, APPLIED.   [J.B.Jo.]

Ecotone   A geographic boundary or transition zone between two different groups of plant or animal distributions. The term has been used to denote transitions at different spatial scales or levels of analysis, and may refer to any one of several attributes of the organisms involved. For example, an ecotone could refer to physiognomy (roughly, the morphology or appearance of the relevant organisms), such as between the boreal forest and grassland biomes; or it could refer to composition, such as between oak-hickory and maple-basswood forest associations; or it could refer to both. Ecotones are generally distinguished from other geographic transitions of biota by their relative sharpness. The ecotone between boreal forest and prairie in central Saskatchewan occurs over a hundred kilometers or so, in contrast to the transition from tropical forest to savanna in South America or Africa that is associated with increasing aridity and is dispersed over hundreds of kilometers. The "tension zone" between broadleaf deciduous forests in south-cental Michigan and mixed forests to the north is similarly sharp. Ecotones are thought to reflect concentrated long-term gradients of one or more current environmental (rather than historical or human) factors. Though often climatic, these factors can also be due to substrate materials, such as glacial sediments or soils. Regardless of their specific environmental basis, most ecotones are thought to be relatively stable.

Ecotones are often reflected in the distributions of many biota besides the biota used to define them. The prairie-forest ecotone, for example, is defined not only by the dominant vegetation components but also by many faunal members of the associated ecosystems, such as insects, reptiles and amphibians, mammals, and birds, that reach their geographic limits there. *See* BIOME; ECOLOGICAL COMMUNITIES; ECOSYSTEM.

[J.R.Har.]

El Niño In general, an invasion of warm water into the central and eastern equatorial Pacific Ocean off the coast of Peru and Ecuador, with a return period of 4–7 years. El Niño events come in various strengths: weak, moderate, strong, very strong, and extraordinary. The size of an El Niño event can be determined using various criteria: the amount of warming of sea surface temperatures in the central and eastern Pacific from their average condition; the areal extent of that warm water anomaly; and the length of time that the warm water lingers before being replaced by colder-than-average sea surface temperatures in this tropical Pacific region.

Under normal conditions the winds blow up the west coast of South America and then near the Equator turn westward to Asia. The surface water is piled up in the western Pacific, and the sea level there is several tens of centimeters above average while the sea level in the eastern Pacific is below average. As the water is pushed toward the west, cold water from the deeper part of the ocean along the Peruvian coast wells up to the surface to replace it. This cold water is rich with nutrients, making the coastal upwelling region along western South America among the most productive fisheries in the world. *See* UPWELLING.

Every 4–7 years those winds tend to die down and sometimes reverse, allowing the warm surface waters that piled up in the west to move back toward the eastern part of the Pacific Basin. With reduced westward winds the surface water also heats up. Sea level drops in the western Pacific and increases in the eastern part of the basin. El Niño condition can last for 12–18 months, sometimes longer, before the westward flowing winds start to pick up again. Occasionally, the opposite also occurs: the eastern Pacific becomes cooler than normal, rainfall decreases still more, atmospheric surface pressure increases, and the westward winds become stronger. This irregular cyclic swing of warm and cold phases in the tropical Pacific is referred to as ENSO (El Niño Southern Oscillation).

El Niño is considered to be the second biggest climate-related influence on human activities, after the natural flow of the seasons. Although the phenomenon is at least thousands of years old, its impacts on global climate have only recently been recognized. Due to improved scientific understanding and forecasting of El Niño's interannual process, societies can prepare for and reduce its impacts considerably. *See* CLIMATOLOGY; MARITIME METEOROLOGY; TROPICAL METEOROLOGY. [M.H.G.]

Numerical models that couple the atmosphere to the ocean have been used to successfully predict the sea surface temperature of the tropical Pacific a year or so in advance. The basic reason that the cycle is predictable is that ENSO evolves slowly and regularly. If the initial state of the atmosphere-ocean system can be characterized accurately, the classification of this state in the ENSO sequence is made (even if it is not completely recognizable in each system separately), and the future evolution of the cycle can be predicted. *See* CLIMATE MODELING.   [E.S.S.]

Elements, cosmic abundance of   The average chemical and isotopic composition of the solar system is appropriately referred to as cosmic, since this elemental abundance distribution is found to be nearly the same for interstellar gas and for young stars associated with gas and dust in the spiral arms of galaxies. The Sun makes up more than 99.9% of the mass of the solar system, so the bulk chemical composition of the solar system is essentially the same as that of the Sun. The cosmic abundances of the nonvolatile elements are determined from chemical analyses of a type of meteorite known as CI chondrites, whereas the relative abundances of the volatile elements are determined from quantitative measurements of the intensities of elemental emission lines from the Sun's photosphere. In most silicate-rich meteorites and the Earth, Moon, Venus, and Mars, the most abundant elements are oxygen, magnesium, silicon, iron, aluminum, and calcium. Average solar-system composition consists of 70.7 wt % hydrogen, 27.4 wt % helium, and only 1.9 wt % of all remaining elements, lithium to uranium. Cosmic abundances are now widely referred to as standard abundances in the astrophysical literature. *See* ELEMENTS, GEOCHEMICAL DISTRIBUTION OF.

Cosmic abundances of elements have several important uses. First, by comparing cosmic abundances to chemical analyses of various types of meteorites, inferences can be made about chemical fractionation processes that occurred in the primitive solar nebula, such as condensation and vaporization. Also, by comparing them to rock compositions, inferences can be made about processes that occurred early in the history of rocky planets, such as separation of a metallic core and differentiation of silicates into mantle and crust. Second, cosmic abundances serve as a standard of comparison for spectroscopic measurements of elemental abundances of the photospheres of other stars and for measurements of elemental and isotopic abundances in cosmic rays. Finally, nucleosynthesis occurs in many different stellar environments. Explanations of nucleosynthesis must account for how elements and isotopes from various astrophysical sources are made and then mixed to form the solar system's average chemical and isotopic composition.   [A.M.D.]

Elements, geochemical distribution of   The distribution of the chemical elements within the Earth in space and time. Knowledge of the geochemical distribution of the elements in the Earth, particularly in the Earth's crust, and of the processes that lead to the observed distributions make it possible to locate and use efficiently essential elements and minerals and to predict their dispersal patterns when they reenter the natural environment after use.

To understand the present-day distribution of the elements in the Earth, it is necessary to go back to the time of Earth formation approximately 4.5 billion years ago. It is generally believed that the Earth and the other planets in the solar system formed by agglomeration of smaller fragments of solid material orbiting around the Sun. This material had precipitated from a cooling hot gas cloud (the solar nebula), with the most refractory materials condensing out first, the most volatile last. The distribution of elements in the solar system in this early phase thus had much to do with volatility, and the solid material that aggregated to form the planets was a mix of volatile and nonvolatile materials. *See* ELEMENTS, COSMIC ABUNDANCE OF.

Although the Earth may have been an approximately homogeneous mixture of accreted materials at the time of its formation, it is now made of many chemically distinct parts. At the fundamental level, these are the core, the mantle, and the crust.

Element distribution among some of the major subdivisions of the Earth*

Element	Continental crust	Oceanic crust	Upper mantle	Core[†]
Oxygen	45.3	43.6	44.2	—
Silicon	26.7	23.1	21.0	—
Aluminium	8.39	8.47	1.75	—
Iron	7.04	8.16	6.22	85.5
Calcium	5.27	8.08	1.86	—
Magnesium	3.19	4.64	24.0	—
Sodium	2.29	2.08	0.25	—
Potassium	0.91	0.13	0.02	—
Titanium	0.68	1.12	0.11	—
Nickel	0.011	0.014	0.20	5.5
Sulfur	NA	NA	NA	9.0

*Estimates of element abundances are in percent by weight and are arranged in order of decreasing abundance in the continental crust. Sulfur contents are not well known and are designated "not applicable."
[†]The estimate for the core is just one of several models. Others substitute light elements such as oxygen, carbon, or silicon for some or most of the sulfur shown here.

While chemical fractionation in the solar nebula depended upon volatility, chemical differentiation within the Earth took place by the separation of molten material from unmelted residue under the influence of gravity. Because large amounts of energy were released from accreting fragments, the early Earth was very hot, and during the accretion stage itself, temperatures in some parts exceeded the melting point of iron metal. Pools of dense molten iron, with dissolved nickel and other elements, aggregated and sank through the Earth under gravity to form the core, leaving behind a mantle of silicate and oxide minerals. The present core constitutes about 32.4% of the Earth's mass. The distinct parts of the Earth possess unique overall compositions (see table).

The large-scale distribution of the elements in the Earth depends on the affinity of each element for specific compounds or phases. Those elements that alloy easily with iron, for example, are mostly sequestered in the Earth's core; those which form oxides and silicate minerals tend to be concentrated in the Earth's crust and mantle. Although many elements display multiple characteristics depending on the chemical environment, a classification according to geochemical affinity is nevertheless useful. The categories in this classification include atmophile (elements that are gases and concentrate in the atmosphere), lithophile (elements that form silicates or oxides and are concentrated in the minerals of the Earth's crust), siderophile (elements that alloy easily with iron and are concentrated in the core), and chalcophile (elements such as copper which commonly form sulfide minerals if sufficient sulfur is available).

Although geochemists have a good general knowledge of the overall distribution of elements in the core and mantle, much more detailed information is available about the chemical composition of the crust, which is accessible. The crust is actually composed of two major parts with quite different compositions, thickness, and average age: the continental crust and the oceanic crust.

Although all elements are present, the crust is made almost entirely of just nine chemical elements: oxygen, silicon, aluminum, iron, magnesium, calcium, sodium, potassium, and titanium. Oxygen and silicon are by far the most abundant. The most common minerals in the crust are those of the silicate family, in which the basic building block is a silicon atom surrounded by four oxygen atoms in the form of a tetrahedron.

The crust is essentially a framework of oxygen atoms bound together by the common cations. *See* SILICATE MINERALS.

A variety of processes act to make the crust chemically heterogeneous on many scales. Many of these processes involve liquid water. Running water physically sorts particles depending on size and density, which are ultimately related to chemical composition. It is also a superb solvent, carrying many elements in solution under different conditions of temperature and pressure, and depositing them when these conditions change. Processes involving water account for many ore deposits, in which extreme concentrations of some elements occur relative to their average abundance in the crust. One example is the circulating hydrothermal solutions in volcanically active parts of the crust, which can leach metals from their normally dispersed state in large volumes of volcanic rocks and deposit them in concentrated zones as the solutions cool and encounter different rock types. Another example is the action of weathering in tropical regions with high rainfall, which can leach away all but the least soluble components from large volumes of rock, leaving behind mineral deposits rich in aluminum or, depending on the original composition of the rocks being weathered, metals such as iron and nickel. *See* GEOCHEMISTRY; ORE AND MINERAL DEPOSITS; WEATHERING PROCESSES.

[G.Fau.; J.D.MacD.]

Emerald   The medium- to dark-green gem variety of the mineral beryl, crystallizing in the hexagonal system. A flawless emerald with good color is one of the most sought after and highly prized of all precious gems. Emerald is restricted in its occurrence, and only infrequently are exceptional stones found; most emeralds are flawed and cloudy, and few stones command high prices.

In contradistinction to beryl and its other gem varieties, emeralds have only been found in mica schists or metasomalized limestones. The most outstanding occurrences include the Muzo and El Chivor mines in Colombia. Noteworthy occurrences in mica schists include Tokovoja in the Ural Mountains, where emerald occurs with the beryllium minerals chrysoberyl (and its gem variety alexandrite) and phenakite; Habachtal, Austria; Transvaal, South Africa; and Kaliguman, India. The ultimate source of an emerald can often be assessed by a study of its inclusions. *See* BERYL; GEM.   [P.B.M.]

Emery   A natural mixture of corundum with magnetite or with hematite and spinel. Because the mixture is very intimate and appears to be quite homogeneous, it was considered to be a single mineral species until the middle of the 19th century. The aggregate has a gray-to-black color and is extremely tough and difficult to break. The specific gravity varies from 3.7 to 4.3, depending upon the relative amounts of the constituent minerals. The hardness is about 8 (Mohs scale), less than that of pure corundum which is 9, and is more dependent upon the physical state of aggregation than on the percentage of corundum. *See* CORUNDUM.

Although synthetic abrasives have replaced emery in many of its earlier uses, it is still used as an abrasive and polishing material by lapidaries and in the manufacture of lenses, prisms, and other optical equipment. Emery wheels, emery paper, and emery cloth are used not only by lapidaries but also by machinists in the grinding and polishing of steel. *See* MAGNETITE.   [C.S.Hu.]

Enargite   A mineral having composition Cu_3AsS_4. In some places enargite is a valuable ore of copper. The mineral has perfect prismatic cleavage, metallic luster, and grayish- black color. The hardness is 3 on Mohs scale, and the specific gravity is 4.44. Enargite is one of the rarer copper ore minerals and it has been mined in former

Yugoslavia, Peru, the Philippines, and the United States at Butte, Montana, and Bingham Canyon, Utah. Probably the largest deposit is at Chuquicamata, Chile. [C.S.Hu.]

Engineering geology The application of education and experience in geology and other geosciences to solve geological problems posed by civil engineering works. The branches of the geosciences most applicable are surficial geology, petrofabrics, rock and soil mechanics, geohydrology, and geophysics, particularly exploration geophysics and earthquake seismology.

The terms engineering geology and environmental geology often seem to be used interchangeably. Specifically, environmental geology is the application of engineering geology in the solution of urban problems; in the prediction and mitigation of natural hazards such as earthquakes, landslides, and subsidence; and in solving problems inherent in disposal of dangerous wastes and in reclaiming mined lands.

Another relevant term is geotechnics, the combination of pertinent geoscience elements with civil engineering elements to formulate the civil engineering system that has the optimal interaction with the natural environment. [W.R.J.]

Enstatite The name given to the magnesian end member ($MgSiO_3$) of the orthorhombic pyroxene solid-solution series. The mineral is usually yellowish-gray, becoming greenish with a little iron present, and transparent in thin sections.

Enstatite is an important mineral in the upper mantle of the Earth, and coexists with clinopyroxene, garnet, olivine, and plagioclase. Enstatite, clinopyroxene, and garnet rocks called eherzolites are common in certain alpine mountain terrains and as nodules in some diamond-bearing rocks called kimberlites. *See* PYROXENE. [G.W.DeV.]

Environment The sum of all external factors, both biotic (living) and abiotic (non-living), to which an organism is exposed. Biotic factors include influences by members of the same and other species on the development and survival of the individual. Primary abiotic factors are light, temperature, water, atmospheric gases, and ionizing radiation, influencing the form and function of the individual.

For each environmental factor, an organism has a tolerance range, in which it is able to survive. The intercept of these ranges constitutes the ecological niche of the organism. Different individuals or species have different tolerance ranges for particular environmental factors—this variation represents the adaptation of the organism to its environment. The ability of an organism to modify its tolerance of certain environmental factors in response to a change in them represents the plasticity of that organism. Alterations in environmental tolerance are termed acclimation. Exposure to environmental conditions at the limit of an individual's tolerance range represents environmental stress. *See* ECOLOGY.

Abiotic factors. The spectrum of electromagnetic radiation reaching the Earth's surface is determined by the absorptive properties of the atmosphere. Biologically, the most important spectral range is 300–800 nanometers, incorporating ultraviolet, visible, and infrared radiation. Visible light provides the energy source for most forms of life. Light absorbed by pigment molecules (chlorophylls, carotenoids, and phycobilins) is converted into chemical energy through photosynthesis. Light availability is especially important in determining the distribution of plants. Photosynthetic organisms can exist within a wide range of light intensities. Full sunlight in the tropics is around 2000 μmol photons $\cdot$ m$^{-2}\cdot$ s^{-1}. Photosynthetic organisms have survived in locations where the mean light is as low as 0.005% of this value. *See* INSOLATION.

In addition to providing energy, light is important in providing an organism with information about its surroundings. The human eye, for example, is able to respond to wavelengths of light between 400 and 700 nm—the visible range. Within this range, sensitivity is greatest in the green part of the spectrum. This is the portion of the spectrum that plants absorb least, and so is the principal part of the spectrum to be reflected.

Temporal variation in light also provides an important stimulus. Life forms from bacteria upward are able to detect and respond to daily light fluctuations. Such a response may be directly controlled by the presence or absence of light (diurnal rhythms) or may persist when the variation in light is removed (circadian rhythms). In the latter case, regulation is through an internal molecular clock, which is able to predict the daily cycle. Such circadian clocks are normally reset by light on a daily basis. Processes controlled by circadian clocks range from the molecular (gene expression) to the behavioral (for example, sleep patterns in animals or leaf movements in plants).

Ultraviolet radiation has the ability to break chemical bonds and so may lead to damage to proteins, lipids, and nucleic acids. Damage to DNA may result in genetic mutations. The ozone layer in the stratosphere is responsible for absorbing a large proportion of ultraviolet radiation reaching the outer atmosphere. As ozone is destroyed by the action of pollutants such as chloroflurocarbons, the proportion of ultraviolet radiation reaching the surface of the Earth rises.

Water is ubiquitous in living systems, as the universal solvent for life, and is essential for biological activity. Many organisms have evolved the ability to survive prolonged periods in the total absence of water, but this is achieved only through the maintenance of an inactive state. Water availability remains a primary environmental factor limiting survival on land. Primitive land organisms possess little or no ability to conserve water within their cells and are termed poikilohydric. Examples include amphibians and primitive plants such as most mosses and liverworts. These are confined to places where water is in plentiful supply or they must be able to tolerate periods of desiccation. Lichens can survive total water loss and rapidly regain activity upon rewetting. Such organisms must be able to minimize the damage caused to cellular structures when water is lost. Dehydration causes irreversible damage to membranes and proteins. This damage can be prevented by the accumulation of protective molecules termed compatible solutes.

Homeohydric organisms possess a waterproof layer that restricts the loss of water from the cells. Such waterproofing is never absolute, as there is still a requirement to exchange gas molecules and to absorb organic or mineral nutrients through a water phase. Water conservation allows organisms to live in environments in which the water supply is extremely low. In extremely arid environments, behavioral adaptations may allow the water loss to be minimized. Animals may be nocturnal, emerging when temperatures are lower and hence evaporation minimized. Cacti possess a form of photosynthesis, crassulacean acid metabolism (CAM), that allows them to separate gas exchange and light capture. See GROUND-WATER HYDROLOGY.

Temperature is a determinant of survival in two ways: (1) as temperatures decrease, the movement of molecules slows and the rate of chemical reactions declines; (2) temperature determines the physical state of water.

The slowing of metabolic activity at low temperatures is illustrated in reptiles. Such poikilothermic animals, unable to maintain their internal temperature, are typically inactive in the cold of morning. They bask in the sun to increase their body temperature and so become active. High temperatures will cause the three-dimensional structure of proteins to break down, preventing the organisms from functioning. Organisms adapted to extremely high temperatures need more rigid proteins that maintain their structure. Temperature also affects the behavior of cell membranes, made up of lipids and proteins in a liquid crystalline state. At low temperatures, the membrane structure becomes rigid and liable to break. At high temperatures, it becomes too fluid and again liable to disintegrate. In adapting to different temperatures, organisms alter the composition of the lipids in their membranes, whose melting temperature is thereby changed. This outcome also applies to storage lipids. Hence, cold-water fish are a useful source of oils, whereas mammals, with their higher body temperature, contain fats. The effect of temperature on membranes is thought to be a key factor determining the temperature range that an organism is able to survive.

The effect of temperature on the physical state of water is essential to determining the availability of that water to organisms. Poikilothermic organisms may find that the water in their cells begins to freeze at low temperatures. Certain species can survive total freezing through the prevention of ice crystal formation altogether which would otherwise damage cellular structures. To survive low temperatures, cells must be able to survive desiccation, and so low-temperature tolerance involves the formation of compatible solutes. High temperatures increase the rate of evaporation of water. Hence, where water supply is limiting, an organism's ability to survive high temperatures is impaired.

Mammals and birds, homeothermic organisms, are able to regulate their internal temperature, limiting the effects of external temperature variations. Temperature still acts as an environmental constraint in such organisms, however. Cooling is achieved through sweating and hence loss of water. Heat is produced through the metabolism of food, and hence survival in cold climates requires a high metabolic rate.

The atmosphere on Earth is thought to be determined to a large extent by the presence of life. At the same time, organisms have evolved to survive in the atmosphere as it is. The atmospheric constitutents with the most direct biological importance are oxygen (O_2) and carbon dioxide (CO_2). Oxygen makes up approximately 20% of the atmosphere and is due to the occurrence of oxygenic photosynthesis. This process involves the simultaneous uptake of CO_2 to make sugars. Aerobic respiration involves the reverse of this process, the release of CO_2 and the uptake of O_2 to form water. Hence, the current atmosphere represents the balance of previous biological activity. For most terrestrial organisms, neither CO_2 nor O_2 is limiting in the atmosphere; however, the need to get either or both of these gases to cells may represent a limitation on size or on the ability to tolerate water stress. Limitation of either gas may be important in aquatic environments, where the concentration of each is significantly lower.

Nitrogen is also required by all organisms but cannot be used by most in the gaseous form. Nitrogen fixation, the conversion of N_2 gas into a biologically useful form, occurs in some species of bacteria and cyanobacteria or may be caused by lightning.

Atmospheric gases are important in determining the climate and the light environment. Absorption of electromagnetic radiation by the atmosphere determines the spectrum of light reaching the Earth's surface. Absorption and reflectance of infrared radiation by greenhouse gases such as CO_2 and water vapor regulate temperature.

Among other environmental factors determining the range and distribution and form of organisms are mechanical stimuli such as wind or water movement, and the presence of metals, inorganic nutrients, and toxins in the air, soil, or food.

Biotic factors. The biotic environment of an individual is made up of members of the same or other species. Intraspecific interactions involve the need to breed with other individuals, to gain protection through living in a group, and to compete for resources such as food, light, nutrients, and space. The optimal population density depends on the availability of resources and on the behavior, size, and structure of the organism. Interspecific interactions may also be positive or negative. For example, symbiotic relationships involve the mutual benefit of the individuals involved, whereas competition for resources is deleterious to both. Although predation exerts a negative influence on the population as a whole, the success of an individual may be enhanced if a predator removes one of its conspecific competitors.

Humans alter their environment in ways that exceed the impact of all other organisms. For example, the release of greenhouse gases into the atmosphere contributes to climate alterations over the entire planet. This in turn has impacts on the distribution of all other species. The release of pollutants into the environment brings organisms into contact with stresses to which they were not previously exposed. This causes the evolution of new varieties, eventually perhaps new species, adapted to the polluted environments. *See* AIR POLLUTION; BIOSPHERE.

For any given organism, it is often possible to identify a factor in the environment that limits survival and growth. The limiting factor may change through time. Such a change may cause the organism to be at the limit of or outside its tolerance range for that or another environmental factor. In such cases, the organism is said to suffer stress. If the stress to which an individual is exposed is extreme, it may result in irreversible damage and death. Exposure to moderate stress, however, results in a period of acclimation within the organism that allows it to adjust to the new conditions. Organisms exposed gradually to new conditions usually have a higher chance of survival than those exposed suddenly. *See* POPULATION ECOLOGY.

Where a particular environmental factor (or combination of factors) dominates the growth and development of organisms, it is often found that the adaptations and gross features of the landscape will be the same, even when the actual species are different. Thus, mediterranean vegetation is found not only around the Mediterranean Sea but also in California and South Africa, where the conditions of hot dry summers and warm wet winters occur. Regions with similar environmental conditions are classed as biomes. The occurrence of such global vegetation types clearly illustrates the role played by the environment in determining the form and function of individual species.

[G.J.]

Environmental fluid mechanics   The study of the flows of air and water, of the species carried by them, and of their interactions with geological, biological, social, and engineering systems in the vicinity of a planet's surface. The environment on the Earth is intimately tied to the fluid motion of air (atmosphere), water (oceans), and species concentrations (air quality). In fact, the very existence of the human race depends upon its abilities to cope within the Earth's environmental fluid systems.

Meteorologists, oceanologists, geologists, and engineers study environmental fluid motion. Weather and ocean-current forecasts are of major concern, and fluid motion within the environment is the main carrier of pollutants. Biologists and engineers examine the effects of pollutants on humans and the environment, and the means for environmental restoration. Air quality in cities is directly related to the airborne spread of dust particles and of exhaust gases from automobiles. The impact of pollutants on drinking-water quality is especially important in the study of ground-water flow. Likewise, flows in porous media are important in oil recovery and cleanup. Lake levels are significantly influenced by climatic change, a relationship that has become of some concern in view of the global climatic changes that may result from the greenhouse effect (whereby the Earth's average temperature increases because of increasing concentrations of carbon dioxide in the atmosphere). *See* AIR POLLUTION; GREENHOUSE EFFECT; WEATHER FORECASTING AND PREDICTION.

Scales of motion. Environmental fluid mechanics deals with the study of the atmosphere, the oceans, lakes, streams, surface and subsurface water flows (hydrology), building exterior and interior airflows, and pollution transport within all these categories. Such motions occur over a wide range of scales, from eddies on the order of centimeters to large recirculation zones the size of continents. This range accounts in large part for the difficulties associated with understanding fluid motion within the environment. In order to impart motion (or inertia) to the atmosphere and oceans, internal and external forces must develop. Global external forces consist of gravity, Coriolis, and centrifugal forces, and electric and magnetic fields (to a lesser extent). The internal forces of pressure and friction are created at the local level, that is, on a much smaller spatial scale; likewise, these influences have different time scales. The winds and currents arise as a result of the sum of all these external and internal forces.

Governing equations. The foundations of environmental fluid mechanics lie in the same conservation principles as those for fluid mechanics, that is, the conservation of mass, momentum (velocity), energy (heat), and species concentration (for example, water, humidity, other gases, and aerosols). The differences lie principally in the formulations of the source and sink terms within the governing equations, and the scales of motion. These conservation principles form a coupled set of relations, or governing equations, which must be satisfied simultaneously. The governing equations consist of nonlinear, independent partial differential equations that describe the advection and diffusion of velocity, temperature, and species concentration, plus one scalar equation for the conservation of mass. In general, environmental fluids are approximately newtonian, and the momentum equation takes the form of the Navier-Stokes equation. An important added term, neglected in small-scale flow analysis, is the Coriolis acceleration, $2\Omega \times V$, where Ω is the angular velocity of the Earth and V is the flow velocity. *See* Coriolis acceleration.

Fortunately, not every term in the Navier-Stokes equation is important in all layers of the environment. The key to being able to obtain solutions to the Navier-Stokes equation lies in determining which terms can be neglected in specific applications. For convenience, problems can be classified on the basis of the order of importance of the terms in the equations utilizing nondimensional numbers based on various ratios of values.

Measurements. Because of the scales of motion and time associated with the environment, and the somewhat random nature of the fluid motion, it is difficult to conduct full-scale, extensive experimentation. Likewise, some quantities (such as vorticity or vertical velocity) resist direct observations. It is necessary to rely on the availability of past measurements and reports (as sparse as they may be) to establish patterns, especially for climate studies. However, some properties can be measured with confidence.

Modeling. There are two types of modeling strategies: physical and mathematical. Physical models are small-scale (laboratory) mockups that can be measured under variable conditions with precise instrumentation. Such modeling techniques are effective in examining wind effects on buildings and species concentrations within city canyons (flow over buildings). Generally, a large wind tunnel is needed to produce correct atmospheric parameters (such as Reynolds number) and velocity profiles. Mathematical models (algebra- and calculus-based) can be broken down further into either analytical models, in which an exact solution exists, or numerical models, whereby approximate numerical solutions are obtained using computers.   [D.W.P.]

Environmental geology   The branch of geology that deals with the ways in which geology affects people. Examples of the effect of geology on human civilizations include (1) the ways that fertile soils develop from rocks and how these soils can become polluted by human activities; (2) how rocks and soils move down-slope to destroy roads, houses, and other human constructions; (3) sources of surface and subsurface water supplies and how they become polluted; (4) why floods occur where they do and how human activities affect floods; (5) locations of earthquakes and volcanic eruptions and the dangers they pose; (6) location of mineral resources such as copper, oil and gas, and uranium, and how mining these resources can pollute the environment; (7) how human activities can pollute the atmosphere and cause global warming, sea-level rise, and ozone depletion. *See* AIR POLLUTION; EARTHQUAKE; VOLCANO.   [H.Bl.]

Environmental management   The development of strategies to allocate and conserve resources, with the ultimate goal of regulating the impact of human activities on the surrounding environment. "Environment" here usually means the natural surroundings, both living and inanimate, of human lives and activities. However, it can also mean the artificial landscape of cities, or occasionally even the conceptual field of the noosphere, the realm of communicating human minds.

Environmental management is a mixture of science, policy, and socioeconomic applications. It focuses on the solution of the practical problems that humans encounter in cohabitation with nature, exploitation of resources, and production of waste. In a purely anthropocentric sense, the central problem is how to permit technology to evolve continuously while limiting the degree to which this process alters natural ecosystems. Environmental management is thus intimately intertwined with questions regarding economic growth, equitable distribution of consumable goods, and conserving resources for future generations. Environmental managers fall within a broad spectrum, from those who would limit human interference in nature to those who would increase it in order to guide natural processes along benign paths. Participants in the process of environmental management fall into seven main groups: (1) governmental organizations at the local, regional, national, and international levels, including world bodies such as the United Nations Environment Programme and the U.N. Conference on Environment and Development; (2) research institutions, such as universities, academies, and national laboratories; (3) bodies charged with the enforcement of regulations, such as the U.S. Environmental Protection Agency; (4) businesses of all sizes and multinational corporations; (5) international financial institutions, such as the World Bank and International Monetary Fund; (6) environmental nongovernmental organizations, such as the World Wildlife Fund for Nature; and (7) representatives of the users of the environment, including tribes, fishermen, and hunters. The agents of environmental management include foresters, soil conservationists, policy-makers, engineers, and resource planners.

Some common themes of environmental management are bilateral and multilateral environmental treaties; design and use of decision-support systems; environmental policy formulation, enactment, and policing of compliance; estimation, analysis, and management of environmental risk; management of recreation and tourism; natural resource evaluation and conservation; positive environmental economics; promotion of positive environmental values by education, debate, and information dissemination; and strategies for the rehabilitation of damaged environments.

The need to improve management of the environment has given rise to several new techniques. There is environmental impact analysis, which was first formulated in California and is codified in the U.S. National Environmental Policy Act (NEPA). Through the environmental impact statement, it prescribes the investigatory and remedial measures that must be taken in order to mitigate the adverse effects of new development. It is intended to act in favor of both prudent conservation and participatory democracy. Another technique is environmental auditing, which uses the model of the financial audit to examine the processes and outcomes of environmental impacts. It requires value judgments, which are usually set by public preference, ideology, and policy, to define what are regarded as acceptable outcomes. Audits use techniques such as life-cycle analysis and environmental burden analysis to assess the impact of, for example, manufacturing processes that consume resources and create waste.   [D.Ale.]

Environmental radioactivity Radioactivity that originates from natural and anthropogenic sources, including radioactive materials in food, housing, and air, radioactive materials used in medicine, nuclear weapon tests in the open atmosphere, and radioactive materials used in industry and power generation.

Natural radioactivity, which is by far the largest component to which humans are exposed, is of both terrestrial and extraterrestrial (cosmic) origin. About 340 nuclides are known in nature, of which 70 are radioactive and are found mainly among the heavy elements. Three nuclides which are responsible for most of the terrestrial component are potassium-40, uranium-238, and thorium-232.

The average person in the United States receives 80–180 mrem/year (0.8–1.8 millisieverts/year) from natural sources of ionizing radiation, depending on the organ considered. Most of this dose originates from radioactive materials in the Earth's crust. The external dose due to cosmic rays is an average of about 28 mrem/year (0.28 mSv/year), a value that increases with altitude due to reduced shielding of cosmic radiation by the atmosphere. The human body is also exposed to radionuclides in food and water. Potassium-40 is the most important of these, with radium-226 and radium-228 of perhaps less importance from the point of view of the dose delivered.

There are wide deviations from the average doses. Thus, at one extreme, miners working underground in the presence of radioactive ore can be exposed to such high levels of atmospheric radon that they develop lung cancer. There are also geographical areas where the levels of natural radioactivity are unusually high. Six types of anomalies that can be important from the point of view of population exposure are: monazite sands and other placers, alkaline intrusives and granites of the Conway type in New Hampshire, bauxites and intensely weathered soils, uraniferous phosphate rock (and soils), ground waters enriched in radium and radon, and black shales and related organic accumulations. The natural radioactive environment can also be altered by human activities, such as building construction, combustion of fossil fuels, aircraft travel, medical procedures, nuclear weapons testing, and nuclear power plants.

Although various national and international regulatory organizations have proposed guidelines that limit the per capita dose received by individuals in the general population to 170 mrem/year (1.7 mSv/year), it has become evident that nuclear power plants can be routinely operated so that the general population will not be exposed to more than 1% of this limit.   [M.E.]

Eocene The second oldest of the five major worldwide divisions (epochs) of the Tertiary Period (Cenozoic Era), the interval of time (epoch) extending from the end of the Paleocene Epoch to the beginning of the Oligocene Epoch; the middle epoch of the older Tertiary (Paleogene of some authors, Nummulitic of earlier French authors). See CENOZOIC; OLIGOCENE; PALEOCENE; TERTIARY.

The Paleocene/Eocene boundary has been formally defined at the 5.2-ft (1.6-m) level in the Dababiya Quarry section approximately 22 mi (35 km) south of Luxor in the Upper Nile Valley, Egypt. This level coincides with a global carbon isotope excursion associated with significant climatic warming and biotic changes and is about 1 million years older than the base of the classic Ypresian Stage, normally considered the oldest stage of the Eocene. There are varying opinions regarding what to do with the associated stage boundaries. The most prevalent proposes retaining the Ypresian Stage in its present position, with an estimated age for its base of about 54 Ma, and insert the Sparnacian Stage as the lowest stage of the newly redefined Eocene.

Eocene strata are widespread throughout the world and on the deep ocean floor. They include the common sedimentary types and vary from terrestrial, to marginal (estuarine), to normal marine pelagic origin. Igneous activity, while not as extensive as in the later part of the Cenozoic, was notable in some areas such as East Greenland, Oregon, Washington, and British Columbia.

Early Paleogene temperatures, including those of high latitudes, were the warmest of the Cenozoic; peak warming occurred in the early Eocene. The Earth was in a greenhouse state, with partial pressure of carbon dioxide (pCO_2) levels in the early Eocene estimated to have been six times higher than present-day values. During the late Paleocene to the early Eocene, deep-sea temperatures at high southern latitudes warmed by some 7–9°F (4–5°C), from about 50–52°F (10–11°C) to about 57–61°F (14–16°C), while surface temperatures increased by some 9–11°F (5–6°C), with maximum temperatures in excess of 68°F (20°C). At low latitudes, surface water temperatures remained relatively constant and comparable to values of the present-day ocean. Superimposed on this long-term trend was a relatively abrupt (<10,000 years) 2.5–3% drop in $\delta^{13}C$ (the difference in isotopic ratios $^{12}C/^{13}C$ between a sample and a standard) and concomitant marine productivity that has been associated, in turn, with a major turnover (extinction of almost 50%) of the deep-sea benthic (bottom-dwelling) forminiferal fauna. This drop in $\delta^{13}C$ has been identified both in marine organisms and in mammalian bone enamel and paleosol carbonates in terrestrial sections in the Big Horn Basin of the western interior of North America and in the type Sparnacian (that is earliest Eocene) in the Paris Basin. *See* EXTINCTION (BIOLOGY); GEOLOGIC THERMOMETRY; PALEOSOL.

The diversification of life seen in the Paleocene continued in the Eocene, a reflection of the poleward expansion of the tropics, particularly during early Eocene time. In the oceanic realm, microplanktonic animals (foraminiferans) and plants (calcareous nannoplankton) flourished and diversified, as did true bony fishes and siphonate gastropods. In shallow, tropical waters the so-called larger foraminiferans extended their geographic range to latitude 50° north, but the latter group disappeared at the end of the Eocene owing to cooling temperatures. Indeed, microplanktonic animals and plants experienced a gradual but inexorable decline in diversity starting in the late middle Eocene. Succeeding Oligocene faunas and floras were much reduced in diversity and much more uniformly distributed.

On land, subtropical floras extended as far north as southern England and the North American Pacific coast of Puget Sound and southern Alaska. Indeed, the floras of southern England resembled those of modern-day China, Malaysia, and Australia. In the humid interior, thick and extensive mud deposits (the Green River Shale) in Colorado contain a beautifully preserved fresh-water fish fauna eagerly sought after by fossil collectors.

Europe was separated from the Eurasian land mass east of the Urals by a north-south seaway extending from the Arctic to the Tethys Sea—the Turgai Straits. Following the elimination of the elevated corridor that allowed transatlantic poleward migration between Europe and North America in late early Eocene time, middle and late Eocene time witnessed the development of extensive endemic animal evolution. Bats, flying lemurs, creodont carnivores, artiodactyls (cloven-hoof mammals, such as cattle, deer, and camels) and perissodactyls (odd-toed, hoofed mammals, such as rhinoceroses and horses), notoungulates (predominantly South American), and edentates reflect the diversification of primitive placental forms. The massive, rhinoceroslike herbivores called titanotheres and uintatheres appeared alongside the small early progenitors of the modern horse, *Hyracotherium* (known popularly as *Eohippus*).

In the Eocene, some mammals turned toward life in the sea; sea cows appeared in the middle Eocene, while the earliest whales (zeuglodonts) appeared in the North Atlantic–Gulf of Mexico region and the aquatic ancestors of the proboscideans appeared in the late Eocene.   [W.A.Ber.]

Eolian landforms   Topographic features generated by the wind. The most commonly seen eolian landforms are sand dunes created by transportation and accumulation of windblown sand. Blankets of wind-deposited loess, consisting of fine-grained silt, are less obvious than dunes, but cover extensive areas in some part of the world.

Where abundant loose sand is available for the wind to carry, sand dunes develop. As soon as enough sand accumulates in one place, it interferes with the movement of air and a wind shadow is produced which contributes to the shaping of the pile of sand. Dunes advance downwind by erosion of sand on the windward side and redeposition on the slip face. Dunes may have a variety of shapes, depending on wind conditions, vegetation, and sand supply. The fine silt and clay winnowed out from coarser sand is often blown longer distances before coming to rest as a blanket of loess mantling the preexisting topography. Thick deposits of loess are most often found in regions downwind from glacial outwash plains or alluvial valleys. See DUNE; LOESS; SAND.   [D.J.E.]

Epidote   The group name for a family of minerals of general composition $Ca_2(Fe^{3+}$, Al, $Mn^{3+})Al_2O[SiO_4][Si_2O](OH)$ that occur widely in metamorphic and igneous rocks. Epidote [octahedral ferric iron (Fe^{3+}) dominant] and clinozoisite [aluminum (Al) dominant] represent the most common compositions among the epidote group; a third composition, piemontite [manganese (Mn^{3+}) dominant], is less abundant. Allanite refers to compositions displaying significant rare-earth (such as lanthanum or cerium) substitution for calcium (Ca^{2+}), with corresponding replacement of Fe^{3+} by ferrous iron (Fe^{2+}). A fifth member, zoisite, is equivalent to clinozoisite, but it has a different crystalline system. Rare epidote-clinozoisites abundant in chromium (Cr), vanadium (V), and lead (Pb) and allanites rich in fluorine (F), beryllium (Be), and phosphorus (P) also exist. See SOLID SOLUTION.

Epidote group minerals, particularly epidote and clinozoisite, are common and widespread in regional- and contact-metamorphic rocks, both as primary and secondary (that is, alteration) minerals. They occur together as individual grains, as intergrowths, or as zoned crystals. Epidote and (clino)zoisite are found in aluminous limestones with grossularite, anorthite, microcline, quartz, and calcite; in mafic schists and gneisses with hornblende, albite, and chloritoid; in actinolite greenschists with chlorite, sphene, albite, quartz, calcite, and magnetite; in hornfels with diopside, actinolite, grossularite, and albite; in glaucophane schists; in quartzites; and in slates. Approximate depth-temperature conditions of their formation range from 5–25 km (3–15 mi) and 300–500°C (570–1020°F; low-grade) to 5–25 km (3–15 mi) and 450–650°C (840–1200°F; medium-grade). However, their stabilities are also sensitive to the pressure of oxygen in the rock during metamorphism. See METAMORPHIC ROCKS.   [P.S.D.]

Epsomite   A mineral with the chemical composition $MgSO \cdot 7H_2O$. Epsomite, or epsom salt, occurs in clear, needle-like, orthorhombic crystals. More commonly it is massive or fibrous, although crystals from salt lakes on Kruger Mountain near Orville, Washington, are reported to be several feet long. Fracture is conchoidal. Luster varies from vitreous to silky. Hardness is 2–2.5 on Mohs scale and specific gravity is 1.68. The mineral has a salty bitter taste and is soluble in water. Epsomite is found as a capillary coating in limestone caves and in coal or metal mine galleries. It is also found associated with gypsum and in thin layers in salt deposits of oceanic origin or from salt lakes.   [E.C.T.C.]

Equator   The great circle around the Earth, equally distant from the North and South poles, which divides the Earth into Northern and Southern hemispheres. It is the greatest circumference of the Earth because of centrifugal force from rotation, and resultant flattening of the polar areas.

The Earth's rotational axis is vertical to the plane of the Equator, and because the inclination of the axis is $66^{1}/_{2}°$ from the plane of the ecliptic, the plane of the Equator is always inclined $23^{1}/_{2}°$ from the ecliptic.

The celestial equator in astronomy is equally distant from the celestial poles and is the great circle in which the plane of the terrestrial Equator intersects the celestial sphere. *See* MATHEMATICAL GEOGRAPHY.   [V.H.E.]

Equatorial currents   Ocean currents near the Equator. The westward trade winds that prevail over the tropical Atlantic and Pacific oceans drive complex oceanic circulations characterized by alternating bands of eastward and westward currents. The intense currents are confined to the surface layers of the ocean; below a depth of approximately 100 m (330 ft) the temperature is much lower, and the speed of ocean currents is much slower. The westward surface currents tend to be divergent—they are associated with a parting of the surface waters—and therefore entrain cold water from below. The water temperature rises as the currents flow westward, so that temperatures are low in the east and high in the west, except between 3 and 10°N where eastward surface currents create a band of warm water across the Pacific and Atlantic oceans. The distinctive sea surface temperature pattern in which surface waters are warm in the west and cold in the east, except for the warm band just north of the Equator, reflects the oceanic circulation. A dramatic change in this pattern every few years during El Niño episodes, when the temperature of the eastern tropical Pacific Ocean rises, is associated with an intensification of the eastward currents and a weakening (sometimes reversal) of the westward currents. *See* EL NIÑO.

The South Equatorial Current flows westward in the upper ocean, has its northern boundary at approximately 3°N, and attains speeds in excess of 1 m/s (3.3 ft/s) near the Equator. It is directly driven by the westward trade winds and has its origins in the cold, northwestward-flowing Peruvian coastal current. Because the Coriolis force deflects water parcels to their right in the Northern Hemisphere and to their left in the Southern Hemisphere, this current is divergent at the Equator. As a consequence, cold water from below wells up along the Equator. *See* CORIOLIS ACCELERATION; UPWELLING.

The North Equatorial Countercurrent flows eastward immediately to the north of the South Equatorial Current. The boundary between these two currents is a sharp thermal front that is clearly evident in satellite photographs. The front can literally be a green line, hundreds of yards wide, because of the abundance of phytoplankton. This current, which is counter to the wind, is driven by the torque (curl) that the wind exerts on the ocean. To its north is a colder westward current known as the North Equatorial Current. *See* OCEAN WAVES.

The Equatorial Undercurrent, which in the Pacific Ocean was originally known as the Cromwell Current, is an intense, narrow, eastward, subsurface jet that flows precisely along the Equator across the width of the Pacific. Its core, where speeds can be in excess of 1.5 m/s (5 ft/s), is at an approximate depth of 100 m (330 ft); its width is approximately 200 km (120 mi). A similar current exists in the Atlantic Ocean. In the Indian Ocean it is often present along the Equator, in the western part of the basin during March and April when westward winds prevail over that region. Such winds (including the trade winds over the Pacific and Atlantic oceans) pile up warm surface waters in the west while exposing cold waters to the surface in the east. *See* ATLANTIC OCEAN; INDIAN OCEAN; PACIFIC OCEAN. [S.G.P.]

Erosion   The result of processes that entrain and transport earth materials along coastlines, in streams, and on hillslopes. Wind and water are common agents through which forces are applied to resistant rocks, soils, or other unconsolidated materials. Erosion types often are designated on the basis of the agent: wind erosion, fluvial (water) erosion, and glacial erosion. Fluvial erosion usually has been regarded as the most effective type in shaping the land surface during recent geologic time. Under certain environmental conditions, however, wind erosion moves considerable quantities of earth materials, as demonstrated during the "dust bowl" years in the United States. Glacial erosion shaped much of the land surface during the Quaternary Period of geologic time. Each type of erosion produces distinctive landforms, contributing to the diversity of terrestrial landscapes. *See* DESERT EROSION FEATURES; EOLIAN LANDFORMS; FLUVIAL EROSION LANDFORMS; GEOMORPHOLOGY; GLACIOLOGY; MASS WASTING; QUATERNARY; STREAM TRANSPORT AND DEPOSITION.

Forces exerted by erosion processes must exceed resistances of earth materials for entrainment and transportation to occur. Environmental conditions determine the magnitude of the forces, the resistances, and the relations among them. Erosion rates are highly variable in time and space due to changing relations between forces and resistances. The major factors governing wind-erosion rates are wind velocity, topography, surface roughness, soil properties and soil moisture, vegetation cover, and land use. The major factors governing fluvial-erosion rates on hillslopes are rainfall energy, topography, soil properties, vegetation cover, and land use. The major factors governing fluvial-erosion rates in stream channels are depth and velocity of water flow, together with the size and cohesiveness of the bed and bank materials. The major factors governing glacial-erosion rates are the depth and velocity of ice flow, together with the hardness of the bed and side-wall materials.

Accelerated erosion by fluvial processes may be the most important environmental problem worldwide because of its spatial and temporal ubiquity. Erosion rates commonly exceed soil-formation rates, causing depletion of soil resources. The effects of erosion are insidious due to the removal of the fertile topsoil horizon, compromising food production. Sediment frequently is transported well beyond the source area to degrade water quality in streams and lakes, harm aquatic life, reduce the water-storage capacity of reservoirs, and increase channel-maintenance costs. *See* SOIL.   [T.J.T.]

Escarpment   A long line of cliffs or steep slopes that break the general continuity of the land by separating it into two level or sloping surfaces. Some very high escarpments, or scarps, may form by vertical movement along faults. Often a whole block of land may be forced upward while the adjacent block is downfaulted. *See* FAULT AND FAULT STRUCTURES.

Other types of escarpments form by differential weathering and erosion of contrasted rock types. Less resistant rocks, such as clay or shale, are often eroded from beneath resistant cap rocks, such as sandstone and limestone. With support removed from below, the cap rock fails and the escarpment retreats. Escarpments are often very prominent in arid regions, where hardened weathering products may form extensive cap rocks known as duricrusts.

Some of the largest known escarpments occur on the planet Mars, where erosion has presumably been much slower than on the Earth in reducing primary structural relief.   [V.R.B.]

Esker   A sinuous ridge composed predominantly of sand and gravel deposited by glacial meltwater. Eskers vary in degree of continuity, and range in size from a few meters (1 m = 3.3 ft) to tens of meters high and from a few meters to a hundred or more kilometers long. They have steep ice-contact slopes and were deposited in channels confined by ice. Most eskers generally parallel the direction of ice flow, and while most follow valleys and have a normal down-drainage slope, some trend up a regional or local slope.   [W.H.J.]

Estuarine oceanography   The study of the physical, chemical, biological, and geological characteristics of estuaries. An estuary is a semienclosed coastal body of water which has a free connection with the sea and within which the sea water is measurably diluted by fresh water derived from land drainage. Many characteristic features of estuaries extend into the coastal areas beyond their mouths, and because the techniques of measurement and analysis are similar, the field of estuarine oceanography is often considered to include the study of some coastal waters which are not strictly, by the above definition, estuaries. Also, semienclosed bays and lagoons exist in which evaporation is equal to or exceeds freshwater inflow, so that the salt content is either equal to that of the sea or exceeds it. Hypersaline lagoons have been termed negative estuaries, whereas those with precipitation and river inflow equaling evaporation have been called neutral estuaries. Positive estuaries, in which river inflow and precipitation exceed evaporation, form the majority, however.

Within estuaries, the river discharge interacts with the sea water, and river water and sea water are mixed by the action of tidal motion, by wind stress on the surface, and by the river discharge forcing its way toward the sea. There is a small difference in salinity between river water and sea water, but it is sufficient to cause horizontal pressure gradients within the water which affect the way it flows. Salinity is consequently a good indicator of estuarine mixing and the patterns of water circulation.

Estuarine ecological environments are complex and highly variable when compared with other marine environments. They are richly productive, however. Because of the variability, fewer species can exist as permanent residents in this environment than in some other marine environments, and many of these species are shellfish that can easily tolerate short periods of extreme conditions. Motile species can escape the extremes. A number of commercially important marine torms are indigenous to the estuary, and the environment serves as a spawning or nursery ground for many other species. *See* MARINE ECOLOGY.

The patterns of sediment distribution and movement depend on the type of estuary and on the estuarine topography. The type of sediment brought into the estuary by the rivers, by erosion of the banks, and from the sea is also important; and the relative importance of each of these sources may change along the estuary. Fine-grained material will move in suspension and will follow the residual water flow, although there may be deposition and re-erosion during times of locally low velocities. The coarser-grained material will travel along the bed and will be affected most by high velocities and, consequently, in estuarine areas, will normally tend to move in the direction of the maximum current.   [K.R.D.]

Europe   Although long called a continent, in many physical ways Europe is but a great western peninsula of the Eurasian landmass. Its eastern limits are arbitrary and are conventionally drawn along the water divide of the Ural Mountains, the Ural River, the Caspian Sea, and the Caucasus watershed to the Black Sea. On all other sides Europe is surrounded by salt water. Of the oceanic islands of Franz Josef Land, Spitsbergen (Svalbard), Iceland, and the Azores, only Iceland is regarded as an integral part of Europe; thus the northwestern boundary is drawn along the Danish Strait.

Europe is not only peninsular but has a large ratio of shoreline to land area reflecting a notable interfingering of land and sea. Excluding Iceland, the maximum north-south distance is (3529 mi) (5680 km); and the greatest east-west extent is 2398 mi (3860 km). Of Europe's area of 3,881,000 mi^2 (10,050,000 km^2) 73% is mainland, 19% peninsulas, and 8% islands. Also, 51% of the land is less than 155 mi (250 km) from shores and another 23% lies closer than 310 mi (500 km). This situation is caused by the inland seas that enter, like arms of the ocean, deep into the northern and southern regions of Europe, which thus becomes a peninsula of peninsulas. The most notable of these branching arms of salt water are the White Sea, the North Sea, the Baltic Sea with the Gulf of Bothnia, the English Channel (La Manche), the Mediterranean Sea with its secondary branches, and finally, the Black Sea. Even the Caspian Sea, presently the largest saltwater lake of the world, formed part of the southern seas before the folding of the Caucasus. The penetration of the landmass by these seas brings marine influences deep into the continent and provides Europe with a balanced climate favorable for human evolution and settlement.

Europe has a unique diversity of land forms and natural resources. The relief, as varied as that of other continents, has an average elevation of 980 ft (300 m) as compared with North America's 1440 ft (440 m). The shape and the overall physiographic aspect of the great peninsula are controlled by geologic structure which delimits the major regional units.

Climate is determined by a number of factors. Probably the most important are a favorable location between 35° and 71°N latitudes on the western or more maritime side of the world's largest continental mass; the west-to-east trend (rather than north-south) of the lofty southern ranges and the Central Lowlands, as well as of the inland seas, which permit the prevailing westerly winds of these latitudes to carry marine influences deep into the continent; the beneficial influence of the North Atlantic Drift, which makes possible ice-free coasts far within the Arctic Circle; and the low elevation of the northwestern mountain ranges and the Urals, which allows the free shifting of air masses over their crests.

The intricate relief and the climates of Europe are well reflected in the drainage system. Extensive drainage basins with large slow-flowing rivers are developed only in the Central Lowlands, especially in the eastern part. Streams with the greatest discharge empty into the Black Sea and the North Sea, although Europe's longest river, the Volga, feeds the Caspian Sea. Second in dimension is the Danube, which crosses the Carpathian Basin and cuts its way twice through mountain ranges at the Gate of Bratislava and at the Iron Gate. The Rhine and Rhone are the two major Alpine rivers with headwater sources close to each other but feeding the North Sea and the Western Mediterranean Basin, respectively. Abundant precipitation throughout the year, as well as the permeable soils and the dense vegetation which temporarily store the water, provides the streams of Europe north of the Southern Highlands with ample water throughout the seasons. The combined effects of poor vegetation, rocky and desolate limestone karstlands, and slight annual precipitation result in intermittent flow of the rivers along the Mediterranean coast, especially on the eastern side of peninsulas. Only the Alpine rivers carry enough water, and if it were not for the Danube and Rhone, both originating in regions north of the Alps, the only major river of the

Mediterranean basin would be the Po. *See* ATLANTIC OCEAN; BALTIC SEA; BLACK SEA; CONTINENT; MEDITERRANEAN SEA.   [G.T.]

Eutrophication   The deterioration of the esthetic and life-supporting qualities of lakes and estuaries, caused by excessive fertilization from effluents high in phosphorus, nitrogen, and organic growth substances. Algae and aquatic plants become excessive, and, when they decompose, a sequence of objectionable features arise. Water supplies drawn from such lakes must be filtered and treated. Diversion of sewage, better utilization of manure, erosion control, improved sewage treatment and harvesting of the surplus aquatic crops alleviate the symptoms. Prompt public action is essential.
[A.D.H.]

Extinction (biology)   The death and disappearance of a species. The fossil record shows that extinctions have been frequent in the history of life. Mass extinctions refer to the loss of a large number of species in a relatively short period of time. Episodes of mass extinction occur at times of rapid global environmental change; five such events are known from the fossil record of the past 600 million years. Human activity is causing extinctions on a scale comparable to the mass extinctions in the fossil record.

Record. An extinction may be of two types; phyletic or terminal. Phyletic extinction occurs when one species evolves into another with time; in this case, the ancestral species can be called extinct. However, because the evolutionary lineage has continued, such extinctions are really pseudoextinctions. In contrast, terminal extinction marks the end of an evolutionary lineage, termination of a species without any descendants. Most extinctions recorded in the fossil record and those occurring today are terminal. It has been estimated that 99% of all species that have ever lived are now extinct.

The fossil record is best known for marine organisms. The mass extinctions of the marine fossil record occurred during the Late Ordovician, Late Devonian, Late Permian, Late Triassic, and Late Cretaceous. These mass extinctions affected a variety of organisms in many different ecological settings. Terrestrial and marine mass extinctions seem to occur at about the same time. The Late Permian, Late Triassic, and Late Cretaceous are also times of extinction for terrestrial vertebrates; the most dramatic extinction of terrestrial vertebrates took place at the end of the Cretaceous, when the last dinosaurs died off.

The best record of terrestrial vertebrate extinction is that of the Pleistocene. Late Pleistocene extinctions in North America are especially well known—33 genera of mammals vanished during the last 100,000 years. These extinctions were concentrated among the large mammals—those over 100 lb (44 kg) in weight—and most occurred during a short time interval approximately 11,000 years ago.

Causes. Ever since the work of Georges Cuvier, the French naturalist who demonstrated the reality of extinction, explanations have fascinated both scientists and the general public. Cuvier invoked sudden catastrophic events, whereas his contemporaries favored more gradual processes. These two themes, catastrophism and gradualism, are still debated.

In 1980 high concentrations of iridium were reported precisely at the Cretaceous-Tertiary boundary. Iridium is rare in most rocks but more abundant in meteorites. It was proposed, therefore, that an asteroid struck the Earth 65 million years ago. The impact darkened the atmosphere with dust, caused a catastrophic short-term cooling of the climate, and thus led to the extinction of dinosaurs and many other Cretaceous species. The iridium-rich layer at the boundary marks this terminal Cretaceous event.

Astronomical theories have been put forward to explain the Late Cretacous extinctions as well as the 26-million-year periodicity. In one theory, the Sun has a distant companion star that would pass in orbit near the solar system's cloud of comets every 26 million years. This might perturb many comets, sending a few into the Earth. A comet would produce the same effects as an asteroid.

Other explanations for mass extinctions include lowered sea level, climatic cooling, and changes in oceanic circulation. Biotic processes such as disease, predation, and competition may also cause the extinction of species but are difficult to prove from the fossil record because they leave little evidence. Biotic factors usually affect only one or a few interdependent species. Predation and competition are important causes of more recent extinctions, which continue today. Human activities such as hunting and fishing (predation), habitat alteration (competition for space), and pollution have probably destroyed thousands of species. These activities, together with continued tropical deforestation and resulting changes in climate, are likely to cause extinctions that will be comparable to the mass extinctions seen in the fossil record. *See* Fossil.

[K.W.F.]

F

Facies (geology) Any observable attribute of rocks, such as overall appearance, composition, or conditions of formation, and changes that may occur in these attributes over a geographic area. The term facies is widely used in connection with sedimentary rock bodies, but is not restricted to them. In general, facies are not defined for sedimentary rocks by features produced during weathering, metamorphism, or structural disturbance. In metamorphic rocks specifically, however, facies may be identified by the presence of minerals that denote degrees of metamorphic change.

Sedimentary facies. The term sedimentary facies is applied to bodies of sedimentary rock on the basis of descriptive or interpretive characteristics. Descriptive facies are based on lithologic features such as composition, grain size, bedding characteristics, and sedimentary structures (lithofacies) or on biological (fossil) components (biofacies), or on both. Individual lithofacies or biofacies may be single beds a few millimeters thick or a succession of beds tens to hundreds of meters thick. For example, a river deposit may consist of decimeters-thick beds of a conglomerate lithofacies interbedded with a cross-bedded sandstone lithofacies. The fill of certain major Paleozoic basins may be divided into units hundreds of meters thick comprising a shelly facies, containing such fossils as brachiopods and trilobites, and graptolitic facies.

The term facies can be used also in an interpretive sense for groups of rocks that are thought to have been formed under similar conditions. This usage may emphasize specific depositional processes, such as a turbidite facies, or a particular depositional environment such as a shelf carbonate facies, encompassing a range of depositional processes.

Groups of facies (usually lithofacies) that are commonly found together in the sedimentary record are known as facies assemblages or facies associations. These groupings provide the basis for defining broader, interpretive facies for the purpose of paleogeographic reconstruction. *See* PALEOGEOGRAPHY. [A.D.M.]

A metamorphic facies is a collection of rocks containing characteristic mineral assemblages developed in response to burial and heating to similar depths and temperatures. It can represent either the diagnostic mineral assemblages that indicate the physical conditions of metamorphism or the pressure-temperature conditions that produce a particular assemblage in a rock of a specific composition.

The metamorphic facies to which a rock belongs can be identified from the mineral assemblage present in the rock; the pressure and temperature conditions represented by each facies are broadly known from experimental laboratory work on mineral stabilities. These facies names are based on the mineral assemblages that develop during metamorphism of a rock with the composition of a basalt, which is a volcanic rock rich in iron and magnesium and with relatively little silica. For example, the dominant mineral of the blueschist facies (in a rock of basaltic composition) is a sodium- and magnesium-bearing silicate called glaucophane, which is dark blue in outcrop and blue or violet when viewed under the microscope. Characteristic minerals of the greenschist facies include chlorite and actinolite, both of which are green in outcrop and under the microscope. Basaltic rocks metamorphosed in the amphibolite facies are largely composed of an amphibole called hornblende. The granulite facies takes its name from a texture rather than a specific mineral: the pyroxenes and plagioclase that are common minerals in granulite facies rocks typically form rounded crystals of similar size that give the rock a granular fabric. *See* AMPHIBOLITE; BASALT; BLUESCHIST; GLAUCOPHANE; GRANULITE; METAMORPHIC ROCKS; METAMORPHISM; PYROXENE.   [J.Sel.]

Fault and fault structures   Products of fracturing and differential movements along fractures in continental and oceanic crustal rocks. Faults range in length and magnitude of displacement from small structures visible in hand specimens, displaying offsets of a centimeter or less, to long, continuous crustal breaks, extending hundreds of kilometers in length and accommodating displacements of tens or hundreds of kilometers. Faults exist in deformed rocks at the microscopic scale, but these are generally ignored or go unrecognized in most geological studies. Alternatively, where microfaults systematically pervade rock bodies as sets of very closely spaced subparallel, planar fractures, they are recognized and interpreted as a type of cleavage which permitted flow of the rock body. Fractures along which there is no visible displacement are known as joints. Large fractures which have accommodated major dilational opening (a meter or more) perpendicular to the fracture surfaces are known as fissures. Formation of fissures is restricted to near-surface conditions, for example, in areas of crustal stretching of subsidence.

In addition to describing the physical and geometric nature of faults and interpreting time of formation, it has been found to be especially important to determine the orientations of minor fault structures (such as striae and drag folds) which record the sense of relative movement. Evaluating the movement of faulting can be difficult, for the apparent relative movement (separation) of fault blocks as seen in map or outcrop

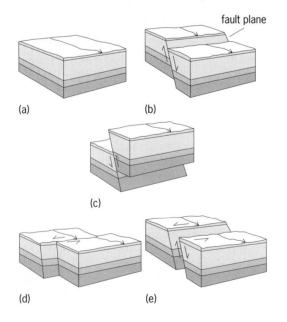

Slip on faults. (*a*) Block before faulting; (*b*) normal-slip; (*c*) reverse-slip; (*d*) strike-slip; (*e*) oblique-slip. (*After F. Press and R. Siever, Earth, 2d ed., 1978*)

may bear little or no relation to the actual relative movement (slip). The slip of the fault is the actual relative movement between two points or two markers in the rock that were coincident before faulting (see illustration). Strike-slip faults have resulted in horizontal movements between adjacent blocks; dip-slip faults are marked by translations directly up or down the dip of the fault surface; in oblique-slip faults the path of actual relative movement is inclined somewhere between horizontal and dip slip.

Recognizing even the simplest translational fault movements in nature is often enormously difficult because of complicated and deceptive patterns created by the interference of structure and topography, and by the absence of specific fault structures which define the slip path. While mapping, the geologist mainly documents apparent relative movement (separation) along a fault, based on what is observed in plan-view or cross-sectional exposures. *See* TRANSFORM FAULT.   [G.H.D.]

Feldspar   Any of a group of aluminosilicate minerals whose crystal structures are composed of corner-sharing $[AlO_4]$ and $[SiO_4]$ tetrahedra linked in an infinite three-dimensional array, with charge-balancing cations primarily sodium (Na), potassium (K), and calcium (Ca)] occupying large, irregular cavities in the framework of the tetrahedra. Collectively, the feldspars constitute about 60% of the outer 8–10 mi (13–17 km) of the Earth's crust. They are nearly ubiquitous igneous and metamorphic rocks, and are a primary constituent of arkosic sediments derived from them. The importance of the many feldspars that occur so widely in igneous, metamorphic, and some sedimentary rocks cannot be underestimated, especially from the viewpoint of a petrologist attempting to unravel earth history. *See* ARKOSE; MINERALOGY; PETROLOGY; SILICATE MINERALS.

With weathering, feldspars form commercially important clay materials. Economically, feldspars are valued as raw material for the ceramic and glass industries, as fluxes in iron smelting, and as constituents of scouring powders. Occasionally their luster or colors qualify them as semiprecious gemstones. Some decorative building and monument stones are predominantly composed of weather-resistant feldspars. *See* CLAY MINERALS; IGNEOUS ROCKS; METAMORPHIC ROCKS.

The general formula AT_4O_8 characterizes the chemistry of feldspars, where T (for tetrahedrally coordinated atom) represents aluminum (Al) or silicon (Si). The A atom is Ca^{2+} or barium (Ba^{2+}) for the $[Al_2Si_2O_8]^{2-}$ alkaline-earth feldspars and Na^+ or K^+ for the $[AlSi_3O_8]^-$ alkali feldspar series of solid solutions and mixed crystals.

Knowledge of a feldspar's composition and its crystal structure is indispensable to an understanding of its properties. However, it is the distribution of the Al and Si atoms among the available tetrahedral sites in each chemical species that is essential to a complete classification scheme, and is of great importance in unraveling clues to the crystallization and thermal history of many igneous and metamorphic rocks.

Alkali feldspars are assigned to the polymorphs of $KAlSi_3O_8$ and $NaAlSi_3O_8$ in accordance with their symmetry and the Al content of their tetrahedral sites. *See* ALBITE; MICROCLINE; ORTHOCLASE.

Anorthoclase is a triclinic solid solution of composition $Or_{37}Ab_{63}$-Or_0Ab_{100} containing up to 10 mol % anorthite, or more. *See* ANORTHOCLASE.

Plagioclase feldspars containing significant amounts of exsolved K-rich feldspar are called antiperthites. It is only in once-molten rocks quenched at very high temperatures that the full range of so-called high plagioclases exist as simple solid solutions. With very slow cooling over millions of years, complex textures develop in most feldspar crystals as a coupled NaSi, CaAl ordering. *See* ANDESINE; BYTOWNITE; LABRADORITE; OLIGOCLASE; SOLID SOLUTION.

The variable properties of feldspars are determined by their structure, symmetry, chemical composition, and crystallization and subsequent history of phase transformation, exsolution, and alternation or deformation. Very few feldspars are transparent and colorless; many are white or milky due to internal reflections of light from inclusions, exsolution interfaces, and fracture or cleavage surfaces. Plagioclases are slightly harder (6–6.5) on Mohs scale than K-rich feldspars (6). Feldspars are brittle and, when broken, cleave along the (001) and (010) crystallographic planes.   [P.H.R.]

Feldspathoid   A member of the feldspathoid group of minerals. Members of this group are characterized by the following related features: (1) All are aluminosilicates with one or more of the large alkali ions (for example, sodium, potassium) or alkaline-earth ions (for example, calcium, barium). (2) The proportion of aluminum relative to silicon, both of which are tetrahedrally coordinated by oxygen, is high. (3) Although the crystal structures of many members are different, they are all classed as tektosilicates. (4) They occur principally in igneous rocks, but only in silica-poor rocks, and do not coexist with quartz (SiO_2). Feldspathoids react with silica to yield feldspars, which also are alkali-alkaline-earth aluminosilicates. Feldspathoids commonly occur with feldspars. *See* FELDSPAR; SILICATE MINERALS.

The principal species of this group are the following:

Nepheline	$KNa_3[AlSiO_4]_4$
Leucite	$K[AlSi_2O_6]$
Cancrinite	$Na_6Ca[CO_3 \mid (AlSiO_4)_6] \cdot 2H_2O$
Sodalite	$Na_8[Cl_2 \mid (AlSiO_4)_6]$
Nosean	$Na_8[SO_4 \mid (AlSiO_4)_6]$
Haüyne	$(Na,Ca)_{8-4}(SO_4)_{2-1}(AlSiO_4)_6]$
Lazurite	$(Na,Ca)_8[(SO_4,S,Cl)_2 \mid (AlSiO_4)_6]$

The last four species (sodalite group) are isostructural, and extensive solid solution occurs between end members; but members of the sodalite group, cancrinite, leucite, and nepheline have different crystal structures. *See* CANCRINITE; LAZURITE; LEUCITE ROCK; SODALITE.   [D.R.P.]

Fiord   A segment of a troughlike glaciated valley partly filled by an arm of the sea. It differs from other glaciated valleys only in the fact of submergence. The floors of many fiords are elongate basins excavated in bedrock, and in consequence are shallower at the fiord mouths than in the inland direction. The seaward rims of such basins represent lessening of glacial erosion at the coastline, where the former glacier ceased to be confined by valley walls and could spread laterally. Some rims are heightened by glacial drift deposited upon them in the form of an end moraine.

Fiords occur conspicuously in British Columbia and southern Alaska, Greenland, Arctic islands, Norway, Chile, New Zealand, and Antarctica—all of which are areas of rock resistant to erosion, with deep valleys, and with strong former glaciation.   [R.F.FI.]

Fission track dating   A method of dating geological and archeological specimens by counting the radiation-damage tracks produced by spontaneous fission of uranium impurities in minerals and glasses. During fission two fragments of the uranium nucleus fly apart with high energy, traveling a total distance of about 25 micrometers (0.001 in.) and creating a single, narrow but continuous, submicroscopic trail of altered material, where atoms have been ejected from their normal positions. Such a trail, or track, can be revealed by using a chemical reagent to dissolve the altered material, and the trail can then be seen in an ordinary microscope. The holes produced in this way can be enlarged by continued chemical attack until they are visible to the unaided eye.

Track dating is possible because most natural materials contain some uranium in trace amounts and because the most abundant isotope of uranium, ^{238}U, fissions spontaneously. Over the lifetime of a rock substantial numbers of fissions occur; their tracks are stored and thus leave a record of the time elapsed since track preservation began. The number of tracks produced in a given volume of material depends on the uranium content as well as the age, so that it is necessary to measure the uranium content before an age can be determined.

One feature unique to this dating technique is the time span to which it is applicable. It ranges from less than 100 years for certain synthetic, decorative glasses to approximately 4,500,000,000 years, the age of the solar system. A second useful feature is that measurements can sometimes be made on extremely minute specimens, such as chips of meteoritic minerals or fragments of glass from the ocean bottom. A third useful feature is that each mineral dates the last cooling through the temperature below which tracks are retained permanently. Since this temperature is different for each mineral, it is possible to measure the cooling rate of a rock by dating several minerals—each with a different track-retention temperature. *See* AMINO ACID DATING; ROCK AGE DETERMINATION.

[R.L.F.; P.B.P.; R.M.W.]

Floodplain The relatively broad and smooth valley floor that is constructed by an active river and periodically covered with floodwater from that river during intervals of overbank flow. Engineers consider the floodplain to be any part of the valley floor subject to occasional floods that threaten life and property. Various channel improvements or impoundments may be used to restrict the natural process of overbank flow. Geomorphologist consider the floodplain to be a surface that develops by the active erosional and depositional processes of a river. Floodplains are underlain by a variety of sediments, reflecting the fluvial history of the valley.

Most floodplains consist of the following types of deposits: colluvium—slope wash and mass-wasting products from the valley sides, as is common in small, narrow floodplains; channel lag—coarse debris marking the bottoms of former channels; lateral accretion deposits—sand and gravel deposited as the meandering river migrates laterally; vertical accretion deposits—clay and silt deposited by overbank flooding of the river; crevassesplay deposits—relatively coarse sediment carried through breaks in the natural river levees and deposited in areas that usually receive overbank deposition; and channel-fill deposits—fills of former river channels. Channel fills may be coarse for sandy rivers. The noncohesive character of coarse sediments allows these rivers to easily erode laterally. *See* STREAM TRANSPORT AND DEPOSITION.

[V.R.B.]

Fluorite A mineral of composition CaF_2. It is the most abundant fluorine-bearing mineral, and occurs as cubes or compact masses and more rarely as octahedra with complex modifications. Fluorite has a perfect octahedral cleavage, hardness 4 (Mohs scale), and specific gravity 3.18. The color is extremely variable, the most common being green and purple; but fluorite may also be colorless, white, yellow, blue, or brown. Colors may result from the presence of impurity ions. Fluorite frequently emits a blue-to-green fluorescence under ultraviolet radiation, especially if rare-earth or hydrocarbon material is present. Some fluorites are thermoluminescent; that is, they emit light when heated.

Fluorite occurs as a typical hydrothermal vein mineral with quartz, barite, calcite, sphalerite, and galena. Crystals of great beauty from Cumberland, England, and

Rosiclare, Illinois, are highly prized by mineral fanciers. It also occurs as a metasomatic replacement mineral in limestones and marbles.   [P.B.M.]

Fluvial erosion landforms   Landforms that result from erosion by water flowing on land surfaces. This water may concentrate in channels as streams and rivers or flow in thin sheets and rills down slopes. Essentially all land surfaces are subjected to modification by running water, and it is among the most important surface processes. Valleys are cut, areas become dissected, and sediment is moved from land areas to ocean basins. With increasing dissection and lowering of the landscape, the land area may pass through a series of stages known as the fluvial erosion cycle.

The most distinctive fluvial landform is the stream valley. Valleys range greatly in size and shape, as do the streams that flow in them. They enlarge both through down and lateral cutting by the stream and mass wasting processes acting on the valley sides.

Waterfalls occur where there is a sudden drop in the stream bed. This is often the case where a resistant rock unit crosses the channel and the stream is not able to erode through it at the same rate as the adjacent less resistant rock. Waterfalls also occur where a main valley has eroded down at a faster rate than its tributary valleys which are left hanging above the main stream. With time, waterfalls migrate upstream and are reduced to rapids.

Many streams flow in a sinuous or meandering channel, and stream velocity is greatest around the outside of meander bends. Erosion is concentrated in this area, and a steep, cut bank forms. If the river meander impinges against a valley wall, the valley will be widened actively.

A stream terrace represents a former floodplain which has been abandoned as a result of rejuvenation or downcutting by the stream. It is a relatively flat surface with a scarp slope that separates it from the current floodplain or from a lower terrace. Terraces are common features in valleys and are the result of significant changes in the stream system through time. *See* FLOODPLAIN.

Fluvial erosion also has regional effects. Streams and their valleys form a drainage network which reflects the original topography and geologic conditions in the drainage basin. A dendritic drainage pattern, like that of a branching tree, is the most common and reflects little or no control by underlying earth materials. Where the underlying earth materials are not uniform in resistance, streams develop in the least resistant areas, and the drainage pattern reflects the geology. If the rocks contain a rectangular joint pattern, a rectangular drainage pattern develops; if the rock units are tilted or folded, a trellis pattern of drainage is common. Topography also controls drainage development; parallel and subparallel patterns are common on steep slopes, and a radial pattern develops when streams radiate from a central high area. *See* EROSION; RIVER; STREAM TRANSPORT AND DEPOSITION.   [W.H.J.]

Fluvial sediments   Deposits formed by rivers. A river accumulates deposits because its capacity to carry sediment has been exceeded, and some of the sediment load is deposited. Such accumulations range from temporary bars deposited on the insides of meander bends as a result of a loss of transport energy within a local eddy, to deposits tens to hundreds of meters thick formed within major valleys or on coastal plains as a result of the response of rivers to a long-term rise in base level or to the uplift of sediment source areas relative to the alluvial plain. The same processes control the style of rivers and the range of deposits that are formed, so that a study of the deposits may enable the geologist to reconstruct the changes in controlling factors during the accumulation of the deposits. *See* DEPOSITIONAL SYSTEMS AND ENVIRONMENTS; RIVER; STREAM TRANSPORT AND DEPOSITION.

Coarse debris generated by mechanical weathering, including boulders, pebbles, and sand, is rolled or bounced along the river bed and is called bedload. The larger particles may be moved only infrequently during major floods. Finer material, of silt and clay grade, is transported as a suspended load, and there may also be a dissolved load generated by chemical weathering. Whereas the volume of sediment tends to increase downstream within a drainage system, as tributaries run together, the grain size generally decreases as a result of abrasion and selective transport. This downstream grain-size decrease may assist in the reconstruction of transport directions in ancient deposits where other evidence of paleogeography has been obscured by erosion or tectonic change. *See* Mass wasting.

River deposits of sediment occur as four main types. (1) Channel-floor sediments consist of the coarsest bedload, such as gravel, waterlogged vegetation, or fragments of caved bank material. (2) Bar sediments are accumulations of gravel, sand, or silt which occur along river banks and are deposited within channels, forming bars that may be of temporary duration, or may last for many years, eventually becoming vegetated and semipermanent. (3) Channel-top and bar-top sediments are typically composed of fine-grained sand and silt, and are formed in the shallow-water regions on top of bars, in the shallows at the edges of channels, and in abandoned channels. (4) Floodplain deposits are formed when the water level rises above the confines of the channel and overflows the banks. Much of the coarser floodplain sediment is deposited close to the channel, in the form of levees; silt and mud may be carried considerable distances from the channel, forming blanketlike deposits. *See* Floodplain.

The thickest (up to 6 mi or 10 km) and most extensive fluvial deposits occur in convergent plate-tectonic settings, including regions of plate collision, because this is where the highest surface relief and consequently the most energetic rivers and most abundant debris are present. Some of the most important accumulations occur in foreland basins, which are formed where the continental margin is depressed by the mass of thickened crust formed by convergent tectonism. *See* Basin.

Thick fluvial deposits also occur in rift basins, where continents are undergoing stretching and separation. The famous hominid-bearing sediments of Olduvai Gorge and Lake Rudolf are fluvial and lacustrine deposits formed in the East Africa Rift System. Fluvial deposits are also common in wrench-fault basins, such as those in California.

Significant volumes of oil and gas are trapped in fluvial sandstones. Placer gold, uranium, and diamond deposits of considerable economic importance occur in the ancient rock record in South Africa and Ontario, Canada, and in Quaternary deposits in California and Yukon Territory. Fluvial deposits are also essential aquifers, especially the postglacial valley-fill complexes of urban Europe and North America.   [A.D.M.]

Fog   A cloud comprising waterdroplets or (less commonly) ice crystals formed near the ground and resulting in a reduction in visibility to below 0.6 mi (1 km). This is lower than that occurring in mist, comprising lower concentration of waterdroplets, and haze, comprising smaller-diameter aerosol particles.

Fog results from the cooling of moist air below its saturation (dew) point. Droplets form on hygroscopic nuclei originating from ocean spray, combustion, or reactions involving trace chemicals in the atmosphere. Visibility is reduced even more when such nuclei are present in high concentrations and faster cooling rates activate a larger fraction of such nuclei. Thus, polluted fog, with more numerous smaller droplets, results in lower visibility for a given water content. *See* Dew point.

Haze, the precursor to fog and mist, forms at relative humidity below 100% to about 80%. It is composed of hygroscopic aerosol particles grown by absorption of water vapor to a diameter of about 0.5 micrometer, concentration 1000 to 10,000 per cubic centimeter. Fog and mist form as the relative humidity increases just beyond saturation (100%), so that larger haze particles grow into cloud droplets with a diameter of 10 μm and a concentration of several hundred per cubic centimeter. Fog and mist are a mix of lower-concentration cloud droplets and higher-concentration haze particles. By contrast, smog is formed of particles of 0.5–1-μm diameter, produced by photochemical reactions with organic vapors from automobile exhaust. *See* ATMOSPHERIC CHEMISTRY; HUMIDITY; SMOG.   [J.Hal.]

Fold and fold systems   Layered rocks that have been distorted into wavelike forms. Some folds are fractions of an inch across and have lengths measured in inches, whereas others are a few miles wide and tens of miles long.

Some terms used to describe folds are shown in the illustration. The axial surface divides a fold into two symmetrical parts, and the intersection of the axial surface with any bed is an axis. In general, an axis is undulatory, its height changing along the trend of the fold. Relatively high points on an axis are culminations; low points are depressions. The plunge of a fold is the angle between an axis and its horizontal projection. The limbs or flanks are the sides. A limb extends from the axial surface of one fold to the axial surface of the adjacent fold. Generally, the radius of curvature of a fold is small compared to its wavelength and amplitude, so that much of the limb is planar. The region of curvature is the hinge. *See* ANTICLINE; SYNCLINE.

The geometry of folds is described by the inclination of their axial surfaces and their plunges. Upright folds have axial surfaces that dip from 81° to 90°; inclined folds have axial surfaces that dip from 10° to 80°; and recumbent folds have axial surfaces that dip less than 10°. Vertical folds plunge from 81° to 90°; plunging folds plunge from 10° to 80°; and horizontal folds plunge less than 10°. Auxiliary descriptive terms depend on the attitude or the relative lengths of the limbs. Overturned folds are inclined folds in which both limbs dip in the same direction; isoclinal folds are those in which both limbs are parallel; symmetrical folds have limbs of equal length; and asymmetrical folds have limbs of unequal length. The descriptions of folds consist of combinations of the above terms, for example, isoclinal upright horizontal fold, overturned plunging fold, asymmetrical inclined horizontal fold.

In a section of folded rocks the layers possess different rheological properties; some have apparent stiffness (competency), whereas others behave less stiffly (incompetency). The most competent layers or group of layers control the folding, and the less competent units tend to conform to the fold-form of the most competent units.

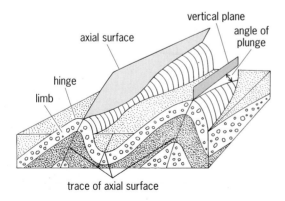

Elements of folds.

Folds generally do not occur singly but are arranged in festoons in mobile belts with lengths of thousands of miles and widths of hundreds of miles. The folds of these belts, or fold systems, commonly consist of great complex structures composed of many smaller folds. The development of fold systems is closely tied to concepts of global tectonics. The favored hypothesis is that of plate tectonics, in which fold systems are formed at converging margins where continents are underriden by oceanic crust, or where collision occurs between a continent and an island arc or between two continents.

[P.H.O.]

Food web   A diagram depicting those organisms that eat other organisms in the same ecosystem. In some cases, the organisms may already be dead. Thus, a food web is a network of energy flows in and out of the ecosystem of interest. Such flows can be very large, and some ecosystems depend almost entirely on energy that is imported. A food chain is one particular route through a food web.

A food web helps depict how an ecosystem is structured and functions. Most published food webs omit predation on minor species, the quantities of food consumed, the temporal variation of the flows, and many other details.

Along a simple food chain, A eats B, B eats C, and so on. For example, the energy that plants capture from the sun during photosynthesis may end up in the tissues of a hawk. It gets there via a bird that the hawk has eaten, the insects that were eaten by the bird, and the plants on which the insects fed. Each stage of the food chain is called a trophic level. More generally, the trophic levels are separated into producers (the plants), herbivores or primary consumers (the insects), carnivores or secondary consumers (the bird), and top carnivores or tertiary consumers (the hawk).

Food chains may involve parasites as well as predators. The lice feeding in the feathers of the hawk are yet another trophic level. When decaying vegetation, dead animals, or both are the energy sources, the food chains are described as detrital. Food chains are usually short; the shortest have two levels. One way to describe and simplify various food chains is to count the most common number of levels from the top to the bottom of the web. Most food chains are three or four trophic levels long (if parasites are excluded), though there are longer ones.

There are several possible explanations for why food chains are generally short. Between each trophic level, much of the energy is lost as heat. As the energy passes up the food chain, there is less and less to go around. There may not be enough energy to support a viable population of a species at trophic level five or higher.

This energy flow hypothesis is widely supported, but it is also criticized because it predicts that food chains should be shorter in energetically poor ecosystems such as a bleak arctic tundra or extreme deserts. These systems often have food chains similar in length to energetically more productive systems.

Another hypothesis about the shortness of food chains has to do with how quickly particular species recover from environmental disasters. For example, in a lake with phytoplankton, zooplankton, and fish, when the phytoplankton decline the zooplankton will also decline, followed by the fish. The phytoplankton may recover but will remain at low levels, kept there by the zooplankton. At least transiently, the zooplankton may reach higher than normal levels because the fish, their predators, are still scarce. The phytoplankton will not completely recover until all the species in the food chain have recovered. Mathematical models can show that the longer the food chain, the longer it will take its constituent species to recover from perturbations. Species atop very long food chains may not recover before the next disaster. Such arguments predict that food chains will be longer when environmental disasters are rare, short when they are common, and will not necessarily be related to the amount of energy entering the system.

The number of trophic levels a food web contains will determine what happens when an ecosystem is subjected to a short, sharp shock—for example, when a large number of individuals of one species are killed by a natural disaster or an incident of human-made pollution and how quickly the system will recover. The food web will also influence what happens if the abundance of a species is permanently reduced (perhaps because of harvesting) or increased (perhaps by increasing an essential nutrient for a plant).

Some species have redundant roles in an ecosystem so that their loss will not seriously impair the system's dynamics. Therefore, the loss of such species from an ecosystem will not have a substantial effect on ecosystem function. The alternative hypothesis is that more diverse ecosystems could have a greater chance of containing species that survive or that can even thrive during a disturbance that kills off other species. Highly connected and simple food webs differ in their responses to disturbances, so once again the structure of food webs makes a difference. *See* ECOLOGICAL COMMUNITIES; ECOSYSTEM; POPULATION ECOLOGY.   [S.Pi.]

Formation   A fundamental geological unit used in the description and interpretation of layered sediments, sedimentary rocks, and extrusive igneous rocks. A formation is defined on the basis of lithic characteristics and position within a stratigraphic succession. It is usually tabular or sheetlike, and is mappable at the Earth's surface or traceable in the subsurface (for example, between boreholes or in mines). Examples are readily recognized in the walls of the Grand Canyon of northern Arizona. Each formation is referred to a section or locality where it is well developed (a type section), and assigned an appropriate geographic name combined with the word formation or a descriptive lithic term such as limestone, sandstone, or shale (for example, Temple Butte Formation, Hermit Shale). This usage of "formation" by geologists differs from its informal lay usage for stalactites, stalagmites, and other mineral buildups in caves. *See* STRATIGRAPHY.

Distinctive lithic characteristics used to designate formations include chemical and mineralogical composition, particle size and other textural features, primary sedimentary or volcanic structures related to processes of accumulation, fossils or other organic content, and color. Contacts or boundaries between formations are chosen at surfaces of abrupt lithic change or within zones of gradational lithic character. Commonly, these contacts correspond with recognizable changes in topographic expression, related to variations in resistance to weathering. *See* SEDIMENTARY ROCKS; SEDIMENTOLOGY.

Mappability is an essential characteristic of a formation because such units are used to delineate geological structure (faults and folds), and it is useful to be able to recognize individual formations in isolated outcrops or areas of poor exposure. Well-established formations are commonly divisible into two or more smaller-scale units termed members and beds (for example, subdivisions of the Redwall Limestone). In other cases, formations of similar lithic character or related genesis are combined into composite units termed groups and supergroups (for example, Supai Group). The rank of a named unit may vary from one area to another (for example, from group to formation) according to whether or not subunits are readily mappable. Changes in rank are also justified in the light of new geological knowledge.

Although commonly used as a framework for interpreting geological history, formations and related units are conceptually independent of geological time. They may represent either comparatively short or comparatively long intervals of time. Accumulation of a particular unit may have begun earlier in some places than in others, and the time span represented by a unit may be influenced by later erosion. In some cases, a formation cropping out at one locality may be entirely older or younger than the same lithic unit at another locality. Although the concept of time plays no role in the definition of a formation, evidence of age is useful in the recognition of lithologically similar units far from their type localities.   [N.C.B.]

Fossil   A record of earlier life buried in rock. Originally meaning any distinctive object that has been dug up (from Latin *fodio*, dig), the term "fossil" soon came to refer particularly to things resembling animals and plants.

Fossilization. A widespread conception is that a fossil is a shell or skeleton turned into stone. This picture represents only a few of the ways in which life can leave its trace in Earth's accreting skin of sedimentary rocks. Any part of an organism can get preserved, not only hard parts but also soft tissues, cells, and organelles. Even when nothing of the original organisms is preserved, impressions and traces in the sediment give important information about their former presence, activities, and ecological roles. Also, fossilization can imply everything from preservation of the almost unaltered original tissues to their complete replacement with sediment or minerals growing in place. Organic molecules can be preserved, though in a more or less degraded state.

The biosphere normally recycles all organic and inorganic matter produced by organisms; fossils represent dead individuals that to some degree escaped that process. Most decomposition is by aerobic scavengers, fungi, and bacteria, and so a prerequisite for fossilization is that the dead body is quickly and permanently subjected to an environment in which decomposers cannot be active. A combination of anoxic water and rapid sedimentation is a typical condition favorable to fossilization, though other conditions, such as extreme temperatures, salinity, poisonous environment, desiccation, or rapid mineralization, are also known to promote fossilization. Some kind of microbial activity, however, seems to be a prerequisite for many types of fossilization, particularly of soft tissue.

Shells and skeletons. Mineralized hard parts, such as shells, spicules, and bones, are by far the most common type of fossil. Although they typically contain a substantial proportion of organic material, their mineral phase usually ensures that they are more resistant than soft tissues to biological decomposers. The most common skeletal minerals are opal (hydrated silica), apatite (calcium phosphates), and calcite and aragonite (calcium carbonates). Even when none of the original hard tissue is preserved, its former presence may promote fossilization through its initial resistance to degradation. Shells are frequently preserved as molds or casts; for example, lithified infillings (internal molds) of mollusk shells are common fossils. Some hard or protective tissues may not contain any appreciable mineral phase but are nevertheless resistant to degradation and may commonly be fossilized. Arthropod cuticle, cnidarian perisarc, hemichordate periderm, leaf cuticle, and spore exine are examples of such outer protective tissues. What gives them their resistance is usually tanned proteins, polysaccharides, or waxes.

Exceptional preservation. Paleontology is increasingly dependent on sites with unusual conditions of preservation, allowing for the fossilization of soft as well as hard parts and of more complete samples of the total biota. Early silicification of sediments may trap and preserve biota; this is the main process responsible for knowledge of the microbially dominated biosphere of the Archean and Proterozoic eons, up to about 550 million years ago. Another process known to promote exquisite preservation is impregnation with calcium phosphate during early diagenesis (physical and chemical changes occurring in sediments between deposition and solidification); this has been known to preserve soft tissues, even to cellular detail. Seemingly destructive forest fires may result in excellently preserved plant tissues through coalification. Amber,

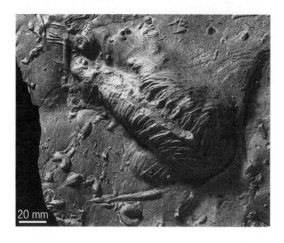

Trace fossil, *Cruziana*, formed by an arthropod (probably a trilobite) digging for a wormlike animal in soft mud during the Cambrian Period. The picture shows the lower side of a sandstone bed casting of the original markings in the mud (now vanished). (*Swedish Geological Survey; from S. Jensen, Trace fossils from the Lower Cambrian Mickwitzia sandstone, south-central Sweden, Fossils and Strata, vol. 42, 1997*)

fossilized tree resin, is well known for its capacity to trap and fossilize insects and other small animals and plant parts. Freezing has yielded spectacular finds of soft-tissue preservation of, for example, mammoths. The dependence on permanent low temperatures for maintaining the fossils, however, limits this kind of preservation to the most recent fossil biotas.

Various types of fine-grained shales and mudstones are more or less compressed by the weight of overlying sediment, and so the fossils are not preserved in as full relief as in the other types of extraordinary preservation mentioned earlier. However, the shaley deposits are capable of preserving much larger fossils than most of the other processes. *See* BURGESS SHALE.

Other types of exceptional fossil preservation are known, though they are more incidental and may be restricted to a short stratigraphic interval.

Trace fossils (marks of animal activities in sediment) and coprolites (fossilized feces) generally give less information than body fossils about the anatomy of the ancient organisms, but they are important sources of ecological and behavioral information (see illustration). *See* TRACE FOSSILS.   [J.J.Se.]

Franklinite   A natural member of the spinel structure type, with composition $Zn^{2+}Fe_2^{3+}O_4$. The habit is octahedral, often modified by the cube and dodecahedron, but the mineral usually occurs as bands of isolated rounded grains, blebs, or compact masses. It is black with a metallic luster and red internal reflections; its hardness is 6 (Mohs scale); specific gravity is 5.3; and it is weakly magnetic.

Franklinite is confined in its occurrence to the unique ore bodies at Franklin and Sterling Hill (Ogdensburg), Sussex County, New Jersey. Franklinite is a major ore mineral and is still mined at Sterling Hill for spiegeleisen and zinc. *See* WILLEMITE.   [P.B.M.]

Fresh-water ecosystem   Fresh water is best defined, in contrast to the oceans, as water that contains a relatively small amount of dissolved chemical compounds. Some studies of fresh-water ecosystems focus on water bodies themselves, while others include the surrounding land that interacts with a lake or stream. *See* ECOLOGY; ECOSYSTEM.

Fresh-water ecosystems are often categorized by two basic criteria: water movement and size. In lotic or flowing-water ecosystems the water moves steadily in a uniform direction, while in lentic or standing-water systems the water tends to remain in the same general area for a longer period of time. Size varies dramatically in each category. Lotic systems range from a tiny rivulet dripping off a rock to large rivers. Lentic systems range from the water borne within a cup formed by small plants or tree holes to very large water bodies such as the Laurentian Great Lakes. Fresh-water studies also consider the interactions of the geological, physical, and chemical features along with the biota, the organisms that occur in an area.

Physical environment. The quantity and spectral quality of light have major influences on the distribution of the biota and also play a central role in the thermal structure of lakes. The light that reaches the surface of a lake or stream is controlled by latitude, season, time of day, weather, and the conditions that surround a water body. Light penetration is controlled by the nature of water itself and by dissolved and particulate material in a water column.

Water exhibits a number of unusual thermal properties, including its existence in liquid state at normal earth surface temperatures, a remarkable ability to absorb heat, and a maximum density at $39.09°F$ ($3.94°C$), which leads to a complex annual cycle in the temperature structure of fresh-water ecosystems.

As water is warmed at the surface of a lake, a stable condition is reached in which a physically distinct upper layer of water, the epilimnion, is maintained over a deeper, cooler stratum, the hypolimnion. The region of sharp temperature changes between these two layers is called the metalimnion. The characteristic establishment of two layers is of major importance in the chemical cycling within lakes and consequently for the biota.

As the surface waters of a lake cool, the density of epilimnetic waters increases, which decreases their resistance to mixing with the hypolimnion. If cooling continues, the entire water column will mix, an event known as turnover. At temperatures below $39.09°F$ ($3.94°C$), water again becomes less dense; ice and very cold water float above slightly warmer water, maintaining liquid water below ice cover even in lakes in the Antarctic. Many lakes in the temperate zones undergo two distinct periods of mixing annually, one in the spring and the other in the fall, that separate periods of stratification in the summer and winter.

Water movement is more extensive in lotic than in standing-water ecosystems, but water motion has important effects in both types. Turbulence occurs ubiquitously and affects the distribution of organisms, particles, dissolved substances, and heat. Turbulence increases with the velocity of flowing water, and the amount of material transported by water increases with turbulence. Flowing-water ecosystems are characterized by large fluctuations in the velocity and amount of water. Aside from surface waves on large lakes, most water movement in lentic systems is not conspicuous. *See* LAKE; LIMNOLOGY; RIVER.

Chemical environment. For an element, three basic parameters are of importance: the forms in which it occurs, its source, and its concentration in water relative to its biological demand or effect. Most elements are derived from dissolved gases in the atmosphere or from minerals in geological materials surrounding a lake. In some cases the presence of elements is strongly mediated by biological activities. *See* BIOGEOCHEMISTRY.

Oxygen occurs as dissolved O_2 and in combination with other elements resulting from chemical or biological reactions. It enters water primarily from the atmosphere through a combination of diffusion and turbulent mixing. When biological demands for oxygen exceed supply rates, it can be depleted from fresh-water ecosystems. Anoxic conditions occur in hypolimnia during summer and under ice cover in winter when lake strata are isolated from the atmosphere. Oxygen depletion may also occur in rivers that receive heavy organic loading. Aside from specialized bacteria, few organisms can occur under anoxic conditions.

Carbon dioxide is derived primarily from the atmosphere, with additional sources from plant and animal respiration and carbonate minerals. Its chemical species exert a major control on the hydrogen ion concentration of water (the acidity or pH).

Phosphorus occurs primarily as a phosphate ion or in a number of complex organic forms. It is the element which is most commonly in the shortest supply relative to biological demand. Phosphorus is thus a limiting nutrient, and its addition to fresh-water ecosystems through human activities can lead to major problems due to increased growth of aquatic plants.

Nitrogen occurs in water as N_2, NO_2, NO_3, NH_4, and in diverse organic forms. It may be derived from precipitation and soils, but its availability is usually regulated by bacterial processes. Nitrogen occurs in relatively short supply relative to biological demand. It may also limit growth in some fresh-water systems, particularly when phosphorus levels have been increased because of human activity.

A variety of other elements also help determine the occurrence of fresh-water organisms either directly or by the elements' effects on water chemistry.

Biota. In addition to taxonomy, fresh-water organisms are classified by the areas in which they occur, the manner in which they move, and the roles that they occupy in trophic webs. Major distinctions are made between organisms that occur in bottom areas and those within the water column, the limnetic zone. Production is the most difficult variable to measure, but it provides the greatest information on the role of organisms in an ecosystem. *See* BIOMASS.

Plankton organisms occur in open water and move primarily with general water motion. Planktonic communities occur in all lentic ecosystems. In lotic systems they are important only in slow-moving areas.

Phytoplankton (plant plankton) comprise at least eight major taxonomic groups of algae, most of which are microscopic. They exhibit a diversity of forms ranging from one-celled organisms to complex colonies.

Zooplankton (animal plankton) comprise protozoans and three major groups of eukaryotic organisms: rotifers, cladocerans, and copepods. Most are microscopic but some are clearly visible to the naked eye. *See* POPULATION ECOLOGY.

Animals, such as fishes and swimming insects, that occur in the water column and can control their position independently of water movement are termed nekton. In addition to their importance as a human food source, fishes may affect zooplankton, benthic invertebrates, vegetation, and lake sediments.

Benthic organisms are a diverse group associated with the bottoms of lakes and streams. The phytobenthos ranges from microscopic algae to higher plants. Benthic animals range from microscopic protozoans and crustaceans to large aquatic insects and fishes. *See* FOOD WEB.

Bacteria occur throughout fresh-water ecosystems in planktonic and benthic areas and play a major role in biogeochemical cycling. Most bacteria are heterotrophic, using reduced carbon as an energy source; others are photosynthetic or derive energy from reduced compounds other than carbon.

Interactions. Ultimately the conditions in a fresh-water ecosystem are controlled by numerous interactions among biotic and abiotic components. Primary production in a fresh-water ecosystem is controlled by light and nutrient availability. As light diminishes with depth in a column or water, a point is reached where energy for photosynthesis balances respiratory energy demands. In benthic areas, the region where light is sufficient for plant growth is termed the littoral zone; deeper areas are labeled profundal.

Nutrient availability generally controls the total amount of primary production that occurs in fresh-water ecosystems. One classification scheme for lakes ranks them according to total production, ranging from oligotrophic lakes, where water is clear and production is low, to eutrophic systems, characterized by high nutrient concentrations, high standing algal biomass, high production, low water clarity, and low concentrations of oxygen in the hypolimnion. Eutrophic conditions are more likely to occur as a lake ages. This aging process, termed eutrophication, occurs naturally but can be greatly accelerated by anthropogenic additions of nutrients. A third major lake category, termed dystrophy, occurs when large amounts of organic materials that are resistant to decomposition wash into a lake basin. These organic materials stain the lake water and have a major influence on water chemistry which results in low production. *See* BOG; EUTROPHICATION.   [T.M.F.]

Front   An elongated, sloping zone in the troposphere, within which changes of temperature and wind velocity are large compared to changes outside the zone. Thus the passage of a front at a fixed location is marked by rather sudden changes in temperature and wind and also by rapid variations in other weather elements such as moisture and sky condition.

In its idealized sense, a front can be regarded as a sloping surface of discontinuity separating air masses of different density or temperature. In practice, the temperature change from warm to cold air occurs mainly within a zone of finite width, called a transition or frontal zone. The three-dimensional structure of the frontal zone is shown in the illustration. In typical cases, the zone is about 1 km (0.6 mi) in depth and 100–200 km (60–120 mi) in width, with a slope of approximately 1/100. The cold air lies beneath the warm in the form of a shallow wedge. Temperature contrasts generally are strongest at or near the earth's surface, the frontal zone usually being narrowest near the ground and becoming wider and more diffuse with height.

The surface separating the frontal zone from the adjacent warm air mass is referred to as the frontal surface, and it is the line of intersection of this surface with a second surface, usually horizontal or vertical, that strictly speaking constitutes the front. According to this more precise definition, the front represents a discontinuity in temperature gradient rather than in temperature itself. The boundary on the cold air side is often ill-defined, especially near the earth's surface, and for this reason is not represented in routine analysis of weather maps. *See* WEATHER MAP.

The wind gradient, or shear, like the temperature gradient, is larger within the frontal zone than on either size of it. In well-developed fronts the shift in wind direction often

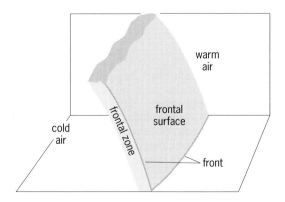

Schematic diagram of the frontal zone, angle with Earth's surface much exaggerated

is concentrated along the frontal surface, while a more gradual change in wind speed may occur throughout the frontal zone. An upper-level jet stream normally is situated above the frontal zone. *See* JET STREAM. [S.E.M.]

Frost   A covering of ice in one of several forms produced by the freezing of supercooled water droplets on objects colder than 32°F (0°C). The partial or complete killing of vegetation, by freezing or by temperatures somewhat above freezing for certain sensitive plants, also is called frost. Air temperatures below 32°F (0°C) sometimes are reported as "degrees of frost"; thus, 10°F (−12°C) is 22 degrees of frost (this usage is confined to the Fahrenheit scale and is not applied to Celsius temperatures).

Frost forms in exactly the same manner as dew except that the individual droplets that condense in the air a fraction of an inch from a subfreezing object are themselves supercooled, that is, colder than 32°F (0°C). When the droplets touch the cold object, they freeze immediately into individual crystals. When additional droplets freeze as soon as the previous ones are frozen, and hence are still close to the melting point because all the heat of fusion has not been dissipated, amorphous frost or rime results.

At more rapid rates of condensation, the drops form a thin layer of liquid before freezing, and glaze or glazed frost ("window ice" on house windows, "clear ice" on aircraft) generally follows. Glaze formation on plants, buildings and other structures, and especially on wires sometimes is called an ice storm, a silver frost storm, or thaw.

At slower deposition rates, such that each crystal cools well below the melting point before the next joins it, true crystalline or hoar frosts form. These include fernlike assemblages on snow surfaces, called surface hoar; similar feathery plumes in cold buildings, caves, and crevasses, called depth hoar; and the common window frost or ice flowers on house windows.

Killing frosts or freezes damage or kill vegetation depending on their duration and their intensity, that is, how far the plant temperatures go below 32°F (0°C). Such conditions result from advection of much colder air, which then cools the plants, as in the infamous cold waves of the north-central United States; or from radiational cooling of the plants themselves, by long-wave radiation to clear skies at night. In either case, the extent to which plant fluids freeze determines the severity of the frost. *See* AIR TEMPERATURE; DEW; DEW POINT. [A.Cou.]

Fuller's earth   Any natural earthy material that decolorizes mineral and vegetable oils and has high sorbent capacity for water and oil. The term fuller's earth has no genetic or mineralogic significance. However, the most common earthy materials classed as fuller's earth are calcium montmorillonites and palygorskites (attapulgites) and sepiolites. The term originated in England, where in ancient times raw wool was cleaned by kneading it in water with clay materials that adsorbed dirt and lanolin. The process was known as fulling, and the clay or earth became known as fuller's earth. *See* MONTMORILLONITE; SEPIOLITE.

Several clay deposits in the world are mined and processed for their absorbent, adsorbent, and decolorizing or bleaching properties. Some clays have a high natural decolorizing ability; however, in most instances a clay, normally a calcium montmorillonite, is acid-activated to enhance its bleaching or decolorizing properties. Sulfuric acid is commonly used, and in the treatment process sodium, calcium, magnesium, and iron that occupy the cation exchange sites on the clay surface are removed by the acid and replaced by hydrogen. Also, some aluminum, iron, or magnesium is removed from the mineral structure, increasing the negative charge on the clay surface. These highly charged surfaces covered with hydrogen ions selectively absorb the color bodies and other impurities in the oil. *See* CLAY; CLAY MINERALS.

The largest applications for fuller's earth are as sorbents, and by far the biggest market is pet-litter production. Other large sorbent applications are as carriers for insecticides, pesticides, and fertilizers used in agriculture and as absorbers of oil and water spills on the floors of machine shops, factories, service stations, and other manufacturing plants for safety purposes.   [H.H.Mu.]

Fused-salt phase equilibria   Conditions in which two or more phases of fused-salt mixtures can coexist in thermodynamic equilibrium. Phase diagrams of these equilibrium conditions summarize basic knowledge about fused salts. Numerous advances in the technologies which are based on high-temperature chemistry have become possible through the increase in knowledge about fused salts. The increasingly significant role of fused salts in industrial processes is evident in the widening application of these materials as heat-transfer media, in extractive metallurgy, in nonaqueous reprocessing of nuclear reactor fuels, and in the development of nuclear reactors which create more fuel than they consume (breeder reactors).

Fused-salt mixtures find application in technology when the need arises for liquids which are stable at high temperatures. For most applications, suitably low melting temperatures and low vapor pressures are primary considerations. To some extent these requirements are conflicting, because salts which are useful in obtaining low freezing temperatures often tend to have appreciable covalent character and therefore to exhibit unfavorably high vapor pressures.

As a special class of liquids, one which is composed entirely of positively and negatively charged ions undiluted by weak-electrolyte supporting media, fused salts are used in many different types of research. For example, advances in solution theory, thermodynamics, and crystal chemistry have come about through studies of fused-salt systems.

A close connection between fused-salt phase diagrams and geochemistry stems from the model principle developed by V. M. Goldschmidt, who noted that isomorphic structures are assumed by ions of the same proportionate size and stoichiometric relations but of different charge. Thus the fluorides of beryllium, calcium, and magnesium, for example, are structural models for silicon dioxide (SiO_2), titanium dioxide (TiO_2), and zirconium dioxide (ZrO_2). The fluoride structures are referred to as weakened models because of the smaller electrostatic forces resulting from smaller ionic charges; they have been useful for comparisons with oxide and silicate systems. According to Goldschmidt's interpretation, saltlike materials were derived from components such as water (H_2O), carbon dioxide (CO_2), sulfur trioxide (SO_3), chlorine (Cl_2), and fluorine (F_2), which were volatilized from molten magmas as they crystallized. Crystallization equilibria in fused-salt systems therefore provide a convenient way to study the mechanisms occurring in the formation of igneous rocks. *See* IGNEOUS ROCKS.   [R.E.Th.]

G

Gabbro The plutonic equivalent of its more abundant extrusive equivalent, basalt. Because it crystallized from a magma intruded deep within the crust, gabbro has a grain size visible to the naked eye with approximately equal amounts of calcic plagioclase (with 50% or more anorthite, the calcium aluminium feldspar) and pyroxene. Olivine is common as an early crystallized mineral, but either nepheline or quartz could be a late-stage crystallization product found in the matrix. Hornblende or biotite is commonly formed as an alteration product of pyroxene during the late stages of the magmatic crystallization, when water becomes enriched in the residual magma. *See* BIOTITE; HORNBLENDE.

Gabbro is found in diverse tectonic environments, ranging through oceanic ridges, convergent plate boundaries, stable continents, and rifts. The forms of the intrusive gabbro bodies include dikes, sills, pipes, laccoliths, stocks, batholiths, and large layered intrusive complexes.

The grain size of gabbroic rocks ranges from a millimeter to centimeters. The finer-gained gabbro is commonly referred to as a diabase (or dolerite in the United Kingdom) that usually has small granular pyroxene interstitially enclosed by randomly oriented laths of calcic plagioclase. This diabasic texture results from the faster cooling of the magma due to its injection as small dikes and sills in shallow crust. In coarser-grained gabbro found in large plutons injected in deeper crust, the pyroxenes are larger and enclose partially (a texture called subophitic) or fully (a texture called ophitic) randomly oriented labradorite. Gabbros with mineral gains larger than 2 cm (0.8 in.) are rare, but such rocks are referred to as gabbro pegmatites. At contacts with country rock, the gabbro is commonly very fine grained or glassy because of the fast chilling of the magma. In some plutons, especially near their margins, gabbroic minerals are aligned perpendicular to the walls, yielding comb layering. *See* DOLERITE; MAGMA; PLUTON.

Unlike basalts, which are found in surface or near-surface environments, gabbros are found as shallow-to-deep intrusive bodies. Large blocks measuring more than 10 km (6 mi) thick, with a suite of rocks including serpentinite, pillow lavas, and chert, contain gabbro. This suite, called ophiolites, is commonly found on continents along convergent plate boundaries; it is thought to have been tectonically emplaced by thrusting onto the continental margins. It consists of a suite of rocks that are believed to represent the oceanic crust and upper mantle. *See* CHERT; OPHIOLITE; SERPENTINITE; STRATIGRAPHY.

[A.M.K.]

Galena A mineral with composition PbS (lead sulfide) and belonging to the rock salt (NaCl) structure type. Galena usually occurs as cubes, sometimes modified by the octahedral form, with perfect cubic cleavage, brilliant metallic luster, color lead gray, specific gravity 7.5, and hardness $2\frac{1}{2}$ on Mohs scale.

Galena is widely distributed and constitutes by far the most important ore for lead. Silver, antimony, arsenic, copper, and zinc minerals often occur in intimate association with galena; consequently, galena ores mined for lead also include many valuable by-products. Important localities include Broken Hill, Australia; the tristate district of Missouri, Kansas, and Oklahoma; and numerous occurrences in Colorado, Montana, and Idaho.   [P.B.M.]

Garnet   A hard, dense silicate mineral which occurs as crystals of cubic symmetry in a wide range of geologic environments. The general chemical formula of the silicate garnet group is $A_3B_2(SiO_4)_3$ where, in natural occurrences, the A cations are dominantly Fe^2, Mn^2, Mg, and Ca, and the B cations are Al, Fe^3, and Cr^3.

The garnet mineral group is generally divided into a number of individual species on the basis of chemical composition. The more common of these species are pyrope, almandine, spessartine, grossular, andradite, and uvarovite.

Garnets are substantially denser than most chemically analogous silicates, with specific gravities ranging between 3.58 (pyrope) and 4.32 (almandine). They also have high refractive indices (1.71–1.89) and hardness, on Mohs scale, of $6^1/_2$ to $7^1/_2$. The relative hardness, coupled with the absence of cleavage, has led to the use of garnet as an abrasive. The color of garnet is primarily controlled by its chemical composition. Uvarovite is emerald green; gem varieties of garnet are generally clear, deep red pyrope. See GEM; SILICATE MINERALS.

Garnets are widespread in their occurrence, particularly in rocks which formed at high temperatures and pressures. Because of the large, readily identifiable crystals which form, the first appearance of garnet is commonly used by geologists as an index of the intensity, or grade, of metamorphism. Garnets are strongly resistant to weathering and alteration and are hence widespread constituents of sands and sediments in areas of garnetiferous primary rocks. See METAMORPHIC ROCKS.   [B.J.Wo.]

Gem   A mineral or other material that has sufficient beauty for use as personal adornment and has the durability to make this feasible. With the exception of a few materials of organic origin, such as pearl, amber, coral, and jet, and inorganic substances of variable composition, such as natural glass, gems are lovely varieties of minerals.

Natural gems. Each distinct mineral is called a species by the gemologist. Two stones that have the same essential composition and crystal structure but that differ in color are considered varieties of the same species. Thus ruby and sapphire are distinct varieties of the mineral species corundum, and emerald and aquamarine are varieties of beryl. See MINERAL; MINERALOGY.

Most gemstones are crystalline (that is, they have a definite atomic structure) and have characteristic properties, most of which are related directly to either beauty or durability. Each mineral has a characteristic hardness (resistance to being scratched) and toughness (resistance to cleavage and fracture). With few exceptions, the most important gemstones are those at the top of the Mohs hardness scale; for example, diamond is 10, ruby and sapphire are 9, chrysoberyl is $8^1/_2$, and topaz, beryl (emerald and aquamarine), and spinel are 8.

Optical properties are particularly important to the beauty of the various gem materials. The important optical properties include color; dispersion (or "fire"); refractive index (relating the breaking up of white light into colors—a rough measure of brilliancy); and pleochroism (the property of some doubly refractive materials of absorbing light unequally in the different directions of transmission, resulting in color, differences). Gemstones usually are cherished for their color, brilliancy, fire, or one of the several optical phenomena, such as asterism (the star effect caused by certain reflections of

Hardness, specific gravity, and refractive indices of gem materials

Gem material	Hardness	Specific gravity	Refractive index
Amber	2–$2\frac{1}{2}$	1.05	1.54
Beryl	$7\frac{1}{2}$–8	2.67–2.85	1.57–1.58
Chrysoberyl	$8\frac{1}{2}$	3.73	1.746–1.755
Corundum	9	4.0	1.76–1.77
Diamond	10	3.52	2.42
Feldspar	6–$6\frac{1}{2}$	2.55–2.75	1.5–1.57
Garnet			
Almandite	$7\frac{1}{2}$	4.05	1.79
Pyrope	7–$7\frac{1}{2}$	3.78	1.745
Rhodolite	7–$7\frac{1}{2}$	3.84	1.76
Andradite	$6\frac{1}{2}$–7	3.84	1.875
Grossularite	7	3.61	1.735
Spessartite	7–$7\frac{1}{2}$	4.15	1.80
Hematite	$5\frac{1}{2}$–$6\frac{1}{2}$	5.20	
Jade			
Jadeite	$6\frac{1}{2}$–7	3.34	1.66–1.68
Nephrite	6–$6\frac{1}{2}$	2.95	1.61–1.63
Lapis lazuli	5–6	2.4–3.05	1.50
Malachite	$3\frac{1}{2}$–4	3.34–3.95	1.66–1.91
Opal	5–$6\frac{1}{2}$	2.15	1.45
Pearl	4	2.7	
Peridot	$6\frac{1}{2}$–7	3.34	1.654–1.690
Quartz			
Crystalline	7	2.65	1.54–1.55
Chalcedonic	$6\frac{1}{2}$–7	2.60	1.535–1.539
Spinel	8	3.60	1.72
Spodumene	6–7	3.18	1.66–1.676
Topaz	8	3.53	1.61–1.62
Tourmaline	7–$7\frac{1}{2}$	3.06	1.624–1.644
Turquois	5–6	2.76	1.61–1.65
Zircon			
Blue and colorless	$7\frac{1}{2}$	4.7	1.92–1.98
Green	6	4.0	1.81

light); chatoyancy, or a cat's-eye effect; play of color, such as displayed by an opal; and adularescence (the billowy light effect seen in adularia or moonstone varieties of orthoclase feldspar).

Gemstones are commonly designated as precious or semiprecious. This is a somewhat meaningless practice, however, and often misleading, since many of the so-called precious gem varieties are inexpensive and many of the more attractive varieties of the semiprecious stones are exceedingly expensive and valuable. For example, a piece of fine-quality jadeite may be valued at approximately 100 times the price per carat of a low-quality star ruby. Fine black opals, chrysoberyl cat's-eyes, and alexandrites are often much more expensive than many sapphires of certain colors. *See* PRECIOUS STONES.

More than 100 natural materials have been fashioned at one time or another for ornamental purposes. Of these, however, only a relatively small number are likely to be encountered in jewelry articles.

The table lists the important gem minerals and the properties most useful in identification. For further information on the individual species or groups *see* AMBER; AMETHYST; AZURITE; BERYL; CAMEO; CHRYSOBERYL; CORUNDUM; DIAMOND; EMERALD; FELDSPAR; GARNET; INTAGLIO (GEMOLOGY); JADE; JET (GEMOLOGY); LABRADORITE; LAZURITE; MALACHITE; OLIVINE; ONYX; OPAL; ORTHOCLASE; PEARL; QUARTZ; RUBY; SAPPHIRE; SPINEL; SPODUMENE; TOPAZ; TOURMALINE; TURQUOISE; ZIRCON.

Manufactured gems. A mineral or other material that has sufficient beauty and durability for use as a personal adornment can be manufactured. The term "manufactured," as used here, does not include such processes as shaping, faceting, and polishing, but only the processes that affect the material from which the finished gem is produced. These processes are (1) those that change the mineral in some fundamental characteristic, such as color, called a treated gem; (2) those by which a material is made that is identical with the naturally occurring mineral, called a synthetic gem; and (3) those that produce a simulated material with the appearance but not both the composition and structure of the natural gem, called an imitation gem.

Treated gems. There are four basic methods of treatment: (1) dyeing and staining, (2) plastic or other impregnation, (3) heat treatment, and (4) radiation. When the process to which a gem material is subjected changes its structure or adds something, such as a dye or a plastic binder, an effect on value takes place. When such changes are made, the nature of the alteration must be disclosed. An example of the second category of treatment, wherein there is no obvious effect on value, is gentle heating of amethyst to even its color, or stronger heating to change it to yellow or brown citrine. The change is permanent and nothing but temporary heat has been added. Many colored stones, including most green tourmaline, aquamarine, and colorless and flame-colored zircon and all pink topaz and blue zircon have been heated to improve their color.

Synthetic gems. The U.S. Federal Trade Commission has restricted the term synthetic gems to manufactured materials that have the same chemical, physical, and optical properties as their naturally occurring counterparts. Many gem materials, including diamond, have been synthesized, but in such small crystals or poor quality that they are unsatisfactory as gemstones. Some attempts to make gemstones have resulted in producing substances hitherto not known, many of which are of great importance industrially. Others have resulted in significant improvements in existing processes.

Imitation gems. Since prehistoric times glass has been the most widely used gem imitation. Since World War II colored plastics have replaced glass to a great extent in the least expensive costume jewelry.

Identification. Materials made by a flame-fusion process almost always contain gas bubbles, which are usually spherical or nearly so. Those with medium to dark tones of color often show color banding, or striae, with a curvature corresponding to that of the top of the boule. Natural gem materials are characterized by angular inclusions and straight color bands, if any are present.

Flux-fusion synthetic emeralds have distinctly lower refractive indices and specific gravities than natural emeralds and are characterized by wisplike or veillike flux inclusions. They show a red fluorescence under ultraviolet light, whereas most natural emeralds are inert.

Hydrothermally made synthetic emeralds have properties similar to many natural emeralds, but their inclusions differ and they are characterized by a very strong red fluorescence under ultraviolet.

The cheaper forms of glass are cast in molds and, under a hand lens, show rounded edges at the intersections of facets; the facets are often concave. The better grades, known as cut glass, have been cut and polished after first being molded approximately into the desired form. Cut glass has facets that intersect in sharp edges. Both types may contain gas bubbles or have a roiled appearance in the heart or have both, in contrast to most of the colored stones they imitate.   [R.T.L.]

Gemology   The science of those minerals and other materials which possess sufficient beauty and durability to make them desirable as gemstones. It is concerned with the identification, grading, evaluation, fashioning, and other aspects of gemstones. *See* GEM. [R.T.L.]

Geochemical prospecting   The use of chemical properties of naturally occurring substances (including rocks, glacial debris, soils, stream sediments, waters, vegetation, and air) as aids in a search for economic deposits of metallic minerals or hydrocarbons. In exploration programs, geochemical techniques are generally integrated with geological and geophysical surveys. *See* GEOCHEMISTRY; GEOPHYSICAL EXPLORATION.

General principles. Mineral deposits represent anomalous concentrations of specific elements, usually within a relatively confined volume of the Earth's crust. Most mineral deposits include a central zone, or core, in which the valuable elements or minerals are concentrated, often in percentage quantities, to a degree sufficient to permit economic exploitation. The valuable elements surrounding this core generally decrease in concentration until they reach levels, measured in parts per million (ppm) or parts per billion (ppb), which appreciably exceed the normal background level of the enclosing rocks. These zones or halos afford means by which mineral deposits can be detected and traced; they are the geochemical anomalies being sought by all geochemical prospectors.

The zone surrounding the core deposit is known as a primary halo or anomaly, and it represents the distribution patterns of elements which formed as a result of primary dispersion. Primary dispersion halos vary greatly in size and shape as a result of the numerous physical and chemical variables that affect fluid movements in rocks. Some halos can be detected at distances of hundreds of meters from their related ore bodies; others are no more than a few centimeters in width.

Abnormal chemical concentrations in weathering products are known as secondary dispersion halos or anomalies and are more widespread. They are sometimes referred to as dispersion trains. The shape and extent of secondary dispersion trains depend on a host of factors, of which topography and ground-water movement are perhaps most important. Ground waters frequently dissolve some of the constituents of mineralized bodies and may transport these for considerable distances before eventually emerging in springs or streams. Further dispersion may ensue in stream sediments when soil or weathering debris that has anomalous metal content becomes incorporated through erosion in stream sediment. Analysis of the fine sand arid silt of stream sediment can be a particularly effective method for detection of mineralized bodies within the area drained by the stream.

Survey design. The degree of success of a geochemical survey in a mineral exploration program is often a reflection of the amount of care taken with initial planning and survey design. This phase of activity is often referred to as an orientation survey; its practical importance cannot be overstressed.

When a geochemical prospecting survey is contemplated, four basic considerations must be addressed: the nature of the mineral deposits being sought; the geochemical properties of the elements likely to be present in the target mineral deposit; geological factors likely to cause variations in geochemical background; and environmental, or landscape, factors likely to influence the geochemical expression of the target mineral deposit. Elucidation of these factors in an orientation survey will permit design of a geochemical prospecting survey that is most likely to prove effective under the prevailing conditions.

Geochemical prospecting surveys fall into two broad categories, strategic or tactical, which may be further subdivided according to the material sampled. Strategic surveys imply coverage of a large area (generally several thousands of square kilometers) where the primary objective is to identify districts of enhanced mineral potential; tactical surveys comprise the more detailed follow-up to strategic reconnaissance. Typically the area covered by a tactical survey is divided into discrete areas of high mineral potential within the general anomalous district.

Soil and glacial till surveys have been used extensively in geochemical prospecting and have resulted in the discovery of a number of ore bodies. Generally, such surveys are of a detailed nature and are run over a closely spaced grid.

Biogeochemical surveys are of two types. One type utilizes the trace-element content of plants to outline dispersion halos, trains, and fans related to mineralization; the other uses specific plants or the deleterious effects of an excess of elements in soils on plants as indicators of mineralization. The latter type of survey is often referred to as a geobotanical survey.

Rock geochemical surveys are reconnaissance surveys carried out on a grid or on traverses of an area, with samples taken of all available rock outcrops or at some specific interval. One or several rock types may be selected for sampling and analyzed for various elements. Geochemical maps are compiled from the analyses, and contours of equal elemental values are drawn. These are then interpreted, often by using statistical methods. Under favorable conditions, mineralized zones or belts may be outlined in which more detailed work can be concentrated. If the survey is executed over a large expanse of territory, geochemical provinces may be outlined.

Isotopic surveys are applicable to elements which exist in two or more isotopic forms. They employ the ratios between isotopes such as ^{204}Pb, ^{206}Pb, ^{207}Pb, ^{208}Pb, or ^{32}S and ^{34}S to "fingerprint" or indicate certain types of mineral deposits which may share a common origin. Isotopic ratios may also be used to determine the ages of minerals or given rock types and may, thus, assist in elucidating questions of ore formation.

Geochemistry applied to hydrocarbon exploration differs from that in the search for metallic mineral deposits; the former chiefly involves detection and study of organic substances found during drilling; the latter, detection and study of inorganic substances at the surface. Once hydrocarbon accumulations have been discovered, their classification into geochemical families is important. The final stages of detailed exploration may involve complex multivariate computer-aided modeling of all available geological, geochemical, geophysical, and hydrological data—to determine the ultimate hydrocarbon potential of a given basin. [R.F.H.]

Geochemistry A field that encompasses the investigation of the chemical composition of the Earth, other planets, and the solar system and universe as a whole, as well as the chemical processes that occur within them. The discipline is large and very important because basic knowledge about the chemical processes involved is critical for understanding subjects as diverse as the formation of economically valuable ore deposits, safe disposal of toxic wastes, and variations in the Earth's climate.

Isotope geochemistry is based on the fact that the isotopic compositions of various chemical elements may reveal information about the age, history, and origin of terrestrial and extraterrestrial materials. Isotopes of an element share the same chemical properties but have slightly different nuclear makeups and therefore different masses. Some naturally occurring isotopes are radioactive and decay at known rates to form daughter isotopes of another element; for example, radioactive uranium isotopes decay to stable isotopes of lead. Radioactive decay is the basis of geochronology, or age determination: the age of a sample can be found by measuring its content of the daughter isotope. Both radioactive decay and the processes that enrich or deplete materials in certain isotopes cause different parts of the Earth and solar system to have different, characteristic isotopic compositions for some elements. These differences serve as fingerprints for tracing the origins of, and characterizing the interactions between, various geochemical reservoirs. *See* DATING METHODS; ELEMENTS, GEOCHEMICAL DISTRIBUTION OF; GEOCHRONOMETRY; LEAD ISOTOPES (GEOCHEMISTRY).

Cosmochemistry deals with nonearthly materials. Typically, cosmochemists use the same kinds of analytical and theoretical approaches as other geochemists but apply them to problems involving the origin and history of meteorites, the formation of the solar system, the chemical processes on other planets, and the ultimate origin of the elements themselves in stars.

Organic geochemistry deals with carbon-containing compounds, largely those produced by living organisms. These are widely dispersed in the outer part of the Earth—in the oceans, the atmosphere, soil, and sedimentary rocks. Organic geochemistry is important for understanding many of the chemical cycles that occur on Earth because biology often plays a major role. Organic geochemists are also active in investigating such areas as the origin of life, the formation of some types of ore deposits that may be biologically mediated, and the origin of coal, petroleum, and natural gas. *See* BIOGEOCHEMISTRY; COAL; NATURAL GAS; ORGANIC GEOCHEMISTRY; PETROLEUM.

In recent years there has been widespread application of geochemical techniques to problems in paleoclimatology and paleoceanography. In this approach, ocean sediments, sedimentary rocks on land, ice cores, and other continuous records of the Earth's history are analyzed for fossil chemical evidence of past climates or seawater composition. As in most areas of geochemistry, precise and accurate analytical methods for determining the isotopic and elemental composition of the samples are critical. *See* EARTH SCIENCES; PALEOCEANOGRAPHY; PALEOCLIMATOLOGY.   [J.D.MacD.]

Geochronometry   The measurement of the age of rocks, minerals, water, and biological materials. Measurements are based primarily on the radioactive decay or fission of such naturally occurring isotopes as ^{238}U, ^{235}U, ^{232}Th, ^{187}Re, ^{176}Lu, ^{147}Sm, ^{87}Rb, ^{40}K, ^{129}I, ^{36}Cl, ^{26}Al, ^{14}C, and ^{10}Be. These radioactive isotopes can be divided into two groups: primordial isotopes that are residual from early nucleosynthesis, and cosmogenic isotopes that are continuously produced by cosmic-ray-induced spallation reactions primarily within the Earth's atmosphere or on the surfaces of meteorites. For the first group, the relative amounts of the radioactive parent and radiogenic daughter are used as a measure of age. Age is determined for the second group by the amount of radioactive isotope remaining after the object is isolated from further intake—for example, by death of an organism participating in the carbon-oxygen cycle, or by trapping of the cosmogenic isotope in sediment or ice. Tree-ring dating (dendrochronology), which is based on the counting of annual rings, may also be used and provides a very precise measure of age of the last eight millennia. *See* DENDROCHRONOLOGY; LEAD ISOTOPES (GEOCHEMISTRY); RADIOCARBON DATING.

There are also methods of establishing the relative sequence of events in time, most importantly, the use of unidirectional biologic evolution upon which the boundaries of the Phanerozoic time scale are based (5.5×10^8 years to the present). The virtues of isotopic dating are its applicability to the full range of geologic time, including the Precambrian for which an adequate paleontologic time scale does not exist; better resolution of events during the Cenozoic (6.5×10^7 years to present); and provision of the fourth physical dimension of astronomic time to quantify rates and energies involved in geologic processes. These isotopic chronometers have been used to measure the age of the Earth, Moon, and meteorites (4.5×10^9 years), the age of the oldest datable rocks (3.7×10^9 years), and many other significant geologic events such as the advance and retreat of continental glaciers. They have also been used to establish a Precambrian time scale, to calibrate the Phanerozoic time scale in solar years, and to provide a chronology for significant biologic, cultural, and environmental events related to the evolution of the human race. On a much shorter time scale, these methods have been used to determine rates of flow of water through aquifers and rates of material (aerosols) transport through the atmosphere. *See* AMINO ACID DATING; DATING METHODS; GEOLOGIC TIME SCALE; ROCK AGE DETERMINATION. [P.E.Da.]

Geode A roughly spheroidal hollow body, lined on the inside with inward-projecting small crystals (see illustration). Geodes are found most frequently in lime-

Geode, lined with quartz crystals, keokuk, Iowa. (*Brooks Museum, University of Virginia*)

stone beds but may occur in some shales. Typically, a geode consists of a thin outer shell of dense chalcedonic silica and an inner shell of quartz crystals. Many geodes are filled with water; others, having been exposed for some time at the surface, are dry. Calcite or dolomite crystals line the interior of some geodes, and a host of other minerals are less commonly found. In some geodes there is an alternation of layers of silica and calcite, but almost all geodes show some banding suggestive of rhythmic precipitation. *See* CHALCEDONY. [R.Si.]

Geodesy The science of measuring the size, shape, and gravity field of the Earth. Geodesy supplies positioning information about locations on the Earth, and this information is used in a variety of applications, including civil engineering, boundary demarcations, navigation, resource management and exploration, and geophysical studies of the dynamics of the Earth. *See* EARTH.

The conventional measurement systems in geodesy are triangulation and trilateration for determining horizontal positions, and leveling for determining heights. These techniques depend on the Earth's gravity field, and so a major part of geodesy has been not only position determination but also the measurement of the Earth's gravity field. *See* EARTH, GRAVITY FIELD OF.

Two major measurement systems were developed in the late 1970s and early 1980s: satellite laser ranging (SLR) systems, which could measure the distance from the ground to a satellite equipped with special corner-cube mirrors; and very long baseline interferometry (VLBI), which could measure the difference in arrival times between radio signals from extragalactic radio sources. With these systems it is possible to measure accurately (within a few centimeters) the distances between points located on different continents, making possible the creation of truly global coordinate systems. Both systems were deployed around the world to measure not only the positions of locations but also the changes in those positions; and thus it was confirmed that the Earth is not a static but a highly dynamic body, with much of this dynamism causing catastrophic events such as earthquakes and volcanic eruptions. The more recent Global Positioning System (GPS) offers much of the capability of SLR and VLBI.

The most recent development in geodetic techniques is interferometric synthetic aperture radar (InSAR). This technique is used to measure heights of the topography or, if the topography is already known, the changes in the topography between two synthetic aperture radar (SAR) images. Heights measured with InSAR are far less accurate than normal geodetic height measurements, but since InSAR is an imaging system, large areas can be measured easily. If the InSAR instrument is on an orbiting spacecraft, global topography can be measured. The measurement of changes in topography with InSAR has been widely used to measure the surface displacements after earthquakes (by comparing before and after SAR images) and for monitoring volcanic deformations.

Some of the major impacts of modern geodetic measurements have been in the study of the dynamics of the Earth. The measurement systems enable the observation of many of the minute motions of the Earth, such as those associated with plate tectonics and other geophysical processes, and changes in the rotation of the Earth.   [T.He.]

Geodynamics   The branch of geophysics that studies the processes leading to deformation of planetary mantle and crust and the related earthquakes and volcanism that shape the structure of the Earth and other planets. On the largest scale, these processes are a consequence of the transfer of heat out of planetary interiors due to cooling at their surfaces. Rock contracts as it cools, so that its density increases. The cool surface layer is heavier than the interior and has a tendency to sink into it. At the same time, cooling and solidification of the metallic core heats the deepest portion of the surrounding rocky mantle, causing it to become buoyant. The resulting flow of the mantle causes deformation at the surface. Volcanism arises from the partial melting of hot mantle that rises toward the surface from the deeper interior, in response either to buoyancy or to surface deformation. Surface deformation also results from external loads, such as the distribution of ice and water, tidal loads due to the gravitational attraction of nearby planetary bodies, and meteor impacts. See EARTH, HEAT FLOW IN; GEOPHYSICS; VOLCANO.

A planet's response to its internal heat flow depends largely on the rheology of deforming rock. At low temperatures, near the surface, rock behaves as a brittle-elastic material, allowing the propagation of seismic waves and the support of surface loads by elastic stresses. Deformation occurs by the formation of cracks or faults. On geologic time scales and at the higher temperatures of the deeper interior, thermally activated creep allows the solid, rocky mantle to flow like a viscous fluid. But even at these high temperatures, rock behaves elastically on short time scales so that elastic shear waves propagate through the slowly flowing mantle. In the case of the Earth in its current stage of evolution, plate tectonics describes how the surface behaves: large, cold, relatively rigid plates move laterally across the surface while the deeper mantle flows by creep. In the cold plates, deformation is largely confined to boundary faults between the plates. Faults slip with a stick-slip behavior, giving rise to large earthquakes that occur primarily on the plate boundaries. See EARTHQUAKE; PLATE TECTONICS; ROCK MECHANICS.

Given the difficulty of direct observation and the wide range of scales involved in phenomena of interest, multiple approaches are needed to understand geodynamic processes. Laboratory experiments on relatively small samples of rock are used to characterize the rock's physical properties, such as its rheology, at high pressures and temperatures. The rate of deformation due to creep in nature is much too slow to measure directly in the laboratory. Field studies of rocks once deep in the interior and brought to the surface by uplift and erosion provide evidence of the processes that have affected them. But the interior of the Earth where the processes of interest are actually occurring is not directly accessible for study. Thus geodynamicists must design large-scale observational experiments that allow them to create conceptual and physical images of the interior, using combinations of seismic, gravitational, electromagnetic, and heat-flow measurements. Variations in global gravity and corresponding surface topography can be remotely sensed from orbiting spacecraft. *See* EARTH CRUST; EARTH INTERIOR; GEODESY; GEOMAGNETISM; SEISMOLOGY.   [E.M.P.]

Geodynamo   The mechanism thought to be responsible for the generation of the Earth's magnetic field through the convection of conducting fluids in the Earth's core.

Paleomagnetic measurements suggest that the Earth has possessed a magnetic field for at least 3.5 billion years. Geophysicists generally accept that the ambient magnetic field measured at the Earth's surface is due to electric currents flowing in its liquid iron core (see illustration). In the absence of electromotive forces, like those of chemical batteries, electric currents will decay as magnetic energy is converted to heat. Without some regenerative process to offset such natural ohmic dissipation in the Earth's core, any electric currents and the associated magnetic field would vanish in about 15,000 years. Regeneration of the field is necessary. In the Earth it is thought that the magnetic field is maintained by dynamo action, whereby the kinetic energy of convective motion in the Earth's liquid core is converted into magnetic energy. Since this process operates without an external energy source, the geodynamo is said to be self-sustaining. *See* GEOELECTRICITY; GEOMAGNETISM; PALEOMAGNETISM.

It is not obvious how a simply connected conducting fluid body, like the Earth's core, functions as a dynamo without the induced currents simply short-circuiting and eliminating field generation. In fact, the electric current in a dynamo and the magnetic field that it sustains cannot be too simple; a theorem, due to T. G. Cowling, says that

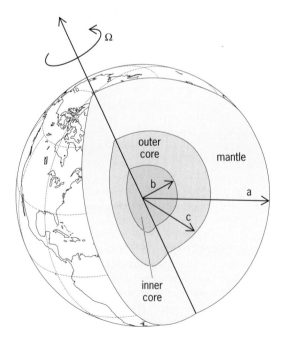

Anatomy of the Earth. The rocky mantle has a radius $a = 6371$ km (3959 mi), the liquid iron outer core has a radius $c = 3485$ km (2165 mi), and the solid inner core has a radius $b = 1215$ km (755 mi). The Earth's rotational vector is Ω.

no axisymmetric, or even two-dimensional, dynamo magnetic field can exist. Although the magnetic north and south poles usually are nearly coincident with the geographic poles, indicating that the rotation arising from the Coriolis force plays an important role in the core's dynamics, it is no accident that the compass does not point toward true north everywhere on the Earth's surface—an inherent lack of symmetry. As a result, theoretical progress has been slow since scientists often take advantage of symmetry, should it be present, when solving mathematical equations. *See* CORIOLIS ACCELERATION.

Geophysicists do, however, have a good qualitative understanding of how the geodynamo works. In the 1940s and 1950s, W. M. Elsasser and E. N. Parker first elucidated the so-called α-ω (alpha-omega) mechanism, by which core fluid motion can act as a dynamo if it consists of a combination of differential rotation and convective helical motion. Since then it has been shown mathematically that dynamo regeneration can arise from the turbulent motion of a rotating fluid. Although the α-ω mechanism probably describes how the field is amplified, it is the dynamics that ultimately governs the strength of the field.

There are two possible sources of the energy sustaining the fluid convection in the outer core—thermal and compositional. Thermal convection is perhaps most familiar, with heat sources, such as radioactive potassium, distributed over the volume of the outer core. With sufficient internal heating, the fluid is gravitationally unstable and, as a result, convection is sustained. Compositional convection is currently favored by most geophysicists as the energy source of the geodynamo. Although the core is primarily of iron, there are probably light impurities, such as sulfur. Due to the effects of pressure, as the Earth slowly cools, iron solidifies at the inner-core boundary. This causes the inner core to grow and leaves the lighter constituents behind in the fluid at the base of the outer core, supplying the buoyancy that drives the convection.

The Sun is a familiar dynamo, and it reverses regularly almost every 11 years. So, why does the Earth's magnetic field not display such regularity? The difference is thought to be due to the presence in the Earth of a solid electrically conducting inner core, where the magnetic field can change only rather slowly by diffusion. Recent calculations suggest that because the inner core is electromagnetically coupled to the outer core, its presence acts to stabilize the magnetic field, so that only particularly large fluctuations of the field in the outer core are sufficient to overcome the damping effect of the inner core. *See* GEOPHYSICS; MAGNETIC REVERSALS.   [J.J.Lo.]

Geoelectricity   Electromagnetic phenomena and electric currents, mostly of natural origin, that are associated with the Earth. Geophysical methods utilize natural and artificial electric currents to explore the properties of the Earth's interior and to search for natural resources (for example, petroleum, water, and minerals). Geoelectricity is sometimes known as terrestrial electricity. All electric currents (natural or artificial, local or worldwide) in the solid Earth are characterized as Earth currents. The term telluric currents is reserved for the natural, worldwide electric currents whose origins are almost entirely outside the atmosphere. Geoelectromagnetism is a more comprehensive term than geoelectricity. Time variations of any magnetic field are associated with an electric field that induces electric currents in conducting media such as the Earth.

Magnetic fields, electric fields, and electric currents are the constituents of electromagnetism, and are related by Maxwell's equations. For instance, the illustration shows the time variations of the natural magnetic and electric fields simultaneously measured at one location at the surface of the Earth. These two traces are related to each other, not only by Maxwell's equations but also by the physical properties of the subsurface rocks in the vicinity of the measuring site. Either one of the two traces may be computed

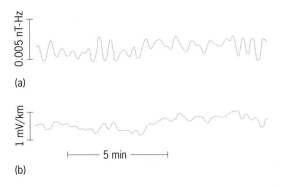

(a)

(b)

├────── 5 min ──────┤

Time variations of the horizontal orthogonal components of the natural (*a*) magnetic and (*b*) electric fields, simultaneously measured at one site at the surface.

synthetically from the other if the properties of the subsurface rocks are known. Conversely, the two traces together can yield geologic information; this is a form of geophysical exploration or prospecting. Thus, the terms geoelectricity, geomagnetism, and geoelectromagnetism are essentially interchangeable, although each one may have a somewhat different emphasis. For example, the term geomagnetism is sometimes used for the study of the Earth's quasistationary main magnetic field. *See* GEOMAGNETISM; GEOPHYSICAL EXPLORATION.   [S.H.Y.]

Geographic information systems
Computer-based technologies for the storage, manipulation, and analysis of geographically referenced information. Attribute and spatial information is integrated in geographic information systems (GIS) through the notion of a data layer, which is realized in two basic data models: raster and vector. The major categories of applications comprise urban and environmental inventory and management, policy decision support, and planning; engineering and defense applications; and scientific analysis and modeling.

A geographic information system differs from other computerized information systems in two major respects. First, the information in this type of system is geographically referenced (geocoded). Second, a geographic information system has considerable capabilities for data analysis and scientific modeling, in addition to the usual data input, storage, retrieval, and output functions.

A geographic information system is composed of software, hardware, and data. The notion of data layer (or coverage) and overlay operation lies at the heart of most software designed for geographic information systems.

Two fundamental data models, the vector and raster models, embody the overlay idea in geographic information systems. In a vector geographic information system, the geometrical configuration of a coverage is stored in the form of points, arcs (line segments), and polygons, which constitute identifiable objects in the database. In a raster geographic information system, a layer is composed of an array of elementary cells of pixels, each holding an attribute value without explicit reference to the geographic feature of which the pixel is a part.

A data layer or coverage integrates two kinds of information: attribute and spatial (geographic). The functionality of a geographic information system consists of the ways in which that information may be captured, stored, manipulated, analyzed, and presented to the user. Spatial data capture (input) may be from primary sources such as remote sensing scanners, radars, or global positioning systems, or from scanning or digitizing images and maps derived from remote sensing. Output (whether as a display on a cathode-ray tube or as hard copy) is usually in map or graph form, accompanied by tables and reports linking spatial and attribute data. The critical data management and analysis functions fall into four categories: retrieval, classification, and measurement; overlay functions; neighborhood operations; and connectivity functions. *See* REMOTE SENSING. [H.Co.]

Business applications of geographic information systems are increasingly widespread and include market analysis, store location, and agribusiness (for example, determining the correct amount of fertilizers or pesticides needed at each point of a cultivated field). Engineers use geographic information systems when modeling terrain, building roads and bridges, maintaining cadastral maps, routing vehicles, drilling for water, determining what is visible from any point on the terrain, integrating intelligence information on enemy targets, and so forth. Such applications have been facilitated through the integration of geographic information systems with global positioning systems.

Among the earliest and still most widespread applications of the technology are land information and resource management systems (for example, forest and utility management). Other common uses of geographic information systems in an urban policy context include emergency planning, determination of optimal locations for fire stations and other public services, assistance in crime control and documentation, and electoral and school redistricting. Uses of geographic information systems have spread well beyond geography, the source discipline, and now involve most applied sciences, both social and physical, that deal with spatial data. The nature of the applications of geographic information systems in these areas ranges from simple thematic mapping for illustration purposes to complex statistical and mathematical modeling for the exploration of hypotheses or the representation of dynamic processes. [H.Co.]

Geography   The study of physical and human landscapes, the processes that affect them, how and why they change over time, and how and why they vary spatially. Geographers consider, to varying degrees, both natural and human influences on the landscape, although a common division separates human and physical geography. Physical geographers may study landforms (geomorphology), water (hydrology), climate and meteorology (climatology), the biotic environment (biogeography), or soils (pedology). Human geographers include urban, regional, and environmental planners; cultural geographers; regional and area specialists; economic geographers; political geographers; transportation analysts; location analysts; and specialists in the spatial nature of ethnic or gender issues. *See* BIOGEOGRAPHY; CLIMATOLOGY; GEOMORPHOLOGY; HYDROLOGY; PEDOLOGY.

Many geographers are involved with the development of techniques and applications that support spatial analytical studies or the display of spatial information and data. Maps, whether printed, digital, or conceptual, are the basic tools of geography. Geographers are involved in map interpretation and use, as well as map production and design. Cartographers supervise the compilation, design, and development of maps, globes, and other graphic representations. *See* CARTOGRAPHY.

A geographic information system (GIS) is a relatively new technology that combines the advantages of computer-assisted cartography with those of spatial database management. It facilitates the storage, retrieval, and analysis of spatial information in the form of digital map "overlays," each representing a different landscape component (terrain, hydrologic features, roads, vegetation, soil types, or any mappable factor). Each of these data layers can be fitted digitally to the same map scale and map projection—in any combination—permitting the analysis of relationships among any combination of environmental variables for which data have been input into the geographic information systems. *See* GEOGRAPHIC INFORMATION SYSTEMS.

Many geographers are applied practitioners, solving problems using a variety of tools, including computer-assisted cartography, statistical methods, remotely sensed imagery, the Global Positioning System (GPS), and geographic information systems. Today, nearly all geographers, regardless of their subdisciplinary emphases, employ some or all of these techniques in their professional endeavors. *See* PHYSICAL GEOGRAPHY; TERRAIN AREAS.   [J.F.P.]

Geologic thermometry   The measurement or estimation of temperatures at which geologic processes take place. Methods used can be divided into two groups, nonisotopic and isotopic. Nonisotopic methods involve measurements of earth temperatures either directly by surface and near-surface features or indirectly from various properties of minerals and fossils. The isotopic methods involve the determination of distribution of isotopes of the lighter elements between pairs of compounds in equilibrium at various temperatures, and application of these data to problems of the temperature at which these compounds (commonly minerals) form in nature.   [E.I.]

Geologic time scale   An ordered, internally consistent, internationally recognized sequence of time intervals, each distinct in its own history and record of life on Earth, including the assignment of absolute time in years to each geologic period.

Geologic time scale		

Eon
Era
    Period [system]
      Epoch [series]
	Age at
	beginning of
	interval,
	10^6 years
		Interval
		length,
		10^6 years

Phanerozoic		
Cenozoic		65
Quaternary (Q)*		1.8
Recent	0.01	0.01
Pleistocene	1.8	1.79
Tertiary (T)	65	63.2
Pliocene (Tpl)	5.3	3.5
Miocene	23.8	18.5
Oligocene (To)	33.7	9.9
Eocene (Te)	54.8	21.1
Paleocene (Tp)	65	10.2
Mesozoic	250	185
Cretaceous (K)	144	79
Jurassic (J)	206	62
Triassic (Tr)	250	44
Paleozoic	543	297
Permian (P)	290	40
Carboniferous (M, P)	354	64
Devonian	417	63
Silurian	443	26
Ordovician	490	47
Cambrian	543	53
Precambrian		
Proterozoic	2500	1957
Late (Z)† (Neoproterozoic)	900	357
Middle (Y) (Mesoproterozoic)	1600	700
Early (X) (Paleoproterozoic)	~2500	900
Archean	3800	
Late (W)	3000	500
Middle (U)	3400	400
Early (V)	>3800	>400

*In parentheses are the symbols for the periods and epochs used on geologic maps and figures in North America, as well as other parts of the world.
†Letter designations of Precambrian age intervals are used by the U.S. Geological Survey.

The geologic time scale (*see* table) has a relative scale, consisting of named intervals of geologic history arranged in historical sequence; and a numerical (or absolute) time scale, providing absolute ages for the boundaries of these intervals.

In order to establish a geologic time scale, an independent means of dating rocks is required. Before the discovery of radioactivity, crude estimates of the length of a geologic history were made based on the total thicknesses of sedimentary rock and assumed rates of erosion and sedimentation. These estimates varied by as much as a factor of 10.

The modern geologic time scale is based on many measurements of various rock types by quantitative isotopic chronometers such as uranium-lead (U-Pb) and potassium-argon (K-Ar). *See* ROCK AGE DETERMINATION.　　　　　　　　　　[A.R.P.; J.W.Gei.; J.L.K.]

Geology　　The science of the Earth. The study of the Earth's materials and of the processes that shape them is known as physical geology. Historical geology is the record of past events. *See* EARTH; EARTH SCIENCES.

Geology is an interdisciplinary subject that overlaps and depends on other scientific disciplines. Physical geology is concerned primarily with the Earth's materials (minerals, rocks, soils, water, ice, and so forth) and the processes of their origin and alteration. Chemistry and physics are the two scientific disciplines most closely related—study of the chemistry of the Earth's materials is geochemistry, and study of the physical properties of the Earth is geophysics. *See* GEOCHEMISTRY; GEOPHYSICS; STRUCTURAL GEOLOGY.

Historical geology is based on two complementary disciplines, stratigraphy and paleontology. Stratigraphy is the systematic study of stratified rocks through geologic time. The stratigraphic record reveals the sequence of events that have affected the Earth through eons of time. Absolute dates for the stratigraphic record are provided from geochemical studies of naturally occurring radioactive isotopes. Paleontology is the study of fossilized plants and animals with regard to their distribution in space and time. Paleontology is closely related to biology. The distinctions between physical and historical geology are more matters of convenience than substance, because it is increasingly clear, within the framework of plate tectonics, that all aspects of geology are interrelated. *See* PALEONTOLOGY; PLATE TECTONICS; STRATIGRAPHY.

Mineralogy concerns the study of natural inorganic substances (minerals), the basic building blocks of rocks. About 3600 minerals have been identified, but fewer than 50 are common constituents in the types of rocks that are abundant in the Earth. The most common minerals in the crust are feldspars, quartz, micas, amphiboles, pyroxenes, olivine, and calcite. Modern laboratories have effective devices for resolving the mineral content of rock materials; even the ultramicroscopic particles in clays are clearly defined under the electron microscope. *See* MINERAL; MINERALOGY.

Petrology is the study of rocks, their physical and chemical properties, and their modes of origin. The primary families are igneous rocks, which have solidified from molten matter (magma); sedimentary rocks, made of fragments derived by weathering of preexisting rocks, of chemical precipitates from sea or lake water, and of organic remains; and metamorphic rocks derived from igneous or sedimentary rocks under conditions that brought about changes in mineral composition, texture, and internal structure (fabric). The secondary rock families are pyroclastic rocks, which are partly igneous and partly sedimentary rocks because they are composed largely or entirely of fragments of igneous matter erupted explosively from a volcano; diagenetic rocks are transitional between sedimentary and metamorphic rocks because their textures or compositions were affected by low-temperature, postsedimentation processes below conditions of metamorphism; migmatites are transitional between metamorphic and igneous rocks because they form when metamorphic rocks are raised to temperatures and pressures so that small localized fractions of the rock start to melt but the melting is insufficient for a large body of magma to develop. *See* PETROGRAPHY; PETROLOGY; ROCK.

A general knowledge of geology has many practical applications, and large numbers of geologists receive special training for service in solving problems met in the mining of metals and nonmetals, in discovering and producing petroleum and natural gas, and in engineering projects of many kinds. Human use of materials has become so great that waste materials are influencing natural geological processes. As a result, a new discipline, environmental geology, is starting to emerge. *See* ENGINEERING GEOLOGY; PETROLEUM GEOLOGY.   [B.J.S.]

Geomagnetic variations   Variations in the natural magnetic field measured at the Earth's surface. This field changes with periodicities from about 0.3 s to thousands of years. Many of these variations of observed field—the very short-period, daily, seasonal, semiannual, and solar-cycle (11-year) variations—arise from sources that are external to the Earth but are superposed upon the larger main dipolar field of the Earth by the typical measuring instruments. The daily and seasonal atmospheric motions cause field variations that are smooth in form and relatively predictable, given the time and location of the observation. During occasions of high solar-terrestrial disturbance activity that give rise to auroras at high latitudes, very large geomagnetic variations occur that mask the quiet daily changes. These geomagnetic variations are so spectacular in size and global extent that they have been named geomagnetic storms. *See* AURORA; GEOMAGNETISM; IONOSPHERE; MAGNETOSPHERE.   [W.H.Ca.]

Geomagnetism   The magnetism of the Earth; also, the branch of science that deals with the Earth's magnetism. Formerly called terrestrial magnetism, geomagnetism involves any topic pertaining to the magnetic field observed near the Earth's surface, within the Earth, and extending upward to the magnetospheric boundary. Modern usage of the term is generally confined to historically recorded observations to distinguish it from the sciences of archeomagnetism and paleomagnetism, which deal with the ancient magnetic field frozen respectively in archeological artifacts and geologic structures. *See* PALEOMAGNETISM; ROCK MAGNETISM.

The primary component of the magnetic field observed at the Earth's surface is caused by electric currents flowing in its liquid core, and is called the main field. Vectorially added to this component are the crustal field of magnetized rocks, transient variations imposed from external sources, and the field from electric currents induced in the Earth from these variations.

The geomagnetic field is specified at any point by its vector **F**. Its direction is that of a magnetized needle perfectly balanced before it is magnetized, and freely pivoted about that point, when in equilibrium. The north pole of such a needle is the one that at most places on the Earth takes the more northerly position. Over most of the Northern Hemisphere, that pole dips downward (see illustration). The elements used to describe the vector **F** are H, the component of the vector projected onto a horizontal plane; its north and east components X and Y, respectively; Z the vertical component; F the magnitude of the vector **F**; the angles I, the dip of the field vector below the horizontal; and D the magnetic declination or deviation of the compass from geographic north. By convention, Z and I are positive downward, and D is positive eastward (or may be indicated as east or west of north). These elements can be related to each other by trigonometric equations.

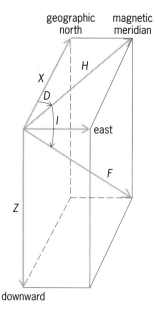

Elements of the geomagnetic field. D = declination, I = inclination, H = horizontal intensity, X = north intensity, Y = east intensity, Z = vertical intensity, F = total intensity.

A magnetic pole is a location where the field is vertically aligned, $H = 0$. Due to the presence of sometimes strong (for example, > 1000 nanoteslas) magnetic anomalies at the Earth's surface, there are a number of locations where the field is locally vertical. However, those field components that extend to sufficient altitude to control charged particles can be accurately located by using the computations from a spherical harmonic expansion using degrees up to only about $n = 10$. Indeed, a pole can be defined by using only the main dipole ($n = 1$), or many terms. See AURORA.

The $n = 1$ poles are sometimes referred to as the geomagnetic poles, and those computed using higher terms as dip poles. The term geomagnetic could also refer to the eccentric geomagnetic pole, which can be computed from $n = 1$ and $n = 2$ harmonics so as to be the best representation of a dipole offset from the center of the Earth. The latter has been used as a simplified field model at distances of 3 or 4 earth radii. Due to the more rapid fall-off of the higher terms with distance from the Earth, the two principal poles approach those of the $n = 1$ term with increasing altitude, until the distortions due to external effects begin to predominate. See MAGNETOSPHERE.

The distribution of the dip angle I over the Earth's surface can be indicated on a globe or map by contours called isoclines, along which I is constant. The isocline for which $I = 0$ (where a balanced magnetized needle rests horizontal) is called the dip equator. The dip equator is geophysically important because there is a region in the ionospheric E layer in which small electric fields can produce a large electric current called the equatorial electrojet. See GEOMAGNETIC VARIATIONS; IONOSPHERE.

A magnetized compass needle can be weighted so as to rest and move in a horizontal plane at the latitudes for which it is designed, thus measuring the declination D. The lines on the Earth's surface along which D is constant are called isogonic lines or isogones. The compass points true geographic north on the agonic lines where $D = 0$. At nonpolar latitudes, D is a useful tool for marine and aircraft navigational reference. Indeed, isogones appear on navigation charts, electronic navigational aids are referenced to D, and airport runways are marked with D/10. A runway painted with the number 11 indicates that its direction has a compass heading of $110°$. The compass needle becomes less reliable in polar regions because the horizontal component H becomes smaller as the magnetic poles are approached.

The intensity of the field can also be represented by maps, and the lines of equal intensity are called isodynamic lines. The dipole dominates the patterns of magnetic intensity on Earth in that the intensity is about double at the two poles compared to the value near the Equator. However, it can also be seen that the next terms of the spherical harmonic expansion also have a significant effect, in that there is a second maximum in Siberia, and an area near Brazil that is weaker than any other. This so-called Brazilian anomaly allows charged particles trapped in the magnetic field to reach a low altitude and be lost by collisions with atmospheric gases. The highest intensity of this smooth field is about 70 microteslas near the south magnetic pole in Antarctica, and the weakest is about 23 μT near the coast of Brazil.

The term magnetic anomaly has become clearer than it was previously because it is recognized that the geomagnetic field has a continuous spectrum but with two distinct contributors. Originally, the term meant a field pattern that was very local in extent; the modern definition is that portion of the field whose origin is the Earth's crust. The sizes of the strong and easily observable features are generally up to only a few tens of kilometers. Their intensity ranges typically from a few hundred nanoteslas up to several thousand, and they are highly variable depending on the geology of the region.

The main or core component of the geomagnetic field undergoes slow changes that necessitate continual adjustment of the model coefficients and redrawing of the isomagnetic maps. In any magnetic element at a particular place, the variation may be an increase or a decrease and is not constant in either magnitude or sign. This distribution of the rate for any element can be indicated on isoporic maps by lines (isopors) along which the rate is constant. Typically, the pattern of isopors is more complex than that of the isomagnetic lines for the same element, partly because the spectrum of such change is not dominated by the dipole as is the case of the static field.

Studies indicate that the dipole component of the field 2000 years ago was about 50% stronger than the present. Its average decay rate has averaged about 0.05% per year (15 nanoteslas per year) since about 1840 when absolute measurements were first begun, but accelerated from 1970 to its 1994 value of 0.08% per year (24 nT/yr). However, there is also evidence that the decade of the 1940s showed a rate of only about 10 nT per year. A linear projection of the present rate would have the dipole decreasing to zero in less than 1500 years. Although archeomagnetic evidence indicates that the field has indeed decayed to near zero level within the last 50,000 years with a subsequent return to the present polarity, and paleomagnetic results show that the field has reversed its polarity many times since the Earth's formation (the last time, about a million years ago), there is no model that can predict the future course of field change.

Deriving a suitable model that explains the source of the Earth's magnetic field has been one of the most frustrating problems that theoreticians have faced. Starting with the physical laws that should govern the behavior of a highly conducting, rotating, spherical fluid and coming up with a model of the geomagnetic field is exceedingly difficult. Dynamo means that a current is generated as an electrical conductor is moved through a magnetic field, as in a dynamo supplying electrical power. *See* GEODYNAMO.

The main source of data for magnetic maps and models before the advent of satellites was fixed magnetic observatories. These stations, numbering about 140, provided the continuous record of changes in the magnetic field at their location. Their data are generally accurate and an excellent indicator of both secular change and the transient variations, but their global coverage is too sparse for a determination of the whole field. Spherical harmonic analyses based only on such data produce distorted results because of the large gaps in coverage, especially because of the sparseness of observing locations in southern oceanic regions. [J.Cai.]

Geomorphology   The study of landforms, including the description, classification, origin, development, and history of planetary surface features. Emphasis is placed on the genetic interpretation of the erosional and depositional features of the Earth's surface. However, geomorphologists also study primary relief elements formed by movements of the Earth's crust, topography on the sea floor and on other planets, and applications of geomorphic information to problems in environmental engineering.

Geomorphologists analyze the landscape. Their purview includes the structural framework of landscape, weathering and soils, mass movement and hillslopes, fluvial features, eolian features, glacial and periglacial phenomena, coastlines, and karst landscapes. Processes and landforms are analyzed for their adjustment through time, especially the most recent portions of Earth history. Geomorphologists consider processes from the perspectives of pedology, soil mechanics, sedimentology, geochemistry, hydrology, fluid mechanics, remote sensing, and other sciences. The complexity of geomorphic processes has required this interdisciplinary approach, but it has also led to a theoretical vacuum in the science. At present many geomorphologists are organizing their studies through a form of systems analysis. The landscape is conceived of as a series of elements linked by flows of mass and energy. Process studies measure the inputs, outputs, and transfers for these systems. Systems analysis provides an organizational framework within which geomorphologists are developing models to predict selected phenomena. [V.R.B.]

Geophysical exploration   Making, processing, and interpreting measurements of the physical properties of the Earth with the objective of practical application of the findings. Most exploration geophysics is conducted to find commercial accumulations of oil, gas, or other minerals, but geophysical investigations are also employed with engineering objectives, in studies aimed at predicting the nature of the Earth for the foundations of roads, buildings, dams, tunnels, nuclear power plants, and other structures, and in the search for geothermal areas, water resources, archeological ruins, and so on.

Geophysical exploration is also called applied geophysics or geophysical prospecting. The physical properties and effects of subsurface rocks and minerals that can be measured at a distance include density, electrical conductivity, thermal conductivity, magnetism, radioactivity, elasticity, and other properties. Exploration geophysics is often divided into subsidiary fields according to the property being measured, such as magnetic, gravity, seismic, electrical, thermal, or radioactive properties.

Magnetic exploration. Rocks and ores containing magnetic minerals become magnetized by induction in the Earth's magnetic field so that their induced field adds to the Earth's field. Magnetic exploration involves mapping variations in the magnetic field to determine the location, size, and shape of such bodies. The magnetic susceptibility of sedimentary rock is generally orders of magnitude less than that of igneous or metamorphic rock. Consequently, the major magnetic anomalies observed in surveys of sedimentary basins usually result from the underlying basement rocks. Determining the depths of the tops of magnetic bodies is thus a way of estimating the thickness of the sediments. *See* GEOMAGNETISM; MAGNETOMETER; ROCK MAGNETISM.

Except for magnetite and a very few other minerals, mineral ores are only slightly magnetic. However, they are often associated with bodies such as dikes that have magnetic expression so that magnetic anomalies may be associated with minerals empirically. For example, placer gold is often concentrated in stream channels where magnetite is also concentrated.

Gravity exploration. Gravity exploration is based on the law of universal gravitation: the gravitational force between two bodies varies in direct proportion to the product of their masses and in inverse proportion to the square of the distance between them. Because the Earth's density varies from one location to another, the force of gravity varies from place to place. Gravity exploration is concerned with measuring these variations to deduce something about rock masses in the immediate vicinity. Gravity surveys are used more extensively for petroleum exploration than for metallic mineral prospecting. The size of ore bodies is generally small; therefore, the gravity effects are quite small and local despite the fact that there may be large density differences between the ore and its surroundings. *See* GRAVITY METER.

Seismic exploration. Seismic exploration is the predominant geophysical activity. Seismic waves are generated by one of several types of energy sources and detected by arrays of sensitive devices called geophones or hydrophones. The most common measurement made is of the travel times of seismic waves, although attention is being directed increasingly to the amplitude of seismic waves or changes in their frequency content or wave shape. *See* SEISMOLOGY.

Electrical and electromagnetic exploration. Variations in the conductivity or capacitance of rocks form the basis of a variety of electrical and electromagnetic exploration methods, which are used primarily in metallic mineral prospecting. Both natural and induced electrical currents are measured. Direct currents and low-frequency alternating currents are measured in ground surveys, and ground and airborne electromagnetic surveys involving the lower radio frequencies are made. *See* GEOELECTRICITY.

Radioactivity exploration. Natural radiation from the Earth, especially of gamma rays, is measured both in land surveys and airborne surveys. Natural types of radiation are usually absorbed by a few feet of soil cover, so that the observation is often of diffuse equilibrium radiation. The principal radioactive elements are uranium, thorium, and potassium; radioactive exploration has been used primarily in the search for uranium and other ores, such as columbium, which are often associated with them. The Geiger counter and scintillation counter are instruments generally used to detect and measure the radiation.

Remote sensing. Measurements of natural and induced electromagnetic radiation made from high-flying aircraft and earth satellites are referred to collectively as remote sensing. This comprises both the observation of natural radiation in various spectral bands, including both visible and infrared radiation, such as by photography and measurements of the reflectivity of infrared and radar radiation. *See* REMOTE SENSING.

Well logging. A variety of types of geophysical measurements are made in boreholes, including self-potential, electrical conductivity, velocity of seismic waves, natural and induced radioactivity, and temperature variations. Borehole logging is used extensively in petroleum exploration to determine the characteristics of the rocks which the borehole has penetrated, and to a lesser extent in mineral exploration.   [R.E.Sh.]

Geophysical fluid dynamics   The branch of physics that studies the dynamics of naturally occurring large-scale flows in the atmosphere and oceans. Examples of such flows are weather patterns, atmospheric fronts, and ocean currents. The fluids are either air or water in a moderate range of temperatures and pressures.

Because of their large scale (from tens of kilometers up to the size of the planet), geophysical flows are strongly influenced by the diurnal rotation of the Earth, which is manifested in the equations of motion as the Coriolis force. Another fundamental characteristic is stratification, that is, density heterogeneity within the fluid in the presence of the Earth's gravitational field, which is responsible for buoyancy forces. Thus, geophysical fluid dynamics may be considered to be the study of rotating and stratified fluids. It is the common denominator of dynamical meteorology and physical oceanography. *See* CORIOLIS ACCELERATION; EARTH; METEOROLOGY; OCEANOGRAPHY.

The first of the two distinguishing attributes of geophysical fluid dynamics is the effect of the Earth's rotation. Because geophysical flows are relatively slow and spread over long distances, the time taken by a fluid particle (be it a parcel of air in the atmosphere or water in the ocean) to traverse the region occupied by a certain flow structure is comparable to, and often longer than, a day. Thus, the Earth rotates significantly during the travel time of the fluid, and rotational effects enter the dynamics. Fluid flows viewed in a rotating framework of reference are subject to two additional types of forces, namely the centrifugal force and the Coriolis force. (Properly speaking, these originate not as actual forces but as acceleration terms to correct for the fact that viewing the flow from a rotating frame—the rotating Earth in the case of geophysical fluid dynamics— demands a special transformation of coordinates.) Contrary to intuition, the centrifugal force plays no role on fluid motion because it is statically compensated by the tilting of the gravitational force caused by the departure of the Earth's shape from sphericity. Thus, of the two, only the Coriolis force acts on fluid parcels.

Variations of moisture in the atmosphere, of salinity in the ocean, and of temperature in either can modify the density of the fluid to such an extent that buoyancy forces become comparable to other existing forces. The fluid then has a strong tendency to arrange itself vertically so that the denser fluid sinks under the lighter fluid. The resulting arrangement is called stratification, the second distinguishing attribute of geophysical fluid dynamics. The greater the stratification in the fluid, the greater the resistance to vertical motions, and the more potential energy can affect the amount of kinetic energy available to the horizontal flow.

A quantity central to the understanding of geophysical flows, which are simultaneously rotating and stratified, is the potential vorticity, q. This quantity incorporates both rotation and stratification. Geophysical flows are replete with vortices, resulting from baroclinic instability. Their interactions generate highly complex flows not unlike those commonly associated with turbulence. Unlike classical fluid turbulence, however, geophysical flows are wide and thin (with, furthermore, a high degree of vertical rigidity as a result of rotational effects), and their turbulence is nearly two-dimensional.

In meteorology, geophysical fluid dynamics has been the key to understanding the essential properties of midlatitude weather systems, including the formation of cyclones and fronts. Geophysical fluid dynamics also explains the dynamical features of hurricanes and tornadoes, sea and land breezes, the seasonal formation and break-up of the polar vortex that is associated with high-latitude stratospheric ozone holes, and a host of other wind-related phenomena in the lower atmosphere. *See* CYCLONE; HURRICANE; TORNADO.

In oceanography, successes of geophysical fluid dynamics include the explanation of major oceanic currents, such as the Gulf Stream. Coastal river plumes, coastal upwelling, shelf-break fronts, and open-ocean variability on scales ranging from tens of kilometers to the size of the basin are among the many other marine applications. The El Niño phenomenon in the tropical Pacific is rooted in processes that fall under the scope of geophysical fluid dynamics. *See* EL NIÑO; GULF STREAM. [B.C.R.]

Geophysics   Those branches of earth sciences in which the principles and practices of physics are used to study the Earth. Geophysics is considered by some to be a branch of geology, by others to be of equal rank. It is distinguished from the other earth sciences largely by its use of instruments to make direct or indirect measurements of the parts of the Earth being studied, in contrast to the more direct observations which are typical of geology. *See* GEOLOGY.

Geophysics consists of several principal fields plus parallel and subsidiary divisions. These are commonly considered to include plutology, with geodesy, geothermometry, seismology, and tectonophysics as subdivisions; hydrospheric studies, hydrology (groundwater studies), oceanography, and glaciology; atmospheric studies, meteorology, and aeronomy; and several fields of geophysics which overlap one another, including geomagnetism and geoelectricity, geochronology, geocosmogony, and geophysical exploration and prospecting. Planetary sciences, the study of the planets and satellites aside from the Earth, are usually considered a branch of geophysics because the techniques used have been, until the first landings on each, entirely instrumental rather than directly observational. *See* GEOCHRONOMETRY; GEODESY; GEOELECTRICITY; GEOMAGNETISM; GLACIOLOGY; HYDROLOGY; METEOROLOGY; OCEANOGRAPHY; SEISMOLOGY.

[B.F.H.]

Geostrophic wind   A hypothetical wind based upon the assumption that a perfect balance exists between the horizontal components of the Coriolis force and the horizontal pressure gradient force per unit mass, with the implication that viscous forces and accelerations are negligible. The geostrophic wind blows parallel to the isobars (lines of equal pressure) with lower pressure to the left of the direction of the wind in the Northern Hemisphere and to the right in the Southern Hemisphere. It represents a good approximation to the actual wind at elevations greater than about 3000 ft (900 m), except in instances of strongly curved flow and in the vicinity of the Equator.

The term thermal wind denotes the net change in the geostrophic wind over some specific vertical distance. This change arises because the rate of change of pressure in the vertical is different in two air columns of different air density, so that the horizontal component of the pressure gradient force per unit mass varies in the vertical. The thermal wind is directed approximately parallel to the isotherms of air temperature with cold air to the left and warm air to the right in the Northern Hemisphere, and vice versa in the Southern Hemisphere. Thus, for example, the increasing predominance of westerly winds aloft may be viewed as a consequence of the warmth of tropical latitudes and the coldness of polar regions. *See* CORIOLIS ACCELERATION; GRADIENT WIND; WIND; WIND STRESS.

[F.S.]

Geosyncline   A linear part of the crust of the Earth that sagged deeply through time, that is, a great trough hundreds of kilometers long and tens of kilometers wide that subsided as it received thousands of meters of sedimentary and volcanic rocks through millions of years. Thicknesses are roughly 10 times greater than synchronous strata in adjacent stable regions. Linear geosynclinal belts were great subsiding tectonic divisions of the crust that lay between more stable parts of continents (cratons) or the deep abyssal ocean basins of their time, or both. Geosynclines generally contain volcanic (eugeosynclinal) and nonvolcanic (miogeosynclinal) zones (see illustration).

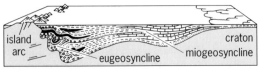

Diagrammatic section of Cordilleran geosyncline in southeastern Alaska and British Columbia at the close of Permian time (about 230 million years ago). (*After A. J. Eardley, J. Geol., 55:319–342, 1947*)

Key: ■ volcanic rocks

Although defined on the basis of thickness and types of rocks, the geosyncline has always been so closely linked with the origin of mountains that it has little meaning in any other context. *See* CORDILLERAN BELT; OROGENY; SYNCLINE.   [R.H.Dot.]

Geyser   A natural spring or fountain which discharges a column of water or steam into the air at more or less regular intervals. Perhaps the best-known area of geysers is in Yellowstone Park, Wyoming, where there are more than 100 active geysers and more than 3000 noneruptive hot springs.

The eruptive action of geysers is believed to result from the existence of very hot rock not far below the surface. The neck of the geyser is usually an irregularly shaped tube partly filled with water which has seeped in from the surrounding rock. Far down the pipe the water is at a temperature much above the boiling point at the surface, because of the pressure of the column of water above it. Its temperature is constantly increasing, because of the volcanic heat source below. Eventually the superheated water changes into steam, lifting the column of water out of the hole.   [A.N.S./R.K.Li.]

Glacial epoch   An informal reference to a time during the history of the Earth when there were larger ice sheets (continental size) and mountain glaciers than today. The most recent glacial epoch, better known as the Pleistocene glacial epoch, and also by the older term Quaternary period, encompassed at least the last 3,000,000 years.

Many side effects resulted from the existence of these ice sheets and glaciers, including climate changes, sea-level rise and fall, depressions of the Earth's crust, and large-scale migrations of plants, animals, and humans as well as mass extinctions. Mountain landscapes were sculptured by glaciers, and erosional and depositional landforms were formed. Ocean temperatures were cold during glaciations and warm at times of interglacials. Early human evolution, development, and migrations resulted from the ever-changing climates closely related to glacier advances and retreats.

Glacial epochs seem to recur at intervals of 200,000,000 to 250,000,000 years. In overall occurrence, all the glacial epochs that have ever occurred occupy only 5 to 10% of all geologic time. During major glacial epochs, great ice sheets formed in the high latitudes and spread out to cover as much as 40% of the Earth's land surface. Accompanying drops in temperature during some glacial epochs may have been as much as $25°F$ ($14°C$) in the mid-latitudes. During a glacial epoch, major glaciations are short-lived, each lasting less than 10,000 years, with the interglacials persisting for only about 10,000 years, so that for most of an epoch, the ice sheets either grow or diminish in size. The Pleistocene glacial epoch was distinguished by seven or eight glacial advances within the last 700,000 years. Its last glaciation ended about 9000 years ago in Fennoscandia, and less than 8000 years ago in north-central Canada. *See* GLACIAL GEOLOGY; GLACIOLOGY; PLEISTOCENE.   [S.E.Wh.]

Glacial geology   The scientific study of the effects of glaciers on the broad land areas, on the oceans, and on climate, of their erosion and deposition, and of their modification of the Earth's surface in detail. Included in the realm of glacial geology is the history of glacial theory, consideration of the origin of glacial ages, extent and times of past glaciations, erosion and sculpturing of plains and mountains, deposition of ice-contact and meltwater sediments, and the consequences of glaciers on worldwide climate, and also on local climate around their edges. Quite distinct from glacial geology, however, is the separate, growing subscience of glaciology, the study of glaciers themselves. *See* GLACIOLOGY.

Features on the Earth's surface explained by former worldwide glaciation are numerous, embracing, for example, glacially eroded and molded valleys and mountains;

ice-transported and deposited sediments and nonglacial sediments; abandoned stream channels with associated floodwater deposits; elevated silts and clays that collected around continental edges when sea level was higher; valleys eroded across and into continental shelves and slopes when sea level was much lower; communities of plants and animals similar to each other but separated by shallow seaways where land bridges once existed; fossil shells and microorganisms in deep-sea sediments reflecting colder or warmer water temperatures than today; vegetated sand dunes aligned to wind systems no longer operating; ancient shorelines and beach ridges ringing dry empty lake basins far inland; and orderly patterns of stones and fine sediments next to glacier margins in polar regions and high mountains. *See* CIRQUE; DRUMLIN; GLACIAL EPOCH; MORAINE.

[S.E.Wh.]

Glaciology A broad field encompassing all aspects of the study of ice, specifically glaciers, the largest ice masses on Earth.

Glaciers are classified principally on the basis of size, shape, and temperature. Cirque glaciers occupy spectacular steep-walled, overdeepened basins a few square kilometers ($1 \text{ km}^2 = 0.36 \text{ mi}^2$) in area, called cirques. Most cirques are in high mountain areas that have been repeatedly inundated by ice. The cirques and the deep valleys leading away from them were, in fact, eroded by larger glaciers over the past 3 million years. *See* CIRQUE.

As a cirque glacier expands, it is usually constrained, at least initially, to move down such a valley. It then becomes a valley glacier (see illustration). Where such a valley ends in a deep fiord in the sea, the glacier is called a tidewater glacier. *See* FIORD.

In contrast, some glaciers are situated on relatively flat topography. Such glaciers can spread out in all directions from a central dome. When small, on the order of a few tens of kilometers across, these are called ice caps. Large ones, like those in Antarctica and Greenland, are ice sheets. *See* ANTARCTICA; ICE FIELD.

Thermally, glaciers are usually classified as either temperate or polar. In the simplest terms, a temperate glacier is one that is at the melting point throughout. The term melting point is used in this context rather than 0°C (32°F), because the temperature at

Storglaciären, a small valley glacier in northern Sweden.

which ice melts decreases as the pressure increases. Thus, the temperature at the base of a temperate glacier that is 500 m (1700 ft) thick will be about $-0.4°C$ ($31.2°F$), but if heat energy is added to the ice, it can melt without an increase in temperature. Most valley glaciers are temperate.

In polar glaciers the temperature is below the melting point nearly everywhere. The temperature of a polar glacier increases with depth, however, because the deeper ice is warmed by heat escaping from within the Earth and by frictional heat generated by deformation of the ice. Thus, at its base, a polar glacier may be frozen to the substrate or may be at the melting point. Ice caps and ice sheets are normally polar, as are some valley glaciers in high latitudes.

As was the case with the classification based on size and shape, there is a continuum of thermal regimes in glaciers. The most common intermediate type has a surficial layer of cold ice, a few tens of meters (1 m = 3.28 ft) thick in its lower reaches, but is temperate elsewhere. Such glaciers are sometimes called subpolar or polythermal.

Glaciers exist because there are places where the climate is so cold that some or all of the winter snow does not melt during the following summer. The next winter's snow then buries that remaining from the previous winter, and over a period of years a thick snow pack or snowfield develops. Deep in such a snow pack the snow is compacted by the weight of the overlying snow. In addition, evaporation of water molecules from the tips of snowflakes and condensation of this water in intervening hollows results in rounding of grains. These processes of compaction and metamorphism gradually transform the deeper snow, normally known as firn, into ice. Melt water percolating downward into this firn may refreeze, accelerating the transformation. *See* SNOWFIELD AND NÉVÉ.

When, during a given year, the mass of snow added in the accumulation area of a glacier exceeds the mass of ice lost from the ablation area, the glacier is said to have had a positive mass balance. If such a situation persists for several years, the glacier will advance to lower elevations or more temperate latitudes, thus increasing the size of its ablation area and the mass loss. Conversely, persistent negative mass balances lead to retreat. Contrary to one implication of the word retreat, a retreating glacier does not flow backward. Rather, a glacier retreats when the ice flow toward the terminus is less than the melt rate at the terminus.

Under certain rather rare circumstances, the high subglacial water pressures that develop in the spring do not dissipate quickly but persist for weeks. This occurs under glaciers that have been thickening for several years or decades but have not advanced appreciably as a result of the thickening. On these occasions, the increase in sliding speed resulting from the increased water pressure inhibits development of an integrated subglacial conduit system, so water pressures remain high. The glacier then may advance at speeds of meters to tens of meters per day, in what is known as a surge.

Ice stream flow occurs in some parts of the Antarctic Ice Sheet. These high flow rates occur in streams of ice tens of kilometers wide and hundreds of kilometers long, and are sustained for centuries. These streams are bounded not by valley sides but by ice that is moving much more slowly. Ice Stream B, for example, which drains to the Ross Ice Shelf in West Antarctica, has a maximum speed of 825 m/yr, while ice on either side of it is moving at only 10–20 m/yr. The high speeds of these ice streams are attributed to slippery conditions at the bed, where high water pressures reduce friction between the ice and the bed. Changes in paths of water flow at the bed are believed to be responsible for the changes in ice stream activity.

Among the hazards associated with glaciers are jökulhlaups, or sudden releases of water from lakes dammed by glaciers. Jökulhlaup is an Icelandic word; it has entered the vocabulary of geology because such floods are common in Iceland where localized volcanic heat is responsible for the presence of deep lakes surrounded by ice. In other regions, the lakes are more commonly formed where a glacier in a trunk valley extends across the open mouth of a tributary valley. *See* GLACIAL GEOLOGY.   [R.LeB.H.]

Glauconite   The term glauconite as currently used has a two-fold meaning. It is used as both a mineralogic and morphologic term. The mineral glauconite is defined as an illite type of clay mineral. A fundamental characteristic of glauconite is that the unit cell is composed of a single silicate layer rather than the double layer of most other dioctahedral micas. *See* CLAY MINERALS; ILLITE.

Glauconite is known to occur in flakes and as pigmentary materials. When used in the morphological sense, the term glauconite often refers to small, green, spherical, earthy pellets. Some of these pelletal varieties are composed solely of the mineral described above, others are a mixed-layer association of this mineral and other three-layer structures.

Glauconite forms during marine diagenesis, in relatively shallow water, and at times of slow or negative deposition. Glauconite has been identified in both recent and ancient sediments. It is a major component in some "greensand" deposits and has been used commercially for the extraction of potassium from such sources. *See* AUTHIGENIC MINERALS; DIAGENESIS; MARINE SEDIMENTS.   [F.M.W.; R.E.Gr.]

Glaucophane   A monoclinic sodic amphibole with composition close to $Na_2(Mg_3Al_2)Si_8O_{22}(OH)_2$. This mineral exhibits a characteristic blue color with distinct pleochroism from colorless to lavender blue when viewed in thin section by plane-polarized light. Outcrops of glaucophane-rich metamorphic rocks are commonly blue and tend to have good foliation; these rocks are called blueschists. *See* BLUESCHIST.

Glaucophane is an index mineral of blueschist, which is generated under unusually high pressures at low temperatures in a tectonic environment exclusively associated with a subducted lithospheric slab or related tectonic loading. The glaucophane-bearing assemblages occur in recrystallized graywackes and pelitic rocks and in metabasites and metacherts of oceanic affinity; they are typically found in subduction zone complexes at plate boundaries, a setting first recognized in the Jurassic and Cretaceous Franciscan Complex of northern California. Blueschists are most common and best developed in Mesozoic and Cenozoic terranes: some Paleozoic and even latest Precambrian blueschists have been described in Russia and China. Blueschists formed earlier in geologic time may have been eroded or been recrystallized under normal geothermal conditions. The preservation of glaucophane in blueschists of continental or island arc margins indicates either rapid uplift or maintenance of low geothermal gradients by steady-state subduction for tens of million years. *See* GRAYWACKE; SUBDUCTION ZONES.
                    [J.G.L.; R.Y.Z.; S.Maru.]

Global climate change   The periodic fluctuations in global temperatures and precipitation, such as the glacial (cold) and interglacial (warm) cycles of the Pleistocene (a geological period from 1.8 million to 10,000 years ago). Presently, the increase in global temperatures since 1900 is of great interest. Many atmospheric scientists and meteorologists believe it is linked to human-produced carbon dioxide (CO_2) in the atmosphere.

Greenhouse effect. The greenhouse effect is a process by which certain gases (water vapor, carbon dioxide, methane, nitrous oxide) trap heat within the Earth's atmosphere and thereby produce warmer air temperatures. These gases act like the glass of a greenhouse: they allow short (ultraviolet; UV) energy waves from the Sun to penetrate into the atmosphere, but prevent the escape of long (infrared) energy waves that are emitted from the Earth's surface. *See* ATMOSPHERE; GREENHOUSE EFFECT.

Human-induced changes in global climate caused by release of greenhouse gases into the atmosphere, largely from the burning of fossil fuels, have been correlated with global warming. Since 1900, the amount of two main greenhouse gases (carbon dioxide and methane) in the Earth's atmosphere has increased by 25%. Over the same period, mean global temperatures have increased by about 0.5°C (0.9°F). The most concern centers on carbon dioxide. Not only is carbon dioxide produced in much greater quantities than any other pollutant, but it remains stable in the atmosphere for over 100 years. Methane, produced in the low-oxygen conditions of rice fields and as a by-product of coal mining and natural gas use, is 100 times stronger than carbon dioxide in its greenhouse effects but is broken down within 10 years.

Chloroflurocarbon (CFC) pollution, from aerosol propellants and coolant systems, affects the Earth's climate because CFCs act as greenhouse gases and they break down the protective ozone (O_3) layer. Other pollutants released into the atmosphere are also likely to influence global climate. Sulfur dioxide (SO_2) from car exhaust and industrial processes, such as electrical generation from coal, cool the Earth's surface air temperatures and counteract the effect of greenhouse gases. Nevertheless, there have been attempts in industrialized nations to reduce sulfur dioxide pollution because it also causes acid rain. *See* AIR POLLUTION.

Possible impact. A rise in mean global temperatures is expected to cause changes in global air and ocean circulation patterns, which in turn will alter climates in different regions. Changes in temperature and precipitation have already been detected. In the United States, total precipitation has increased, but it is being delivered in fewer, more extreme events, making floods (and possibly droughts) more likely. *See* OCEAN CIRCULATION.

Global warming has caused changes in the distribution of a species throughout the world. By analyzing preserved remains of plants, insects, mammals, and other organisms which were deposited during the most recent glacial and interglacial cycles, scientists have been able to track where different species lived at times when global temperatures were either much warmer or much cooler than today's climate. Several studies have documented poleward and upward shifts of many plant and insect species during the current warming trend.

Changes in the timing of growth and breeding events in the life of an individual organism, called phenological shifts, have resulted from global warming. For example, almost one-third of British birds are nesting earlier (by 9 days) than they did 25 years ago, and five out of six species of British frog are laying eggs 2–3 weeks earlier.

Community reassembly, changes in the species composition of communities, has resulted from climate change because not all species have the same response to environmental change.

To date, there have been no extinctions of species directly attributable to climate change. However, there is mounting evidence for drastic regional declines. For example, the abundance of zooplankton (microscopic animals and immature stages of many species) has declined by 80% off the California coast. This decline has been related to gradual warming of sea surface temperatures. *See* CLIMATE HISTORY; CLIMATE MODIFICATION; EXTINCTION (BIOLOGY).   [C.Pa.]

Globe (Earth)   A sphere on the surface of which is a map of the world. The map may be drawn, engraved, or painted directly on the surface but is more commonly prepared as a series of gores, or segments in other designs, to be affixed to the globe ball (see illustration).

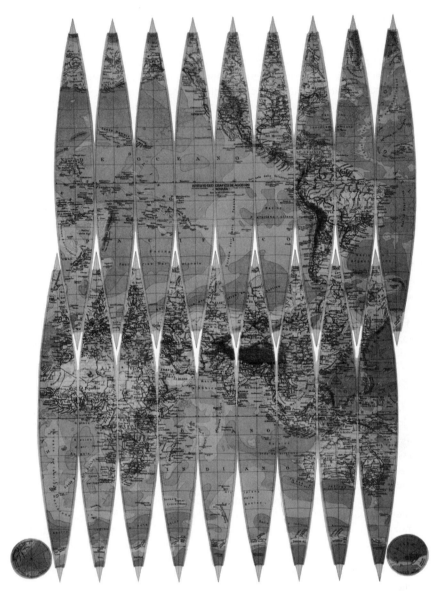

Globe gores from collections of Library of Congress. (*Istituto Geografico de Agostini, Novara, Italy*)

Globes are both artistically interesting and scientifically useful. Their principal value is in stimulating sound concepts of worldwide patterns and in rectifying errors induced by the limitations of flat maps. All flat maps distort the Earth's surface patterns, but carefully made globes constitute truer scale models of the Earth, with correct areas, shapes, and distances as well as continuity of surface. Globes have long been used as aids in navigation, in the teaching of earth sciences, and as room ornaments.

Many modern globes have special attachments to improve their utility. A meridian ring, extending from pole to pole, may be calibrated in degrees to measure latitude.

The longitude of points directly beneath that ring will be indicated at the intersection of the ring with the equatorial scale. A horizon ring at right angles to the meridian ring may be calibrated in miles or in meters, degrees, and hours to expedite distance and time measurement. A hinged horizon ring may be lifted to serve as a meridian ring, or placed in an oblique position to show great circle routes and distances.   [A.C.G.]

Gneiss   Coarse-grained, banded crystalline rock. Gneiss is composed of mineral grains large enough to be seen with the naked eye (see illustration). Banding arises from segregation of the various minerals present, typically into dark- and light-colored layers. Individual bands are commonly 0.04 to 0.4 in. (1 mm to 1 cm) thick. Although individual mineral grains are often flattened parallel to banding, such shape orientation is not present in many gneisses. Sheetlike minerals such as micas may be present but form only a subordinate amount of the rock. Banded rock of coarse grain containing substantial amounts of such minerals is named schist. Crystalline rock which has flattened grains but lacks obvious banding is generally called leptite. *See* SCHIST.

1 cm

Gneiss formed by metamorphism of preexisting granite. Dark minerals are mica; light-colored minerals are quartz and feldspar. The streaky nature of banding is typical of gneisses. The sample is from the Great Smoky Mountains of North Carolina.

Gneiss is defined by its texture, or arrangement of mineral grains, rather than by its mineral composition. However, the term gneiss is often taken to imply a mineral composition of granitic type, dominated by quartz and feldspar. Gneisses of other compositions are identified by qualifying terms such as compositional rock names, as in diorite gneiss and amphibolite gneiss, or a partial list of minerals present, as in biotite-plagioclase gneiss and hornblende-plagioclase gneiss. *See* FELDSPAR; QUARTZ.

Most gneisses are formed by recrystallization of preexisting rock during intense regional metamorphism. Shear stress present during such metamorphism causes formation of gneissic banding, although the exact mechanisms of this process are not well understood. Gneisses typically occupy large areas within the high-grade cores of regional metamorphic belts. Such terranes are often difficult to understand, because the processes which cause formation of gneissic texture are also sufficient to obscure preexisting rock structures. High temperature and shear are sufficient to cause plastic flow of gneissic rock on a gigantic scale. Such conditions of metamorphism are probably brought about by deep tectonic burial and major regional compression. Thus gneissic terranes may be expected to form in areas of convergent plate tectonics. *See* METAMORPHIC ROCKS; METAMORPHISM; METASOMATISM; PLATE TECTONICS.   [D.W.Mo.]

Goethite   A mineral of composition FeO · OH, crystallizing in the orthorhombic system. Crystals are rare, and the mineral is usually in reniform or stalactitic masses which have a radiating fibrous internal structure. The luster is adamantine to dull, and the color light to dark brown. The Mohs hardness is 5.0–5.5, and the density is 4.28 for crystals and 3.3–4.3 for massive material. Most of the common, yellow-brown, earthy ferric oxides known as limonite are mixtures composed largely of cryptocrystalline goethite.

Goethite is one of the most common minerals. It is the major constituent of the gossan at the surface of metalliferous deposits rich in iron-bearing sulfides, as at Bisbee, Arizona, and of laterites, as in Cuba. Well-formed crystals are found at Pribram, Bohemia, and Cornwall, England. It is an important iron ore in Alsace-Lorraine, in the Lake Superior hematite deposits, and in the southern Appalachians. *See* LIMONITE.

[L.Gr.]

Graben   A block of the Earth's crust, generally with a length much greater than its width, that has been dropped relative to the blocks on either side (see illustration). The

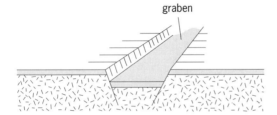

Diagram of simple graben. (*After A. K. Lobeck, Geomorphology, McGraw-Hill, 1939*)

size of a graben may vary. The faults that separate a graben from the adjacent rocks are inclined from 50 to 70° toward the down-thrown block and have displacements ranging from inches to thousands of feet. The direction of slip on these indicates that they are gravity faults. *See* FAULT AND FAULT STRUCTURES; HORST; RIFT VALLEY.   [P.H.O.]

Gradient wind   A hypothetical wind based upon the assumption that the sum of the horizontal components of the Coriolis force and the atmospheric pressure gradient force per unit mass is equivalent to a wind acceleration which is normal to the direction of the wind itself (centripetal acceleration), with the implication that there are no viscous forces acting. The direction of the gradient wind is the same as that of the geostrophic wind. The gradient wind speed is less than the geostrophic speed when the air moves in a cyclonically curved path and greater when the air moves in an anticyclonically curved path. The gradient wind is a good approximation of the actual wind and is often superior to the geostrophic wind, particularly when the flow is strongly curved in the cyclonic sense. *See* CORIOLIS ACCELERATION; GEOSTROPHIC WIND.   [F.S.]

Granite   A crystalline igneous rock that consists largely of alkali feldspar (typically perthitic microcline or orthoclase), quartz, and plagioclase (commonly calcic albite or oligoclase). Its average grain size is 0.04–1.0 in. (1–25 mm); finer-grained rocks of this composition include rhyolite and aplite, and coarser-grained ones are granite pegmatite. *See* APLITE; PEGMATITE; RHYOLITE.

The revised nomenclature of the International Union of Geological Sciences (IUGS) subcommission defines granite as containing 80–100% by volume quartz, alkali feldspar, and plagioclase in the proportions given in the illustration, and 20–0% accessory minerals. The three essential minerals must include 20–60% quartz, and alkali

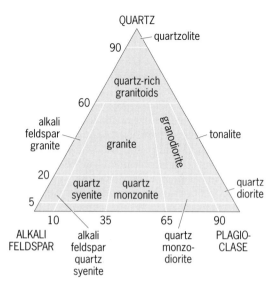

Classification of granitic rocks by the IUGS scheme. All proportions are by volume percentages.

feldspar must constitute 65–90% of the total feldspar. The variety alkali feldspar granite is similar except that alkali feldspar constitutes 90–100% of its total feldspar. The term granitic rocks includes granodiorite and tonalite as well as granite, and as used by some geologists may include quartz syenite to quartz diorite (see illustration). See FELDSPAR; GRANODIORITE; HORNBLENDE; MUSCOVITE.

Granites may be divided into three major types: calc-alkaline, peraluminous, and alkaline. Calc-alkaline granites typically are biotite or biotite-hornblende granites, some contain augite, and sphene is a common accessory.

Peraluminous granites, also known as S-type granites, contain aluminum in excess of that contained in feldspars and biotite; thus muscovite is an accessory mineral. Other aluminous minerals such as andalusite, sillimanite, cordierite, or garnet also may be accessory. Subalkaline to alkaline granites are characterized by iron-rich mafic minerals and relatively sodic alkali feldspar. The subalkaline type typically contains ferruginous biotite or hornblende, or both, but varieties containing ferrohedenbergite or fayalite are not uncommon. Allanite and zircon are common accessories. The alkaline type contains the Na-Fe minerals aegirine or riebeckite-arfvedsonite, or all three, and may also contain ferruginous biotite or even astrophyllite, eudialyte, or other rare minerals. See IGNEOUS ROCKS.   [F.B.]

Granodiorite   A phaneritic (visibly crystalline) plutonic rock composed chiefly of sodic plagioclase (oligoclase or andesine), alkali feldspar (microcline or orthoclase, usually perthitic), quartz, and subordinate dark-colored (mafic) minerals (biotite, amphibole, or pyroxene). Granodiorite is intermediate between granite and quartz diorite (tonalite). For convenience granite and granodiorite are commonly grouped and referred to as granite. See GRANITE; IGNEOUS ROCKS.   [C.A.C.]

Granulite   An important class of metamorphic rocks exposed at the surface of the Earth's crust, and inferred to make up a large portion of the deeper crust. Granulites are known to have formed at higher temperatures, and in many cases, higher pressures, than most other crustal rock assemblages. Thus, they are believed to have formed at considerable depths in the crust. See METAMORPHIC ROCKS.

Granulites may be of many different bulk compositions, inherited from precursor sedimentary, igneous, or lower-grade metamorphic rocks. The high temperatures of crystallization have resulted in very low water content, reflected in nearly anhydrous mineralogy. Characteristic minerals of granulite metabasalts are plagioclase, orthopyroxene, clinopyroxene, hornblende, and garnet. These minerals are also characteristic of granulites of intermediate to granitic composition, together with progressively greater amounts of quartz and potassium feldspar. The association of potassium feldspar with orthopyroxene is definitive for charnockite, a granulite of approximately granitic composition characteristic of ancient high-grade terrains. *See* BASALT; GRANITE.

Granulites characteristically contain CO_2-rich fluid inclusions in the mineral grains, in contrast to the more aqueous fluid inclusions of other kinds of rock. This has suggested that action of volatiles low in H_2O and rich in CO_2, probably of subcrustal origin, were important in crustal metamorphism early in the Earth's history. *See* METAMORPHISM.

[R.C.N.]

Gravel   An unconsolidated sedimentary aggregate containing more than 50% by weight of gravel-sized particles (mean diameter greater than 0.08 in. or 2 mm). The gravel-sized particles are termed the framework; those less than 0.08 in. in diameter are the matrix. There is an important distinction between framework-supported and matrix-supported gravels. The latter may possess a muddy matrix, in which case they are termed diamictons. Typically, diamictons are unstratified internally, contain subangular framework clasts, and are deposited by mass-flow processes such as debris flow or glacial-ice transport. Water-laid gravels are typically stratified or cross-stratified and are framework-supported, with subangular to rounded clasts in a sand matrix. Less commonly, they may be sand-matrix-supported, or they may lack matrix and then are termed openwork gravels. Water-worn gravel clasts tend to conform to the shape of triaxial ellipsoids and develop preferred orientation, with long axes normal to stream flow and intermediate axes dipping gently upstream.

The consolidated equivalents of gravels are conglomerates and breccias, the latter including only angular particles. Paleoenvironmental indicators for conglomerates include stratification, size grading, particle roundness, particle orientation, and matrix–framework relations. *See* BRECCIA; CONGLOMERATE; SEDIMENTARY ROCKS.   [B.R.R.]

Gravity meter   A device that measures local acceleration due to the Earth's gravity; it is also called a gravimeter. Such instruments fall into two categories: relative gravity meters, which are used to determine gravity differences among a number of geographic locations or changes in gravity that occur at a single location over time; and absolute gravity meters, which can measure the true value of the acceleration due to gravity at a given location and time.

The local value of gravity is the acceleration undergone by a freely falling mass upon which gravity is the only force acting. Because the value of gravity at any particular position depends on the distribution of mass throughout the Earth (and also slightly on the Earth's rotation), measurements of the gravity field can yield information on the density of underlying rock. Thus, gravity meters are used for geologic studies and for oil and mineral exploration. Local gravity also depends on the shape of the Earth; the observation of gravity over time, then, provides a measure of deformations in the Earth that can be caused by a wide variety of phenomena, including tides, tectonic activity, and volcanism.

Nowhere on the surface of the Earth does the value of gravity differ from the nominal value of 980 Gal by more than about 0.5%. (The SI unit for gravity is the meter/second^2; the more commonly used unit is the Gal, defined as 1 cm/s^2, or the milliGal, which equals 0.001 Gal.) Values of gravity predicted with a latitude-and-height-dependent Earth model usually agree with observed values to within about 30 mGal. The gravitational acceleration produced by the mass of a 1-m-thick (3-ft) sheet of water (having infinite lateral extent) is 0.043 mGal.

A number of different instruments are available. One of two methods is used in all gravity meters. The first, employed by absolute gravity meters, is the direct determination of the acceleration of a test mass falling inside a vacuum chamber by using optical interferometry. The second is the observation of variations in the position of a mass supported by a mechanical or magnetic spring. This method, applied in relative gravity meters and shipboard gravity meters, is usually used in conjunction with an additional applied force (nulling force) that maintains the mass at a null position. The small nulling force is a relative measure of gravity.   [M.A.Z.]

Graywacke   A well-indurated dark gray sandstone that is characterized by abundant dark-colored detrital rock fragments and more than 15% clay matrix minerals between sand grains. Graywacke sands were deposited chiefly in marine basins near the edge of continental margins where plate subduction was taking place. Subsequent compressional deformation and uplift of rocks in the sedimentary basins results in the occurrence of most graywackes in Alpine-type (compressional) mountain ranges. *See* CLAY MINERALS; CONTINENTAL MARGIN.

Graywackes have a wide range in mineral composition, which reflects the varied source rocks from which the detritus in them was derived. They tend to be quartz-poor (10–50%), to be rich in both feldspar and unstable rock fragments, and to contain several percent of unstable accessory minerals such as micas, pyroxenes, and amphiboles. Feldspathic graywackes (those in which feldspar exceeds rock fragments) are derived chiefly from plutonic cores of denuded island arcs. Lithic graywackes (those in which rock fragments exceed feldspar) are derived either from volcanic island arcs or from sedimentary rocks in adjacent basins that were deformed and uplifted. Volcanic rock fragments characterize the former type of lithic graywackes, whereas sandstone, shale, and their weakly metamorphosed equivalents characterize the latter type.

Most graywackes were deposited in submarine fans and adjacent basin-plain environments by turbidity currents. They commonly display graded bedding, Bouma sequences, and current-formed and biogenic sole marks. The term Bouma sequence refers to five divisions of a single, ideal turbidity current deposit. Graywackes are interbedded with shale beds that were deposited by dilute turbidity currents and other marine processes. Thicknesses of several miles of interbedded turbidite graywacke and shale accumulated in many basins. Burial and subsequent compressional deformation of these sequences resulted in the generation of clay matrix, loss of porosity, and strong induration. The gray color of the sandstone is derived from rock fragments and organic-stained clay minerals. *See* ARKOSE; SANDSTONE; SEDIMENTARY ROCKS; SHALE; TURBIDITE; TURBIDITY CURRENT.   [E.F.McB.]

Great circle, terrestrial   A circle or near-circle representing a trace on the Earth's surface of a plane that passes through the center of the Earth and divides it into equal halves (see illustration). The Equator is a great circle, the trace of the plane that bisects and is perpendicular to the Earth's axis. Planes through the Poles cut the Earth along meridians. All meridians are great circles; actually, they are not quite circular because of the slightly flattened Earth. The equatorial diameter is 1.0034 times the size of the polar diameter. All parallels other than the Equator are called small circles, being smaller than a great circle.

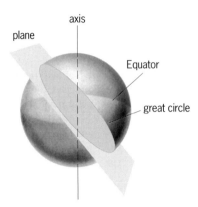

Diagram of a great circle described by a plane through the center of the Earth.

Two common methods can be used to calculate the distance of a great circle arc. One method uses trigonometric functions: $\cos D = \sin a \sin b + \cos a \cos b \cos c$. Here D is the arc distance between points A and B in degrees, a is the latitude of A, b is the latitude of B, and c is the difference in longitude between A and B. After D is calculated, it can be converted to a linear distance measure by multiplying D by the length of one degree of the Equator, which is 111.32 km or 69.17 mi.

The other method uses the azimuthal equidistant projection. Unlike the gnomonic projection, the azimuthal equidistant projection can be centered at any point on the Earth's surface and can show the entire sphere. More importantly, a straight line from the center of the projection to any other point is a great circle route and the distances are at a comparable (consistent) scale between the two points. The azimuthal equidistant projection is therefore useful in showing any movement directed toward or away from a center, such as seismic waves, radio transmissions, missiles, and aircraft flights.   [K.t.C.]

Greenhouse effect   The ability of a planetary atmosphere to inhibit heat loss from the planet's surface, thereby enhancing the surface warming that is produced by the absorption of solar radiation. For the greenhouse effect to work efficiently, the planet's atmosphere must be relatively transparent to sunlight at visible wavelengths so that significant amounts of solar radiation can penetrate to the ground. Also, the atmosphere must be opaque at thermal wavelengths to prevent thermal radiation emitted by the ground from escaping directly to space. The principle is similar to a thermal blanket, which also limits heat loss by conduction and convection. In recent decades the term has also become associated with the issues of global warming and climate change induced by human activity. *See* ATMOSPHERE.

Basic understanding of the greenhouse effect dates back to the 1820s, when the French mathematician and physicist Joseph Fourier performed experiments on atmospheric heat flow and pondered the question of how the Earth stays warm enough for plant and animal life to thrive; and to the 1860s, when the Irish physicist John Tyndall demonstrated by means of quantitative spectroscopy that common atmospheric trace gases, such as water vapor, ozone, and carbon dioxide, are strong absorbers and emitters of thermal radiant energy but are transparent to visible sunlight. It was clear to Tyndall that water vapor was the strongest absorber of thermal radiation and, therefore, the most influential atmospheric gas controlling the Earth's surface temperature. The principal components of air, nitrogen and oxygen, were found to be radiatively inactive, serving instead as the atmospheric framework where water vapor and carbon dioxide can exert their influence.

The impact of water vapor behavior was noted by the American geologist Thomas Chamberlin who, in 1905, described the greenhouse contribution by water vapor as a positive feedback mechanism. Surface heating due to another agent, such as carbon dioxide or solar radiation, raises the surface temperature and evaporates more water vapor which, in turn, produces additional heating and further evaporation. When the heat source is taken away, excess water vapor precipitates from the atmosphere, reducing its contribution to the greenhouse effect to produce further cooling. This feedback interaction converges and, in the process, achieves a significantly larger temperature change than would be the case if the amount of atmospheric water vapor had remained constant. The net result is that carbon dioxide becomes the controlling factor of long-term change in the terrestrial greenhouse effect, but the resulting change in temperature is magnified by the positive feedback action of water vapor.

Besides water vapor, many other feedback mechanisms operate in the Earth's climate system and impact the sensitivity of the climate response to an applied radiative forcing. Determining the relative strengths of feedback interactions between clouds, aerosols, snow, ice, and vegetation, including the effects of energy exchange between the atmosphere and ocean, is an actively pursued research topic in current climate modeling. *See* CLIMATE MODIFICATION.   [A.A.L.]

Greenockite   A mineral having composition CdS (cadmium sulfide). Greenockite usually occurs as earthy coatings with resinous luster and yellow-to-orange color. There is good prismatic cleavage; the hardness is 3 (Mohs scale) and specific gravity is 4.9. Greenockite and wurtzite, ZnS, are isostructural, and a complete solid-solution series exists between the two minerals. Although greenockite is the most common cadmium mineral, no deposits of it are sufficiently large to warrant mining it solely as a source of cadmium.   [C.S.Hu.]

Greisen   A type of hydrothermal wall-rock alteration and a class of tin-tungsten deposits (greisen deposits). Hydrothermal wall-rock alteration is the process whereby rocks on the margins of hydrothermal flow channels are changed from an original assemblage of minerals to a different one. This change occurs because of heat and mass exchange between water and rock.

Granitic rocks altered to greisen are known as apogranites. They are composed mainly of quartz, topaz (fluor-aluminosilicate), and muscovite (white mica), accompanied by accessory minerals such as tourmaline and fluorite. Abundant veins of quartz-topaz are characteristic of intensely greisenized zones. Skarn and limestone on the margins of apogranites may also be altered to greisen (aposkarn greisen and apocarbonate greisen, respectively) with abundant fluorite. Apogranite greisen commonly is accompanied by other types of hydrothermal wall-rock alteration, including early feldspathic and late sericitic and lesser argillic.

Tin-tungsten-(beryllium-molybdenum) deposits in peraluminous granites commonly are accompanied by greisen. Ore minerals may include cassiterite, wolframite, scheelite, molybdenite, bismuth, and bismuthinite, accompanied in some deposits by pyrrhotite and sphalerite, in addition to chalcopyrite and other sulfides. *See* GRANITE.

Tin greisens represent the dominant world source of lode tin, with examples in Southeast Asia (Malaya, Indonesia, Burma, Thailand); southeast China; Tasmania, Australia; Zinnwald and Altenberg (Erzegebirge), Germany; and Cornwall-Devon, southwest England. Greisenized skarn and apocarbonate greisen, also mostly tin deposits (but including beryllium and tungsten), are found in western Tasmania, Australia; Seward Peninsula, Alaska; and Yunnan (tungsten), China. Many of the tungsten deposits of southeast China, the richest tungsten province in the world, occur in greisenized granite. *See* METASOMATISM; ORE AND MINERAL DEPOSITS; PNEUMATOLYSIS .   [M.T.E.]

Ground-water hydrology   The occurrence, circulation, distribution, and properties of any liquid water residing beneath the surface of the earth. Generally ground water is that fraction of precipitation which infiltrates the land surface and subsequently moves, in response to various hydrodynamic forces, to reappear once again as seeps or in a more obvious fashion as springs. Most of ground-water discharge is not evident because it occurs through the bottoms of surface water bodies. *See* SPRING (HYDROLOGY).

Ground water can be found, at least in theory, in any geological horizon containing interconnected pore space. Thus a ground-water reservoir (an analogy to an oil reservoir) can be a classical porous medium, such as sand or sandstone; a fractured, relatively impermeable rock, such as granite; or a cavernous geologic horizon, such as certain limestone beds. Ground-water reservoirs which readily yield water to wells are known as aquifers; in contrast, aquitards are formations which do not normally provide adequate water supplies, and aquicludes are considered, for all practical purposes, to be impermeable. These terms are, of course, subjective descriptions; the flow of water which constitutes an economically viable supply depends upon the intended use and the availability of alternative sources. *See* AQUIFER.

To effectively utilize ground water as a natural resource, it is necessary to be able to forecast the impact of exploitation on water availability. When ground water is used for water supply, a concern is the potential energy in the aquifer as reflected in the water levels in the producing well or neighboring wells. When a ground-water reservoir which does not readily transmit water is tapped, the energy loss associated with flow to the well can be such that the well must be drilled to prohibitively great depths to provide adequate supplies. On the other hand, in a formation able to transmit fluid easily, water levels may drop because the reservoir is being depleted of water. This is generally encountered in reservoirs of limited areal extent or those in which natural infiltration has been reduced either naturally or through human activities.

Problems involving ground-water quantity were once the primary concern of hydrologists; interest is now focused on ground-water quality. Ground-water contamination is a serious problem, particularly in the highly urbanized areas of the United States. *See* HYDROLOGY .   [G.F.P.]

Gulf of California   A young, elongate ocean basin on the west coast of Mexico. It is flanked on the west by the narrow mountainous peninsula and continental shelf of Baja California, while the eastern margin has a wide continental shelf and coastal plain. The floor of the gulf consists of a series of basins 3300–12,000 ft (1000–3600 m) deep, whereas the northern gulf is dominated by a broad shelf which is the result of deltaic deposition from the Colorado River. The structural depression of the gulf continues northward into the Imperial Valley of California, which is cut off from the ocean by the delta of the Colorado River. *See* CONTINENTAL MARGIN.

Most of the gulf lies within an arid climate, with 4–6 in. (10–15 cm) of annual rainfall over Baja California and ranging on the eastern side from 4 in. (10 cm) in the north to about 34 in. (85 cm) in the southeast. No year-round streams enter the gulf on the west; a series of intermediate-size rivers flow in on the east side; and the major source of fresh-water sediment came from the Colorado River at the north prior to damming it upstream in the United States.

Water circulation is driven by seasonal wind patterns. Surface water is blown into the gulf in the summer by the southwesterly wind regime. In the winter, surface water is driven out of the gulf by the northwesterly wind regime, and upwelling occurs along the eastern margin, resulting in high organic productivity. Bottom sediments of the gulf range from deltaic sediments of the Colorado River at the north and coalesced deltas of the intermediate-size rivers on the east. A strong oxygen minimum occurs between 990 and 3000 ft (300 and 900 m) water depth, where seasonal influx of terrigenous sediments and blooms of diatoms due to upwelling produce varved sediments consisting of alternating diatom-rich and clay-rich layers. Rates of sediment accumulation are high, and total sediment fill beneath the Colorado River delta at the north may attain thicknesses of greater than 6 mi (10 km), even though the structural depression and the underlying crust are geologically young. *See* DELTA; MARINE SEDIMENTS; OCEAN CIRCULATION; UPWELLING; VARVE.   [J.R.C.]

Gulf of Mexico   A subtropical semienclosed sea bordering the western North Atlantic Ocean. It connects to the Caribbean Sea on the south through the Yucatan Channel and with the Atlantic on the east through the Straits of Florida. To the north, it is bounded by North America, to the west and south by Mexico and Central America, and on the east and southeast by Florida and Cuba respectively. *See* INTRA-AMERICAS SEA.

The continental shelves surrounding the gulf are very broad along the eastern (Florida), northern (Texas, Louisiana, Mississippi, Alabama), and southern (Campeche) area, averaging 125–186 mi (200–300 km) wide. The continental shelves along the western and southwestern (Mexico) and southeastern (Cuba) boundaries of the gulf are narrow, often being less than 12 mi (20 km) wide. Between the continental shelves and the Sigsbee Abyssal Plain are three steep continental slopes: the Florida Escarpment off west Florida, the Campeche Escarpment off Yucatan, and the Sigsbee Escarpment south of Texas and Louisiana. Two major submarine canyons crease the gulf's shelf areas: the De Soto Canyon near the Florida-Alabama border, and the Campeche Canyon west of the Yucatan Peninsula. *See* CONTINENTAL MARGIN; ESCARPMENT; GLACIAL EPOCH; HOLOCENE; MARINE GEOLOGY.

Compared with the North American rivers, the Mexican rivers are short, but they still provide approximately 20% of the fresh-water input to the gulf because of extensive orographic rainfall from the trade winds that dominate the southern flank of the basin. Meteorologically, the Gulf of Mexico is a transition zone between the tropical wind system (easterlies) and the westerly frontal-passage-dominated weather (in winter particularly) to the north, punctuated with intense tropical storms in summer/autumn called the West Indian Hurricane. Much of the atmospheric moisture supplied to the North American heartland during spring and summer has its origin over the gulf, and thus it is a vital element in the so-called North American Monsoon. *See* HURRICANE; MONSOON METEOROLOGY; STORM SURGE; TROPICAL METEOROLOGY.

The Gulf Stream System dominates the oceanic circulation in the Gulf of Mexico. The Yucatan Current, flowing northward into the eastern Gulf of Mexico, is the first recognizable western boundary current in the Gulf Stream System. North of the Yucatan Peninsula, the flow penetrates into the eastern gulf (where it is called the Gulf Loop Current) at varying distances with a distinctive chronology, loops around clockwise, and finally exits through the Straits of Florida, where it is called the Florida Current. This intense current reaches to more than 3300 ft (1000 m) depth, and transports 1.1×10^9 ft^3/s (3×10^7 m^3/s) of water, an amount 1800 times that of the Mississippi River. *See* GULF STREAM; MEDITERRANEAN SEA; OCEAN CIRCULATION.

Surrounding the Gulf of Mexico are many population centers that exploit the numerous estuaries, lagoons, and oil and gas fields. Coral reefs off Yucatan, Cuba, and Florida provide important fishery and recreational activities. There are extensive wetlands along most coastal boundaries with ecological connections to many seagrass beds nearshore and coastal mangrove forests of Mexico, Cuba, and Florida. This biogeographic confluence creates one of the most productive marine areas on Earth, providing the food web for commercially important species such as lobster, demersal (bottom-dwelling) fish, and shrimp; this same ecology supports large populations of sea turtles and marine mammals. The coastal and nearshore waters also support large phytoplankton populations. The juxtaposition of these enormous marine resources and human activities has led to a distinctive anthropogenic impact on the health of the marine ecosystem. *See* BIOGEOGRAPHY; ESTUARINE OCEANOGRAPHY; FOOD WEB; MARINE ECOLOGY; REEF; WETLANDS. [G.A.Ma.]

Gulf Stream A great ocean current transporting about 70,000,000 tons (63,000,000 metric tons) of water per second (1000 times the discharge of the Mississippi River) northward from the latitude of Florida to the Grand Banks off Newfoundland. The Gulf Stream is thought of as a portion of a great horizontal circulation in the ocean, where each particle of water executes a closed circuit, sometimes moving slowly in midocean regions and other times rapidly in strong currents like the Gulf Stream. Thus the beginning and end of the Stream have arbitrary geographical limits. *See* ATLANTIC OCEAN.

The Gulf Stream is a narrow (62 mi or 100 km) and swift (up to 5 knots or 250 cm/s) eastward-flowing current jet which is embedded in a weaker and broader mean westward flow and which is surrounded by intense eddies. As it leaves the coast at Cape Hatteras, the Stream meanders from side to side like a river.

The near-surface Gulf Stream transports warm water from southern latitudes eastward to the Grand Banks, where the flow becomes broader and weaker, separating into several branches and eddies. About half the near-surface flow continues eastward across the Mid-Atlantic Ridge, and half recirculates southwestward, with part of the recirculation consisting of a countercurrent located south of the Stream.

The Gulf Stream is predominantly driven by the large-scale wind pattern, the westerlies in the north and the trades in the south. The winds exert a torque on the ocean that, due to the shape and rotation of the Earth, causes a large western-intensified gyre. Cold, deep water is formed in northern seas and flows southward as a western boundary current; warm water flows northward and replaces it. *See* OCEAN CIRCULATION. [P.R.]

Gypsum The most common sulfate mineral, characterized by the chemical formula $CaSO_4 \cdot 2H_2O$; it shows little variation from this composition.

Gypsum is one of the several evaporite minerals. This mineral group includes chlorides, carbonates, borates, nitrates, and sulfates. These minerals precipitate in seas, lakes, caves, and salt flats due to concentration of ions by evaporation. When heated or subjected to solutions with very large salinities, gypsum converts to bassanite ($CaSO_4 \cdot H_2O$) or anhydrite ($CaSO_4$). Under equilibrium conditions, this conversion to anhydrite is direct. The conversion occurs above 108°F (42°C) in pure water. The presence of halite (NaCl) or other sulfates in the solution lowers this temperature, although metastable gypsum exists at higher temperatures. *See* ANHYDRITE; HALITE.

Crystals of gypsum are commonly tabular, diamond-shaped, or lenticular; swallow-tailed twins are also common. The mineral is monoclinic with symmetry $2/m$. The common colors displayed are white, gray, brown, yellow, and clear. Cleavage surfaces show a pearly to vitreous luster. Gypsum is the index mineral chosen for hardness 2 on Mohs scale with a specific gravity of 2.32. In addition to free crystals, the common forms of gypsum are satin spar (fibrous), alabaster (finely crystalline), and selenite (massive crystalline).

Gypsum is used for a variety of purposes, but chiefly in the manufacture of plaster of paris, in the production of wallboard, in agriculture to loosen clay-rich soils, and in the manufacture of fertilizer. Plaster of paris is made by heating gypsum to 392°F (200°C) in air. A hemihydrate is formed as part of the water of crystallization is driven off. Later, when water is added, rehydration occurs. The interlocking, finely crystalline texture that results forms a uniform hardened mass. The slightly increased volume of the set plaster serves to fill the mold into which it has been poured.

Gypsum deposits are mined throughout the world, with the United States being a world leader in gypsum production. The majority of United States gypsum is mined in Michigan, Iowa, Texas, California, and Oklahoma. Canada is the world's second largest producer. Most Canadian production is in the province of Nova Scotia. Among the other leading producers are France, Japan, Iran, Russia, Italy, Spain, and the United Kingdom.   [M.L.H.; C.Sc.]

H

Hadean The eon of geological time extends for several hundred million years from the end of the accretion of the Earth to the formation of the oldest recognized rocks. According to current models, the inner planets formed by the accretion of planetesimals in an environment where gas and volatiles had been swept away by early intense solar activity. The accretion of the Earth appears to have been completed between 50 and 100 million years (m.y.) after the beginning of the solar system (T_0) as recorded in the oldest refractory inclusions in the Allende Meteorite, whose age of 4566 ± 2 m.y., ascertained by lead isotope dating, is taken as T_0. Core formation on the Earth appears to have been coeval with accretion and so preceded the Hadean. Any primitive atmosphere was removed by early collisional events, and the present atmosphere has arisen by a combination of degassing and additions from comets. *See* LEAD ISOTOPES (GEOCHEMISTRY).

The Acasta Gneiss in the Northwest Territories of Canada, dated at 3960 m.y., is often regarded as the oldest rock. However, that date refers to relict zircon crystals in the rock rather than the age of formation of the rock itself. The oldest definitely dated rocks are at Isua, Greenland, with an age of 3650–3700 m.y. Thus the Hadean Eon begins around 4500–4450 m.y. ago and extends to between 3900 and 3650 m.y. ago depending on the age assigned to the oldest rock.

Conditions on the Hadean Earth bore little resemblance to more recent times. A picture dimly appears of a hot young Earth with a thick basaltic crust, covered by an ocean. Dry land was rare. Plate tectonics had not yet begun. A few remnant zircon crystals indicate the formation of an occasional felsic rock, produced by remelting of the basalt. Sporadic disruption of the surface was caused by the collisions of basin-forming impactors that probably culminated in a spike or cataclysm around 3850–4000 m.y. ago. Such events must have frustrated the origin and development of life, which emerged in post-Hadean time.   [S.R.T.]

Hail Precipitation composed of chunks or lumps of ice formed in strong updrafts in cumulonimbus clouds. Individual lumps are called hailstones. Most hailstones are spherical or oblong, some are conical, and some are bumpy and irregular. Diameters range from 0.2 to 6 in. (5 to 150 mm) or more. That is, the largest stones are grapefruit or softball size, and the smallest are pea size.

Very often hailstones are observed to be made of alternating rings of clear and white ice (see illustration). These rings indicate the growth processes of the hail. The milky or white portion of the growth occurs when small cloud droplets are collected by the hailstone and freeze almost instantaneously, trapping bubbles of air between the droplets and creating a milky appearance. The clear portion is formed when many droplets are collected so rapidly that a film of water spreads over the stone and freezes gradually, giving time for any trapped air bubbles to escape from the liquid.

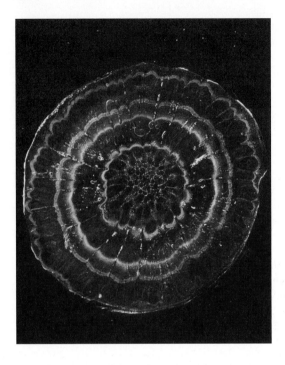

Cross section of a large hailstone showing the structure of alternating rings of clear and white ice. (*Alberta Research Council, Edmonton*)

The most favorable conditions for hail formation occur in the mountainous, high plains regions of the world. Hailstorms normally have relatively high, cool cloud bases and very strong updrafts within the clouds to carry the hailstones into the cooler regions of the cloud, where maximum growth occurs. Both small ice particles and supercooled liquid water (liquid water at temperatures below 32°F or 0°C) are needed for the ice particles to grow into hailstones. *See* CLOUD PHYSICS; PRECIPITATION (METEOROLOGY).

[H.D.O.]

Halite   One of the group of minerals referred to as evaporites, halite is commonly known as salt. Halite is one of many substances that are essential for human life. Evaporite minerals form when ions are concentrated to their saturation point by the progressive evaporation of seawater or saline lake water. Halite precipitates after calcium sulfate, but before the highly soluble salts of potassium and magnesium. *See* HALOGEN MINERALS; SALINE EVAPORITES.

Halite (chemical formula NaCl) is composed of sodium cations and chlorine anions in equal proportion. It is the most common chloride mineral in natural sequences which proceed beyond the precipitation of sulfates. Even in sequences which contain a high percentage of potassium and magnesium salts, halite is often the most common chloride present.

Crystals of halite are generally cubic or hopper-shaped (skeletal). Although the mineral is colorless generally, impurities can color it gray, red, orange, or brown. Blue or violet halite results from exposure to radioactivity, which produces dislocations and defects in the crystal structure. Halite is characterized by a hardness of 2.5 on Mohs scale and a specific gravity of 2.16.

The deformation of bedded halite deposits is of importance to the petroleum industry. Salt rises, in part as a result of density contrasts, to form domelike structures.

Hydrocarbons (oil and gas) are commonly associated with salt domes. Exploration for these structures by geophysical techniques often results in major discoveries by the petroleum industry. *See* GEOPHYSICAL EXPLORATION; PETROLEUM GEOLOGY; SALT DOME.

[M.L.H.; B.C.Sc.]

Halloysite   A clay mineral similar in structure to kaolinite, having a 1:1 structure in which a silica tetrahedral sheet is joined to an alumina octahedral sheet. Unlike kaolinite, however, the structure is disordered in both the a and the b axis directions in successive layers, and it frequently contains water between the layers. *See* KAOLINITE.

Two principal modifications exist: a less hydrous form with a composition and structure near to that of kaolinite, $Al_2Si_2O_5(OH)_4$; and a hydrous form with the composition $Al_2Si_2O_5(OH)_4 \cdot 2H_2O$. The less hydrous form has a c-dimension of about 0.72 nanometer, whereas the hydrous form has a c-dimension of about 1.01 nm, the difference between them being roughly the thickness of a single sheet of water molecules. The hydrated form converts spontaneously and irreversibly into the less hydrous form when dried.

Electron microscopy reveals that the morphology of halloysite is usually tubular. Because the 1:1 layers in halloysite generally are separated from each other by water, halloysite has a larger cation exchange capacity, surface area, and catalytic activity than does kaolinite.

Halloysite is formed in nature from the weathering of feldspar under intense leaching conditions, and may also form in low-temperature hydrothermal systems. It has not been synthesized in the laboratory beyond doubt, although products resembling halloysite have been obtained by the artificial weathering of feldspar, and by the intercalation of kaolinite. Halloysite may precede kaolinite as a weathering product, and the transformation of halloysite into kaolinite may explain why halloysite is not common in sediments. *See* FELDSPAR; WEATHERING PROCESSES.

Halloysite is used as a catalyst and in the manufacture of ceramic products. *See* CLAY MINERALS.   [D.Eb.]

Halo   Either of two large circles of light surrounding the Sun or Moon that result from the refraction of sunlight by small, hexagonal ice crystals falling slowly through the air. Light passing through the side faces of a hexagonal prism is refracted by an amount that depends on the orientation of the crystal; but a collection of many crystals refracts light passing through two side faces by an average angle of about $22°$. If such crystals tumble randomly as they fall, they will produce the $22°$ halo, a circle around the Sun with an angular radius of $22°$. Rays that pass through a side face and an end face of the prism similarly produce the larger and fainter $46°$ halo. The halos sometimes have a red inner edge and otherwise appear nearly white.

Many similar effects result from rays passing through ice crystals that assume special orientations as they fall, and from rays undergoing combination of reflection and refraction in an ice crystal. Usually, all of these effects are referred to as halo effects. *See* METEOROLOGICAL OPTICS; SUN DOG.   [R.Gr.]

Halogen minerals   Naturally occurring compounds containing a halogen as the sole or principal anionic constituent. There are over 70 such minerals, but only a few are common and can be grouped according to the following methods of formation.

1. Saline deposition by evaporation of seawater or salt lakes. Halite (rock salt), NaCl, is the most important of this type. Of the other minerals associated with halite, sylvite, KCl, and carnallite, $KMgCl_3 \cdot 6H_2O$, are the most important.

2. Hydrothermal deposition. Fluorite, CaF_2, is the chief representative of this type. Cryolite, Na_3AlF_6, may be of primary deposition or may result from the action of fluorine-bearing solutions on preexisting silicates. *See* CRYOLITE; FLUORITE.

3. Secondary alteration. Chlorides, iodides, or bromides of silver, copper, lead, or mercury may form as surface alterations of ore bodies carrying these metals. The most common are cerargyrite, $AgCl$, and atacamite, $CU_2(OH)_3Cl$. *See* CERARGYRITE.

4. Deposition by sublimation. Halides formed as sublimation products about volcanic fumaroles include sal ammoniac, NH_4Cl; malysite, $FeCl_3$; and cotunnite, $PbCl_2$. At Mount Vesuvius, Italy, is the most noted occurrence of such minerals.

5. Meteorites. Lawrencite, $FeCl_2$, has been found in iron meteorites.   [C.S.H.]

Hatchettite   Mountain tallow, a yellow-white to yellow-green hydrocarbon occurring in Belgian coal seams; it is also called hatchettine. Hatchettite is translucent but darkens on exposure to air. It is soft, has no odor, is greasy to the touch, and consists of 85.5% carbon and 14.5% hydrogen. Its index of refraction is 1.47–1.50; it melts at 46–47°C (115–117°F), is sparingly soluble in alcohol or ether, decomposes in concentrated sulfuric acid, and has a specific gravity of 0.89–0.98.   [I.A.B.]

Hausmannite   A mineral with composition $Mn^{2+}Mn_2^{3+}O_4$. Hausmannite is most frequently massive-granular and possesses one perfect basal cleavage. The color is black, and streak dark brown. Hardness is 5.5 on Mohs scale, and the specific gravity is 4.81. It is an occasional ore of manganese, and it most frequently occurs in metamorphosed sedimentary manganese ore deposits, such as some small deposits in central Sweden and in the Central Provinces, India.   [P.B.M.]

Heat balance, terrestrial atmospheric   The balance of various types of energy in the atmosphere and at the Earth's surface. At the top of the atmosphere, the incoming solar radiation that is absorbed by the Earth-atmosphere system is approximately balanced by the terrestrial radiation emitted from this system over long periods of time. The flux of solar energy (energy per time) across a surface of unit area normal to the solar beam at the mean distance between the Sun and the Earth is referred to as the solar constant. Based on recent satellite measurements, a value of 1365 watts per square meter (W/m^2) for the solar constant has been suggested. Because the area of the spherical Earth is four times that of its cross section facing the parallel solar beam, the top of the Earth's atmosphere receives an average of about 341 W/m^2. Based on measurements from satellite radiation budget experiments, about 30% of this is reflected back to space, and is referred to as the global albedo. The reflecting power of the Earth-atmosphere system includes the scattering of molecules, aerosols, and various types of clouds, as well as reflection by different surfaces. Thus, only about 70% of the incoming solar flux, that is, about 239 W/m^2, is absorbed within the Earth-atmosphere system. For this system to be in thermodynamic equilibrium or balance, it must emit the same amount of thermal infrared radiation.

For the presentation of internal heat balance components, the effective solar constant of 341 W/m^2 may be arbitrarily represented by 100 units (see illustration). Of these units, roughly 26 are absorbed within the atmosphere, including 22 by clear column and 4 by clouds. A total of 30 units are reflected back to space, including about 7 from clear column, 17 from cloudy atmospheres, and 6 directly from the Earth's surface. The remaining 44 units are absorbed by the surface. The Earth-atmosphere system emits terrestrial radiation according to its temperature and composition distributions. The upward flux from the warmer surface accounts for about 115 units. The colder troposphere emits both upward and downward fluxes, with about 70 and 100 units

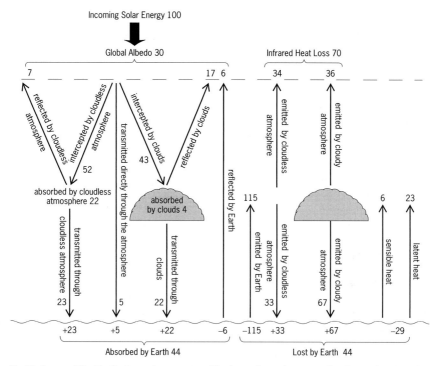

Heat balance of the Earth-atmosphere system. The incoming solar energy is taken to be 100 units. On a climatic scale, the incoming solar energy at the top of the atmosphere is approximately balanced by the reflected solar energy and thermal infrared heat loss. At the surface, the heat balance involves sensible and latent heat components, in addition to net radiative energy.

at the top and surface, respectively. The clear and cloudy portions are 34 and 36 at the top and 33 and 67 at the surface, respectively. The net upward flux at the surface, representing the difference between the flux emitted by the surface and the downward flux from the atmosphere reaching the surface, is about 15 units. *See* TERRESTRIAL RADIATION.

As a result of thermal emission, the atmosphere loses 55 units. With absorption of the incoming solar flux contributing only 26 units, the net radiative loss from the atmosphere amounts to about 29 units. This deficit is balanced by convective fluxes of sensible and latent heat associated with temperature gradient and evaporation. Based on statistical analyses, the average annual ratio of sensible to latent heat loss at the surface has a global value of about 0.27. It follows that the latent and sensible heat components are about 23 and 6 units, respectively, in order to produce an overall heat balance at the surface (see illustration). The atmosphere experiences a net radiative cooling that must be balanced by the latent heat of condensation released in precipitation processes and by the convection and conduction of sensible heat from the underlying surface.

[K.N.L.]

Heavy minerals Minerals with a density greater than 2.9 g/cm^3. The term is most commonly used to denote high-density components of siliciclastic sediments. Most heavy mineral studies are undertaken to determine sediment provenance, because heavy mineral suites provide important information on the mineralogical composition of source areas. Since heavy minerals rarely constitute more than 1% of sandstones, their study normally requires them to be concentrated. This is achieved by disaggregation of the sandstone, followed by mineral separation using dense liquids such as bromoform, tetrabromoethane, or the more recently developed nontoxic polytungstate liquids. See PROVENANCE (GEOLOGY); SANDSTONE.

Geographic and stratigraphic variations in heavy mineral suites within a sedimentary basin can be used to infer differences in sediment provenance. Such differences result either from the interplay between a number of sediment transport systems draining different source regions, or from erosional unroofing within a single source area. Heavy mineral data therefore play an important role in the understanding of depositional history and paleogeography. In some cases, sophisticated mathematical and statistical treatment of heavy mineral data may be required to elucidate the interplay between multiple sediment transport systems. See DEPOSITIONAL SYSTEMS AND ENVIRONMENTS; PALEOGEOGRAPHY; SEDIMENTOLOGY; STRATIGRAPHY.

Heavy minerals have important economic applications. Their use in paleogeographic reconstructions, especially in elucidating sediment transport pathways, is of particular value in hydrocarbon exploration, and their use in correlation has important applications in hydrocarbon reservoir evaluation and production. Recent advances have made it possible to utilize the technique on a real-time basis at the well site, where it is used to help steer high-angle wells within the most productive reservoir horizons. Heavy minerals may become concentrated naturally by hydrodynamic sorting, usually in shallow marine or fluvial depositional settings. Naturally occurring concentrates of economically valuable minerals are known as placers, and such deposits have considerable commercial significance. Cassiterite, gold, diamonds, chromite, monazite, and rutile are among the minerals that are widely exploited from placer deposits. See DATING METHODS; ZIRCON.   [A.Mor.]

Hematite   The most important ore of iron, with composition α-Fe_2O_3. The crystals are thick tabular, usually flattened parallel to the base, and are frequently platy in habit. Hematite usually occurs as rouge-red earthy masses of finely divided particles. It is the major red-coloring agent in rocks and is a common interstitial cement in sediments. When mixed with quartzite or finely divided quartz, the mixture is called jasper, jaspilite, or taconite. Botryoidal masses are called "kidney ore," and splinters of these masses are "pencil ore."

The color is steel gray, blood red in thin fragments, and streak and powder are rouge red; hardness is 6 on Mohs scale and specific gravity is 5.25. The mineral is only weakly magnetic.

Hematite is the most widespread iron mineral. The most important ores are in low-to-medium-grade metamorphic rocks of sedimentary origin. Enormous beds occur in the Great Lakes region of the United States. Hematite also occurs in contact metamorphic and metasomatic deposits, often derived from the oxidation of magnetite and frequently associated with limestones.

Nearly every country in the world mines some hematite ore; the most important occurrences outside of the United States include India, Cuba, China, Chile, north African nations, and Russia. See REDBEDS.   [P.B.M.]

Hemimorphite   A mineral sorosilicate with the composition $Zn_4Si_2O_7(OH)_2 \cdot H_2O$; an ore of zinc. Crystals are usually colorless and the aggregates white, but in some cases there are faint shades of green, yellow, and blue. The mineral has a vitreous luster, a hardness of $4\frac{1}{2}$ to 5 on Mohs scale, and a specific gravity of 3.45.

Hemimorphite has a wide distribution and has been mined in Belgium, Germany, Romania, England, Algeria, and Mexico. In the United States it is found at Sterling Hill, New Jersey; Friedensville, Pennsylvania; and Elkhorn Mountains, Montana. *See* SILICATE MINERALS.   [C.S.Hu.]

Heulandite   A mineral belonging to the zeolite family of silicates. It usually occurs in crystals with the prominent side being pinacoid, often having a diamond shape. There is perfect side pinacoid cleavage on which the luster is pearly; elsewhere the luster is vitreous. The hardness is $3^1/_2$ to 4 on Mohs scale; specific gravity is 2.18–2.20. The mineral is usually white or colorless but may be yellow or red. *See* ZEOLITE.

Heulandite is essentially a hydrous calcium aluminum silicate, $Ca(Al_2Si_7O_{18}) \cdot 6H_2O$. Heulandite is a secondary mineral found in cavities in basalts associated with other zeolites and calcite. Notable localities are in the Faeroe Islands, India, Nova Scotia, and West Paterson, New Jersey.   [C.Fr.; C.S.Hu.]

High-pressure mineral synthesis   A laboratory technique for studying the behavior of minerals under high-pressure conditions.

The nature of minerals as they exist at atmospheric pressure represents only a very limited part of their real nature. The range of pressure and temperature prevailing at the surface of the Earth is very limited compared to the ranges that exist in the other planets of the solar system. The bottom of the ocean, which is at the highest pressure that can be observed directly, is only 0.1 GPa (1 kilobar), while the pressure at the center of the Earth is 390 GPa (3900 kilobars). Pressures at the centers of large planets such as Saturn and Jupiter exceed 1000 GPa (10,000 kilobars). Therefore, to study the formation and structure of the Earth and other planets, it is essential to study the behavior of minerals under high pressure. It has become clear through high-pressure experiments that the minerals constituting the Earth's lower mantle (which extends from 650 to 2900 km or 400 to 1800 mi from the surface and occupies more than 50% of the entire volume of the Earth) are mostly so-called silicate perovskites that can never be formed on the surface of the Earth. *See* EARTH INTERIOR.

Pressure is defined as a force per unit area; therefore, in order to apply a high pressure, it is necessary to concentrate a large force in a small area. Because of the limited strength of materials used to produce sample chambers, many different techniques are required, depending on the pressure range (see illustration).

A large number of phase transformations has been found in minerals under high pressure, but most of these structures have already been observed in other minerals existing under atmospheric pressure. For example, rutile-type SiO_2 (stishovite) is formed only above 10 GPa (100 kilobars), but the same structure is obtained at atmospheric pressure when the Si ion is replaced by the larger germanium (Ge) ion. This implies that crystal structure is determined mainly by the ratio of the cation radius to that of the anion.

When the very dense structure is compressed further, the bond length becomes shorter and shorter, and the orbitals of the electrons around the ions begin to overlap. This means that the orbital electrons can move freely in the material, which changes into a so-called metallic state. This metallic transition is believed to occur in all materials when they are subjected to high enough pressure. Even hydrogen, helium, and ice are believed to exist in the metallic state in the interiors of Jupiter and Saturn. In the laboratory, however, this transformation into the metallic state under pressure has been confirmed in only a limited number of materials such as Si, Ge, and gallium arsenide (GaAs).

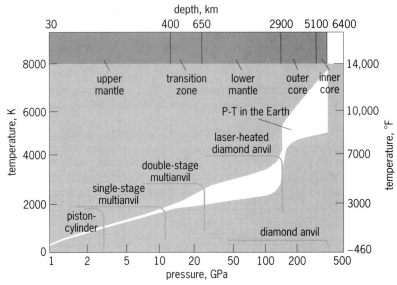

Diagram of pressure (P) and temperature (T) within the Earth, showing capabilities of various types of high-pressure apparatus. 1 GPa = 10 kilobars. 1 km = 0.6 mi.

It has become clear that many of the major phases of silicate transform into the perovskite structure above 25 GPa (250 kilobars). Therefore it is believed that silicate perovskite is the most abundant mineral within the Earth, although it is an exotic mineral on the surface. In order to clarify the nature of the high-pressure minerals believed to be present in the interior of the Earth many studies have been made using various techniques. For this type of study, it is important to obtain a single crystal. The multianvil apparatus has been widely used for such experiments, and various single crystals of high-pressure minerals such as silicate perovskite, spinel, and stishovite have been synthesized.

High-pressure synthesis is a powerful method not only for use in earth and planetary sciences but also for the creation of new materials. Many industrial diamonds are synthesized by using high-pressure techniques, and some new high-pressure materials, such as cubic boron nitride, have found wide application. Pressure is one of the most fundamental parameters that can alter the state of materials, and research in this field is also expected to expand in the future. *See* DIAMOND; SILICATE PHASE EQUILIBRIA.   [T.Ya.]

Hill and mountain terrain   Land surfaces characterized by roughness and strong relief. The distinction between hills and mountains is usually one of relative size or height, but the terms are loosely and inconsistently used.

Uplift of the Earth's crust is necessary to give mountain and hill lands their distinctive elevation and relief, but most of their characteristic features—peaks, ridges, valleys, and so on—have been carved out of the uplifted masses by streams and glaciers. Hill lands, with their lesser relief, indicate only lesser uplift, not a fundamentally different course of development. The features of hill and mountain lands are chiefly valleys and divides produced by sculpturing agents, especially running water and glacier ice. Local peculiarities in the form and pattern of these features reflect the arrangement and character of the rock materials within the upraised crustal mass that is being dissected.

Hill and mountain terrain occupies about 36% of the Earth's land area. The greater portion of that amount is concentrated in the great cordilleran belts that surround the Pacific Ocean, the Indian Ocean, and the Mediterranean Sea. Additional rough terrain, generally low mountains and hills, occurs outside the cordilleran systems in eastern North and South America, northwestern Europe, Africa, and western Australia. Eurasia is the roughest continent, more than half of its total area and most of its eastern portion being hilly or mountainous. Africa and Australia lack true cordilleran belts. The broad-scale pattern of crustal disturbance, and hence of rough lands, is now known to be related to the relative movements of a worldwide system of immense crustal plates. *See* MOUNTAIN; PLATE TECTONICS.   [E.H.Ha.]

Holocene   That portion of geologic time that postdates the latest episode of continental glaciation. The Holocene Epoch is synonymous with the Recent or Postglacial interval of Earth's geologic history and extends from 10,000 years ago to the present day. It was preceded by the Pleistocene Epoch and is part of the Quaternary Period, a time characterized by dramatic climatic oscillations from warm (interglacial) to cold (glacial) conditions that began about 1.6 million years ago. The term Holocene is also applied to the sediments, processes, events, and environments of the epoch.

As the interval of time closest to us, the Holocene Epoch is very convenient to study. Holocene sediments cover virtually every part of the Earth's surface and represent almost every environment of deposition. With the development of ^{14}C dating (a method of age determination based on the measurement of radioactive carbon decay), Holocene sediments are relatively easy to date. From a scientific standpoint, the Holocene Epoch is of great interest because it provides a recent analog for past environments and processes. Its sediments and landforms provide important clues to changes that occurred as a result of the last shift from the glacial to the nonglacial climatic mode. *See* DEPOSITIONAL SYSTEMS AND ENVIRONMENTS; RADIOCARBON DATING.

The Pleistocene/Holocene transition was a time of dramatic environmental change. The huge ice sheets that had developed over the northern and western parts of North America (Laurentide and Cordilleran, respectively) and most of Scandinavia were at their maximum geographic extent about 18,000 ^{14}C years B.P. (before present, where present is defined as the year 1950) and in full retreat by 14,000 ^{14}C years B.P. By 10,000 ^{14}C years B.P., the Laurentide ice sheet had withdrawn from the Great Lakes. The ice sheets survived in the northern latitudes for another 3000 ^{14}C years or so. The progress of deglaciation was complex, because the overall glacial meltback was interrupted by intervals of glacier readvance. It remains unclear whether these readvances were synchronous on a hemispheric or global scale and what role ice sheet/oceanic interactions played in the deglaciation. *See* PLEISTOCENE.

The early phase of the Holocene was geologically the most eventful. The periglacial (near the edge of the ice) landscape was unstable and very dynamic. As the Pleistocene ice sheets melted, enormous volumes of water, stored as glacier ice for many thousands of years, returned to the oceans via meltwater streams or by way of ice streams that flowed directly to the ocean.

As the ice sheets shrank, sea level rose an average of 130 m (426 ft), drowning the continental margins and closing many land bridges, including the land bridge across the Bering Strait between Asia and North America that had enabled humans to migrate to the Americas. In parts of Canada and Scandinavia, temporary marine invasions occurred when the ice melted from low areas where the Earth's crust had been depressed by the weight of the ice sheets.

As the ice sheets waned, the Earth's crust rose, rebounding from the release of the weight of thousands of meters of glacier ice and creating uplifted shoreline features and sediments. Parts of Hudson Bay and Scandinavia were uplifted several hundred meters. Maximum uplift occurred in the early Holocene, but uplift continues even today although at much slower rates.

The middle phase of the Holocene has been called the hypsithermal, a name for the warmest interval of the present interglacial episode. It has also been referred to as the climatic optimum, a term which is more appropriately applied to the peak warmth of the hypsithermal phase. At the climatic optimum, world temperature was probably 2 or 3°C (3.6 or 5.4°F) higher than today. The climate was warm enough to melt much of the sea ice in the Arctic Ocean, as indicated by the occurrence of fossil driftwood (dated at 4000–6000 ^{14}C years B.P.) on uplifted beaches.

After the climatic optimum, the Earth experienced climatic cooling. The shift to a cooler, moister climate began about 5000–4000 ^{14}C years B.P. in the midcontinent. In western North America at about 5000 ^{14}C years B.P., the mountain glaciers began to expand again. This renewed glacial activity is called Neoglaciation. At least three intervals of glacial expansion have occurred in the late Holocene. The glacial advances are cyclic. In the mountains of the western United States, the three advances have been dated at about 5000, 2800, and 300 ^{14}C years B.P. The most impressive of the three glacial intervals is the last, called the Little Ice Age. It is well documented because it occurred in historic time. Between the intervals of glacier expansion were times of climatic warming. One, called the Little Climatic Optimum to differentiate it from the hypsithermal of the middle Holocene, peaked about 1800 ^{14}C years B.P.

During the late Holocene, human populations expanded and human culture developed into the complex agricultural, industrial, and technological society of today. The result is that humans have become significant factors in altering the Earth's surface environment, including, most believe, Holocene climate. *See* GEOLOGIC TIME SCALE; GLACIAL EPOCH; QUATERNARY.   [A.K.H.]

Hornblende   The name that was traditionally assigned to common calcic amphiboles of metamorphic and igneous rocks. However, a nomenclature scheme for amphiboles was introduced in 1997 in which the names now carry strict compositional restrictions. Magnesiohornblende (contains magnesium) and ferrohornblende (contains iron) are monoclinic amphiboles with end-member compositions $Ca_2(Mg_4Al)(Si_7Al)O_{22}(OH)_2$ and $Ca_2(Fe^{2+}_4Al)(Si_7Al)O_{22}(OH)_2$, respectively (Ca = calcium, MG = magnesium, Al = aluminum, Si = silicon, O = oxygen, Fe = iron, OH = hydroxyl). Most natural compositions differ significantly from these ideal end members. Significant deviations from these compositions are denoted by the addition and replacement of prefixes and adjectival modifiers characteristic of the compositions involved. Thus fluorohornblende (contains fluorine, F) has the end-member composition $Ca_2(Mg_4Al)(Si_7Al)-O_{22}F_2$, in which all of the OH in hornblende has been replaced by F. When used to denote an amphibole of known chemical composition, the term hornblende is never used without a prefix or adjectival modifier. The unmodified term hornblende specifically refers to a calcic amphibole identified by physical or optical properties without characterization of the chemical composition.

Hornblende is a common rock-forming mineral in medium- and high-grade metamorphic rocks, particularly those of mafic and ultramafic composition. In mafic rocks, it first appears in the upper part of the low grade by a chemical reaction involving the disappearance of actinolite, a nonaluminous calcic amphibole. This change is extremely noticeable in thin sections, very pale-green actinolite giving way to blue-green hornblende. With prograde metamorphism, the composition of the hornblende gradually changes in a highly complex manner that is a function of temperature, pressure, oxygen fugacity (a measure of the activity of oxygen), and the chemical composition of the rock. This causes a gradual color change from blue-green through various shades of green to olive green and brown. At the middle of the high grade, hornblende becomes unstable and breaks down to form pyroxene (plus other minerals). The prominence of hornblende in medium-grade metabasic rocks has led to these rocks being called amphibolites. *See* PYROXENE.

Hornblende is commonly found as a minor phase in a wide variety of igneous rocks. Magnesium-rich hornblendes do occur as primary phases in basic and ultrabasic rocks, but this is not common. Igneous amphiboles are most abundant in calcic-alkaline diorites, granodiorites, and granites, becoming more iron-rich with increasing acidity of the host rock. This compositional trend is also characterized by a progressive increase in the alkali content of the amphibole, and hornblende grades into hastingsite, riebeckite, and arfvedsonite in granitic rocks. *See* GRANITE; GRANODIORITE; IGNEOUS ROCKS.

Due to its complex structure and chemistry, hornblende contains much information on its formation. Its behavior is understood reasonably well, and hornblende is of considerable use in interpreting the geological history of the rocks in which it occurs. *See* AMPHIBOLITE; METAMORPHISM.   [F.C.Ha.]

Hornfels   A metamorphic rock that has been subjected to heating during contact metamorphism around intrusive igneous rocks. Hornfels is typically fine-grained, although where it is subjected to high temperatures, large crystals called porphyroblasts can form. Mineral grains in hornfels are randomly oriented, with no preferred alignment of crystals to form foliation or cleavage planes. This texture indicates that the hornfels was not subjected to significant stresses during contact metamorphism.

Hornfels generally originates from sediments that undergo mineralogical changes, the nature of which depend on the magnitude of heating. The types of minerals that form are strongly dependent on the bulk composition. Minerals in hornfels formed from metamorphism of limestones, which are rich in calcium oxide (CaO), carbon dioxide (CO_2), and various amounts of magnesium oxide (MgO), iron oxide (FeO), and aluminum oxide (Al_2O_3), include (from high to low temperature) fosterite, diopside, tremolite, talc, and brucite. Other minerals that may be present include wollastonite, vesuvianite, anorthite, and grossular garnet, depending on the bulk composition of the rock. *See* LIMESTONE.

Pelitic sediments are rich in chemical constituents such as silicon dioxide (SiO_2), Al_2O_3, MgO, FeO, potassium oxide (K_2O), and water (H_2O), with relatively minor amounts of CaO, sodium oxide (Na_2O), manganese oxide (MnO), and titanium dioxide (TiO_2). Metamorphism of these sediments to form hornfels results in formation of minerals such as chlorite, muscovite, biotite, andalusite, sillimanite, cordierite, garnet, staurolite, and K-feldspar. At extremely high temperatures ($>800°C$ or $1470°F$) aluminum-rich minerals such as sapphirine, spinel, and corundum form. Deposits of emery, utilized for abrasives, are aluminum-rich hornfels that are products of high-temperature contact metamorphism. Chemical study of emeries indicates a general lack of alkali elements (K, Na, and Ca), which has been used to argue that they form as a result of extraction of a melt phase during high-temperature contact metamorphism. *See* EMERY.

During contact metamorphism, hornfels typically forms in the highest-temperature part of aureoles adjacent to the pluton. Further away from the pluton, metamorphism of sediments results in development of schists and phyllites. Around the pluton, low-grade chlorite-bearing slates are progressively metamorphosed, resulting in the systematic appearance from low to higher temperature of cordierite + biotite + muscovite phyllite to cordierite + K-feldspar + biotite hornfels. In hornfels of a slightly different composition, muscovite is preserved, resulting in a hornfels with the composition andalusite + K-feldspar + cordierite + biotite + muscovite. Adjacent to the contact with the pluton, these muscovite-bearing hornfels undergo partial melting, resulting in the segregation of K-feldspar + plagioclase + quartz from the metamorphosed sediment as a result of partial melting. *See* PLUTON; SLATE.

Metamorphic studies of hornfels provide an important avenue to documenting the temperature and, in particular, the pressure (that is, the depth) during emplacement of the intrusive igneous rock that provides the heat. *See* METAMORPHIC ROCKS; METAMORPHISM; MINERALOGY.   [M.W.N.]

Horst   A segment of the Earth's crust, generally long as compared to its width, that has been upthrown relative to the adjacent rocks (see illustration). Horsts range in size

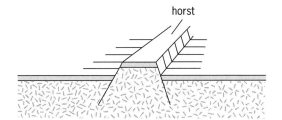

A simple horst with associated faults. (*After A.K. Lobeck, Geomorphology*, McGraw-Hill, *1939*)

from those that have lengths and upward displacement of a few inches to those that are tens of miles long with upward displacements of thousands of feet. The faults bounding a horst on either side commonly have inclinations of 50–70° toward the down-thrown blocks, and the direction of movement on these displacements indicates that they are gravity faults. These relationships suggest that horsts develop in regions where the crust has undergone extension. They may form in the crests of anticlines or domes, or may be related to broad regional warpings. *See* FAULT AND FAULT STRUCTURES.   [P.H.O.]

Hot spots (geology)   The surface manifestations of plumes, that is, columns of hot material, that rise from deep in the Earth's mantle. Hot spots are widely distributed around the Earth. One of their characteristics is an abundance of volcanic activity which persists for long time periods (greater than 1 million years). When the lithosphere (the rigid outer layer of the Earth) moves over a plume, a chain of volcanoes is left behind that progressively increases in age along its length. Hot spots are believed to be fixed with respect to each other and the deep mantle so that the age and orientation of these chains provide information on the absolute motions of the tectonic plates. *See* LITHOSPHERE; PLATE TECTONICS.

The Hawaiian-Emperor seamount chain in the central Pacific Ocean is a good example of a volcanic chain that was generated at a hot spot. The 3400-mi-long (5700-km) chain is made up mainly of tholeiitic lavas and ash tuff and pumice deposits. The lavas may have evolved from an initial submarine shield-building stage, through an explosive stage as they build up to sea level, and finally to a subaerial post-erosional stage. *See* LAVA; SEAMOUNT AND GUYOT.

Not all hot-spot volcanism is expressed in terms of highly lineated, multistage, volcanic chains. Aseismic ridges that extend up to or close to the axes of mid-oceanic ridges are another example of hot-spot volcanism. When a hot spot (for example, Iceland) is centered on the axis, pairs of ridges such as the Iceland-Faeroes Rise and the Greenland

Rise are formed. Sometimes the plate (for example, Africa) has migrated off the hot spot (such as Tristan da Cunha), leaving behind ridge systems that no longer extend to the ridge axis (such as Rio Grande Rise and Western Walvis). *See* MID-OCEANIC RIDGE; VOLCANO; VOLCANOLOGY.

Another characteristic of hot spots is their association with broad swells in the Earth's topography. The Hawaiian hot-spot swell is believed to have been formed in response to either thermal or dynamic effects in an underlying mantle plume. The crustal and upper-mantle structure, which is constrained by seismic refraction data, shows that the oceanic crust is of uniform thickness beneath the swell. The long-wavelength correlation that is observed between the gravity anomaly and the topography (about 37 mGal mi^{-1} or 22 mGal km^{-1}) indicates that the mass excess of the swell is compensated by a low-density, high-temperature region below the crust. The uplift of hot-spot swells is believed to result from thermal perturbations in the underlying plume. The excess heights of swells suggest, on isostatic grounds, that temperature differences of about 450°F (250°C) occur between the plume and the surrounding mantle. Hot ascending plumes may raise the temperature of the overlying lithosphere, thereby thinning it.

Two classes of models have been proposed to explain hot-spot swells. In the reheating model, uplift is produced by thermal expansion that is confined to the conducting portion of the lithosphere (the thermal boundary layer). In the dynamic model, however, there is a contribution to the uplift that is produced by vertical normal stresses exerted to the seismically defined base of the lithosphere (the mechanical boundary layer) by convection.

The main distinguishing feature between the uplift models is that the reheating model predicts a higher heat flow than the dynamic model. Discrimination between these models therefore depends on how the subsidence history, heat flow, and long-term strength (which is controlled mainly by the temperature) differ from those for unperturbed lithosphere of the same age.   [A.B.W.]

Huebnerite   A mineral with the chemical composition $MnWO_4$. Huebnerite is the manganese member of the wolframite solid-solution series. It commonly contains small amounts of iron. It occurs in monoclinic, short, prismatic crystals. Fracture is uneven. Luster varies from adamantine to resinous. Hardness is 4 on Mohs scale and specific gravity is 7.2. Huebnerite is transparent and yellowish to reddish-brown in color; streak is brown. *See* WOLFRAMITE.   [E.C.T.C.]

Humidity   Atmospheric water-vapor content, expressed in any of several measures, especially relative humidity, absolute humidity, humidity mixing ratio, and specific humidity.

Relative humidity is the ratio, in percent, of the moisture actually in the air to the moisture it would hold if it were saturated at the same temperature and pressure. It is a useful index of dryness or dampness for determining evaporation, or absorption of moisture. *See* PSYCHROMETRICS.

Absolute humidity is the weight of water vapor in a unit volume of air expressed, for example, as grams per cubic meter or grains per cubic foot.

Humidity mixing ratio is the weight of water vapor mixed with unit mass of dry air, usually expressed as grams per kilogram. Specific humidity is the weight per unit mass of moist air and has nearly the same values as mixing ratio.   [J.R.F.]

Humite   A homologous series of magnesium nesosilicate minerals having the general composition $Mg_{2n+1}(SiO_4)_n(F, OH)_2$. The known species include norbergite ($n = 1$), chondrodite ($n = 2$), humite ($n = 3$), and clinohumite ($n = 4$). All are based on hexagonal close-packed oxygen and fluorine atoms, the Mg atoms occupying octahedral interstices and the Si atoms occupying tetrahedral interstices. Forsterite, norbergite, and humite are orthorhombic; chondrodite and clinohumite are monoclinic; brucite is trigonal. Manganese analogs of these minerals occur as pink grains in metamorphosed manganese ores derived from preexisting siliceous carbonate and sedimentary manganese oxides. Other cations which can occur as substituents are Fe^{2+}, Ca^{2+}, Al^{3+}, and Ti^{4+}.

The minerals of the humite group have similar physical properties. The luster is resinous, and the color usually light yellow, brown, orange, or red. The pure synthetic Mg end members are colorless. Hardness is $6–6^1/_2$ on Mohs scale, specific gravity is 3.1–3.2. They are very difficult to distinguish visually, and x-ray diffraction, electron microprobe, or optical techniques are required. They are found in regionally crystallized marbles, usually the skarn minerals associated with iron ores. *See* SILICATE MINERALS.

[P.B.M.]

Humus   The amorphous, ordinarily dark-colored, colloidal matter in soil, representing a complex of the fractions of organic matter of plant, animal, and microbial origin that are most resistant to decomposition.

Humus consists of the combined residues of organic materials which have lost their original structure following the rapid decomposition of the simpler ingredients and includes synthesized cell substance as well as by-products of microorganisms. It is not a definite substance and is in a continual state of flux, disappearing by slow decomposition, and being constantly renewed by incorporation of residual matter. With a balance between these processes, humus, though not static, remains relatively uniform in nature and amount in a given soil. It constitutes a reservoir of stabilizing material which imparts beneficial physical, chemical, and biological properties to soil. Fertile soils are rich in humus.

Humus improves the texture of soils. It exerts a binding effect on sandy soils, and loosens the harder, clayey soils, thus increasing their porosity and permeability. It increases the moisture-holding capacity and improves the granular structure by cementing mineral particles into stable crumbs. This helps soils resist the pulverizing and eroding action of wind, water, and cultivation. As a storehouse of elements important to plants, humus functions as a regulator of soil processes by liberating gradually nutrients that would otherwise drain away. A soil rich in humus provides optimum conditions for the development of beneficial microorganisms and constitutes the best medium for growth of plants.

Peat is a type of humus that results from the decomposition of plant material under conditions of excessive moisture or in areas submerged in water. It is an organic deposit formed in marshes and swamps by the partial decomposition of countless generations of a variety of plants. *See* BOG; PEAT.

[A.G.L.]

Hurricane   A tropical cyclone whose maximum sustained winds reach or exceed a threshold of 119 km/h (74 mi/h). In the western North Pacific ocean it is known as a typhoon. Many tropical cyclones do not reach this wind strength. *See* CYCLONE.

Maximum surface winds in hurricanes range up to about 200 mi (320 km) per hour. However, much greater losses of life and property are attributable to inundation from hurricane tidal surges and riverine or flash flooding than from the direct impact of winds on structures.

Tropical cyclones of hurricane strength occur in lower latitudes of all oceans except the South Atlantic and the eastern South Pacific, where combinations of cooler sea temperatures and prevailing winds whose velocities vary sharply with height prevent the establishment of a central warm core through a deep enough layer to sustain the hurricane wind system.

In the United States, property losses resulting from hurricanes have climbed steadily because of the increasing number of seashore structures. However, the loss of life, which has been huge in many storms, has decreased markedly. This is due mainly to the fact that warnings, aided by a more complete surveillance from aircraft and satellite, and extensive programs of public education, have become more accurate and more effective. Improvements in methodology for hurricane prediction have reduced the error in pinpointing hurricane landfall and have greatly reduced the probability of larger errors in prediction. *See* STORM; TROPICAL METEOROLOGY.   [R.Sim.; J.Sim.]

Hydrography   The measurement and description of the physical features and conditions of navigable waters and adjoining coastal areas, including oceans, rivers, and lakes. It involves geodesy, physical oceanography, marine geology, geophysics, photogrammetry (in coastal areas), remote sensing, and marine cartography. Basic parameters observed during a hydrographic survey are time, geographic position, depth of water, and bottom type. However, observation, analysis, and prediction of tides and currents area are also normally included in order to reduce depth measurements to a common vertical datum. *See* GEODESY.

A principal objective of hydrography is to provide for safe navigation and protection of the marine environment through the production of up-to-date nautical charts and related publications. In addition, hydrographic data are essential to a multitude of other activities such as global studies, for example, shoreline erosion and sediment transport studies; coastal construction; delimitation of maritime boundaries; environmental protection and pollution control; exploration and exploitation of marine resources, both living and nonliving; and development of marine geographic information systems (GIS). *See* GEOGRAPHIC INFORMATION SYSTEMS.

Modern depth information is achieved with sonar measurements. Dual-frequency echo sounders are used, with a high-frequency, narrow beam to measure the depth below the vessel, and a lower-frequency, wider beam to obtain larger coverage of the terrain. Side-scan sonar, an instrument that transmits acoustic signals obliquely through the water, is normally towed behind the survey vessel and displays the returning echoes via an onboard graphic recorder. Although this technique does not allow exact determination of position and depth (both can be approximated), it provides excellent resolution with a depiction with what lies to either side of the vessel. Multibeam hydrographic survey systems consist of hull-mounted arrays such that a fan-shaped array of sound beams is transmitted perpendicular to the direction of the ship%s track. This provides for the possibility of 100% coverage of the sea floor. *See* ECHO SOUNDER.

Laser airborne systems mounted in fixed-wing aircraft or helicopters are also available for hydrographic surveys. The system emits a two-color laser beam, usually green and red, such that a return is received from the surface of the water by the red laser and from the bottom by the lower-frequency green laser, allowing the depth to be determined from the time difference. They can be operated in depths down to 165 ft (50 m), but more normally to 66 ft (20 m), depending on water clarity. Hydrographers use tide-coordinated aerial photography to delineate the high and low water lines for charting, which in turn is used for base-line determination of offshore boundaries. Satellite positioning of the aircraft using the Global Positioning System with carrier phase measurement and postprocessing of the data provides for determination of the position of the aircraft of the decimeter level.   [C.An.; G.An.]

Hydrology The study of the waters of the Earth: their occurrence, circulation, and distribution; their chemical and physical properties; and their reaction with the environment, including their relation to living things. *See* TERRESTRIAL WATER.

Water in liquid and solid form covers most of the crust of the Earth. By a complex process powered by gravity and the action of solar energy, an endless exchange of water, in vapor, liquid, and solid forms, takes place between the atmosphere, the oceans, and the crust. This is known as the hydrologic cycle. Water circulates in the air and in the oceans, as well as over and below the surface of landmasses. The distribution of water in the planet is uneven. General patterns of circulation are present in the atmosphere, the oceans, and the landmasses, but regional features are very irregular and seemingly random in detail. Therefore, while causal relations underlie the overall process, it is believed that important elements of chance affect local hydrological events. *See* ATMOSPHERIC GENERAL CIRCULATION.

Whereas the global linkages of the hydrologic cycle are recognized, the science of hydrology has traditionally confined its direct concern to the detailed study of the portion of the cycle limited by the physical boundaries of the land; thus, it has generally excluded specialized investigations of the ocean (which is the subject of the science of oceanography) and the atmosphere (which is the subject of the science of meteorology). The heightened interest in anthropogenically induced environmental impacts has, however, underlined the critical role of the hydrologic cycle in the global transport and budgeting of mass, heat, and energy. Hydrology has become recognized as a science concerned with processes at the local, regional, and global scales. This enhanced status has strengthened its links to meteorology, climatology, and oceanography. *See* CLIMATOLOGY; METEOROLOGY; OCEANOGRAPHY.

A number of field measurements are performed for hydrologic studies. Among them are the amount and intensity of precipitation; the quantities of water stored as snow and ice, and their changes in time; discharge of streams; rates and quantities of infiltration into the soil, and movement of soil moisture; rates of production from wells and changes in their water levels as indicators of ground-water storage; concentration of chemical elements, compounds, and biological constituents in surface and ground waters; amounts of water transferred by evaporation and evapotranspiration to the atmosphere from snow, lakes, streams, soils, and vegetation; and sediment lost from the land and transported by streams.

In addition, hydrology is concerned with research on the phenomena and mechanisms involved in all physical and biological components of the hydrologic cycle, with the purpose of understanding them sufficiently to permit quantitative predictions and forecasting. The field investigations and measurements not only provide the data whereby the behavior of each component may be evaluated in detail, permitting formulation in quantitative terms, but also give a record of the historical performance of the entire system. Thus, two principal vehicles for hydrological forecasting and prediction become available: a set of elemental processes, whose operations are expressible in mathematical terms, linked to form deterministic models that permit the prediction of hydrologic events for given conditions; and a group of records or time series of measured hydrologic variables, such as precipitation or runoff, which can be analyzed by statistical methods to formulate stochastic models that permit inferences to be made on the future likelihood of hydrologic events. *See* GROUND-WATER HYDROLOGY; HYDROSPHERE. [M.A.Ma.]

Hydrometeorology The study of the occurrence, movement, and changes in the state of water in the atmosphere. The term is also used in a more restricted sense, especially by hydrologists, to mean the study of the exchange of water between the atmosphere and continental surfaces. This includes the processes of precipitation and direct condensation, and of evaporation and transpiration from natural surfaces. Considerable emphasis is placed on the statistics of precipitation as a function of area and time for given locations or geographic regions.

Water occurs in the atmosphere primarily in vapor or gaseous form. The average amount of vapor present tends to decrease with increasing elevation and latitude and also varies strongly with season and type of surface. Precipitable water, the mass of vapor per unit area contained in a column of air extending from the surface of the Earth to the outer extremity of the atmosphere, varies from almost zero in continental arctic air to about 6 g/cm^2 in very humid, tropical air.

Although a trivial proportion of the water of the globe is found in the atmosphere at any one instant, the rate of exchange of water between the atmosphere and the continents and oceans is high. Evaporation from the ocean surface and evaporation and transpiration from the land are the sources of water vapor for the atmosphere. Water vapor is removed from the atmosphere by condensation and subsequent precipitation in the form of rain, snow, sleet, and so on. The amount of water vapor removed by direct condensation at the Earth's surface (dew) is relatively small. *See* HYDROLOGY; METEOROLOGY; PRECIPITATION (METEOROLOGY). [E.M.R.]

Hydrosphere The water portion of the Earth as distinguished from the solid part and from the gaseous outer envelope (atmosphere). Approximately 74% of the Earth's surface is covered by water, in either the liquid or solid state. These waters, combined with minor contributions from ground waters, constitute the hydrosphere.

The oceans account for about 97% of the weight of the hydrosphere, while the amount of ice reflects the Earth's climate, being higher during periods of glaciation. There is a considerable amount of water vapor in the atmosphere. The circulation of the waters of the hydrosphere results in the weathering of the landmasses. The annual evaporation from the world oceans and from land areas results in an annual precipitation of 320,000 km^3 (76,000 mi^3) on the world oceans and 100,000 km^3 (24,000 mi^3) on land areas. The rainwater falling on the continents, partly taken up by the ground and partly by the streams, acts as an erosive agent before returning to the seas.

The unique chemical properties of water make it an effective solvent for many gases, salts, and organic compounds. Circulation of water and the dissolved material it contains is a highly dynamic process driven by energy from the Sun and the interior of the Earth. Each component has its own geochemical cycle or pathway through the hydrosphere, reflecting the component's relative abundance, chemical properties, and utilization by organisms. The introduction of materials by humans has significantly altered the composition and environmental properties of many natural waters. *See* GROUND-WATER HYDROLOGY; HYDROLOGY; LAKE; TERRESTRIAL WATER. [J.S.H.]

Hydrothermal vent A hot spring on the ocean floor, where heated fluids exit from cracks in the Earth's crust. Most hydrothermal vents occur along the central axes of mid-oceanic ridges, which are underwater mountain ranges that wind through all of the deep oceans. The best-studied vents are at tectonic spreading centers on the East Pacific Rise and at the Mid-Atlantic Ridge. However, vents are also found over hot spots such as the Hawaiian Islands and Iceland, in back-arc basins such as those in the western Pacific, in shallow geothermal systems such as those off the Kamchatka Peninsula, and on the flanks of some underwater volcanoes and seamounts. Hydrothermal vent sites, or closely grouped clusters of vent deposits and exit ports, may cover areas from hundreds to thousands of square feet (tens to hundreds of square meters). Individual vent sites may be separated along mid-ocean ridges by more than 1000 mi (1600 km). *See* MID-OCEANIC RIDGE; SEAMOUNT AND GUYOT.

All of the hydrothermal vent sites occur in areas where quantities of magma exist below the sea floor. Cold seawater is drawn down into the oceanic crust toward the heat source. As the seawater is heated and reacts with surrounding rock, its composition changes. Sulfate and magnesium are major components of seawater lost during the reactions; sulfide, metals, and gases such as helium and methane are major components gained. This modified seawater is known as hydrothermal fluid. Buoyant, hot hydrothermal fluid rises toward the sea floor in a concentrated zone of upflow to exit from the sea floor at temperatures ranging from 50°F (10°C) to greater than 750°F (400°C), depending on the degree of cooling and of mixing with seawater during the ascent. If the sea floor is shallow enough and the fluid hot enough, the solution may boil; but it usually does not because of the pressure of overlying seawater. *See* MAGMA.

Hydrothermal fluid that mixes extensively with seawater below the sea floor surface may reach the sea floor as warm springs, with temperatures of 50–86°F (10–30°C). This outflow is usually detectable as cloudy or milky water, but the flow is slow and no mineral deposits accumulate except for some hydrothermal staining or oxidation of sea floor basalts. When hotter, relatively undiluted hydrothermal fluid reaches the sea floor, it is still buoyant with respect to seawater, so that the hot solution rises out of cracks in the sea floor at velocities up to about 6 ft (2 m) per second, mixing turbulently with seawater as it rises. Mixing of hydrothermal fluid with seawater leads to precipitation of minerals from solution, forming mineral deposits at the exit from the sea floor and so-called smoke, tiny mineral particles suspended in the rising plume of fluid. Black smoker vents are distinguished by the presence of such large quantities of minute mineral particles that the plumes become virtually opaque.

Formation and outflow of hydrothermal fluid makes a major contribution to the concentration and balance of elements in the oceans by changing the composition of seawater. The quantities of elements added or removed from the oceans by hydrothermal venting around the world are comparable to quantities contributed by the worldwide flow of rivers into the oceans. Hydrothermal venting also represents a major flow of heat from the Earth's crust and a major mechanism for cooling of new oceanic lithosphere. *See* LITHOSPHERE; SEAWATER.

Perhaps the most striking feature of sea-floor hydrothermal vents is their dense biologic communities. Vent faunas tend to be dominated by mollusks, annelids, and crustaceans, whereas faunas on nonvent hard-bottom habitats consist predominantly of cnidarians, sponges, and echinoderms. Biologically, vents are among the most productive ecosystems on Earth. Sulfide from hydrothermal fluids provides the energy to drive these productive systems. Whereas most animal life depends on food of photosynthetic origin (inorganic carbon converted to useful sugars by plants using energy from the Sun), the animals at hydrothermal vents obtain most or all of their food by a process of chemosynthesis. Chemosynthesis is accomplished by specialized bacteria residing in hydrothermal fluids, in mats on the sea floor, or in symbiotic relationships with other organisms. The bacteria convert inorganic carbon to sugars by mediating the oxidation of hydrogen sulfide, thereby exploiting the energy stored in chemical bonds. A few vent animals are also known to use methane gas as a source of energy and carbon. The physical and chemical conditions at hydrothermal vents would be lethal to most marine animals, but vent species have adapted to the conditions there. *See* DEEP-SEA FAUNA.

In a remarkable discovery, it was shown that chemosynthetic microbes known as Archaea are flushed from cavities deep within the Earth's crust by hydrothermal and volcanic activity. These microbes are hyperthermophilic (hot-water-loving) and thrive at temperatures exceeding 90°C (194°F). It is now suspected that an entire community of such microbes inhabits the rocks deep within the water-saturated portions of the Earth's crust. *See* MARINE GEOLOGY.   [M.Go.]

Ice field   A network of interconnected glaciers or ice streams, with common source area or areas, in contrast to ice sheets and ice caps. (An ice sheet is a broad, cakelike glacial mass with a relatively flat surface and gentle relief. Ice caps are properly defined as domelike glacial masses, usually at high elevation.) Being generally associated with terrane of substantial relief, ice-field glaciers are mostly of the broad-basin, cirque, and mountain-valley type. Thus, different sections of an ice field are often separated by linear ranges, bedrock ridges, and nunataks.   [M.M.Mi.]

Iceberg   A large mass of glacial ice broken off and drifted from parent glaciers or ice shelves along polar seas. Icebergs should be distinguished from polar pack ice which is sea ice, or frozen sea water, though rafted or hummocked fragments of the latter may resemble small bergs. *See* GLACIOLOGY; SEA ICE.

Icebergs are classified by shape and size. The terms used are arched, blocky, dome, pinnacled, tabular, valley, and weathered for berg description, and bergy-bit and growler for berg fragments ranging smaller than cottage size above water. The lifespan of an iceberg may be indefinite while the berg remains in cold polar waters, eroding only slightly during summer months. But under the influence of ocean currents, an iceberg that drifts into warmer water will disintegrate rapidly.

In the Arctic, icebergs (see illustration) originate chiefly from glaciers along Greenland coasts. It is estimated that a total of about 16,000 bergs are calved annually in the Northern Hemisphere, of which over 90% are of Greenland origin; but only about half of these have a size or source location to enable them to achieve any significant drift. No icebergs are discharged or drift into the North Pacific Ocean or its adjacent seas, except a few small bergs each year that calve from the piedmont glaciers along the Gulf of Alaska.

Arctic iceberg, eroded to form a valley or dry-dock type: grotesque shapes are common to the glacially produced icebergs of the North.

In the Southern Ocean, bergs originate from the giant ice shelves all along the Antarctic continent. These result in huge, tabular bergs or ice islands several hundred feet high and often over a hundred miles in length, which frequent the entire waters of the Antarctic seas.   [R.P.D.]

Igneous rocks   Those rocks which have congealed from a molten mass. They may be composed of crystals or glass or both depending on the conditions of formation. The molten matter from which they come is called magma; where erupted to the surface, it is commonly known as lava. Solidification of the hot rock melt occurs in response to loss of heat. Generated at depth the magma tends to rise. It commonly breaks through the Earth's crust and spills out on the Earth's surface or ocean floor to form volcanic or extrusive rocks. At the surface where cooling is rapid, fine-grained or glassy rocks are formed.

Where unable to reach the surface, magma cools more slowly, insulated by the overlying rocks; and a coarser texture develops. The resulting igneous rocks appear intrusive relative to adjacent rocks. In general, deeply formed (plutonic) rocks display the coarsest texture. Igneous rocks formed at shallow depths (hypabyssal) display features somewhat intermediate between those of volcanic and plutonic types. *See* MAGMA; PLUTON; VOLCANO; VOLCANOLOGY.

Texture refers to the mutual relation of the rock constituents within a uniform aggregate. It is dependent upon the relative amounts of crystalline and amorphous (glassy) matter as well as the size, shape, and arrangement of the constituents. Rock textures are highly significant; they shed light on the problem of rock genesis, and tell much about the conditions and environment under which the rock formed.

Schemes for classifying igneous rocks are numerous. Three principal methods of classification are used. (1) Megascopic schemes are based on the appearance of the rock-in-hand specimen or as seen with a magnifying glass (hand lens). Such schemes are useful in the field study of rocks. (2) Microscopic schemes (largely mineralogical) are employed in laboratory investigations where more detailed information is needed. (3) Chemical schemes are very useful but have more limited application. The mineral content and texture of a rock generally tell much more about the rock's origin than does a bulk chemical analysis. Igneous rocks show great variations chemically, mineralogically, texturally, and structurally with few if any natural boundaries.

Plutonic rocks occur in large intrusive masses (batholiths, stocks, and other large plutons). They form at great depth and are often referred to as abyssal rocks. They are generated from large bodies of magma which cooled slowly.

Volcanic rocks are formed as lava flows or as pyroclastic rocks (heterogeneous accumulations of volcanic ash and coarser fragmental matter). They have solidified rapidly, and expanding gas bubbles formed by escaping volatiles frequently create highly porous rocks.

Hypabyssal rocks exhibit characteristics more or less intermediate between those of volcanic and plutonic types. They differ from volcanic rocks in that they are intrusive and generally free from glass and vesicular structures. They differ from plutonic rocks in that they occur in small bodies (dikes and sills) or in larger bodies formed at shallow depths (laccoliths) and they have textures characteristically resulting from more rapid cooling. *See* ANORTHOSITE; BASALT; FELDSPAR; FELDSPATHOID; GABBRO; GRANITE; GRANODIORITE; LABRADORITE; LEUCITE ROCK; PYROXENITE; QUARTZ; RHYOLITE; SYENITE; TRACHYTE.

Igneous rock-forming minerals may be classed as primary or secondary. The primary minerals are those formed by direct crystallization from the magma. Secondary minerals may form at any subsequent time. The principal primary minerals are relatively few and may be classed as light-colored (felsic) or dark-colored (mafic) varieties. Felsic is a mnemonic term for feldspathic minerals (feldspar and feldspathoids) and silica (quartz, tridymite, and cristobalite). Mafic is mnemonic for magnesium and iron-rich minerals (biotite, amphibole, pyroxene, and olivine). Felsic minerals are composed largely of silica, alumina, and alkalies. Mafics are rich in iron, magnesium, and calcium.

Secondary minerals include minerals formed by addition of material subsequent to solidification of the rock or by alteration of minerals already present in the rock. Alteration in which certain minerals become more or less reconstituted is common and widespread. The common alteration products derived from the essential primary minerals are as follows.

Primary mineral	Secondary mineral
Quartz	Not altered
Potash feldspar	Kaolinite, sericite
Plagioclase	Kaolinite, sericite (paragonite), epidote, zoisite, calcite
Nepheline	Cancrinite, analcite, natrolite
Leucite	Nepheline and potash feldspar
Sodalite	Analcite, cancrinite
Biotite	Chlorite, sphene, epidote, rutile, iron oxide
Hornblende	Actinolite, biotite, chlorite, epidote, calcite
Orthopyroxene	Antigorite, actinolite, talc
Clinopyroxene	Hornblende, actinolite, biotite, chlorite, epidote, antigorite
Olivine	Serpentine, magnetite, talc, magnesite

See PETROLOGY.   [C.A.C.]

Ignimbrite   A pyroclastic rock deposit formed by one or more ground-hugging flows of hot volcanic fragments and particles, essentially synonymous with pyroclastic-flow deposit, ash-flow tuff, flood tuff, or welded tuff.

Ignimbrites are commonly produced during explosive eruptions and are associated with most of the world's volcanic systems. They vary in size by orders of magnitude (10^{-3} to 10^3 km^3 of erupted material) and have chemical compositions that span the entire range commonly exhibited by igneous rocks (basaltic to rhyolitic). An ignimbrite can be of any form and size, but most deposits have sheetlike shapes and cover many thousands of square kilometers.

Ignimbrite deposits are characterized by a poorly sorted aggregate of ash (crystals and glass shards) and pumice. In the larger deposits, the pumice fragments may be flattened and stretched to yield ovoid-to-lenticular shapes, reflecting the compaction and welding of the deposit after or during emplacement. *See* IGNEOUS ROCKS; PUMICE; PYROCLASTIC ROCKS; TUFF; VOLCANIC GLASS; VOLCANO.   [R.I.T.]

Illite   A clay-size, micaceous mineral; a common component of soil, sediments, sedimentary rocks, and hydrothermal deposits. Illite is considered to possess a smaller layer charge and potassium (K) content than muscovite. It is characterized by the ideal formula $Al_2(Si_{3.2}Al_{0.8})O_{10}(OH)_2K_{0.8}$.

Illite frequently is interlayered with smectite, and can be considered the nonswelling end member in an illite-smectite mixed-layer clay series. Pure end-member illite, with no interlayer smectite, is rare.

Illitic clays are used for manufacturing structural clay products such as brick and tile. Some degraded illites (vermiculites) are used for molding sands. Illite may also be useful for storing certain types of radioactive wastes, because it is less subject to transformation by heat than are other common clays and because it is highly specific for the sorption of cesium. *See* CLAY, COMMERCIAL; CLAY MINERALS.   [D.D.E.]

Ilmenite   A rhombohedral mineral with composition $Fe^{2+}Ti^{4+}O_3$. The hardness is $5\frac{1}{2}$ on Mohs scale, specific gravity 4.72. Color is black, and there is no cleavage. The mineral usually occurs massive or in thin plates. Two other minerals belonging to the ilmenite structure type are geikielite, $MgTiO_3$, and pyrophanite, $MnTiO_3$.

Ilmenite is the most abundant titanium mineral in igneous rocks and the most important ore of titanium. Important occurrences include the Ilmen Mountains, Russia (whence the name); Kragerø, Norway; and Allard Lake, Quebec, Canada.   [P.B.M.]

Impsonite   A black, naturally occurring carbonaceous material having specific gravity 1.10–1.25 and fixed carbon 50–85%. The origin of impsonite is not well understood, but it appears to be derived from a fluid bitumen that polymerized after it filled the vein in which it is found. *See* ASPHALT AND ASPHALTITE.   [I.A.B.]

Index fossil   The ancient remains and traces of a plant or animal that lived during a particular span of geologic time and that geologically dates the containing rocks. Index fossils are almost exclusively confined to sedimentary rocks which originated in such diverse environments as open oceans, tropical lagoons, coral reefs, beaches, lakes, and rivers.

The choice of a fossil as an index depends on several criteria. In general, the fossil represents a group that evolved rapidly. The greater the rate of evolution, the shorter the period of time represented by any given index fossil and the narrower the limits of relative age assigned to the rocks containing the index. Commonly, the span of geologic time during which a fossil lived is referred to as its range, and the thickness of rocks through which a particular index fossil or selected group of fossils occurs is referred to as a faunal zone. An index fossil also must be present in the rocks in sufficient numbers to be found with reasonable effort, must be relatively easy to collect or identify, and must be geographically extensive so that the zone it defines is widely applicable.

The fossil groups most useful as index fossils are generally marine and either floaters or open ocean swimmers, such as cephalopods, or bottom dwellers that had a floating or swimming stage in their life cycles, such as the medusa stage in the brachiopods. Such characteristics are necessary for rapid dispersal of newly evolved forms. On land, such mobile forms as the horses or wind-borne pollen and spores were relatively unrestricted by environmental barriers and became widely dispersed. All of these groups have provided biochronological zones of worldwide extent.

During the Cambrian Period (5.7×10^8 years before present) the oldest highly developed animals appeared; among them the trilobites provide the first important group of index fossils. Small plantlike floating colonial animals called graptolites have proved useful in correlating Ordovician (4.75×10^8 years B.P.) and Silurian (4.25×10^8 years B.P.) rocks. Ammonoids are a classic example of the internationally useful index fossil and are important beginning in the Devonian Period (4.13×10^8 years B.P.) and extending to the end of the Cretaceous Period (6.5×10^8 years B.P.). From the Pennsylvanian Period (3.1×10^8 years B.P.), fusulinids, a family of Foraminiferida, and pollen and spores from the coal forests are important indices. Small phosphatic teethlike fossils known as conodonts have been useful for detailed zonation throughout the Paleozoic Era as well as the early part of the Mesozoic. Closer to present time, the bones and teeth of vertebrate animals serve as index fossils for the Tertiary Era, while the remains of primitive humans have been used to date the Recent past. *See* FOSSIL; GEOLOGIC TIME SCALE; STRATIGRAPHY.   [C.C.]

Indian Ocean The smallest and geologically the most youthful of the three oceans. It differs from the Pacific and Atlantic oceans in two important aspects. First, it is landlocked in the north, does not extend into the cold climatic regions of the Northern Hemisphere, and consequently is asymmetrical with regard to its circulation. Second, the wind systems over its equatorial and northern portions change twice each year, causing an almost complete reversal of its circulation.

The eastern and western boundaries of the Indian Ocean are 147 and 20°E, respectively. In the southeastern Asian waters the boundary is usually placed across Torres Strait, and then from New Guinea along the Lesser Sunda Islands, across Sunda Strait and Singapore Strait.

The ocean floor is divided into a number of basins by a system of ridges. The largest is the Mid-Ocean Ridge, the greater part of which has a rather deep rift valley along its center. It lies like an inverted Y in the central portions of the ocean and ends in the Gulf of Aden. The Sunda Trench, stretching along Java and Sumatra, is the only deep-sea trench in the Indian Ocean. East of the Mid-Ocean Ridge, deep-sea sediments are chiefly red clay; in the western half of the ocean, globigerina ooze prevails and, near the Antarctic continent, diatom ooze.

Atmospheric circulation over the northern and equatorial Indian Ocean is characterized by the changing monsoons. In the southern Indian Ocean atmospheric circulation undergoes only a slight meridional shift during the year. The surface circulation is caused largely by winds and changes in response to the wind systems. In addition, strong boundary currents are formed, especially along the western coastline, as an effect of the Earth's rotation and of the boundaries created by the landmasses.

North of 10°S the changing monsoons cause a complete reversal of surface circulation twice a year. In February, during the Northeast Monsoon, flow north of the Equator is mostly to the west and the North Equatorial Current is well developed. Its water turns south along the coast of Somaliland and returns to the east as the Equatorial Countercurrent between about 2 and 10°S. In August, during the Southwest Monsoon, the South Equatorial Current extends to the north of 10°S; most of its water turns north along the coast of Somaliland, forming the strong Somali Current. North of the Equator flow is from west to east and is called the Monsoon Current. Parts of this current turn south along the coast of Sumatra and return to the South Equatorial Current. During the two transition periods between the Northeast and the Southwest monsoons in April–May and in October, a strong jetlike surface current flows along the Equator from west to east in response to the westerly winds during these months. *See* OCEAN CIRCULATION.

Both semidiurnal and diurnal tides occur in the Indian Ocean. The semidiurnal tides rotate around three amphidromic points situated in the Arabian Sea, southeast of Madagascar, and west of Perth. The diurnal tide also has three amphidromic points: south of India, in the Mozambique Channel, and between Africa and Antarctica. It has more the character of a standing wave, oscillating between the central portions of the Indian Ocean, the Arabian Sea, and the waters between Australia and Antarctica. *See* TIDE. [K.W.]

Industrial meteorology The application of meteorological information to industrial, business, or commercial problems. Generally, industrial meteorology is a branch of applied meteorology, which is the broad field where weather data, analyses, and forecasts are put to practical use. The term "private sector meteorology" has taken on the broader context of traditional industrial meteorology, expanding to include the provision of weather instrumentation/remote sensing devices, systems development and integration, and various consulting services to government and academia as well as value-added products and services to markets in industry (such as media, aviation, and utilities). Some areas in which industrial meteorology may be applied include environmental health and air-pollution control, weather modification, agricultural and forest management, and surface and air transportation. *See* AERONAUTICAL METEOROLOGY; AGRICULTURAL METEOROLOGY; METEOROLOGICAL INSTRUMENTATION.

Specific examples of the uses of industrial meteorology include many in the public sphere. For example, electric utilities need hourly predictions of temperature, humidity, and wind to estimate system load. In addition, they need to know when and where thunderstorms will impact their service area, so that crews can be deployed to minimize or correct disruptions to their transmission and distribution systems. Highway departments need to know when and where frozen precipitation will affect their service areas so that crews can be alerted, trucks loaded with sand and salt, and, if necessary, contractors hired to assist. Since a few degrees' change in temperature, or a slight change in intensity of snow or ice, determines the type of treatment required, early prediction and close monitoring of these parameters are critical.

Agricultural enterprises, from farmers to cooperatives to food manufacturers, rely on precise weather information and forecasts. Weather is the single most important factor in determining crop growth and production. Thus, monitoring and prediction of drought, floods, heat waves, and freezes are of extreme importance. *See* WEATHER FORECASTING AND PREDICTION.

Professionals involved with the meteorological aspects of air pollution are generally concerned with the atmospheric transport, distribution, transformation, and removal mechanisms of air pollutants. They are often called upon to evaluate the effectiveness of pollution control technologies or regulatory (policy) actions used to achieve and maintain air-quality goals. *See* AIR POLLUTION.   [T.S.G.]

Insolation   The incident radiant energy emitted by the Sun, which reaches a unit horizontal area of the Earth's surface. The term is a contraction of incoming solar radiation. About 99.9% of the Sun's energy is in the spectral range of 0.15–4.0 micrometers. About 95% of this energy is in the range of 0.3–2.4 μm; 1.2% is below 0.3 μm and 3.6% is above 2.4 μm. The bulk of the insolation (99%) is in the spectral region of 0.25–4.0 μm. About 40% is found in the visible region of 0.4–0.7 μm and only 10% is in wavelengths shorter than the visible. Energy of wavelengths shorter than 0.29 μm is absorbed high in the atmosphere by nitrogen, oxygen, and ozone.

Insolation depends on several factors: (1) the solar constant—that is, the amount of energy that in a unit time reaches a unit plane surface perpendicular to the Sun's rays outside the Earth's atmosphere, when the Earth is at its mean distance from the Sun; (2) the Sun's elevation in the sky; (3) the amount of solar radiation returned to space at the Earth-atmosphere boundary; and (4) the amount of solar radiation absorbed by the atmosphere and the amount of solar radiation reflected at the lower boundary of the Earth. Insolation is commonly expressed in units of watts per square meter, or calories per square centimeter per minute, also known as langley/min. For instance, the mean value of the solar constant has been estimated as 1368 W/m^2 ($\sim$1.96 ly/min), and the average insolation in summer for a midlatitude clear region could be 340 W/m^2 (700 ly/day), while for a cloudy region it is only about 120 W/m^2 (250 ly/day). *See* ATMOSPHERE; TERRESTRIAL RADIATION.   [R.T.P.]

Instrumented buoys   Unattended, floating structures equipped with systems for collecting, processing, and transmitting meteorological and oceanographic data. Such information is useful for many purposes, including storm warnings and forecasts, coastal engineering, climatology, and oceanographic and atmospheric research.

Moored systems, anchored to the ocean bottom, record currents and water properties as the flow passes the buoy (the eulerian framework). Drifting buoys move with the waters, indicating where they go (the lagrangian framework). In both cases, additional sensors may record a variety of parameters, such as temperature, pressure, acceleration, and water properties.

Surface moorings come in many shapes and sizes depending upon a number of factors, such as research versus operational requirements, duration of deployment, surface versus subsurface measurements, strength of currents, and weather considerations. To operate for a long time, moored buoys require highly reliable components, including a strong mooring line to prevent failure due to wear and tear in heavy seas. This in turn requires large buoys with substantial surface flotation to support the weight of the line. These buoys support a suite of atmospheric sensors and power systems for their operation and data telemetry.

Studies of the subsurface ocean typically employ moorings with flotation below the surface. Subsurface moorings experience much less fatigue of the hardware and mooring line due to the absence of wave motion, permitting the use of lighter-weight, less expensive hardware. More importantly, the absence of surface wave motion results in a much calmer mooring line, greatly facilitating accurate current measurements. Subsurface moorings also eliminate the risk of piracy and entanglement with fishing nets, a major source of equipment and data loss for surface moorings. However, these moorings require costly subsurface flotation devices because they must have the strength to withstand the higher hydrostatic pressure. Large steel spheres are used for near-surface flotation applications, and glass balls are used at greater pressures. Syntactic foam, consisting of a matrix of glass microspheres embedded in epoxy, has proven very effective in applications where the flotation must have a special shape to support instruments, yet remain streamlined to minimize drag forces in strong currents.

Drifting buoys drift with the waters, either as surface drifters or as subsurface floats. In order to follow the waters while minimizing the effect of winds, surface drifters have very little exposure at the surface but have large drag elements hanging beneath. Today, the nearly universal drifter design includes a spherical flotation element and a long, large-diameter canvas tube with large holes, known as a holey sock. Its large size guarantees that the drifter moves with the waters around the sock, readily pulling the small surface sphere with it. Depending upon the application, the sock might hang just to 10 m or as deep as 100 m. The movement of the drifters are tracked by satellites. These drifters have made major contributions to our knowledge of the surface circulation of all the oceans. Some new designs come equipped with temperature sensors, and increasingly, with barometric pressure gauges. They also telemeter valuable sea surface weather information, including acoustic measurements of winds. These buoys cost-effectively complement the functions of the surface buoys.

Buoy sensors can be categorized according to whether they measure scalar or vector properties. Scalar sensors measure the state of a fluid, such as temperature, pressure, humidity, salinity, and light intensity. Vector sensors measure speed and direction of winds and currents. In the past, precipitation measurement presented formidable problems, but the tropical moorings across the Pacific have successfully measured rainfall. An acoustic method that measures the noise generated by raindrops as they hit the surface also shows promise. *See* OCEAN CIRCULATION.   [H.T.R.]

Intaglio (gemology)   The name given to the type of carved gemstone in which the figure is engraved into the surface of the stone, rather than left in relief by cutting away the background, as in a cameo. Intaglios are almost as old as recorded history, for this type of carving was popular in ancient Egypt in the form of cylinders. Intaglios have been carved in a variety of gem materials, including emerald, crystalline quartz, hematite, and the various forms of chalcedony. *See* CAMEO.   [R.T.L.]

International Date Line   The $180°$ meridian, where each day officially begins and ends. As a person travels eastward, against the apparent movement of the Sun, 1 h is gained for every $15°$ of longitude; traveling westward, time is lost at the same rate. Two people starting from any meridian and traveling around the world in opposite

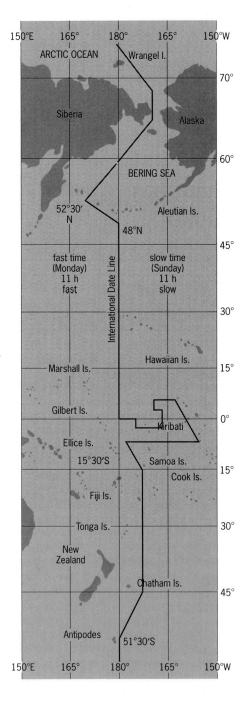

The International Date Line.

directions at the same speed would have the same time when they meet, but would be 1 day apart in date. When a traveler goes west across the line, a day is lost; if it is Monday to the east, it will be Tuesday immediately as the traveler crosses the International Date Line.

The 180° meridian is ideal for serving as the International Date Line (see illustration). It is exactly halfway around the world from the zero, or Greenwich, meridian, from which all longitude is reckoned. It also falls almost in the center of the largest ocean; consequently there is the least amount of inconvenience as regards population centers. A few deviations in the alignment have been made, such as swinging the line east around Siberia to keep that area all in the same day, and westward around the Aleutian Islands so that they will be within the same day as the rest of Alaska. Other variations for the same purpose have been made near Kiribati, at the Equator, and the Fiji Islands, in the South Pacific. *See* MATHEMATICAL GEOGRAPHY. [V.H.E.]

Intra-Americas Sea   That area of the tropical and subtropical western North Atlantic Ocean encompassing the Gulf of Mexico, the Caribbean Sea, the Bahamas and Florida, the northeast coast of South America, and the juxtaposed coastal regions, including the Antillean Islands.

Meteorologically, the Intra-Americas Sea is a transition zone between truly tropical conditions in the south and a subtropical climate in the north. The Sea is also the region that either spawns or interacts with the intense tropical storms known locally as the West Indian Hurricane. Air flowing over the Sea acquires moisture that is the source of much of the precipitation over the central plains of North America. *See* HURRICANE.

Ocean currents of the Intra-Americas Sea are dominated by the Gulf Stream system. Surface waters flow into the Sea through the passages of the Lesser Antilles, and to a lesser extent through the Windward Passage between Cuba and Haiti, and the Anegada Passage between Puerto Rico and Anguilla. These inflowing waters form the Caribbean Current, which flows westward and northward into the Gulf of Mexico through the Yucatán Channel. *See* OCEAN CIRCULATION.

River discharge from several South American rivers, notably the Orinoco and Amazon, drifts through the Intra-Americas Sea and carries materials thousands of kilometers from the deltas. The large deltas are heavily impacted by anthropogenic activities, but they remain the source of rich fisheries and plankton communities. Because of the small tidal range in the Sea, most deltas are wind-dominated geological features. *See* DELTA; EARTHQUAKE; MARINE FISHERIES; PLATE TECTONICS; TIDE. [G.A.Ma.]

Ionosphere   The part of the upper atmosphere that is sufficiently ionized that the concentration of free electrons affects the propagation of radio waves. Existence of the ionosphere was suggested simultaneously in 1902 by O. Heaviside in England and A. E. Kennelly in the United States to explain the transatlantic radio communication that was demonstrated the previous year by G. Marconi; and for many years it was commonly referred to as the Kennelly-Heaviside layer. The existence of the ionosphere as an electronically conducting region had been postulated earlier by B. Steward to explain the daily variations in the geomagnetic field. *See* ATMOSPHERE.

The ionosphere is highly structured in the vertical direction. It was first thought that discrete layers were involved, referred to as the D, E, F_1, and F_2 layers; however, the layers actually merge with one another to such an extent that they are now referred to as regions rather than layers. The very high temperatures in the Earth's upper atmosphere are colocated with the upper ionosphere, since both are related to the effect of x-rays from the Sun. That is, the x-rays both ionize and heat the very uppermost portion of the Earth's atmosphere. Tremendous variations occur in the ionosphere at high latitudes because of the dynamical effects of electrical forces and because of the additional sources of plasma production. The most notable is the visual aurora, one of the most spectacular natural sights.

The aurora has a poleward and equatorward limit during times of magnetic storms. A resident of the arctic regions of the Northern Hemisphere see the "northern" lights in their southern sky, for example. The aurora forms two rings around the poles of the Earth. The size of the rings waxes and wanes while wavelike disturbances propagate along its extent. See AURORA; UPPER-ATMOSPHERE DYNAMICS.   [M.C.K.; F.S.J.]

Isentropic surfaces   Surfaces along which the entropy and potential temperature of air are constant. Potential temperature, in meteorological usage, is defined by the relationship

$$\Theta = T \left(\frac{1000}{P} \right)^{(c_p - c_v)/c_p}$$

in which T is the air temperature, P is atmospheric pressure expressed in millibars, C_P is the heat capacity of air at constant pressure, and C_V is the heat capacity at constant volume. Since the potential temperature of an air parcel does not change if the processes acting on it are adiabatic (no exchange of heat between the parcel and its environment), a surface of constant potential temperature is also a surface of constant entropy. The slope of isentropic surfaces in the atmosphere is of the order of 1/100 to 1/1000. An advantage of representing meteorological conditions on isentropic surfaces is that there is usually little air motion through such surfaces, since thermodynamic processes in the atmosphere are approximately adiabatic. See ATMOSPHERIC GENERAL CIRCULATION.   [F.S.]

Island biogeography   The distribution of plants and animals on islands. Islands harbor the greatest number of endemic species. The relative isolation of many islands has allowed populations to evolve in the absence of competitors and predators, leading to the evolution of unique species that can differ dramatically from their mainland ancestors.

Plant species produce seeds, spores, and fruits that are carried by wind or water currents, or by the feet, feathers, and digestive tracts of birds and other animals. The dispersal of animal species is more improbable, but animals can also be carried long distances by wind and water currents, or rafted on vegetation and oceanic debris. Long-distance dispersal acts as a selective filter that determines the initial composition of an island community. Many species of continental origin may never reach islands unless humans accidentally or deliberately introduce them. Consequently, although islands harbor the greatest number of unique species, the density of species on islands (number of species per area) is typically lower than the density of species in mainland areas of comparable habitat.

Once a species reaches an island and establishes a viable population, it may undergo evolutionary change because of genetic drift, climatic differences between the mainland and the island, or the absence of predators and competitors from the mainland. Consequently, body size, coloration, and morphology of island species often evolve rapidly, producing forms unlike any related species elsewhere. Examples include the giant land tortoises of the Galápagos, and the Komodo dragon, a species of monitor lizard from Indonesia.

If enough morphological change occurs, the island population becomes reproductively isolated from its mainland ancestor, and it is recognized as a unique species.

Because long-distance dispersal is relatively infrequent, repeated speciation may occur as populations of the same species successively colonize an island and differentiate. The most celebrated example is Darwin's finches, a group of related species that inhabit the Galápagos Islands and were derived from South American ancestors. The island species have evolved different body and bill sizes, and in some cases occupy unique ecological niches that are normally filled by mainland bird species. The morphology of these finches was first studied by Charles Darwin and constituted important evidence for his theory of natural selection.

Island biogeography theory has been extended to describe the persistence of single-species metapopulations. A metapopulation is a set of connected local populations in a fragmented landscape that does not include a persistent source pool region. Instead, the fragments themselves serve as stepping stones for local colonization and extinction. The most successful application of the metapopulation model has been to spotted owl populations of old-growth forest fragments in the northwestern United States. *See* BIOGEOGRAPHY; ECOLOGICAL COMMUNITIES; ECOSYSTEM.   [N.J.Go.]

Isobar (meteorology)   A curve along which pressure is constant. Leading examples of its uses are in weather forecasting and meteorology. The most common weather maps are charts of weather conditions at the Earth's surface and mean sea level, and they contain isobars as principal information. Areas of bad or unsettled weather are readily defined by roughly circular isobars around low-pressure centers at mean sea level. Likewise, closed isobars around high-pressure centers define areas of generally fair weather. *See* AIR PRESSURE.

A principal use of isobars stems from the so-called geostrophic wind, which approximates the actual wind on a large scale. The direction of the geostrophic wind is parallel to the isobars, in the sense that if an observer stands facing away from the wind, higher pressures are to the person's right if in the Northern Hemisphere and to the left if in the Southern. Thus, in the Northern Hemisphere, flow is counterclockwise about low-pressure centers and clockwise about high-pressure centers, with the direction of the flow reversed in the Southern Hemisphere. *See* GEOSTROPHIC WIND; WEATHER MAP.
                                                                    [F.B.Sh.]

Isopycnic   The line of intersection of an atmospheric isopycnic surface with some other surface, for instance, a surface of constant elevation or pressure. An isopycnic surface is a surface in which the density of the air is constant. Since specific volume is the reciprocal of density, isosteric surfaces coincide with isopycnic surfaces. On a surface of constant pressure, isopycnics coincide with isotherms, because on such a surface, density is a function solely of temperature. On a constant-pressure surface, isopycnics lie close together when the field is strongly baroclinic and are absent when the field is barotropic. *See* BAROCLINIC FIELD; BAROTROPIC FIELD; SOLENOID (METEOROLOGY).
                                                                    [F.S.; H.B.B.]

Isostasy   The application of Archimedes' principle to the layered structure of the Earth. The elevated topography of Earth is roughly equivalent to an iceberg that floats in the surrounding, denser water. Just as an iceberg extends beneath the exposed ice, the concept of isostasy proposes that topography is supported, or compensated, by a deep root. The buoyant outer shell of the Earth, the crust, displaces the denser, viscous mantle in proportion to the surface elevation. Isostasy implies the existence of a level surface of constant pressure within the mantle, the depth of compensation. Above this surface the mass of any vertical column is equal. Equal pressure at depth can also be

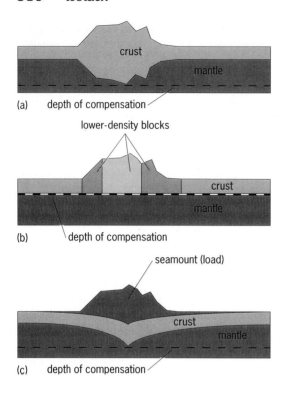

(a) depth of compensation

lower-density blocks

(b) depth of compensation

seamount (load)

(c) depth of compensation

Three major modes of isostatic compensation. (*a*) Airy isostasy, where the crustal density is constant beneath both the elevated topography and the level region; a large root extends beneath the elevated topography, and the depth of compensation is at the base of the crust where the pressure is constant. (*b*) Pratt isostasy, where the density of the crust varies inversely with the height of the topography; the depth of compensation is at the base of the horizontal crust-mantle boundary. (*c*) Flexural or regional isostasy, where the crust has some strength and is deflected beneath the elevated topography; the depth of compensation is a horizontal surface beneath the lowest extent of the crust.

achieved by varying density structure or by the regional deflection of the lithosphere. *See* EARTH CRUST.

Local isostasy achieves equilibrium directly beneath a load by varying either the density or thickness of that mass column. This model attributes no inherent strength to the crust and assumes that the mantle is a simple fluid, redistributing mass to minimize pressure differences at depth. From studies of seamounts, oceanic trenches, foreland basins, and glacial rebound, it has become known that the outer shell of the Earth is rigid, responding to loads over a region broader than the load itself, and that the mantle is a viscous fluid with a time-dependent response to loads.

The simplest method of examining the response of the Earth is to study an area influenced by a discrete load such as a seamount or a continental glacier. If local isostasy (illus. *a*, *b*) is applicable, the region surrounding the load will be horizontal, unaffected by the load. In contrast, if the lithosphere has finite strength and regional or flexural isostasy (illus. *c*) is applicable, the surrounding regions will be deflected down toward the load. Gravity, bathymetry, and seismic studies of the crust surrounding Hawaii and other seamounts have demonstrated that the crust is downwarped beneath seamounts. The implication of this regional response is that the oceanic lithosphere has some strength and that the Earth's outer shell behaves elastically. *See* EARTH; EARTH, GRAVITY FIELD OF; GEODESY; LITHOSPHERE.　　　　　　　　　　　　　　　[R.E.Bel.; B.J.C.]

Isotach　A line along which the speed of the wind is constant. Isotachs are customarily represented on surfaces of constant elevation or atmospheric pressure, or in vertical cross sections. The closeness of spacing of the isotachs is indicative of the intensity of the wind shear on such surfaces. In the region of a jet stream the isotachs are approximately parallel to the streamlines of wind direction and are closely spaced on either side of the core of maximum speed. *See* JET STREAM; WIND.   [F.S.; H.B.B.]

Isothermal chart   A map showing the distribution of air temperature (or sometimes sea surface or soil temperature) over a portion of the Earth's surface, or at some level in the atmosphere. On it, isotherms are lines connecting places of equal temperature. The temperatures thus displayed may all refer to the same instant, may be averages for a day, month, season, or year, or may be the hottest or coldest temperatures reported during some interval.

Isothermal charts are drawn daily in major weather forecasting centers; 5-day, 2-week, and monthly charts are used regularly in long-range forecasting; mean monthly and mean annual charts are compiled and published by most national weather services, and are presented in standard books on, for example, climate, geography, and agriculture. *See* AIR TEMPERATURE; WEATHER MAP.   [A.Cou.]

J

Jade A name that may be applied correctly to two distinct minerals. The two true jades are jadeite and nephrite. In addition, a variety of other minerals are incorrectly called jade. Idocrase is called California jade, dyed calcite is called Mexican jade, and green grossularite garnet is called Transvaal or South African jade. The most commonly encountered jade substitute is the mineral serpentine. It is often called "new jade" or "Korean jade." The most widely distributed and earliest known true type is nephrite, the less valuable of the two. Jadeite, the most precious of gemstones to the Chinese, is much rarer and more expensive. *See* JADEITE.   [R.T.L.]

Jadeite The monoclinic sodium aluminum pyroxene, $NaAlSi_2O_6$. Free crystals are rare. Jadeite usually occurs as dense, felted masses of elongated blades or as fine-grained granular aggregates. It has a Mohs hardness of 6.5 and a density of 3.25–3.35. It has a vitreous or waxy luster, and is commonly green but may also be white, violet, or brown.

Jadeite is always found in metamorphic rocks. It is associated with serpentine at Taw-maw, Burma; Kotaki, Japan; and San Benito County, California. It occurs in metasedimentary rocks of the Franciscan group in California and in Celebes. It is found in Tibet; Yunan Province, China; and Guatemala.

Jadeite is the more cherished of the two jade minerals, because of the more intense colors it displays. It is best known in the lovely intense green color resembling that of emerald. *See* JADE; PYROXENE.   [L.Gr.]

Jasper An opaque, impure type of massive fine-grained quartz that typically has a tile-red, dark-brownish-red, brown, or brownish-yellow color. Jasper has been used since ancient times as an ornamental stone, chiefly of inlay work, and as a semiprecious gem material. *See* GEM; QUARTZ.

Jasper has a smooth conchoidal fracture with a dull luster. The specific gravity and hardness are variable; both values approach those of quartz. The color of jasper often is variegated in banded, spotted, or orbicular types.   [C.Fr.]

Jet (gemology) A black, opaque material that takes a high polish. Jet has been used for many centuries for ornamental purposes. It is a compact variety of lignite coal. It has a refractive index of 1.66, a hardness of 3–4 on Mohs scale, and a specific gravity of 1.30–1.35. Jet is compact and durable, and can be carved or even turned on a lathe. The main source is Whitby, England, in hard shales. *See* GEM; LIGNITE.   [R.T.L.]

Jet stream A relatively narrow, fast-moving wind current flanked by more slowly moving currents. Jet streams are observed principally in the zone of prevailing westerlies above the lower troposphere and in most cases reach maximum intensity, with regard both to speed and to concentration, near the tropopause. At a given time, the position and intensity of the jet stream may significantly influence aircraft operations because of the great speed of the wind at the jet core and the rapid spatial variation of wind speed in its vicinity. Lying in the zone of maximum temperature contrast between cold air masses to the north and warm air masses to the south, the position of the jet stream on a given day usually coincides in part with the regions of greatest storminess in the lower troposphere, though portions of the jet stream occur over regions which are entirely devoid of cloud. The jet stream is often called the polar jet, because of the importance of cold, polar air. The subtropical jet is not associated with surface temperature contrasts, like the polar jet. Maxima in wind speed within the jet stream are called jet streaks. *See* ATMOSPHERIC GENERAL CIRCULATION. [F.S.; H.B.B.]

Jurassic The system of rocks deposited during the middle part of the Mesozoic Era, and encompassing an interval of time between about 208 and 145 million years ago, based on radiometric dating. It takes its name from the Jura Mountains, which run along the border of France and Switzerland.

The main continental masses were grouped together as the supercontinent Pangaea, with a northern component, Laurasia, separated from a southern component, Gondwana, by a major seaway, Tethys, which expanded in width eastward (see illustration). From about Middle Jurassic times onward, this supercontinent began to split up, with a narrow ocean being created between eastern North America and northwestern Africa, corresponding to the central sector of the present Atlantic Ocean. At about the same time, and continuing into the Late Jurassic, separation began between the continents that now surround the Indian Ocean, namely Africa, India, Australia, and Antarctica. As North America moved westward, it collided with a number of oceanic islands in the eastern part of the Paleo-Pacific.

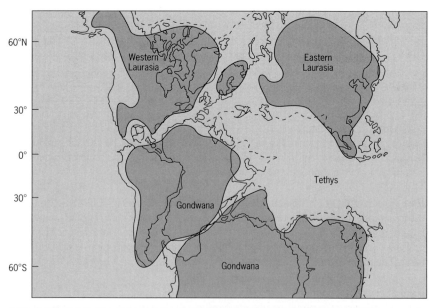

Approximate distribution of land and sea in the Oxfordian stage, the first stage of the late Jurassic. Small islands are excluded, but boundaries of modern continents are included as a reference.

The climate of Jurassic times was clearly more equable than at present. A number of ferns whose living relatives cannot tolerate frost are distributed over a wide range of paleolatitudes, sometimes as far as 60° N and S. Similarly, coral reefs, which are at present confined to the tropics, occur in Jurassic strata in western and central Europe, beyond the paleotropical zone. Many other groups of organisms had wide latitudinal distribution, and there was much less endemism (restriction to a particular area) with respect to latitude than there is today. In addition, there is a lack of evidence for polar icecaps.

The vertebrate terrestrial life of the Jurassic Period was dominated by the reptiles. The dinosaurs had first appeared late in the Triassic from a thecodont stock, which also gave rise to pterosaurs and, later, birds. From small bipedal animals such as *Coelophysis*, there evolved huge, spectacular creatures. These include the herbivorous *Apatosaurus*, *Brontosaurus*, *Brachiosaurus*, *Diplodocus*, and *Stegosaurus* as well as the carnivorous, bipedal *Allosaurus*.

Flying animals include the truly reptilian pterosaurs and the first animals that could be called birds as distinct from reptiles, as represented by the pigeon-sized *Archaeopteryx*. There were two important groups of reptiles that lived in the sea, the dolphinlike ichthyosaurs and the long-necked plesiosaurs. Both of these groups had streamlined bodies and limbs beautifully adapted to marine life. Turtles and crocodiles are also found as fossils in Jurassic deposits.

Jurassic mammals, known mainly from their teeth alone, were small and obviously did not compete directly with the dinosaurs. The fish faunas were dominated by the holosteans, characterized by heavy rhombic scales. Their evolutionary successors, the teleosts, probably appeared shortly before the end of the period.

Because they are far more abundant, the invertebrate fossil faunas of the sea are of more importance to stratigraphers and paleoecologists than are the vertebrates. By far the most useful for stratigraphic correlation are the ammonites, a group of fossil mollusks related to squids. They were swimmers that lived in the open sea, only rarely braving the fluctuating salinity and temperature of inshore waters. They are characteristically more abundant in marine shales and associated fine-grained limestones. From a solitary family that recovered from near extinction at the close of the Triassic, there radiated an enormous diversity of genera. Many of these were worldwide in distribution, but increasingly throughout the period these was a geographic differentiation into two major realms. The Boreal Realm occupied a northern region embracing the Arctic, northern Europe, and northern North America. The Tethyan Realm, with more diverse faunas, occupied the rest of the world. *See* LIMESTONE; SHALE.

With regard to the plant kingdom, the Jurassic might well be called the age of gymnosperms, the nonflowering "naked seed" plants, forests of which covered much of the land. They included the conifers, gingkos, and their relatives, the cycads. Ferns and horsetails made up much of the remainder of the land flora. These and others of the Jurassic flora are still extant in much the same forms.

Jurassic source rocks in the form of organic-rich marine shale and associated rocks contain a significant proportion of the world's petroleum reserves. A familiar example is the Upper Jurassic Kimmeridge Clay of the North Sea, and its stratigraphic equivalents in western Siberia. Some of the source rocks of the greatest petroleum field of all, in the Middle East, are also of Late Jurassic age. *See* MESOZOIC; PETROLEUM GEOLOGY.   [A.Ha.]

K

Kaliophilite A rare mineral tectosilicate found in volcanic rocks high in potassium and low in silica. Kaliophilite is one of three polymorphic forms of $KAlSiO_4$. It crystallizes in the hexagonal system in prismatic crystals. The hardness is 6 on Mohs scale, and the specific gravity is 2.61. The principal occurrence of kaliophilite is at Monte Somma, Italy. *See* KALSILITE; SILICATE MINERALS. [C.S.Hu.]

Kalsilite A rare mineral found in volcanic rocks at Mafuru, in southwest Uganda. It is one of the three polymorphic forms of $KAlSiO_4$. The mineral is hexagonal. The specific gravity is 2.59. In index of refraction and general appearance in thin section it resembles nepheline and is difficult to distinguish from it. *See* KALIOPHILITE; SILICATE MINERALS. [C.S.Hu.]

Kame A round hill or small knoll of sand and gravel, a few meters to more than 100 m high (330 ft). It is one of a family of stratified glacial sediments formed by meltwater in contact with a disintegrating ice sheet. Melting stagnant ice produces large volumes of water carrying boulders, sand, silt, and clay into holes melted in the wasting glacier. When the ice melts completely, the sediment is left standing as small hills. Mixtures of loose rock debris (till) carried above in the melting ice may be lowered onto the kame. Kames are common wherever ice sheets melted, as in New England, New York, the midwestern United States, the British Isles, and Sweden. [S.E.Wh.]

Kaolinite A common hydrous aluminum silicate mineral found in sediments, soils, hydrothermal deposits, and sedimentary rocks. It is a member of a group of clay minerals called the kaolin group minerals, which include dickite, halloysite, nacrite, ordered kaolinite, and disordered kaolinite. These minerals have a theoretical chemical composition of 39.8% alumina, 46.3% silica, and 13.9% water $[Al_2Si_2O_5(OH)_4]$, and they generally do not deviate from this ideal composition. They are sheet silicates comprising a single silica tetrahedral layer joined to a single alumina octahedral layer. Although the kaolin group minerals are chemically the same, each is structurally unique as a result of how these layers are stacked on top of one another. Kaolinite is the most common kaolin group mineral and is an important industrial commodity used in ceramics, paper coating and filler, paint, plastics, fiberglass, catalysts, and other specialty applications. *See* CLAY MINERALS; SILICATE MINERALS. [J.E.K.]

Karst topography Distinctive associations of third-order, erosional landforms indented into second-order structural forms such as plains and plateaus. They are produced by aqueous dissolution, either acting alone or in conjunction with (and as the trigger for) other erosion processes. Karst is largely restricted to the most soluble rocks, which are salt, gypsum and anhydrite, and limestone and dolomite. *See* DOLOMITE ROCK; GYPSUM; LIMESTONE.

The essence of the karst dynamic system is that meteoric water (rain or snow) is routed underground, because the rocks are soluble, rather than flowing off in surface river channels. It follows that dissolutional caves develop in fracture systems, resurging as springs at the margins of the soluble rocks or in the lowest places. A consequence is that most karst topography is "swallowing topography," assemblages of landforms created to deliver meteoric water down to the caves.

Karst landforms develop at small, intermediate, and large scales. Karren is the general name given to small-scale forms—varieties of dissolutional pits, grooves, and runnels. Individuals are rarely greater than 10 m (30 ft) in length or depth, but assemblages of them can cover hundreds of square kilometers. On bare rock, karren display sharp edges; circular pits or runnels extending downslope predominate. Beneath soil, edges are rounded and forms more varied and intricate.

Sinkholes, also known as dolines or closed depressions, are the diagnostic karst (and pseudokarst) landform. They range from shallow, bowllike forms, through steep-sided funnels, to vertical-walled cylinders. Asymmetry is common. Individual sinkholes range from about 1 to 1000 m (3 to 3300 ft) in diameter and are up to 300 m (1000 ft) deep. Many may become partly or largely merged.

Dry valleys and gorges are carved by normal rivers, but progressively lose their water underground (via sinkholes) as the floors become entrenched into karst strata. Many gradations exist, from valleys that dry up only during dry seasons (initial stage) to those that are without any surface channel flow even in the greatest flood periods (paleo-valleys). They are found in most plateau and mountain karst terrains and are greatest where river water can collect on insoluble rocks before penetrating the karst (allogenic rivers).

Poljes, a Serbo-Croatian term for a field, is the generic name adopted for the largest individual karst landform. This is a topographically closed depression with a floor of alluvium masking an underlying limestone floor beveled flat by planar corrosion.

Karst plains and towers are the end stage of karst topographic development in some regions, produced by long-sustained dissolution or by tectonic lowering. The plains are of alluvium, with residual hills (unconsumed intersinkhole limestone) protruding through. Where strata are massively bedded and the hills are vigorously undercut by seasonal floods or allogenic rivers, they may be steepened into vertical towers.   [D.Fo.]

Kernite   A hydrated borate mineral with chemical composition $Na_2B_4O_6(OH)_2 \cdot 3H_2O$. It occurs only very rarely in crystals but is found most commonly in coarse, cleavable masses and aggregates. It is colorless to white; colorless and transparent specimens tend to become chalky white on exposure to air.

Boron compounds are used in the manufacture of glass, especially in glass wool used for insulation purposes. They are also used in soap, in porcelain enamels for coating metal surfaces, and in the preparation of fertilizers and herbicides. *See* BORATE MINERALS.   [C.K.]

Kerogen   The complex, disseminated organic matter present in sedimentary rocks that remains undissolved by sequential treatment with common organic solvents followed by treatment with nonoxidizing mineral hydrochloric acid and hydrofluoric acid. *See* SEDIMENTARY ROCKS.

Kerogen is considered to be the major starting material for most oil and gas generation as sediments are subjected to geothermal heating in the subsurface. It is the most abundant form of organic carbon on Earth—about 1000 times more abundant than coal, which forms primarily from terrigenous remains of higher plants. Kerogen is formed from the remains of marine and lacustrine microorganisms, plants and animals, and variable amounts of terrigenous debris in sediments. The terrestrial portions of kerogen have elemental compositions similar to coal. *See* COAL.

Kerogens are classified according to their atomic ratios of hydrogen to carbon (H/C) and oxygen to carbon (O/C), with oil-prone kerogens being generally higher in H/C and lower in O/C than the gas-prone kerogens. With increasing length of exposure to subsurface temperatures, all kerogens show decreases in the O/C and H/C ratios as they generate preferentially carbon dioxide and water, then oil, and finally only gas (methane) at progressively higher subsurface depths and temperatures. *See* NATURAL GAS; PETROLEUM.   [J.K.W.]

Kimberlite   A variety of peridotite, an igneous rock containing at least 35% olivine. Kimberlite is richer in carbon dioxide than most peridotites, and has crystals larger in diameter than 0.5 mm of olivine, garnet, clinopyroxene, phlogopite, and orthopyroxene. All of these silicate minerals have high Mg/Fe ratios in kimberlites. Diamonds are the only economically significant mineral extracted from kimberlite. They form deeper than 150 km (93 mi) in the Earth's mantle and are carried upward as "accidental tourists" in kimberlite.

The magmatic liquid that forms kimberlite is generated by the melting of small amounts of the Earth's upper mantle containing water and carbonate. The liquid moves upward, gathering crystals (including diamond) and rock fragments along the way. During the violent injection of kimberlites into the upper crust, some detached fragments from the crust move downward and others from the lower crust and mantle move upward. Kimberlite bodies are important scientifically because they contain fragments of rocks that were once above the present-day erosion surface as well as fragments of the Earth's mantle.

Kimberlites usually occur in regions of thick and stable continental crust, in southern Africa (including the Kimberley district), India, Siberia, Canada, Colorado-Wyoming, Venezuela, and Brazil. Most kimberlite outcrops appear on the surface as small, roughly circular areas less than 1 km (0.6 mi) in diameter; they are usually not well exposed because kimberlite weathers rapidly. In three dimensions, kimberlite bodies are dikes or, more commonly, downward-tapering cylinders (pipes). *See* DIAMOND; IGNEOUS ROCKS; OLIVINE; PERIDOTITE.   [D.S.Ba.]

Kuroshio   A swift, intense current flowing northeastward off the coasts of China and Japan in the upper waters of the North Pacific Ocean. The Kuroshio is the western portion of a giant clockwise, horizontal circulation known as the North Pacific subtropical gyre. This circulation extends from 15° to 45°N across the entire width of the Pacific Ocean. It is driven by the large-scale winds—the trades in the south and the westerlies in the north. As with all other western boundary currents, such as the Gulf Stream, the effect of the Earth's rotation and its spherical shape is to concentrate the Kuroshio flow into a current that is only about 100 km (62 mi) wide with speeds up to 2 m/s (4 mi/h). *See* CORIOLIS ACCELERATION; GULF STREAM; PACIFIC OCEAN.

The Kuroshio (Japanese, meaning "Black Current") has an apparent blackness resulting from the water clarity, which is a consequence of the low biological productivity of seawater in the area. It originates off the southeast coast of Luzon, the main island of the Philippines. For the first 1000 km (620 mi), the Kuroshio flows northward along the east coasts of Luzon and Taiwan, until it enters the East China Sea. For the next 1000 km, it flows northeastward near the edge of the continental shelf off eastern China, until it exits the East China Sea through the Tokara Strait. During its final 1000 km, it flows east-northeastward off the southern coast of Japan (where it is sometimes called the Japan Current). Finally, it leaves the Asian coast near Tokyo and travels into the interior of the North Pacific Ocean as a slowly expanding jetlike current known as the Kuroshio Extension. Here it merges with the Oyashio, a cold current with high biological productivity, and becomes the North Pacific Current.

Like the Gulf Stream in the North Atlantic, the Kuroshio rapidly carries large quantities of warm water from the tropics into midlatitude regions. It is consequently an important agent in redistributing global heat. North of 30°N, where prevailing winds are westerlies, the North American climate is strongly affected by the warmth of these waters.   [M.Wi.]

Kyanite   A nesosilicate mineral, Al_2SiO_5, crystallizing in the triclinic system and occurring in metamorphic rocks. It is essentially a pure phase, but minor amounts of iron (Fe^{3+}), chromium (Cr^{3+}), and titanium (Ti^{4+}) may substitute for aluminum (Al). The structure of kyanite is based on cubic close-packed oxygens (O). Ten percent of the tetrahedral (fourfold) interstices are filled with silicon (Si), and 40 percent of the octahedral (sixfold) interstices are filled with Al. The Al, with O at the corners, occurs in zigzag edge-sharing chains of Al-O octahedra. Si-O tetrahedra share corners with Al-O octahedra along the sides. This structure is about 10 percent denser than that of the other two Al_2SiO_5 polymorphs, sillimanite and andalusite, making kyanite the high-pressure polymorph.

Kyanite occurs in well-formed bladed crystals and aggregates. Luster is vitreous to pearly. It is usually light blue because of minor Fe and Ti, and, rarely, light green because of Fe only. Kyanite may also be white or gray. Hardness is 5 (Mohs scale) along the length of crystals and 7 at right angles to the length. Kyanite has a single perfect cleavage parallel to the bladed face of crystals.

Kyanite is a source of material for the manufacture of highly refractory porcelains such as those used for spark plugs. *See* ANDALUSITE; MINERALOGY; SILICATE MINERALS; SILLIMANITE.   [M.J.H.]

L

Labradorite A plagioclase feldspar with composition range $Ab_{50}An_{50}$ to $Ab_{30}An_{70}$ ($Ab = NaAlSi_3O_8$; $An = CaAl_2Si_2O_8$). In some labradorite samples brilliant colors, much like those seen in oil films on water, result from the interference of light reflected at successive lamellar interfaces. *See* FELDSPAR; IGNEOUS ROCKS.   [P.H.R.]

Lake An inland body of water, small to moderately large in size, with its surface exposed to the atmosphere. Most lakes fill depressions below the zone of saturation in the surrounding soil and rock materials. Generically speaking, all bodies of water of this type are lakes, although small lakes usually are called ponds, tarns (in mountains), and less frequently pools or meres. The great majority of lakes have a surface area of less than 100 mi^2 (259 km^2). More than 30 well-known lakes, however, exceed 1500 mi^2 (3885 km^2) in extent, and the largest fresh-water body, Lake Superior, North America, covers 31,180 mi^2 (80,756 km^2). Most lakes are relatively shallow features of the Earth's surface. Because of their shallowness, lakes in general may be considered evanescent features of the Earth's surface, with a relatively short life in geological time.

Lakes differ as to the salt content of the water and as to whether they are intermittent or permanent. Most lakes are composed of fresh water, but some are more salty than the oceans. Generally speaking, a number of water bodies which are called seas are actually salt lakes; examples are the Dead, Caspian, and Aral seas. All salt lakes are found under desert or semiarid climates, where the rate of evaporation is high enough to prevent an outflow and therefore a discharge of salts into the sea.

Lakes with fresh waters also differ greatly in the composition of their waters. Because of the balance between inflow and outflow, fresh lake water composition tends to assume the composite dissolved solids characteristics of the waters of the inflowing streams—with the lake's age having very little influence. Under a few special situations, as crater lakes in volcanic areas, sulfur or other gases may be present in lake water, influencing color, taste, and chemical reaction of the water. *See* FRESH-WATER ECOSYSTEM; HYDROSPHERE; MEROMICTIC LAKE; SURFACE WATER.

Both natural and artificial lakes are economically significant for their storage of water, regulation of stream flow, adaptability to navigation, and recreational attractiveness. A few salt lakes are significant sources of minerals. *See* EUTROPHICATION.   [E.A.A.]

Lamprophyre Any of a heterogeneous group of gray to black, mafic igneous rocks characterized by a distinctive panidiomorphic and porphyritic texture in which abundant euhedral, dark-colored ferromagnesian (femic) minerals (dark mica, amphibole, pyroxene, olivine) occur in two generations—both early as phenocrysts and later in the matrix or groundmass—while felsic minerals (potassium feldspar, plagioclase, analcime, melilite) are restricted to the groundmass.

Many varieties of lamprophyre are known. Minettes, kersantites, vogesites, and spessartites are the most common and are sometimes collectively called calcalkaline lamprophyres. Camptonites and monchiquites are less common; alnoites are rare. These varieties, along with some others, are referred to as alkaline lamprophyres.

Lamprophyres are widespread but volumetrically minor rocks that apparently are restricted to the continents and are the last manifestation of igneous activity in a given area. They usually occur as subparallel or radial swarms of thin ($\sim$1.6–160 ft or $\sim$0.5–50 m) dikes or, less commonly, sills, volcanic neck fillings, or diatremes, or, rarely, lava flows. *See* IGNEOUS ROCKS.   [S.W.B.]

Landscape ecology   The study of the distribution and abundance of elements within landscapes, the origins of these elements, and their impacts on organisms and processes. A landscape may be thought of as a heterogeneous assemblage or mosaic of internally uniform elements or patches, such as blocks of forest, agricultural fields, and housing subdivisions. Biogeographers, land-use planners, hydrologists, and ecosystem ecologists are concerned with patterns and processes at large scale. Landscape ecologists bridge these disciplines in order to understand the interplay between the natural and human factors that influence the development of landscapes, and the impacts of landscape patterns on humans, other organisms, and the flows of materials and energy among patches. Much of landscape ecology is founded on the notion that many observations, such as the persistence of a small mammal population within a forest patch, may be fully understood only by accounting for regional as well as local factors.

Factors that lead to the development of a landscape pattern include a combination of human and nonhuman agents. The geology of a region, including the topography and soils along with the regional climate, is strongly linked to the distribution of surface water and the types of vegetation that can exist on a site. These factors influence the pattern of human settlement and the array of past and present uses of land and water. One prevalent effect of humans is habitat fragmentation, which arises because humans tend to reduce the size and increase the isolation among patches of native habitat.

The pattern of patches on a landscape can in turn have direct effects on many different processes. The structure and arrangement of patches can affect the physical movement of materials such as nutrients or pollutants and the fate of populations of plants and animals. Many of these impacts can be traced to two factors, the role of patch edges and the connectedness among patches.

The boundary between two patches often act as filters or barriers to the transport of biological and physical elements. As an example, leaving buffer strips of native vegetation along stream courses during logging activities can greatly reduce the amount of sediment and nutrients that reach the stream from the logged area. Edge effects can result when forests are logged and there is a flux of light and wind into areas formerly located in the interior of a forest. In this example, edges can be a less suitable habitat for plants and animals not able to cope with drier, high-light conditions. When habitats are fragmented, patches eventually can become so small that they are all edge. When this happens, forest interior dwellers may become extinct. When patch boundaries act as barriers to movement, they can have pronounced effects on the dynamics of populations within and among patches. In the extreme, low connectivity can result in regional extinction even when a suitable habitat remains. This can occur if populations depend on dispersal from neighboring populations. When a population becomes extinct within a patch, there is no way for a colonist to reach the vacant habitat and reestablish the population. This process is repeated until all of the populations within a region disappear. Landscape ecologists have promoted the use of corridors of native habitat between patches to preserve connectivity despite the fragmentation of a landscape.   [D.Sk.]

Landslide   The perceptible downward sliding, falling, or flowing of masses of soil, rock, and debris (mixtures of soil and weathered rock fragments). Landslides range in size from a few cubic meters to over 10^9 m^3 (3.5×10^{10} ft^3), their velocities range from a few centimeters per day to over 100 m/s (330 ft/s), and their displacements may be several centimeters to several kilometers. *See* MASS WASTING.

The U.S. Highway Research Board classification divides landsliding of rock, soil, and debris, on the basis of the types of movement, into falls, slides, and flows. Other classifications consider flows, along with creep and other kinds of landslides, as general forms of mass wasting.

Falls occur when soil or rock masses free-fall through air. Falls are usually the result of collapse of cliff overhangs which result from undercutting by rivers or simply from differential erosion. Slides invariably involve shear displacement or failure along one or more narrow zones or planes. Internal deformation of the sliding mass after initial failure depends on the kinetic energy of the moving mass (size and velocity), the distance traveled, and the internal strength of the mass. Flows have internal displacement and a shape that resemble those of viscous fluids. Relatively weak and wet masses of shale, weathered rock, and soil may move in the form of debris flows and earthflows; water-soaked soils or weathered rock may displace as mudflows.

Mining and civil engineering works have induced myriads of landslides, a few of them of a catastrophic nature. Open-pit mines and road cuts create very high and steep slopes, often quite close to their stability limit. Local factors (weak joints, fault planes) or temporary ones (surges of water pressure inside the slopes, earthquake shocks) induce the failure of some of these slopes. The filling of reservoirs submerges the lower portion of natural, marginally stable slopes or old landslides. Water lowers slope stability by softening clays and by buoying the lowermost portion, or toe, of the slope.

Advances in soil and rock engineering have improved the knowledge of slope stability and the mechanics of landsliding. Small and medium-sized slopes in soil and rock can be made more stable. Remedial measures include lowering the slope angle, draining the slope, using retaining structures, compressing the slope with rock bolts or steel tendons, and grouting. *See* ENGINEERING GEOLOGY; EROSION.   [A.S.N.]

Langmuir circulation   A form of motion found in the near-surface water of lakes and oceans under windy conditions. When the wind is stronger than 5–8 m/s (10–15 knots), streaks of bubbles, seaweed, or flotsam form lines running roughly parallel to the wind, called windrows. Windrows are seen at one time or another on all bodies of water, from ponds to oceans. In the 1920s, Irving Langmuir hypothesized that they are produced by convergences in the water rather than by a direct action of the wind. Langmuir proposed that as the surface water is blown downwind it moves in a spiral fashion, first angling toward the streaks along the surface, next sinking to some depth, then diverging out from under the streaks, and finally rising again in between the streaks (see illustration). In a series of observations and experiments conducted in the North Atlantic and on Lake George in New York, he was able to confirm this basic form of the circulation.

In the ocean, the downwelling under windrows can be strong enough to pull bubbles, seaweed, and other buoyant particles down tens to hundreds of meters below the surface. The downward motion is eventually halted by a subtle increase in the water density with depth, associated with colder temperatures and/or higher salinity. The mixed surface layer typically spans most or all of the depth that light penetrates (the

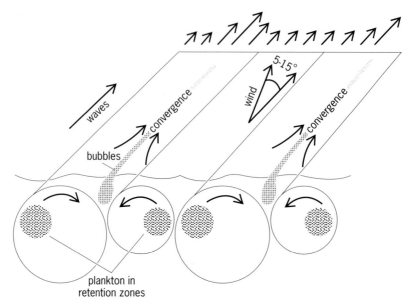

Langmuir circulation consists of a set of alternating rolls of water nearly aligned with the wind. Downwelling zones tend to be narrow and intense compared to the broader, gentler upwelling in between. The rolls may be asymmetric, with stronger flows to the right of the wind at the surface (in the Northern Hemisphere). Details of the lower part of the motion are not yet clear.

euphotic zone). This layer acts like the skin of the sea, through which heat, water vapor, oxygen, carbon dioxide, and all other materials must pass as they enter and leave the air and the sea. The mixing at the bottom of this layer also brings up nutrients and colder water from below.   [J.A.Smi.]

Larnite   The alpha polymorph of calcium silicate (Ca_2SiO_4). Larnite is a mineral which crystallizes at high temperature. Its occurrences are practically confined to limestone or chalk zones in contact with semimolten basalts. At room temperature, larnite is metastable and inverts to its low-temperature polymorph calcio-olivine through shock. This leads to "fall," or disintegration of slags with time, and presents problems in the cement industry. The mineral is very rare, known from its type locality at Scawt Hill, County Antrim, Ireland, and from Crestmore, near Riverside, California. See SILICATE MINERALS.   [P.B.M.]

Laterite   Originally the name for the iron-rich weathering product of basalt in southern India. The term is now used in a compositional sense for weathering products composed principally of the oxides and hydrous oxides of iron, aluminum, titanium, and manganese. Clay minerals of the kaolin group are typically associated with, and are genetically related to, laterite. Laterites range from soft, earthy, porous material to hard, dense rock. Concretionary forms of varying size and shape commonly are developed. The color depends on the content of iron oxides and ranges from white to dark red or brown, commonly variegated. See BAUXITE; CLAY MINERALS; KAOLINITE; WEATHERING PROCESSES.

Mature lateritic soils lack fertility for most systems of agriculture. Savannas or parklike grasslands are typical on laterite. Clay, not laterite, is found beneath rainforests and jungle vegetation.   [S.S.G.]

Lava   Molten rock material that is erupted by volcanoes through openings (volcanic vents) in the Earth's surface. Volcanic rock is formed by the cooling and solidification of erupted lava. Beneath the Earth's surface, molten rock material is called magma. All magmas and lavas consist mainly of a liquid, along with much smaller amounts of solid and gaseous matter. The liquid is molten rock that contains some dissolved gases or gas bubbles; the solids are suspended crystals of minerals or incorporated fragments of preexisting rock. Rapid cooling (quenching) of this liquid upon eruption forms a natural volcanic glass, whereas slower cooling allows more minerals to crystallize from the liquid and preexisting minerals to grow in size. The dissolved gases, a large proportion of which are lost on eruption, are mostly water vapor, together with lesser amounts of carbon, sulfur, chlorine, and fluorine gases. With very rare exception, the chemical composition of the liquid part of magmas and lavas is dominated by silicon and oxygen, which form polymers or compounds with other common rock-forming elements, such as aluminum, iron, magnesium, calcium, sodium, potassium, and titanium.

Viscosity is the principal property which determines the form of erupted lava. It is mainly dependent on chemical composition, temperature, gas content, and the amount of crystals in the magma. Liquid lava with basaltic composition (such as in Hawaii)— relatively low in silicon and aluminum and high in iron, magnesium, and calcium—has higher fluidity (lower viscosity) compared with lava of rhyolitic or dacitic composition (such as at Mount St. Helens, Washington), with higher abundance of silicon and aluminum but lower amounts of iron, magnesium, and calcium. High temperature and gas content of the liquid lava, combined with low crystal abundance, also contribute to increased lava fluidity. Measured maximum temperatures of basaltic lava (1150–1200°C; 2100–2190°F) are higher than those for andesitic and more silicic lavas (720–850°C; 1330–1560°F). Very fluid basaltic lavas can flow great distances, tens to hundreds of kilometers, from the eruptive vents; in contrast, more silicic lavas travel much shorter distances, forming stubby flows or piling up around the vent to form lava domes.

Volcanic products formed by erupted lava vary greatly in size and appearance, depending on volcano type, lava composition, and eruptive style. Most lava products are either lava flows, formed during nonexplosive eruptions by cooling and hardening of flowing lava; or fragmental (pyroclastic) products, formed during explosive eruptions by the shredding apart and ejection into the air of liquid lava. *See* ANDESITE; DACITE; MAGMA; PYROCLASTIC ROCKS; RHYOLITE; VOLCANIC GLASS; VOLCANO.   [R.I.T.]

Lawsonite   A metamorphic silicate mineral related chemically and structurally to the epidote group of minerals. Its composition is $CaAl_2(H_2O)(OH)_2[Si_2O_7]$. It possesses two perfect cleavages; crystals are orthorhombic prismatic to tabular, and colorless to pale blue; specific gravity is 3.1, and hardness is 6.5 on Mohs scale. *See* EPIDOTE; SILICATE MINERALS.   [P.B.M.]

Layered intrusion   In geology, an igneous rock body of large dimensions, 5–300 mi (8–480 km) across and as much as 23,000 ft (7000 m) thick, within which distinct subhorizontal stratification, or layering, is apparent and may be continuous over great distances, in some cases more than 60 mi (100 km). Although conspicuous layering may be found in other rocks of syenitic to granitic composition that are richer in silica, the great layered complexes of the world are, in an overall sense, of tholeiitic basaltic composition. (They may be viewed as intrusive analogs to continental flood basalts.) Indeed, their basaltic composition is of paramount significance to their origin. Only basaltic melts, originating in the mantle beneath the crust of the Earth, are both voluminous enough to occupy vast magma chambers and fluid enough for mineral layering to develop readily. The relatively low viscosity of basaltic melt is a consequence of its high temperature, 2100–2200°F (1150–1200°C), derived from the mantle source region, and its silica-poor, magnesium- and iron-rich (mafic) composition. *See* BASALT; EARTH; MAGMA.

Layered mafic complexes develop upon intrusion of large volumes of basaltic magma (120–24,000 mi^3 or 500–100,000 km^3) into more or less funnel-shaped (smaller complexes) or dish-shaped (larger complexes) chambers 3–5 mi (5–8 km) beneath the Earth's surface. It is widely held that such layering is dominantly produced by gravitational settling of early-formed (cumulus) crystals. These crystals begin to grow as the magma cools and, on reaching a critical size, begin to sink because of their greater density relative to that of the hot silicate melt. Although the sequential order of mineral crystallization can vary depending on subtle differences in magma chemistry, a classic sequence of crystallization from basaltic magma is olivine, $(Mg,Fe)SiO_4$; orthopyroxene, $(Mg,Fe)SiO_3$; clinopyroxene, $(Ca,Mg,Fe)SiO_3$; plagioclase $(Ca,Na)(Si,Al)_4O_8$.

While layers containing only one cumulus mineral may form under special circumstances, coprecipitation of two or three cumulus minerals—for example, olivine + orthopyroxene; orthopyroxene + plagioclase; or orthopyroxene + clinopyroxene + plagioclase—is more common. Under the influence of gravity and current movements in the cooling, tabular magma chamber, the cumulus minerals accumulate on the ever-rising floor of the chamber. The solid rock formed is known as a cumulate, a term that emphasizes its mode of origin and predominant content of cumulus minerals.

Lithologic layering within a complex is typically displayed on a variety of scales. On the broadest scale, a layered mafic complex may contain ultramafic cumulates rich in olivine and orthopyroxene at its base; mafic pyroxene- and plagioclase-rich cumulates at intermediate levels; and more evolved plagioclase-rich cumulates, or even granitic (granophyric) rocks, near its top.

Study of layered mafic complexes is of far more than academic interest because many of them host important deposits of chromium, copper, nickel, titanium, vanadium, and the platinum-group elements platinum, palladium, iridium, osmium, and rhodium. Each of these elements has a relatively high initial concentration in mafic magmas, but all must be dramatically concentrated within restricted layers to be recoverable economically. *See* IGNEOUS ROCKS. [G.K.Cz.]

Lazurite   The chief mineral constituent in the ornamental stone lapis lazuli. Lazurite is a feldspathoid. It crystallizes in the isometric system, but well-formed crystals, usually dodecahedral, are rare. Most commonly, it is granular or in compact masses. The hardness is 5–5.5 on Mohs scale, and the specific gravity is 2.4–2.5. There is vitreous luster and the color is a deep azure, more rarely a greenish-blue. Lazurite is a tectosilicate, the composition of which is expressed by the formula $Na_4Al_3Si_3O_{12}S$.

Lapis lazuli is a mixture of lazurite with other silicates and calcite and usually contains disseminated pyrite. It has long been valued as an ornamental material. Localities of occurrence are in Afghanistan; Lake Baikal, Siberia; Chile; and San Bernardino County, California. *See* FELDSPATHOID; SILICATE MINERALS. [C.S.Hu.]

Lead isotopes (geochemistry)   The study of the isotopic composition of stable and radioactive lead in geological and environmental materials to determine their ages or origins.

Lead isotope geochemistry provides the principal method for determining the ages of old rocks and the Earth itself, as well as the sources of metals in mineral deposits and the evolution of the mantle. Lead (Pb) has four stable isotopes of mass 204, 206, 207, and 208. Three are produced by the radioactive decay of uranium (U) and thorium (Th) [reactions (1)–(3), where $t_{1/2}$ is the half-life of the isotope and α and β denote

$$^{238}U\,(t_{1/2} = 4.5 \times 10^9 \text{ years}) \rightarrow {}^{206}Pb + 8\alpha + 6\beta \tag{1}$$

$$^{235}U\,(t_{1/2} = 0.71 \times 10^9 \text{ years}) \rightarrow {}^{207}Pb + 7\alpha + 4\beta \tag{2}$$

$$^{232}Th\,(t_{1/2} = 13.9 \times 10^9 \text{ years}) \rightarrow {}^{208}Pb + 6\alpha + 4\beta \tag{3}$$

alpha and beta particles, respectively].

The lead produced by the decay of uranium and thorium is termed radiogenic. Since ^{204}Pb is not produced by the decay of any naturally occurring radionuclide, it can be used as a monitor of the amount of initial (nonradiogenic) lead in a system. This will include all of the ^{204}Pb and variable amounts of ^{206}Pb, ^{207}Pb, and ^{208}Pb.

It is possible to calculate the isotopic composition of lead at any time t in the past by calculating and deducting the amount of radiogenic lead that will have accumulated, provided a mineral or rock represents a closed system. A closed system is one in which there has been no chemical transfer of uranium, thorium, or lead in or out of the mineral or rock since it formed. All calculations for uranium-lead dating should yield the same age; this is a unique and powerful property. The ratio of radiogenic ^{207}Pb to ^{206}Pb is simply a function of age, not the U/Pb ratio. Certain minerals such as zircon, monazite, and uraninite are particularly well suited for dating because of extremely high concentrations of uranium or thorium relative to initial lead. However, the degree to which they behave as closed systems can vary. *See* ROCK AGE DETERMINATION.

Even if a rock or mineral contains appreciable initial lead, it may still be dated by using isochron methods. Since the amount of radiogenic lead relative to nonradiogenic lead is a function of the U/Pb ratio and time, the slope on a plot of $^{206}Pb/^{204}Pb$ against $^{238}U/^{204}Pb$ is proportional to age. An isochron is a line on a graph defined by data for rocks of the same age with the same initial lead isotopic composition, the slope of which is proportional to the age. In practice, the $^{238}U/^{204}Pb$ ratio may well have been disturbed by recent alteration of the rock because uranium is highly mobile in near-surface environments. For this reason it is more common to combine the two uranium decay schemes and plot $^{207}Pb/^{204}Pb$ against $^{206}Pb/^{204}Pb$; the slope of an isochron on this plot is a function of age.

Isochron dating has been used to determine an age of 4.55 billion years for the Earth and the solar system by dating iron and stony meteorites. The position of data along the isochron is a function of the U/Pb ratio. The iron meteorites are particularly important for defining the initial lead isotopic composition of the solar system since they contain negligible uranium. The meteorite isochron is commonly termed the geochron. *See* GEOCHRONOMETRY.

Lead isotopes can serve as tracers in the lithosphere, atmosphere, and hydrosphere. Lead isotopes are commonly used to trace the sources of constituents in continental terranes, granites, ore deposits, and pollutants. For example, some granites such as those of the Isle of Skye in northwest Scotland have very unradiogenic lead, indicating that the magmas were derived by melting portions of the lower continental crust that were depleted in uranium about 3 billion years ago. *See* ORE AND MINERAL DEPOSITS.

While there are at least 11 known radioactive isotopes of lead, only ^{212}Pb, ^{214}Pb, and especially ^{210}Pb have been of interest geochemically. Unlike their noble-gas parents, the radioactive lead isotopes as well as other daughter products have a strong affinity for atmospheric aerosols. Both ^{212}Pb and ^{214}Pb have been used to study the process of diffusion of ions in gases and the mechanism of attachment of small ions to aerosols. Measurement of the distribution of radon (Rn) daughter product activities with respect to aerosol size has been important in the development of theoretical models of ion-aerosol interactions. The short half-lives of ^{212}Pb and ^{214}Pb also make these isotopes suitable for studies of near-ground atmospheric transport processes. ^{210}Pb, because of its longer half-life, is removed mainly by precipitation and dry deposition. Its horizontal and vertical distributions are the result of the integrated effects of the distribution and intensity of sources, the large-scale motions of the atmosphere, and the distribution and intensity of removal processes. *See* AIR MASS; ATMOSPHERIC CHEMISTRY.

One of the most important uses of ^{210}Pb is for dating recent coastal marine and lake sediments. As the isotope is rapidly removed from water to underlying deposits, surface sediments often have a considerable excess of ^{210}Pb. The excess is defined as that present in addition to the amount produced by the decay of radium in the sediments. When the sedimentation rate is constant and the sediments are physically undisturbed, the excess ^{210}Pb decreases exponentially with sediment depth as a result of radioactive decay during burial. The reduction in activity at a given depth, compared with that at the surface, provides a measure of the age of the sediments at that depth. Typically, excess ^{210}Pb can be measured for up to about five half-lives or about 100 years, and it is therefore ideally suited for dating sediments that hold records of human impact on the environment. *See* SEDIMENTOLOGY. [A.N.H.; J.A.R.]

Lepidolite   A mineral of variable composition which is also known as lithium mica and lithionite, $K_2(Li,Al)_{5-6}(Si_{6-7}, Al_{2-1})O_{20-21}(F,OH)_{3-4}$. Rubidium (Rb) and cesium (Cs) may replace potassium (K); small amounts of Mn, Mg, Fe(II), and Fe(III) normally are present; and the OH/F ratio varies considerably. Polithionite is a silicon- and lithium-rich, and thus aluminum-poor, variety of lepidolite.

Lepidolite usually forms small scales or fine-grained aggregates. Its colors, pink, lilac, and gray, are a function of the Mn/Fe ratio. Hardness is 2.5–4.0 on Mohs scale; specific gravity is 2.8–3.0. *See* MICA; SILICATE MINERALS.

Lepidolite is uncommon, occurring almost exclusively in structurally complex granitic pegmatites, commonly in replacement units. Common associates are quartz, cleavelandite, alkali beryl, and alkali tourmaline. It is a commercial source of lithium, commonly used directly in lithium glasses and other ceramic products. [E.W.H.]

Leucite rock   Igneous rocks rich in leucite but lacking or poor in alkali feldspar. Those types with essential alkali feldspar are classed as phonolites, feldspathoidal syenite, and feldspathoidal monzonite. The group includes an extremely wide assortment both chemically and mineralogically.

The rocks are generally dark-colored and aphanitic types of volcanic origin. They consist principally of pyroxene and leucite and may or may not contain calcic plagioclase or olivine. Leucite rocks are rare. They occur principally as lava flows and small intrusives (dikes and volcanic plugs). *See* IGNEOUS ROCKS. [C.A.C.]

Lightning   An abrupt, high-current electric discharge that occurs in the atmospheres of the Earth and other planets and that has a path length ranging from hundreds of feet to tens of miles. Lightning occurs in thunderstorms because vertical air motions and interactions between cloud particles cause a separation of positive and negative charges. *See* ATMOSPHERIC ELECTRICITY.

The vast majority of lightning flashes between cloud and ground begin in the cloud with a process known as the preliminary breakdown. After perhaps a tenth of a second, a highly branched discharge, the stepped leader, appears below the cloud base and propagates downward in a succession of intermittent steps. The leader channel is usually negatively charged, and when the tip of a branch of the leader gets to within about 30 m (100 ft) of the ground, the electric field becomes large enough to initiate one or more upward connecting discharges, usually from the tallest objects in the local vicinity of the leader. When contact occurs between an upward discharge and the stepped leader, the first return stroke begins. The return stroke is basically a very intense, positive wave of ionization that propagates up the partially ionized leader channel into the cloud at a speed close to the speed of light. After a pause of 40–80 milliseconds, another leader, the dart leader, forms in the cloud and propagates down the previous return-stroke channel without stepping. When the dart leader makes contact with the ground, a subsequent return stroke propagates back to the cloud. A typical cloud-to-ground flash lasts 0.2–0.3 s and contains about four return strokes; lightning often appears to flicker because the human eye is capable of just resolving the interval between these strokes.

Lightning between cloud and ground is usually classified according to the direction of propagation and polarity of the initial leader. For example, in the most frequent type of cloud-to-ground lightning a negative discharge is initiated by a downward propagating leader as described above (illus. *a*). In this case, the total discharge will effectively lower negative charge to ground or, equivalently, will deposit positive charge in the cloud.

A discharge can be initiated by a downward-propagating positive leader (illus. *b*). Positive discharges occur less frequently than negative ones, but positive discharges are often quite deleterious. Another type of lightning is a ground-to-cloud discharge that begins with a positive leader propagating upward (illus. *c*); this type is relatively rare and is usually initiated by a tall structure or a mountain peak. The rarest form of lightning is a discharge that begins with a negative leader propagating upward (illus. *d*).

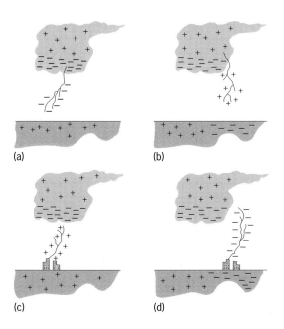

(a)   (b)   (c)   (d)

Sketches of the different types of lightning between an idealized cloud and the ground: (*a*) type 1, (*b*) type 2, (*c*) type 3, and (*d*) type 4. Channel development within the cloud is not shown. Type 1 is the most common form of cloud-to-ground lightning, and type 4 is very rare. (**After M. A. Uman, The Lightning Discharge, Academic Press, 1987**)

The electric currents that flow in return strokes have been measured during direct strikes to instrumented towers. The peak current in a negative first stroke is typically 30 kiloamperes, with a zero-to-peak rise time of just a few microseconds. This current decreases to half-peak value in about 50 microseconds, and then low-level currents of hundreds of amperes may flow for a few to hundreds of milliseconds. The long-continuing currents produce charge transfers on the order of tens of coulombs and are frequently the cause of fires. Subsequent return strokes have peak currents that are typically 10–15 kA, and somewhat faster current rise times. Five percent of the negative discharges to ground generate peak currents that exceed 80 kA, and 5% of the positive discharges exceed 250 kA. Positive flashes frequently produce very large charge transfers, with 50% exceeding 80 coulombs and 5% exceeding 350 coulombs.

[E.P.K.]

Red sprites, elves, and blue jets are upper atmospheric optical phenomena associated with thunderstorms and have only recently been documented using low-light-level television technology. Sprites are massive but weak luminous flashes appearing directly above active thunderstorms coincident with cloud-to-ground or intracloud lightning. They extend from the cloud tops to about 95 km (59 mi) and are predominantly red. High-speed photometer measurements show that the duration of sprites is only a few milliseconds. Their brightness is comparable to a moderately bright auroral arc. Elves are associated with sprites. They are optical emissions of approximately 1 millisecond, with a fast lateral, horizontal expansion that emits more red than blue light. They occur at altitudes of 75–95 km (47–59 mi). Blue jets are optical ejections from the top of the electrically active core regions of thunderstorms. Following the emergence from the top of the thundercloud, they typically propagate upward in narrow cones of about 15° full width at vertical speeds of roughly 100 km/s (60 mi/s), fanning out and disappearing at heights of about 40–50 km (25–30 mi).

[R.E.O.]

Lignite   A brownish-black, low-rank coal, with a heating value of less than 8300 Btu/lb (4611 kcal/kg) on a moist, mineral-matter-free basis. Lignite occurs in two subclasses: lignite A [8300–6300 Btu/lb (4611–3500 kcal/kg)] and lignite B [less than 6300 Btu/lb (3500 kcal/kg)]. Outside North America, low-rank coal is classified as brown coal, which includes lignite, subbituminous, and most high-volatile C bituminous coal of the North American classification system. Brown coal is divided into soft and hard coal; hard coal is subdivided into dull and bright coal. *See* COAL.

Because lignite has undergone less coalification than higher-rank coals, the organic precursor constituents are more easily recognized than those in high-rank coals, and lignite serves as an invaluable link between peat and high-rank coals in studying the coal origin. *See* COAL PALEOBOTANY; PEAT.

Lignite is used primarily to generate electricity at mine-mouth power plants. Lignite has been successfully used as a feedstock for gasification, liquefaction, and pyrolysis. Minor uses of lignite are montan wax, activated carbon, firing kilns, and home heating.

[W.B.A.]

Limestone   A common sedimentary rock composed predominantly of carbonates of calcium and magnesium. Limestones are the most voluminous of the nonsiliciclastic sedimentary rocks. In the strict sense, limestones refer to sedimentary rocks composed of the calcium carbonate mineral calcite ($CaCO_3$). Those rocks, dominated by the magnesium-calcium carbonate mineral dolomite [$CaMg(CO_3)_2$], are known as dolomites or dolostones. Although most limestones are similar in chemical and mineralogical composition, the complex organic and chemical origins of carbonate sediments lead to a wide range of textures and fabrics in the resulting limestones. These textures and fabrics share significant parallels with those found in siliciclastic rocks, and they are quite useful for the classification and determination of depositional environments for limestones. Limestones and dolomites are used commercially as building materials and as a source for industrial and agricultural lime. In addition, limestones and dolomites are important reservoirs for oil and gas and are the hosts for important mineral deposits, including lead, zinc, silver, and fluorite. See ORE AND MINERAL DEPOSITS; PETROLEUM GEOLOGY; STONE AND STONE PRODUCTS.

Most marine limestones (perhaps 90% or more) originate as calcium carbonate skeletal elements of various organisms, including both plants (marine algae such as *Lithothamnion* and phytoplankton such as coccoliths) and animals (such as corals, clams, snails, and oysters). The larger organisms are broken down into cobble-to-silt-sized sediments by biological processes, such as boring, browsing, and grazing, in the environment. Once formed, these sediments react to environmental processes as do their siliciclastic counterparts. See CHALK; DEPOSITIONAL SYSTEMS AND ENVIRONMENTS; MARINE SEDIMENTS; STRATIGRAPHY.

Some limestones and limestone components are formed by direct chemical precipitation from marine and meteoric waters. Most modern, tropical, marine surface water is supersaturated with respect to calcium carbonate. If carbon dioxide is removed from this water by warming, agitation, or photosynthesis, there is a tendency for calcium carbonate to be precipitated. This precipitation can take several forms: an aragonite or magnesian calcite cement, which lithifies carbonate sediment, such as the beach rock commonly found along tropical beaches; an aragonite precipitate on a moving nucleus in a high-energy environment, forming highly polished, round, sand-sized particles termed ooids; or clouds of spontaneously precipitated, clay-sized aragonite, forming on shallow carbonate platforms or in restricted bays. See OOLITE.

Finally, some limestones are formed in fresh-water environments associated with caves (speleothems, such as stalactites and stalagmites), springs (tufa and travertine), and lakes (almost always chemically precipitated fine muds of calcite, dolomite, or alkali-carbonates). See CAVE; STALACTITES AND STALAGMITES; TRAVERTINE; TUFA.

Textures and fabrics in limestones are much more difficult to interpret than in siliciclastics, because of the organic origin of most carbonate grains. While grain size distribution in siliciclastics is controlled by the flow velocity at the site of deposition, grain size distribution in carbonates may be controlled by the types of organisms present in the environment that furnishes the grains. As an example, an environment dominated by large mollusks will tend to produce a sediment characterized by coarse grain sizes, whereas a benthic foraminiferal community will tend to produce grain sizes that are much finer. Roundness in siliciclastic deposits may be used to infer transport and depositional processes. Roundness in the individual grains of a limestone, however, may reflect only the original shape of the organism or the architecture of its skeleton.   [C.H.Mo.]

Limnology   The study of lakes, ponds, rivers, streams, swamps, and reservoirs that make up inland water systems. Each of these inland aquatic environments is physically and chemically connected with its surroundings by meteorologic and hydrogeologic processes (see illustration).

Aquatic systems with excellent physical conditions for production of organisms and high nutrient levels may show signs of eutrophication. Eutrophic lakes are generally identified by large numbers of phytoplankton and aquatic macrophytes and by low

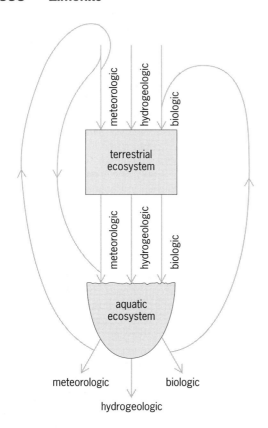

Diagrammatic model of the functional linkages between terrestrial and aquatic ecosystems. Vectors may be meteorologic, hydrogeologic, or biologic components moving nutrients or energy along the pathway shown.

oxygen concentrations in the profundal zone. *See* ECOLOGY; EUTROPHICATION; FRESH-WATER ECOSYSTEM; HYDROLOGY; LAKE; RIVER.   [J.E.S.]

Limonite   A field or generic term for natural hydrous iron oxides, the most common phase being the mineral goethite, α-FeO(OH). Limonite includes the so-called bog iron ores. It is the characteristic brown stain which coats rocks containing sulfide ores, such as pyrite and pyrrhotite, in the zone of weathering of these ores referred to as a gossan. It is formed by biogenic or inorganic precipitation in bog, spring, lacustrine, or marine deposits.   [P.B.M.]

Lithosphere   The rigid or mechanically strong outer layer of the Earth that can support stress. The lithosphere is divided into 12 major plates, the boundaries of which are zones of intense activity that produce many of the large-scale geological features that characterize the Earth. These plates move as coherent units with velocities of up to several centimeters per year, and their relative movement and interaction form the foundation for the theory of plate tectonics.

The lithosphere comprises the crust (either continental or oceanic) and a portion of the upper mantle that together overlie a zone of relative weakness termed the astheno-sphere. The boundary between the crust and the mantle is known as the Mohorovičić discontinuity (Moho), and is compositional in origin—that is, the crust and mantle are distinguished by fundamental differences in rock chemistry. In contrast, the boundary between the lithosphere and the asthenosphere represents an isotherm that separates a conductively cooling lithosphere from a quasi-isothermal convecting asthenosphere. The asthenosphere differs from the overlying lithosphere principally in its ability to flow on geological time scales. These differences arise from the fact that temperature (and thus the fluid or flow properties of rocks) increases as a function of depth in the Earth. Whereas the lithosphere tends to be resistant to deformation, the asthenosphere deforms by flowing. The lithosphere is either oceanic or continental, each type being fundamentally different in terms of the formation and composition of the rocks that constitute the crust and upper mantle. While the most common definition of the lithosphere is in terms of its temperature structure, there exist a whole range of alternative definitions that consider the seismic, mechanical, rheological, and chemical characteristics of the crust and mantle. See ASTHENOSPHERE; EARTH CRUST; MOHO (MOHOROVIČIĆ DISCONTINUITY); PLATE TECTONICS.

The chemical lithosphere is defined as a chemical boundary layer between the surface of the Earth and the asthenosphere that cools by conduction and contains both the material differentiated or extracted from the mantle (for example, oceanic and continental crust) and mantle material modified by various degrees of depletion.   [G.D.K.]

Loess   Silt-dominated sediment of eolian (windblown) origin. Loess is a common deposit in and near areas that were glaciated during the Quaternary Period, and most loess deposits are indirectly related to glaciation. See EOLIAN LANDFORMS; GLACIAL EPOCH; QUATERNARY.

Loess is a well-sorted clastic deposit which is unconsolidated, relatively homogeneous, seemingly nonstratified, and extremely porous. Colors range from buff to shades of pink, gray, yellow, or brown. Silt-sized particles, most of which are 0.0002–0.002 in. (0.005–0.05 mm) in diameter, usually make up 60–90% of the deposit, with small amounts of fine sand and small to moderate amounts of clay-sized material. The particles are generally angular to subangular.

Quartz is the dominant mineral, with subordinate amounts of feldspar, calcite, dolomite, clay minerals, and small amounts of other minerals. Clay minerals are primarily smectite, illite, and chlorite. They occur as silt-sized aggregates and, along with calcite, as coatings or fillings on silt grains, in interstices, and in vertical tubes left from the decay of grass roots. These latter characteristics partially bind the particles together and give loess with relatively large dry strength. As a result, many loess deposits maintain near-vertical slopes in both natural and artificial cuts. See CALCITE; CLAY MINERALS; DOLOMITE; FELDSPAR; QUARTZ.

Loess occurs as a relatively thin (generally <90 ft or 30 m), blanket-type deposit which drapes over an irregular landscape. It is common in many areas of the world, but is particularly thick near valleys that served as meltwater drainageways during Quaternary glaciation. Loess also may be derived from desert areas, in which case the particles must be produced by either weathering processes or eolian abrasion. See SEDIMENTOLOGY; SOIL.   [W.H.J.]

M

Magma   The hot material, partly or wholly liquid, from which igneous rocks form. Besides liquids, solids and gas may be present in magma. Most observed magmas are silicate melts with associated crystals and gas, but some inferred magmas are carbonate, phosphate, oxide, sulfide, and sulfur melts.

Strictly, any natural material which contains a finite proportion of melt (hot liquid) is a magma. However, magmas which contain more than about 60% by volume of solids generally have finite strength and fracture like solids.

Hypothetical, wholly liquid magmas which develop by partial melting of previously solid rock and segregation of the liquid into a volume free of suspended solids and gas are called primary magmas. Hypothetical, wholly liquid magmas which develop by crystallization of a primary magma and isolation of rest liquid free of suspended solids are called parental (or secondary) magmas. Although no unquestioned natural examples of either primary or parental magmas are known, the concepts implied by the definitions are useful in discussing the origins of magmas.

Bodies of flowing lava and natural volcanic glass prove the existence of magmas. Such proven magmas include the silicate magmas corresponding to such rocks as basalt, andesite, dacite, and rhyolite as well as rare carbonate-rich magmas and sulfur melts. Oxide-rich and sulfide-rich magmas are inferred from textural and structural evidence of fluidity as well as mineralogical evidence of high temperature, together with the results of experiments on the equilibrium relations of melts and crystals. *See* IGNEOUS ROCKS; LAVA.

Magma is presumed to underlie regions of active volcanism and to occupy volumes comparable in size and shape to plutons of eroded igneous rocks. However, it is not certain that individual plutons existed wholly as magma at one time. Magma may underlie some regions where no volcanic activity exists, because many plutons appear not to have vented to the surface. *See* PLUTON.

Diverse origins are probable for various magmas. Basaltic magmas because of their high temperatures probably originate within the mantle several tens of kilometers beneath the surface of the Earth. Rhyolitic magmas may originate through crystallization of basaltic magmas or by melting of crustal rock. Intermediate magmas may originate within the mantle or by crystallization of basaltic magmas, by melting of appropriate crustal rock, and also by mixing of magmas or by assimilation of an appropriate rock by an appropriate magma. *See* IGNEOUS ROCKS; VOLCANO.   [A.T.A.]

Magnesite   A member of the calcite-type carbonates having the formula $MgCO_3$. It forms dolomite $[CaMg(CO_3)_2]$ with calcite $(CaCO_3)$ in the system $CaCO_3$—$MgCO_3$. Pure magnesite is not common in nature because there exists a complete series of solid solutions between $MgCO_3$ and $FeCO_3$, which is constantly present in magnesite in its natural occurrence. *See* CARBONATE MINERALS.

Magnesite is usually white, but it may be light to dark brown if iron-bearing. The hardness of magnesite is $3^1/_2$ to $4^1/_2$ on the Mohs scale, and the specific gravity is 3.00.

Magnesite deposits are of two general types: massive and crystalline. Massive magnesite is an alteration product of serpentine which has been subjected to the action of carbonate waters. Crystalline magnesite is usually found in association with dolomite. It is generally thought to be a secondary replacement of magnesite in preexisting dolomite by magnesium-rich fluids.

Magnesite is an important industrial mineral. Various types of magnesite or magnesia (MgO) are produced by different thermal treatments. The caustic-calcined magnesite or magnesia is used in the chemical industry for the production of magnesium compounds, while dead-burned or sintered magnesite or magnesia is used in refractory materials. Fused magnesia is used as an insulating material in the electrical industry because of its high electrical resistance and high thermal conductivity.   [L.L.Y.C.]

Magnetic reversals   The Earth's magnetic field has reversed polarity hundreds of times. That is, at different times in Earth's past, a compass would have pointed south instead of north. Recognition that the geomagnetic field has repeatedly reversed polarity played a key role in the revolution that transformed the geological sciences in the 1960s—the acceptance of the theory of plate tectonics. It is generally accepted that the geomagnetic field is generated by motion of electrically conducting molten metal in Earth's outer core. However, the mechanism by which the field decays and reverses polarity remains one of the great unknowns in geophysics. *See* GEODYNAMO; GEOMAGNETISM; PLATE TECTONICS.

The last magnetic field reversal occurred long before humans were aware of the geomagnetic field (780,000 years ago), so it is necessary to study geological records to understand the process by which the field reverses. The ability of rocks to act as fossilized compasses, which record a permanent "memory" of Earth's magnetic field at the time of rock formation, makes them suitable for detailed studies of ancient geomagnetic field behavior. *See* PALEOMAGNETISM; ROCK MAGNETISM.   [A.P.R.]

Magnetite   A cubic mineral and member of the spinel structure type with composition $[Fe^{3+}]^{IV}[Fe^{2+}Fe^{3+}]^{VI}O_4$. The color is opaque iron-black and streak black, the hardness is 6 (Mohs scale), and the specific gravity is 5.20. The habit is octahedral, but the mineral usually occurs in granular to massive form, sometimes of enormous dimensions. Magnetite is a natural ferrimagnet, but heated above 1072°F (578°C; the Curie temperature) it becomes paramagnetic.

The major magnetic ore of iron, magnetite may be economically important if it occurs in sufficient quantities. The most spectacular ore body occurs at Kiruna in northern Sweden. Other important occurrences are in Norway, Russia, and Canada. *See* SPINEL.   [P.B.M.]

Magnetometer   A device used to measure the intensity and direction of a magnetic field. Magnetometers may be classified as either scalar or vector instruments. A scalar magnetometer measures the strength of the total magnetic field, whereas a vector magnetometer measures one or more vector components of the magnetic field. Most magnetometers are relative instruments that must be calibrated with respect to a known magnetic field. A few magnetometers are absolute instruments that yield accurate magnetic field values without the need for calibration. Three modern devices in regular use are the nuclear magnetometer, fluxgate magnetometer, and SQUID magnetometer.

Two general classes of nuclear magnetometers are the proton precession magnetometer and the optically pumped magnetometer. Both are absolute instruments that measure total field strength without the need for calibration using a known magnetic field.

The fluxgate (saturable-core) magnetometer employs a sensor constructed from an identical pair of cores made from high-magnetic-permeability material. All fluxgate magnetometers are relative vector instruments that require calibration in a known magnetic field to produce accurate results. Orthogonal sets of fluxgate sensors can be used to measure all three field components and thereby the total field vector. Like proton magnetometers, fluxgate pairs can be configured as vector field gradiometers.

The cryogenic or SQUID magnetometer uses one or more Josephson junctions as a magnetic field sensor. A Josephson junction is a zone of weak magnetic coupling (a weak link) between two regions of superconducting material in which current will flow without resistance. A change in the magnetic field applied to the weak link produces a proportional change in magnetic flux within the Josephson junction. The SQUID is the most sensitive magnetometer in use, capable of measuring flux changes only a small fraction of a flux quantum Φ_0 (2.07×10^{-15} weber).

Two types of SQUID are in common use. The radio-frequency SQUID employs a single weak link, whereas the direct-current SQUID uses a pair of Josephson junctions. All SQUID magnetometers are relative, vector instruments. The principal advantages of the SQUID magnetometer over proton, optically pumped, and fluxgate magnetometers are sensitivity and frequency response. The principal disadvantage of the SQUID magnetometer is that it must be kept in a superconducting state. [M.O.McW.]

Magnetosphere   A comet-shaped cavity or bubble around the Earth, carved in the solar wind. This cavity is formed because the Earth's magnetic field represents an obstacle to the solar wind, which is a supersonic flow of plasma blowing away from the Sun. As a result, the solar wind flows around the Earth, confining the Earth and its magnetic field into a long cylindrical cavity with a blunt nose. Since the solar wind is a supersonic flow, it also forms a bow shock a few earth radii away from the front of the cavity. The boundary of the cavity is called the magnetopause. The region between the bow shock and the magnetopause is called the magnetosheath. The Earth is located about 10 earth radii from the blunt-nosed front of the magnetopause. The long cylindrical section of the cavity is called the magnetotail, which is on the order of a few thousand earth radii in length, extending approximately radially away from the Sun.

The magnetosphere has been extensively explored by a number of satellites carrying sophisticated instruments. The satellite observations have indicated that the cavity is not an empty one, but is filled with plasmas of different characteristics. The Earth's dipolar magnetic field is considerably deformed by these plasmas and the electric currents generated by them. *See* VAN ALLEN RADIATION.

All other magnetic planets, such as Mercury, Jupiter, and Saturn, have magnetospheres which are similar in many respects to the magnetosphere of the Earth. [S.I.A.]

Malachite   A bright-green, basic carbonate of copper [$Cu_2CO_3(OH)_2$]. Malachite is the most stable copper mineral in natural environments in contact with the atmosphere and hydrosphere. It occurs as an ore mineral in oxidized copper sulfide deposits; as a stain on fractures in rock outcrops; as a corrosion product of copper and its alloys (except in industrial-urban environments, where the basic copper sulfate dominates); as suspended particles in streams and in alluvial sediments; and as encrustations on bronze artifacts in seawater and on coccoliths floating in the oceans. It can be distinguished from other green copper minerals by its effervescence in acid. The combination of hardness (3.5–4 on Mohs scale) ideal for carving, color variation in concentric layers, and adamantine-to-silky luster has made malachite a highly prized ornamental stone. Its rare blocky-tabular crystals up to 5 mm (0.2 in.), its pseudomorphs after azurite crystals to 2 cm (0.8 in.), and its more common felty tufts perched on bright blue azurite are eagerly sought by mineral collectors. Malachite is an important copper ore mineral in supergene copper oxide deposits formed by weathering of primary copper sulfide deposits. *See* Azurite; Carbonate minerals.   [M.T.E.]

Manganese nodules   Concentrations of manganese and iron oxides found on the floors of many oceans. The origin of these potato-shaped metal-rich deposits has been elucidated; their complex growth histories are revealed by the textures of nodule interiors shown in the illustration.

Marine manganese nodules from certain regions are significantly enriched in nickel, copper, cobalt, zinc, molybdenum, and other elements so as to make them important reserves for these strategic metals.

Although manganiferous nodules and crusts have been sampled or observed on most sea floors, attention has focused on the nickel-plus-copper-rich nodules (2–3 wt% metals) from the north equatorial Pacific in a belt stretching from southeast Hawaii to Baja California, as well as the high-cobalt nodules from seamounts in the Pacific Ocean. Manganese nodules from the Atlantic Ocean and from higher latitudes in the Pacific Ocean have significantly lower concentrations of the minor strategic metals. However, surveys of the Indian Ocean have revealed metal-enrichment trends comparable to those found in the Pacific Ocean nodules; high Ni + Cu-bearing nodules are found adjacent to the Equator.

Microchemical analyses have revealed that chemical differences exist between the outermost top (exposed to sea water) and bottom (immersed in sediment) layers of manganese nodules. Surfaces buried in underlying sediments are generally higher in

Reflected-light photograph of the polished surface of a sectioned manganese nodule showing the complex growth history of the concretionary deposit (diameter 1.6 in. or 4 cm).

Mn, Ni, and Cu contents, compared to the more Fe + Co-rich surfaces exposed to sea water. Episodic rolling-over of a nodule accounts for fluctuating concentrations of Mn, Fe, Ni, Cu, Co, and other metals across sectioned manganese nodules.   [R.G.Bu.]

Manganese oxide minerals   Minerals that contain manganese (Mn) and oxygen (O) or the hydroxyl ion (OH) as principal components. Over 20 manganese oxide minerals have been identified. They can be broadly categorized by the primary oxidation state of manganese in the mineral as tetravalent (Mn^{4+}), trivalent (Mn^{3+}), or divalent (Mn^{2-}) manganese oxides. Tetravalent and trivalent manganese oxides occur in widespread continental and marine environments. They are the primary constituents of manganese nodules and crusts that occur in vast quantities on the ocean floors. Manganese oxide minerals are of economic importance as a source of manganese for the manufacture of steel, and some are used as the cathodic material in dry-cell batteries.

The basic structural unit for many of the manganese oxide minerals is an octahedron formed from a manganese cation surrounded by six oxygens (MnO_6). Octahedra in these structures are linked to one another by sharing either corners (one oxygen atom) or edges (two oxygen atoms). The octahedra are then linked in some minerals to form chains of different widths, and in other minerals to form layers or sheets of octahedra.

The manganese oxides are black and opaque with the exception of manganosite (emerald green) and pyrochroite (colorless to pale green or blue). Both manganosite and pyrochroite become black on exposure to air. Most of the manganese oxides have a specific gravity of 4–5 and a hardness of 5–7 or less. Some of the minerals have a wide range of hardness. Many of the manganese oxides commonly are poorly crystalline and occur in irregular masses or grains. Many manganese oxides are intimately intergrown with other manganese oxides or other minerals on a fine scale. *See* HAUSMANNITE; MANGANITE; PYROLUSITE.   [S.T.]

Manganite   A mineral having composition MnO(OH) and crystallizing in the orthorhombic system in prismatic crystals with deep vertical striations. The hardness is 4 on Mohs scale, and the specific gravity is 4.3. The luster is metallic and the color iron black. Fine crystals have been found in the Harz Mountains; in Cornwall, England; and in the United States at Negaunee, Michigan. Manganite is a minor ore of manganese.   [C.S.Hu.]

Map design   The systematic process of arranging and assigning meaning to elements on a map for the purpose of communicating geographic knowledge in a pleasing format. Careful design is crucial to map effectiveness to avoid distorted or inaccurately represented information.

The first design stage involves determining the type of map to be created for the problem at hand. Decisions must be made about the map's spatial format in terms of size and shape, the basic layout, and the data to be represented. In this step, the experience, cultural background, and educational attainment of the intended audience must be considered. The second stage involves the exploration of preliminary ideas through the manipulation of design parameters such as symbols, color, typography, and line weight. In the third step, alternatives are evaluated and may be accepted or rejected. Under some circumstances, prototype maps may be developed for sample readers as a means of evaluating design scenarios. The last step involves the selection of a final design.

Design considerations include the selection of scale (the relationship between the mapping media format and the area being mapped), symbols to represent geographic features, the system of projection (the method used to translate Earth coordinates to flat media), titles, legends, text, borders, and credits. The process of arranging each map element is referred to as map composition. Success in map composition is achieved when design principles are applied to create a pleasing image with a high degree of information content and readability.

Computers now duplicate all capabilities of manual map design. Design iterations can be explored and evaluated faster and with lower cost, compared to manual drafting methods. The digital environment facilitates the independent storage of map elements that can be combined to form composite images. Other computer innovations such as interactive mapping allow a map user to act as map creator in exploring geographic relationships by selecting and tailoring data sets available through software or the World Wide Web. In addition to its benefits to the design process, mapping software has brought challenges to the mapping sciences, including the proliferation of poorly designed maps constructed by persons untrained in cartography. *See* CARTOGRAPHY; MAP PROJECTIONS.   [T.A.Wi.]

Map projections   Systematic methods of transforming the spherical representation of parallels, meridians, and geographic features of the Earth's surface to a nonspherical surface, usually a plane. Map projections have been of concern to cartographers, mathematicians, and geographers for centuries because globes and curved-surface reproductions of the Earth are cumbersome, expensive, and difficult to use for making measurements. Although the term "projection" implies that transformation is accomplished by projecting surface features of a sphere to a flat piece of paper using a light source, most projections are devised mathematically and are drawn with computer assistance. The task can be complex because the sphere and plane are not applicable surfaces. As a result, each of the infinite number of possible projections deforms the geometric relationships among the points on a sphere in some way, with directions, distances, areas, and angular relationships on the Earth never being completely recreated on a flat map.

It is impossible to transfer spherical coordinates to a flat surface without distortion caused by compression, tearing, or shearing of the surface (see illustration). Conceptually, the transformation may be accomplished in two ways: (1) by geometric transfer to some other surface, such as a tangent or intersecting cylinder, cone, or plane, which can then be developed, that is, cut apart and laid out flat; or (2) by direct mathematical transfer to a plane of the directions and distances among points on the sphere. Patterns of deformation can be evaluated by looking at different projection families. Whether a projection is geometrically or mathematically derived, if its pattern of scale variation is like that which results from geometric transfer, it is classed as cylindrical, conic, or in the case of a plane, azimuthal or zenithal. *See* CARTOGRAPHY; TERRESTRIAL COORDINATE SYSTEM.

Cylindrical projections result from symmetrical transfer of the spherical surface to a tangent or intersecting cylinder. True or correct scale can be obtained along the great circle of tangency or the two homothetic small circles of intersection. If the axis of the cylinder is made parallel to the axis of the Earth, the parallels and meridians appear as perpendicular lines. Points on the Earth equally distant from the tangent great circle (Equator) or small circles of intersection (parallels equally spaced on either side of the Equator) have equal scale departure. The pattern of deformation therefore parallel the parallels, as change in scale occurs in a direction perpendicular to the parallels. A cylinder turned $90°$ with respect to the Earth's axis creates a transverse projection with a pattern of deformation that is symmetric with respect to a great circle through the

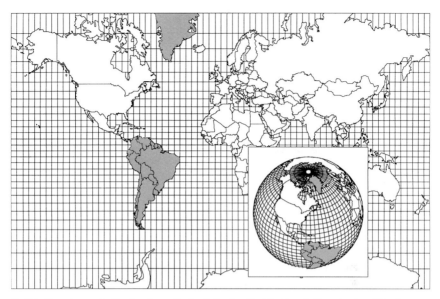

On this Mercator projection (mathematically derived, cylindrical type), Greenland and South America appear similar in size. The inset map shows that South America is actually about 15 times larger than Greenland.

Poles. Transverse projections based on the Universal Transverse Mercator grid system are commonly used to represent satellite images, topographic maps, and other digital databases requiring high levels of precision. If the turn of the cylinder is less than 90°, an oblique projection results. All cylindrical projections, whether geometrically or mathematically derived, have similar patterns of deformation. *See* GREAT CIRCLE, TERRESTRIAL.

Transfer to a tangent or intersecting cone is the basis of conic projections. For these projections, true scale can be found along one or two small circles in the same hemisphere. Conic projections are usually arranged with the axis of the cone parallel to the Earth's axis. Consequently, meridians appear as radiating straight lines and parallels as concentric angles. Conical patterns of deformation parallel the parallels; that is, scale departure is uniform along any parallel. Several important conical projections are not true conics in that their derivation either is based upon more than one cone (polyconic) or is based upon one cone with a subsequent rearrangement of scale variation. Because conic projections can be designed to have low levels of distortion in the midlatitudes, they are often preferred for representing countries such as the United States.

Azimuthal projections result from the transfer to a tangent or intersecting plane established perpendicular to a right line passing through the center of the Earth. All geometrically developed azimuthal projections are transferred from some point on this line. Points on the Earth equidistant from the point of tangency or the center of the circle of intersection have equal scale departure. Hence the pattern of deformation is circular and concentric to the Earth's center. All azimuthal projections, whether geometrically or mathematically derived, have two aspects in common: (1) all great circles that pass through the center of the projection appear as straight lines; and (2) all azimuths from the center are truly displayed.   [A.H.Ro.; T.A.Wi.]

Marble   A term applied commercially to any limestone or dolomite taking polish. Marble is extensively used for building and ornamental purposes. *See* DOLOMITE; LIME-STONE.

In petrography the term marble is applied to metamorphic rocks composed of re-crystallized calcite or dolomite. Schistosity, often controlled by the original bedding, is usually weak except in impure micaceous or tremolite-bearing types. Calcite (marble) deforms readily by plastic flow even at low temperatures. Therefore, granulation is rare, and instead of schistosity there develops a flow structure characterized by elongation and bending of the grains concomitant with a strong development of twin lamellae. *See* METAMORPHIC ROCKS; MINERALOGY; SCHIST.

Pure marbles attaining 99% calcium carbonate, $CaCO_3$, are often formed by simple recrystallization of sedimentary limestone. Dolomite marbles are usually formed by metasomatism. *See* CALCITE; DOLOMITE; METASOMATISM.   [T.F.W.B.]

Marcasite   A mineral having composition FeS_2 and crystallizing in the orthorhom-bic system. Marcasite frequently has a radiating structure and may be globular or sta-lactitic. There is poor prismatic cleavage. The hardness is 6–6.5 on Mohs scale and the specific gravity is 4.89. The luster is metallic and the color pale bronze-yellow to nearly white on a fresh fracture. Marcasite and pyrite are dimorphous; both have the composition FeS_2. Because marcasite is whiter, it is called white iron pyrite.

Marcasite is found in metalliferous deposits associated with lead and zinc ores, as replacement deposits in limestone, and in concretions in clays and shales. The nodular and lenticular masses in coal known as brasses are in part marcasite and in part pyrite. *See* PYRITE.   [C.S.Hu.]

Marine biological sampling   The collection and observation of living or-ganisms in the sea, including the quantitative determination of their abundance in time and space. The biological survey of the ocean depends to a large extent on specially equipped vessels. Sampling in intertidal regions at low tide is one of the few instances where it is possible to observe and collect marine organisms without special apparatus.

A primary aim of marine biology is to discover how ocean phenomena control the distribution of organisms. Sampling is the means by which this aim is accomplished. Traditional techniques employ the use of samplers attached to wires lowered over the side of a ship by means of hydraulic winches. These samplers include bottles designed for enclosing seawater samples from particular depths, fine-meshed nets that are towed behind the ship to sieve out plankton and fish, and grabs or dredges that are used to collect animals inhabiting the ocean bottom. These types of gear are relied upon in many circumstances; however, they illustrate some of the problems common to all methods by which the ocean is sampled. First, sampling is never synoptic, which means that it is not possible to sample an area of ocean so that conditions can be considered equivalent at each point. Usually, it is assumed that this is so. Second, there are marine organisms for which there exists no sampling methodology. For example, knowledge of the larger species of squid is confined to the few animals that have been washed ashore. A third problem concerns the representativeness of the samples collected. The open ocean has no easily definable boundaries, and organisms are not uniformly distributed. The actual sampling is regularly done out of view of the observer; thus, sampling effectiveness is often difficult to determine. Furthermore, navigational systems are not error-free, and therefore the position of the sample is never precisely known. All developments in methods for sampling the ocean try to resolve one or more of these difficulties by improving synopticity, devising more efficient sampling gear, or devising methods for observation such that more meaningful samples can be obtained.

Direct and remote observation methods provide valuable information on the undersea environment and thus on the representativeness of various sampling techniques. Personnel-operated deep-submergence research vessels (DSRVs) are increasingly being employed to observe ocean life at depth and on the bottom, and for determining appropriate sampling schemes. The deep-submergence research vessels are used with cameras and television recording equipment and are also fitted with coring devices, seawater samplers, and sensors of various types. Cameras are deployed from surface ships on a wire. Other cameras are operated unattended at the bottom for months at a time, recording changes occurring there. Scuba diving is playing a larger role, especially in open-ocean areas, and is used to observe marine organisms in their natural habitat as well as to collect the more fragile marine planktonic forms such as foraminifera, radiolaria, and jellyfish. Remotely piloted vehicles (RPVs) will continue to assume greater importance in sampling programs since they can go to greater depths than can divers, and they overcome a limitation in diving in that remotely piloted vehicles can be operated at night. Optical sensors carried aboard Earth-orbiting satellites can provide images of ocean color over wide areas. Ocean color is related to the turbidity and also to the amount of plant material in the seawater. This thus establishes a means by which sampling programs carried out from ships can be optimized. *See* DIVING; SEAWATER FERTILITY; UNDERWATER PHOTOGRAPHY; UNDERWATER TELEVISION; UNDERWATER VEHICLE. [J.Marr.]

Marine ecology An integrative science that studies the basic structural and functional relationships within and among living populations and their physical-chemical environments in marine ecosystems. Marine ecology draws on all the major fields within the biological sciences as well as oceanography, physics, geology, and chemistry. Emphasis has evolved toward understanding the rates and controls on ecological processes that govern both short- and long-term events, including population growth and survival, primary and secondary productivity, and community dynamics and stability. Marine ecology focuses on specific organisms as well as on particular environments or physical settings. *See* ENVIRONMENT.

Marine environments. Classification of marine environments for ecological purposes is based very generally on two criteria, the dominant community or ecosystem type and the physical-geological setting. Those ecosystems identified by their dominant community type include mangrove forests, coastal salt marshes, submersed seagrasses and seaweeds, and tropical coral reefs. Marine environments identified by their physical-geological setting include estuaries, coastal marine and nearshore zones, and open-ocean-deep-sea regions. *See* DEEP-SEA FAUNA; ECOLOGICAL COMMUNITIES; HYDROTHERMAL VENT.

An estuary is a semienclosed area or basin with an open outlet to the sea where fresh water from the land mixes with seawater. The ecological consequences of freshwater input and mixing create strong gradients in physical-chemical characteristics, biological activity and diversity, and the potential for major adverse impacts associated with human activities. Because of the physical forces of tides, wind, waves, and freshwater input, estuaries are perhaps the most ecologically complex marine environment. They are also the most productive of all marine ecosystems on an area basis and contain within their physical boundaries many of the principal marine ecosystems defined by community type. *See* ESTUARINE OCEANOGRAPHY.

Coastal and nearshore marine ecosystems are generally considered to be marine environments bounded by the coastal land margin (seashore) and the continental shelf 300–600 ft (100–200 m) below sea level. The continental shelf, which occupies the greater area of the two and varies in width from a few to several hundred kilometers, is strongly influenced by physical oceanographic processes that govern general patterns of circulation and the energy associated with waves and currents. Ecologically, the coastal and nearshore zones grade from shallow water depths, influenced by the adjacent landmass and input from coastal rivers and estuaries, to the continental shelf break, where oceanic processes predominate. Biological productivity and species diversity and abundance tend to decrease in an offshore direction as the food web becomes supported only by planktonic production. Among the unique marine ecosystems associated with coastal and nearshore water bodies are seaweed-dominated communities (for example, kelp "forests"), coral reefs, and upwellings. *See* CONTINENTAL MARGIN; REEF; UPWELLING.

Approximately 70% of the Earth's surface is covered by oceans, and more than 80% of the ocean's surface overlies water depths greater than 600 ft (200 m), making open-ocean–deep-sea environments the largest, yet the least ecologically studied and understood, of all marine environments. The major oceans of the world differ in their extent of landmass influence, circulation patterns, and other physical-chemical properties. Other major water bodies included in open-ocean–deep-sea environments are the areas of the oceans that are referred to as seas. A sea is a water body that is smaller than an ocean and has unique physical oceanographic features defined by basin morphology. Because of their circulation patterns and geomorphology, seas are more strongly influenced by the continental landmass and island chain structures than are oceanic environments.

Within the major oceans, as well as seas, various oceanographic environments can be defined. A simple classification would include water column depths receiving sufficient light to support photosynthesis (photic zone); water depths at which light penetration cannot support photosynthesis and which for all ecological purposes are without light (aphotic zone); and the benthos or bottom-dwelling organisms. Classical oceanography defines four depth zones; epipelagic, 0–450 ft (0–150 m), which is variable; mesopelagic, 450–3000 ft (150–1000 m); bathypelagic, 3000–12,000 ft (1000–4000 m); and abyssopelagic, greater than 12,000 ft (4000 m). These depth strata correspond approximately to the depth of sufficient light penetration to support photosynthesis; the zone in which all light is attenuated; the truly aphotic zone; and the deepest oceanic environments.

Marine ecological processes. Fundamental to marine ecology is the discovery and understanding of the principles that underlie the organization of marine communities and govern their behavior, such as controls on population growth and stability, quantifying interactions among populations that lead to persistent communities, and coupling of communities to form viable ecosystems. The basis of this organization is the flow of energy and cycling of materials, beginning with the capture of radiant solar energy through the processes of photosynthesis and ending with the remineralization of organic matter and nutrients.

Photosynthesis in seawater is carried out by various marine organisms that range in size from the microscopic, single-celled marine algae to multicellular vascular plants. The rate of photosynthesis, and thus the growth and primary production of marine plants, is dependent on a number of factors, the more important of which are availability and uptake of nutrients, temperature, and intensity and quality of light. Of these three, the last probably is the single most important in governing primary production and the distribution and abundance of marine plants. Considering the high attenuation of light in water and the relationships between light intensity and photosynthesis, net autotrophic production is confined to relatively shallow water depths. The major primary producers in marine environments are intertidal salt marshes and mangroves, submersed seagrasses and seaweeds, phytoplankton, benthic and attached microalgae, and—for coral reefs—symbiotic algae (zooxanthellae). On an areal basis, estuaries and nearshore marine ecosystems have the highest annual rates of primary production. From a global perspective, the open oceans are the greatest contributors to total marine primary production because of their overwhelming size.

The two other principal factors that influence photosynthesis and primary production are temperature and nutrient supply. Temperature affects the rate of metabolic reactions, and marine plants show specific optima and tolerance ranges relative to photosynthesis. Nutrients, particularly nitrogen, phosphorus, and silica, are essential for marine plants and influence both the rate of photosynthesis and plant growth. For many phytoplankton-based marine ecosystems, dissolved inorganic nitrogen is considered the principal limiting nutrient for autotrophic production, both in its limiting behavior and in its role in the eutrophication of estuarine and coastal waters.

Marine food webs and the processes leading to secondary production of marine populations can be divided into plankton-based and detritus-based food webs. They approximate phytoplankton-based systems and macrophyte-based systems. For planktonic food webs, current evidence suggests that primary production is partitioned among groups of variously sized organisms, with small organisms, such as cyanobacteria, playing an equal if not dominant role at times in aquatic productivity. The smaller autotrophs—both through excretion of dissolved organic compounds to provide a substrate for bacterial growth and by direct grazing by protozoa (microflagellates and ciliates)—create a microbially based food web in aquatic ecosystems, the major portion of autotrophic production and secondary utilization in marine food webs may be controlled, not by the larger organisms typically described as supporting marine food webs, but by microscopic populations.

Macrophyte-based food webs, such as those associated with salt marsh, mangrove, and seagrass ecosystems, are not supported by direct grazing of the dominant vascular plant but by the production of detrital matter through plant mortality. The classic example is the detritus-based food webs of coastal salt marsh ecosystems. These ecosystems, which have very high rates of primary production, enter the marine food web as decomposed and fragmented particulate organics. The particulate organics of vascular plant origin support a diverse microbial community that includes bacteria, flagellates, ciliates, and other protozoa. These organisms in turn support higher-level consumers.

Both pelagic (water column) and benthic food webs in deep ocean environments depend on primary production in the overlying water column. For benthic communities, organic matter must reach the bottom by sinking through a deep water column, a process that further reduces its energy content. Thus, in the open ocean, high rates of secondary production, such as fish yields, are associated with areas in which physical-chemical conditions permit and sustain high rates of primary production over long periods of time, as is found in upwelling regions.

Regardless of specific marine environment, microbial processes provide fundamental links in marine food webs that directly or indirectly govern flows of organic matter and nutrients that in turn control ecosystem productivity and stability. *See* ECOLOGY; ECOSYSTEM; SEAWATER FERTILITY. [R.We.]

Marine fisheries The harvest of animals and plants from the ocean to provide food and recreation for people, food for animals, and a variety of organic materials for industry. It is now generally agreed that the world catch is approaching a maximum, which may be less than 100,000,000 metric tons per year. If methods can be devised to harvest smaller organisms not heretofore used because they have been too costly to catch and process, it has been estimated that the yield could perhaps be increased severalfold. Russia is said to have succeeded in developing an acceptable human food product from Antarctic krill. [J.L.McH.]

Marine geology   The study of the portion of the Earth beneath the oceans. Approximately 70% of the Earth's surface is covered with water. Marine geology involves the study of the sea floor; of the sediments, rocks, and structures beneath the sea floor; and of the processes that are responsible for their formation. The average depth of the ocean is about 3800 m (12,500 ft), and the greatest depths are in excess of 11,000 m (36,000 ft; the Marianas Trench). Hence, the study of the sea floor necessitates employing a complex suite of techniques to measure the characteristic properties of the Earth's surface beneath the oceans. Contrary to popular views, only a minority of marine geological investigations involve the direct observation of the sea floor by scuba diving or in submersibles. Rather, most of the ocean floor has been investigated by surface ships using remote-sensing geophysical techniques, and more recently by the use of satellite observations.

The oceanic crust is relatively young, having been formed entirely within the last 200 million years (m.y.), a small fraction of the nearly 5-billion-year history of the Earth. The process of renewing or recycling the oceanic crust is the direct consequence of plate tectonics and sea-floor-spreading processes. It is therefore logical that the geologic history of the sea floor be outlined within the framework of plate tectonic tenets. Where plates move apart, molten lava reaches the surface to fill the voids, creating new oceanic crust. Where the plates come together, oceanic crust is thrust back within the interior of the Earth, creating the deep oceanic trenches. These trenches are located primarily around the rim of the Pacific Ocean. The material can be traced by using the distribution of earthquakes to depths of about 700 km (420 mi). At that level, the character of the subducted lithosphere is lost, and this material is presumably remelted and assimilated with the surrounding upper-mantle material. *See* EARTHQUAKE; GEODYNAMICS; LITHOSPHERE; PLATE TECTONICS.

Mid-oceanic ridges. Most of the ocean floor can be classified into three broad physiographic regions, one grading into the other (see illustration). The approximate

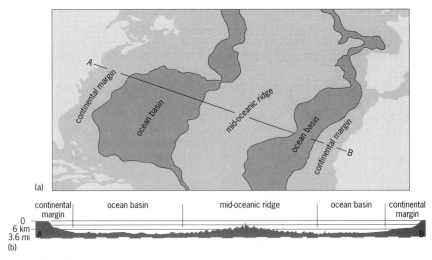

Geology of the North Atlantic Ocean. (*a***) Physiographic divisions of the ocean floor. (***b***) Principal morphologic features along the profile between North America and Africa.**

centers of the ocean basin are characterized by spectacular, globally encircling mountain ranges, the mid-oceanic ridge (MOR) system, which formed as the direct consequence of the splitting apart of oceanic lithosphere. The detailed morphologic characteristics of these mountain ranges depend somewhat upon the rate of separation of the plates involved. Abyssal hill relief, especially within 500 km (300 mi) of ridge crest, is noticeably rougher on the slow-spreading Mid-Atlantic Ridge than on the fast-spreading East Pacific Rise. The profile of the East Pacific Rise is also broader and shallower than for the Mid-Atlantic Ridge. If the entire mid-oceanic ridge system were spreading rapidly, the expanded volume of the ridge system would displace water from the ocean basins onto the continents.

The broad cross-sectional shape of this mid-ocean mountain range can be related directly and simply to its age. The depth of the mid-oceanic ridge at any place is a consequence of the steady conduction of heat to the surface and the associated cooling of the oceanic crust and lithosphere. As it cools, contracts, and becomes more dense, the oceanic crust plus the oceanic lithosphere sink isostatically (under its own weight) into the more fluid substrate (the asthenosphere). Hence, the depth to the top of the oceanic crust is a predictable function of the age of that crust; departures from such depth predictions represent oceanic depth anomalies. These depth anomalies are presumably formed as a consequence of processes other than lithospheric cooling, such as intraplate volcanism. The Hawaiian island chain and the Polynesian island groups are examples of this type of volcanism. *See* ASTHENOSPHERE; MID-OCEANIC RIDGE; OCEANIC ISLANDS; VOLCANOLOGY.

Basins. The deep ocean basins, which lie adjacent to the flanks of the mid-oceanic ridge, represent the older portions of the sea floor that were once the shallower flanks of the ridge (see illustration). The bulk of sediments found on the ocean floor can be broadly classified as terrigenous or biogenic. Terrigenous sediments are derived from adjacent landmasses and are brought to the sea through river systems. This sediment load is sometimes transported across the continental shelves, often utilizing, as pathways, submarine canyons that dissect the shelves, the continental slope, and the continental rise. Biogenic sediments are found in all parts of the ocean, either intermixed with terrigenous sediments or in near "pure form" in those areas inaccessible to terrigenous sedimentation.

Biogenic sediments are composed mostly of the undissolved tests of siliceous and calcareous microorganisms, which settle slowly to the sea floor. This steady so-called pelagic rain typically accumulates at rates of a few centimeters per thousand years. The composition and extent of the input to the biogenic sediment depend upon the composition and abundances of the organisms, which in turn are largely reflective of the water temperature and the available supply of nutrients. The Pacific equatorial zones and certain other regions of deep ocean upwelling are rich in nutrients and correspondingly rich in the microfauna and flora of the surface waters. Such regions are characterized by atypically high pelagic sediment rates. *See* UPWELLING.

Continental margins. The continental margins lie at the transition zone between the continents and the ocean basins and mark a major change from deep water to shallow water and from thin oceanic crust to thick continental crust. Rifted margins are found bounding the Atlantic Ocean (see illustration). These margins represent sections of the South American and North American continents that were once contiguous to west Africa and northwest Africa, respectively. These supercontinents were rifted apart 160–200 m.y. ago as the initial stages of sea-floor spreading and the birth of the present Atlantic Ocean sea floor. *See* CONTINENTS, EVOLUTION OF.

Continental margins are proximal to large sources of terrestrial sediments that are the products of continental erosion. The margins are also the regions of very large vertical motions through time. This vertical motion is a consequence of cooling of the rifted continental lithosphere and subsidence. During initial rifting of the continents, fault-bound rift basins are formed that serve as sites of deposition for large quantities of sediment. These sedimented basins constitute significant loads onto the underlying crust, giving rise to an additional component of margin subsidence. The continental margins are of particular importance also because, as sites of thick sediment accumulations (including organic detritus), they hold considerable potential for the eventual formation and concentration of hydrocarbons. As relatively shallow areas, they are also accessible to offshore exploratory drilling and oil and gas production wells.

Many sedimentary aprons or submarine fans are found seaward of prominent submarine canyons that incise the continental margins. Studies of these sedimentary deposits have revealed a number of unusual surface features that include a complex system of submarine distributary channels, some with levees. The channel systems control and influence sediment distribution by depositional or erosional interchannel flows. Fans are also effected by major instantaneous sediment inputs caused by large submarine mass slumping and extrachannel turbidity flows. *See* SUBMARINE CANYON.

In contrast to the rifted margins, the continental margins that typically surround the Pacific Ocean represent areas where plates are colliding. As a consequence of these collisions, the oceanic lithosphere is thrust back into the interior of the Earth; the loci of underthrusting are manifest as atypically deep ocean sites known as oceanic trenches. The processes of subducting the oceanic lithosphere give rise to a suite of tectonic and morphologic features characteristically found in association with the oceanic trenches. An upward bulge of the crust is created seaward of the trench that represents the flexing of the rigid oceanic crust as it is bent downward at the trench. The broad zone landward of most trenches is known as the accretionary prism and represents the accumulation of large quantities of sediment that was carried on the oceanic crust to the trench. Because the sediments have relatively little strength, they are not underthrust with the more rigid oceanic crust, but they are scraped off. In effect, they are plastered along the inner wall of the trench system, giving rise to a zone of highly deformed sediments. These sediments derived from the ocean floor are intermixed with sediments transported downslope from the adjacent landmass, thus creating a classic sedimentary melange. *See* CONTINENTAL MARGIN; SEDIMENTOLOGY.

Anomalous features. In addition to the major morphologic and sediment provinces, parts of the sea floor consist of anomalous features that obviously were not formed by fundamental processes of sea-floor spreading, plate collisions, or sedimentation. Examples are long, linear chains of seamounts and islands. Many of these chains are thought to reflect the motion of the oceanic plates over hot spots that are fixed within the mantle. *See* MAGMA; SEAMOUNT AND GUYOT.

The presence of large, anomalously shallow regions known as oceanic plateaus may also represent long periods of anomalous regional magmatic activity that may have occurred either near divergent plate boundaries or within the plate. Alternatively, many oceanic plateaus are thought to be small fragments of continental blocks that have been dispersed through the processes of rifting and spreading, and have subsequently subsided below sea level to become part of the submarine terrain.

Other important features of the ocean floor are the so-called scars represented by fracture zone traces that were formed as part of the mid-oceanic ridge system, where the ridge axis was initially offset. Oceanic crusts on opposite sides of such offsets have different ages and hence they have different crustal depths. A structural-tectonic discontinuity exists across this zone of ridge axis offset known as a transform zone. Although relative plate motion does not occur outside the transform zone, the contrasting properties represented by the crustal age differences create contrasting topographic and subsurface structural discontinuities, which can sometimes be traced for great distances. Fracture zone traces define the paths of relative motion between the two plates involved. Those mapped by conventional methods of marine survey have provided fundamental information that allows rough reconstructions of the relative positions of the continents and oceans throughout the last 150–200 m.y. The study that deals with the relative motions of the plates is known as plate kinematics. *See* TRANSFORM FAULT.

Marginal seas. The sea-floor features described so far are representative of the main ocean basins and reflect their evolution mostly through processes of plate tectonics. Other, more complicated oceanic regions, typically found in the western Pacific, include a variety of small, marginal seas (back-arc basins) that were formed by the same general processes as the main ocean basins. These regions define a number of small plates whose interaction is also more or less governed by the normal tenets of plate tectonics. One difficulty in studying these small basins is that they are typified by only short-lived phases of evolution. Frequent changes in plate motions interrupt the process, creating tectonic overprints and a new suite of ocean-floor features. Furthermore, conventional methods of analyzing rock magnetism or depths of the sea floor to date the underlying crust do not work well in these small regions. The small dimensions of these seas bring into play relatively large effects of nearby tectonic boundaries and render invalid key assumptions of these analytical techniques. The number of small plates that actually behave as rigid pieces is not well known, but it is probably only 10–20 for the entire world.

[D.E.H.]

Marine microbiology   An independent discipline applying the principles and methods of general microbiology to research in marine biology and biogeochemistry. Marine microbiology focuses primarily on prokaryotic organisms, mainly bacteria. Because of their small size and easy dispersability, bacteria are virtually ubiquitous in the marine environment. Furthermore, natural populations of marine bacteria comprise a large variety of physiological types, can survive long periods of starvation, and are able to start their metabolic activity as soon as a substrate becomes available. As a result, the marine environment, similar to soil, possesses the potential of a large variety of microbial processes that degrade (heterotrophy) but also produce (autotrophy) organic matter. Considering the fact that the marine environment represents about 99% of the biosphere, marine microbial transformations are of tremendous global importance. *See* BIOSPHERE.

Heterotrophic transformations. Quantitatively, the most important role of microorganisms in the marine environment is heterotrophic decomposition and remineralization of organic matter. It is estimated that about 95% of the photosynthetically produced organic matter is recycled in the upper 300–400 m (1000–1300 ft) of water, while the remaining 5%, largely particulate matter, is further decomposed during sedimentation. Only about 1% of the total organic matter produced in surface waters arrives at the deep-sea floor in particulate form. In other words, the major source of energy and carbon for all marine heterotrophic organisms is distributed over the huge volume of pelagic water mass with an average depth of about 3800 m (2.5 mi). In this highly dilute medium, particulate organic matter is partly replenished from dissolved organic carbon by microbial growth, the so-called microbial loop.

Of the large variety of organic material decomposed by marine heterotrophic bacteria, oil and related hydrocarbons are of special interest. Other environmentally detrimental pollutants that are directly dumped or reach the ocean as the ultimate sink by land runoff are microbiologically degraded at varying rates. Techniques of molecular genetics are aimed at encoding genes of desirable enzymes into organisms for use as degraders of particular pollutants.

A specifically marine microbiological phenomenon is bacterial bioluminescence, which may function as a respiratory bypass of the electron transport chain. Free-living luminescent bacteria are distinguished from those that live in symbiotic fashion in light organelles of fishes or invertebrates.

Photoautotrophs and chemoautotrophs. The type of photosynthesis carried out by purple sulfur bacteria uses hydrogen sulfide (instead of water) as a source of electrons and thus produces sulfur, not oxygen. Photoautotrophic bacteria are therefore limited to environments where light and hydrogen sulfide occur simultaneously, mostly in lagoons and estuaries. In the presence of sufficient amounts of organic substrates, heterotrophic sulfate-reducing bacteria provide the necessary hydrogen sulfide where oxygen is depleted by decomposition processes. Anoxygenic photosynthesis is also carried out by some blue-green algae, which are now classified as cyanobacteria.

Chemoautotrophic bacteria are able to reduce inorganic carbon to organic carbon (chemosynthesis) by using the chemical energy liberated during the oxidation of inorganic compounds. Their occurrence, therefore, is not light-limited but depends on the availability of oxygen and the suitable inorganic electron source. Their role as producers of organic carbon is insignificant in comparison with that of photosynthetic producers (exempting the processes found at deep-sea hydrothermal vents). The oxidation of ammonia and nitrite to nitrate (nitrification) furnishes the chemically stable and biologically most available form of inorganic nitrogen for photosynthesis. *See* NITROGEN CYCLE.

The generation of methane and acetic acid from hydrogen and carbon dioxide stems from anaerobic bacterial chemosynthesis, and is common in anoxic marine sediments.

Marine microbial sulfur cycle. Sulfate is quantitatively the most prominent anion in seawater. Since it can be used by a number of heterotrophic bacteria as an electron acceptor in respiration following the depletion of dissolved oxygen, the resulting sulfate reduction and the further recycling of the reduced sulfur compounds make the marine environment microbiologically distinctly different from fresh water and most soils. The marine anaerobic, heterotrophic sulfate-reducing bacteria are classified in three genera; *Desulfovibrio, Desulfotomaculum*, and *Clostridium*.

The marine aerobic sulfur-oxidizing bacteria fall into two groups: the thiobacilli and the filamentous or unicellular organisms. While the former comprise a wide range from obligately to facultatively chemoautotrophic species (requiring none or some organic compounds), few of the latter have been isolated in pure culture, and chemoautotrophy has been demonstrated in only a few.

Hydrothermal vent bacteria. Two types of hydrothermal vents have been investigated: warm vents (8–25°C or 46–77°F) with flow rates of 1–2 cm (0.4–0.8 in.) per second, and hot vents (260–360°C or 500–600°F) with flow rates of 2 m (6.5 ft) per second. In their immediate vicinity, dense communities of benthic invertebrates are found with a biomass that is orders of magnitude higher than that normally found at these depths and dependent on photosynthetic food sources. This phenomenon has been explained by the bacterial primary production of organic carbon through the chemosynthetic oxidation of reduced inorganic compounds. The chemical energy required for this process is analogous to the light energy used in photosynthesis and is provided by the geothermal reduction of inorganic chemical species. The specific compounds contained in the emitted vent waters and suitable for bacterial chemosynthesis are mainly hydrogen sulfide, hydrogen, methane, and reduced iron and manganese. The extremely thermophilic microorganisms isolated from hydrothermal vents belong, with the exception of the genus *Thermotoga*, to the Archaebacteria. Of eight archaeal genera, growing within a temperature range of about 75–110°C (165–230°F), three are able to grow beyond the boiling point of water, if the necessary pressure is applied to prevent boiling. These organisms are strictly anaerobic. However, unlike mesophilic bacteria, hyperthermophilic marine isolates tolerate oxygen when cooled below their minimum growth temperature. *See* HYDROTHERMAL VENT. [H.W.J.]

Marine sediments   The accumulation of minerals and organic remains on the sea floor. Marine sediments vary widely in composition and physical characteristics as a function of water depth, distance from land, variations in sediment source, and the physical, chemical, and biological characteristics of their environments. The study of marine sediments is an important phase of oceanographic research and, together with the study of sediments and sedimentation processes on land, constitutes the subdivision of geology known as sedimentology. *See* OCEANOGRAPHY.

Traditionally, marine sediments are subdivided on the basis of their depth of deposition into littoral 0–66 ft (0–20 m), neritic 66–660 ft (20–200 m), and bathyal 660–6600 ft (200–2000 m) deposits. This division overemphasizes depth. More meaningful, although less rigorous, is a distinction between sediments mainly composed of materials derived from land, and sediments composed of biological and mineral material originating in the sea. Moreover, there are significant and general differences between deposits formed along the margins of the continents and large islands, which are influenced strongly by the nearness of land and occur mostly in fairly shallow water, and the pelagic sediments of the deep ocean far from land.

Sediments of continental margins. These include the deposits of the coastal zone, the sediments of the continental shelf, conventionally limited by a maximum depth of 330–660 ft (100–200 m), and those of the continental slope. Because of large differences in sedimentation processes, a useful distinction can be made between the coastal deposits on one hand (littoral), and the open shelf and slope sediments on the other (neritic and bathyal). Furthermore, significant differences in sediment characteristics and sedimentation patterns exist between areas receiving substantial detrital material from land, and areas where most of the sediment is organic or chemical in origin.

Coastal sediments include the deposits of deltas, lagoons, and bays, barrier islands and beaches, and the surf zone. The zone of coastal sediments is limited on the seaward side by the depth to which normal wave action can stir and transport sand, which depends on the exposure of the coast to waves and does not usually exceed 66–100 ft (20–30 m); the width of this zone is normally a few miles. The sediments in the coastal zone are usually land-derived. The material supplied by streams is sorted in the surf zone; the sand fraction is transported along the shore in the surf zone, often over long distances, while the silt and clay fractions are carried offshore into deeper water by currents. Consequently, the beaches and barrier islands are constructed by wave action mainly from material from fairly far away, although local erosion may make a contribution, while the lagoons and bays behind them receive their sediment from local rivers.

The types and patterns of distribution of the sediments are controlled by three factors and their interaction: (1) the rate of continental runoff and sediment supply; (2) the intensity and direction of marine transporting agents, such as waves, tidal currents, and wind; and (3) the rate and direction of sea level changes. The balance between these three determines the types of sediment to be found. *See* DELTA; ESTUARINE OCEANOGRAPHY.

On most continental shelves, equilibrium has not yet been fully established and the sediments reflect to a large extent the recent rise of sea level. Only on narrow shelves with active sedimentation are present environmental conditions alone responsible for the sediment distribution. Sediments of the continental shelf and slope belong to one or more of the following types: (1) biogenic (derived from organisms and consisting mostly of calcareous material); (2) authigenic (precipitated from sea water or formed by chemical replacement of other particles, for example, glauconite, salt, and phosphorite); (3) residual (locally weathered from underlying rocks); (4) relict (remnants of earlier environments of deposition, for example, deposits formed during the transgression leading to the present high sea level stand); (5) detrital (products of weathering and erosion of land, supplied by streams and coastal erosion, such as gravels, sand, silt, and clay).

Much of the fine-grained sediment transported into the sea by rivers is not permanently deposited on the self but kept in suspension by waves. This material is slowly carried across the shelf by currents and by gravity flow down its gentle slope, and is finally deposited either on the continental slope or in the deep sea. If submarine canyons occur in the area, they may intercept these clouds, or suspended material, channel them, and transport them far into the deep ocean as turbidity currents. If the canyons intersect the nearshore zone where sand is transported, they can carry this material also out into deep water over great distances. *See* CONTINENTAL MARGIN; REEF.

[T.H.V.A.]

Deep-sea sediments. In general, classifications are difficult to apply because so many deep-sea sediments are widely ranging mixtures of two or more end-member sediment types. However, they can be divided into biogenic and nonbiogenic sediments.

Biogenic sediments, those formed from the skeletal remains of various kinds of marine organisms, may be distinguished according to the composition of the skeletal material, principally either calcium carbonate or opaline silica. The most abundant contributors of calcium carbonate to the deep-sea sediments are the planktonic foraminiferids, coccolithophorids and pteropods. Organisms which extract silica from the sea water and whose hard parts eventually are added to the sediment are radiolaria, diatoms, and to a lesser degree, dilicoflagellates and sponges. The degree to which deep-sea sediments in any area are composed of one or more of these biogenic types depends on the organic productivity of the various organisms in the surface water, the degree to which the skeletal remains are redissolved by sea water while setting to the bottom, and the rate of sedimentation of other types of sediment material. Where sediments are composed largely of a single type of biogenic material, it is often referred to as an ooze, after its consistency in place on the ocean floor.

The nonbiogenic sediment constituents are principally silicate materials and, locally, certain oxides. These may be broadly divided into materials which originate on the continents and are transported to the deep sea (detrital constituents) and those which originate in place in the deep sea, either precipitating from solution (authigenic minerals) or forming from the alteration of volcanic or other materials. The coarser constituents of detrital sediments include quartz, feldspars, amphiboles, and a wide spectrum of other common rock-forming minerals. The finer-grained components also include some quartz and feldspars, but belong principally to a group of sheet-silicate minerals known as the clay minerals, the most common of which are illite, montmorillonite, kaolinite, and chlorite. The distributions of several of these clay minerals have yielded information about their origins on the continents and, in several cases, clues to their modes of transport to the oceans. [P.E.Bi.]

Maritime meteorology Those aspects of meteorology that occur over, or are influenced by, ocean areas. Maritime meteorology serves the practical needs of surface and air navigation over the oceans. Phenomena such as heavy weather, high seas, tropical storms, fog, ice accretion, sea ice, and icebergs are especially important because they seriously threaten the safety of ships and personnel. The weather and ocean conditions near the air-ocean interface are also influenced by the atmospheric planetary boundary layer, the ocean mixed layer, and ocean fronts and eddies.

To support the analysis and forecasting of many meteorological and oceanographic elements over the globe, observations are needed from a depth of roughly 1 km (0.6 mi) in the ocean to a height of 30 km (18 mi) in the atmosphere. In addition, the observations must be plentiful enough in space and time to keep track of the major features of interest, that is, tropical and extratropical weather systems in the atmosphere and fronts and eddies in the ocean. Over populated land areas, there is a fairly dense meteorological network; however, over oceans and uninhabited lands, meteorological observations are scarce and expensive to make, except over the major sea lanes and air routes. Direct observations in the ocean, especially below the sea surface, are insufficient to make a synoptic analysis of the ocean except in very limited regions. Fortunately, remotely sensed data from meteorological and oceanographic satellites are helping to fill in some of these gaps in data. Satellite data can provide useful information on the type and height of clouds, the temperature and humidity structure in the atmosphere, wind velocity at cloud level and at the sea surface, the ocean surface temperature, the height of the sea, and the location of sea ice. Although satellite-borne sensors cannot penetrate below the sea surface, the height of the sea can be used to infer useful information about the density structure of the ocean interior. *See* REMOTE SENSING.

The motion of the atmosphere and the ocean is governed by the laws of fluid dynamics and thermodynamics. These laws can be expressed in terms of mathematical equations that can be put on a computer in the form of a numerical model and used to help analyze the present state of the fluid system and to forecast its future state. This is the science of numerical prediction, and it plays a very central role in marine meteorology and physical oceanography.

The first step in numerical prediction is known as data assimilation. This is the procedure by which observations are combined with the most recent numerical prediction valid at the time that the observations are taken. This combination produces an analysis of the present state of the atmosphere and ocean that is better than can be obtained from the observations alone. Data assimilation with a numerical model increases the value of a piece of data, because it spreads the influence of the data in space and time in a dynamically consistent way.

The second step is the numerical forecast itself, in which the model is integrated forward in time to predict the state of the atmosphere and ocean at a future time. Models of the global atmosphere and world ocean, as well as regional models with higher spatial resolution covering limited geographical areas, are used for this purpose. In meteorology and oceanography the success of numerical prediction depends on collecting sufficient data to keep track of meteorological and oceanographic features of interest (including those in the earliest stages of development), having access to physically complete and accurate numerical models of the atmosphere and ocean, and having computer systems powerful enough to run the models and make timely forecasts. *See* WEATHER FORECASTING AND PREDICTION.   [R.L.Han.]

Marl   A sediment that consists of a mixture of calcium carbonate ($CaCO_3$) and any other constituents in varying proportions. Marls usually are fine-grained, so that most consist of $CaCO_3$ mixed with silt and clay. The dominant carbonate mineral in most marls is calcite ($CaCO_3$), but other carbonate minerals such as aragonite (another form of $CaCO_3$), dolomite [$Ca, Mg(CO_3)_2$], and siderite ($FeCO_3$) may be present. *See* ARAGONITE; CALCITE; CARBONATE MINERALS; DOLOMITE; SIDERITE.

In North America, the name marl may be limited to a lake deposit that is rich in $CaCO_3$, but usually the term is extended to include marine deposits. Deep-sea marls consist of mixtures of clay and the $CaCO_3$ skeletons of microscopic planktonic animals (foraminiferans) and plants (coccoliths and discoasters). Marls deposited in the deeper parts of lakes consist of fine-grained $CaCO_3$, but marls deposited in shallow water may contain $CaCO_3$ in the form of shells of mollusks and fragments of calcareous algae such as *Chara* (stonewort) mixed with fine-grained $CaCO_3$ that is precipitated on the leaves of rooted aquatic vegetation. The $CaCO_3$ from these various sources is then mixed with sand, silt, or clay brought in by streams from the surrounding drainage basin. *See* MARINE SEDIMENTS.

Although the term marl has been applied to sediments with a greatly variable content of $CaCO_3$, strictly speaking the $CaCO_3$ content should range between 30 and 70%. Because of the high $CaCO_3$ content, most marls are light to medium gray, although they can be almost any color. The high $CaCO_3$ content also tends to make dried marl earthy and crumbly.

The indurated rock equivalent of marl is marlstone. Equivalent terms range from calcareous claystone to argillaceous or impure limestone, depending on the amount of $CaCO_3$ that is present. Marlstone is common in marine sequences of all ages, and is particularly common in ancient lake sequences of Tertiary age in the western United States. For example, the famous "oil shale" of the Green River Formation of Wyoming, Colorado, and Utah is not a shale at all but a marlstone in which the dominant carbonate mineral is dolomite. *See* LIMESTONE; OIL SHALE; SEDIMENTARY ROCKS.   [W.E.De.]

Mass wasting   A generic term for downslope movement of soil and rock, primarily in response to gravitational body forces. Mass wasting is distinct from other erosive processes in which particles or fragments are carried down by the internal energy of wind, running water, or moving ice and snow.

The stability of slope-making materials is lost when their shear strength (or sometimes their tensile strength) is overcome by shear (or tensile) stresses, or when individual particles, fragments, and blocks are induced to topple or tumble. The shear and tensile strength of earth materials depends on their mineralogy and structure. Processes that generally decrease the strength of earth materials include one or more of the following: structural changes, weathering, groundwater, and meteorological changes. Stresses in slopes are increased by steepening, heightening, and external loading due to static and dynamic forces. Processes that increase stresses can be natural or result from human activities. Although other classifications exist, these movements can be conveniently classified according to their velocity into two types: creep and landsliding.

Geologically, creep is the imperceptible downslope movement at rates as slow as a fraction of millimeter per year; its cumulative effects are ubiquitously expressed in slopes as the downhill bending of bedded and foliated rock, bent tree trunks, broken retaining walls, and tilted structures. There are two varieties of geologic creep. Seasonal creep is the slow, episodic movement of the uppermost several centimeters of soil, or fractured and weathered rock. It is especially important in regions of permanently frozen ground. Rheologic creep, sometimes called continuous creep, is a time-dependent deformation at relatively constant shear stresses of masses of rock, soil, ice, and snow. This type of creep affects rock slopes down to depths of a few hundred meters, as well as the surficial layer disturbed by seasonal creep. Continuous creep is most conspicuous in weak rocks and in regions where high horizontal stresses (several tens of bars or several megapascals) are known to exist in rock masses at depths of 330 to 660 ft (100 to 200 m).

Landsliding includes all perceptible mass movements. Three types are generally recognized on the basis of the type of movement: falls, slides, and flows. Falls involve free-falling material; in slides the moving mass displaces along one or more narrow shear zones; and in flows the distribution of velocities within the moving mass resembles that of a viscous flow. *See* LANDSLIDE.

Mass wasting is an important consideration in the interaction between humans and the environment. Deforestation accelerates soil creep. Engineering activities such as damming and open-pit mining are known to increase landsliding. On the other hand, enormous natural rock avalanches have buried entire villages and claimed tens of thousands of lives. [A.S.N.]

Massif A block of the Earth's crust commonly consisting of crystalline gneisses and schists, the textural appearance of which is generally markedly different from that of the surrounding rocks. Common usage indicates that a massif has limited areal extent and considerable topographic relief. Structurally, a massif may form the core of an anticline or may be a block bounded by faults or even unconformities. In any case, during the final stages of its development a massif acts as a relatively homogeneous tectonic unit which to some extent controls the structures that surround it. Numerous complex internal structures may be present; many of these are not related to its development as a massif but are the mark of previous deformations. [P.H.O.]

Mathematical geography The branch of geography that examines human and physical activities on the Earth's surface using models and statistical analysis. The primary areas in which mathematical methods are used include the analysis of spatial patterns, the processes that are responsible for creating and modifying these patterns, and the interactions among spatially separated entities.

What sets geographic methods apart from other quantitative disciplines is geography's focus on place and relative location. Latitude and longitude provide an absolute system of recording spatial data, but geographic databases also typically contain large amounts of relative and relational data about places. Thus, geographers have devoted much effort to accounting for spatial interrelations while maintaining consistency with the assumptions of mathematical models and statistical theory. *See* GEOGRAPHY.

Spatial pattern methodologies attempt to describe the arrangement of phenomena over space. In most cases these phenomena are either point or area features, though computers now allow for advanced three-dimensional modeling as well. Point and area analyses use randomness (or lack of pattern) as a dividing point between two opposite pattern types—dispersed or clustered.

An important innovation in geographic modeling has been the development of spatial autocorrelation techniques. Unlike conventional statistics, in which many tests assume that observations are independent and unrelated, very little spatial data can truly be considered independent. Soil moisture or acidity in one location, for example, is a function of many factors, including the moisture or acidity of nearby points. Because most physical and human phenomena exhibit some form of spatial interrelationships, several statistical methods, primarily based on the Moran Index, have been developed to measure this spatial autocorrelation. Once identified, the presence and extent of spatial autocorrelation can be built into the specification of geographical models to more accurately reflect the behavior of spatial phenomena. [J.C.Co.]

Mediterranean Sea The Mediterranean Sea lies between Europe, Asia Minor, and Africa. It is completely landlocked except for the Strait of Gibraltar, the Bosporus, and the Suez Canal. The Mediterranean is conveniently divided into an eastern basin and a western basin, which are joined by the Strait of Sicily and the Strait of Messina.

The total water area of the Mediterranean is 965,900 mi^2 (2,501,000 km^2), and its average depth is 5040 ft (1536 m). The greatest depth in the western basin is 12,200 ft (3719 m), in the Tyrrhenian Sea. The eastern basin is deeper, with a greatest depth of 18,140 ft (5530 m) in the Ionian Sea about 34 mi (55 km) off the Greek mainland. The Atlantic tide disappears in the Strait of Gibraltar. The tides of the Mediterranean are predominantly semidiurnal.   [J.Ly.]

Melanterite   A mineral having composition $FeSO_4 \cdot 7H_2O$. Melanterite occurs mainly in green, fibrous or concretionary masses, or in short, monoclinic, prismatic crystals. Luster is vitreous, hardness is 2 on Mohs scale, and specific gravity is 1.90.

Melanterite is a common secondary mineral derived from oxidation and hydration of iron sulfide minerals such as pyrite and marcasite. Its occurrence is widespread. It is not an ore mineral. *See* MARCASITE.   [E.C.T.C.]

Melilite   A complete solid solution series ranging from gehlenite, $Ca_2Al_2SiO_7$, to akermanite, $Ca_2MgSi_2O_7$, often containing appreciable Na and Fe. The Mohs hardness is 5–6, and the density increases progressively from 2.94 for akermanite to 3.05 for gehlenite. The luster is vitreous to resinous, and the color is white, yellow, greenish, reddish, or brown. Akermanite-rich varieties occur in thermally metamorphosed siliceous limestones and dolomites, but more gehlenite-rich ones result if Al is present. Melilites are found instead of plagioclase in silica-deficient, feldspathoid-bearing basalts. *See* SILICATE MINERALS.   [L.Gr.]

Meromictic lake   A lake whose water is permanently stratified and therefore does not circulate completely throughout the basin at any time during the year. Normally lakes in the temperate zone mix completely during the spring and autumn when water temperatures are approximately the same from top to bottom. In meromictic lakes there are no periods of overturn or complete mixing because seasonal changes in the thermal gradient are either small or overridden by the stability of a chemical gradient, or the deeper waters are physically inaccessible to the mixing energy of the wind. Most commonly, the vertical stratification is stabilized by a chemical gradient in meromictic lakes.

The upper stratum of water in a meromictic lake is mixed by the wind and is called the mixolimnion. The bottom, denser stratum, which does not mix with the water above, is referred to as the monimolimnion. The transition layer between these strata is called the chemocline.

Of the hundreds of thousands of lakes on the Earth, only about 120 are known to be meromictic. In general, meromictic lakes in North America are restricted to: sheltered basins that are proportionally very small in relation to depth and that often contain colored water, basins in arid regions, and isolated basins in fiords. *See* FRESH-WATER ECOSYSTEM; LAKE; LIMNOLOGY.   [G.E.Li.]

Mesometeorology   That portion of meteorology comprising the knowledge of intermediate-scale atmospheric phenomena, that is, in the size range of approximately 1–1200 mi (2–2000 km) and with time periods typically, but not always, less than 1 day. Unlike the larger weather systems on synoptic scales (the scales resolved by current weather reporting station networks) which typically produce significant changes over periods of days, most mesoscale phenomena have interdiurnal periods (less than

1 day), and consequently their changes are often more startling. In addition to time and space criteria for defining mesoscale, dynamical considerations can be used.

For observing mesoscale phenomena over midlatitude land masses, the average spacing between atmospheric sounding stations is about 180–360 mi (300–600 km) and soundings are taken twice each day. Consequently, only the largest mesoscale phenomena, with wavelengths greater than about 600 mi (1000 km), are routinely observed (resolved) by this network. Information with higher time-and-space resolution is available from aircraft observations and networks of radar stations, profilers, and satellite imagery. The profiler is a ground-based hybrid observing system of vertically pointing radar and microwave radiometry. The remote-sensing platforms often show that mesoscale weather systems are distinct components of larger synoptic-scale cyclones (low-pressure systems) and anticyclones (high-pressure systems).

Although the satellite imagery and radar data provide extensive areal coverage and clearly reveal the presence of the mesoscale systems, they do not provide measurements of certain atmospheric parameters (such as temperature, moisture, and pressure) in a form in which mesoscale structures and circulations can be readily quantified and understood. Thus, while mesoscale phenomena can be "observed," they cannot be studied and predicted (in the conventional manner) as easily as synoptic-scale systems. *See* METEOROLOGICAL SATELLITES; RADAR METEOROLOGY; WEATHER FORECASTING AND PREDICTION.

Mesoanalysis is the analysis of meteorological data in a manner that reveals the presence and characteristics of mesoscale phenomena. Because sounding stations are so widely spaced, only the largest of mesoscale systems can be resolved by the free-air (sounding) data. On the other hand, the density of surface observing stations is often satisfactory for identifying mesoscale features or circulations. Probably the most common application of mesoanalysis is for forecasting convective (thunderstorm) weather systems. *See* THUNDERSTORM.

Modern weather-forecasting techniques rely heavily upon predictions made from computers. These predictions, commonly called numerical model forecasts, require as input the three-dimensional initial state of the atmosphere. Normally, this initial condition is produced from the previous forecast and the most recent observations from the network of atmospheric soundings. For many mesoscale phenomena, it has been possible to develop relationships between the large-scale (synoptic) environment and the occurrence of particular types of mesoscale events. By using these relationships, the prediction of the synoptic-scale environment by the numerical models is then used to infer the likelihood of specific mesoscale events. Using satellite, radar, and conventional surface observations, the onset of an event is readily detected and appropriate adjustments to local forecasts are implemented. These adjustments usually come in the form of very short-term forecasts, commonly called nowcasts, and typically are valid for only about 3 h. However, depending upon the particular mesoscale phenomena, longer-term (3–12 h) forecasts sometimes are possible. *See* METEOROLOGY; STORM DETECTION.

[J.Mi.F.]

Mesosphere   A layer within the Earth's atmosphere that extends from about 50 to 85 km (31 to 53 mi) above the surface. The mesosphere is predominantly characterized by its thermal structure. On average, mesospheric temperature decreases with increasing height.

Temperatures range from as high as 12°C (53°F) at the bottom of the mesosphere to as low as −133°C (−208°F) at its top. The top of the mesosphere, called the mesopause, is the coldest area of the Earth's atmosphere. Temperature increases with increasing altitude above the mesopause in the layer known as the thermosphere, which absorbs the Sun's extreme ultraviolet radiation. In the stratosphere, the atmospheric layer immediately below the mesosphere, the temperature also increases with height. The stratosphere is where ozone, which also absorbs ultraviolet radiation from the Sun, is most abundant. The transition zone between the mesosphere and the stratosphere is called the stratopause. Mesospheric temperatures are comparatively cold because very little solar radiation is absorbed in this layer. Meteorologists who predict weather conditions or study the lowest level of the Earth's atmosphere, the troposphere, often refer to the stratosphere, mesosphere, and thermosphere collectively as the upper atmosphere. However, scientists who study these layers distinguish between them; they also refer to the stratosphere and mesosphere as the middle atmosphere. See ATMOSPHERE; METEOROLOGY; STRATOSPHERE; THERMOSPHERE; TROPOSPHERE.

In the lower part of the mesosphere, the difference between the temperature at the summer and winter poles is of order 35°C (63°F). This large temperature gradient produces the north-south or meridional winds that blow from summer to winter. Temperatures in the upper mesosphere are colder in summer and warmer in winter, resulting in return meridional flow from the summer to the winter hemisphere. Although the temperature gradient in the upper part of the mesosphere remains large, additional complications result in wind speeds that are much slower than they are in the lower part of the mesosphere. Winds in the east-west or zonal direction are greatest at mesospheric middle latitudes. Zonal winds blow toward the west in summer and toward the east in winter. Like their meridional counterparts, zonal winds are comparatively strong near the bottom of the mesosphere and comparatively weak near the top. Thus, on average both temperature and wind speed decrease with increasing height in the mesosphere. See ATMOSPHERIC GENERAL CIRCULATION.

Meteors which enter the Earth's atmosphere vaporize in the upper mesosphere. These meteors contain significant amounts of metallic atoms and molecules which may ionize. Metallic ions combined with ionized water clusters make up a large part of the D-region ionosphere that is embedded in the upper mesosphere. See IONOSPHERE.

The upper mesosphere is also where iridescent blue clouds can be seen with the naked eye and photographed in twilight at high summer latitudes when the Sun lights them up in the otherwise darkening sky. These clouds are called noctilucent clouds (NLC). Noctilucent clouds are believed to be tiny ice crystals that grow on bits of meteoric dust.

Large-scale atmospheric circulation patterns transport tropospheric air containing methane and carbon dioxide from the lower atmosphere into the middle atmosphere. While carbon dioxide warms the lower atmosphere, it cools the middle and upper atmosphere by releasing heat to space. Methane breaks down and contributes to water formation when it reaches the middle atmosphere. If the air is sufficiently cold, the water can freeze and form noctilucent clouds. Temperatures must be below −129°C (−200°F) for noctilucent clouds to form. These conditions are common in the cold summer mesopause region at high latitudes.   [M.Hag.]

Mesozoic   The middle era of the three major divisions of the Phanerozoic Eon (Paleozoic, Mesozoic, and Cenozoic eras) of geologic time, encompassing an interval from 245 to 65 million years ago (Ma) based on various isotopic age dates. The Mesozoic Era is also known as the Age of the Dinosaurs and the interval of middle life. The Mesozoic Erathem (the largest recognized time-stratigraphic unit) encompasses all sedimentary rocks, body and trace fossils of organisms preserved, metamorphic rocks, and intrusive and extrusive igneous rocks formed during the Mesozoic Era. See GEOCHRONOMETRY.

The Mesozoic Era records dramatic changes in the geologic and biologic history of the Earth. At the beginning of the Mesozoic Era, all the continents were amassed into one large supercontinent (Pangaea), with both the marine and continental biotas impoverished from the mass extinction that marked the end of the Paleozoic Era. During the Mesozoic Era, many significant events were recorded in the geologic and fossil record of the Earth, including the breakup of Pangaea and the evolution of modern ocean basins by continental drift, the rise of the dinosaurs, the ascension of the angiosperms (flowering plants), and the appearance of the mammals. The end of the Mesozoic Era is marked by a major mass extinction (at the Cretaceous-Tertiary boundary) that records numerous meteorite impacts, the extinction of the dinosaurs, the rise to dominance of the mammals, and the beginning of the Cenozoic Era and the advanced life forms dominant today. *See* CONTINENTAL DRIFT; PLATE TECTONICS.

The Mesozoic Era comprises three periods of geologic time: the Triassic Period (245–208 Ma), the Jurassic Period (208–146 Ma), and the Cretaceous Period (146–65 Ma). These periods are each subdivided into epochs, formal designations of geologic time designated as Early, Middle, and Late (except for the Cretaceous, which has no middle epoch). The packages of rock themselves are subdivided into series designated Lower, Middle, and Upper (except for Cretaceous). Each epoch is subdivided into ages. Likewise, each series is subdivided into stages, which are time-stratigraphic units whose boundaries are based on unconformities (erosional surfaces), on correlations to a type section (place were rocks are first described), or preferably on changes in the biota that depict true measurable time (for example, evolutionary changes). *See* CRETACEOUS; JURASSIC; TRIASSIC; UNCONFORMITY. [S.T.H.; R.F.Du.]

Metamict state   The state of a special class of amorphous materials that were initially crystalline. W. C. Broegger first used the term *metamikte* in 1893 to describe minerals that were optically isotropic with a "glasslike" fracture but still retained well-formed crystal faces. In 1914 A. Hamburg correctly attributed the transition from the periodic, crystalline state to the aperiodic, metamict state as induced by alpha-decay damage. In minerals, this damage is the result of the decay of naturally occurring radionuclides and their daughter products in the uranium and thorium (^{238}U, ^{235}U, and ^{232}Th) decay series. A wide variety of complex oxides, silicates, and phosphates are reported as occurring in the metamict state. All of these structures can accommodate uranium and thorium.

The presence of uranium and thorium distinguishes metamict minerals from other naturally occurring amorphous materials that have not experienced this radiation-induced transformation. Lanthanide elements are also common (in some cases over 50 wt %) and water of hydration may be high (up to 70 mol %).

The radiation damage caused by the alpha-decay event is the result of two separate but simultaneous processes: (1) An alpha particle with an energy of approximately 4.5 MeV and a range of 10,000 nanometers dissipates most of its energy by ionization; however, at low velocities near the end of its track, it displaces several hundred atoms, creating Frenkel defect pairs. (2) The alpha-recoil atom with an energy of approximately 0.09 MeV and a range of 10 to 20 nm produces several thousand atomic displacements, creating tracks of disordered material. These two damaged areas are separated by thousands of unit cell distances and have different effects on the crystalline structure. Local point defects cause an increase in the distortion; therefore, there is an increase in the strain in the structure. Alpha-recoil tracks create regions of aperiodic material that at high enough alpha-decay doses (usually 10^{24} to 10^{25} alpha-decay events/m^3) overlap and finally lead to the metamict state. The former causes broadening of x-ray diffraction maxima and an increase in unit cell volume (and a decrease in the density); the latter causes a decrease in diffraction peak intensities. The radiation-induced transition from the crystalline to the metamict state occurs over a narrow range of alpha-decay dose (10^{24}–10^{25} alpha decays/m^3), which corresponds to 0.1 to 1.0 displacements per atom (dpa).

A renewed interest in the metamict state has been stimulated by concern for the long-term stability of crystalline materials (nuclear waste forms) that will serve as hosts for actinides (for example, plutonium, americium, curium, and neptunium). Various crystalline materials (phases) may appear in a single waste form; each phase may or may not suffer radiation damage. For some nuclear waste-form phases, the radiation-induced transformation to the metamict state has been stimulated by doping phases with highly radioactive plutonium-238 or curium-244.   [R.C.Ew.]

Metamorphic rocks   Preexisting rock masses in which new minerals, textures, or structures are formed at higher temperatures and greater pressures than those normally present at the Earth's surface. *See* IGNEOUS ROCKS; SEDIMENTARY ROCKS.

Two groups of metamorphic rocks may be distinguished; cataclastic rocks, formed by the operation of purely mechanical forces; and recrystallized rocks, or the metamorphic rocks properly so called, formed under the influence of metamorphic pressures and temperatures. Cataclastic rocks are mechanically sheared and crushed. They represent products of dynamometamorphism, or kinetic metamorphism. Chemical and mineralogical changes generally are negligible. The rocks are characterized by their minute mineral grain size. Each mineral grain is broken up along the edges and is surrounded by a corona of debris or strewn fragments. *See* METAMORPHISM.

Metamorphic rocks, properly so called, are recrystallized rocks. The laws of recrystallization are not the same as those of simple crystallization from a liquid, because the crystals can develop freely in a liquid, but during recrystallization the new crystals are encumbered in their growth by the old minerals. Consequently, the structures which develop in metamorphic rocks are distinctive and of great importance, because in many ways they reflect the physiochemical environment of recrystallization and thereby the genesis and history of the metamorphic rock.

The metamorphic minerals may be arranged in an idioblastic series (crystalloblastic series) in their order of decreasing force of crystallization as follows: (1) sphene, rutile, garnet, tourmaline, staurolite, kyanite; (2) epidote, zoisite; (3) pyroxene, hornblende; (4) ferrogmagnesite, dolomite, albite; (5) muscovite, biotite, chlorite; (6) calcite; (7) quartz, plagioclase; and (8) orthoclase, microcline. Crystals of any of the listed minerals tend to assume idioblastic outlines at surfaces of contact with simultaneously developed crystals of all minerals of lower position in the series.

Igneous magma at high temperature may penetrate into sedimentary rocks, it may reach the surface, or it may solidify in the form of intrusive bodies (plutons). Heat from such bodies spreads into the surrounding sediments, and because the mineral assemblages of the sediments are adjusted to low temperatures, the heating-up will result in a mineralogical and textural reconstruction known as contact metamorphism. *See* CONTACT AUREOLE; PLUTON.

The effects produced do not depend only upon the size of the intrusive. Other factors are amount of cover and the closure of the system, composition and texture of the country rock, and abundance of gaseous and hydrothermal magmatic emanations. The heat conductivity of rocks is so low that gases and vaporous emanations become chiefly responsible for transportation and transfer of heat into the country rock.

Crystalline schists, gneisses, and migmatites are typical products of regional metamorphism and mountain building. If sediments accumulate in a slowly subsiding geosynclinal basin, they are subject to down-warping and deep burial, and thus to gradually increasing temperature and pressure. They become sheared and deformed, and a general recrystallization results. However, subsidence into deeper parts of the crust is not the only reason for increasing temperature. It is not known what happens at the deeper levels of a live geosyncline, but obviously heat from the interior of the Earth is introduced regionally and locally, partly associated with magmas, partly in the form of "emanations" following certain main avenues, determined by a variety of factors. From this milieu rose the lofty mountain ranges of the world, with their altered beds of thick sediments intercalated with tuffs, lava, and intrusives, all thrown into enormous series of folds and elevated to thousands of feet. Thus were born the crystalline schists with their variants of gneisses and migmatites. *See* OROGENY.

Well-defined series of mineral facies have been singled out. Sedimentary rocks of the lowest metamorphic grade have recrystallized to give rocks of the zeolite facies. At slightly higher temperatures the greenschist facies develops—chlorite, albite, and epidote being characteristic minerals. A higher degree of metamorphism produces the epidote-amphibolite facies, and a still higher degree the true amphibolite facies in which hornblende and plagioclase mainly take the place of chlorite and epidote. Representative of the highest regional metamorphic grade is the granulite facies, in which most of the stable minerals are water-free, for example, pyroxenes and garnets. Any sedimentary unit will recrystallize according to the rules of the several mineral facies, the complete sequence of events being a progressive change of the sediment by deformation, recrystallization, and alteration in the successive stages: greenschist facies → epidote-amphibolite facies → amphibolite facies → granulite facies. *See* FACIES (GEOLOGY); GRANULITE.

The normal continental crust is entirely made up of metamorphic rocks; where thermal, mechanical, and geochemical equilibrium prevails, there are only metamorphic rocks. Border cases of this normal situation occur in the depths where ultrametamorphism brings about differential melting and local formation of magmas. When equilibrium is restored, these magmas congeal and recrystallize to (metamorphic) rocks. At the surface, weathering processes oxidize and disintegrate the rocks superficially and produce sediments as transient products. Thus the cycle is closed; petrology is without a break. All rocks that are found in the continental crust were once metamorphites. *See* AMPHIBOLITE; GREISEN; MARBLE; MIGMATITE; PHYLLITE; QUARTZITE; SCAPOLITE; SERPENTINITE; SOAPSTONE. [T.F.W.B.; R.C.N.]

Metamorphism   The alterations and transformations in preexisting rock masses effected by temperature and pressure, but excluding changes produced by weathering and sedimentation. The changes may include the production of new minerals, structures, or textures, or all three. They give a distinctive new character to the rock as a whole, but they do not involve the loss of individuality of a rock mass, such as changes brought about by fusion. Quantitatively, the metamorphic rocks, including gneisses and migmatites, are the most important group of rocks in the crust of the continents. *See* GNEISS; METAMORPHIC ROCKS; MIGMATITE.

Different kinds of metamorphism may be defined according to genetic criteria, such as the geologic processes that were assumed to have caused the metamorphism, or the physical and chemical conditions that appear to have been predominant in determining the course of metamorphism. Using these criteria three general kinds of metamorphism are noted below.

1. Dislocation, mechanical, or dynamic metamorphism is the result of pressure (or stress) along dislocations in the Earth's crust. The deformed rocks commonly show marked zones of extremely fine-grained rocks, such as mylonites, whose structures are determined by crushing and movement of the grains without important recrystallization

of old, or growth of new, minerals. This type of metamorphism is local and restricted in occurrence.

2. Contact or thermal metamorphism occurs in response to increased temperature induced by adjacent intrusions of magma. Chemical reconstitution of the rocks is due to magmatic exhalation; other conditions, such as confining pressure, exert subordinate influence.

3. Regional metamorphism, the most widespread type, is brought about by an increase in both temperature and pressure in orogenic regions, which are vast segments of the crust represented by the folded mountain ranges. Heat and pressure are mainly consequences of downwarping and deep burial. Pressure is also generated by shearing stresses accompanying the orogenic movements. See OROGENY.   [T.F.W.B./R.C.N.]

A fluid phase plays an important role during metamorphism as an agent of heat and mass transfer. The presence of a static film of fluid around mineral grains greatly facilitates chemical reactions because the fluid film speeds the movement of matter from reactant to product minerals. Flowing fluid carries substantial quantities of materials in solution that may be precipitated far from their source. Heated rocks are cooled more rapidly, and cool rocks are more quickly heated, by flowing fluids than would otherwise be possible by heat conduction. Fluids of metamorphic rocks consist primarily of water (H_2O), variable amounts of carbon dioxide (CO_2) and methane (CH_4), and minor quantities of hydrogen sulfide (H_2S), carbon monoxide (CO), hydrogen (H_2), and sulfur dioxide (SO_2).   [D.Rum.]

Metasomatism   The process by which the bulk chemical composition of a rock is changed from some previous state by the introduction of components from an external source. In contrast with metamorphism, where rocks are converted to a new set of minerals with little or no change in bulk composition, metasomatism involves the import and export of chemical components through the agency of a chemically active fluid.

Clastic rocks and mafic-to-felsic igneous rocks react in similar ways with metasomatic fluids by exchange of alkalies, alkaline earths, and hydrogen. Hydrolytic alteration dominates lower-temperature metasomatic processes. Hydrolysis involves hydrogen-ion metasomatism—exchange of hydrogen ion for potassium, sodium, and calcium.

Metasomatism of aluminum-poor rocks carbonate and ultramafic rocks generally involves addition of silica, metals, and alumina.

Metasomatism is best developed in environments characterized by extreme physical and chemical gradients and high fluid flux. At the centimeter scale, chemical contrasts along shale-limestone contacts lead to diffusive exchange of components on heating during regional or contact metamorphism. At the kilometer scale in mid-ocean rifts, island arcs, and continental-margin plutonic arcs, metasomatism results from emplacement of magma at depths of a few kilometers and infiltration of hot, saline, aqueous fluids through fractured rocks. At global scales, metasomatism accompanies mass fluxing between the crust and the mantle, such as on emplacement of mantle plumes into the lower crust or subduction of oceanic crust into the mantle. See ASTHENOSPHERE; EARTH CRUST; MAGMA; METAMORPHISM.   [M.T.E.]

Meteorological instrumentation   Devices that measure or estimate properties of the Earth's atmosphere. Meteorological instruments take many forms, from simple mercury thermometers and barometers to complex observing systems that remotely sense winds, thermodynamic properties, and chemical constituents over large volumes of the atmosphere.

Weather station measurements provide a description of conditions near the ground. In addition to the average regional conditions, these measurements also provide local information on mesoscale phenomena such as cold fronts, sea breezes, and disturbed conditions resulting from nearby thunderstorms. Traditional thermodynamic instruments are mechanical or heat-conductive devices relying on the expansion and contraction of metallic and nonmetallic liquids or solid materials as a function of temperature, pressure, and humidity. Among these are the mercury, alcohol, and bimetallic thermometers for measurement of temperature, mercury and metallic bellow (aneroid) barometers for measurement of pressure, human hair hygrometers, and wet/dry-bulb thermometers (called psychrometers) for measurement of relative humidity. Mercury barometers are simply weighing devices that balance the mass of the atmospheric column against the mass of a mercury column. On average, a column of atmosphere weighs the same as 76 cm (29.92 in.) of mercury. Psychrometers measure humidity by means of the wet-bulb depression technique. A moist thermometer is cooled by evaporation when relative humidity is less than 100%. The temperature difference between wet and dry thermometers is referred to as the wet-bulb depression, a well-known function of relative humidity at standard airflow speeds. A related method of humidity measurement is the chilled mirror technique (dewpointer). A polished surface is cooled to the temperature of water vapor saturation, at which point the cooled surface becomes fogged. Dewpoint saturation uniquely defines humidity at a known temperature and pressure. See BAROMETER; DEW POINT; PSYCHROMETER.

Precipitation measurement devices may be described as precision buckets, which measure the depth or weight of that which falls into them. These gages work best for rainfall, but they are also used in an electrically heated mode for weighing snow. Rulers are routinely used for measurement of snow depth. Time-resolved measurements of rainfall are traditionally made by counting quantum amounts (0.01 in. or 0.25 mm) of rain with a small, mechanically controlled tipping bucket located beneath a large collecting orifice. Modern rain measuring is sometimes performed along short paths via drop-induced scintillations of infrared radiation, which is emitted by a laser. When the raindrop size distribution is needed, optical-shadowing spectrometers are employed, as are momentum-measuring impact distrometers, devices that measure the number density versus the size distribution of raindrops or other hydrometeors. See SNOW GAGE.

Wind measurements are performed by anemometers, some of which use wind-driven spinning cups for wind speed determination. Vanes are used in conjunction with cups for indication of wind direction. Alternatively, three-axis propeller anemometers may be employed to provide orthogonal components of the three-dimensional wind vector. Many hybrids of these basic approaches continue to be successfully employed. Fast-response sonic anemometers employ ultrasound transmission, where the apparent propagation speed of sound is measured. The difference between this measured speed and the actual speed for a fluid at rest is the wind speed. Such measurements are made on a time scale of 0.01 s and are used to determine the fluxes of momentum, water vapor, sensible heat, and other scalars in the planetary boundary layer. See WIND MEASUREMENT.

Balloon-borne vertical profiles or soundings of temperature, humidity, and winds are central to computerized (numerical) weather prediction. Such observations are made simultaneously or synoptically worldwide on a daily basis. The temperature and humidity sensors are lightweight expendable versions of traditional surface station instruments. Balloon drift during ascent provides the wind measurement. The preferred method of tracking these rawinsondes is to use global navigation aid systems such as Omega, Loran-C, and the Global Positioning System. Parachute-borne dropsondes are often released from aircraft in data-sparse regions.

Remote sensing, principally via electromagnetic radiation, is a mainstay of modern meteorology. Such devices typically operate in the optical, infrared, millimeter-wave, microwave, and high-frequency radio regions of the electromagnetic spectrum. Passive radiometers typically operate at infrared and microwave frequencies; they are used for estimates of temperature, water vapor, cloud heights, cloud liquid water mass, and trace-gas concentrations. These observations are made from the ground, aircraft, and satellites, usually measuring naturally emitted radiation. Radarlike, active remote-sensing devices are among the most powerful tools available to meteorology. Collectively, these instruments are capable of measuring kinematic, microphysical, chemical, and thermodynamic properties of the troposphere at high spatial and temporal resolution. Active meteorological remote sensors are principally deployed on land, ships, and aircraft platforms, as well as aboard satellites. Unlike passive instruments, active remote sensors can precisely resolve the distance at which a measurement is located.

At optical frequencies, lidars measure conditions in relatively clear air. Capabilities include determining the properties of tenuous clouds; determining concentrations of aerosol, ozone, and water vapor; and measuring winds through the Doppler frequency-shift effect. Millimeter-wave radars are used to probe opaque, nonprecipitating clouds. Polarimetric and Doppler techniques reveal hydrometeor type, water mass, and air motions.

The best-known meteorological remote sensor is the microwave weather radar. In addition to measuring rainfall and tracking movement of storms, powerful and sensitive meteorological radars can measure detailed flow fields in and around storms by using hydrometeors, insects, and blobs of water vapor as reflective targets. These radars can also distinguish between rain, hail, and snow. When Doppler measurements are combined with the atmospheric equations of motion, thermodynamic perturbation fields, such as buoyancy, are revealed inside violent convective storms. At ultrahigh and very high radio frequencies, radars known as wind profilers measure the mean wind as a function of height in the clear and cloudy air. Superior to infrequent weather balloons, radio wind profiling methods permit continuous measurement of winds with regularity and high accuracy. When radio wind profilers are colocated with acoustic transponders, the speed of sound is easily measured through radar tracking of the acoustic wave. This permits the computation of atmospheric density and temperature profiles, on which the speed of sound is strongly dependent. *See* METEOROLOGY; RADAR METEOROLOGY; REMOTE SENSING.   [R.E.Car.]

Meteorological optics

Meteorological optics   The study of optical phenomena occurring in the atmosphere. Many light effects can be seen by looking skyward, and all of them, resulting from the interaction of light with the atmosphere, lie in the province of atmospheric optics or meteorological optics. The subject also includes the effect of light waves too long or too short to be detected by the human eye—light-type radiation in the infrared or ultraviolet regions of the spectrum. Light interacts with the different components of the atmosphere by a variety of physical processes, the most important being scattering, reflection, refraction, diffraction, absorption, and emission. *See* ATMOSPHERE.   [R.Gr.]

Meteorological radar

Meteorological radar   A remote-sensing device that transmits and receives microwave radiation for the purpose of detecting and measuring weather phenomena. Radar is an acronym for radio detection and ranging. Today, many types of sophisticated radars are used in meteorology, ranging from Doppler radars, which are used to determine air motions (for example, to detect tornadoes), to multiparameter radars, which provide information on the phase (ice or liquid), shape, and size of hydrometeors. Airborne Doppler radars play a vital role in meteorological research. Radars are also used to detect hail, estimate rainfall rates, probe the clear-air atmosphere to

Parabolic dish antenna of the CSU-CHILL 11-cm multiparameter Doppler radar operating at Colorado State University. The antenna is housed within a large, inflatable radome. (*P. Kennedy, Colorado State University*)

monitor wind patterns, and study the electrification processes in thunderstorms that generate lightning discharges.

Commonly used, pulsed Doppler radar operates in the microwave region, with standard wavelengths of 10, 5, and 3 cm, referred to as S-, C-, and X-band radars, respectively. The electromagnetic radiation is focused into a narrow beam by illuminating a parabolic dish reflector with microwave energy provided by the radar transmitter. S-band radars require the use of large antennas (see illustration) to generate a narrow beam of microwave energy; transmit high power (peak power of 1 megawatt); and suffer relatively little attenuation as the radar beam passes through regions of heavy rain and hail. X-band radars use much smaller antennas to achieve similar narrow beams, and are highly portable. However, X-band radars suffer from attenuation when used to probe precipitation, which significantly limits their range. Attenuation results when the radar energy is either absorbed by the raindrops or reemitted from the raindrops in directions other than toward the radar.

A pulsed Doppler radar typically emits 1000 electromagnetic pulses per second. These individual pulses are typically 1 microsecond (10^{-6} s) in duration. The Doppler radar provides information on the target's velocity, either toward or away from the radar when viewed along the radar beam. The Doppler shift, which is measured as a small difference between the frequency of the transmitted pulse and the frequency of the energy backscattered to the radar, provides a measure of the scatterer's radial motion. Scatterers in the case of meteorological radar include raindrops, ice particles (snowflakes), hailstones, and even insects, providing clear air returns. A Doppler radar also detects the amplitude of the backscattered signal, which can be used as a measure of storm intensity and as a means of estimating rainfall rates. *See* PRECIPITATION (METEOROLOGY); STORM; STORM DETECTION.

Dual-wavelength radar transmits electromagnetic energy at two wavelengths, and it also receives energy at both wavelengths. Typically, S- and X-band wavelengths are used. Dual-wavelength techniques were originally proposed to detect large hail. At S-band, hail is usually a Rayleigh target, whereas at X-band, hail is considered a Mie scatterer. Since radar energy is scattered in various directions by a Mie target, the power returned at the X-band wavelength is reduced relative to that at S-band. The presence of large hail is interpreted on the basis of the ratio of backscattered power at X-band to that at S-band. Dual-wavelength techniques are also used to estimate rainfall rates by comparing the backscattered power at a nonattenuating wavelength (S-band) to that at attenuating wavelengths (X-band). *See* HAIL.

Dual-polarization radar is able to transmit and receive both horizontally and vertically polarized radiation (the polarization of a radar beam is defined by the orientation of the electric field vector that comprises an electromagnetic wave). These radars are now used in meteorological research and have largely superseded dual-wavelength radars. A suite of multiparameter variables is being used to infer information on particle phase (ice or water), size, orientation, and shape.

Radar operating at a wavelength of 2 cm is in low Earth orbit (350 km or 210 m above the surface) and is used for mapping tropical precipitation. Understanding the amount and distribution of tropical rainfall is crucial for better understanding the Earth's climate. This space-borne radar and associated satellite was jointly developed by the United States (NASA) and Japan (National Space Development Agency); it is known as the *Tropical Rainfall Measuring Mission* (*TRMM*) satellite. Space-borne radar presents many challenging problems, including cost, size constraints, reliability issues, and temporal sampling. It is obviously impossible to continuously sample every precipitating cloud in the tropics from radar orbiting the Earth. But the *TRMM* satellite will help scientists develop a statistical distribution of rain rates within a certain area, and calculate the probability of a specific rain rate occurring. Based on this information, it will be possible to generate monthly mean rain amounts within areas of 10^5 km^2. Such information will be vital for the verification of climate models. *See* SATELLITE METEOROLOGY.

Radars used to probe the clear air, or regions devoid of clouds, are known as profilers. A profiler is essentially a Doppler radar that operates at much longer wavelengths compared to weather radar. Wavelengths of 6 m, 70 cm, and 33 cm are commonly used. In the case of a profiler, the reflected power is not only from hydrometeors but also from gradients in the index of refraction of air, which are caused by turbulent motions in the atmosphere. These turbulent motions in turn cause small fluctuations in air temperature and moisture content, which also change the index of refraction. A profiler can determine the airflow in the cloud-free atmosphere, roughly up to 10 km above the Earth's surface. Optical radars, called lidars, use lasers as the radiation source. At these short wavelengths (0.1–10 μm), the laser beam is scattered by small aerosol particles and air molecules, allowing air motions to be determined, especially in thin, high tropospheric clouds and in the Earth's boundary layer (approximately the lowest 1 km or 0.6 mi of the Earth's atmosphere). *See* HYDROMETEOROLOGY .

In the late 1990s, the weather radars used by the National Weather Service to provide warnings of impending severe weather were updated from antiquated WSR-57 and WSR-74 noncoherent radars to NEXRADs (Next Generation Weather Radars). NEXRAD (WSR-88D) radars are state-of-the-art Doppler radars operating at a wavelength of 10 cm. Using NEXRAD's Doppler capability, weather forecasters are able to warn the public sooner of approaching tornadoes and other severe weather. In severe storms, a mesocyclone first develops within the storm. The mesocyclone may be 10 km

(6 mi) or more wide, and represents a deep rotating column of air within the storm. Severe and long-lasting tornadoes are often associated with mesocyclones. The mesocyclone is readily detected by a Doppler radar such as NEXRAD. The entire continental United States is covered by the NEXRAD network, consisting of more than 100 radars. NEXRADs along the Gulf Coast, in Florida, and along the eastern seaboard provide warning information on land-falling hurricanes. About 60 of the nation's busiest airports are also equipped with Doppler radars. These radars (operating at a wavelength of 5 cm, known as Terminal Doppler Weather Radars) provide weather-related warnings to air-traffic controllers and pilots. One particularly dangerous weather condition is wind shear, which often occurs as a microburst or intense downdraft. Microbursts can severely affect the flight of landing and departing aircraft, and have been identified as a factor in many aircraft accidents. *See* RADAR METEOROLOGY; TORNADO; WEATHER FORECASTING AND PREDICTION. [S.A.Ru.]

Meteorological rocket A small rocket system used for extending observations of the atmosphere above feasible limits for balloon-borne and telemetering instruments. Synoptic exploration of the middle-atmospheric circulation (20–95 km or 12–60 mi altitude) through use of these systems (also known as rocketsondes) matured in the 1960s into a highly productive source of information on atmospheric structure and dynamics. Many thousands of small meteorological rockets have been launched in a coordinated investigation of the wind field and the temperature and ozone structures in the middle atmosphere region at 25–55 km (16–34 mi) altitude. These data produced dramatic changes in the scientific view of this region of the atmosphere, with a resulting alteration of the structural concepts into an atmospheric model that is primarily characterized by intense dynamics.

The development of the small meteorological rocket began in 1959 with a rocketsonde system known as ARCAS. The maximum altitude reached by this system was about 60 km (37 mi). A less wind-sensitive rocketsonde system, the Loki-Datasonde (PWN-8B), replaced the ARCAS system during the early 1970s. It was soon replaced with the Super Loki-Datasonde (PWN-11D). The PWN-11D rocketsonde motor burns for 2 s before separation from its inert dart and payload, which are thereby propelled to about 80 km (50 mi) altitude, where the payload is ejected. The payload consists of a small bead thermistor temperature sensor attached to a radio transmitter that sends the temperature data to a ground receiver, and a Starute parachute. The meteorological measurements are made during payload descent. At launch, the Super Loki-Datasonde has an overall weight of 31 kg (68 lb), and its length is approximately 4 m (13 ft).

Synoptic-scale circulation systems in the upper atmosphere are demonstrated by rocketsonde data to be very obviously keyed to the geographic and orographic structures of the Earth's surface. In winter, oceanic regions characteristically have poleward extensions of ridges of high pressure, and continental regions have shifty troughs of low pressure extending equatorward over them. This intimate relationship between the surface and 50 km (31 mi) is most likely the direct result of turbulent energy transport in the vertical direction, and a total understanding of the entire atmospheric system cannot be realized until these factors are incorporated. *See* ATMOSPHERIC GENERAL CIRCULATION; STRATOSPHERE; UPPER-ATMOSPHERE DYNAMICS. [W.L.W.; F.J.Sc.]

Meteorological satellites Satellites dedicated to the observation of meteorological phenomena and atmospheric or surface properties used for weather forecasting. Operational meteorological satellites provide routine observations of weather conditions as well as an ever expanding range of environmental properties, such as aerosol, dust and ash clouds from volcanic eruptions, ozone, and land vegetation cover. For this reason, they are known in the United States as operational environmental satellites. *See* METEOROLOGY; SATELLITE METEOROLOGY.

Optical imaging sensors. The first recognized application of orbital observation was the visual exploitation of cloud images associated with weather systems. Recent instruments, such as the Advanced Very High Resolution Radiometer (AVHRR) on *NOAA* satellites, use a variety of quantitative applications, such as remote sensing of sea surface temperature, monitoring changes in land vegetation, and discriminating between different kinds of clouds. There is a pervasive trend to increase the number of spectral bands in imaging sensors, from 5 channels in the current AVHRR to 36 channels in the experimental Moderate-resolution Imaging Spectroradiometer (MODIS) developed by NASA. These channels sample the full spectrum of backscattered solar radiation in the visible, near-infrared, and longwave infrared, and a good part of the emitted terrestrial radiation spectrum (thermal infrared). This multiplicity of spectral bands allows the detection of a wide variety of features, from aerosols and smoke in the atmosphere to chlorophyll in the ocean. *See* CLOUD; REMOTE SENSING; TERRESTRIAL RADIATION; WEATHER.

Except for observing polar regions, or providing meteorological support to operations in remote locations worldwide, the ideal platforms for cloud imaging are those in geosynchronous equatorial orbit, also known as geostationary orbit, at the precise altitude (35,900 km) where the orbital period matches the period of rotation of the Earth, so that the satellite appears to hover over a fixed location at the Equator. The international system of four to six geostationary meteorological satellites provides uninterrupted visibility of the global tropics and midlatitudes (up to $60°$ north and south at the satellite longitude) with the ability to monitor fast-developing weather systems that often are the most dangerous. The sharpness of cloud images (1-km picture elements in the visible), as well as the ability to scan the same scene repeatedly at time intervals as short as 5 minutes, allow for tracking the apparent motion of clouds, deducing wind velocity, and instantaneously assessing the strength of developing storms, a valuable capability in warm climate regions. *See* TROPICAL METEOROLOGY.

Imaging microwave radiometers. Also interesting is the detection of diverse atmospheric properties and surface features using multifrequency microwave radiometers with small antenna beams. Water molecule absorption of microwave radiation emitted by the ocean provides an accurate estimation of total precipitable water in the atmospheric column. Microwave radiation emitted by the relatively homogeneous moist atmosphere below is scattered in a recognizable way by waterdrops and ice particles in rain clouds, thus providing an indirect means to estimate precipitation rates. Microwave radiation contrast discriminates ice floes from open ocean water, and wet from dry soil. Microwave radiometry enables diagnostics of sea state and wind strength over the surface of the ocean, or the sea surface temperature. The principal design constraint of imaging microwave radiometers is the diffraction limit of the sensor— large apertures are desirable, but bulky antennas are a problem because mechanical scanning is needed to preserve radiometric accuracy. In order to achieve reasonably small footprints, microwave sensors are currently deployed in low Earth orbit. *See* PRECIPITATION (METEOROLOGY).

Sounding sensors. The retrieval of temperature profile and water vapor information from spectral data is a difficult and not a fully determined mathematical problem. The solutions are highly sensitive to spectral resolution and small errors in radiometric measurements. The latest Atmospheric Infra-Red Sounder (AIRS) instrument developed by NASA is expected to yield temperature profiles as accurate as balloon measurements, $1°C$ within each successive 1-km-thick layer of the lower atmosphere. *See* HYDROMETEOROLOGY.

Atmospheric sounders operate in the thermal infrared, using the absorption bands of carbon dioxide molecules ($3.7–4.9$ μm and $13–15$ μm), and in the microwave spectrum, using the 54-GHz absorption band of oxygen. Emitted radiation is much weaker and atmospheric sounders correspondingly less sensitive in the microwave region. However, nonprecipitating clouds are largely transparent to such relatively long wavelengths, thus allowing all-weather albeit less accurate observations.

Measurements of temperature and moisture are used mainly to update numerical weather prediction computations that forecast the circulation of the global atmosphere several days in advance. For this quantitative application, a delay of a few hours is immaterial but homogeneous global coverage is essential. Thus, atmospheric sounders are principally deployed on Sun-synchronous polar orbits. The parameters of these circular low Earth orbits are selected from a discrete set of altitudes (800–1000 km) and inclinations (retrograde quasi-polar) that allow the orbital plane to drift by about $1°$ of longitude per day and match the change in Sun-Earth direction. Thus, a Sun-synchronous satellite crosses the Equator at (nearly) the same local time on every successive orbit.

Active sensors. Orbital systems are now powerful enough to probe the atmospheric medium or the surface with beams of electromagnetic radiation generated in space. The first operational sensor of this kind was a coarse radar or scatterometer that measured microwave radiation backscattered by the ocean surface. Backscatter is sensitive to surface roughness and thus provides a measurement of vector wind speed over the ocean (as well as a coarse all-weather mapping of sea ice).

Various radar altimeters have been used to map the changing topography of the ocean surface (principally to reconstruct the oceanic circulation from measured altitude gradients). Higher-frequency experimental radar and lidar systems are being tested to profile the distribution and optical properties of aerosol and cloud ice particles and waterdrops. The first demonstration of a space-borne precipitation radar in being conducted with the United States-Japan Tropical Rain Measuring Mission (TRMM) launched in 1997. Rain rate can be deduced from the three-dimensional distribution of ice particles and water in rain clouds, as observed by the *TRMM* satellite. *See* METEOROLOGICAL RADAR; RADAR METEOROLOGY.

Yet another promising technology will determine wind velocity in clear air from direct measurements of the frequency shift (Doppler effect) of multiple laser pulses backscattered by aerosol and other diffusive particles. Global wind measurements will provide an invaluable enhancement of the worldwide meteorological observing network, especially at low latitudes where the wind field cannot be deduced from atmospheric pressure. [P.Mo.]

Meteorology A discipline involving the study of the atmosphere and its phenomena. Meteorology and climatology are rooted in different parent disciplines, the former in physics and the latter in physical geography. They have, in effect, become interwoven to form a single discipline known as the atmospheric sciences, which is devoted to the understanding and prediction of the evolution of planetary atmospheres and the broad range of phenomena that occur within them. The atmospheric sciences comprise a number of interrelated subdisciplines. *See* CLIMATOLOGY.

Atmospheric dynamics (or dynamic meteorology) is concerned with the analysis and interpretation of the three-dimensional, time-varying, macroscale motion field. It is a branch of fluid dynamics, specialized to deal with atmospheric motion systems on scales ranging from the dimensions of clouds up to the scale of the planet itself. The activity within dynamic meteorology that is focused on the description and interpretation of large-scale (greater than 1000 km or 600 mi) tropospheric motion systems such as extratropical cyclones has traditionally been referred to as synoptic meteorology, and that devoted to mesoscale (10–1000 km or 6–600 mi) weather systems such as severe thunderstorm complexes is referred to as mesometeorology. Both synoptic meteorology and mesometeorology are concerned with phenomena of interest in weather forecasting, the former on the day-to-day time scale and the latter on the time scale of minutes to hours. *See* DYNAMIC METEOROLOGY; MESOMETEOROLOGY.

The complementary field of atmospheric physics (or physical meteorology) is concerned with a wide range of processes that are capable of altering the physical properties and the chemical composition of air parcels as they move through the atmosphere. It may be viewed as a branch of physics or chemistry, specializing in processes that are of particular importance within planetary atmospheres. Overlapping subfields within atmospheric physics include cloud physics, which is concerned with the origins, morphology, growth, electrification, and the optical and chemical properties of the droplets within clouds; radiative transfer, which is concerned with the absorption, emission, and scattering of solar and terrestrial radiation by aerosols and radiatively active trace gases within planetary atmospheres; atmospheric chemistry, which deals with a wide range of gas-phase and heterogeneous (that is, involving aerosols or cloud droplets) chemical and photochemical reactions on space scales ranging from individual smokestacks to the global ozone layer; and boundary-layer meteorology or micrometeorology, which is concerned with the vertical transfer of water vapor and other trace constituents, as well as heat and momentum across the interface between the atmosphere and the underlying surfaces and their redistribution within the lowest kilometer of the atmosphere by motions on scales too small to resolve explicitly in global models. Aeronomy is concerned with physical processes in the upper atmosphere (above the 50-km or 30-mi level). *See* AERONOMY; ATMOSPHERIC CHEMISTRY; ATMOSPHERIC ELECTRICITY; ATMOSPHERIC GENERAL CIRCULATION; ATMOSPHERIC WAVES, UPPER SYNOPTIC; CLOUD PHYSICS; METEOROLOGICAL OPTICS; MICROMETEOROLOGY; TERRESTRIAL RADIATION.

Although atmospheric dynamics and atmospheric physics in some circumstances can be successfully pursued as separate disciplines, important problems such as the development of numerical weather prediction models and the understanding of the global climate system require a synthesis. Physical processes such as radiative transfer and the condensation of water vapor onto cloud droplets are ultimately responsible for the temperature gradients that drive atmospheric motions, and the motion field, in turn, determines the evolving, three-dimensional setting in which the physical processes take place.

The atmospheric sciences cannot be completely isolated from related disciplines. On time scales longer than a month, the evolution of the state of the atmosphere is influenced by dynamic and thermodynamic interactions with the other elements of the climate system, that is, the oceans, the cryosphere, and the terrestrial biosphere. A notable example is the El Niño-Southern Oscillation phenomenon in the equatorial Pacific Ocean, in which changes in the distribution of surface winds force anomalous ocean currents; the currents can alter the distribution of sea-surface temperature, which in turn can alter the distribution of tropical rainfall, thereby inducing further changes in the surface wind field. On a time scale of decades or longer, the cycling of chemical species such as carbon, nitrogen, and sulfur between these same global reservoirs also influences the evolution of the climate system. Human activities represent an increasingly significant atmospheric source of some of the radiatively active trace gases that play a role in regulating the temperature of the Earth. *See* BIOSPHERE; MARITIME METEOROLOGY; TROPICAL METEOROLOGY.

Throughout the atmospheric sciences, prediction is a unifying theme that sets the direction for research and technological development. Prediction on the time scale of minutes to hours is concerned with severe weather events such as tornadoes, hail, and flash floods, which are manifestations of intense mesoscale weather systems, and with urban air-pollution episodes; day-to-day prediction is usually concerned with the more ordinary weather events and changes that attend the passage of synoptic-scale weather systems such as extratropical cyclones; and seasonal prediction is concerned with regional climate anomalies such as drought or recurrent and persistent cold air outbreaks. Prediction on still longer time scales involves issues such as the impact of human activity on the temperature of the Earth, regional climate, the ozone layer, and the chemical makeup of precipitation. *See* CLIMATE MODELING; DROUGHT; HAIL; TORNADO.

The evolution of the atmospheric sciences from a largely descriptive field to a mature, quantitative physical science discipline is apparent in the development of vastly improved predictive capabilities based upon the numerical integration of specialized versions of the Navier-Stokes equations, which include sophisticated parametrizations of physical processes such as radiative transfer, latent heat release, and microscale motions. The so-called numerical weather prediction models have largely replaced the subjective and statistical prediction methods that were widely used as a basis for day-to-day weather forecasting. The state-of-the-art numerical models exhibit significant skill for forecast intervals as long as about a week.

A distinction is often made between weather prediction, which is largely restricted to the consideration of dynamic and physical processes internal to the atmosphere, and climate prediction, in which interactions between the atmosphere and other elements of the climate system are taken into account. The importance and complexity of these interactions tend to increase with the time scale of the phenomena of interest in the forecast. Weather prediction involves shorter time frames (days to weeks), in which the information contained in the initial conditions is the dominant factor in determining the evolution of the state of the atmosphere; and climate prediction involves longer time frames (seasons and longer), for boundary forcing is the dominant factor in determining the state of the atmosphere.

Atmospheric prediction has benefited greatly from major advances in remote sensing. Geostationary and polar orbiting satellites provide continuous surveillance of the global distribution of cloudiness, as viewed with both visible and infrared imagery. These images are used in positioning of features such as cyclones and fronts on synoptic charts. Cloud motion vectors derived from consecutive images provide estimates of winds in regions that have no other data. Passive infrared and microwave sensors aboard satellites also provide information on the distribution of sea-surface temperature, sea state, land-surface vegetation, snow and ice cover, as well as vertical profiles of temperature and moisture in cloud-free regions. Improved ground-based radar imagery and vertical profiling devices provide detailed coverage of convective cells and other significant mesoscale features over land areas. Increasingly sophisticated data assimilation schemes are being developed to incorporate this variety of information into numerical weather prediction models on an operational basis. *See* ATMOSPHERE; CYCLONE; FRONT; RADAR METEOROLOGY; SATELLITE METEOROLOGY; WEATHER FORECASTING AND PREDICTION. [J.M.Wa.]

Mica Any one of a group of hydrous aluminum silicate minerals with platy morphology and perfect basal (micaceous) cleavage. The most common micas are muscovite $[KAl_2(AlSi_3O_{10})(OH)_2]$, paragonite $[NaAl_2(AlSi_3O_{10})(OH)_2]$, phlogopite $[K(Mg,Fe)_3 (AlSi_3O_{10})(OH)_2]$, biotite $[K(Fe,Mg)_3(AlSi_3O_{10})(OH)_2]$, and lepidolite

$[K(Li,Al)_{2.5-3.0}(Al_{1.0-0.5}Si_{3.0-3.5}O_{10})(OH)_2]$. Calcium (Ca), barium (Ba), rubidium (Rb), and cesium (Cs) can substitute for sodium (Na) and potassium (K); manganese (Mn), chromium (Cr), and titanium (Ti) for magnesium (Mg), iron (Fe), and lithium (Li); and fluorine (F) for hydroxyl (OH). The three major species, muscovite, biotite, and phlogopite, are widely distributed rock-forming minerals, occurring as essential constituents in a variety of igneous, metamorphic, and sedimentary rocks and in many mineral deposits.

Mica is commonly found as small flakes or lamellar plates without a crystal outline. Muscovite and biotite sometimes occur in thick books, tabular prisms with a hexagonal outline that can be up to several feet across. The prominent basal cleavage is a consequence of the layered crystal structure. Thin cleavage sheets of micas, particularly muscovite and phlogopite, are flexible, elastic, tough, and translucent to transparent (isinglass). They have low electrical and thermal conductivity and high dielectric strength.

Micas have Mohs hardnesses of 2–3 and specific gravities of 2.8–3.2. Upon heating in a closed tube, they evolve water. They have a vitreous-to-pearly luster. Muscovite is colorless to pale shades of brown, green, or gray. Paragonite is colorless to pale yellow. Phlogopite is pale yellow to brown. Biotite is dark green, brown, or black. Lepidolite is most often pale lilac, but it can also be colorless, pale yellow, or pale gray. *See* BIOTITE.

Commercial mica is of two main types: sheet, and scrap or flake. Sheet muscovite, mostly from pegmatites, is used as a dielectric in capacitors and vacuum tubes in electronic equipment. Lower-quality muscovite is used as an insulator in home electrical products such as hot plates, toasters, and irons. Scrap and flake mica is ground for use in coatings on roofing materials and waterproof fabrics, and in paint, wallpaper, joint cement, plastics, cosmetics, well drilling products, and a variety of agricultural products. *See* SILICATE MINERALS.　　　　　　　　　　　　　　　　　　　　　　　　[L.Gr.; S.Sim.]

Microcline　Triclinic potassium feldspar, $KAlSi_3O_8$, that usually contains a few percent sodium feldspar (Ab = $NaAlSi_3O_8$) in solid solution. Its hardness is 6; specific gravity, 2.56; mean refractive index, 1.52; color, white (green varieties are called amazon stone or amazonite). Microcline is found in some relatively high-grade regional metamorphic rocks, but is much more common in pegmatites, granites, and related plutonic igneous rocks. In the last, it often occurs as a microcline perthite, containing exsolved low albite intergrowths. *See* FELDSPAR; PERTHITE.　　　　　　　[P.H.R.]

Micrometeorology　The study of small-scale meteorological processes associated with the interaction of the atmosphere and the Earth's surface. The lower boundary condition for the atmosphere and the upper boundary condition for the underlying soil or water are determined by interactions occurring in the lowest atmospheric layers. Momentum, heat, water vapor, various gases, and particulate matter are transported vertically by turbulence in the atmospheric boundary layer and thus establish the environment of plants and animals at the surface. These exchanges are important in supplying energy and water vapor to the atmosphere, which ultimately determine large-scale weather and climate patterns. Micrometeorology also includes the study of how air pollutants are diffused and transported within the boundary layer and the deposition of pollutants at the surface.

In many situations, atmospheric motions having time scales between 15 min and 1 h are quite weak. This represents a spectral gap that provides justification for distinguishing micrometeorology from other areas of meteorology. Micrometeorology studies phenomena with time scales shorter than the spectral gap (time scales less than 15 min to 1 h and horizontal length scales less than 2–10 km or 1–6 mi). Some phenomena studied by micrometeorology are dust devils, mirages, dew and frost formation, evaporation, and cloud streets. *See* AIR POLLUTION; ATMOSPHERE; MESOMETEOROLOGY.

Much of the early understanding of micrometeorology was obtained by studying conditions in large, flat, uniform areas that are relatively simple situations. Micrometeorologists have turned their attention to more complex situations that represent conditions over more of the Earth's surface. The micrometeorology of complex terrain, that is, hills and mountains, is important for air pollution in many towns and cities and for visibility in national parks and for locating wind generators. Another interest is the study of micrometeorology in areas of widely varied surface conditions. For instance, several different crops, dry unirrigated lands, lakes, and rivers may be located near one another. In these cases it is important to understand how the micrometeorology associated with each of these surfaces interacts to produce the overall heat and moisture fluxes of the region so that these areas can be correctly included in weather and climate forecast computer programs. *See* CLIMATOLOGY; MOUNTAIN METEOROLOGY; WEATHER FORECASTING AND PREDICTION.

Microscale meteorological features are too small to be observed by the standard national and international weather observing network. Generally, micrometeorological phenomena must be studied during specific experiments by using specially designed instruments. Instruments used to study turbulent fluxes must be able to respond to very rapid fluctuations. Special cup anemometers are made from very light materials, and high-quality bearings are used to minimize drag. Other anemometers use the speed of sound waves or measure the temperature of heated wires to measure wind. Tiny thermometers are used, so that time constants are short. Instruments are usually placed on towers or in aircraft, or are suspended in packages from tethered balloons. Instruments have been developed that can measure turbulence remotely. Wind speed and boundary-layer convection can be measured with Doppler radar, lidar devices using lasers, and sodar (sound detection and ranging) using sound waves. *See* ATMOSPHERIC ACOUSTICS; METEOROLOGICAL INSTRUMENTATION; METEOROLOGICAL RADAR; METEOROLOGY.

[S.A.St.]

Micropaleontology   A branch of paleontology dealing with the fossilized microscopic organic remains (microfossils) of the geologic past, their structure, biology, phylogenetic relations, and distribution in space and time. The study of these microfossils has become an independent scientific field largely because: (1) The size of these fossils requires special methods for collection and examination. (2) Their abundance in geologic formations makes it possible to analyze their spatial distribution and the rates of morphological changes during the course of evolution by means of statistical methods which can be used only under exceptional circumstances in the study of larger fossils. (3) Microfossils have become indispensable tools in certain branches of applied geology, especially in the exploration for oil-bearing strata, because countless numbers of these minute fossils may be obtained from small pieces of subsurface rock recovered from drill holes. (4) The diversity of microfossils, their wide spatial distribution in varied environments, and their distinctive steps in evolution and the ease of studying them have contributed to make micropaleontology one of the most actively studied branches of the earth sciences.

The material subjected to micropaleontological studies forms a spectrum from primitive plants to advanced vertebrates (see illustration). The only prerequisite for organisms to become the subject of micropaleontological studies is their possession of resistant skeletal components ensuring their preservation in sedimentary strata as fossilized remains even after biological, chemical, or mechanical processes have destroyed the organisms' soft parts.

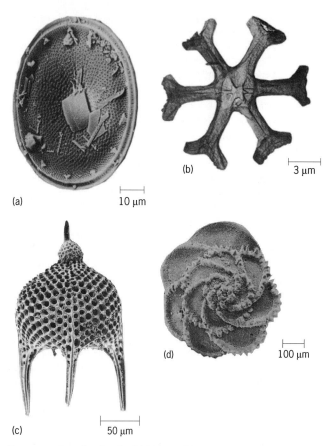

(a)   10 µm

(b)   3 µm

(c)   50 µm

(d)   100 µm

Representative microfossils. (*a*) Diatom; (*b*) asterolith, a calcareous nannoplankton; (*c*) radiolarlan; and (*d*) planktonlc foraminiferan.

Most major groups of organisms incorporate, besides organic compounds, hard resistant materials that serve for structural support or protection. The more common substances found among the microfossils are calcium carbonate, silicon dioxide (or silica), calcium phosphate in the form of the mineral apatite (typical of bones and teeth), sporonine (principal constituent of pollen and spore walls), and various complex organic compounds.   [T.S.]

Middle-atmosphere dynamics   The motion of that portion of the atmosphere that extends in altitude roughly from 10 to 100 km (6 to 60 mi). The Earth's climate is determined by a balance between incoming solar and outgoing Earth thermal radiative energy, both of which must necessarily pass through the middle atmosphere. The lower portion, the stratosphere, contains many greenhouse gases (ozone, water vapor, carbon dioxide, methane, nitrous oxide, chlorofluorocarbons, and others); and it is predicted to cool at the same time as the lower atmosphere is warmed by the greenhouse effect. The middle atmosphere is also a focus for effects of emissions from proposed commercial fleets of stratospheric aircraft. In addition to the chemistry involved, dynamical transport modeling and measurements are needed to predict the widespread transport of these important trace gases and emissions over the globe. *See* ATMOSPHERE; STRATOSPHERE; TERRESTRIAL RADIATION.

The stratosphere constitutes the lower part of the middle atmosphere, from about 10 to 50 km (6 to 30 mi) altitude; from about 50 to 80 km (30 to 48 mi) or so lies the mesosphere. The location of the base of the stratosphere (called the tropopause) depends on meteorological conditions, varying on average from about 10 km (6 mi) in altitude at the poles to about 16 km (10 mi) at the Equator.

Atmospheric gravity waves result from combined gravitational and pressure gradient forces. Typical characteristics are transverse polarization, vertical wavelengths of 0.1–0 km (0.06–6 mi), horizontal wavelengths of 1–100 km (0.6–60 mi) or more, and periods in the range of 5 min to several hours. These waves may be excited by airflow over orography (mountains) as standing lee waves, by growing clouds, and by large-scale storm complexes in the lower atmosphere; and then they propagate up into the middle atmosphere.

Planetary-scale Rossby waves are large and slowly moving waves affected by the Coriolis effect due to the Earth's rotation. Rossby waves are common at middle latitudes in winter, where they can propagate up into the middle atmosphere from excitation regions below. Near the Equator, hybrid Rossby-gravity waves and also Kelvin waves (a special class of eastward-propagating internal gravity waves having no north-south velocity component) have been observed in the middle atmosphere. *See* CORIOLIS ACCELERATION.

A variety of global-scale normal-mode oscillations are also found in the middle atmosphere, prominent examples being wave 1, westward-moving waves with periods of about 5 and 16 days, and wave 3, a westward-moving feature with a period of about 2 days. Another observed oscillation in the middle atmosphere has been found with periods in the range 1–2 months (propagating up from the lower atmosphere).

Waves resulting from fluid dynamical instabilities are also observed: medium-scale (waves 4–7) eastward-moving waves, which are actually the tops of tropospheric storm systems, can dominate the circulation of the summer Southern Hemisphere lower stratosphere. The medium-scale waves have periods of 10–20 days. *See* DYNAMIC INSTABILITY; DYNAMIC METEOROLOGY.   [J.L.Sta.]

Mid-Oceanic Ridge

Mid-Oceanic Ridge   An interconnected system of broad submarine rises totaling about 60,000 km (36,000 mi) in length, the longest mountain range system on the planet. The origin of the Mid-Oceanic Ridge is intimately connected with plate tectonics. Wherever plates move apart sufficiently far and fast for oceanic crust to form in the void between them, a branch of the Mid-Oceanic Ridge will be created. In plan view the plate boundary of the Mid-Oceanic Ridge comprises an alternation of spreading centers (or axes or accreting plate boundaries) interrupted or offset by a range of different discontinuities, the most prominent of which are transform faults. As the plates move apart, new oceanic crust is formed along the spreading axes, and the ideal transform fault zones are lines along which plates slip past each other and where oceanic crust is neither created nor destroyed. *See* PLATE TECTONICS; TRANSFORM FAULT.

Separation of plates causes the hot upper mantle to rise along the spreading axes of the Mid-Oceanic Ridge; partial melting of this rising mantle generates magmas of basaltic composition that segregate from the mantle and rise in a narrow zone at the axis of the Mid-Oceanic Ridge to form the oceanic crust. The partially molten mantle "freezes" to the sides and bottoms of the diverging plates to form the mantle lithosphere that, together with the overlying "rind" of oceanic crust, comprises the lithospheric plate. At the axis of the Mid-Oceanic Ridge the underlying column of crust and mantle is hot and thermally expanded; this thermal expansion explains why the Mid-Oceanic Ridge is a ridge. With time, a column of crust plus mantle lithosphere cools and shrinks as it moves away from the ridge axis as part of the plate. The gentle regional slopes of the Mid-Oceanic Ridge therefore represent the combined effects of sea-floor spreading (divergent plate motion) and thermal contraction. *See* LITHOSPHERE; MAGMA.

The height and thermal contraction rate of the ridge crest are relatively independent of the rate of sea-floor spreading; thus the width and regional slopes of the Mid-Oceanic Ridge depend primarily on the rate of plate separation (spreading rate). Where the plates are separating at 2 cm (0.8 in.) per year, the Mid-Oceanic Ridge has five times the regional slope but only one-fifth the width of a part of the ridge forming where the plates are separating at 10 cm (4 in.) per year. One consequence of the relation between the width and plate separation rate of the Mid-Oceanic Ridge is that more ocean water is displaced, thereby raising sea level, during times of globally faster plate motion.

Although the Mid-Oceanic Ridge exhibits little systematic depth variation along much of its length, there are several bulges (swells) of shallower sea floor. For reasons not well understood, the sea-floor bulges are more prominent along parts of the Mid-Oceanic Ridge where the rate of plate separation (spreading rate) is slower, for example, along the northern Mid-Atlantic Ridge and the Southwest Indian Ridge.

The axis of the Mid-Oceanic Ridge—that is, the active plate boundary between two separating plates—is a narrow zone only a few kilometers wide, characterized by frequent earthquakes, intermittent volcanism, and scattered clusters of hydrothermal vents where seawater, percolating downward and heated by proximity to hot rock, is expelled back into the ocean at temperatures as high as 350°C (660°F). Surrounding such vents are deposits of hydrothermal minerals rich in metals, as well as exotic animal communities including, in some vent fields, giant tubeworms and clams. *See* HYDROTHERMAL VENT; MARINE GEOLOGY; VOLCANO.   [P.R.V.]

Migmatite   Rocks originally defined as of hybrid character due to intimate mixing of older rocks (schist and gneiss) with granitic magma. Now most plutonic rocks of mixed appearance, regardless of how the granitic phase formed, are called migmatites. Commonly they appear as veined gneisses.

Several modes of origin have been proposed. (1) Granitic magma may be intercalated between thin layers of schist (lit-par-lit injection) to form a banded rock called injection gneiss. (2) The granitic magma may form in place by selective melting of the rock components. (3) The granitic layers may develop by metamorphic differentiation (redistribution of minerals in solid rock by recrystallization). (4) The granitic layers may represent selectively replaced or metasomalized portions of the rock. *See* METAMORPHISM; METASOMATISM.   [C.A.C.]

Millerite   A mineral having composition NiS and crystallizing in the hexagonal system. Millerite usually occurs in hair-like tufts and radiating groups of slender to capillary crystals. The hardness is 3–3.5 (Mohs scale) and the specific gravity is 5.5. The luster is metallic and the color pale brass yellow. Millerite is found in many localities in Europe, notably in Germany and Czech Republic. In the United States it is found with pyrrhotite at the Gap Mine, Lancaster County, Pennsylvania; with hematite at Antwerp, New York; and in geodes in limestone at Keokuk, Iowa. In Canada large cleavable masses are mined as a nickel ore in Lamotte Township, Quebec.   [C.S.Hu.]

Mineral   A naturally occurring homogeneous solid with a definite (but generally not fixed) chemical composition and a highly ordered atomic arrangement; it is usually formed by inorganic chemical processes. The fact that a mineral has a definite chemical composition implies that it can be expressed by a specific chemical formula. For example, the chemical composition of quartz (silicon dioxide) is expressed as SiO_2; its formula is definite because quartz contains no chemical elements other than silicon and oxygen. Most minerals, however, do not have such a well-defined composition. Dolomite [$CaMg(CO_3)_2$], for example, is not always a pure calcium-magnesium carbonate. It may contain considerable amounts of iron and manganese in place of magnesium, and because these amounts vary, the composition of dolomite is said to range between certain limits and is, therefore, not fixed. The description of mineral structure as a highly ordered atomic arrangement indicates that a mineral possesses an internal structural framework of atoms (or ions) arranged in a regular geometric pattern. Minerals are crystalline because this is the criterion of a crystalline solid. Under favorable conditions, crystalline materials may express their ordered internal structure by well-developed external form, also known as crystal form, or morphology. *See* APATITE; ARAGONITE.

Classification. Chemical composition has been the basis for the classification of minerals, whereby they are divided into classes depending on the dominant anion or anionic group (such as oxides, halides, sulfides, or silicates). However, it was recognized early in the development of mineralogy that chemistry alone does not adequately characterize a mineral. A full appreciation of the nature of minerals evolved only after x-rays were used to determine internal structures. It has become clear that mineral classification must be based on chemical composition and internal structure, because these together represent the essence of a mineral and determine its physical properties. Major groups in the mineral classification are native elements, sulfides, sulfosalts, oxides, hydroxides, halides, carbonates, nitrates, borates, phosphates, sulfates, tungstates, and silicates. These groups are subdivided on the basis of chemical types, and may be refined further on the basis of structural similarity. *See* BORATE MINERALS; CARBONATE MINERALS; HALOGEN MINERALS ; NATIVE ELEMENTS; NITRATE MINERALS; PHOSPHATE MINERALS; SILICATE MINERALS.

Names. Minerals may be given names on the basis of some physical property or chemical aspect; or they may be named after a locality, a public figure, a mineralogist, or almost any other subject considered appropriate. Some examples of mineral names and their derivations are as follows: albite ($NaAlSi_3O_8$) from the Latin albus (white) in allusion to its color; rhondonite ($MnSiO_3$) from the Greek rhodon (a rose) in allusion to its characteristically pink color; chromite ($FeCr_2O_4$) because of the presence of a large amount of chromium in the mineral; magnetite (Fe_3O_4) because of its magnetic properties; franklinite ($ZnFe_2O_4$) after Franklin, New Jersey, where it occurs as the dominant zinc mineral; sillimanite (Al_2SiO_2) after Professor Benjamin Silliman of Yale University.

Occurrence and formation. Minerals form in all geological environments, reflecting a wide range of chemical and physical conditions, such as temperature and pressure. The main categories of mineral formation are (1) igneous, or magmatic, in which minerals form as crystallization products from a melt; (2) sedimentary, in which minerals are the result of the processes of weathering, erosion, and sedimentation; (3) metamorphic, in which new minerals form at the expense of earlier ones, as a result of changing (usually increasing) temperatures, pressures, or both, on some earlier rock type; metamorphic minerals are the result of new mineral growth in the solid rock, without the intervention of a melt (as in igneous processes); and (4) hydrothermal, in which minerals are chemically precipitated from hot solutions. The first three processes generally lead to rock types in which different mineral grains are closely intergrown in an interlocking fabric. Hydrothermal solutions, and even solutions at very low temperatures such as ground water, tend to follow fracture zones in rocks that may provide open spaces for chemical precipitation of minerals from solution. It is from such open spaces, partially filled by minerals deposited from solutions, that most of the spectacular mineral specimens, seen in mineral museums worldwide, have been collected. If a mineral in the process of its growth (as a result of precipitation) is allowed to grow in a free space, it will commonly exhibit a well-developed crystal form, which adds to a specimen's esthetic beauty. Similarly, geodes, which are rounded, hollow, or partially hollow bodies, commonly found in limestones, may contain very well-formed crystals lining the central cavity. Geodes are the result of mineral deposition from solutions such as ground water. *See* GEODE.

Economic importance. Society today depends on minerals in countless ways—from the construction of skyscrapers to the manufacture of televisions. A few minerals such as talc, asbestos, and sulfur are used essentially as they come from the ground, but most are first processed to obtain a usable material such as bricks, glass, cement, plaster, and a score of metals ranging from iron to gold. Both forests and farms are dependent upon soils, which are composed chiefly of minerals. *See* ASBESTOS; SOIL; TALC.

Metallic ores and industrial minerals are mined on every continent, wherever specific minerals are sufficiently concentrated to be extracted economically. The location of minable metal and industrial mineral deposits, and the study of the origin, size, and ore grade of these deposits are the domain of economic geologists, but a knowledge of the chemistry, occurrence, and physical properties of minerals is basic to pursuits in economic geology. *See* MINERALOGY.   [C.K.]

Mineralogy   The science which concerns the study of natural inorganic substances, whether of terrestrial or extraterrestrial origin, called minerals. Mineralogy is a science that cannot be easily defined. It is most properly a branch of inorganic

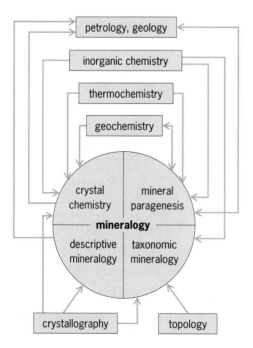

Fig. 1. Diagram showing the transmission of Information between mineralogy and some other sciences. Arrows Imply the direction of Information.

chemistry, but the discipline concentrates on the origin, description, and classification of minerals. *See* MINERAL.

Thus, four main categories may be considered: crystal chemistry (composition and atomic arrangement of minerals); paragenetic mineralogy (the study of mineral association and occurrence both in natural and synthetic systems); descriptive mineralogy (the study of the physical properties of minerals and the means for their identification); and taxonomic mineralogy (mineral classification, systematization, and nomenclature).

Crystal chemistry. This is the most vital aspect of mineralogy because it is the basis for the other studies. The fields of mineralogy, crystallography, inorganic chemistry, geochemistry, petrology, and geology are connected in the domain of crystal chemistry (Fig. 1).

A particular mineral, called a mineral species, is defined on the basis of a specified chemical composition and a specified crystal structure (atomic arrangement). These two criteria provide almost sufficient knowledge for characterization of a mineral since in principle all other properties can be derived from them.

Crystalline substances, that is, periodic arrangements of matter in three dimensions, can be divided into six crystal systems: triclinic, monoclinic, orthorhombic, tetragonal, hexagonal (including the trigonal and rhombohedral subdivisions), and cubic. These crystal systems can be considered as structure cells, each of which contains a certain integral number of atoms of a substance. Figure 2 depicts these systems and offers the criteria for distinguishing them.

Any crystalline substance has a certain integral number of its essential chemical formula units loosely called molecules) in its structure cell. Thus, each unique atomic position in the structural unit can be occupied by a particular kind (or kinds) of atom(s). Consider the ideal cell formula $(A_a)(B_b)\dots(P_p)$. Each of the parentheses specifies a unique atomic position. The capital letters specify the element present and the small subscripts the number of times it occurs in the cell. If the small subscripts have a factor in common, it is factored out and what remains is the formula unit. The ideal formula is further defined in terms of the atomic element which occurs in excess of 50 mole % within each of the parentheses. The structure type, along with the ideal formula unit, defines a mineral species.

This strict definition of a species is required since a particular mineral may have a range of compositions. The range of compositions is called a series. The ideal limiting compositions are called end members and each of the end members has a specific name.

Paragenetic mineralogy. Paragenetic mineralogy is the study of mineral paragenesis, or the association and order of crystallization of minerals. The problem may concern mineral association within a single hand specimen or may embrace a much larger region, such as an entire ore body, in which case many representative specimens are judiciously collected. This study usually accompanies the analysis of the general geological structures within and around the ore body, such as the bedding, folding, and faulting. Included among the important aspects of paragenetic mineralogy are ore mineralogy, the mineralogy of a sequence of phases crystallized from a parent magma, the sequence of minerals crystallized in a vein, and so forth. *See* GEOLOGY; PETROLOGY.

Mineral paragenesis is usually considered in relative time and the absolute difference in time between the oldest and youngest minerals is often not known. Absolute age differences can be obtained in some instances, for example, by lead isotope age dating of a sequence of crystallized lead-bearing minerals.

Descriptive mineralogy. Mineral recognition directly by the senses is very subjective and requires considerable experience. Gross features of a mineral such as color, form, hardness, and specific gravity are important criteria for identification in the field

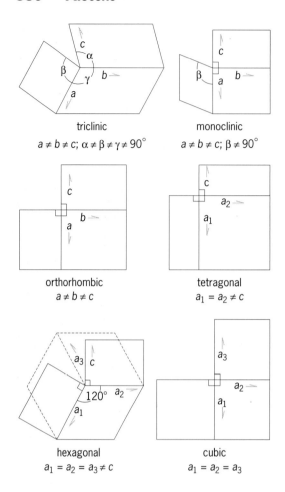

triclinic
$a \neq b \neq c; \alpha \neq \beta \neq \gamma \neq 90°$

monoclinic
$a \neq b \neq c; \beta \neq 90°$

orthorhombic
$a \neq b \neq c$

tetragonal
$a_1 = a_2 \neq c$

hexagonal
$a_1 = a_2 = a_3 \neq c$

cubic
$a_1 = a_2 = a_3$

Fig. 2. The cell shapes of the six crystal systems shown as their principal projections.

where a well-equipped laboratory is usually not available. More objective criteria such as the optical properties and x-ray powder diffraction spectra of a mineral require specialized equipment, but the results are usually certain since these data are known for most mineral species and are extensively tabulated. Older methods such as fusibility, flame tests, and blowpipe analysis have been largely abandoned.

Taxonomic mineralogy. There are approximately 3000 distinct mineral species known to science. About 60 new mineral species are discovered each year. For a new species to be properly defined, the chemical analysis, structure cell and space group, crystal morphology, powder pattern, optical data, all physical properties, and paragenesis must be given as completely as possible.   [P.B.M.]

Miocene   The second subdivision of the Tertiary Period (Eocene, Miocene, and Pliocene) by Charles Lyell in 1833; the fourth in a more modern sevenfold subdivision (epochs) of the Cenozoic Era; and the first epoch of the Neogene Period (which includes in successive order the Miocene, Pliocene, Pleistocene, and Holocene). The Miocene represents the interval of time from the end of the Oligocene to the beginning of the Pliocene and the rocks (series) formed during this epoch. *See* CENOZOIC; HOLOCENE; OLIGOCENE; PLEISTOCENE; PLIOCENE; TERTIARY.

The Miocene spans the time interval between 23.8 and 5.32 million years ago (Ma) based on integrated astronomical and radioisotopic dating. The Miocene/Pliocene boundary is located in Sicily, just above a major unconformity separating the youngest late Miocene (Messinian) deposits (of the Great Terminal Miocene Salinity Crisis) and the overlying white chalks of the Zanclean. *See* UNCONFORMITY.

Major orogenic and volcanic events characterize the Miocene. Plate-tectonic motions, originating in the Mesozoic, resulted in the gradual dismemberment of the Tethyan Ocean and the upthrusting of the Alpine-Himalayan orogenic belt in three major phases: the late Eocene (about 40 Ma) and the early (21–17 Ma) and mid-late Miocene (10–7 Ma). Along the eastern margins of the Pacific Ocean, the ocean crust was subducted under the North and South American continents, giving rise to major orogenic movements stretching from the Aleutians to Tierra del Fuego. The Andes range was thrust up during the later part of the Miocene. The Pacific Coast developed as a result of westward drift of North America over, and partial consumption by, the Farallon plate and collision with the Farallon Ridge. Only two relatively minor plates remain as remnants of the Farallon plate: the Juan de Fuca and Cocos plates between Mexico and Alaska. The plate margin was bounded by transform faults rather than a subduction zone, and northwestern propagation of a major transform fault issuing from the Cocos plate formed the Gulf of California in the late Miocene, and its continued extension northward is familiar to residents of the west coast as the San Andreas Fault System. The latter was responsible for the formation of many of the off- and onshore basins of southern California, some of which contain prolific petroleum resources. Subduction of the Pacific plate at the Middle America Trench during the late Paleogene and Neogene resulted in arc magmatism and eventual uplift of the Central American Isthmus into a series of archipelagos in the late Miocene (about 7 Ma) and eventual fusion into a continuous land bridge in the early Pliocene (about 3 Ma) that resulted in the separation of the Atlantic and Pacific oceans and concomitant disruption in marine faunal communities as well as transcontinental migration of vertebrate animals in the Great American Faunal Interchange. *See* OROGENY; PLATE TECTONICS; SUBDUCTION ZONES; TRANSFORM FAULT.

Ocean circulation essentially assumed its modern form during the Miocene as enhanced refrigeration in the form of growth of the Antarctic Ice Sheet plunged the Earth inexorably deeper into an icehouse state, although there were some details that were completed during the succeeding Pliocene and Pleistocene epochs. An ice cap has been present on Antarctica, at least intermittently, since at least the early Oligocene (about 34 Ma). The opening of the Drake Passage between South America and Antarctica took place during the latest Oligocene–early Miocene (about 25–23 Ma), allowing the unhindered circulation of ocean currents around the Antarctic continent. The development of the Circum-Antarctic Current thermally isolated high southern latitude waters and the continent of Antarctica from the warmer, low-latitude waters and resulted in the replacement of calcareous oozes (comprising planktonic foraminifera and calcareous nannoplankton) by biosiliceous oozes (diatoms and radiolarians). *See* GLACIAL EPOCH.

During the Miocene, life assumed much of its modern aspect. The spread of grasses and weeds throughout this epoch, but particularly in the late Miocene, and concomitant reduction in and thinning of forests reflected the global Neogene cooling as the Earth entered deeper into an ice house state. In this environment, snakes, frogs, and murids (rats, mice) expanded in diversity and habitat; songbirds reflect the expansion of seed-bearing herbs and, like frogs, the concomitant diversification of insects, many of which are found entombed in middle Miocene amber from the Dominican Republic. Grazing animals (elephants, rodents, horses, camelids, and rhinos, for example) developed high crowned teeth to resist significant wear caused by silicon fragments in the developing grasses. Some animals assumed gigantic proportions such as *Baluchitherium*, a Eurasian rhino that stood 16 ft (5 m) at the shoulders, and the tallest camel known, a giraffelike form that was over 12 ft (3.5 m) tall.

The relatively free interchange between Eurasia and Africa between 18 and 12 Ma appears to have come to an end in the late Miocene, after which (about 8 Ma) the hominoids of Eurasia and Africa appear to have followed separate and independent lines of evolution: the pongids in Asia, and the panids and hominoids leading eventually to the true hominids in (predominantly East) Africa. This scenario has been linked, in turn, with the development of the East African Rift (and its northward extension into the Red Sea and Gulf of Suez), which would have served as a geographic barrier allowing independent evolution toward forest (panid) and savannah (hominid) adapted forms. With the late Miocene change in climate (7–5 Ma) to cooler, drier conditions and the spread of open savannah and grasslands, monkeys came to dominate the African forest at the expense of dryomorphs. There is a gap in the terrestrial fossil record during this interval of time, and it is only in the early Pliocene (about 4 Ma) that the story of human evolution resumes with the discovery of the earliest true hominids (australopithecines) in East Africa, about 1 million years older than the australopithecine footprints of Laeotili and the skeletons of Lucy and other australopithecines at Hadar in Ethiopia at about 3 million years.

In the marine realm, major radiation of mammals including walruses, seals, sea lions, and whales occurred during the early Miocene. On the sea floor, large bivalve mollusks of the scallop family thrived in the early Miocene, and a distinct horizon of large pectenids occurs in lower Miocene rocks of Europe and North America and in corresponding levels in the deposits of the Paratethyan Sea in east-central Europe and at least as far to the east as Iran, attesting to an interval of global climatic amelioration. Among the protozoa, planktonic foraminifera experienced a major radiation in the early and middle Miocene following the drastic reduction in diversity during the middle and late Eocene, some 15–20 million years earlier. Mangroves and coral reefs flourished in a circumequatorial belt spanning the Indo-Pacific and Caribbean regions, but the latter were eliminated from the Mediterranean during the terminal Miocene Salinity Crisis, never to return with early Pliocene flushing from the Atlantic. *See* REEF.   [W.A.Ber.]

Mirage   A name for a variety of unusual images of distant objects seen as a result of the bending of light rays in the atmosphere during abnormal vertical distribution of air density. If the air closer to the ground is much warmer than the air above, the rays are bent in such a way that they enter the observer's eyes along a line lower than the direct line of sight. The object is then seen below the horizon, the inferior mirage. If the air closer to the ground is much colder than the air above, the rays are bent in the opposite direction, arriving at the observer's eyes above the line of sight; the object then seems to be elevated or floating in the air, the superior mirage. Mirages can be seen most frequently along an overheated highway surface; the inferior mirage of the sky gives the impression of water reflection over a wet pavement, which disappears upon a closer viewing.   [Z.S.]

Mississippian   The fifth period of the Paleozoic Era. The Mississippian System (referring to rocks) or Period (referring to time during which these rocks were deposited) is employed in North America as the lower (or older) subdivision of the Carboniferous, as used on other continents. The name Mississippian is derived from rock exposures on the banks of the Mississippi River between Illinois and Missouri.

The limits of the Mississippian Period are radiometrically dated. Its start (following the Devonian) is dated as 345–360 million years before the present (Ma). Its end (at the start of the next younger North American period, the Pennsylvanian) is dated as 320–325 Ma. The duration of the Mississippian is generally accepted as 40 million years (m.y.). Biochronologic dating within the Mississippian, based on a combination of conodont (phosphatic teeth of eel- or hagfish-like primitive fish), calcareous foraminiferan, and coral zones permits a relative time resolution to within about 1 m.y. *See* Carboniferous; Pennsylvanian.

The Mississippian is divided, in ascending order, into the Lower Mississippian, comprising the Kinderhookian and Osagean, and the Upper Mississippian, comprising the Meramecian and Chesterian. Kinderhookian, Osagean, Meramecian, and Chesterian are used in North America as series (for rocks) and as stages (for time). In Illinois, Valmeyeran is commonly used for the Osagean and Meramecian combined.

During much of Mississippian time, the central North American craton (stable part of the continent) was the site of an extensive marine carbonate platform on which mainly limestones and some dolostones and evaporites were deposited. This platform extended either from the present Appalachian Mountains or Mississippi Valley to the present Great Basin. The craton was covered by shallow, warm, tropical epicontinental seas that had maximum depths of only about 60 m (200 ft) at the shelf edge.

Mississippian North America was subjected to a number of sea-level rises, associated with transgressions, and sea-level falls, associated with regressions. The two major rises, which were probably eustatic (referring to worldwide change of sea level), took place during the late Kinderhookian and at the start of the middle Osagean zone. These sea-level rises caused progradation or seaward migration of carbonate platforms as organisms produced buildups that maintained their niches relative to sea level. The rises also caused stratification of the water column in deeper basins, so that bottom conditions became deficient to lacking in oxygen. *See* Anoxic zones; Basin.

Crinoids were probably the most abundant biota in Mississippian seas, but are only uncommonly used for correlation because most specific identifications require study of calyxes, which generally disarticulated after death. Corals are probably the most widely preserved Mississippian megafossil group, and they are one of the most useful biochronologic tools for constructing biostratigraphic zones for carbonate-platform rocks. The earliest Mississippian is characterized mainly by solitary corals and tubular colonial corals known as *Syringopora*, which had survived the late Frasnian mass extinction. Other colonial corals began a gradual return late in the Kinderhookian. The Late Mississippian was a heyday for reef-building colonial corals and large solitary corals. *See* Reef.

Forests flourished during the Mississippian, and tree trunks, plant stems, roots, and spores occur commonly in terrestrial and peritidal rocks, particularly in coal beds. *Lepidodendron* trunks and *Stigmaria* roots are among the best-known plant remains. *See* Geologic time scale; Paleontology; Paleozoic.   [C.A.Sa.]

Moho (Mohorovičić discontinuity)

Moho (Mohorovičić discontinuity)   The level in the Earth where the velocity of sonic waves first increases rapidly or discontinuously to a value between 4.7 and 5.3 mi/s (7.6 and 8.6 km/s). A. Mohorovičić discovered this boundary while investigating seismograms of the Zagreb (in former Yugoslavia) earthquake of October 8, 1909. He recognized that low-velocity waves traveling directly from the earthquake source were overtaken at large distances by refracted waves traveling through the deeper, high-velocity layer. Modern determinations of the depth and nature of the Moho are commonly made in seismic refraction studies that use artificial seismic sources, such as explosions, rather than earthquakes. This method allows identification of the wave traveling in the high-velocity medium (P_n) and a wide-angle reflection (P_mP) from the boundary. The Moho is generally assumed to mark the boundary between the crust and mantle, although this need not always be the case. *See* EARTH INTERIOR.   [D.M.Fo.]

While the definition of Moho in the continents is based solely on seismic refraction observations, the term has been utilized to describe a range of observations pertaining to the oceanic environment. In the strictest sense, Moho refers to the depth where material velocities, as determined from seismic refraction methods, exceed 5 mi/s (8 km/s). From geological studies of ocean crust and ophiolites (ancient oceanic sections subsequently emplaced on continents), oceanic crust is understood to originate from partial melting of mantle that upwells and decompresses in response to sea-floor spreading. The eruption and intrusion of this melt, and the formation of Moho, occurs in a very narrow zone at the axis of sea-floor spreading. Different mantle conditions can lead to different volumes of melting, accounting for most of the observed variations in crustal thickness (depth to the Moho). Very slow spreading centers, while rare, are expected to form somewhat thinner crust, largely reflecting the magma-starred nature of these areas. The Moho is a proxy for the transition from crustal materials (for example, basalts, diabase dikes, or gabbros) formed by mafic melts extracted from the mantle to the ultramafic residual mantle (for example, peridotite) that has remained at depth. *See* BASALT; DOLERITE; GABBRO; OPHIOLITE; PLATE TECTONICS.   [C.Z.M.]

Molybdenite   A mineral having composition MoS_2. Molybdenite is the chief ore of molybdenum. It crystallizes in the hexagonal system, but crystals are rare and when found are hexagonal plates. It is commonly in scales or foliated masses. The mineral has a greasy feel. The hardness is 1.5 (Mohs scale) and the specific gravity is 4.7. The luster is metallic and the color lead gray. Molybdenite and graphite have long been confused because of their nearly identical physical properties. They can be distinguished by the streak left on glazed paper, black for graphite and green for molybdenite. Molybdenite has been used as a lubricant.

Molybdenite occurs in various places in Norway, Sweden, Australia, England, China, and Mexico. In the United States, molybdenite is found in small amounts at many localities but the most important occurrence is at Climax, Colorado.   [C.S.Hu.]

Monazite   A rare mineral that incorporates the light rare-earth elements (lanthanum, cerium, praseodymium, neodymium, promethium, samarium, europium, gadolinium) and also yttrium. Monazite has a general formula of $(La,Ce, Nd)PO_4$, but Pr, Sm, Eu, Gd, and Y substitute for La, Ce, and Nd in solid solution in minor amounts. The dominant rare-earth element in a particular monazite is denoted by the atomic suffix, such as monazite-(Ce) in which cerium exists in amounts greater than other rare-earth atoms. Monazite-(Ce), monazite-(La), and monazite-(Nd) are officially recognized by the International Mineralogical Association. *See* MINERAL.

The atomic arrangement of monazite is formed of a packing arrangement of (PO_4) tetrahedra and distorted (REO_9) polyhedra, where RE = the rare-earth elements in the particular monazite mineral. The arrangement is formed of chains of alternating phosphate tetrahedra and RE polyhedra, parallel to the *c* axis. Monazite is similar in structure and chemistry to the tetragonal mineral xenotime, $Y(PO_4)$, that selectively incorporates the heavy rare-earth elements. *See* PHOSPHATE MINERALS.

Monazite is variably green, yellow, brown, or red-brown, and rarely occurs in crystals large enough to discern with the unaided eye. Mohs hardness is 5–5.5, and the specific gravity is 4.6–5.5, varying with substitution of different elements.

Monazite is one of the main ore minerals for the rare-earth elements that are used in the manufacture of television and computer screens, fluorescent light bulbs, and highly efficient batteries, among other industrial applications.   [J.M.Hu.; J.Ra.]

Monsoon meteorology   The study of the structure and behavior of the atmosphere in those areas of the world that have monsoon climates. In lay terminology, monsoon connotes the rains of the wet summer season that follows the dry winter. However, for mariners, the term monsoon has come to mean the seasonal wind reversals.

In true monsoon climates, both the wet summer season that follows the dry winter and the seasonal wind reversals should occur. Winds from cooler oceans blow toward heated continents in summer, bringing warm, unsettled, moisture-laden air and the season of rains, the summer monsoon. In winter, winds from the cold heartlands of the continents blow toward the oceans, bringing dry, cool, and sunny weather, the winter monsoon.

Based on these criteria, monsoon climates of the world include almost all of the Eastern Hemisphere tropics and subtropics, which is about 25% of the surface area of the Earth. The areas of maximum seasonal precipitation straddle or are adjacent to the Equator. Two of the world's areas of maximum precipitation (heavy rainfall) are within the domain of the monsoons: the central and south African region, and the larger south Asia-Australia region. The monsoon surface winds emanate from the cold continents of the winter hemisphere, cross the Equator, and flow toward and over the hot summer-hemisphere landmasses.

India presents the classic example of a monsoon climate region, with an annual cycle that brings southwesterly winds and heavy rains in summer (the Indian southwest monsoon) and northeasterly winds and dry weather in winter (the northeast winter monsoon).

Like all weather systems on Earth, monsoons derive their primary source of energy from the Sun. About 30% of the Sun's energy that enters the top of the atmosphere is transmitted back to space by cloud and surface reflections. Little of the remainder is absorbed directly by the clear atmosphere; it is absorbed at the Earth's surface according to a seasonal cycle. The opposition of seasons in the Northern and Southern hemispheres leads to a slow movement of surface air across the Equator from winter hemisphere to summer hemisphere, forced by horizontal pressure gradients and vertical buoyancy forces resulting from differential seasonal heating. Such a seasonally reversing rhythm is most pronounced in the monsoon regions. *See* ATMOSPHERE; INSOLATION; METEOROLOGY; TROPICAL METEOROLOGY.   [J.S.Fe.]

Montmorillonite   A group name for all clay minerals with an expanding structure, except vermiculite, and also a specific mineral name for the high alumina end member of the group. *See* CLAY MINERALS; VERMICULITE.

Montmorillonite clays have wide commercial use. The high colloidal, plastic, and binding properties make them especially in demand for bonding molding sands and for oil-well drilling muds. They are also widely used to decolorize oils and as a source of petroleum cracking catalysts. *See* CLAY.

Members of the montmorillonite group of clay minerals vary greatly in their modes of formation. Alkaline conditions and the presence of magnesium particularly favor the formation of these minerals. Several important modes of occurrence are in soils, in bentonites, in mineral veins, in marine shales, and as alteration products of other minerals. Recent sediments have a fairly high montmorillonite content. *See* BENTONITE; MARINE SEDIMENTS.   [F.M.W.; R.E.Gr.)

Moraine   An accumulation of glacial debris, usually till, with distinct surface expression related to some former ice front position. End moraine, the most common form, is an uneven ridge of till built in front of or around the terminus of a glacier

End moraine in Pennsylvania
(*U.S. Geological Survey*).

margin, and reflects some degree of equilibrium between rate of ice motion, supply of rock debris at the ice front, temperature of the glacier base, and shape and resistance of underlying bedrock (*see* illustration).

If an end moraine represents the farthest forward position a glacier ever moved, it is a terminal moraine. It demonstrates a steady-state condition for a period of time within the ice body where constant forward motion is balanced by frontal melting; and a continual supply of debris, as on an endless conveyor belt, is brought forward to the glacier terminus. If the ice front then melts farther back than it moves forward, till is spread unevenly over the land as ground moraine. If a retreatal position of steady-state equilibrium is maintained again, a recessional moraine may be constructed.

Drumlins, produced by glacier streamlining of ground moraine, are probably the best-known moraine forms.   [S.E.Wh.]

Mountain   A feature of the Earth's surface that rises high above its base and has generally steep slopes and a relatively small summit area. Commonly the features designated as mountains have local heights measurable in thousands of feet, lesser features of the same type being called hills, but there are many exceptions. *See* HILL AND MOUNTAIN TERRAIN.

Mountains rarely occur as isolated individuals. Instead they are usually found in roughly circular groups or massifs, such as the Olympic Mountains of northwestern Washington, or in elongated ranges, like the Sierra Nevada of California. An array of linked ranges and groups, such as the Rocky Mountains, the Alps, or the Himalayas, is a mountain system. North America, South America, and Eurasia possess extensive cordilleran belts, within which the bulk of their higher mountains occur. *See* CORDILLERAN BELT; MASSIF; MOUNTAIN SYSTEMS.

As a rule, mountains represent portions of the Earth's crust that have been raised above their surroundings by upwarping, folding, or buckling, and have been deeply carved by streams or glaciers into their present surface form. Some individual peaks and massifs have been constructed upon the surface by outpourings of lava or eruptions of volcanic ash. *See* OROGENY.   [E.H.Ha.]

Mountain meteorology   The effects of mountains on the atmosphere, ranging over all scales of motion, including very small (such as turbulence), local (for instance, cloud formations over individual peaks or ridges), and global (such as the monsoons of Asia and North America).

The most readily perceived effects of a mountain, or even of a hill, are related to the blocking of air flow. When there is sufficient wind, the air either goes around the obstacle or over it, causing waves in the flow similar to those in a river washing over a boulder. Since ascending air cools by adiabatic expansion, the saturation point of water vapor may be reached in such waves as they form over an obstacle, and a cloud then forms in the ascending branch of the wave motion. Such a cloud dissipates in the descending branch where adiabatic warming takes place. The shapes and amplitudes of these lee waves (they form over and to the lee of mountains) depend not only on the thermal stability and on the vertical wind shear in the overlying atmosphere but also on the shape of the underlying terrain. *See* CLOUD; CLOUD PHYSICS.

On a grander scale, mountain ranges, such as the Sierras of North and South America, place an obstacle in the path of the westerly winds (that is, winds from the west), which generally prevail in middle latitudes. Such a blockage tends to generate a high-pressure region upwind from the mountains (this may be viewed as air piling up as it prepares to jump the hurdle), and a low-pressure area downwind. Thus, there is a stronger push against the mountains on the high-pressure western side than on the low-pressure eastern side. The net effect is the slowing down of the atmospheric flow (mountain torque).

Less subtle than mountain torque effects are the large-scale meanders that develop in the global flow patterns once they have been perturbed, mainly by the North and South American Andes and by the Plateau of Tibet and its Himalayan mountain ranges. These meanders in the large-scale flow are known as planetary waves. They appear prominently in the pressure patterns of hemispheric or global weather maps. *See* WEATHER MAP.

The major monsoon circulations interact with the global circulation, shaped in part by sea-surface temperature anomalies in the equatorial Pacific. The various aspects of mountain meteorology, therefore, have to be viewed within the larger picture. There is a continuous interaction between the weather effects on all space and time scales generated by the mountains and the weather patterns that prevail elsewhere on the Earth. *See* METEOROLOGY.   [E.R.R.]

Mountain systems   Long, broad, linear to arcuate belts in the Earth's crust where extreme mechanical deformation and thermal activity have been (or are being) concentrated.

Mountain systems in the general sense occur both on continents and in ocean basins, but the geological properties of the systems in continental as opposed to oceanic settings are distinctly different. The mechanical strain in classical, continental mountain systems is expressed in the presence of major folds, faults, and intensive fracturing and cleavage. Thermal effects are in the form of vast volcanic outpourings, intruded bodies of igneous magma, and metamorphism. Uplift and deformation in young mountain systems are conspicuously displayed in the physiographic forms of topographic relief. Where mountain building is presently taking place, the dynamics are partly expressed in warping of the land surface and significant shallow or deep earthquake activity. Locations of ancient mountain systems in continental regions now beveled flat by erosion are clearly disclosed by the presence of highly deformed, intruded, and metamorphosed rocks.

Two basic classes of oceanic mountain systems exist. A world-encircling oceanic rift mountain system has been built along the extensional tectonic boundary between plates diverging at rates of 0.8–2.4 in. or 2–6 cm per year from the mid-oceanic ridges. This rift mountain system is exposed to partial view in Iceland. The second type, island arc mountain systems, occur in oceanic basins where the crust dives downward at trench sites, thus underthrusting adjacent oceanic crust. *See* MARINE GEOLOGY.

The classical, conspicuous mountain systems of the Earth occur at the continent/ocean interface, for this is the site where plate convergence has led to major sedimentation, subduction of oceanic crust under continents, collision of island arc mountain systems with continents, and head-on collision of continents. *See* OROGENY; PLATE TECTONICS.   [G.H.D.]

Muscovite   A mineral of the mica group with an ideal composition of $KAl_2(AlSi_3)O_{10}(OH)_2$. Sometimes it is referred to as a white mica or potash mica.

Physical properties include specific gravity 2.76–2.88, hardness on the Mohs scale 2–2.5, and luster vitreous to pearly. Thin sheets are flexible and may be colorless, with books (thick crystals) translucent, yellow, brown, reddish, or green. Muscovite occurs commonly in all the major rock types, in igneous rocks (granites, pegmatites, and hydrothermal alteration products), in metamorphic rocks (slates, phyllites, schists and gneisses), and in sedimentary rocks (sandstones and other clastic rocks). As larger flakes, muscovite is used as an electrical insulator, both for its dielectric properties and for its resistance to heat. Ground muscovite is used for fireproofing, as an additive to paint to provide a sheen and for durability, as a filler, and for many other applications. *See* MICA; SILICATE MINERALS.   [S.Gu.]

Mylonite   A rock that has undergone significant modification of original textures by predominantly plastic flow due to dynamic recrystallization. Mylonites form at depth beneath brittle faults in continental and oceanic crust, in rocks from quartzo-feldspathic to olivine-pyroxenite composition. Mylonites were once confused with cataclasites, which form by brittle fracturing, crushing, and comminution. Microstructures that develop during mylonitization vary according to original mineralogy and modal compositions, temperature, confining pressure, strain, strain rate, applied stresses, and presence or absence of fluids.

At low to moderate metamorphic grades, mylonitization reduces the grain size of the protolith and commonly produces a very fine-grained, well-foliated rock with a pronounced linear fabric defined by elongate minerals. Lineations may be weak or absent, however, in high-strain zones that lack a significant rotational component. At high metamorphic grades, grain growth during mylonitization can produce a net increase in grain size, and the term mylonitic gneiss is used where there is a preserved or inferred undeformed protolith. *See* METAMORPHIC ROCKS.   [C.Si.]

N

Native elements Those elements which occur in nature uncombined with other elements. Aside from the free gases of the atmosphere there are about 20 elements that are found as minerals in the native state. These are divided into metals, semimetals, and nonmetals. Gold, silver, copper, and platinum are the most important metals and each of these has been found abundantly enough at certain localities to be mined as an ore. Rarer native metals are others of the platinum group, lead, mercury, tantalum, tin, and zinc. Native iron is found sparingly both as terrestrial iron and meteoric iron.

The native semimetals can be divided into (1) the arsenic group, including arsenic, antimony, and bismuth; and (2) the tellurium group, including tellurium and selenium.

The native nonmetals are sulfur, and carbon in the forms of graphite and diamond. Native sulfur is the chief industrial source of that element.   [C.S.Hu.]

Natrolite A fibrous or needlelike mineral belonging to the zeolite family of silicates. Most commonly it is found in radiating fibrous aggregates. The hardness is $5-5^{1}/_{2}$ on Mohs scale, and the specific gravity is 2.25. The mineral is white or colorless with a vitreous luster that inclines to pearly in fibrous varieties. The chemical composition is $Na_2(Al_2Si_3O_{10}) \cdot 2H_2O$, but some potassium is usually present substituting for sodium.

Natrolite is a secondary mineral found lining cavities in basaltic rocks. Its outstanding locality in the United States is at Bergen Hill, New Jersey. See ZEOLITE.   [C.Fr.; C.S.Hu.]

Natural gas A combustible gas that occurs in porous rock of the Earth's crust and is found with or near accumulations of crude oil. Being in gaseous form, it may occur alone in separate reservoirs. More commonly it forms a gas cap, or mass of gas, entrapped between liquid petroleum and impervious capping rock layer in a petroleum reservoir. Under conditions of greater pressure it is intimately mixed with, or dissolved in, crude oil.

Typical natural gas consists of hydrocarbons having a very low boiling point. Methane (CH_4) makes up approximately 85% of the typical gas. Ethane (C_2H_6) may be present in amounts up to 10%; and propane (C_3H_8) up to 3%. Butane (C_4H_{10}); pentane (C_5H_{12}); hexane; heptane; and octane may also be present.

Whereas normal hydrocarbons having 5–10 carbon atoms are liquids at ordinary temperatures, they have a definite vapor pressure and therefore may be present in the vapor form in natural gas. Carbon dioxide, nitrogen, helium, and hydrogen sulfide may also be present.

Types of natural gas vary according to composition and can be dry or lean (mostly methane) gas, wet gas (considerable amounts of so-called higher hydrocarbons), sour gas (much hydrogen sulfide), sweet gas (little hydrogen sulfide), residue gas (higher paraffins having been extracted), and casinghead gas (derived from an oil well by extraction at the surface). Natural gas has no distinct odor. Its main use is for fuel, but it is also used to make carbon black, natural gasoline, certain chemicals, and liquefied petroleum gas. Propane and butane are obtained in processing natural gas.

Gas occurs on every continent. Wherever oil has been found, a certain amount of natural gas is also present. Successful exploitation of these resources involves drilling, producing, gathering, processing, transporting, and metering the use of the gas. Long before supplies of natural gas run out or become expensively scarce, it is expected that some process of coal gasification will produce a gas which is completely interchangeable with natural gas and at a competitive price. This is important because coal makes up a majority of the world's known fossil fuel reserves. But when energy consumers indicated in the marketplace their preference for fluid and gaseous fuels over the solid forms, coal gasification research, already well under way, was given additional impetus.

[M.A.A.; M.T.H.]

Nearshore processes   Processes that shape the shore features of coastlines and begin the mixing, sorting, and transportation of sediments and runoff from land. In particular, the processes include those interactions among waves, winds, tides, currents, and land that relate to the waters, sediments, and organisms of the nearshore portions of the continental shelf. The nearshore extends from the landward limit of storm-wave influence, seaward to depths where wave shoaling begins. *See* COASTAL LANDFORMS.

The energy for nearshore processes comes from the sea and is produced by the force of winds blowing over the ocean by the gravitational attraction of Moon and Sun acting on the mass of the ocean, and by various impulsive disturbances at the atmospheric and terrestrial boundaries of the ocean. These forces produce waves and currents that transport energy toward the coast. The configuration of the landmass and adjacent shelves modifies and focuses the flow of energy and determines the intensity of wave and current action in coastal waters. Rivers and winds transport erosion products from the land to the coast, where they are sorted and dispersed by waves and currents.

In temperate latitudes, the dispersive mechanisms operative in the nearshore waters of oceans, bays, and lakes are all quite similar, differing only in intensity and scale, and are determined primarily by the nature of the wave action and the dimensions of the surf zone. The most important mechanisms are the orbital motion of the waves, the basic mechanism by which wave energy is expended on the shallow sea bottom, and the currents of the nearshore circulation system that produce a continuous interchange of water between the surf zone and offshore areas. The dispersion of water and sediments near the coast and the formation and erosion of sandy beaches are some of the common manifestations of nearshore processes.

Erosional and depositional nearshore processes play an important role in determining the configuration of coastlines. Whether deposition or erosion will be predominant in any particular place depends upon a number of interrelated factors: the amount of available beach sand and the location of its source; the configuration of the coastline and of the adjoining ocean floor; and the effects of wave, current, wind, and tidal action. The establishment and persistence of natural sand beaches are often the result of a delicate balance among a number of these factors, and any changes, natural or anthropogenic, tend to upset this equilibrium. *See* DEPOSITIONAL SYSTEMS AND ENVIRONMENTS; EROSION.

[D.L.I.]

Nepheline   A mineral of variable composition: in its purest state, $NaAlSiO_4$; often nearly $Na_3K(AlSiO_4)_4$; but generally $(Na, K, \square, Ca, Mg, Fe^{2+}, Mn, Ti)_8(Al, Si, Fe^{3+})_{16}O_{32}$, where $\square$ represents vacant crystallographic sites, and Ca, Mg, Fe^{2+}, Mn, Ti, and Fe^{3+} are usually present in only minor or trace amounts. The most important variations in nepheline composition are due to crystalline solution of $KAlSiO_4$ (the mineral kalsilite), and substitution of □ for K. *See* SILICATE MINERALS.

The salient physical properties of nepheline are: a Mohs scale hardness of 5.5–6.0; a specific gravity between 2.56 and 2.67; a typically dark gray, light gray, or white color, but it can also be colorless (nepheline is colorless in petrographic thin section); and a vitreous or greasy luster. Nepheline occurs as simple hexagonal prisms or, more commonly, as isolated shapeless grains or irregular polycrystalline masses.

Nepheline is the most abundant feldspathoid mineral; it occurs in a wide variety of SiO_2-deficient (quartz-free) and alkali-rich volcanic, plutonic, and metamorphic rocks. In volcanic rocks, nepheline occurs chiefly as a primary mineral in phonolites, kenytes, and melilite basalts, and it is the characteristic mineral of nephelinites.

Both "pure" (processed) nepheline and nepheline syenite are used as raw materials for the manufacture of glass, various ceramic materials, alumina, pottery, and tile. *See* FELDSPATHOID; IGNEOUS ROCKS; NEPHELINITE. [J.G.B.]

Nephelinite   A dark-colored, aphanitic (very finely crystalline) rock of volcanic origin, composed essentially of nepheline (a feldspathoid) and pyroxene. *See* KALSILITE.

The texture is usually porphyritic with large crystals (phenocrysts) of augite and nepheline in a very fine-grained matrix. Augite phenocrysts may be diopsidic or titanium-rich and may be rimmed with soda-rich pyroxene (aegirine-augite). Microscopically the matrix is seen to be composed of tiny crystals or grains of nepheline, augite, aegirite, and sodalite with occasional soda-rich amphibole, biotite, and brown glass.

Nephelinite and related rocks are very rare. They occur as lava flows and small, shallow intrusives. A great variety of these feldspathoidal rocks is displayed in Kenya. *See* FELDSPATHOID; IGNEOUS ROCKS. [C.A.C.]

New Zealand   A landmass in the Southern Hemisphere, bounded by the South Pacific Ocean to the north, east, and south and the Tasman Sea to the west, with a total land area of 103,883 mi^2 (269,057 km^2). The exposed landmass represents about one-quarter of a subcontinent, with three-quarters submerged. This long, narrow, mountainous country, oriented northeast to southwest, consists of two main islands, North Island and South Island, surrounded by a much greater area of crust submerged to depths reaching 1.2 mi (2 km).

South Island lowlands are either alluvial plains as in Otago, Southland, and Nelson, or glacial outwash fans as in Westland and Canterbury. North Island lowlands such as Hawke's Bay, Wairarapa, and Manawatu are alluvial; the Waikato, Hauraki, and Bay of Plenty lowlands occupy structural basins that contain large volumes of reworked volcanic debris from the central volcanic region. The alluvial lowlands of both main islands form the most agriculturally productive areas of the country. *See* PLAINS.

The climate of New Zealand is influenced by three main factors: a location in latitudes where the prevailing airflow is westerly; an oceanic environment; and the mountain chains, which modify the weather systems as they pass eastward, causing high rainfalls on windward slopes and sheltering effects to leeward.

Weather is determined mostly by series of anticyclones and troughs of low pressure that produce alternating periods of settled and variable conditions. Westerly air masses are occasionally replaced by southerly airstreams, which bring cold conditions with snow in winter and spring to areas south of 39°S, and northerly tropical maritime air, which brings warm humid weather to the north and east coasts. *See* METEOROLOGY.

Rainfall on land is 16–470 in. (400–12,000 mm) per year, with the highest rainfall being on the western windward slopes of the mountains, and the lowest on the eastern basins in the lee of the Southern Alps in Central Otago and south Canterbury. Annual rain days are at least 130 for most of North Island, but on South Island the totals are far more variable, with over 200 occurring in Fiordland, 180 on the west coast, and fewer than 80 in Central Otago. Summer droughts are relatively common in Northland, and in eastern regions of both islands. *See* DROUGHT; PRECIPITATION (METEOROLOGY).

Droughts, springtime air frosts, and hailstorms are the major common climatic hazards for the farming industry, but floods associated with prolonged intense rainstorms are the major general hazard.

The economy is heavily dependent on the natural resources soil, water, and plants. New Zealand has few exploitable minerals, but possesses a climate generally favorable for agriculture, pastoral farming, renewable forestry, and tourism. With a small population (3.4 million), much of its manufacturing is concerned with processing produce from the land and surrounding seas, and supplying the needs of those industries.

Because of its high relief and its location on an active crustal plate boundary in the zone of convergence between Antarctic air masses and tropical air masses, New Zealand is prone to high-intensity and high-frequency natural hazards—earthquakes, volcanic eruptions, large and small landslides, and floods. [M.J.Se.]

Nickeline   A minor ore of nickel. Nickeline is a mineral having composition NiAs and crystallizing in the hexagonal system. Crystals are rare, and nickeline usually occurs in massive aggregates with metallic luster and pale copper-red color. Because of the color, not the composition, it is called copper nickel. The hardness is 5.5 on Mohs scale and the specific gravity is 7.78. Nickeline is frequently associated with other nickel arsenides and sulfides in massive pyrrhotite. It is also found in vein deposits with cobalt and silver minerals, as in the silver mines of Saxony, Germany, and Cobalt, Ontario, Canada. *See* PYRRHOTITE. [C.S.Hu.]

Niter   A potassium nitrate mineral with chemical composition KNO_3. Niter crystallizes in the orthorhombic system, generally in thin crusts and delicate acicular crystals; it occurs in massive, granular, or earthy forms. It is brittle; hardness is 2 on Mohs scale; specific gravity is 2.109. The luster is vitreous, and the color and streak are colorless to white. *See* NITRATE MINERALS.

Niter is commonly found, usually in small amounts, as a surface efflorescence in arid regions and in caves and other sheltered places. Niter occurs associated with soda niter in the desert regions of northern Chile, and in similar occurrences in Italy, Egypt, Russia, the western United States, and elsewhere. [G.Sw.]

Nitrate minerals   These minerals are few in number and with the exception of soda niter are of rare occurrence. Normal anhydrous and hydrated nitrates occurring as minerals are soda niter, $NaNO_3$; niter, KNO_3: ammonia niter, NH_4NO_3; nitrobarite, $Ba(NO_3)_2$; nitrocalcite, $Ca(NO_3)_2 \cdot 4H_2O$; and nitromagnesite, $Mg(NO_3)_2 \cdot 6H_2O$. In addition there are three known naturally occurring nitrates containing hydroxyl or halogen, or compound nitrates. They are gerhardtite, $Cu_2(NO_3)(OH)_3$; buttgenbachite, $Cu_{19}(NO_3)_2Cl_4(OH)_{32}\dot{m}\, 3H_2O$; and darapskite, $Na_3(NO_3)(SO_4) \cdot H_2O$. *See* NITER; SODA NITER.

The natural nitrates are for the most part readily soluble in water. For this reason they occur most abundantly in arid regions, particularly in South America along the Chilean coast. [G.Sw.]

Nitrogen cycle   The collective term given to the natural biological and chemical processes through which inorganic and organic nitrogen are interconverted. It includes

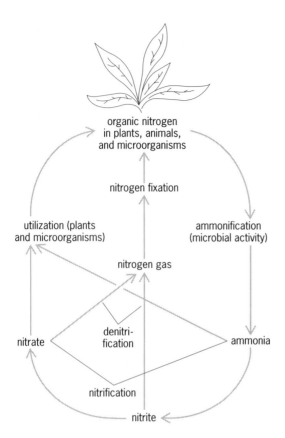

organic nitrogen
in plants, animals,
and microorganisms

nitrogen fixation

utilization (plants
and microorganisms)

ammonification
(microbial activity)

nitrogen gas

denitri-
fication

nitrate

ammonia

nitrification

nitrite

Diagram of the nitrogen cycle.

the process of ammonification, ammonia assimilation, nitrification, nitrate assimilation, nitrogen fixation, and denitrification.

Nitrogen exists in nature in several inorganic compounds, namely N_2, N_2O, NH_3, NO_2^-, and NO_3^-, and in several organic compounds such as amino acids, nucleotides, amino sugars, and vitamins. In the biosphere, biological and chemical reactions continually occur in which these nitrogenous compounds are converted from one form to another. These interconversions are of great importance in maintaining soil fertility and in preventing pollution of soil and water.

An outline showing the general interconversions of nitrogenous compounds in the soil-water pool is presented in the illustration. There are three primary reasons why organisms metabolize nitrogen compounds: (1) to use them as a nitrogen source, which means first converting them to NH_3, (2) to use certain nitrogen compounds as an energy source such as in the oxidation of NH_3 to NO_2^- and of NO_2^- to NO_3^-, and (3) to use certain nitrogen compounds (NO_3^-) as terminal electron acceptors under conditions where oxygen is either absent or in limited supply. The reactions and products involved in these three metabolically different pathways collectively make up the nitrogen cycle.

There are two ways in which organisms obtain ammonia. One is to use nitrogen already in a form easily metabolized to ammonia. Thus, nonviable plant, animal, and microbial residues in soil are enzymatically decomposed by a series of hydrolytic and other reactions to yield biosynthetic monomers such as amino acids and other small-molecular-weight nitrogenous compounds. These amino acids, purines, and pyrimidines are decomposed further to produce NH_3 which is then used by plants and bacteria for biosynthesis, or these biosynthetic monomers can be used directly by some microorganisms. The decomposition process is called ammonification.

The second way in which inorganic nitrogen is made available to biological agents is by nitrogen fixation (this term is maintained even though N_2 is now called dinitrogen), a process in which N_2 is reduced to NH_3. Since the vast majority of nitrogen is in the form of N_2, nitrogen fixation obviously is essential to life. The N_2-fixing process is confined to prokaryotes (certain photosynthetic and nonphotosynthetic bacteria). The major nitrogen fixers (called diazotrophs) are members of the genus *Rhizobium*, bacteria that are found in root nodules of leguminous plants, and of the cyanobacteria (originally called blue-green algae).   [L.E.Mo.]

North America   The third largest continent, extending from the narrow isthmus of Central America to the Arctic Archipelago. The physical environments of North America, like the rest of the world, are a reflection of specific combinations of the natural factors such as climate, vegetation, soils, and landforms. *See* CONTINENT.

Location. North America covers 9,400,000 mi^2 (24,440,000 km^2) and extends north to south for 5000 mi (8000 km) from Central America to the Arctic. It is bounded by the Pacific Ocean on the west and the Atlantic Ocean on the east. The Gulf of Mexico is a source of moist tropical air, and the frozen Arctic Ocean is a source of polar air. With the major mountain ranges stretching north-south, North America is the only continent providing for direct contact of these polar and tropical air masses, leading to frequent climatically induced natural hazards such as violent spring tornadoes, extreme droughts, subcontinental floods, and winter blizzards, which are seldom found on other continents. *See* AIR MASS; ARCTIC OCEAN; ATLANTIC OCEAN; GULF OF MEXICO; PACIFIC OCEAN.

Geologic structure. The North American continent includes (1) a continuous, broad, north-south-trending western cordilleran belt stretching along the entire Pacific coast; (2) a northeast-southwest-trending belt of low Appalachian Mountains paralleling the Atlantic coast; (3) an extensive rolling region of old eroded crystalline rocks in the north-central and northeastern part of the continent called the Canadian Shield; (4) a large, level interior lowland covered by thick sedimentary rocks and extending from the Arctic Ocean to the Gulf of Mexico; and (5) a narrow coastal plain along the Atlantic Ocean and the Gulf of Mexico. These broad structural geologic regions provide the framework for the natural regions of this continent and affect the location and nature of landform, climatic, vegetation, and soil regions.

Canadian Shield. Properly referred to as the geological core of the continent, the exposed Canadian Shield extends about 2500 mi (4000 km) from north to south and almost as much from east to west. The rest of it dips under sedimentary rocks that overlap it on the south and west. The Canadian Shield consists of ancient Precambrian rocks, over 500 million years old, predominantly granite and gneiss, with very complex structures indicating several mountain-building episodes. It has been eroded into a rolling surface of low to moderate relief with elevations generally below 2000 ft (600 m). Its surface has been warped into low domes and basins, such as the Hudson Basin, in which lower Paleozoic rocks, including Ordovician limestones, have been preserved. Since the end of the Paleozoic Era, the Shield has been dominated by erosion. Parts of the higher surface remain at about 1500–2000 ft (450–600 m) above sea level, particularly in the Labrador area. The Shield remained as land throughout the Mesozoic Era, but its western margins were covered by a Cretaceous sea and by Tertiary terrestrial sediments derived from the Western Cordillera. *See* CRETACEOUS; ORDOVICIAN; MESOZOIC; PALEOZOIC; PRECAMBRIAN; TERTIARY.

The entire exposed Shield was glaciated during the Pleistocene Epoch, and its surface was intensely eroded by ice and its meltwaters, erasing major surface irregularities and eastward-trending rivers that were there before. The surface is now covered by glacial till, outwash, moraines, eskers, and lake sediments, as well as drumlins formed by advancing ice. A deranged drainage pattern is evolving on this surface with thousands of lakes of various sizes. *See* DRUMLIN; ESKER; GLACIAL EPOCH; MORAINE; PLEISTOCENE; TILL.

The Canadian Shield extends into the United States as Adirondack Mountains in New York State, and Superior Upland west of Lake Superior.

Southeastern Coastal Plain. The Southeastern Coastal Plain is geologically the youngest part of the continent, and it is covered by the youngest marine sedimentary rocks. This flat plain, which parallels the Atlantic and Gulf coastline, extends for over 3000 mi (4800 km) from Cape Cod, Massachusetts, to the Yucatán Peninsula in Mexico. It is very narrow in the north but increases in width southward along the Atlantic coast and includes the entire peninsula of Florida. As it continues westward along the Gulf, it widens significantly and includes the lower Mississippi River valley. It is very wide in Texas, narrows again southward in coastal Mexico, and then widens in the Yucatán Peninsula and continues as a wide submerged plain, or a continental shelf, into the sea. *See* COASTAL PLAIN.

Extending from Cape Cod, Massachusetts, to Mexico and Central America, the Coastal Plain is affected by a variety of climates and associated vegetation. While a humid, cool climate with four seasons affects its northernmost part, subtropical air masses affect the southeastern part, including Florida, and hot and arid climate dominates Texas and northern Mexico; Central America has hot, tropical climates.

Varied soils characterize the Coastal Plain, including the fertile alluvial soils of the Mississippi Valley. Broadleaf forests are present in the northeast, citrus fruits grow in Florida, grasslands dominate the dry southwest, and tropical vegetation is present on Central American coastal plains.

Eastern Seaboard Highlands. Between the Southeastern Coastal Plain and the extensive interior provinces lies a belt of mountains that, by their height and pattern, create a significant barrier between the eastern seaboard and the interior of North America. These mountains consist of the Adirondack Mountains and the New England Highlands.

The Adirondack Mountains are a domal extension of the Canadian Shield, about 100 mi (160 km) in diameter, composed of complex Precambrian rocks. The New England Highlands consist of a north-south belt of mountains east of the Hudson Valley, including the Taconic mountains in the south and the Green mountains in the north, and continuing as the Notre Dame Mountains along the St. Lawrence Valley and the Chic-Choc Mountains of the Gaspé Peninsula. The large area of New England east of these mountains is an eroded surface of old crystalline rocks culminating in the center as the White Mountains, with their highest peak of the Presidential Range, Mount Washington, reaching over 6200 ft (1880 m). This area has been intensely glaciated, and it meets the sea in a rugged shoreline. Nova Scotia and Newfoundland have a similar terrain.

New England is a hilly to mountainous region carved out of ancient rocks, eroded by glaciers, and covered by glacial moraines, eskers, kames, erratics, and drumlins, with hundreds of lakes scattered everywhere. It has a cool and moist climate with four seasons, thin and acid soils, and mixed coniferous and broadleaf forests.

Appalachian Highlands. The Appalachian Highlands are traditionally considered to consist of four parts: the Piedmont, the Blue Ridge Mountains, the Ridge and Valley

Section, and the Appalachian Plateau. These subregions are all characterized by different geologic structures and rock types, as well as different geomorphologies.

The northern boundary of the entire Appalachian System is an escarpment of Paleozoic rocks trending eastward along Lake Erie, Lake Ontario, and the Mohawk Valley. The boundary then swings south along Hudson River Valley and continues southwestward along the Fall Line to Montgomery, Alabama. The western boundary trends northeastward through Cumberland Plateau in Tennessee, and up to Cleveland, Ohio, where it joins the northern boundary. Together with New England, this region forms the largest mountainous province in eastern United States.

Interior Domes and Basins Province. The southwestern part of the Appalachian Plateau, overlain mainly by the Mississippian and Pennsylvanian sedimentary rocks, has been warped into two low structural domes called the Blue Grass and Nashville Basins, and a structural basin, drained by the Green River; its southern fringe is called the Pennyroyal Region. The Interior Dome and Basin Province is contained roughly between the Tennessee River in the south and west and the Ohio River in the north.

There is no boundary on the east, because the domes are part of the same surface as the Appalachian Plateau. However, erosional escarpments, forming a belt of hills called knobs, clearly mark the topographic domes and basins. The northern dome, called the Blue Grass Basin or Lexington Plain, has been eroded to form a basin surrounded by a series of inward-facing cuesta escarpments. The westernmost cuesta reaches about 600 ft (180 m) elevation while the central part of the basin lies about 1000 ft (300 m) above sea level, which is higher than the surrounding hills. This gently rolling surface with deep and fertile soils exhibits some solutional karst topography. *See* FLUVIAL EROSION LANDFORMS.

Ozark and Ouachita Highlands. The Paleozoic rocks of the Pennyroyal Region continue Westward across southern Illinois to form another dome of predominantly Ordovician rocks, called the Ozark Plateau. This dome, located mainly in Missouri and Arkansas, has an abrupt east side, and a gently sloping west side, called the Springfield Plateau. Its surface is stream eroded into hilly and often rugged topography that is developed mainly on limestones, although shales, sandstone, and chert are present. Much residual chert, eroded out of limestone, is present on the surface. There are some karst features, such as caverns and springs. In the northeast, Precambrian igneous rocks protrude to form the St. Francois Mountains, which reach an elevation of 1700 ft (515 m).

Central Lowlands. One of the largest subdivisions of North America is the Central Lowlands province which is located between the Appalachian Plateau on the east, the Interior Domes and Basins Province and the Ozark Plateau on the south, and the Great Plains on the west. It includes the Great Lakes section and the Manitoba Lowland in Canada. This huge lowland in the heart of the continent (whose elevations vary from about 900 ft or 270 m above sea level in the east and nearly 2000 ft or 600 m in the west) is underlain by Paleozoic rocks that continue from the Appalachian Plateau and dip south under the recent coastal plain sediments; meet the Cretaceous rocks on the west; and overlap the crystalline rocks of the Canadian Shield on the northeast.

The present surface of nearly the entire Central Lowlands, roughly north of the Ohio River and east of the Missouri River, is the creation of the Pleistocene ice sheets. When the ice formed and spread over Canada, and southward to the Ohio and Missouri rivers, it eroded much of the preexisting surface. During deglaciation, it left its deposits over the Canadian Shield and the Central Lowlands.

The Central Lowlands are drained by the third longest river system in the world, the Missouri-Mississippi, which is 3740 mi (6000 km) long. This mighty river system, together with the Ohio and the Tennessee, drains not only the Central Lowlands but also parts of the Appalachian Plateau and the Great Plains, before it crosses the Coastal Plain and ends in the huge delta of the Mississippi. The river carries an enormous amount of water and alluvium and continues to extend its delta into the Gulf. In 1993 it reached a catastrophic level of a hundred-year flood, claimed an enormous extent of land and many lives, and created an unprecedented destruction of property. This flood again alerted the population to the extreme risk of occupying a river floodplain. *See* FLOODPLAIN; RIVER.

Great Plains. The Great Plains, which lie west of the Central Lowlands, extend from the Rio Grande and the Balcones Escarpment in Texas to central Alberta in Canada. On the east, they are bounded by a series of escarpments, such as the Côteau du Missouri in the Dakotas. The dry climate with less than 20 in. (50 cm) of precipitation, and steppe grass vegetation growing on calcareous soils, help to determine the eastern boundary of the Great Plains. On the west, the Great Plains meet the abrupt front of the Rocky Mountains, except where the Colorado Piedmont and the lower Pecos River Valley separate them from the mountains.

The Great Plains region shows distinct differences between its subsections from south to north. The southernmost part, called the High Plains or Llano Estacado, and Edwards Plateaus are the flattest. While Edwards Plateau, underlain by limestones of the Cretaceous age, reveals solutional karst features, the High Plains have the typical Tertiary bare cap rock surface, devoid of relief and streams.

The central part of the Great Plains has a recent depositional surface of loess and sand. The Sand Hills of Nebraska form the most extensive sand dunes area in North America, covering about 24,000 mi^2 (62,400 km^2). They are overgrown by grass and have numerous small lakes. The loess region to the south provides spectacular small canyon topography. *See* DUNE.

The northern Great Plains, stretching north of Pine Ridge and called the Missouri Plateau, have been intensly eroded by the western tributaries of the Missouri River into river breaks and interfluves. In extreme cases, badlands were formed, such as those of the White River and the Little Missouri.

The terrain of the Canadian Great Plains consists of three surfaces rising from east to west: the Manitoba, Saskatchewan, and Alberta Prairies developed on level Creteceous and Tertiary rocks. Climatic differences between the arid and warm southern part and the cold and moist northern part have resulted in regional differences. The eastern boundary of the Saskatchewan Plain is the segmented Manitoba Escarpment, which extends for 500 mi (800 km) northwestward, and in places rises 1500 ft (455 m) above the Manitoba Lowland. Côteau du Missouri marks the eastern edge of the higher Alberta Plain.

Western Cordillera. The mighty and rugged Western Cordilleras stretch along the Pacific coast from Alaska to Mexico. There are three north-south-trending belts: (1) Brooks Range, Mackenzie Mountains, and the Rocky Mountains to the north and Sierra Madre Oriental in Mexico; (2) Interior Plateaus, including the Yukon Plains, Canadian Central Plateaus and Ranges, Columbia Plateau, Colorado Plateau, and Basin and Range Province stretching into central Mexico; and (3) Coastal Mountains from Alaska Range to California, Baja California, and Sierra Madre Occidental in Mexico.

This subcontinental-size mountain belt has the highest mountains, greatest relief, roughest terrain, and most beautiful scenery of the entire continent. It has been formed by earth movements resulting from the westward shift of the North American lithospheric plate. The present movements, and the resulting devastating earthquakes along the San Andreas fault system paralleling the Pacific Ocean, are part of this process. *See* CORDILLERAN BELT; PLATE TECTONICS.

This very high, deeply eroded and rugged Rocky Mountains region comprises several distinct parts: Southern, Middle, and Northern Rockies, plus the Wyoming Basin in the United States, and the Canadian Rockies. The Southern Rockies, extending from Wyoming to New Mexico, include the Laramie Range, the Front Range, and Spanish Peaks with radiating dikes on the east; Medicine Bow, Park, and Sangre de Cristo ranges in the center; and complex granite Sawatch Mountains and volcanic San Juan Mountains of Tertiary age on the west. Most of the ranges are elongated anticlines with exposed Precambrian granite core, and overlapping Paleozoic and younger sedimentary rocks which form spectacular hogbacks along the eastern front. There are about 50 peaks over 14,000 ft (4200 m) high, while the Front Range alone has about 300 peaks over 13,000 ft (3940 m) high. The southern Rocky Mountains, heavily glaciated into a beautiful and rugged scenery with permanent snow and small glaciers, form a major part of the Continental Divide.

The interior Plateaus and Ranges Province of the Western Cordillera lies between the Rocky Mountains and the Coastal Mountains. It is an extensive and complex region. It begins in the north with the wide Yukon Plains and Uplands; narrows into the Canadian Central Plateaus and Ranges; widens again into the Columbia Plateau, Basin and Range Province, and Colorado Plateau; and finally narrows into the Mexican Plateau and the Central American isthmus.

The coastal Lowlands and Ranges extend along the entire length of North America and include Alaskan Coast Ranges, Aleutian Islands, Alaska Range, Canadian Coast Ranges, and a double chain of the Cascade Mountains and Sierra Nevada on the east, and Coast Ranges on the west, separated by Puget Sound, Willamette Valley, and Great Valley of California. These ranges continue southward as Lower California Peninsula, Baja California, and Sierra Madre Occidental in Mexico.

The basin-and-range type of terrain of the southwest United States continues into northern Mexico and forms its largest physiographic region, the Mexican Plateau. This huge tilted block stands more than a mile above sea level—from about 4000 ft (1200 m) in the north, it rises to about 8000 ft (2400 m) in the south. The Mexican Plateau is separated from the Southern Mexican Highlands (Sierra Madre del Sur) by a low, hot and dry Balsas Lowland drained by the Balsas River. To the east of the Southern Highlands lies a lowland, the Isthmus of Tehuantepec, which is considered the divide between North and Central America. Here the Pacific and Gulf coasts are only 125 mi (200 km) apart. The lowlands of Mexico are the coastal plains. The Gulf Coastal Plain trends southward for 850 mi from the Rio Grande to the Yucatán Peninsula. It is about 100 mi (160 km) wide in the north, just a few miles wide in the center, and very wide in the Yucatan Peninsula. Barrier beaches, lagoons, and swamps occur along this coast. The Pacific Coastal Plains are much narrower and more hilly. North-south-trending ridges of granite characterize the northern part, and islands are present offshore. Toward the south, sandbars, lagoons, and deltaic deposits are common.

East of the Isthmus of Tehuantepec begins Central America with its complex physiographic and tectonic regions. This narrow, mountainous isthmus is geologically connected with the large, mountainous islands of the Greater Antilles in the Carribean. They are all characterized by east-west-trending rugged mountain ranges, with deep depressions between them. One such mountain system begins in Mexico and continues in southern Cuba, Puerto Rico, and the Virgin Islands. North of this system, called the Old Antillia, lies the Antillian Foreland, consisting of the Yucatán Peninsula and the Bahama Islands. Central American mountains are bordered on both sides by active volcanic belts. Along the Pacific, a belt of young volcanoes extends for 800 mi (1280 km) from Mexico to Costa Rica. Costa Rica and Panama are mainly a volcanic chain of mountains extending to South America. Nicaragua is dominated by a major crustal fracture trending northwest-southeast.   [B.Z.B.]

North Pole   That end of the Earth's axis which points toward the North Star, Polaris (Alpha Ursae Minoris). It is the geographical pole where all meridians converge, and should not be confused with the north magnetic pole, which is in the Canadian Archipelago. The North Pole's location falls near the center of the Arctic Sea. The North Pole has phenomena unlike any other place except the South Pole. For 6 months the Sun does not appear above the horizon, and for 6 months it does not go below the horizon. As there is a long period (about 7 weeks) of continuous twilight before March 21 and after September 23, the period of light is considerably longer than the period of darkness.   [V.H.E.]

North Sea   A flooded portion of the northwest continental margin of Europe occupying an area of over 200,000 mi^2 (500,000 km^2). The North Sea has extensive marine fisheries and important offshore oil and gas reserves. In the south, its depth is less than 150 ft (50 m), but north of 58° it deepens gradually to 600 ft (200 m) at the top of the continental slope. A band of deep water down to 1200 ft (400 m) extends around the south and west coast of Norway and is known as the Norwegian Trench.

The nontidal residual current circulation of the southern North Sea is mainly determined by wind velocity, but in the north, well-defined non-wind-driven currents have been identified, especially in the summer. Two of these currents bring in water from outside the North Sea; one flows through the channel between Orkney and Shetland (the Fair Isle current), and the other follows the continental slope north of Shetland and merges with the Fair Isle current southwest of Norway before entering the Skagerrak. The north-flowing Norwegian coastal current provides the exit route for North Sea waters, and is formed from the waters of these two major inflows and from other much smaller inputs such as river runoff, the English Channel, and the Baltic Sea.

There is a rich diversity of zooplankton within the North Sea. Copepods are of particular importance in the food web. There are a wide range of fish stocks in the North Sea and adjacent waters and, in terms of species exploited by commercial fisheries, they constitute the richest area in the northeast Atlantic. The commercially important stocks exploited for human consumption include cod, haddock, whiting, pollock, plaice, sole, herring, mackerel, lobster, prawn, and brown shrimp (*Crangon crangon*). A number of stocks are used for fishmeal and oil; these stocks include sand eel, Norway pout, blue whiting, and sprat. *See* MARINE ECOLOGY.   [H.D.D.]

Oasis An isolated fertile area, usually limited in extent and surrounded by desert. The term was initially applied to small areas in Africa and Asia typically supporting trees and cultivated crops with a water supply from springs and from seepage of water originating at some distance. However, the term has been expanded to include areas receiving moisture from intermittent streams or artificial irrigation systems. Thus the floodplains of the Nile and Colorado rivers can be considered vast oases, as can arid areas irrigated by humans. *See* DESERT.

Oases are restricted to climatic regions where precipitation is insufficient to support crop production. Such regions may be classified as extremely arid (annual rainfall less than 2 in. or 50 mm), arid (annual rainfall less than 10 in. or 250 mm), and semiarid (rainfall less than 20 in. or 500 mm). Many African and Asian oases are in extremely arid areas. Most oases are found in warm climates. Oasis soils are weakly developed, high in organic matter but often saline, and have been strongly affected by human occupation.   [W.G.McG.]

Obsidian A volcanic glass, usually of rhyolitic composition, formed by rapid cooling of viscous lava. The color is jet-black because of abundant microscopic, embryonic crystal growths (crystallites) which make the glass opaque except on thin edges. Iron oxide dust may produce red or brown obsidian.

Obsidian usually forms the upper parts of lava flows. Well-known occurrences are Obsidian Cliffs in Yellowstone Park, Wyoming; Mount Hekla, Iceland; and the Lipari Islands off the coast of Italy. *See* IGNEOUS ROCKS; VOLCANIC GLASS.   [C.A.C.]

Ocean One of the major subdivisions of the interconnected body of salt water that occupies almost three-quarters of the Earth's surface. Earth is the only planet in the solar system whose surface is covered with significant quantities of water. Of the nearly 1.4 billion cubic kilometers of water found either on the surface or in relatively accessible underground supplies, more than 97% is in the oceans. *See* OCEANOGRAPHY.

Oceans and the seas that connect them cover some 73% of the surface of the Earth, with a mean depth of 3729 m (12,234 ft) (table). More than 70% of the oceans have a depth between 3000 and 6000 m (10,000 and 20,000 ft). Less than 0.2% of the oceans have depths as great as 7000 m (23,000 ft).

The oceans are cold and salty. Some 50% have a temperature between 0 and 2°C (32 and 36°F) and a salinity between 34.0 and 35.0. To a high degree of approximation, a salinity of 34 is the equivalent of 34 grams of salt in a kilogram of seawater. Water with a temperature above a few degrees Celsius is confined to a relatively thin surface layer of the ocean. *See* SEAWATER.

Ocean salinity is primarily controlled by the balance of precipitation, river runoff, and evaporation of water at the sea surface. The highest salinities are found in major

Ocean basin characteristics			
	Area, km^2	Volume, km^3	Mean depth, m
Pacific	181,344,000	714,410,000	3940
Atlantic	94,314,000	337,210,000	3575
Indian	74,118,000	284,608,000	3840
Arctic	12,257,000	13,702,000	1117
Total	362,033,000	1,349,929,000	3729

evaporation basins with little rainfall or river runoff, such as the Red Sea. The lowest salinities are found near the mouths of major rivers such as the Amazon. *See* RED SEA.

Nearly all elements known to humankind have been found dissolved in seawater, and those that have not are assumed to be present. However, all but a few are found in very small amounts. Sodium chloride accounts for some 85% of the dissolved salts, and an additional four ions (sulfate, magnesium, calcium, and potassium) bring the total to more than 99.3%. The ratio of ions is remarkably constant from one ocean to another and from top to bottom of each.

The oceans are continually transporting excess heat (warm water) from the tropics toward the Poles and returning colder water toward the tropics. This process of moving excess heat from lower (south of 40°) to higher (north of 40°) latitudes is shared approximately equally by the oceans and the atmosphere. A significant part of the ocean heat exchange process is carried out by the major ocean currents, the "named" currents such as the Gulf Stream, Brazil Current, California Current, and Kuroshio. These currents are primarily driven by the winds, and there is considerable similarity in their pattern from one ocean basin to another. *See* GULF STREAM; KUROSHIO.

The average winds over the North and South Atlantic as well as the North and South Pacific oceans come out of the west (westerlies) at the middle latitudes and from the east at the lower latitudes (trade winds). The frictional drag of these winds on the surface of the water imparts a spin or torque to the surface of the ocean, clockwise in the Northern Hemisphere and counterclockwise in the Southern Hemisphere. The major exception is the Indian Ocean north of the Equator, where the circulation is strongly influenced by the winds of the seasonal monsoon. *See* ATLANTIC OCEAN; CORIOLIS ACCELERATION; EQUATORIAL CURRENTS; INDIAN OCEAN; OCEAN CIRCULATION; PACIFIC OCEAN.   [J.A.K.]

Ocean circulation   The general circulation of the ocean. The term is usually understood to include large-scale, nearly steady features, such as the Gulf Stream, as well as current systems that change seasonally but are persistent from one year to the next, such as the Davidson Current, off the northwestern United States coast and the equatorial currents in the Indian Ocean. A great number of energetic motions have periods of a month or two and horizontal scales of a few hundred kilometers—a very low-frequency turbulence, collectively called eddies. Energetic motions are also concentrated near the local inertial period (24 h, at 30° latitude) and at the periods associated with tides (primarily diurnal and semidiurnal). *See* TIDE.

The greatest single driving force for currents, as for waves, is the wind. Furthermore, the ocean absorbs heat at low latitudes and loses it at high latitudes. The resultant effect on the density distribution is coupled into the large-scale wind-driven circulation. Some subsurface flows are caused by the sinking of surface waters made dense by cooling or high evaporation. *See* OCEAN WAVES.

Except in western boundary currents, and in the Antarctic Circumpolar Current, the system of strong surface currents is restricted mainly to the upper 330–660 ft

(100–200 m) of the sea. The mid-latitude anticyclonic gyres, however, are coherent in the mean well below 3300 ft (1000 m). The average speeds of the open-ocean surface currents remain mostly below 0.4 knot (20 cm/s). Exceptions to this are found in the western boundary currents, such as the Gulf Stream, and in the Equatorial Currents of the three oceans, all of which have velocities of 2–4 knots (1–2 m/s).

The deep circulation results in part from the wind stress and in part from the internal pressure forces which are maintained by the budgets of heat, salt, and water. Both groups of forces are dependent upon atmospheric influences. Apart from Coriolis and frictional forces, the topography of the sea bottom exercises a decisive influence on the course of deep circulation.

The deep circulation in marginal seas depends largely on the climate of the region, whether arid or humid. Under the influence of an arid climate, evaporation is greater than precipitation. The marginal sea is therefore filled with relatively salty water of a high density. Its surface lies at a lower level than that of the neighboring ocean. Examples of this type are the Mediterranean Sea, Red Sea, and Persian Gulf. The deep circulation of marginal seas in humid climates shows a different pattern. The level of the sea is higher than in the neighboring ocean. Therefore, the surface water with its lower density and accordingly its lower salinity flows outward, and the relatively salty ocean water of higher density flows over the sill into the marginal sea. Examples of this circulation are the Baltic Sea with the shallow Darsser and Drogden rises, the Norwegian and Greenland fiords, and the Black Sea with its entrance through the Bosporus. *See* BLACK SEA; FIORD; MEDITERRANEAN SEA.

The deep circulation in the oceans is more difficult to perceive than the circulation in the marginal seas. In addition to the internal pressure forces, determined by the distribution of density and the piling up of water by the wind, there are also the influences of Coriolis forces and large-scale turbulence. There are areas in tropical latitudes in which the surface water, as a result of strong evaporation, has a relatively high density. In thermohaline convection, the water sinks while flowing horizontally until it reaches a density corresponding to its own, and then spreads out horizontally. In this way the colder and deeper levels of the oceans take on a layered structure consisting of the so-called bottom water, deep water, and intermediate water. *See* ATLANTIC OCEAN; PACIFIC OCEAN.   [W.Stu.]

Wherever oceanographers have made long-term current and temperature measurements, they have found energetic fluctuations with periods of several weeks to several months. These low-frequency fluctuations (compared to tides) are caused by oceanic mesoscale eddies which are in many respects analogous to the atmospheric mesoscale pressure systems that form weather. Like the weather, mesoscale eddies often dominate the instantaneous current, and are thought to be an integral part of the ocean's general circulation.

Eddies occur in virtually all oceans and seas, but their amplitude varies greatly from place to place. The largest amplitudes are found on the western sides of the oceans in conjunction with the strongest ocean currents (the Gulf Stream in the North Atlantic, the Kuroshio in the North Pacific) and near the Equator. Much weaker eddies are found in the ocean interior, distant from major currents. This consistent pattern of eddy amplitude suggests that instabilities of western boundary currents are an important source of eddy energy. Atmospheric forcing by variable winds can also generate eddies, and is probably most important at low latitudes where the horizontal scales of the oceanic eddies best match the scales of the atmospheric forcing.   [J.F.P.]

Ocean waves   The irregular moving bumps and hollows on the ocean surface. Winds blowing over the ocean, in addition to producing currents, create surface water undulations called waves or a sea. The characteristics of these waves (or the state of the sea) depend on the speed of the wind, the length of time that it has blown, the distance over which it has blown, and the depth of the water. If the wind dies down, the waves that remain are called a dead sea.

Surface waves. Ocean surface waves are propagating disturbances at the atmosphere-ocean interface. They are the most familiar ocean waves. Surface waves are also seen on other bodies of water, including lakes and rivers.

A simple sinusoidal wave train is characterized by three attributes: wave height (H), the vertical distance from trough to crest; wavelength (L), the horizontal crest-to-crest distance; and wave period (T), the time between passage of successive crests past a fixed point. The phase velocity ($C = L/T$) is the speed of propagation of a crest. For a given ocean depth (h), wavelength increases with increasing period. The restoring force for these surface waves is predominantly gravitational. Therefore, they are known as surface gravity waves, unless their wavelength is shorter than 1.8 cm (0.7 in.), in which case surface tension provides the dominant restoring force.

Surface gravity waves may be classified according to the nature of the forces producing them. Tides are ocean waves induced by the varying gravitational influence of the Moon and Sun. They have long periods, usually 12.42 h for the strongest constituent. Storm surges are individual waves produced by the wind and dropping barometric pressure associated with storms; they characteristically last several hours. Earthquakes or other large, sudden movements of the Earth's crust can cause waves, called tsunamis, which typically have periods of less than an hour. Wakes are waves resulting from relative motion of the water and a solid body, such as the motion of a ship through the sea or the rapid flow of water around a rock. Wind-generated waves, having periods from a fraction of a second to tens of seconds, are called wind waves. Like tides, they are ubiquitous in the ocean, and continue to travel well beyond their area of generation. The ocean is never completely calm. *See* STORM SURGE; TIDE; TSUNAMI.

The growth of wind waves by the transfer of energy from the wind is not fully understood. At wind speeds less than 1.1 m/s (2.5 mi/h), a flat water surface remains unruffled by waves. Once generated, waves gain energy from the wind by wave-coupling of pressure fluctuations in the air just above the waves. For waves traveling slower than the wind, secondary, wave-induced airflows shift the wave-induced pressure disturbance downwind so the lowest pressure is ahead of the crests. This results in energy transfer from the wind to the wave, and hence growth of the wave.

If a constant wind blows over a sufficient length of ocean, called the fetch, for a sufficient length of time, a wave field develops whose statistical characteristics depend only on wind velocity.

$$S(f) = A \frac{g^2}{f^5} e^{-1.25(fm/f)^4}$$

Because of viscosity, surface waves lose energy as they propagate, short-period waves being dampened more rapidly than long-period waves. Waves with long periods (typically 10 s or more) can travel thousands of kilometers with little energy loss. Such waves, generated by distant storms, are called swell.

When waves propagate into an opposing current, they grow in height. For example, when swell from a Weddell Sea storm propagates northeastward into the southwestward-flowing Agulhas Current off South Africa, high steep waves are formed. Many large ships in this region have been severely damaged by such waves.

Because actual ocean waves consist of many components with different periods, heights, and directions, occasionally a large number of these components can, by chance, come in phase with one another, creating a freak wave with a height several times the significant wave height of the surrounding sea. According to linear theory, waves with different periods propagate with different speeds in deep water, and hence the wave components remain in phase only briefly. But nonlinear effects are bound to be significant in a large wave. In such a wave, the effects of nonlinearity can compensate for those of dispersion, allowing a solitary wave to propagate almost unchanged. Consequently, a freak wave can have a lifetime of a minute or two.   [M.Wi.]

Internal waves. Internal waves are wave motions of stably stratified fluids in which the maximum vertical motion takes place below the surface of the fluid. The restoring force is mainly due to gravity; when light fluid from upper layers is depressed into the heavy lower layers, buoyancy forces tend to return the layers to their equilibrium positions. In the oceans, internal oscillations have been observed wherever suitable measurements have been made. The observed oscillations can be analyzed into a spectrum with periods ranging from a few minutes to days. At a number of locations in the oceans, internal tides, or internal waves having the same periodicity as oceanic tides, are prominent.

Internal waves are important to the economy of the sea because they provide one of the few processes that can redistribute kinetic energy from near the surface to abyssal depths. When they break, they can cause turbulent mixing despite the normally stable density gradient in the ocean. Internal waves are known to cause time-varying refraction of acoustic waves because the sound velocity profile in the ocean is distorted by the vertical motions of internal waves. Internal waves have been found by recording fluctuating currents in middepths by moored current meters, by acoustic backscatter Doppler methods, and by studies of the fluctuations of the depths of isotherms as recorded by instruments repeatedly lowered from shipboard or by autonomous instruments floating deep in the water.

Internal waves are thought to be generated in the sea by variations of the wind pressure and stress at the sea surface, by the interaction of surface waves with each other, and by the interaction of tidal motions with the rough sea floor.   [C.S.C.]

Oceanic islands   Islands rising from the deep sea floor. Oceanic islands range in size from mere specks of rock or sand above the reach of tides to large masses such as Iceland (39,800 mi^2 or 103,000 km^2). Excluded are islands that have continental crust, such as the Seychelles, Norfolk, or Sardinia, even though surrounded by ocean; all oceanic islands surmount volcanic foundations. A few of these have active volcanoes, such as on Hawaii, the Galápagos islands, Iceland, and the Azores, but most islands are on extinct volcanoes. On some islands, the volcanic foundations have subsided beneath sea level, while coral reefs growing very close to sea level have kept pace with the subsidence, accumulating thicknesses of as much as 5000 ft (1500 m) of limestone deposits between the underlying volcanic rocks and the present-day coral islands. *See* REEF; VOLCANO.

Oceanic islands owe their existence to volcanism that began on the deep sea floor and built the volcanic edifices, flow on flow, up to sea level and above. The highest of the oceanic islands is Hawaii, where the peak of Mauna Kea volcano reaches 14,000 ft (4200 m). Most volcanic islands are probably built from scratch in less than 10^6 years, but minor recurrent volcanism may continue for millions of years after the main construction stage. *See* VOLCANOLOGY.

Islands in regions of high oceanic fertility are commonly host to colonies of sea birds, and the deposits of guano have been an important source of phosphate for fertilizer. On some islands, for example, Nauru in the western equatorial Pacific, the original guano has been dissolved and phosphate minerals reprecipitated in porous host limestone rocks. The principal crop on most tropical oceanic islands is coconuts, exploited for their oil content, but some larger volcanic islands, with rich soils and abundant water supplies, are sites of plantations of sugarcane and pineapple. Atoll and barrier-reef islands have very limited water supplies, depending on small lenses of ground water, augmented by collection of rainwater. *See* ATOLL; ISLAND BIOGEOGRAPHY; REEF.   [E.L.Wi.]

Oceanographic vessels   Research vessels designed to collect quantitative data from the sea's surface, its depths, the sea floor, and the overlying atmosphere. Their primary purpose is to carry scientists and increasingly sophisticated equipment to and from study sites on the ocean's surface, and in some cases below the surface. The ships must have the ability to lower and retrieve instruments by using winches and wires. The ship's equipment and instrumentation must determine precisely the location on the sea surface, and provide suitable communication, data gathering, archiving, and computational facilities for the scientific party.

The requirements list includes seakeeping (sea-kindliness, a measure of a ship's response to severe seas; and station-keeping, the ability of a ship to maintain its fixed location on the sea surface); work environment; endurance (range, days at sea); scientific complement (number of researchers accommodated); operating economy and scientific effectiveness; subdued acoustical characteristics; payload (scientific storage, weight handling); speed; and ship control. These requirements often conflict, necessitating compromise.

Ships typically can be considered in three major groups based on use: general purpose (classical biological, physical, chemical, geological, and ocean engineering research, or a combination); dedicated special purpose (hydrographic survey, mapping, geophysical, or fisheries); or unique (deep-sea drilling, crewed spar buoy, or support of submersible operations). They can be used simply as delivery and support systems for exploratory devices, such as floats and bottom landers, as well as crewless remote operating vehicles (ROVs—tethered, powered, surface-controlled robots), or autonomous underwater vehicles (AUVs—freely operating robots, using computer programmed guidances). *See* OCEANOGRAPHY.   [J.J.Gr.; J.F.Ba.]

Oceanography   The science of the sea; including physical oceanography, marine chemistry, marine geology, and marine biology. The need to know more about the impact of marine pollution and possible effects of the exploitation of marine resources, together with the role of the ocean in possible global warming and climate change, means that oceanography is an important scientific discipline. Improved understanding of the sea has been essential in such diverse fields as fisheries conservation, the exploitation of underwater oil and gas reserves, and coastal protection policy, as well as in national defense strategies. The scientific benefits include not only improved understanding of the oceans and their inhabitants, but important information about the evolution of the Earth and its tectonic processes, and about the global environment and climate, past and present, as well as possible future changes. *See* CLIMATE HISTORY; MARINE SEDIMENTS; MARITIME METEOROLOGY.

The traditional basis of modern oceanography is the hydrographic station. Hydrographic studies are still carried out at regular intervals, with the research vessel in a specific position. Seawater temperature, depth, and salinity can be measured continuously by a probe towed behind the ship. The revolution in electronics has provided not only a new generation of instruments for studying the sea but also new ways of collecting and analyzing the data they produce. Computers are employed in gathering and processing data in all fields, and are also used in the creation of mathematical models to aid in understanding. Much information can also be gained by remote sensing using satellites, which are also a valuable navigational aid. These provide data on sea surface temperature and currents, and on marine productivity. Satellite altimetry gives information on wave height and winds and even bottom topography (because this affects sea level). Deep-sea cameras and submersibles now permit visual evidence of creatures in remote depths. See HYDROGRAPHY; OCEANOGRAPHIC VESSELS; MARINE ECOLOGY; REMOTE SENSING; SEAWATER.

Since the early 1900s, all recorded ocean depths have been incorporated in the General Bathymetric Chart of the Ocean. The amount of data available increased greatly with the introduction of continuous echo sounders; subsequently, side-scan sonar permitted very detailed topographical surveys to be made of the ocean floor. The features thus revealed, in particular the midocean ridges (spreading centers) and deep trenches (subduction zones), are integral to the theory of plate tectonics. An important discovery made toward the end of the twentieth century was the existence of hydrothermal vents, where hot mineral-rich water gushes from the Earth's interior. The deposition of minerals at these sites and the discovery of associated ecosystems make them of potential economic as well as great scientific interest. See ECHO SOUNDER; HYDROTHERMAL VENT; MARINE GEOLOGY; MID-OCEANIC RIDGE; PLATE TECTONICS; SUBDUCTION ZONES.   [M.De.]

Oil field waters   Waters of varying mineral content which are found associated with petroleum and natural gas or have been encountered in the search for oil and gas. They are also called oil field brines, or brines. They include a variety of underground waters, usually deeply buried, and have a relatively high content of dissolved mineral matter. These waters may be (1) present in the pore space of the reservoir rock with the oil or gas, (2) separated by gravity from the oil or gas and thus lying below it, (3) at the edge of the oil or gas accumulation, or (4) in rock formations which are barren of oil and gas. Brines are commonly defined as water containing high concentrations of dissolved salts. Potable or fresh waters usually are not considered oil field waters but may be encountered, generally at shallow depths, in areas where oil and gas are produced.

Probably the most important geological use of oil field water analyses is their application to the quantitative interpretation of electrical and neutron well logs, particularly micrologs. See PETROLEUM GEOLOGY.   [P.McG.]

Oil sand   A loose to consolidated sandstone or a porous carbonate rock, impregnated with a heavy asphaltic crude oil, too viscous to be produced by conventional methods; also known as tar sand or bituminous sand.

Oil sands are distributed throughout the world but the largest proven accumulation occurs in Alberta, Canada (the Athabasca deposit). A large accumulation appears to be present in the Orinoco Basin in Venezuela, and far smaller deposits occur in Russia, the United States, Madagascar, Albania, Trinidad, and Romania.   [G.R.G.]

Oil shale   A sedimentary rock containing solid, combustible organic matter in a mineral matrix. The organic matter, often called kerogen, is largely insoluble in petroleum solvents, but decomposes to yield oil when heated. Although "oil shale" is used as a lithologic term, it is actually an economic term referring to the rock's ability to yield oil; oil shale appears to be the cheapest source after natural petroleum for large amounts of liquid fuels. No real minimum oil yield or content of organic matter can be established to distinguish oil shale from sedimentary rocks. Additional names given to oil shales include black shale, bituminous shale, carbonaceous shale, coaly shale, cannel shale, cannel coal, lignitic shale, torbanite, tasmanite, gas shale, organic shale, kerosine shale, coorongite, maharahu, kukersite, kerogen shale, algal shale, and "the rock that burns." *See* KEROGEN.

The world's oil shale deposits represent a tremendous store of fossil energy. It has been estimated that the organic matter in sedimentary rocks contains 1.2×10^{16} tons (1.1×10^{16} metric tons) of organic carbon, nearly 1000 times that found in coals. Although part of that organic carbon has matured to produce oil and gas, most of it is still oil shale. Unfortunately, most of this tremendous resource is not well known. Oil shales occur on *every* continent in sediments ranging in age from Cambrian to Tertiary. Estimates for the total oil resource in shales of all grades reached 1.75×10^{15} barrels. Just 1% of that total shale oil represents more oil than the world is expected to produce as natural petroleum (2×10^{12} bbl). Oil shale represents a tremendous supply of liquid fuels.

Although the oil potential of the world's oil shales is great, commercial production of this oil has been considered uneconomic. Oil shales are lean ores, producing only limited amounts of oil which historically has been low in price. Mining and heating 1 ton of relatively rich oil shale yielding 25 gal/ton produces only 0.6 bbl of oil.

Shale oil is produced from the organic matter in oil shale when the rock is heated in the absence of oxygen (destructive distillation). This heating process is called retorting, and the equipment that is used to do the heating is known as a retort. The rate at which the oil is produced depends upon the temperature at which the shale is retorted. Most references report retorting temperatures as being about 500°C (930°F).   [J.W.S.; H.B.J.]

Oligocene   The third oldest of the seven geological epochs of the Cenozoic Era. It corresponds to an interval of geological time (and rocks deposited during that time) from the close of the Eocene Epoch to the beginning of the Miocene Epoch. The most recent geological time scales assign an age of 34 to 24 million years before present (m.y. B.P.) to the Oligocene Epoch. *See* CENOZOIC; EOCENE; MIOCENE.

An important event that characterizes the Oligocene Epoch was the development of extensive glaciation on the continent of Antarctica. Prior to that time, the world was largely ice-free through much of the Mesozoic and early Tertiary. A significant amount of ice is now known to have existed on the Antarctic continent since at least the beginning of the Oligocene, when the Earth was ushered into its most recent phase of ice-house conditions. This in turn created revolutions in the global climatic and hydrographic systems, with important repercussions for the marine and terrestrial biota. The changes include steepened latitudinal and vertical thermal gradients affecting major fluctuations in global climates, and the shift in the route of global dispersal of marine biota from an ancestral equatorial Tethys seaway, which had become severely restricted by Oligocene time, to the newly initiated circum-Antarctic circulation. *See* GLACIAL EPOCH.

Worldwide, the epoch represents an overall regressive sequence when there was a drawdown of global sea level, with relatively deeper, marine facies in the Early Oligocene and shallower-water to nonmarine facies in the late Oligocene. *See* FACIES (GEOLOGY).

Due to accentuated thermal gradients and seasonality in the Oligocene, marine biotic provinces became more fragmented. Extreme climates, with greater diurnal and seasonal temperature contrasts, are held responsible for reduced diversities in marine plankton. The Oligocene was characterized by transitional faunal features between the Paleogene and the Neogene.

Rhinocerids, tapirs, and wild boarlike hog species with strong incisors and large canines appeared. *Hyaenodon* was an Oligocene carnivore with strong canines and sharp molars, much like those of the modern cat species. The horses that had first appeared in the Eocene continued to increase in size and became three-toed, typified by

Mesohippus. Elephants made their first appearance near the Eocene-Oligocene boundary and developed a short trunk and two pairs of tusks. An early simian, *Propliopithecus*, made its first appearance in Oligocene, and is considered ancestral to the modern family of gibbons. The general uniformity of mammalian fauna in the Oligocene suggests that the widespread regressions of the sea most likely resulted in land bridges that reconnected some of the Northern Hemispheric landmasses, which may have led to transmigrations of some families of mammals between North America, Asia, and Africa. Birds had achieved some of their modern characteristics, and at least 10 modern genera had already made their appearance by the close of Oligocene time. *See* GEOLOGIC TIME SCALE. [B.U.H.]

Oligoclase A plagioclase feldspar with composition in the range $Ab_{90}An_{10}$ to $Ab_{70}An_{30}$, where Ab represents the composition of albite, $NaAlSi_3O_8$, and An represents the composition of anorthite, $CaAl_2Si_2O_8$. The diagnostic properties are hardness on Mohs scale, 6–6.5; density, 2.65 g/cm^3; and color usually white or colorless, transparent to translucent. The presence of minute, mutually parallel inclusions of hematite (Fe_2O_3) causes a golden play of color in the variety of oligoclase called aventurine or sunstone. Oligoclase is triclinic. The mineral is common in igneous rocks and metamorphic rocks. *See* ALBITE; ANORTHITE; FELDSPAR; IGNEOUS ROCKS; METAMORPHIC ROCKS. [D.T.G.]

Olivine It is generally accepted that the Earth's upper mantle consists mainly of olivine, an orthorhombic silicate with the composition $(Mg_{1.8},Fe_{0.2})SiO_4$, together with some pyroxene and garnet. The natural occurrence of two high-pressure forms (polymorphs) of olivine—orthorhombic wadsleyite, and cubic ringwoodite (with a spinel structure)—was predicted from high-pressure experiments and was later confirmed by meteorite investigations. The names olivine, wadsleyite, and ringwoodite refer only to naturally occurring compositions [$(Mg,Fe)_2SiO_4$].

Because of their abundance in the Earth's mantle, knowledge of physical and chemical properties of olivine, wadsleyite, and ringwoodite is of great geophysical importance. Until recently, many of these properties had to be inferred from theoretical considerations and from experiments on chemical analogs, which transform at lower pressures. With the development of new experimental apparatus capable of generating very high pressures and temperatures (multianvil press and diamond anvil cell), a growing number of experimental studies are being performed on phases of natural composition.

Experimentally determined thermodynamic phase equilibria data indicate that in the Earth's mantle olivine transforms to wadsleyite, then to ringwoodite, and finally to compositions of magnesiowüstite plus perovskite. Estimated transformation pressures correspond closely to the discontinuities of seismic velocities at, respectively, 246, 322, and 417 mi (410, 520, and 670 km) depth in the mantle. [L.Ke.]

Omphacite The pale to bright green monoclinic pyroxene found in eclogites and related rocks. Omphacites are essentially members of the solid solution series between jadeite ($NaAlSi_2O_6$) and diopside ($CaMgSi_2O_6$). The density ranges from 3.16 to 3.43 g/cm^3; hardness is 5–6.

Omphacite is stable only at the relatively high pressures of the blueschist and eclogite facies of metamorphism, where it is associated with minerals such as glaucophane and lawsonite or pyropic garnet, respectively. In such environments it also occurs in veins, either on its own or with quartz. *See* GLAUCOPHANE; PYROXENE. [T.J.B.H.]

Onyx Banded chalcedonic quartz, in which the bands are straight and parallel, rather than curved, as in agate. Unfortunately, in the colored-stone trade, gray chalcedony dyed in various solid colors such as black, blue, and green is called onyx, with the color used as a prefix. Because the color is permanent, the fact that it is the result of dyeing is seldom mentioned.

The natural colors of true onyx are usually red or brown with white, although black is occasionally encountered as one of the colors. When the colors are red-brown with white or black, the material is known as sardonyx; this is the only kind commonly used as a gemstone. Its most familiar gem use is in cameos and intaglios. *See* CAMEO; CHALCEDONY.   [R.T.L]

Oolite A deposit containing spheroidal grains with a mineral cortex, most commonly calcite or aragonite, accreted around a nucleus formed primarily of shell fragments or quartz grains. The term ooid is applied to grains less than 0.08 in. (2 mm) in diameter, and the term pisoid to those greater than 0.08 in. (2 mm). Accretionary layering (growth banding) is usually developed clearly. A flattened or elongate shape may occur if the nucleus shows that form. Ooids formed on nuclei of shells and shell fragments and composed of fine, radial calcite are cemented by coarse, clear calcite. Growth banding is visible in most ooids. The pisoids are composed of many thin layers of very small, tangential (lighter layers) and radial (darker layers) aragonite crystals. These pisoids are cemented with fibrous aragonite.

Ooids are primarily marine, forming in agitated shallow, warm waters. Under those conditions, the ooids are kept intermittently moving, so accretion occurs on all sides. Some ooids and most pisoids form in nonmarine environments, such as hypersaline and fresh-water lakes, hot springs, caves, caliche soils, and some rivers. *See* ARAGONITE; CALCITE.   [P.A.Sa.]

Opal A natural hydrated form of silica. Opal is a relatively common mineral in its nongem form, which is known as common opal and lacks the play of color for which gem, or precious, opal is known. All opal is of relatively simple chemical composition, $SiO_2 \cdot nH_2O$. The hardness of opal on Mohs hardness scale ranges from 5 to 6, the specific gravity from 2.25 to 1.99, and the refractive index from 1.455 to 1.435. *See* SILICATE MINERALS.

The color of common opal ranges from transparent, glassy, and colorless to white and bluish white. Common pigmenting agents, such as iron, produce yellow, brown, red, and green colors, and frequently several colors in a single specimen. Precious opal has a play of color that is the result of white light being diffracted by the relatively regular internal array of silica spheres. Because opal is a hydrous mineral, certain opals from specific geologic occurrences may crack because of water loss. Therefore, considerable care is required in the polishing and handling of opal.

Several trade terms are used to describe the appearance of precious opal based on transparency, body color, and the type of play of color. Some of these terms are black opal, which is translucent to almost opaque, with dark gray to black body color, with play of color; fire opal, which is transparent to semitransparent, with yellow, orange, red, or brown body color and with or without play of color; harlequin or mosaic opal, in which the play of color occurs in distinct, broad, angular patches; and matrix opal, which consists of thin seams of high-quality gem opal in a matrix. *See* GEM.   [P.J.Da.; C.K.]

Ophiolite A distinctive assemblage of mafic plus ultramafic rocks generally considered to be fragments of the oceanic lithosphere that have been tectonically emplaced onto continental margins and island arcs. An ophiolite is a formation made up of an association of typical rocks in a clearly defined sequence. A complete idealized ophiolite sequence from bottom to top includes (1) an ultramafic tectonite complex composed mostly of multilayered, deformed harzburgite, dunite, and minor chromitite; (2) a plutonic complex of layered mafic-ultramafic cumulates at the base, grading upward to massive gabbro, diorite, and possibly plagiogranite; (3) a mafic sheeted-dike complex; (4) an extrusive section of massive and pillow lavas, pillow breccias, and intercalated pelagic sediments; and (5) a top layer of abyssal or bathyal sediments, which may include ribbon chert, red pelagic limestone, metalliferous sediments, volcanic breccias, or pyroclastic deposits. Most ophiolites lack complete sections, and are dismembered and fragmented. Their estimated original thickness is variable, ranging from about 2 km (1.2 mi) to more than 8 km (5 mi). *See* EARTH CRUST; LITHOSPHERE.

Ophiolites typically occur in collisional mountain belts or island arcs and define a suture zone marking the boundary where two plates have welded together. The ophiolite complex is interpreted as evidence for a closed marginal ocean or back-arc basin. Throughout the world, ophiolites occur as long narrow belts, up to 10 km (6 mi) wide, that can extend more than 1000 km (600 mi) in length, in two distinct geographic settings. (1) Those in the Alpine-Mediterranean region, Tethyan ophiolite, were formed in small ocean basins that were surrounded by older, attenuated continental crust. (2) Those in western North America and the Circum-Pacific (Cordilleran) region seem to have formed in inter-arc basins. The Cordilleran ophiolites, such as the Trinity ophiolite and the Coast Range ophiolite of California, are generally incomplete, metamorphosed, or dismembered, but they commonly form the basement rocks for many North American continental margin terranes. *See* BASIN; CONTINENTAL MARGIN; STRUCTURAL GEOLOGY; GEODYNAMICS.

Ophiolites represent new oceanic crust formed in a variety of spreading environments, including oceanic ridge, back-arc basin, and island arcs above a subduction zone, and subsequently emplaced onto the continents. Their occurrence along plate sutures marks the sites of ancient tectonic interaction between oceanic and continental crust. Ophiolites provide the best opportunity for geologists to study the ocean floor on land; they also offer vertical sections in addition to horizontal distributions. Moreover, ophiolite formations record the ages of oceanic fragments that escaped, disappearing into subduction zones. [J.Li.; S.Mar.; Y.O.]

Ordovician The second-oldest period in the Paleozoic Era. The Ordovician is remarkable because not only did one of the most significant Phanerozoic radiations of marine life take place (early Middle Ordovician), but also one of the two or three most severe extinctions of marine life occurred (Late Ordovician). The early Middle Ordovician radiation of life included the initial colonization of land. These first terrestrial organisms were nonvascular plants. Vascular plants appeared in terrestrial settings shortly afterward. *See* GEOLOGIC TIME SCALE.

The rocks deposited during this time interval (these are termed the Ordovician System) overlie those of the Cambrian and underlie those of the Silurian. The Ordovician Period was about 7×10^7 years in duration, and it lasted from about 5.05×10^8 to about 4.35×10^8 years ago.

The Ordovician System is recognized in nearly all parts of the world, including the peak of Mount Everest, because the groups of fossils used to characterize the system are so broadly delineated. Biogeographic provinces limited the distribution of organisms in the past to patterns similar to those of modern biogeographic provinces. Three broadly defined areas of latitude—the tropics, the midlatitudes (approximately 30–60°S), and the Southern Hemisphere high latitudes—constitute the biogeographic regions. Provinces may be distinguished within these three regions based upon organismal associations unique to each province.

Early Ordovician environmental conditions in most areas were similar to those of the Late Cambrian. Accordingly, Early Ordovician life was similar to that of the latter part of the Cambrian. Trilobites were the prominent animal in most shelf sea environments. Long straight-shelled nautiloids, certain snails, a few orthoid brachiopods, sponges, small echinoderms, algae, and bacteria flourished in tropical marine environments. Linguloid brachiopods and certain bivalved mollusks inhabited cool-water, nearshore environments.

Middle Ordovician plate motions were acompanied by significant changes in life. On land, nonvascular, mosslike plants appeared in wetland habitats. Vascular plants appeared slightly later in riverine habitats. The first nonvascular plants occurred in the Middle East on Gondwanan shores. The Middle Ordovician radiation of marine invertebrates is one of the most extensive in the record of Phanerozoic marine life. Corals, bryozoans, several types of brachiopods, a number of crinozoan echniderms, conodonts, bivalved mollusks, new kinds of ostracodes, new types of trilobites, and new kinds of nautiloids suddenly developed in tropical marine environments. As upwelling conditions formed along the plate margins, oxygen minimum zones—habitats preferred by many graptolites—expanded at numerous new sites. Organic walled microfossils (chitinozoans and acritarchs) radiated in mid- to high-latitude environments. Ostracoderms (jawless, armored fish) radiated in tropical marine shallow-shelf environments. These fish were probably bottom detritus feeders.

The latest Ordovician stratigraphic record suggests that glacial ice melted relatively quickly, accompanied by a relatively rapid sea-level rise in many areas. Some organisms—certain conodonts, for example—did not endure significant extinctions until sea levels began to rise and shelf sea environments began to expand. *See* PALEO-CEANOGRAPHY; STRATIGRAPHY.   [W.B.N.B.]

Ore and mineral deposits   Ore deposits are naturally occurring geologic bodies that may be worked for one or more metals. The metals may be present as native elements, or, more commonly, as oxides, sulfides, sulfates, silicates, or other compounds. The term ore is often used loosely to include such nonmetallic minerals as fluorite and gypsum. The broader term, mineral deposits, includes, in addition to metalliferous minerals, any other useful minerals or rocks. Minerals of little or no value which occur with ore minerals are called gangue. Some gangue minerals may not be worthless in that they are used as by-products; for instance, limestone for fertilizer or flux, pyrite for making sulfuric acid, and rock for road material.

Mineral deposits that are essentially as originally formed are called primary or hypogene. The term hypogene also indicates formation by upward movement of material. Deposits that have been altered by weathering or other superficial processes are secondary or supergene deposits. Mineral deposits that formed at the same time as the enclosing rock are called syngenetic, and those that were introduced into preexisting rocks are called epigenetic.

The distinction between metallic and nonmetallic deposits is at times an arbitrary one since some substances classified as nonmetals, such as lepidolite, spodumene, beryl, and rhodochrosite, are the source of metals. The principal reasons for distinguishing nonmetallic from metallic deposits are practical ones, and include such economic factors as recovery methods and uses.

Most mineral deposits are natural enrichments and concentrations of original material produced by different geologic processes. Economic considerations, such as the amount and concentration of metal, the cost of mining and refining, and the market value of the metal, determine whether the ore is of commercial grade. To be of commercial grade, for example, the following metals must be concentrated in the amounts indicated: aluminum, about 30%; copper, 0.7–10%; lead, 2–4%; zinc, 3–8%; and gold, silver, and uranium, only a small fraction of a percent of metal. *See* GEOCHEMICAL PROSPECTING; MINERAL.   [A.F.H.]

Organic geochemistry The study of the abundance and composition of naturally occurring organic substances, their origins and fate, and the processes that affect their distributions on Earth and in extraterrestrial materials. These activities share the common need for identification, measurement, and assessment of organic matter in its myriad forms.

Organic geochemistry was born from a curiosity about the organic pigments extractable from petroleum and black shales. It developed with extensive investigations of the chemical characteristics of petroleum and petroleum source rocks as clues to their occurrence and formation, and now encompasses a broad scope of activities within interdisciplinary areas of earth and environmental science. This range of studies recognizes the potential of geological records of organic matter to help characterize sedimentary depositional environments and to provide evidence of ancient life and indications of evolutionary developments through the Earth's history. Organic geochemistry includes determinations of anthropogenic contaminants amid the natural background of organic molecules and the assessment of their environmental impact and fate. Marine organic geochemistry addresses and interprets aquatic processes involving carbon species. It involves investigations of the chemical character of particulate and dissolved organic matter, evaluation of oceanic primary production including the factors (light, temperature, nutrient availability) that influence the uptake of carbon dioxide (CO_2), the composition of marine organisms, and the subsequent processing of organic constituents through the food web. Organic geochemistry extends to broader biogeochemical issues, such as the carbon cycle, and the effects of changing carbon dioxide levels, especially efforts to use geochemical data and proxies to help constrain global climate models. Examination of the organic chemistry of meteorites and lunar materials also falls within its compass, and as a critical part of the quest for remnants of life on Mars, such extraterrestrial studies are now regaining the prominence they held in the 1970s during lunar exploration. *See* GEOCHEMISTRY.

Global inventories of carbon. Carbon naturally exists as oxidized and reduced forms in carbonate carbon and organic matter. The major reservoir of both forms of carbon on Earth is the geosphere. It contains carbonate minerals deposited as sediments and organic matter accumulated from the remains of dead organisms. Estimates of the size of the geological reservoir of carbon vary within the range of 5 to 7 $\times$ 10^{22} g, of which 75% is carbonate carbon and 25% is organic carbon. The amounts of carbon contained in living biota (5×10^{17} g), dissolved in the ocean (4×10^{19} g), and present in atmospheric gases (7×10^{17} g) are miniscule compared to the quantity of organic carbon buried in the rock record. The importance of buried organic matter extends beyond its sheer magnitude; it includes the fossil fuels—coal, natural gas, and petroleum—that supply 85% of the world's energy. *See* BIOGEOCHEMISTRY; CARBONATE MINERALS; COAL; NATURAL GAS; PETROLEUM; SEDIMENTARY ROCKS.

Sedimentary organic matter. The vast amounts of organic matter contained in geological materials represent the accumulated vestiges of organisms amassed over the expanse of geological time. Yet, survival of organic cellular constituents of biota into the rock record is the exception rather than the norm. Only a small portion of the carbon fixed by organisms during primary production, especially by photosynthesis, escapes degradation as it settles through the water column and eludes microbial alteration during subsequent incorporation and assimilation into sedimentary detritus. *See* BIODEGRADATION.

Sedimentary organic matter can be divided operationally into solvent-extractable bitumen and insoluble kerogen. Bitumens contain a myriad of structurally distinct molecules, especially hydrocarbons, which can be individually identified (such as by gas chromatography-mass spectrometry) although they may be present in only minute quantities (nanograms or picograms). The range of components includes many biomarkers that retain structural remnants inherited from their source organisms, which attest to their biological origins and subsequent geological fate. *See* BITUMEN; KEROGEN.

Biomarkers are individual compounds whose chemical structures carry evidence of their origins and history. Recognition of the specificity of biomarker structures initially helped confirm that petroleum was derived from organic matter produced by biological processes. Of the thousands of individual petroleum components, hundreds reflect precise biological sources of organic matter, which distinguish and differentiate their disparate origins. The diagnostic suites of components may derive from individual families of organisms, but contributions at a species level can occasionally be recognized. Biomarker abundances and distributions help to elucidate sedimentary environments, providing evidence of depositional settings and conditions. They also reflect sediment maturity, attesting to the progress of the successive, sequential transformations that convert biological precursors into geologically occurring products. Thus, specific biomarker characteristics permit assessment of the thermal history of individual rocks or entire sedimentary basins. *See* BASIN.

Carbon isotopes. Carbon naturally occurs as three isotopes: carbon-12 (^{12}C), carbon-13 (^{13}C), and radiocarbon (^{14}C). Temporal excursions in the ^{13}C values of sediment sequences can reflect perturbations of the global carbon cycle. Radiocarbon is widely employed to date archeological artifacts, but the sensitivity of its measurement also permits its use in exploration of the rates of biogeochemical cycling in the oceans. This approach permits assessment of the ages of components in sediments, demonstrating that bacterial organic matter is of greater antiquity than components derived from phytoplankton sources. *See* MARINE SEDIMENTS; PALEOCEANOGRAPHY; RADIOCARBON DATING.   [S.C.B.]

Orogeny   The process of mountain building. As traditionally used, the term orogeny refers to the development of long, mountainous belts on the continents that are called orogenic belts or orogens. These include the Appalachian and Cordilleran orogens of North America, the Andean orogen of western South America, the Caledonian orogen of northern Europe and eastern Greenland, and the Alpine-Himalayan orogen that stretches from western Europe to eastern China. It is important to recognize that these systems represent only the most recent orogenic belts that retain the high relief characteristic of mountainous regions. In fact, the continents can be viewed as a collage of ancient orogenic belts, most of which are so deeply eroded that no trace of their original mountainous topography remains. By comparing characteristic rock assemblages from more recent orogens with their deeply eroded counterparts, geologists surmise that the processes responsible for mountain building today extended back through most (if not all) of geologic time and played a major role in the growth of the continents. *See* CONTINENTS, EVOLUTION OF.

The construction of mountain belts is best understood in the context of plate tectonics theory. Orogenic belts form at convergent boundaries, where lithosphere plates collide. *See* LITHOSPHERE; PLATE TECTONICS.

There are two basic kinds of convergent plate boundaries, leading to the development of two end-member classes of orogenic belts. Oceanic subduction boundaries are those at which oceanic lithosphere is thrust (subducted) beneath either continental or oceanic lithosphere. The process of subduction leads to partial melting near the plate boundary at depth, which is manifested by volcanic and intrusive igneous activity in the overriding plate. Where the overriding plate consists of oceanic lithosphere, the result is an intraoceanic island arc, such as the Japanese islands. Where the overriding plate is continental, a continental arc is formed. The Andes of western South America is an example. *See* MARINE GEOLOGY; OCEANIC ISLANDS; SUBDUCTION ZONES.

The second kind of convergent plate boundary forms when an ocean basin between two continental masses has been completely consumed at an oceanic subduction boundary and the continents collide. Continent collisional orogeny has resulted in some of the most dramatic mountain ranges on Earth; a good example is the Himalayan orogen, which began forming roughly 50 million years ago when India collided with the Asian continent. Because the destruction of oceanic lithosphere at subduction boundaries is a prerequisite for continental collision, continent collisional orogens contain deformational features and rock associations developed during arc formation as well as those produced by continental collision. [K.Ho.]

Orpiment   A mineral having composition As_2S_3. Crystals are small, tabular, and rarely distinct; the mineral occurs more commonly in foliated or columnar masses. The hardness is 1.5–2 (Mohs scale) and the specific gravity is 3.49. The luster is resinous and pearly on the cleavage surface; the color is lemon yellow. Orpiment is found in Romania, Peru, Japan, and Russia. In the United States it occurs at Mercer, Utah; Manhattan, Nevada; and in deposits from geyser waters in Yellowstone National Park.

[C.S.Hu.]

Orthoclase   Potassium feldspar (Or $=$ $KAlSi_3O_8$) that usually contains up to 30 mole % albite (Ab $=$ $NaAlSi_3O_8$) in solid solution. Its hardness is 6; specific gravity, 2.57–2.5, depending on Ab content; mean refractive index, 1.52; color, white to dull pink or orange-brown. Some orthoclases may be intergrown with relatively pure albite which exsolved during cooling from a high temperature in pegmatites, granites, or granodiorites. This usually is ordered low albite, but in rare cases it may show some degree of Al,Si disorder, requiring it to be classified as analbite or high albite. If exsolution is detectable by eye, the Or-Ab composite mineral is called perthite; if microscopic examination is required to distinguish the phases, it is called microperthite; and if exsolution is detectable only by x-ray diffraction or electron optical methods, it is called cryptoperthite. Orthoclase is optically monoclinic. Its structure averaged over hundreds of nanometers may be monoclinic, but its true symmetry is triclinic. *See* ALBITE; PERTHITE.

[P.H.R.]

Orthorhombic pyroxene   A group of minerals having the general chemical formula $XYSi_2O_6$, in which the Y site contains Fe or Mg and the X site contains Fe, Mg, Mn, or a small amount of Ca (up to about 3%). The end members of this solid solution series are enstatite ($Mg_2Si_2O_6$) and ferrosilite ($Fe_2Si_2O_6$). Names used for intermediate members of the series are enstatite, bronzite, hypersthene, and orthoferrosilite.

Many of the physical and optical properties of orthopyroxene are strongly dependent upon composition, and especially upon the Fe-Mg ratio. In hand specimens, orthopyroxene can be distinguished from amphibole by its characteristic $88°$ cleavage angles, and from augite by color—augite is typically green to black, while orthopyroxene is more commonly brown, especially on slightly weathered surfaces.

Orthopyroxene is a widespread mineral in metamorphic rocks. It is characteristic of granulite facies metamorphism, in both mafic and feldspathic gneisses. Orthopyroxene occurs in many basalts and gabbros, particularly those of tholeiitic composition, and many meteorites, but is notably absent from most alkaline igneous rocks. The greatest abundance of orthopyroxene is in ultramafic rocks, especially those in large layered intrusions. *See* PYROXENE. [R.J.Tr.]

P

Pacific islands A geographic designation that includes thousands of mainly small coral and volcanic islands scattered across the Pacific Ocean from Palau in the west to Easter Island in the east. Island archipelagos off the coast of the Asian mainland, such as Japan, Philippines, and Indonesia, are not included even though they are located within the Pacific Basin. The large island constituting the mainland of Papua New Guinea and Irian Jaya is also excluded, along with the continent of Australia and the islands that make up Aotearoa or New Zealand. The latter, together with the Asian Pacific archipelagos, contain much larger landmasses, with a greater diversity of resources and ecosystems, than the oceanic islands, commonly labelled Melanesia, Micronesia, and Polynesia. *See* AUSTRALIA; NEW ZEALAND; OCEANIC ISLANDS.

The great majority of these islands are between 4 and 4000 mi^2 (10 and 10,000 km^2) in land surface area. The three largest islands include the main island of New Caledonia (6220 mi^2 or 16,100 km^2), Viti Levu (4053 mi^2 or 10,497 km^2) in Fiji, and Hawaii (4031 mi^2 or 10,440 km^2) the big island in the Hawaiian chain. When the 80-mi (200-km) Exclusive Economic Zones are included in the calculation of surface area, some Pacific island states have very large territories. These land and sea domains, far more than the small, fragmented land areas per se, capture the essence of the island world that has meaning for Pacific peoples. *See* EAST INDIES.

Oceanic islands are often classified on the basis of the nature of their surface lithologies. A distinction is commonly made between the larger continental islands of the western Pacific, the volcanic basalt island chains and clusters of the eastern Pacific, and the scattered coral limestone atolls and reef islands of the central and northern Pacific.

It has been suggested that a more useful distinction can be drawn between plate boundary islands and intraplate islands. The former are associated with movements along the boundaries of the great tectonic plates that make up the Earth's surface. Islands of the plate boundary type form along the convergent, divergent, or tranverse plate boundaries, and they characterize most of the larger island groups in the western Pacific. These islands are often volcanically and tectonically active and form part of the Pacific so-called Ring of Fire, which extends from Antarctica in a sweeping arc through New Zealand, Vanuatu, Bougainville, and the Philippines to Japan.

The intraplate islands comprise the linear groups and clusters of islands that are thought to be associated with volcanism, either at a fixed point or along a linear fissure. Volcanic island chains such as the Hawaii, Marquesas, and Tuamotu groups are classic examples. Others, which have their volcanic origins covered by great thickness of coral, include the atoll territories of Kiribati, Tuvalu, and the Marshall Islands. Another type of intraplate island is isolated Easter Island, possibly a detached piece of a mid-ocean ridge. The various types of small islands in the Pacific are all linked geologically to much larger structures that lie below the surface of the sea. These structures contain the answers to some puzzles about island origins and locations, especially when considered in terms of the plate tectonic theory of crustal evolution. *See* MARINE GEOLOGY; MID-OCEANIC RIDGE; PLATE TECTONICS; SEAMOUNT AND GUYOT; VOLCANO.

The climate of most islands in the Pacific is dominated by two main forces: ocean circulation and atmospheric circulation. Oceanic island climates are fundamentally distinct from those of continents and islands close to continents, because of the small size of the island relative to the vastness of the ocean surrounding it. Because of oceanic influences, the climates of most small, tropical Pacific islands are characterized by little variation through the year compared with climates in continental areas.

The major natural hazards in the Pacific are associated either with seasonal climatic variability (especially cyclones and droughts) or with volcanic and tectonic activity. *See* CLIMATE HISTORY.   [R.D.Be.]

Pacific Ocean   The Pacific Ocean has an area of 6.37×10^7 mi^2 (1.65×10^8 km^2) and a mean depth of 14,000 ft (4280 m). It covers 32% of the Earth's surface and 46% of the surface of all oceans and seas, and its area is greater than that of all land areas combined. Its mean depth is the greatest of the three oceans and its volume is 53% of the total of all oceans. Its greatest depths in the Marianas and Japan trenches are the world's deepest, more than 6 mi (10 km).

The two major wind systems driving the waters of the ocean are the westerlies which lie about $40-50°$ lat in both hemispheres (the "roaring forties") and the trade winds from the east which dominate in the region between 20°N and 20°S. These give momentum directly to the west wind drift (flow to the east) in high latitudes and to the equatorial currents which flow to the west. At the continents there is flow of water from one system to the other and huge circulatory systems result. *See* OCEAN CIRCULATION; SOUTHEAST ASIAN WATERS.

The swiftest flow (greater than 2 knots) is found in the Kuroshio Current near Japan. It forms the northwestern part of a huge clockwise gyre whose north edge lies in the west wind drift centered at about 40°N, whose eastern part is the south-flowing California Current, and whose southern part is the North Equatorial Current.

Equatorward of 30° lat heat received from the Sun exceeds that lost by reflection and back radiation, and surface waters flowing into these latitudes from higher latitudes (California and Peru currents) increase in temperature as they flow equatorward and turn west with the Equatorial Current System. They carry heat poleward and transfer part of it to the high-latitude cyclones along the west wind drift. The temperature of the equatorward currents along the eastern boundaries of the subtropical anticyclones is thus much lower than that of the currents of their western boundaries at the same latitudes. The highest temperatures (more than 82°F or 28°C) are found at the western end of the equatorial region. Along the Equator itself somewhat lower temperatures are found. The cold Peru Current contributes to its eastern end, and there is apparent upwelling of deeper, colder water at the Equator.

Upwelling also occurs at the edge of the eastern boundary currents of the subtropical anticyclones. When the winds blow strongly equatorward (in summer) the surface waters are driven offshore, and the deeper colder waters rise to the surface and further reduce the low temperatures of these equatorward-flowing currents. *See* UPWELLING.

The limiting temperature in high latitudes is that of freezing. Ice is formed at the surface at temperatures slightly less than 30°F (−1°C) depending upon the salinity; further loss of heat is retarded by its insulating effect. The ice field covers the northern and eastern parts of the Bering Sea in winter, and most of the Sea of Okhotsk, including that part adjacent to Hokkaido (the north island of Japan). Summer temperatures, however, reach as high as 43°F (6°C) in the northern Bering Sea and as high as 50°F (10°C) in the northern part of the Sea of Okhotsk. *See* BERING SEA.

Pack ice reaches to about 62°S from Antarctica in October and to about 70°S in March, with icebergs reaching as far as 50°S. *See* ICEBERG; SEA ICE.

Surface waters in high latitudes are colder and heavier than those in low latitudes. As a result, some of the high-latitude waters sink below the surface and spread equatorward, mixing mostly with water of their own density as they move, and eventually become the dominant water type in terms of salinity and temperature of that density over vast regions.

The most conspicuous water masses formed in the Pacific are the Intermediate Waters of the North and of the South Pacific, which on the vertical sections include the two huge tongues of low salinity extending equatorward beneath the surface from about 55°S and from about 45°N. The southern tongue is higher in salinity and density and lies at a greater depth.   [J.L.Re.]

Paleoceanography   The study of the history of the ocean with regard to circulation, chemistry, biology, and patterns of sedimentation. The source of information is largely the biogenous deep-ocean sediments, so the field may be considered a branch of sedimentology or paleontology. However, there are also strong links to geophysics, marine geochemistry, and mathematical modeling. Geophysical sciences are called upon for reconstruction of geography (position of continents, horizontal motion of the ocean floor), topography (changing depth patterns in the ocean, general subsidence of any given piece of sea floor), and dating (radioisotopes, magnetic patterns on the sea floor and in sediments). Geochemical analyses deliver information on sediment composition (stable and unstable isotopes; major and minor components such as carbonate, opal, and trace elements). Such information is useful in correlating sedimentary sequences and, when combined with geochemical arguments, yields insights about the dynamics of carbon and nutrient cycles. Mathematical modeling introduces a strong quantitative element. It draws upon the knowledge reservoir of modern oceanography and climatology. *See* CLIMATOLOGY; GEOCHEMISTRY; GEOPHYSICS; PALEONTOLOGY; SEDIMENTOLOGY.

The study of marine sediments on land is as old as geology itself. Modern paleoceanography is set apart by the study of sediments recovered from the ocean, especially the deep ocean, and by the use of concepts developed by oceanographers (controls on ocean currents and upwelling, vertical stratification, heat budget, nutrient and carbon cycles, pelagic biogeography, and water masses). Important early studies were on cores raised by the United States cable ship *Lord Kelvin* (1936) in the North Atlantic. The glacial debris zones noted in the *Kelvin* cores are now commonly referred to as Heinrich layers; they are witness to sporadic input of iceberg armadas during the last glacial epoch. *See* GLACIAL EPOCH.

Comparisons of climate-related changes between major ocean basins became possible through the systematic recovery of long cores by the circumglobal Swedish Deep Sea Expedition (1947–1949) on the research vessel *Albatross*. The expedition retrieved cores up to 15 m (45 ft) long, with records reaching back 500,000–1,000,000 years. Many fundamental paleoceanographic concepts were established by the geologists who analyzed these cores, including the role of trade winds in promoting glacial-age upwelling, the changing supply of North Atlantic Deep Water, the cyclicity of climatic change in the late Quaternary, and large-scale shifts in biogeographic boundaries. *See* CLIMATE MODELING; PALEOCLIMATOLOGY; QUATERNARY.

A quantum jump in paleoceanographic research resulted from the initiation of deep-sea drilling using the research vessel *Glomar Challenger* (1968). Enormous blank regions on the world's map suddenly became accessible for detailed exploration far back into geologic time, that is, into the Early Cretaceous. Highlights of the first decade of drilling results include the documentation of cooling steps as the planet moved into the present ice age; the reconstruction of long-term fluctuations in the carbonate compensation depth; the documentation of large-scale salt deposition in an isolated Mediterranean basin; and the discovery of temporary anoxic conditions in the Cretaceous deep sea. *See* CRETACEOUS. [W.H.B.; G.We.]

Paleocene   The oldest of the seven geological epochs of the Cenozoic Era, and the oldest of the five epochs that make up the Tertiary Period. The Paleocene Epoch represents an interval of geological time (and rocks deposited during that time) from the end of the Cretaceous Period to the beginning of the Eocene Epoch. Recent revisions of the geological time scales place the Paleocene Epoch between 65 to 55 million years before present. *See* CENOZOIC; EOCENE; GEOLOGIC TIME SCALE; TERTIARY.

The close of the Cretaceous Period was characterized by the disappearance of many terrestrial and marine animals and plants. The dawn of the Cenozoic in the Paleocene Epoch saw the establishment of new fauna and flora that have evolved into modern biota.

Modern schemes of the Paleocene subdivide it into Lower and Upper series, and their formal equivalents, the Danian and Selandian stages. Some authors prefer to use a threefold subdivision of the Paleocene, adding the Thanetian at the top. The older, Danian lithofacies generally tend to be calcium carbonate-rich (pure chalk in the Danian type area), whereas the younger, Selandian and Thanetian facies have greater land-derived components and are more siliciclastic (sand, sandstone, marl). *See* CHALK; FACIES (GEOLOGY); MARL; SAND; SANDSTONE.

Several major tectonic events that began in the Mesozoic continued into the Paleocene. For example, the Laramide Orogeny that influenced deformation and uplift in the North American Rocky Mountains in the Mesozoic continued into the Paleocene. *See* OROGENY.

The establishment of deeper connections between the North and South Atlantic in the Paleocene facilitated enhanced deep-water flow from the northern to the southern basin. In the south, the Drake Passage between South America and Antarctica was still closed, although Australia had already separated from Antarctica by Paleocene time. The lack of circum-Antarctic flow precluded the geographic isolation of Antarctica and the development of cold deep water from a southern source. *See* PALEOCEANOGRAPHY; PALEOGEOGRAPHY.

Terrestrial floras and faunas corroborate the peak warming in the latest Paleocene and early Eocene and suggest that the warm tropical-temperate belt may have been twice its modern latitudinal extent. The temperate floral and faunal elements extended to 60°N, which has been used as an argument to invoke a very low angle of inclination of the Earth's rotational axis in the Paleocene-Eocene. Alternatively, the mild, equable polar climates and well-adapted physiological responses of plants and animals of those times to local conditions may be enough to explain the presence of a rich vertebrate fauna on Ellesmere Island in arctic Canada. *See* CLIMATE HISTORY; PALEOCLIMATOLOGY.

The Paleocene Epoch began after a meteorite struck the Earth, causing massive extinctions at the end of Cretaceous and decimating a large percentage of the terrestrial and marine biota. In the oceans, all ammonites, genuine belemnites, rudistids, most species of planktonic foraminifera and nannoplankton, and marine reptiles disappeared at the close of the Cretaceous Period. Even though some groups, such as squids, octopus, nautilus, and a few species of marine plankton, survived, the genetic pool was relatively small at the dawn of the Tertiary Period. The recovery of the marine biota was, however, fairly rapid after the mid-Paleocene due to overall transgressing seas and ameliorating climates. By the late Paleocene, the biota was well on its way to the high diversification of the Eocene. The end of the Paleocene Epoch saw marked changes in deep-water circulation of the world ocean that resulted in a massive extinction of the benthic marine species. *See* EXTINCTION (BIOLOGY).

On land the large dinosaurs, which had been on the decline for over 20 million years, died out at the close of the Cretaceous Period. However, smaller reptiles, including alligators and crocodiles, and some of the land flora escaped extinction and continued into the Paleocene. The Paleocene saw the first true radiation of mammals. The mammals of this epoch were characteristically primitive and small in size (50 cm or 20 in. or less). As the continent of Australia became more isolated geographically, its mammalian fauna, such as the marsupials, became sequestered and more specialized. *See* PALEONTOLOGY.   [B.U.H.]

Paleoclimatology   The study of ancient climates. Climate is the long-term expression of weather; in the modern world, climate is most noticeably expressed in vegetation and soil types and characteristics of the land surface. To study ancient climates, paleoclimatologists must be familiar with various disciplines of geology, such as sedimentology and paleontology, and with climate dynamics, which includes aspects of geography and atmospheric and oceanic physics. Understanding the history of the Earth's climate system greatly enhances the ability to predict how it might behave in the future. *See* CLIMATOLOGY.

Information about ancient climates comes principally from three sources: sedimentary deposits, including ancient soils; the past distribution of plants and animals; and the chemical composition of certain marine fossils. These are all known as proxy indicators of climate (as opposed to direct indicators, such as temperature, which cannot be measured in the past). In addition, paleoclimatologists use computer models of climate that have been modified for application to ancient conditions. *See* GEOLOGY; PALEONTOLOGY.

Like modern climatologists, paleoclimatologists are concerned with boundary conditions, forcing, and response. Boundary conditions are the limits within which the climate system operates. The boundary conditions considered by paleoclimatologists depend on the part of Earth history that is being studied. For the recent past, that is, the last few million years, boundary conditions that can change on short time scales are considered, for example, atmospheric chemistry. For the more distant past, paleoclimatologists must also consider boundary conditions that change on long time scales. Geographic features—that is, the positions of the continents, the location and orientation of major mountain ranges, the positions of shorelines, and the presence or absence of epicontinental seaways—are important for understanding paleoclimatic patterns. Forcing is a change in boundary conditions, such as continental drift, and response is how forcing changes the climate system. Forcing and response are cause and effect in paleoclimatic change. *See* CONTINENTAL DRIFT; CONTINENTS, EVOLUTION OF; PALEOGEOGRAPHY; PLATE TECTONICS.

Proxy indicators of paleoclimate are abundant in the geologic record. Important sedimentary indicators forming on land are coal, eolian sandstone (ancient sand dunes), evaporites (salt), tillites (ancient glacial deposits), and various types of paleosols (ancient soils), such as bauxite (aluminum ore) and laterite (some iron ores). Coals may form where conditions are favorable for growth of plants and accumulation and preservation of peat, conditions that are partly controlled by climate, especially seasonality of rainfall. *See* BAUXITE; COAL; PALEOSOL.

Fossil indicators provide information about climate mostly by their distribution (paleobiogeography), although a few specific types of fossils may be indicative of certain climatic conditions. The latter are usually fossils from the younger part of the geologic record and are closely related to modern species that have narrow environmental tolerances. Another type of information available for documenting paleoclimatic patterns and change is stable isotope geochemistry of fossils and certain types of sedimentary rock. Many elements that are used by organisms to make shells, teeth, and stems occur naturally in several different forms, known as isotopes. The most climatically useful isotopes are those of oxygen (O). Although the effects of temperature change and ice volume change can be difficult to distinguish, the analysis of oxygen isotopes has provided a powerful quantitative tool for the study of both long-term temperature change and the history of the polar ice caps. *See* FOSSIL.

A great deal of research in paleoclimatology has been devoted to understanding the causes of climatic change, and the overriding conclusion is that any given shift in the paleoclimatic history of the Earth was brought about by multiple factors operating in concert. The most important forcing factors for paleoclimatic variation are changes in paleogeography and atmospheric chemistry and variations in the Earth's orbital parameters. *See* ATMOSPHERIC CHEMISTRY; BIOGEOCHEMISTRY.   [J.T.P.; E.J.Ba.]

Paleogeography   The geography of the ancient past. Paleogeographers study the changing positions of the continents and the ancient extent of land, mountains, and shallow-sea and deep-ocean basins. The Earth's geography changes because its surface is in constant motion due to plate tectonics. The continents move at rates of 2–10 cm/yr (0.75–4 in./yr). Though this may seem slow, over millions of years continents can travel across the globe. As the continents move, new ocean basins form, mountains rise and erode, and sea level rises and falls. Paleogeographic maps are necessary in order to understand global climatic change, migration routes, oceanic circulation, mountain building, and the formation of many of the Earth's natural resources, including oil and gas. *See* BASIN; CONTINENTS, EVOLUTION OF; GEOGRAPHY; MID-OCEANIC RIDGE; PALEOCLIMATOLOGY; PLATE TECTONICS; SUBDUCTION ZONES.

In the late Precambrian the continents were colliding to form supercontinents, and the Earth was locked in a major ice age. About 1100 million years ago (Ma), the supercontinent of Rodinia was assembled. Rodinia split into halves approximately 750 Ma, opening the Panthalassic Ocean. By the end of the Precambrian three continents came together to form the supercontinent of Gondwana(land). This major continent-continent collision is known as the Pan-African orogeny. *See* OROGENY; PRECAMBRIAN; PROTEROZOIC; SUPERCONTINENT.

The supercontinent that formed at the end of the Precambrian Era, approximately 600 Ma, had already begun to break apart by the beginning of the Paleozoic Era. Gondwana, which was considerably larger than any of the other continents, stretched from the Equator to the south. *See* ORDOVICIAN; PALEOZOIC.

By the end of the Paleozoic Era, the continents had collided to form the supercontinent of Pangea. Centered on the Equator, Pangea stretched from the South Pole to the North Pole. Though the supercontinent that formed at the end of the Paleozoic Era is called Pangea (literally, "all land,"), this supercontinent probably did not include all the landmasses that existed at that time.

The supercontinent of Pangea did not rift apart all at once, but in three main episodes. The first episode of rifting began in the Middle Jurassic, about 180 Ma when North America rifted away from northwest Africa, opening the Central Atlantic. *See* JURASSIC.

The second phase in the breakup of Pangea began in the Early Cretaceous, about 140 Ma. Gondwana continued to fragment as South America separated from Africa, opening the South Atlantic, and India together with Madagascar rifted away from Antarctica and the western margin of Australia, opening the Eastern Indian Ocean. *See* CRETACEOUS.

The third and final phase in the breakup of Pangea took place during the early Cenozoic. North America and Greenland split away from Europe, and Antarctica released Australia. Australia, like India some 50 million years earlier, moved rapidly northward on a collision course with Southeast Asia. *See* CENOZOIC.

About 18,000 years ago, all of Antarctica and much of North America, northern Europe, and the mountainous regions of the world were covered by glaciers and great sheets of ice. These ice sheets melted approximately 10,000 years ago, giving rise to familiar geographic features such as Hudson's Bay, the Great Lakes, the English Channel, and the fiords of Norway.   [C.R.Sc.]

Paleomagnetism   The study of the direction and intensity of the Earth's magnetic field through geologic time. Paleomagnetism has been, and continues to be, an important tool in unraveling the past movements of the Earth's tectonic plates. By studying the records of the ancient magnetic field left in rocks, earth scientists are able to learn how the continental and oceanic plates have moved relative to the Earth's spin axis and relative to one another. In addition, the global reference frame of the Earth's magnetic field provides a very useful basis for temporal correlation of rocks on a local or global geographic scale (magnetostratigraphy). *See* GEOMAGNETISM; PLATE TECTONICS.

Many rocks acquire remanent magnetizations at or about the time they are formed. These magnetizations are nearly always parallel to the direction of the Earth's magnetic field at the locality where the rock formed. *See* ROCK MAGNETISM.

In paleomagnetic studies a suite of carefully oriented samples spanning a time interval long enough to average magnetic secular variations is collected. For magnetostratigraphy, an ordered suite of samples spanning the stratigraphic section of interest are collected. The samples are taken to the laboratory, where they are cut into small upright cylinders and their magnetization is measured by using a sensitive magnetometer.

The end product of the laboratory experiments is a suite of magnetization vector directions from the collected samples. These directions are specified by the inclination I, the angle that the magnetization vector makes with the horizontal, and the declination D, the angle that the projection of the magnetization vector upon a horizontal plane makes with true north, reckoned positive clockwise from north. Provided that the sample collection represents a sufficiently long time span to average out secular variation, representative mean D and I values and an associated uncertainty in direction may be calculated by using statistical techniques. The mean declination and inclination, together with the inclination-latitude relationship mentioned earlier and some elementary spherical trigonometry, allow the calculation of a representative paleomagnetic pole from the rock unit. By connecting paleomagnetic poles of different ages in an ordered time sequence, an apparent polar wander path (APWP) may be constructed for a particular tectonic plate (see illustration). The APWP specifies the displacement history of a plate or continent with respect to the spin axis, and can be directly compared with APWPs from other plates or continents to determine whether relative movements have occurred.

The end product of a magnetostratigraphic study is a set of normal (N) and reversed (R) magnetizations from the stratigraphic section under investigation. The positioning and frequency of occurrence of these N-to-R and R-to-N transitions is highly diagnostic in many cases, and by using these data together with other local geologic information, such as the position of major unconformities, one stratigraphic section can be

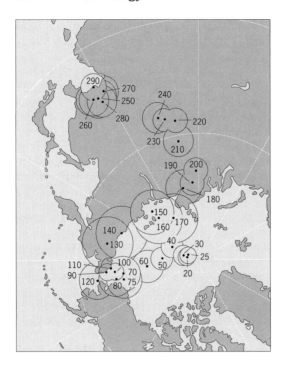

Apparent polar wander path for North America for Late Carboniferous time to the present. The numbers represent time in millions of years before present. Circles encompass standard error. (*After E. Irving, Paleopoles and paleolatitudes of North America and speculations about displaced terranes, Can. J. Earth Sci., 16:669–694, 1979*)

correlated with another over considerable distances. The method can also be used over intracontinental and intercontinental distances. However, because the field has only two possible states (N or R), correlation over longer distances where tectonics and sedimentation rates may vary is correspondingly less accurate.   [M.O.McW.]

Paleontology   The study of animal history as recorded by fossil remains. The fossil record includes a very diverse class of objects ranging from molds of microscopic bacteria in rocks more than 3×10^9 years old to unaltered bones of fossil humans in ice-age gravel beds formed only a few thousand years ago. Quality of preservation ranges from the occasional occurrence of soft parts (skin and feathers, for example) to barely decipherable impressions made by shells in soft mud that later hardened to rock. *See* Fossil; Micropaleontology.

The most common fossils are hard parts of various animal groups. Thus the fossil record is not an accurate account of the complete spectrum of ancient life but is biased in overrepresenting those forms with shells or skeletons. Fossilized worms are extremely rare, but it is not valid to make the supposition that worms were any less common in the geologic past than they are now.

The data of paleontology consist not only of the parts of organisms but also of records of their activities: tracks, trails, and burrows. Even chemical compounds formed only by organisms can, if extracted from ancient rocks, be considered as part of the fossil record. Artifacts made by people, however, are not termed fossils, for these constitute the data of the related science of archeology, the study of human civilizations. *See* Archeology.

Paleontology lies on the boundary between two disciplines, biology and geology. *See* Geology.

Geological aspects. A major task of any historical science, such as geology, is to arrange events in a time sequence and to describe them as fully as possible.

Fossils only tell that a rock is older or younger than another; they do not give absolute age. The decay of radioactive minerals may provide an age in years, but this method is expensive and time-consuming, and cannot always be applied since most rocks lack suitable radioactive minerals. Correlation by fossils remains the standard method for comparing ages of events in different areas. See INDEX FOSSIL; STRATIGRAPHY.

The physical appearance and climate of the Earth during a given period of the geologic past can be described from compilation and analysis of the data which is obtained through studies of the habitats of extant fauna, the geographic distribution of fossils, and the climatic preferences of ancient forms of life. See PALEOCLIMATOLOGY; PALEOGEOGRAPHY.

Biological aspects. The most fundamental fact of paleontology is that organisms have changed throughout earth history and that each geological period has had its characteristic forms of life. An evolutionist has two major interests: first, to know how the process of evolution works; this is accomplished by studying the genetics and population structure of modern organisms; second, to reconstruct the events produced by this process, that is, to trace the history of life. Any modern animal group is merely a stage, frozen at one moment in time, of a dynamic, evolving lineage. Fossils give the only direct evidence of previous stages in these lineages. Horses and rhinoceroses, for example, are very different animals today, but the fossil history of both groups is traced to a single ancestral species that lived early in the Cenozoic Era. From such evidence, a tree of life can be constructed whereby the relationships among organisms can be understood.   [S.J.G.]

Paleoseismology   The study of geological evidence for past earthquakes. This scientific discipline has contributed greatly to modern understanding of the nature of earthquakes. The patterns of earthquakes, in both space and time, evolve over centuries and millennia and cannot be discovered by modern instruments. Knowledge of these patterns is important for understanding the physics of earthquakes and for forecasting future destructive earthquakes.

In certain natural environments, the features related to ancient earthquakes are preserved in the landforms and superficial layers of the Earth's surface. Geologists use this paleoseismological evidence to extend the short historical and instrumental record of earthquakes into ancient centuries and millennia. Such paleoseismological studies have clarified the earthquake record of many parts of the world, including the midcontinent and east coast of the United States, northern Africa, southern Europe, China, Japan, Indonesia, and New Zealand.

The geological preservation of ancient earthquakes has enabled scientists to compare modern earthquakes with those of the past. In 1983, for example, a sparsely populated region of Idaho was struck by a magnitude-7.3 earthquake. Subsequent investigations revealed a fresh, 30-km-long (18-mi) fault scarp running along the western base of the lofty Lost River Range. Inspection of the fresh escarpment revealed that it is surmounted by a more subdued, vegetated escarpment of nearly identical length and height. Excavations across this ancient fault scarp showed that it had formed about 5000 years earlier during an event very similar to the 1983 earthquake. This is one of several examples of what paleoseismologists call a characteristic earthquake: apparently, some earthquakes are nearly identical repetitions of their predecessors. See EARTHQUAKE; SEISMOLOGY.   [K.Si.]

Paleosol   A soil of the past, that is, a fossil soil. Paleosols are most easily recognized when they are buried by sediments. They also include surface profiles that are thought to have formed under very different conditions from those now prevailing, such as the deeply weathered tropical soils of Tertiary geological age that are widely exposed in desert regions of Africa and Australia. Such profiles are generally known as relict paleosols. Those that can be shown to have been buried and then uncovered by erosion are known as exhumed paleosols. The main problem in defining the term paleosol comes from defining what is meant by soil, a term that has very different meanings for agronomists, engineers, geologists, and soil scientists. Soil can be considered distinct from sediment in that it forms in place, but soil need not necessarily include traces of life. At its most general level, soil is material forming the surface of a planet or similar body and altered in place from its parent material by physical, chemical, or biological processes. *See* SOIL.

Paleosols are especially abundant in volcanic, alluvial, and eolian sedimentary sequences. Along with the fossils, sedimentary structures, and volcanic rocks found in such deposits, paleosols provide an additional line of evidence for ancient environments during times between eruptions and depositional events. *See* PALEOCLIMATOLOGY; SEDIMENTOLOGY.   [G.J.R.]

Paleozoic   A major division of time in geologic history, extending from about 540 to 250 million years ago (Ma). It is the earliest era in which significant numbers of shelly fossils are found, and Paleozoic strata were among the first to be studied in detail for their biostratigraphic significance.

The Paleozoic Era is divided into six systems; from oldest to youngest they are Cambrian, Ordovician, Silurian, Devonian, Carboniferous, and Permian. The Carboniferous is subdivided into two subsystems, the Mississippian and the Pennsylvanian which, in North America, are considered systems by many geologists. The Silurian and Devonian systems are closer to international standardization than others; all the series and stage names and lower boundaries have been agreed upon, and most have been accepted. *See* CAMBRIAN; CARBONIFEROUS; DEVONIAN; ORDOVICIAN; PERMIAN; SILURIAN.

Because Alpine and Appalachian mountain chains were among the first studied in detail, orogenies were first named there. In eastern North America, mountain-building effects during the early Paleozoic were ascribed to the Taconic orogeny (Middle and Late Ordovician); middle Paleozoic events were assigned to the Acadian orogeny (Middle and Late Devonian); and late Paleozoic movements were called Appalachian (more accurately Alleghenian) for Permian and, perhaps, Triassic events. *See* DATING METHODS; OROGENY; PLATE TECTONICS; UNCONFORMITY.

The major changes in lithofacies during the Paleozoic were also effected by biotic evolution through the era. Limestone facies became more abundant and more diversified in the shallow warm seas as calcium-fixing organisms became more diverse and more widespread. Sediment input from the land was modified as plants moved from the seas to the low coastal plains and, eventually, to the higher ground during the Devonian. Primitive vertebrates evolved during the Cambro-Ordovician, but true fishes and sharks did not flourish until the Devonian. Amphibians invaded the land during the Late Devonian and early Carboniferous at about the same time that major forests began to populate the terrestrial realm. These changes produced an entirely new suite of nonmarine facies related to coal formation, and the Carboniferous was a time of formation of major coal basins on all continental plates.

Major cycles of cold and warm climates were overlaid on depositional and evolutionary patterns, producing periods of continental glaciation when large amounts of the Earth's water were tied up in ice during the Late Ordovician, the Late Devonian, and the late Permian. During the earliest and latest of these periods, icesheets were concentrated in the Southern Hemisphere on a single large Paleozoic continental mass—Gondwana. *See* DEPOSITIONAL SYSTEMS AND ENVIRONMENTS; FACIES (GEOLOGY); PALEOCLIMATOLOGY.

The Paleozoic featured a single southern landmass (Gondwana) for most of the era. This megaplate moved relatively sedately northward during this entire time interval (540–250 Ma) and always contained the magnetic and geographic south poles. Consequently, many of the facies and biologic provinces in the Gondwanan region were influenced by the cooler marine realms and continental and mountain glaciers in nearly every Paleozoic period. Most of the tectonic action that produced major periods of collision, mountain building, carbonate platform building, back-arc fringing troughs with their distinctive faunas and lithofaces, and formation of coal basins and evaporites took place in the Northern Hemisphere. These pulsations produced combinations of Laurentian (North American), Euro-Baltic, Uralian, Siberian, and Chinese plates a various times during the Paleozoic; and these combined units, in turn, moved slowly across the latitudes, producing climatic change; lithofacies changed in response to both the climate and the plate tectonics. *See* PALEOGEOGRAPHY.

There were fewer and simpler life forms in the Cambrian—often termed the Age of Trilobites. All groups of invertebrates and plants became more numerous through geologic time. For example, 7 major invertebrate animal groups at the beginning of the Cambrian doubled to 14 by the end of the period, 20 by the end of the Ordovician, 23 at the end of the Devonian, and 25 at the end of the Paleozoic. The pattern for plant diversification, although starting later, is similar. Three simple plant groups became 5 by the end of the Silurian, 7 at the end of the Devonian, and 13 at the end of the Paleozoic. The vertebrates also diversified very slowly. From one or two groups in the Cambro-Ordovician (conodonts are now considered primitive vertebrates), the number of major kinds rose to 6 at the end of the Devonian and 8 at the end of the Paleozoic. *See* BIOGEOGRAPHY; GEOLOGIC TIME SCALE; INDEX FOSSIL; STRATIGRAPHY.   [J.T.D.]

Pearl Any mollusk-formed calcareous concretion that displays an orient and is lustrous. There are two major groups of bivalved mollusks in which gem pearls may form: the saltwater pearl oyster (*Pinctada*), and a number of genera of fresh-water clams. Usually, jewelers refer to salt-water pearls as Oriental pearls, regardless of their place of discovery, and to those from fresh-water bivalves as fresh-water pearls.

Between the body mass and the valves of the mollusk extends a curtainlike tissue called the mantle. In order for a pearl to form, a tiny object such as a parasite or a grain of sand must work through the mantle. When this happens, secretion of nacre around the invading object builds a pearl within the body of the mollusk. Whole pearls form within the body mass of the mollusk, in contrast to blister pearls, which form as protrusions on the inner surface of the shell. Edible oysters produce lusterless concretions, but never pearls.

The substitute for natural pearls, to which the name cultured pearl has been given, is usually made by inserting a large bead into a mollusk to be coated with nacre.   [R.T.L.]

Peat A dark-brown or black residuum produced by the partial decomposition and disintegration of mosses, sedges, trees, and other plants that grow in marshes and other wet places. Forest-type peat, when buried and subjected to geological influences of pressure and heat, is the natural forerunner of most coal. Moor peat is formed in relatively elevated, poorly drained moss-covered areas, as in parts of Northern Europe. *See* COAL; HUMUS.   [G.H.C.]

Pectolite   A mineral inosilicate with composition $Ca_2Na-Si_3O_8(OH)$. The hardness is 5 on Mohs scale, and the specific gravity is 2.75. The mineral is colorless, white, or gray with a vitreous to silky luster. Pectolite is found in the United States at Paterson, Bergen Hill, and Great Notch, New Jersey. *See* SILICATE MINERALS.   [C.S.Hu.]

Pedology   Defined narrowly, a science that is concerned with the nature and arrangement of horizons in soil profiles; the physical constitution and chemical composition of soils; the occurrence of soils in relation to one another and to other elements of the environment such as climate, natural vegetation, topography, and rocks; and the modes of origin of soils. Pedology so defined does not include soil technology, which is concerned with uses of soils.

Broadly, pedology is the science of the nature, properties, formation, distribution, and function of soils, and of their response to use, management, and manipulation. The first definition is widely used in the United States and less so in other countries. The second definition is worldwide. *See* SOIL.   [R.W.S.]

Pegmatite   Exceptionally coarse-grained and relatively light-colored crystalline rock composed chiefly of minerals found in ordinary igneous rocks. Extreme variations in grain size also are characteristic, and close associations with dominantly fine-grained aplites are common. Pegmatites are widespread and very abundant where they occur, especially in host rocks of Precambrian age, but their aggregate volume in the Earth's crust is small. Many pegmatites have been economically valuable as sources of clays, feldspars, gem materials, industrial crystals, micas, silica, and special fluxes, as well as beryllium, bismuth, lithium, molybdenum, rare-earth, tantalumniobium, thorium, tin, tungsten, and uranium minerals. *See* APLITE; IGNEOUS ROCKS.

Essential minerals (1) in granitic pegmatites are quartz, potash feldspar, and sodic plagioclase; (2) in syenitic pegmatites, alkali feldspars with or without feldspathoids; and (3) in diorite and gabbro pegmatites, soda-lime or lime-soda plagioclase. Varietal minerals such as micas, amphiboles, pyroxenes, black tourmaline, fluorite, and calcite further characterize the pegmatites of specific districts. Accessory minerals include allanite, apatite, beryl, garnet, magnetite, monazite, tantalite-columbite, lithium tourmaline, zircon, and a host of rarer species.   [R.H.J.]

Pennsylvanian   A major division of late Paleozoic time, considered either as an independent period or as the younger subperiod of the Carboniferous. In North America, the Pennsylvanian has been widely recognized as a geologic period and derives its name from a thick succession of mostly nonmarine, coal-bearing strata in Pennsylvania. Radiometric ages place the beginning of the period at approximately 320 million years ago and its end at about 290 million years ago. In northwestern Europe, strata of nearly equivalent age are commonly designated as Upper Carboniferous and in eastern Europe as Middle and Upper Carboniferous. *See* CARBONIFEROUS.

In North America, the Pennsylvanian Period was characterized by the progressive growth and enlargement of the Alleghenian-Ouachita-Marathon orogenic belt, which formed as the northwestern parts of the large continent Gondwana (mainly northwestern Africa, the area that is now Florida, and northern South America) collided against and deformed the eastern and southern parts of the North American continent. *See* OROGENY.

Much of North America remained a stable, low-lying cratonic platform during the Pennsylvanian and was covered by a relatively thin veneer of shallow-water marine carbonates and marine and nonmarine clastic sediments. These were deposited as sea level repeatedly rose and fell as the polar glaciers of southern Gondwana contracted and expanded. In Pennsylvania and along the western part of the present Allegheny Plateau, Pennsylvanian strata are predominantly nonmarine deposits made up of channel sandstones, floodplain shales, siltstones, sandstones, and coals. *See* CARBONATE MINERALS; COAL; CRATON; MARINE SEDIMENTS.

During Pennsylvanian time, the paleoequator extended across North America from southern California to Newfoundland, through northwestern Europe into the Ukrainian region of eastern Europe, and across parts of northern China. This was a time of extensive coal deposition in a tropical belt that appears to have included areas from 15 to 20° north and south of the paleoequator. Coal of this age is abundant and relatively widespread and has great economic importance.

Petroleum is commonly trapped in nearshore marine deposits of Pennsylvanian age, particularly in carbonate banks near the edge of shelves, in longshore bars and beaches, in reefs and mounds, and at unconformities associated with transgressive-regressive shore lines. Many of these traps contribute significantly to petroleum production. *See* PETROLEUM.

Pennsylvanian paleogeography changed significantly during the period as the supercontinent Pangaea gradually was formed by the joining together of Gondwana and Laurasia. North America and northern Europe, which had been combined into the continent Laurasia since the late Silurian, and South America and northwestern Africa, which formed the northern part of the continent of Gondwana, came together along the Ouachita–Southern Appalachian–Hercynian geosyncline. The result was an extensive orogeny, or mountain-building episode, which supplied the vast amounts of sediments that make up most of the Pennsylvanian strata in the eastern and midwestern parts of the United States. *See* PALEOGEOGRAPHY.

Evidence in the form of well-developed tree rings, less diverse fossil floras and faunas, and glacial deposits indicates that temperate and glacial conditions were common in nonequatorial climatic belts during Pennsylvanian time. Climatic fluctuations during the period caused significant increases and decreases in the amount of water that was temporarily stored in the glaciers in Gondwana and contributed to eustatic changes of sea level. *See* PALEOCLIMATOLOGY.   [C.A.R.; J.R.P.R.]

Pentlandite   A mineral having composition $(Fe,Ni)_9S_8$. Pentlandite is the major ore of nickel. It is usually massive, showing a well-defined octahedral parting. The hardness is 3.5–4 (Mohs scale) and the specific gravity varies from 4.6 to 5.0, depending on the ratio of iron to nickel; greater amounts of iron cause an increase in the specific gravity. The luster is metallic and the color yellowish bronze. Pentlandite is found at many localities in small amounts, but its chief occurrence is at Sudbury, Ontario, where it is mined on a large scale.   [C.S.Hu.]

Peridotite   A rock consisting of more than 90% of millimeter-to-centimeter-sized crystals of olivine, pyroxene, and hornblende, with more than 40% olivine. Other minerals are mainly plagioclase, chromite, and garnet. Much of the volume of the Earth's mantle probably is peridotite.

Peridotites have three principal modes of occurrence corresponding approximately to their textures: (1) Peridotites with well-formed olivine crystals occur mainly as layers in gabbroic complexes. (2) Peridotite nodules in alkaline basalts and diamond pipes generally have equigranular textures, but some have irregular grains. (3) Peridotite also occurs on the walls of rifts in the deep sea floor and as hills on the sea floor, some of which reach the surface. *See* GABBRO.

Peridotites are rich in magnesium, reflecting the high proportions of magnesium-rich olivine. The compositions of peridotites from layered igneous complexes vary widely, reflecting the relative proportions of pyroxenes, chromite, plagioclase, and amphibole.

Peridotite is an important rock economically. Where granites have intruded peridotite, asbestos and talc are common. Pure olivine rock (dunite) is quarried for use as refractory foundry sand and refractory bricks used in steelmaking. Serpentinized peridotite is locally quarried for ornamental stone. Tropical soils developed on peridotite are locally ores of nickel. The sulfides associated with peridotites are common ores of nickel and platinoid metals. The chromite bands commonly associated with peridotites are the world's major ores of chromium. *See* IGNEOUS ROCKS.   [A.T.A.]

Permafrost   Perennially frozen ground, occurring wherever the temperature remains below 32°F (0°C) for several years, whether the ground is actually consolidated by ice or not and regardless of the nature of the rock and soil particles of which the earth is composed. Perhaps 25% of the total land area of the Earth contains permafrost; it is continuous in the polar regions and becomes discontinuous and sporadic toward the Equator. During glacial times permafrost extended hundreds of miles south of its present limits in the Northern Hemisphere.

Temperature of permafrost at the depth of no annual change, about 30–100 ft (10–30 m), crudely approximates mean annual air temperature. It is below 23°F (−5°C) in the continuous zone, between 23–30°F (−5 and −1°C) in the discontinuous zone, and above 30°F (−1°C) in the sporadic zone. Temperature gradients vary horizontally and vertically from place to place and from time to time.

Ice is one of the most important components of permafrost, being especially important where it exceeds pore space. Physical properties of permafrost vary widely from those of ice to those of normal rock types and soil. The cold reserve, that is, the number of calories required to bring the material to the melting point and melt the contained ice, is determined largely by moisture content.

Permafrost develops today where the net heat balance of the surface of the Earth is negative for several years. Much permafrost was formed thousands of years ago but remains in equilibrium with present climates. Permafrost eliminates most groundwater movement, preserves organic remains, restricts or inhibits plant growth, and aids frost action. It is one of the primary factors in engineering and transportation in the polar regions.   [R.F.B.]

Permian   The name applied to the last period of geologic time in the Paleozoic Era and to the corresponding system of rock formations that originated during that period. The Permian Period commenced approximately 290 million years ago and ceased about 250 million years ago. The system of rocks that originated during this interval of time is widely distributed on all the continents of the world. The Permian Period was a time of variable and changing climates, and during much of this time latitudinal climatic belts were well developed. During the latter half of Permian time, many long-established lineages of marine invertebrates became extinct and were not immediately replaced by new fossil-forming lineages. Rocks of Permian age contain many resources, including petroleum, coal, salts, and metallic ores.

During the Permian Period, several important changes took place in the paleogeography of the world. The joining of Gondwana to western Laurasia, which had started during the Carboniferous, was completed during Wolfcampian time (earliest Permian). The addition of eastern Laurasia (Angara) to the eastern edge of western Laurasia finished during Artinskian time (middle to latest early Permian) and completed the assembly of the supercontinent Pangaea. The climatic effects of these changes were dramatic. Instead of having a circumequatorial tropical ocean, such as during the middle Paleozoic, a large landmass with several high chains of mountains extended from the South Pole across the southern temperate, the tropical, and into the north temperate climatic belts. One very large world ocean, Panthalassa and its western tropical branch, the Tethys, occupied the remaining 75% of the Earth's surface, with a few much smaller cratonic blocks, island arcs, and atolls. *See* CONTINENTAL DRIFT; CONTINENTS, EVOLUTION OF; PALEOGEOGRAPHY.

Most marine invertebrates of the Early Permian were continuations of well-established phylogenetic lines of middle and late Carboniferous ancestry. During early Permian time, these faunas were dominated by brachiopods, bryozoans, conodonts, corals, fusulinaceans, and ammonoids. The Siberian traps, an extensive outflow of very late Permian basalts and other basic igneous rocks (dated at about 250 million years ago), are considered by many geologists as contributing to climatic stress that resulted in major extinctions of many animal groups, particularly the shallow-water marine invertebrates. The end of the Permian is also associated with unusually sharp excursions in values of the carbon-12 isotope (^{12}C) in organic material trapped in marine sediments, suggesting major disruption of the ocean chemistry system.

Terrestrial faunas included insects which showed great advances over those of the Carboniferous Coal Measures. Several modern orders emerged, among them the Mecoptera, Odonata, Hemiptera, Trichoptera, Hymenoptera, and Coleoptera.

Of the vertebrates, labyrinthodont amphibians were common and varied; however, reptiles showed the greatest evolutionary radiation and the most significant advances. Reptiles are found in abundance in the lower half of the system in Texas and throughout most of the upper part of the system in Russia and also are common in Gondwana sediments. Of the several Permian reptilian orders, the most significant was the Theriodonta. These reptiles carried their bodies off the ground and walked or ran like mammals. Unlike most reptiles, their teeth were varied—incisors, canines, and jaw teeth as in the mammals—and all the elements of the lower jaw except the mandibles showed progressive reduction. Most of the known theriodonts are from South Africa and Russia. *See* PALEOZOIC.   [C.A.R.; J.R.P.R.]

Perovskite   A minor accessory mineral, formula $CaTiO_3$, occurring in basic rocks. Perovskite has given its name to a large family of materials, synthetic and natural, crystallizing in similar structures. The crystal structure is ideally cubic, with a framework of corner-sharing octahedra, containing titanium (Ti) or other relatively small cations surrounded by six oxygen (O) or fluorine (F) anions. Within this framework are placed calcium (Ca) or other large cations, surrounded by twelve anions. Tilting of the octahedra and other distortions often lower the symmetry from cubic, giving the materials important ferroelectric properties and decreasing the coordination of the central cation. This flexibility gives the structure the ability to incorporate ions of different sizes and charges. Substitution of niobium (Nb), cerium (Ce), and other rare-earth elements in natural calcium titanate ($CaTiO_3$) is common and can make perovskite an ore for these elements.

A number of synthetic perovskites are of major technological importance. Barium titanate ($BaTiO_3$) and lead zirconate-titanate (PZT) ceramics form the basis of a sizable industry in ferroelectric and piezoelectric materials crucial to transducers, capacitors, and electronics. Lanthanum chromate ($LaCrO_3$) and related materials find applications in fuel cells and high-temperature electric heaters.   [A.Na.]

Perthite   Any of the oriented intergrowths of potassium- and sodium-rich feldspars, $(K,Na)AlSi_3O_8$, whose proportions are determined in part by the initial composition of the alkali feldspar from which they exsolved and whose physical properties are thus somewhat variable. The early stages of perthite formation from homogeneous, usually monoclinic (K,Na)AlSi$_3$O$_8$ may be observed experimentally by high-magnification electron microscopy.

If the final K- and Na-rich lamellae or particles are submicroscopic, the composite feldspar is called cryptoperthite. If the particles are small enough and the feldspar relatively clear, Rayleigh-type scattering of light may occur, giving rise to the beautiful blue-to-whitish luster of the semiprecious gem called moonstone. If coarsening has progressed to the micrometer scale and can be seen on a polarizing microscope, the composite is called microperthite; and if the two feldspars are visible to the eye in hand specimen, it is called perthite or macroperthite. Often the albite phase will appear as white veins or blotchy patches against a colored K-rich phase, which may be green to blue microcline (amazonite) or dull pink to orange-brown orthoclase. *See* ALBITE; ANORTHITE; ANORTHOCLASE; FELDSPAR; MICROCLINE; ORTHOCLASE.   [P.H.R.]

Petalite   A rare pegmatitic mineral with composition LiAlSi$_4$O$_{10}$. Its economic significance is markedly disproportionate to the number of its occurrences. It is the only basic raw material suitable for production of a group of materials known as crystallized glass ceramics (melt-formed ceramics). These extremely fine-grained substances are based on a keatite-type structure (stuffed silica derivative). Among the desirable properties of such submicroscopic aggregates are their exceedingly low thermal expansion and high strength, making them suitable for use in cooking utensils and telescopic mirror blanks.

The color is white, pink, pale green, or gray to black. Hardness is $6\frac{1}{2}$ on Mohs scale. Petalite is mined from a large lithium-rich pegmatite at Bikita, Zimbabwe, the only major world source.   [E.W.H.]

Petrifaction   A mechanism by which the remains of extinct organisms are preserved in the fossil record. In petrifactions (though chiefly in plants rather than animals) the original shape and topography of the tissues, and occasionally even minute cytological details, are retained relatively undeformed.

The term petrifaction was adopted as a scientific term before knowledge existed of the geochemical mechanism or processes involved. It was formerly widely believed that in the formation of a petrifaction the organic matter of the organism or tissue was replaced molecule by molecule with mineral material entering in solution in percolating groundwater. It is now evident that what actually happens is that the mineral fills cell lumena and the intermicellar interstices of cell walls with insoluble salts depositing from solution. Petrifaction is hence a form of mineral emplacement or embedding, by which the organic residues are filled with solid substance which infiltrates in solution. The most common substances involved in petrifactions are silica, SiO$_2$, and calcium carbonate, CaCO$_3$ (calcite). Occasionally phosphate minerals, pyrite, hematite, and other less common minerals make up all or part of the petrifaction matrix. *See* FOSSIL
.   [E.S.B.]

Petrofabric analysis   The systematic study of the fabrics of rocks, generally involving statistical study of the orientations and distribution of large numbers of fabric elements. The term fabric denotes collectively all the structural or spatial characteristics of a rock mass. The fabric elements are classified into two groups: (1) megascopic features, including bedding, schistosity, foliation, cleavage, faults, joints, folds, and mineral lineations; and (2) microscopic features, including the shapes, orientations, and mutual arrangement of the constituent mineral crystals (texture) and of internal structures (twin lamellae, deformation bands, and so on) inside the crystals.

The aim of fabric analysis is to obtain as complete and accurate a description as possible of the structural makeup of the rock mass with a view to elucidating its kinematic history. The fabric of a sedimentary rock, for example, may retain evidence of the mode of transport, deposition, and compaction of the sediment in the size, shape, and disposition of the particles; similarly, that of an igneous rock may reflect the nature of the flow or of gravitational segregation of crystals and melt during crystallization. The fabrics of deformed metamorphic rocks (tectonites) have been most extensively studied by petrofabric techniques with the objective of determining the details of the history of deformation and recrystallization. See STRUCTURAL GEOLOGY; STRUCTURAL PETROLOGY.

[J.M.Ch.]

Petrography   The description of rocks with goals of classification and interpretation of origin. Most schemes for the classification of rocks are based on the size of grains and the proportions of various minerals. Interpretations of origin rely on field relations, structure, texture, and chemical composition as well as sizes and proportions of different kinds of grains. The names of rocks are based on the sizes and relative proportions of different minerals; boundaries between the names are arbitrary. The conditions of formation of a rock can be estimated from the types and textures of its constituent minerals.

The description of rocks begins in the field with observation of the shape and structure of bodies of rock at the scale of centimeters to kilometers. The geometrical relations between and structures within mappable rock units are generally the domain of field geology, but are simply rock descriptions at a reduced scale.

A petrographer can correctly name most rocks in which most crystals are larger than about 0.04 in. (1 mm) simply by examining the rock with a 10-power magnifying lens. Rocks with smaller grains require either microscopical examination or chemical analysis for proper classification.

Sizes, shapes, and orientations of grains and voids are the most important features of a rock relevant to its origin. The same features also affect density, porosity, permeability, strength, and magnetic behavior. It is also essential to know the identity, abundance, and compositions of minerals constituting the grains in order to name a rock and infer its conditions of formation.

Petrographers study organic as well as inorganic objects, and petrographic analyses are useful to both paleontologists and petroleum geologists. The quality of coal is revealed with polarizing and reflecting light microscopes. Inclusions of petroleum and brine in crystals of silicates and salt in rocks help scientists infer how petroleum formation is connected with cementation and other modifications of buried sediments. See PALEONTOLOGY; PETROLEUM GEOLOGY.

Petrographers also study synthetic objects. The textures of metals and alloys are scrutinized by petrographers in order to understand what makes these materials strong and resistant to corrosion. Flaws in glasses and ceramics are revealed by microscopical and polarizing techniques. Fragments of minerals and rocks in some pottery can help point to its source and help trace prehistoric routes of trade. The industrial, agricultural, and natural sources of particles in the air and water may be established from petrographic study. See MINERALOGY; PETROLOGY.   [A.T.A.]

Petroleum   Unrefined, or crude, oil is found underground and under the sea floor, in the interstices between grains of sandstone and limestone or dolomite (not in caves). Petroleum is a mixture of liquids varying in color from nearly colorless to jet black, in viscosity from thinner than water to thicker than molasses, and in density from light gases to asphalts heavier than water. It can be separated by distillation into fractions that range from light color, low density, and low viscosity to the opposite extreme. In places where it has oozed from the ground, its volatile fractions have vaporized, leaving the dense, black parts of the oil as a pool of tar or asphalt (such as the Brea Tar Pits in California). Much of the world's crude oil is today produced from drilled wells.

Petroleum consists mostly of hydrocarbon molecules. The four main classes of hydrocarbons are paraffins (also called alkanes), olefins (alkenes), cycloparaffins (cycloalkanes), and aromatics. Olefins are absent in crude oil but can be formed in certain refining processes. The simplest hydrocarbon is one carbon atom bonded to four hydrogen atoms (chemical formula CH_4), and is called methane.

Petroleum usually contains all of the possible hydrocarbon structures except alkenes, with the number of carbon atoms per molecule going up to a hundred or more. These fractions include compounds that contain sulfur, nitrogen, oxygen, and metal atoms. The proportion of compounds containing these atoms increases with increasing size of the molecule.

Asphaltic molecules contain many cyclic compounds in which the rings contain sulfur, nitrogen, or oxygen atoms; these are called heterocyclic compounds. An example is pyridine. *See* ASPHALT AND ASPHALTITE.

It is generally agreed that petroleum formed by processes similar to those which yielded coal, but was derived from small animals rather than from plants. Dead organisms have been buried in mud over millions of years. Further layers deposited over these mud layers have in some cases reached a thickness of thousands of feet, and compacted the layers beneath them, until the mud has become shale rock. The mud layers were heated and compressed by the layers above. The bodies of the organisms in the mud were decomposed and converted into fatty liquids and solids. Heating these fatty materials over a very long time caused their molecules to break into smaller fragments and combine into larger ones, so the original range of molecular size was spread greatly into the range found in crude oil. Bacteria were usually present, and helped remove oxygen from the molecules and turned them into hydrocarbon compounds. The great pressure of the overlying rock layers helped to force the oil out of the compacted mud (shale) layers into less compacted limestone, dolomite, or sandstone layers next to the shale layers. *See* DOLOMITE; LIMESTONE; ORGANIC GEOCHEMISTRY; PETROLEUM GEOLOGY; SANDSTONE; SEDIMENTOLOGY; SHALE.

At depths greater than about 25,000 ft (7620 m), the temperature is so high that the oil conversion processes go all the way to natural gas and soot. Natural gas formed by the conversion processes is now also found over a variety of depths which do not indicate the depth and temperature of their origin. *See* NATURAL GAS.

The oil formed by the natural thermal and bacterial processes was squeezed out of the compacting mud layers into sandstone or limestone layers and migrated upward in tilted layers. Tectonic processes caused such uptilting and bulging of layers to form ridges and domes. When the ridges and domes were covered by shale already formed, the pores of the shale were too tiny to let the oil through, so the shale acted as a sealing cap. When the oil could not rise farther, it was trapped. Porous rock in such a structure that contains oil or gas is called an oil or gas reservoir.

The recovery from typical reservoirs is not as high as might be thought. Multiple-layer reservoirs will typically contain oil-bearing layers with a wide range of permeability. When recovery from the highest-permeability layers is as complete as it can be, the low-permeability layers will usually have been only slightly depleted, despite all efforts to improve the recovery. Despite recovery efforts, half or more of the oil originally present in oil reservoirs is still in them.   [E.L.Cl.]

Heavy oil and tar sand oil (bitumen) are petroleum hydrocarbons found in sedimentary rocks. They are formed by the oxidation and biodegradation of crude oil, and occur in the liquid or semiliquid state in limestones, sandstones, or sands. *See* BITUMEN.

These oils are characterized by their viscosity; however, density (or API gravity) is also used when viscosity measurements are not available. Heavy oils contain 3 wt % or more sulfur and as much as 200 ppm vanadium. Titanium, zinc, zirconium, magnesium, manganese, copper, iron, and aluminum are other trace elements that can be found in these deposits. Their high naphthenic acid content makes refinery processing equipment vulnerable to corrosion.   [E.Ok.]

Petroleum geology   The practice of utilizing geological principles and applying geological concepts to the discovery and recovery of petroleum. Related fields in petroleum discovery include geochemistry and geophysics. The related areas in petroleum recovery are petroleum and chemical engineering. *See* GEOCHEMISTRY; GEOPHYSICS.

Petroleum occurs in a liquid phase as crude oil and condensate, and in a gaseous phase as natural gas. The phase is dependent on the kind of source rock from which the petroleum was formed and the physical and thermal environment in which it exists. Most petroleum occurs at varying depths below the ground surface, but generally petroleum existing as a liquid (crude oil) is found at depths of less than 20,000 ft (6100 m) while natural gas is found both at shallow depths and at depths exceeding 30,000 ft (9200 m). In some cases, oil may seep to the surface, forming massive deposits of oil or tar sands. Natural gas also seeps to the surface but escapes into the atmosphere, leaving little or no surface trace. *See* NATURAL GAS; OIL SAND; PETROLEUM.

Most petroleum is found in sedimentary basins in sedimentary rocks, although many of the 700 or so sedimentary basins of the world contain no known significant accumulations. Several conditions must exist for the accumulation of petroleum: (1) There must be a source rock, usually high in organic matter, from which petroleum can be generated. (2) There must be a mechanism for the petroleum to move, or migrate. (3) A reservoir rock with voids to hold petroleum fluids must exist. (4) The reservoir must be in a configuration to constitute a trap and be covered by a seal—any kind of low-permeability or dense rock formation that prevents further migration. If any of these conditions do not exist, petroleum either will not form or will not accumulate in commercially extractable form. *See* BASIN; SEDIMENTARY ROCKS.

The aim of petroleum geologists is to find traps or accumulations of petroleum. The trap not only must be defined but must exist where other conditions such as source and reservoir rocks occur.

To locate these traps, the geologist must rely on subsurface information and data gathered by drilling exploratory wells and data obtained by geophysical surveying. These data, once interpreted, are used to construct maps, cross sections, and models that are used to infer or to actually depict subsurface configurations that might contain petroleum. Such depictions are prospects for drilling. *See* GEOPHYSICAL EXPLORATION.

Oil and gas must be trapped in an individual reservoir in sufficient quantities to be commercially producible. Worldwide, 25% of all oil discovered so far is contained in only ten fields, seven of which are in the Middle East. Fifty percent of all oil discovered to date is found in only 50 fields.

Most of the large and fairly obvious fields in the United States have been discovered, except those possibly existing in frontier or lightly explored areas such as Alaska and the deep waters offshore. Few areas of the world remain entirely untested, but many areas outside the United States are only partly explored, and advanced techniques have yet to be deployed in the recovery of oil and gas found so far.

Greater efforts in petroleum geology along with petroleum engineering are being made to increase recovery from existing fields. Of all oil discovered so far, it is estimated that there will be recovery of only 35% on the average. Recovering some part of this huge oil resource will require geological reconstruction of reservoirs, a kind of very detailed and small-scale exploration. These reconstructions and models have allowed additional recovery of oil that is naturally movable in the reservoir. If the remaining oil is immobile because it is too viscous or because it is locked in very small pores or is held by capillary forces, techniques must be used by the petroleum geologist and the petroleum engineer to render the oil movable.   [W.L.Fi.]

Petrology   The study of rocks, their occurrence, composition, and origin. Petrography is concerned primarily with the detailed description and classification of rocks, whereas petrology deals primarily with rock formation, or petrogenesis. A petrological description includes definition of the unit in which the rock occurs, its attitude and structure, its mineralogy and chemical composition, and conclusions regarding its origin. *See* MINERALOGY; PETROGRAPHY; ROCK.   [W.I.R.]

One aim of mineralogy and petrology is to decipher the history of igneous and metamorphic rocks. Detailed study of the field geology, the structures, the petrography, the mineralogy, and the geochemistry of the rocks is used as a basis for hypotheses of origin. The conditions at depth within the Earth's crust and mantle, the processes occurring at depth, and the whole history of rocks once deeply buried are deduced from the study of rocks now exposed at the Earth's surface. One approach used to test hypotheses so developed is experimental petrology; the term experimental minerals refers to similar studies involving minerals rather than rocks (mineral aggregates). *See* IGNEOUS ROCKS; METAMORPHIC ROCKS.

The experimental petrologist reproduces in the laboratory the conditions of high pressure and high temperature encountered at various depths within the Earth's crust and mantle where the minerals and rocks were formed. By suitable selection of materials the petrologist studies the chemical reactions that actually occur under these conditions and attempts to relate these to the processes involved in petrogenesis.   [P.J.Wy.]

Phenocryst   A relatively large crystal embedded in a finer-grained or glassy igneous rock. The presence of phe-nocrysts gives the rock a porphyritic texture. Phenocrysts are represented most commonly by feldspar, quartz, biotite, hornblende, pyroxene, and olivine. Strictly speaking, phenocrysts crystallize from molten rock material (lava or magma). They commonly represent an earlier and slower stage of crystallization than does the matrix in which they are embedded. *See* IGNEOUS ROCKS; PORPHYROBLAST.
                                   [C.A.C.]

Phlogopite   A mineral with an ideal composition of $KMg_3(AlSi_3)O_{10}(OH)_2$. Phlogopite belongs to the mica mineral group. It has been occasionally called bronze mica. Phlogopite is a trioctahedral mica, where all three possible octahedral cation sites are occupied by magnesium (Mg). The magnesium octahedra, $Mg(O,OH)_6$, form a sheet by sharing edges. As in all micas, tetrahedra are located on either side of the octahedral sheet, which may be occupied by aluminum (Al) or silicon (Si). Adjacent tetrahedra share corners to form a two-dimensional network of sixfold rings, thus producing a tetrahedral sheet. Two opposing tetrahedral sheets and the included octahedral sheet form a 2:1 layer. Potassium (K) ions are located between adjacent tetrahedral sheets in the interlayer region.

Specific gravity is 2.86, hardness on the Mohs scale is 2.5–3.0, and luster is vitreous to pearly. Thin sheets are flexible. Color is yellow brown, reddish brown, or green, and thin sheets are transparent. Thermal stability varies greatly with composition, with iron or fluorine substitutions reducing or increasing stability, respectively. Weathering of phlogopite may produce vermiculite. See VERMICULITE; WEATHERING PROCESSES.

Phlogopite occurs in marbles produced by the metamorphism of siliceous magnesium-rich limestones or dolomites and in ultrabasic rocks, such as peridotites and kimberlites. See DOLOMITE; LIMESTONE; PERIDOTITE.

Phlogopite is used chiefly as an insulating material and for fireproofing. It has high dielectric properties and high thermal stability. See MICA; SILICATE MINERALS.   [S.Gu.]

Phonolite   A light-colored, aphanitic (not visibly crystalline) rock of volcanic origin, composed largely of alkali feldspar, feldspathoids (nepheline, leucite, sodalite), and smaller amounts of dark-colored (mafic) minerals (biotite, soda amphibole, and soda pyroxene). Phonolite is chemically the effusive equivalent of nepheline syenite and similar rocks. Rocks in which plagioclase (oligoclase or andesine) exceeds alkali feldspar are rare and may be called feldspathoidal latite. See FELDSPATHOID; MAGMA.

Phonolites are rare and highly variable rocks. They occur as volcanic flows and tuffs and as small intrusive bodies (dikes and sills). They are associated with trachytes and a wide variety of feldspathoidal rocks. See IGNEOUS ROCKS; TRACHYTE.   [C.A.C.]

Phosphate minerals   Any naturally occurring inorganic salts of phosphoric acid, $H_3[PO_4]$. All known phosphate minerals are orthophosphates. There are over 150 species of phosphate minerals, and their crystal chemistry is often very complicated. Phosphate mineral paragenesis can be divided into three categories: primary phosphates (crystallized directly from a melt or fluid), secondary phosphates (derived from the primary phosphates by hydrothermal activity), and rock phosphates (derived from the action of water upon buried bone material, skeletons of small organisms, and so forth). See MINERAL.   [P.B.M.]

Phyllite   A type of metamorphic rock formed during low-grade metamorphism of clay-rich sediments called pelites. Phyllites are very fine grained rocks with a grain size barely visible in a hand specimen. They have a well-developed planar element called cleavage defined by alignment of mica grains and interlayering of quartz-rich and mica-rich domains. Typically, mica grains show the greater alignment, although other mineral components (quartz, carbonate, and feldspars) may show a preferred shape orientation. Where all minerals of a particular type show the same degree of alignment and the fabric is well developed throughout the rock, the fabric is termed a penetrative fabric. Cleavage surfaces in phyllites have a glittery, lustrous sheen due to light reflecting off grains of chlorite and muscovite. The mineralogy of phyllites is dependent on chemical composition; typical minerals in phyllites are chlorite, muscovite, and quartz. Other minerals that may be present in phyllites formed during low-grade metamorphism include chlorotoid, garnet (rarely), sodium-mica, and sulfide minerals. See CHLORITE; MUSCOVITE; QUARTZ.

Phyllite is found in most regionally metamorphosed terranes in the world, including the Appalachians of eastern North America, the Scottish Highlands, and the Alps. See METAMORPHIC ROCKS.   [M.W.N.]

Physical geography   The study of the Earth's surface features and associated processes. Physical geography aims to explain the geographic patterns of climate, vegetation, soils, hydrology, and landforms, and the physical environments that result from their interactions. Physical geography merges with human geography to provide a synthesis of the complex interactions between nature and society.

The basic content of physical geography comprises a number of areas of specialization. Climatology, the scientific study of climates, concerns the total complex of weather conditions at a given location over an extended time period; it deals not only with average conditions but with extremes and variations. Geomorphology is the interpretive description and explanation of landforms and the fluvial, glacial, coastal, and eolian process that operate on them. The forms, processes, and patterns within the biosphere, including vegetation and animal distributions, are studied as biogeography. With strong ties to fluvial geomorphology, geographic hydrology concerns the scientific study of water from the aspects of distribution, movement, and utilization. Soil geography, with emphasis on the origin, characteristics, classification, and utilization potential of soils, provides an area of specialization with links to land use. Ultimately, the physical geography of a region is understood through an integration of the multiple aspects.   [J.E.O.]

Picrite   The term picrite has been used with several different meanings. It is generally considered to include certain medium- to fine-grained igneous rocks composed chiefly of olivine with smaller amounts of pyroxene, hornblende, and plagioclase feldspar (labradorite). Its feldspar content is slightly higher than that of peridotite and lower than that of gabbro. Certain analcite-bearing types, associated with teschenite, have also been included under the term picrite. Picrite is rare and is found in small intrusives (sills and dikes). *See* GABBRO; IGNEOUS ROCKS; PERIDOTITE.   [C.A.C.]

Pigeonite   Monoclinic pyroxenes of the general formula $(Mg,Fe)SiO_3$ having some augite in solid solution. Pigeonite bears the same relation to the orthorhombic pyroxenes as augite does to the diopside-hedenbergite series. Pigeonite is the orthorhombic pyroxene equivalent in the volcanic rocks. *See* AUGITE; DIOPSIDE; ORTHORHOMBIC PYROXENE; PYROXENE.   [G.W.DeV.]

Pitchstone   A natural glass with dull or pitchy luster and generally brown, green or gray color. It is extremely rich in microscopic, embryonic crystal growths (crystallites) which may cause its dull appearance. The water content of pitchstone is high and generally ranges from 4 to 10% by weight. Pitchstone is formed by rapid cooling of molten rock material (lava or magma) and occurs most commonly as small dikes or as marginal portions of larger dikes. *See* IGNEOUS ROCKS; VOLCANIC GLASS.   [C.A.C.]

Plains   The relatively smooth sections of the continental surfaces, occupied largely by gentle rather than steep slopes and exhibiting only small local differences in elevation. Because of their smoothness, plains lands, if other conditions are favorable, are especially amenable to many human activities. Thus it is not surprising that the majority of the world's principal agricultural regions, close-meshed transportation networks, and concentrations of population are found on plains. Large parts of the Earth's plains, however, are hindered for human use by dryness, shortness of frost-free season, infertile soils, or poor drainage. Because of the absence of major differences in elevation or exposure or of obstacles to the free movement of air masses, extensive plains usually exhibit broad uniformity or gradual transition of climatic characteristics.

Somewhat more than one-third of the Earth's land area is occupied by plains. With the exception of ice-sheathed Antarctica, each continent contains at least one major expanse of smooth land in addition to numerous smaller areas. The largest plains of North America, South America, and Eurasia lie in the continental interiors, with broad extensions reaching to the Atlantic (and Arctic) Coast. The most extensive plains of Africa occupy much of the Sahara and reach south into the Congo and Kalahari basins. Much of Australia is smooth, with only the eastern margin lacking extensive plains. *See* TERRAIN AREAS.

Surfaces that approach true flatness, while not rare, constitute a minor portion of the world's plains. Most commonly they occur along low-lying coastal margins, the lower sections of major river systems, or the floors of inland basins. Nearly all are the products of extensive deposition by streams or in lakes or shallow seas. The majority of plains, however, are distinctly irregular in surface form, as a result of valley-cutting by streams or of irregular erosion and deposition by continental glaciers.   [E.H.Ha.]

Plate tectonics   The theory that provides an explanation for the behavior of the Earth's crust, particularly the global distribution of mountain building, earthquake activity, and volcanism in a series of linear belts. Numerous other geological phenomena such as lateral variations in surface heat flow, the physiography and geology of ocean basins, and various associations of igneous, metamorphic, and sedimentary rocks can also be logically related by plate tectonics theory.

The theory is based on a simple model of the Earth in which a rigid outer shell 30–90 mi (50–150 km) thick, the lithosphere, consisting of both oceanic and continental crust as well as the upper mantle, is considered to lie above a hotter, weaker semiplastic asthenosphere. The asthenosphere, or low-velocity zone, extends from the base of the lithosphere to a depth of about 400 mi (700 km). The brittle lithosphere is broken into a mosaic of internally rigid plates which move horizontally across the Earth's surface relative to one another. Only a small number of major lithospheric plates exist, which grind and scrape against each other as they move independently like rafts of ice on water. Most dynamic activity such as seismicity, deformation, and the generation of magma occur only along plate boundaries, and it is on the basis of the global distribution of such tectonic phenomena that plates are delineated. *See* ASTHENOSPHERE; EARTHQUAKE; LITHOSPHERE.

The plate tectonics model for the Earth is consistent with the occurrence of sea-floor spreading and continental drift. Convincing evidence exists that both these processes have been occurring for at least the last 6×10^8 years. This evidence includes the magnetic anomaly patterns of the sea floor, the paucity and youthful age of marine sediment in the ocean basins, the topographic features of the sea floor, and the indications of shifts in the position of continental blocks which can be inferred from paleomagnetic data on paleopole positions, paleontological and paleoclimatological observations, the match-up of continental margins and geological provinces across present-day oceans, and the structural style and rock types found in ancient mountain belts. *See* PALEOCLIMATOLOGY; PALEOMAGNETISM.

Geological observations, geophysical data, and theoretical considerations support the existence of three fundamentally distinct types of plate boundaries, named and classified on the basis of whether immediately adjacent plates move apart from one another (divergent plate margins), toward one another (convergent plate margins), or slip past one another in a direction parallel to their common boundary (transform plate margins). The boundaries of plates can, but need not, coincide with the contact between continental and oceanic crust. The velocity at which plates move varies from plate to plate and within portions of the same plate, ranging between 0.8 and 8 in. (2 and 20 cm) per year. *See* CONTINENTS, EVOLUTION OF; MID-OCEANIC RIDGE.

Not only does plate tectonics theory explain the present-day distribution of seismic and volcanic activity around the globe and physiographic features of the ocean basins such as trenches and mid-oceanic rises, but most Mesozoic and Cenozoic mountain belts appear to be related to the convergence of lithospheric plates. Two different varieties of modern mobile belts have been recognized, cordilleran type and collision type. The Cordilleran range, which forms the western rim of North and South America (the Rocky Mountains, Pacific Coast ranges, and the Andes) have for the most part been created by the underthrusting of an ocean lithospheric plate beneath a continental plate. Underthrusting along the Pacific margin of South America is causing the continued formation of the Andes. The Alpine-Himalayan belt, formed where the collision of continental blocks buckled intervening volcanic belts and sedimentary strata into tight folds and faults, is an analog of the present tectonic situation in the Mediterranean, where the collision of Africa and Europe has begun. *See* Cordilleran belt; Orogeny.

Plate tectonics is considered to have been operative as far back as 2.5×10^9 years. Prior to that interval, evidence suggests that plate tectonics may have occurred, although in a markedly different manner, with higher rates of global heat flow producing smaller convective cells or more densely distributed mantle plumes which fragmented the Earth's surface into numerous small, rapidly moving plates. *See* Continental drift; Geodynamics.   [W.C.P.]

Plateau   Any elevated area of relatively smooth land. Usually the term is used more specifically to denote an upland of subdued relief that on at least one side drops off abruptly to adjacent lower lands. In most instances the upland is cut by deep but widely separated valleys or canyons. Small plateaus that stand above their surroundings on all sides are often called tables, tablelands, or mesas. The abrupt edge of a plateau is an escarpment or, especially in the western United States, a rim.   [E.H.Ha.]

Playa   A nearly level, generally dry surface in the lowest part of a desert basin with internal drainage (see illustration). When its surface is covered by a shallow sheet of water, it is a playa lake. Playas and playa lakes are also called dry lakes, alkali flats, mud flats, saline lakes, salt pans, inland sabkhas, ephemeral lakes, salinas, and sinks.

A playa surface is built up by sandy mud that settles from floodwater when a playa is inundated by downslope runoff during a rainstorm. A smooth, hard playa occurs where ground-water discharge is small or lacking and the surface is flooded frequently. These mud surfaces are cut by extensive desiccation polygons caused by shrinkage of the drying clay. Puffy-ground playas form by crystallization of minerals as ground water evaporates in muds near the surface.

Light-colored playa in lowest part of Sarcobatus Flat in southern Nevada.

Subsurface brine is present beneath many playas. The type of brine depends on the original composition of the surface water and reflects the lithology of the rocks weathered in the surrounding mountains. *See* GROUND-WATER HYDROLOGY.

Numerous playas in the southwestern United States yield commercial quantities of evaporite minerals, commonly at shallow depths. Important are salt ($NaCl$) and the borates, particularly borax ($Na_2B_4O_7 \cdot 10H_2O$), kernite ($Na_2B_4O_7 \cdot 4H_2O$), ulexite ($NaCaB_5O_9 \cdot 8H_2O$), probertite ($NaCaB_5O_9 \cdot 5H_2O$), and colemanite ($Ca_2B_6O_{11} \cdot 5H_2O$). Soda ash (sodium carbonate; Na_2CO_3) is obtained from trona ($Na_3H(CO_3)_2 \cdot 2H_2O$) and gaylussite ($Na_2Ca(CO_3)_2 \cdot 5H_2O$). Lithium and bromine are produced from brine waters. *See* DESERT EROSION FEATURES; SALINE EVAPORITES. [J.F.Hu.]

Pleistocene   The older of the two epochs of the Quaternary Period. The Pleistocene Epoch represents the interval of geological time (and rocks accumulated during that time) extending from the end of the Pliocene Epoch (and the end of the Tertiary Period) to the start of the Holocene Epoch. Most recent time scales show the Pleistocene Epoch spanning the interval from 1.8 million years before present (m.y. B.P.) to 10,000 years B.P. The Pleistocene is commonly characterized as an epoch when the Earth entered its most recent phase of widespread glaciation. *See* HOLOCENE; PLIOCENE; QUATERNARY; TERTIARY.

In modern geological time scales, the Pleistocene is subdivided into a lower and an upper series. In Europe the lower series is considered equivalent to the Calabrian Stage, while the upper series is equated with the Sicilian and Tyrrehenian stages.

The onset of the Pleistocene brought glaciations that were more widespread than those in the Pliocene. Mountain glaciers expanded and continental ice fields covered large areas of the temperate latitudes. Sea ice also became more widespread. As evidence has accumulated in recent decades from both the land and the sea, it clearly shows at least 17 glacial events occurred during the Pleistocene. *See* CLIMATE HISTORY; GLACIAL EPOCH.

The expansion and decay of the ice sheets had a direct effect on the global sea level. Global sea-level fluctuations of 50–150 m (170–500 ft) have been estimated for various glacial-interglacial episodes during the Pleistocene. Since the last deglaciation, which began some 17,000 years ago, the sea level has risen by about 110 m (360 ft) worldwide, drowning all of the ancient lowstand shorelines. One important product of the sea-level drops was the migration of large river deltas to the edges of the continental shelves and to the deeper parts of the basins. Conversely, the last marine transgression that started in the late Pleistocene after rapid deglaciation and ended in the Holocene (6000 to 7000 years ago) resulted in new deltas that formed at the mouths of modern rivers. *See* DELTA; PALEOCEANOGRAPHY.

The onset of cooler climate and Pleistocene glaciation is also approximated with a wave of mammalian migration from the east to the west. A relatively modern-looking fauna that included the first true oxen, elephants, and the first one-toed horse appeared at the beginning of Pleistocene. The modern horse, *E. caballus*, made its first appearance some 250,000 years ago in the late Pleistocene in North America. From North America it migrated to Asia and then west to Europe. However, during the last glacial maximum, some 18,000 years ago, it became extinct in North America when it was unable to cross the deserts to migrate to South America. Oxen, deer and reindeer, large cats, mammoth, great elk, wolf, hyena, and woolly rhinos proliferated during the middle and late Pleistocene. Mammoths, which have been found preserved nearly intact in frozen soils in Siberia, ranged over much of Europe during the glacial times. *See* GEOLOGIC TIME SCALE.

The unique Pleistocene mammalian faunas of some of the isolated islands, such as Madagascar, the Philippines, Taiwan, and the Japanese Archipelago, indicate restriction in the dispersal of species during the Pleistocene. The Pliocene Epoch had given rise to the human precursor, *Homo habilis*, around 2 m.y. B.P. The appearance of *H. erectus* came on the scene almost at the Plio-Pleistocene boundary around 1.8 m.y. B.P. The first archaic *H. sapiens* are now considered to have arrived on the scene around about a million years ago. The appearance of *H. neanderthalensis* or the Neanderthal Man, is now dated at least as far back as 250,000 years B.P. Recent datings have the appearance of the first true modern *H. sapiens* (the Cro-Magnon Man) to around 100,000 years B.P. *See* GEOLOGIC TIME SCALE.   [B.U.H.]

Pleochroic halos   Spherical or elliptical regions up to 40 micrometers in diameter in which there is a change in color from the surrounding mineral when viewed with a petrographic microscope. Pleochroic halos are found around small inclusions of radioactive minerals—for example, zircon, monazite, allanite, xenotime, and apatite—and in rock-forming minerals, principally quartz, micas, amphiboles, and pyroxenes. Halos have also been identified in coalified wood preserved in deposits on the Colorado Plateau. *See* PETROFABRIC ANALYSIS; RADIOACTIVE MINERALS.

The change in color is a result of radiation damage caused by alpha particles emitted during the radioactive decay of nuclides in the decay chains of uranium-238, uranium-235, and thorium-232. The range of the alpha particle and ionization effects account for the size and color of the halos. The halos have a distinctive ring structure with varying degrees of discoloration between the rings: the coloration in the halos increases, saturates, and finally diminishes with increasing ion dose. *See* METAMICT STATE.

The early interest in pleochroic halos was in their use for geologic age dating. Careful attempts were made to correlate the halo color with the alpha-irradiation dose in order to estimate the age of the enclosing mineral. Additionally, the constant size of the rings of the uranium and thorium halos for minerals of different ages was taken as evidence that the decay constants for radionuclides used in age dating had remained constant throughout geologic time. Thermal annealing of the halos has been used to model the thermal histories of rock units. *See* FISSION TRACK DATING; GEOCHRONOMETRY.
                [R.Ew.]

Pliocene   The youngest of the five geological epochs of the Tertiary Period. The Pliocene represents the interval of geological time (and rocks deposited during that time) extending from the end of the Miocene Epoch to the beginning of the Pleistocene Epoch of the Quaternary Period. Modern time scales assign the duration of 5.0 to 1.8 million years ago (Ma) to the Pliocene Epoch. *See* MIOCENE; PLEISTOCENE; QUATERNARY; TERTIARY.

Pliocene marine sediments are commonly distributed along relatively restricted areas of the continental margins and in the deep-sea basins. Continental margin sediments are most often terrigenous and range from coarser-grained sandstone to finer-grained mudstone and clay. Major rivers of the world, such as the Amazon, Indus, and Ganges, contain thick piles of Pliocene terrigenous sediments in their offshore fans. The Pliocene deep-sea sediments are carbonate-rich (commonly biogenic oozes) and are often very thick (up to 5000 m or 16,400 ft). *See* BASIN; CONTINENTAL MARGIN.

Modern stratigraphic usage subdivides the Pliocene Epoch into two standard stages, the lower, Zanclean stage and the upper, Piacenzian stage.

The most notable tectonic events in the Pliocene include the beginning of the third and last phase of the Himalayan uplift, the Attican orogeny that began in the late

Miocene and continued into the Pliocene, and the Rhodanian and Walachian orogenies that occurred during the later Pliocene. *See* OROGENY.

The latest Miocene is marked by a global cooling period that continued into the earliest Pliocene, and there is evidence that the East Antarctic ice sheet had reached the continental margins at this time. The global sea level had been falling through the late Miocene, and with the exception of a marked rise in the mid-Zanclean, the trend toward lowered sea levels continued through the Pliocene and Pleistocene. The mid-Zanclean sea-level rise (3.5–3 Ma) was also accompanied by a significant global warming event. The oxygen isotopic data, which record the prevailing sea surface temperatures and total ice volume on the ice caps, show little variations in the Equatorial Pacific during the middle Pliocene. By early Pliocene time, the major surface circulation patterns of the world ocean and the sources of supply of bottom waters were essentially similar to their modern counterparts. *See* GEOLOGIC THERMOMETRY.

By Pliocene time, much of the marine and terrestrial biota had essentially evolved its modern characteristics. The late Pliocene cooling led to the expansion of cooler-water marine assemblages of the higher latitudes into lower latitudes, particularly the foraminifers, bivalves, and gastropods. At the onset of cooling, the warm-water-preferring calcareous nannoplankton group of discoasters began waning in the late Pliocene and became extinct at the close of the epoch.

The widespread grasslands of the Pliocene were conducive to the proliferation of mammals and increase in their average size. The mid-Zanclean sea-level rise led to the geographic isolation of many groups of mammals and the increase in endemism. But the late Pliocene-Pleistocene lowering of sea level facilitated land connections and allowed extensive mammalian migration between continents with interchanges between North and South America. The arrival of the North American mammals led to increased competitive pressure and extinction of many typically South American groups. Horses evolved and spread widely in the Pliocene.

The Pliocene Epoch also saw the appearance of several hominid species that are considered to be directly related to modern human ancestry. The earliest hominid bones have been discovered from Baringo, Kenya, in sediments that are dated to be of earliest Pliocene age. After this first occurrence, a whole suite of australopithicine species made their appearance in the Pliocene. *See* PALEONTOLOGY.   [B.U.H.]

Pluton   A solid rock body that formed by cooling and crystallization of molten rock (magma) within the Earth. Most plutons, or plutonic bodies, are regarded as the product of crystallization of magma intruded into surrounding "country rocks" within the Earth (principally within the crust). Igneous rock bodies are referred to generally as either extrusive or volcanic on one hand, or as intrusive or plutonic on the other, although the term volcanic is sometimes also used to refer to small, shallow intrusive bodies associated with volcanoes. *See* IGNEOUS ROCKS; MAGMA.

Plutons occur in a nearly infinite variety of shapes and sizes, so that definition of types is arbitrary in many cases. In general, two modes of emplacement can be recognized with regard to the country rock. Concordant plutons are intruded between layers of stratified rock, whereas the more common discordant plutons are characterized by boundaries that cut across preexisting structures or layers in the country rock. The principal types of concordant plutons are sills, laccoliths, and lopoliths; the principal types of discordant plutons are dikes, volcanic necks or plugs, stocks, and batholiths.

Several mechanisms of magma intrusion are known or proposed. The most simple ones, pertaining to smaller plutonic bodies, are forceful injection or passive migration into fractures. Larger plutons may form by several processes. For example, less dense magma may migrate upward along a myriad of channelways to accumulate as a large molten body within the upper crust. Further migration could occur by forceful injection, by stoping (a process where the magma rises as blocks of the roof of the magma chamber break off and sink), and by diapiric rise, where country rocks flow around the upward-moving magma body. *See* PETROLOGY.   [W.R.V.S.]

Pneumatolysis   The alteration of rocks or crystallization of minerals by gases or supercritical fluids (generically termed magmatic fluids) derived from solidifying magma. At surface conditions, magmatic fluids contain steam with lesser amounts of carbon dioxide, sulfur dioxide, hydrogen sulfide, hydrogen chloride, and hydrogen fluoride, and trace amounts of many other volatile constituents. Magmatic fluids may contain relatively high concentrations of light and heavy elements, particularly metals, that do not crystallize readily in common rock-forming silicates constituting most of the solidifying magma; thus, valuable rare minerals and ores are sometimes deposited in rocks subjected to pneumatolysis. Magmatic fluids are acidic and may react extensively with rocks in the volcanic edifice or with wall rocks surrounding intrusions. Penetration of magmatic fluids into adjacent rocks is greatly aided by faults, fractures, and cracks developed during intrusion and eruption or created by earlier geologic events. *See* MAGMA; METAMORPHISM; METASOMATISM; ORE AND MINERAL DEPOSITS; VOLCANO.

Pneumatolysis describes specific mechanisms of mineral deposition, hydrothermal alteration, or metasomatism in which magmatic fluids play an extremely significant role. For example, lavas and ejecta at volcanoes may contain blocks (xenoliths) of wall rock that react with magmatic fluids to form pneumatolytic minerals such as vesuvianite (idocrase). Gases streaming from volcanic fumaroles deposit sublimates of sulfur, sulfates, chlorides, fluorides, and oxides of many metals. Wall rocks surrounding volcanic conduits may be thoroughly altered to mixtures of quartz, alunite, anhydrite, pyrite, diaspore, kaolin, as well as other minerals by acidic fluids degassed from magma. Rarely, gold, silver, base-metal sulfides, arsenides, and tellurides are deposited by the fluids, making valuable ores. *See* LAVA; METASOMATISM; PYROCLASTIC ROCKS; VESUVIAN-ITE; XENOLITH.   [F.Go.]

Polar meteorology   The science of weather and climate in the high latitudes of the Earth. In the polar regions the Sun never rises far above the horizon and remains below it continuously for part of the year, so that snow and ice can persist for long periods even at low elevations. The meteorological processes that result have distinctive local and large-scale characteristics in both polar regions.   [U.R.]

Population dispersal   The process by which groups of living organisms expand the space or range within which they live. Dispersal operates when individual organisms leave the space that they have occupied previously, or in which they were born, and settle in new areas. Natal dispersal is the first movement of an organism from its birth site to the site in which it first attempts to breed. Adult dispersal is a subsequent movement when an adult organism changes its location in space. As individuals move across space and settle into new locations, the population to which they belong expands or contracts its overall distribution. Thus, dispersal is the process by which populations change the area they occupy.

Migration is the regular movement of organisms during different seasons. Many species migrate between wintering and breeding ranges. Such migratory movement is marked by a regular return in future seasons to previously occupied regions, and so usually does not involve an expansion of population range. Some migratory species show astounding abilities to return to the exact locations used in previous seasons. Other species show no regular movements, but wander aimlessly without settling permanently into a new space. Wandering (called nomadism) is typical of species in regions where the availibility of food resources are unpredictable from year to year. Neither migration nor nomadism is considered an example of true dispersal.

Virtually all forms of animals and plants disperse. In most higher vertebrates, the dispersal unit is an entire organism, often a juvenile or a member of another young age class. In other vertebrates and many plants, especially those that are sessile (permanently attached to a surface), the dispersal unit is a specialized structure (disseminule). Seeds, spores, and fruits are disseminules of plants and fungi; trochophores and planula larvae are disseminules of sea worms and corals, respectively. Many disseminules are highly evolved structures specialized for movement by specific dispersal agents such as wind, water, or other animals.

A special case of zoochory (dispersal using animal agents) involves transport by humans. The movement of people and cargo by cart, car, train, plane, and boat has increased the potential dispersal of weedy species worldwide. Many foreign aquatic species have been introduced to coastal areas by accidental dispersal of disseminules in ship ballast water. The zebra mussel is one exotic species that arrive in this manner and is now a major economic problem throughout the Great Lakes region of North America. Some organisms have been deliberately introduced by humans into new areas. Domestic animals and plants have been released throughout the world by farmers. A few pest species were deliberately released by humans; European starlings, for example.

Some of the most highly coevolved dispersal systems are those in which the disseminule must be eaten by an animal. Such systems have often evolved a complex series of signals and investments by both the plant and the animal to ensure that the seeds are dispersed at an appropriate time and that the animal is a dependable dispersal agent. Such highly evolved systems are common in fruiting plants and their dispersal agents, which are animals called frugivores. Fruiting plants cover their seeds with an attractive, edible package (the fruit) to get the frugivore to eat the seed. To ensure that fruits are not eaten until the seeds are mature, plants change the color of their fruits as a signal to show that the fruits are ready for eating.

Many plants in the tropical rainforests are coevolved to have their seeds dispersed by specific animal vectors, including birds, mammals, and ants. Many tropical trees, shrubs, and herbaceous plants are specialized to have their seeds dispersed by a single animal species. Temperate forest trees, in contrast, often depend on wind dispersal of both pollen and seeds.

Dispersal barriers are physical structures that prevent organisms from crossing into new space. Oceans, rivers, roads, and mountains are examples of barriers for species whose disseminules cannot cross such features. It is believed that the creation of physical barriers is the primary factor responsible for the evolution of new species. A widespread species can be broken into isolated fragments by the creation of a new physical barrier. With no dispersal linking the newly isolated populations, genetic differences that evolve in each population cannot be shared between populations. Eventually, the populations may become so different that no interbreeding occurs even if dispersal pathways are reconnected. The populations are then considered separate species.

Dispersal is of major concern for scientists who work with rare and endangered animals. Extinction is known to be more prevalent in small, isolated populations. Conservation biologists believe that many species exist as a metapopulation, that is, a group of populations interconnected by the dispersal of individuals or disseminules between subpopulations. The interruption of dispersal in this system of isolated populations can increase the possibility of extinction of the whole metapopulation. Conservation plans sometimes propose the creation of corridors to link isolated patches of habitat as a way of increasing the probability of successful dispersal. *See* EXTINCTION (BIOLOGY).   [J.B.D.]

Population ecology   The study of spatial and temporal patterns in the abundance and distribution of organisms and of the mechanisms that produce those patterns. Species differ dramatically in their average abundance and geographical distributions, and they display a remarkable range of dynamical patterns of abundance over time, including relative constancy, cycles, irregular fluctuations, violent outbreaks, and extinctions. The aims of population ecology are threefold: (1) to elucidate general principles explaining these dynamic patterns; (2) to integrate these principles with mechanistic models and evolutionary interpretations of individual life-history tactics, physiology, and behavior as well as with theories of community and ecosystem dynamics; and (3) to apply these principles to the management and conservation of natural populations.

In addition to its intrinsic conceptual appeal, population ecology has great practical utility. Control programs for agricultural pests or human diseases ideally attempt to reduce the intrinsic rate of increase of those organisms to very low values. Analyses of the population dynamics of infectious diseases have successfully guided the development of vaccination programs. In the exploitation of renewable resources, such as in forestry or fisheries biology, population models are required in order to devise sensible harvesting strategies that maximize the sustainable yield extracted from exploited populations. Conservation biology is increasingly concerned with the consequences of habitat fragmentation for species preservation. Population models can help characterize minimum viable population sizes below which a species is vulnerable to rapid extinction, and can help guide the development of interventionist policies to save endangered species. Finally, population ecology must be an integral part of any attempt to bring the world's burgeoning human population into harmonious balance with the environment. *See* ECOLOGY; THEORETICAL ECOLOGY.   [R.Hol.]

Porphyroblast   A relatively large crystal formed in a metamorphic rock. The presence of abundant porphyroblasts gives the rock a porphyroblastic texture. Minerals found commonly as porphyroblasts include biotite, garnet, chloritoid, staurolite, kyanite, sillimanite, andalusite, cordierite, and feldspar. Porphyroblasts are generally a few millimeters or centimeters across, but some attain a diameter of over 1 ft (30 cm). They may be bounded by well-defined crystal faces, or their outlines may be highly irregular or ragged. Very commonly they are crowded with tiny grains of other minerals that occur in the rock.

Most commonly, porphyroblasts develop in schist and gneiss during the late stages of recrystallization. As the rock becomes reconstituted, certain components migrate to favored sites and combine there to develop the large crystals. *See* GNEISS; METAMORPHIC ROCKS; SCHIST.   [C.A.C.]

Porphyry   An igneous rock characterized by porphyritic texture, in which large crystals (phenocrysts) are enclosed in a matrix of very fine-grained to aphanitic (not visibly crystalline) material. Porphyries are generally distinguished from other porphyritic rocks by their abundance of phenocrysts and by their occurrence in small intrusive bodies (dikes and sills) formed at shallow depth within the earth. In this sense porphyries are hypabyssal rocks. *See* IGNEOUS ROCKS; PHENOCRYST.

Porphyries occur as marginal phases of medium-sized igneous bodies (stocks, laccoliths) or as apophyses (offshoots) projecting from such bodies into the surrounding rocks. They are also abundant as dikes cutting compositionally equivalent plutonic rock, or as dikes, sills, and laccoliths injected into the adjacent older rocks.   [W.I.R.]

Precambrian   A major interval of geologic time between about 540 million years (Ma) and 3.8 billion years (Ga) ago, comprising the Archean and Proterozoic eons and encompassing most of Earth history. The Earth probably formed around 4.6 Ga and was then subjected to a period of intense bombardment by meteorites so that there are few surviving rocks older than about 3.8 billion years. Ancient rocks are preserved exclusively in continental areas. All existing oceanic crust is younger than about 200 million years, for it is constantly being recycled by the processes of sea-floor spreading and subduction. Development of techniques for accurate determination of the ages of rocks and minerals that are billions of years old has revolutionized the understanding of the early history of the Earth. *See* DATING METHODS; GEOCHRONOMETRY; GEOLOGIC TIME SCALE; ROCK AGE DETERMINATION.

Detailed sedimentological and geochemical investigations of Precambrian sedimentary rocks and the study of organic remains have facilitated understanding of conditions on the ancient Earth. Microorganisms are known to have been abundant in the early part of Earth history. The metabolic activities of such organisms played a critical role in the evolution of the atmosphere and oceans. There have been attempts to apply the concepts of plate tectonics to Precambrian rocks. These diverse lines of investigation have led to a great leap in understanding the early history of the planet. *See* PLATE TECTONICS.

Rocks of the Archean Eon (2.5–3.8 Ga) are preserved as scattered small "nuclei" in shield areas on various continents. The Canadian shield contains perhaps the biggest region of Archean rocks in the world, comprising the Superior province. Much of the Archean crust is typified by greenstone belts, which are elongate masses of volcanic and sedimentary rocks that are separated and intruded by greater areas of granitic rocks. The greenstones are generally slightly metamorphosed volcanic rocks, commonly extruded under water, as indicated by their characteristic pillow structures. These structures develop when lava is extruded under water and small sac-like bodies form as the lava surface cools and they are expanded by pressure from lava within. Such structures are common in Archean greenstone assemblages in many parts of the world. *See* ARCHEAN; METAMORPHIC ROCKS.

The Proterozoic Eon extends from 2.5 Ga until 540 Ma, the beginning of the Cambrian Period and Phanerozoic Eon. Proterozoic successions include new kinds of sedimentary rocks, display proliferation of primitive life forms such as stromatolites, and contain the first remains of complex organisms, including metazoans (the Ediacaran fauna). Sedimentary rocks of the Proterozoic Eon contain evidence of gradual oxidation of the atmosphere. Abundant and widespread chemical deposits known as banded iron formations (BIF) make their appearance in Paleo-Proterozoic sedimentary basins. *See* BANDED IRON FORMATION; PROTEROZOIC.   [G.M.Y.]

Precious stones   The materials found in nature that are used frequently as gemstones, including amber, beryl (emerald and aquamarine), chrysoberyl (cat's-eye and alexandrite), coral, corundum (ruby and sapphire), diamond, feldspar (moonstone and amazonite), garnet (almandite, demantoid, and pyrope), jade (jadeite and nephrite), jet, lapis lazuli, malachite, opal, pearl, peridot, quartz (amethyst, citrine, and agate), spinel, spodumene (kunzite), topaz, tourmaline, turquois and zircon. *See* GEM.

The terms precious and semiprecious have been used to differentiate between gemstones on a basis of relative value. Because there is a continuous gradation of values from materials sold by the pound to those valued at many thousands of dollars per carat, and because the same mineral may furnish both, a division is essentially meaningless.   [R.T.L.]

Precipitation (meteorology)   The fallout of water drops or frozen particles from the atmosphere. Liquid types are rain or drizzle, and frozen types are snow, hail, small hail, ice pellets (also called ice grains; in the United States, sleet), snow pellets (graupel, soft hail), snow grains, ice needles, and ice crystals. In England sleet is defined as a mixture of rain and snow, or melting snow. Deposits of dew, frost, or rime, and moisture collected from fog are occasionally also classed as precipitation. *See* HAIL; SNOW.

All precipitation types are called hydrometeors, of which additional forms are clouds, fog, wet haze, mist, blowing snow, and spray. Whenever rain or drizzle freezes on contact with the ground to form a solid coating of ice, it is called freezing rain, freezing drizzle, or glazed frost; it is also called an ice storm or a glaze storm, and sometimes is popularly known as silver thaw or erroneously as a sleet storm. *See* CLOUD; FOG.

Rain, snow, or ice pellets may fall steadily or in showers. Steady precipitation may be intermittent though lacking sudden bursts of intensity. Hail, small hail, and snow pellets occur only in showers; drizzle, snow grains, and ice crystals occur as steady precipitation. Showers originate from instability clouds of the cumulus family, whereas steady precipitation originates from stratiform clouds.

The amount of precipitation, often referred to as precipitation or simply as rainfall, is measured in a collection gage. It is the actual depth of liquid water which has fallen on the ground, after frozen forms have been melted, and is recorded in millimeters or inches and hundredths. A separate measurement is made of the depth of unmelted snow, hail, or other frozen forms. *See* SNOW GAGE. For discussions of other topics related to precipitation *see* CLOUD PHYSICS; DEW; DEW POINT; HUMIDITY; HYDROLOGY; HYDROMETEOROLOGY; RAIN SHADOW; WEATHER MODIFICATION.   [J.R.F.]

Precipitation measurement   Instruments used to measure the amount of rain or snow that falls on a level surface. Such measurements are made with instruments known as precipitation gages. A precipitation gage can be as simple as an open container on the ground to collect rain, snow, and hail; it is usually more complex, however, because of the need to avoid wind effects, enhance accuracy and resolution, and make a measurement representative of a large area. Precipitation is measured as the depth to which a flat horizontal surface would have been covered per unit time if no water were lost by runoff, evaporation, or percolation. Depth is expressed in inches or millimeters, typically per day. The unit of time is often understood and not stated explicitly. Snow and hail are converted to equivalent depth of liquid water. *See* METEOROLOGICAL INSTRUMENTATION; PRECIPITATION (METEOROLOGY); SNOW SURVEYING.   [F.V.B.]

Accurate quantitative precipitation measurement is probably the most important weather radar application. It is extremely valuable for hydrological applications such as watershed management and flash flood warnings. Radar can make rapid and spatially contiguous measurements over vast areas of a watershed at relatively low cost. *See* METEOROLOGICAL RADAR; PRECIPITATION (METEOROLOGY); RADAR METEOROLOGY.   [R.J.Do.]

Prehnite   A mineral with the formula $Ca_2(Al,Fe^{3+})(OH)_2$-$[Si_3AlO_{10}]$, with Al in parentheses in octahedral and Al in brackets in tetrahedral coordination by oxygens. The mineral usually occurs as stalactitic aggregates or as curved crystals, has a vitreous luster, and is yellowish green to pale green in color. Hardness is $6–6^1/_2$ on Mohs scale; specific gravity 2.8–2.9. Common occurrences include vesicular basalts such as the Keweenaw basalts in the Upper Peninsula of Michigan, and the Watchung basalts in New Jersey. *See* SILICATE MINERALS.   [P.B.M.]

Proterozoic   A major division of geologic time spanning from 2500 to 543 million years before present (Ma). The beginning of Proterozoic time is an arbitrary boundary that roughly coincides with the transition from a tectonic style dominated by extensive recycling of the Earth's continental crust to a style characterized by preservation of the crust as stable continental platforms. The end of the Proterozoic coincides with the Precambrian-Cambrian boundary, which is formally defined on the basis of the first appearance of diverse coelomate invertebrate animals. Proterozoic Earth history testifies to several remarkable biogeochemical events, including the formation and dispersal of the first supercontinent, the maturation of life and evolution of animals, the rise of atmospheric oxygen, and the decline of oceanic carbonate saturation. Tremendous iron and lead-zinc mineral deposits occur in Proterozoic rocks, as do the first preserved accumulations of oil and gas. *See* CAMBRIAN; PRECAMBRIAN.

Many of the Earth's Archean cratons are blanketed by little-deformed sequences of Proterozoic sedimentary rocks, which indicate that vigorous recycling of the Earth's crust, characteristic of Archean time, had slowed markedly by the beginning of Proterozoic time. This decrease in crustal recycling is attributed to the development of thick continental roots, which stabilized the cratons, and the decrease in heat that was escaping from the Earth's interior, believed to drive thermal convection in the Earth's mantle and recycling of the crust. Most of the Earth's Archean cratons appear to have participated in the formation of a supercontinent in Mesoproterozoic time, about 1200 Ma. This supercontinent, called Rodinia, seems to have assembled with the North American craton (Laurentia) at its center. Rodinia persisted until the latest part of the Neoproterozoic, about 600 Ma. *See* ARCHEAN; CONTINENTS, EVOLUTION OF; EARTH, HEAT FLOW IN; EARTH CRUST; EARTH INTERIOR; PLATE TECTONICS.

Giant iron oxide deposits were formed by precipitation from seawater about 2000 Ma, whereupon oxygen was free to accumulate in the atmosphere and shallow ocean. During most of Paleoproterozoic time the oceans and atmosphere were reducing and ferrous iron was abundant in seawater.

The partial pressure of carbon dioxide on the early Earth was very high. During Proterozoic time, much of the mass of carbon shifted from the ocean and atmosphere to the solid Earth. Enormous volumes of limestone [$CaCO_3$] and dolostone [$CaMg(CO_3)_2$] were deposited and testify to this shift. *See* DOLOMITE ROCK; LIMESTONE; SEDIMENTARY ROCKS.

Glaciers covered significant parts of the Earth during two widely separated times in Proterozoic history. The first episode occurred about 2200 Ma, and glacial deposits of that age cover various parts of North America and Scandinavia. The second episode consisted of at least two different pulses spanning from 750 to 600 Ma during Neoproterozoic time. Glaciers formed at that time were of almost global extent, and were thought to have extended from the poles to the Equator, according to the snowball Earth hypothesis. *See* GLACIAL EPOCH.

A number of significant events in the evolution of life occurred during Proterozoic time. The record of biological activity is rich, consisting of actual body fossils, in addition to organism traces and impressions, and complex chemical biomarkers. Eukaryotic microbes appear to have evolved by about 1900 Ma, when they became major players in ecosystems present at that time. By the beginning of Neoproterozoic time, about 1000 Ma, multicellular eukaryotic algae are present in numerous sedimentary basins around the world.

The evolution of animals did not take place until the close of Neoproterozoic time. Why these organisms evolved at this particular time in Earth history remains unanswered. General opinion proposes that it was likely the result of the confluence of a number of environmental factors, such as the rise in oxygen. Whatever the cause of their origin, these existed until at least 543 Ma, when another major evolutionary adaptive radiation began which marks the onset of Cambrian time and the end of the Proterozoic Eon. *See* EXTINCTION (BIOLOGY); GEOLOGIC TIME SCALE. [J.P.Gr.]

Proustite   A mineral having composition Ag_3AsS_3. It occurs in prismatic crystals terminated by steep ditrigonal pyramids, but is more commonly massive or in disseminated grains. Hardness is 2–2.5 (Mohs scale) and specific gravity is 5.55. The luster is adamantine and the color ruby red. It is called light ruby silver in contrast to pyrargyrite, dark ruby silver. Proustite and pyrargyrite are found together in silver veins. Noted localities are at Chañiarcillo, Chile; Freiberg, Germany; Guanajuato, Mexico; and Cobalt, Ontario, Canada. *See* PYRARGYRITE.   [C.S.Hu.]

Provenance (geology)   In sedimentary geology, all characteristics of the source area from which clastic (detrital) sediments and sedimentary rocks are derived, including relief, weathering, and source rocks. *See* WEATHERING PROCESSES.

The goal of most provenance studies of sedimentary rocks is the determination of source characteristics of the mountains or hills from which the constituent sediment was derived. Such determinations are difficult to make because sediment composition and texture are continually modified during erosion, transport, deposition, and diagenesis (postdepositional modification). It is most straightforward to determine provenance in situations in which these modifying effects are minimal; provenance may be indeterminate or ambiguous in situations involving extensive modification of sediment composition and texture. The former situation is most common in tectonically active areas, resulting in rapid uplift and erosion of mountains, rapid transport and deposition, and slight diagenetic modification after deposition. In contrast, stable continental areas (for example, cratons) provide ample opportunity for intense weathering so that chemical, mineralogical, and textural characteristics of sediment are intensely modified. *See* DIAGENESIS; EROSION; SEDIMENTARY ROCKS.

Clastic sediment is commonly recycled during multiple episodes of mountain building, erosion, sedimentation, lithification, and renewed mountain building. This process constitutes the rock cycle, within which igneous, sedimentary, and metamorphic rocks are created and modified. During this process, it is common for older sedimentary rocks to be uplifted and eroded, so that individual sedimentary particles (clasts) are recycled to form new sediment, which may be lithified to form new sedimentary rock. Provenance studies must determine the proportion of a sedimentary rock derived directly from indicated source rocks versus the proportion derived directly from another sedimentary rock (that is, rocks exposed during previous cycles of sedimentation). This determination is essential, but commonly difficult to accomplish because the multicyclic nature of sediment may be difficult to recognize. *See* METAMORPHIC ROCKS.

Fundamentally different methods of study are utilized, depending on what aspect of provenance is emphasized, what type of sediment is studied, and what scale of sampling and study is attempted. Grain size of detrital sediment is a dominant control over what methods may be employed.

Methods of determining provenance include direct determination of rock types (primarily used for coarse to medium grains); direct determination of mineralogy (used for all grain sizes); whole-rock geochemistry (used for medium to fine grains); geochemistry of individual mineral species (used for all grain sizes, but especially for medium grains); and radiometric dating of individual mineral species (primarily for medium grains). *See* GEOCHEMISTRY; MINERALOGY; ROCK AGE DETERMINATION.   [R.V.I.]

Psychrometer   An instrument consisting of two thermometers which is used in the measurement of the moisture content of air or other gases. The bulb or sensing area of one of the thermometers either is covered by a thin piece of clean muslin cloth wetted uniformly with distilled water or is otherwise coated with a film of distilled water. The temperatures of both the bulb and the air contacting the bulb are lowered by the evaporation which takes place when unsaturated air moves past the wetted bulb. An equilibrium temperature, termed the wet-bulb temperature (T_W), will be reached; it closely approaches the lowest temperature to which air can be cooled by the evaporation of water into that air. The water-vapor content of the air surrounding the wet bulb can be determined from this wet-bulb temperature and from the air temperature measured by the thermometer with the dry bulb (T_D by using an expression of the form $e = e_{SW} - aP\,(T_D - T_W)$. Here e is the water-vapor pressure of the air, e_{SW} is the saturation water-vapor pressure at the wet-bulb temperature, P is atmospheric pressure, and a is the psychrometric constant, which depends upon properties of air and water, as well as on speed of ventilation of air passing the wet bulb. *See* PSYCHROMETRICS.   [R.M.Sch.]

Psychrometrics

A study of the physical and thermody-namic properties of the atmosphere. The properties of primary concern in air conditioning are (1) dry-bulb temperature, (2) wet-bulb temperature, (3) dew-point temperature, (4) absolute humidity, (5) percent humidity, (6) sensible heat, (7) latent heat, (8) total heat, (9) density, and (10) pressure.

The dry-bulb temperature is the ambient temperature of the air and water vapor as measured by a thermometer or other temperature-measuring device in which the thermal element is dry and shielded from radiation. *See* AIR TEMPERATURE.

If the bulb of a dry-bulb thermometer is covered with a silk or cotton wick saturated with distilled water and the air is drawn over it at a velocity not less than 1000 ft/min (5 m/s), the resultant temperature will be the wet-bulb temperature. Where the dry-bulb and wet-bulb temperatures are the same, the atmosphere is saturated.

The dew-point temperature is the temperature at which the water vapor in the atmosphere begins to condense. This is also the temperature of saturation at which the dry-bulb, wet-bulb, and dew-point temperatures are all the same. *See* DEW POINT.

The actual quantity of water vapor in the atmosphere is designated as the absolute humidity. Percentage or relative humidity is the ratio of the actual water vapor in the atmosphere to the quantity of water vapor the atmosphere could hold if it were saturated at the same temperature. *See* HUMIDITY.

Sensible heat, or enthalpy of dry air, is heat which manifests itself as a change in temperature.

Latent heat, or enthalpy of vaporization, is the heat required to change a liquid into a vapor without change in temperature. Latent heat is sometimes referred to as the latent heat of vaporization and varies inversely as the pressure.

The total heat, or enthalpy, of the atmosphere is the sum of the sensible heat, latent heat, and superheat of the vapor above the saturation or dew-point temperature. Total heat is relatively constant for a constant wet-bulb temperature, deviating only about 1.5–2% low at relative humidities below 30%.

The density of the atmosphere varies with both altitude and percentage humidity. The higher the altitude the lower the density, and the higher the moisture content the lower the density.

Atmospheric pressure is usually referred to as barometric pressure. Pressure varies inversely as elevation, as temperature, and as percentage saturation. *See* PSYCHROMETER.
[J.Ev.]

Pumice

A rock froth, formed by the extreme puffing up of liquid lava by expanding gases liberated from solution in the lava prior to and during solidification. Some varieties will float in water for many weeks before becoming waterlogged. Typical pumice is siliceous (rhyolite or dacite) in composition, but the lightest and most vesicular pumice (known also as reticulite and thread-lace scoria) is of basaltic composition. *See* LAVA; VOLCANIC GLASS.   [G.A.M.]

Pyrargyrite   A mineral having composition Ag_3SbS_3. The mineral occurs as prismatic crystals and in massive form and in disseminated grains. The hardness is 2.5 on Mohs scale and specific gravity is 5.85. The luster is adamantine and the color a deep ruby red to black, giving it the name dark ruby silver. Pyrargyrite is an important silver ore when it is found in veins associated with proustite and other silver minerals. It has been mined as silver ore at Chanarcillo, Chile; Freiberg, Germany; Guanajuato, Mexico; and Cobalt, Ontario, Canada. *See* PROUSTITE.   [C.S.Hu.]

Pyrite   A mineral having composition FeS_2. Pyrite has a Mohs hardness of 6–6.5 and a density of 5.02. The luster is metallic, the color brass yellow, and the streak greenish black or brownish black.

Pyrite, or iron pyrites, is the most common "fool's gold," but it is hard and brittle whereas gold is soft and sectile. Its hardness also distinguishes it from softer chalcopyrite. Marcasite was once thought to be polymorphous with pyrite, but precise analyses show that marcasite contains excess iron, whereas pyrite is stoichiometric FeS_2. *See* MARCASITE.

Pyrite is the most common and most widespread sulfide mineral. It forms under almost all known conditions of mineral deposition. Under oxidizing conditions, pyrite readily alters to iron sulfates and eventually to limonite, forming gossan, the surface expression of pyrite-rich mineral deposits. *See* LIMONITE.

Because of its high sulfur content (53.4%), pyrite has become a source of sulfur for the production of sulfuric acid. In some places, it is mined for sulfur alone.   [L.Gr.]

Pyroclastic rocks   Rocks of extrusive (volcanic) origin, composed of rock fragments produced directly by explosive eruptions. Pyroclastic fragments may represent shattered and comminuted older rocks (volcanic, plutonic, sedimentary, or metamorphic) or solidified lava droplets formed by violent explosion. *See* TUFF; VOLCANO.
                                                                                         [C.A.C.]

Pyroelectricity   The property of certain crystals to produce a state of electric polarity by a change of temperature. Certain dielectric (electrically nonconducting) crystals develop an electric polarization (dipole moment per unit volume) when they are subjected to a uniform temperature change. This pyroelectric effect occurs only in crystals which lack a center of symmetry and also have polar directions (that is, a polar axis). These conditions are fulfilled for 10 of the 32 crystal classes. Typical examples of pyroelectric crystals are tourmaline, lithium sulfate monohydrate, cane sugar, and ferroelectric barium titanate.

Pyroelectric crystals can be regarded as having a built-in or permanent electric polarization. When the crystal is held at constant temperature, this polarization does not manifest itself because it is compensated by free charge carriers that have reached the surface of the crystal by conduction through the crystal and from the surroundings. However, when the temperature of the crystal is raised or lowered, the permanent polarization changes, and this change manifests itself as pyroelectricity.

The magnitude of the pyroelectric effect depends upon whether the thermal expansion of the crystal is prevented by clamping or whether the crystal is mechanically unconstrained. In the clamped crystal, the primary pyroelectric effect is observed, whereas in the free crystal, a secondary pyroelectric effect is superposed upon the primary effect. The secondary effect may be regarded as the piezoelectric polarization arising from thermal expansion, and is generally much larger than the primary effect.

[H.Gr.]

Pyroelectrics have a broad spectrum of potential scientific and technical applications. The most developed is the detection of infrared radiation. In addition, pyroelectric detectors can be used to measure the power generated by a radiation source (in radiometry), or the temperature of a remote hot body (in pyrometry, with corrections due to deviations from the blackbody emission).

An infrared image can be projected on a pyroelectric plate and transformed into a relief of polarization on the surface. Other potential applications of pyroelectricity include solar energy conversion, refrigeration, information storage, and solid-state science.

[A.Had.]

Pyrolusite A mineral having composition MnO_2. Well-developed crystals (polianite) are rare; it is usually in radiating fibers or reniform coatings. The hardness is 1–2 on the Mohs scale (often soiling the fingers) and the specific gravity is 4.75. The luster is metallic and the color iron-black. It frequently forms pseudomorphs after other manganese minerals, notably manganite.

Pyrolusite is extensively mined as a manganese ore in many countries, chiefly in Russia, Ghana, India, the Republic of South Africa, Morocco, Brazil, and Cuba. *See* MANGANITE.

[C.S.Hu.]

Pyromorphite A mineral series in the apatite group, or in the larger grouping of phosphate, arsenate, and vanadate-type minerals. In this series lead (Pb) substitutes for calcium (Ca) of the apatite formula $Ca_5(PO_4)_3(F,OH,Cl)$, and little fluorine (F) or hydroxide (OH) is present. *See* APATITE.

The pyromorphite series crystallizes in the hexagonal system. Crystals are prismatic. Other forms are granular, globular, and botryoidal. Pyromorphite colors range through green, yellow, and brown; vanadinite occurs in shades of yellow, brown, and red.

Pyromorphites are widely distributed as secondary minerals in oxidized lead deposits. Pyromorphite is a minor ore of lead; vanadinite is a source of vanadium and minor ore of lead.

[W.R.Lo.]

Pyrophyllite A hydrated aluminum silicate with composition $Al_2Si_4O_{10}(OH)_2$. The mineral is commonly white, grayish, greenish, or brownish, with a pearly to waxy

Specimen of pyrophyllite.
(*Pennsylvania State University*)

appearance and greasy feel. It occurs as compact masses, as radiating aggregates (see illustration), and as foliated masses. Pyrophyllite belongs to the layer silicate (phyllosilicate) group of minerals. The mineral is soft (hardness $1–1^1/_2$ on Mohs scale) and has easy cleavage parallel to the structural layers. The mineral is highly stable to acids.

Pyrophyllite is used principally for refractory materials and in other ceramic applications. The main sources for pyrophyllite in the United States are in North Carolina. An unusual form from the Transvaal is called African wonderstone. *See* SILICATE MINERALS.

[G.W.Br.]

Pyroxene   A large, geologically significant group of dark, rock-forming silicate minerals. Pyroxene is found in abundance in a wide variety of igneous and metamorphic rocks. Because of their structural complexity and their diversity of chemical composition and geologic occurrence, these minerals have been intensively studied by using a wide variety of modern analytical techniques. Knowledge of pyroxene compositions, crystal structures, phase relations, and detailed microstructures provides important information about the origin and thermal history of rocks in which they occur.

The general chemical formula for pyroxenes is $M2M1T_2O_6$, where T represents the tetrahedrally coordinated sites, occupied primarily by silicon cations (Si^{4+}). Names of specific end-member pyroxenes are assigned based on composition and structure type. Those pyroxenes containing primarily calcium (Ca^{2+}) or sodium (Na^+) cations in the M2 site are monoclinic. Pyroxenes containing primarily Mg^{2+} or iron(II) (Fe^{2+}) cations in the M2 site are orthorhombic at low temperatures, but they may transform to monoclinic at higher temperature.

Common pyroxenes have specific gravity ranging from about 3.2 (enstatite, diopside) to 4.0 (ferrosilite). Hardnesses on the Mohs scale range from 5 to 6. Iron-free pyroxenes may be colorless (enstatite, diopside, jadeite); as iron content increases, colors range from light green or yellow through dark green or greenish brown, to brown, greenish black, or black (orthopyroxene, pigeonite, augite, hedenbergite, aegirine). Spodumene may be colorless, yellowish emerald green (hiddenite), or lilac pink (kunzite). *See* SPODUMENE.

Pyroxenes in the rock-forming quadrilateral are essential constituents of ferromagnesian igneous rocks such as gabbros and their extrusive equivalents, basalts, as well as most peridotites. Pyroxenes may also be present as the dark constituents of more silicic diorites and andesites. *See* ANDESITE; BASALT; DIORITE; GABBRO; IGNEOUS ROCKS; PERIDOTITE.

Pyroxenes, especially those of the diopside-hedenbergite series, are found in medium- to high-grade metamorphic rocks of the amphibolite and granulite facies. *See* METAMORPHISM.

The peridotites found in the Earth's upper mantle contain Mg-rich, Ca-poor pyroxenes, in addition to olivine and other minor minerals. At successively greater depths in the mantle, these Mg-rich pyroxenes will transform sequentially to spinel (Mg_2SiO_4) plus stishovite (SiO_2), an ilmenite structure, or a garnet structure, depending on temperature; and finally at depths of around 360–420 mi (600–700 km) to an $MgSiO_3$ perovskite structure. *See* ASTHENOSPHERE; EARTH INTERIOR; GARNET; ILMENITE; OLIVINE; PEROVSKITE; SILICATE MINERALS; SPINEL; STISHOVITE.

[C.W.Bu.]

Pyroxenite   A heavy, dark-colored, phaneritic (visibly crystalline) igneous rock composed largely of pyroxene with smaller amounts of olivine or hornblende. Pyroxenite composed largely of orthopyroxene occurs with anorthosite and peridotite in large, banded gabbro bodies. Some of these pyroxenite masses are rich sources of chromium. Certain pyroxenites composed largely of clinopyroxene are also of magmatic origin, but many probably represent products of reaction between magma and limestone. Other pyroxene-rich rocks have formed through the processes of metamorphism and metasomatism. *See* GABBRO; IGNEOUS ROCKS; PERIDOTITE; PYROXENE.   [C.A.C.]

Pyroxenoid   A group of silicate minerals whose physical properties resemble those of pyroxenes. In contrast with the two-tetrahedra periodicity of pyroxene single silicate chains, the pyroxenoid crystal structures contain single chains of $(SiO_4)^{4-}$ silicate tetrahedra having repeat periodicities ranging from three to nine (see illustration). The tetrahedron is a widely used geometric representation for the basic building block of most silicate minerals, in which all silicon cations (Si^{4+}) are bonded to four oxygen anions arranged as if they were at the corners of a tetrahedron. In pyroxenoids, as in other single-chain silicates, two of the four oxygen anions in each tetrahedron are shared between two Si^{4+} cations to form the single chains, and the other two oxygen anions of each tetrahedron are bonded to divalent cations, such as calcium (Ca^{2+}), iron (Fe^{2+}), or manganese (Mn^{2+}). These divalent cations bond to six (or sometimes seven or eight) oxygen anions, forming octahedral (or irregular seven- or eight-cornered) coordination polyhedra. *See* PYROXENE; SILICATE MINERALS.

The pyroxenoid structures have composite structural units consisting of strips of octahedra (or larger polyhedra) two or more units wide formed by sharing of polyhedral edges, to which the silicate tetrahedral chains are attached on both top and bottom. The repeat periodicity of the octahedral strips is the same as that of the silicate chains to which they are attached. These composite units are cross-linked to form the three-dimensional crystal structures. The pyroxenoid minerals are triclinic, with either C$\overline{1}$ or I$\overline{1}$ space-group symmetry depending on the stacking of the composite units, and c-axis lengths ranging from about 0.71 nanometer for three-repeat silicate tetrahedral chains to about 2.3 nm for nine-repeat chains.

There are two series of pyroxenoid minerals, one anhydrous and one hydrous. The anhydrous pyroxenoids are significantly more abundant. A general formula for anhydrous pyroxenoids is $(Ca, Mn, Fe^{2+})SiO_3$. Silicate-chain repeat length is inversely proportional to mean divalent cation size. The hydrogen in hydrous pyroxenoids is

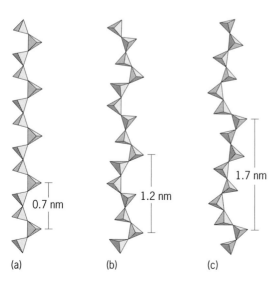

(a)   (b)   (c)

Tetrahedral silicate chains in three pyroxenoid structures: (a) wollastonite with three-tetrahedra periodicity, (b) rhodonite with five-tetrahedra periodicity, and (c) pyroxmangite-pyrox ferroite with seven-tetrahedra periodicity.

0.7 nm

1.2 nm

1.7 nm

hydrogen-bonded between two oxygen atoms; additional hydrogen in santaclaraite is bound as hydroxyl (OH) and as a water molecule (H_2O).   [C.W.Bu.]

Pyrrhotite A mineral with composition $Fe_{1-x}S$ ($x = 0$ to 0.2). Eskebornite, $Fe_{1-x}Se$, is the selenium analog. The iron-deficient pyrrhotites are ferrimagnetic at room temperature, but at some higher temperature they become paramagnetic, presumably because of vacancy disorder.

The mineral occurs as rounded grains to large masses, more rarely as tabular pseudohexagonal crystals and rosettes. Color is brownish bronze-yellow with dark grayish-black streaks. Hardness is 4 on Mohs scale and specific gravity 4.6 (for the composition Fe_7S_8).

Pyrrhotite occurs in basic igneous rocks as a late-stage fractional differentiate, particularly in norites and gabbros, and sufficient quantities may constitute an ore of iron. Pyrrhotite also occurs with magnetite and chondrodite in contact metamorphic marbles, and in low-temperature veins with calcite and other sulfides and sulfosalts.   [P.B.M.]

Q

Quantum mineralogy   Quantum mechanics applied to mineralogical systems. A theoretical understanding of chemical bonding and the electronic structure of minerals is fundamental to understanding the behavior of minerals. The principal goal of quantum mineralogy is to calculate from first principles the properties and transformations of solid earth materials. One of the advantages of this theoretical tool is the ability to predict the behavior of materials at conditions which are not readily accessible to experiment. More importantly, though, experimental observations may be interpreted in terms of quantum-mechanical theory.

The methods used in quantum mineralogy are no different from those used for other chemical systems. They range from simple conceptions of the chemical bond in terms of quantum-mechanical expressions to rigorous calculation of wave functions for model systems. Too often, the more exact the quantum-mechanical calculation, the less understandable are the results for the nonspecialist. Despite such limitations, quantum mineralogy is a powerful theoretical probe into the nature, of structure, bonding, and properties of mineral systems. The size and complexity of model systems that can be considered in rigorous calculations continue to grow rapidly with increasing efficiency of computer hardware and program algorithms. *See* MINERALOGY.

[B.C.C.]

Quartz   The most common oxide on the Earth's surface, constituting 12% of the crust by volume. Quartz is a crystalline form of silicon dioxide (SiO_2). Among the igneous rocks, quartz is especially common within granites, granodiorites, pegmatites, and rhyolites. In addition, quartz can be observed in low- to high-grade metamorphic rocks, including phyllites, quartzites, schists, granulites, and eclogites. Because hydrothermal fluids are enriched in dissolved silica, the passage of fluids through rock fractures results in the emplacement of quartz veins. *See* GRANITE; GRANODIORITE; IGNEOUS ROCKS; METAMORPHIC ROCKS; PEGMATITE; RHYOLITE.

Once quartz has formed, it persists through erosional reworking because of its low solubility in water (parts per million) and its high mechanical hardness (7 on Mohs scale). Consequently, quartz becomes increasingly concentrated in beach sands as they mature, and it is a major component of sandstone. In sedimentary environments, quartz also forms as the final crystallization product during silica diagenesis; amorphous silica on the sea floor that derives from the skeletons of diatoms, radiolarians, and sponges will transform to quartz upon prolonged exposure to increased temperatures ($\leq 300°C$ or 572°F) and pressures (≤ 2 kilobars or 200 pascals) after burial. *See* DIAGENESIS; SANDSTONE.

As with virtually all silicates, the atomic framework of the quartz structure consists of Si^{4+} cations that are tetrahedrally coordinated by oxygen anions (O^{2-}). Every oxygen anion is bonded to two silicon cations, so that the tetrahedral units are corner-linked to form continuous chains. In low-temperature quartz (or α-quartz), two distinct tetrahedral chains spiral about the crystallographic c axis.

Although the silica tetrahedra can be depicted as spirals about the c axis in a left-handed sense, right-handed quartz crystals are found in nature as abundantly as are left-handed crystals. These enantiomorphic varieties are known as the Brazil twins of quartz, and they may be distinguished by crystal shape (corresponding crystal faces occur in different orientations) and by opposite optical activities.

Impurity concentrations in natural α-quartz crystals usually fall below 1000 parts per million. The violet and yellow hues observed in amethyst and citrine are associated with Fe, and black smoky quartz contains Al. The white coloration of milky quartz reflects light scattering off minute fluid inclusions, and the pink tint in rose quartz is believed to arise from fine-scale intergrowths of a pegmatitic mineral called dumortierite [$Al_{27}B_4Si_{12}O_{69}(OH)_3$]. *See* AMETHYST; DUMORTIERITE.

Quartz is used predominantly by the construction industry as gravel and as aggregate in concrete. In addition, quartz is important in advanced technologies. Quartz is piezoelectric and has an extremely high quality factor. The high quality factor means that a bell made of quartz would resonate (ring) for a very long time. This property, combined with its piezoelectric behavior, makes quartz the perfect crystal for oscillators in watches.

Compression of α-quartz perpendicular to the c axis creates an electrostatic charge, and this property is exploited in oscillator plates in electronic components. Large flawless crystals of quartz are routinely synthesized for oscillators and for prisms in laser optic systems. Quartz also is employed in abrasives, fluxes, porcelains, and paints.

[P.J.H.]

Quartzite A metamorphic rock consisting largely or entirely of quartz. Most quartzites are formed by metamorphism of sandstone; but some have developed by metasomatic introduction of quartz, SiO_2, often accompanied by other chemical elements, for example, metals and sulfur (ore quartzites). *See* METAMORPHIC ROCKS; METASOMATISM; SANDSTONE.

Pure sandstones yield pure quartzites. Impure sandstones yield a variety of quartzite types. The cement of the original sandstone is in quartzite recrystallized into characteristic silicate minerals, whose composition often reflects the mode of development. Even the Precambrian quartzites correspond to types that are parallel to present-day deposits. *See* QUARTZ.

[T.F.W.B.]

Quaternary A period that encompasses at least the last 3,000,000 years of the Cenozoic Era, and is concerned with major worldwide glaciations and their effect on land and sea, on worldwide climate, and on the plants and animals that lived then. The Quaternary is divided into the Pleistocene Epoch and Holocene. The universal

CENOZOIC	QUATERNARY	
	TERTIARY	
MESOZOIC	CRETACEOUS	
	JURASSIC	
	TRIASSIC	
PALEOZOIC	PERMIAN	
	CARBONIFEROUS	Pennsylvanian
		Mississippian
	DEVONIAN	
	SILURIAN	
	ORDOVICIAN	
	CAMBRIAN	
PRECAMBRIAN		

term Pleistocene is gradually re-placing Quaternary; Holocene involves the last 7000 years since the Pleistocene. *See* CENOZOIC; GLACIAL EPOCH; HOLOCENE; PLEISTOCENE. [S.E.Wh.]

R

Radar meteorology   The application of radar to the study of the atmosphere and to the observation and forecasting of weather. Meteorological radars transmit electromagnetic waves at microwave and radio-wave frequencies. Water and ice particles, inhomogeneities in the radio refractive index associated with atmospheric turbulence and humidity variations, insects, and birds scatter radar waves. The backscattered energy received at the radar constitutes the returned signal. Meteorologists use the amplitude, phase, and polarization state of the backscattered energy to deduce the location and intensity of precipitation, the wind speed along the direction of the radar beam, and precipitation type (for example, rain or hail). *See* METEOROLOGICAL RADAR; WEATHER FORECASTING AND PREDICTION.

Much of the understanding of the structure of storms derives from measurements made with networks of Doppler radars. They are used to investigate the complete three-dimensional wind fields associated with storms, fronts, and other meteorological phenomena. *See* PRECIPITATION (METEOROLOGY); STORM; STORM DETECTION; THUNDERSTORM.   [R.M.Ra.]

Radioactive minerals   Minerals that contain uranium (U) or thorium (Th) as an essential component of their chemical composition. Examples are uraninite (UO_2) or thorite ($ThSiO_4$). There are radioactive minerals in which uranium and thorium substitute for ions of similar size and charge. There are approximately 200 minerals in which uranium or thorium are essential elements, although many of these phases are rare and poorly described. These minerals are important, as they are found in ores mined for uranium and thorium, most commonly uraninite and its fine-grained variety, pitchblende, for uranium. Thorite and thorogummite are the principal ore minerals of thorium. Minerals in which uranium and thorium occur in trace amounts, such as zircon ($ZrSiO_4$), are important because of their use in geologic age dating. The isotope uranium-238 (^{238}U) decays to lead-206 (^{206}Pb); ^{235}U decays to ^{207}Pb; ^{232}Th decays to ^{208}Pb; thus, the ratios of the isotopes of uranium, thorium, and lead can be used to determine the ages of minerals that contain these elements. *See* DATING METHODS; GEOCHRONOMETRY; LEAD ISOTOPES (GEOCHEMISTRY); THORITE; URANINITE.   [R.Ew.]

Radiocarbon dating   A method of obtaining age estimates on organic materials which has been used to date samples as old as 75,000 years. The method has provided age determinations in archeology, geology, geophysics, and other branches of science.

Radiocarbon (^{14}C) determinations can be obtained on wood; charcoal; marine and fresh-water shell; bone and antler; peat and organic-bearing sediments; carbonate deposits such as tufa, caliche, and marl; and dissolved carbon dioxide (CO_2) and carbonates in ocean, lake, and ground-water sources. Each sample type has specific problems associated with its use for dating purposes, including contamination and special environmental effects. While the impact of ^{14}C dating has been most profound in archeological research and particularly in prehistoric studies, extremely significant contributions have also been made in hydrology and oceanography. In addition, beginning in the 1950s the testing of thermonuclear weapons injected large amounts of artificial ^{14}C ("bomb ^{14}C") into the atmosphere, permitting it to be used as a geochemical tracer.

Carbon (C) has three naturally occurring isotopes. Both ^{12}C and ^{13}C are stable, but ^{14}C decays by very weak beta decay (electron emission) to nitrogen-14 (^{14}N) with a half-life of approximately 5700 years. Naturally occurring ^{14}C is produced as a secondary effect of cosmic-ray bombardment of the upper atmosphere. As $^{14}CO_2$, it is distributed on a worldwide basis into various atmospheric, biospheric, and hydrospheric reservoirs on a time scale much shorter than its half-life. Metabolic processes in living organisms and relatively rapid turnover of carbonates in surface ocean waters maintain ^{14}C levels at approximately constant levels in most of the biosphere. The natural ^{14}C activity in the geologically recent contemporary "prebomb" biosphere was approximately 13.5 disintegrations per minute per gram of carbon. *See* COSMOGENIC NUCLIDE.

To the degree that ^{14}C production has proceeded long enough without significant variation to produce an equilibrium or steady-state condition, ^{14}C levels observed in contemporary materials may be used to characterize the original ^{14}C activity in the corresponding carbon reservoirs. Once a sample has been removed from exchange with its reservoir, as at the death of an organism, the amount of ^{14}C begins to decrease as a function of its half-life. A ^{14}C age determination is based on a measurement of the residual ^{14}C activity in a sample compared to the activity of a sample of assumed zero age (a contemporary standard) from the same reservoir. The relationship between the ^{14}C age and the ^{14}C activity of a sample is given by the equation below, where t is

$$t = \frac{1}{\lambda} \ln \frac{A_o}{A_s}$$

radiocarbon years B.P. (before the present), λ is the decay constant of ^{14}C (related to the half-life $t_{1/2}$ by the expression $t_{1/2} = 0.693/\lambda$), A_o is the activity of the contemporary standards, and A_s is the activity of the unknown age samples. Conventional radiocarbon dates are calculated by using this formula, an internationally agreed half-life value of 5568 ± 30 years, and a specific contemporary standard.

The naturally occurring isotopes of carbon occur in the proportion of approximately 98.9% ^{12}C, 1.1% ^{13}C, and 10^{-10}% ^{14}C. The extremely small amount of radiocarbon in natural materials was one reason why ^{14}C was one of the isotopes which had been produced artificially in the laboratory before being detected in natural concentrations. A measurement of the ^{14}C content of an organic sample will provide an accurate determination of the sample's age if it is assumed that (1) the production of ^{14}C by cosmic rays has remained essentially constant long enough to establish a steady state in the $^{14}C/^{12}C$ ratio in the atmosphere, (2) there has been a complete and rapid mixing of ^{14}C throughout the various carbon reservoirs, (3) the carbon isotope ratio in the sample has not been altered except by ^{14}C decay, and (4) the total amount of carbon in any reservoir has not been altered. In addition, the half-life of ^{14}C must be known with sufficient accuracy, and it must be possible to measure natural levels of ^{14}C to appropriate levels of accuracy and precision.   [R.E.T.]

Radioisotope geochemistry   A branch of environmental geochemistry and isotope geology concerned with the occurrence of radioactive nuclides in sediment, water, air, biological tissues, and rocks. The nuclides have relatively short half-lives ranging from a few days to about 10^6 years, and occur only because they are being produced by natural or anthropogenic nuclear reactions or because they are the intermediate unstable daughters of long-lived naturally occurring radioactive isotopes of uranium and thorium. The nuclear radiation, consisting of alpha particles, beta particles, and gamma rays, emitted by these nuclides constitutes a potential health hazard to humans. However, their presence also provides opportunities for measurements of the rates of natural processes in the atmosphere and on the surface of the Earth.

The unstable daughters of uranium and thorium consist of a group of 43 radioactive isotopes of 13 chemical elements, including all of the naturally occurring isotopes of the chemical elements radium, radon, polonium, and several others. A second group of radionuclides is produced by the interaction of cosmic rays with the chemical elements of the Earth's surface and atmosphere. This group includes hydrogen-3 (tritium), beryllium-10, carbon-14, aluminum-26, silicon-32, chlorine-36, iron-55, and others. A third group of radionuclides is produced artificially by the explosion of nuclear devices, by the operation of nuclear reactors, and by various particle accelerators used for research in nuclear physics. Some of the radionuclides produced in nuclear reactors decay sufficiently slowly to be useful for geochemical research, including strontium-90, cesium-137, iodine-129, and isotopes of plutonium. The explosion of nuclear devices in the atmosphere has also contributed to the abundances of certain radionuclides that are produced by cosmic rays such as tritium and carbon-14. *See* Cosmogenic nuclide; Dating methods; Environmental radioactivity.   [G.Fau.]

Rain shadow   An area of diminished precipitation on the lee side of mountains. There are marked rain shadows, for example, east of the coastal ranges of Washington, Oregon, and California, and over a larger region, much of it arid, east of the Cascade Range and Sierra Nevadas. All mountains decrease precipitation on their lee; but rain shadows are sometimes not marked if moist air often comes from different directions, as in the Appalachian region.

The causes of rain shadow are (1) precipitation of much of the moisture when air is forced upward on the windward side of the mountains, (2) deflection or damming of moist air flow, and (3) downward flow on the lee slopes, which warms the air and lowers its relative humidity.   [J.R.F.]

Rainbow   An optical effect of the sky formed by sunlight falling on the spherical droplets of water associated with a rain shower. The circular arc of colors in the rainbow is seen on the side of the sky away from the Sun. The bright, primary rainbow shows the spectrum of colors running from red, on the outside of the bow, to blue on the inside. Sometimes a fainter, secondary bow is seen outside the primary bow with the colors reversed from their order in the primary bow. The shape of each bow is that of a circle, centered on the antisolar point, a point in the direction exactly opposite to that of the Sun, which is marked by the shadow of the observer's head.

As a light ray from the Sun strikes the surface of a water drop, some light is reflected and some passes through the surface into the drop. The primary bow results from light that enters the drop, reflects once inside the drop, and then leaves the drop headed toward the observer's eye. Light that is reflected twice inside the drop produces the secondary bow. The change of direction that occurs when a light ray enters or leaves the waterdrop (refraction) is different for the different colors that make up white sunlight. As a result, the size of the circle is different for each color, thereby separating the colors into the rainbow sequence. *See* Meteorological optics.   [R.Gr.]

Rare-earth minerals   Naturally occurring solids, formed by geological processes, that contain the rare-earth elements—the lanthanides (atomic numbers 57–71) and yttrium (atomic number 39)—as essential constituents. In a rare-earth mineral, at least one crystallographic site contains a total atomic ratio of lanthanides and yttrium that is greater than that of any other element. The mineral name generally has a suffix, called a Levinson modifier, indicating the dominant rare-earth element; for example, monazite-(La) [LaPO$_4$] contains predominantly lanthanum, and monazite-(Ce) [CePO$_4$] contains predominantly cerium. *See* MINERAL.

So far, about 170 distinct species of rare-earth minerals have been described. A large number of carbonates, phosphates, silicates, niobates, and fluorides are known as rare-earth minerals. It is necessary to obtain structural as well as chemical information about a mineral to judge the essentiality of its rare-earth elements (that is, whether the rare-earth element is part of the mineral's ideal formula or is an impurity). Sometimes, minerals with significant rare-earth content are treated as rare-earth minerals, even if the rare-earth element content appears unessential to the mineral. More than 60 mineral species, including the apatite group minerals, garnet group minerals, and fluorite, are in this category. *See* APATITE; FLUORITE; GARNET.

Rare-earth minerals can be observed as accessory minerals in igneous rocks, such as monazite-(Ce) in granite. Carbonatite is the typical host rock of rare-earth minerals such as bastnäsite-(Ce) [CeCO$_3$F] and monazite-(Ce). Rare-earth minerals also often occur in pegmatite. In both carbonatite and pegmatite, rare-earth elements are concentrated by primary crystallization from melt and by hydrothermal reactions. Carbonatite deposits containing rare-earth elements are found throughout the world. Chemically stable, rare-earth minerals are not weathered easily. As a result, they have been deposited as heavy minerals in beach sand. Such deposits are found in Southeast Asia and Western Australia. *See* CARBONATITE; PEGMATITE.

Among the rare-earth minerals, bastnäsite-(Ce) is the most important source of rare-earth elements. Monazite-(Ce), synchysite-(Ce) [CaCe(CO$_3$)$_2$F], xenotime-(Y), britholite-(Ce) [(Ce$_3$Ca$_2$)(SiO$_4$)$_3$(OH)], and allanite-(Ce) [CaCeAl$_2$Fe(Si$_2$O$_7$)(SiO$_4$)O(OH)] are also sources. Rare-earth elements have been leached with acid from the surface of clay minerals. Rare-earth minerals containing radioactive nuclear species, such as thorium and uranium, are not used as source materials. *See* CLAY MINERALS.

[R.Miy.]

Realgar   A mineral having composition AsS and crystallizing in the monoclinic system. Realgar can occur in short, vertically striated crystals, but more frequently is granular and in crusts. The hardness is 1.5–2 (Mohs scale) and the specific gravity is 3.48. The luster is resinous and the color red to orange. Realgar is found in ores of lead, silver, and gold associated with orpiment and stibnite. It occurs with the silver and lead ores in Hungary, Czechoslovakia, and Germany. Good crystals have come from Binnenthal, Switzerland, and Allchar, Macedonia. In the United States it is found at Manhattan, Nevada; Mercer, Utah; and as deposits from geyser waters in Yellowstone National Park. *See* ORPIMENT; STIBNITE.

[C.S.Hu.]

Red Sea   A body of water that separates northeastern Africa from the Arabian Peninsula. The Red Sea forms part of the African Rift System, which also includes the Gulf of Aden and a complex series of continental rifts in East Africa extending as far south as Malawi. The Red Sea extends for 1920 km (1190 mi) from Ras (Cape) Muhammed at the southern tip of the Sinai Peninsula to the Straits of Bab el Mandab at the entrance to the Gulf of Aden. At Sinai the Red Sea splits into the Gulf of Suez, which extends for an additional 300 km (180 mi) along the northwest trend of the Red Sea and the nearly northward-trending Gulf of Aqaba. The 175-km-long (109-mi) Gulf of Aqaba forms the southern end of the Levant transform, a primarily strike-slip fault system extending north into southern Turkey. The Levant transform also includes the Dead Sea and Sea of Galilee and forms the northwestern boundary of the Arabian plate. *See* ESCARPMENT; FAULT AND FAULT STRUCTURES.

The Red Sea consists of narrow marginal shelves and coastal plains and a broad main trough with depths ranging from about 400 to 1200 m (1300 to 3900 ft). The main trough is bisected by a narrow (<60 km or 37 mi wide) axial trough with a very rough bottom morphology and depths of greater than 2000 m (6600 ft). The maximum recorded depth is 2920 m (9580 ft). *See* REEF.

Water circulation in the Red Sea is driven by monsoonal wind patterns and changes in water density due to evaporation. Evaporation in the Red Sea is sufficient to lower the sea level by over 2 m (6.6 ft) per year. No permanent rivers flow into the sea, and there is very little rainfall. As a result, there must be a net inflow of water from the Gulf of Aden to compensate for evaporative losses. During the winter monsoon, prevailing winds in the Red Sea are from the south, and there is a surface current from the Gulf of Aden into the Red Sea. During the summer monsoon, the wind in the Red Sea blows strongly from the north, causing a surface current out of the Red Sea. *See* MONSOON METEOROLOGY; SEAWATER.   [J.R.Co.]

Redbeds Clastic sediments and sedimentary rocks that are pigmented by red ferric oxide which coats grains, fills pores as cement, or is dispersed as a muddy matrix. These conspicuously colored rocks commonly constitute thick sequences of nonmarine, paralic (marginal marine), and less commonly shallow marine deposits. Clastic redbeds accumulated in many parts of the globe during the past 10^9 years of Earth history. Ferric oxides also pigment marine chert, limestone, and cherty iron formations and ooidal ironstones, but these chemical deposits are not usually included among redbeds.

Some redbeds contain abundant grains of sedimentary and low-grade metamorphic rocks and relatively few grains of iron-bearing minerals. Most of them, however, contain feldspar and relatively abundant grains of opaque black oxides derived from igneous and high-grade metamorphic source rocks. Clay minerals in older redbeds, as in most other ancient clastic deposits, are predominantly illite and chlorite, thus providing no specific clue to the climate in the source area or at the place of deposition.

In many of the younger redbeds the pigmenting ferric oxide mineral cannot be identified specifically because of its poor crystallinity. In most of the older ones, however, hematite is the pigment. As seen under the scanning electron microscope, the hematite is in the form of hexagonal crystals scattered over the surface of grains and clay mineral platelets. In red mudstones most of the pigment is associated with the clay fraction.

Redbeds do not contain significantly more total iron than nonred sedimentary rocks. Normally, iron increases with decreasing grain size of redbeds. Moreover, the amount of iron in the grain-coating pigments is small compared with that in opaque oxides, dark silicates, and clay minerals.

On a global scale, paleomagnetic evidence of the distribution of redbeds relative to their pole position corroborates paleogeographic data, suggesting that most redbeds, evaporites, and eolian sandstones accumulated less than 30° north and south of a paleo equator where hot, dry climate generally prevailed. But diagenetic development of red hematite may be acquired long after deposition. Moreover, continental drift reconstructions reveal that the most widespread redbeds in the geologic record developed near the Equator in late Paleozoic and early Mesozoic time when the continents were assembled in a great landmass, Pangaea. *See* CONTINENTS, EVOLUTION OF; PALEOGEOGRAPHY; PALEOMAGNETISM.   [F.B.V.H.]

Reef   A mass or ridge of rock or rock-forming organisms in a water body, a rock trend on land or in a mine, or a rocky trend in soil. Usually the term reef means a rocky menace to navigation, within 6 fathoms (11 m) of the water surface. Various kinds of calcium carbonate–secreting animals and plants create biogenic, or organic, reefs throughout the warmer seas. Most biogenic reefs are made of corals and associated organisms, but some entire reefs and important parts of others consist mainly of lime-secreting algae, hydrozoans, annelids, oysters, or sponges.

The term fringing reef refers to a coral or other biogenic reef that fringes the edge of the land. A barrier reef ordinarily made of corals or other organisms parallels the shore at the seaward side of a natural lagoon. An atoll is an annular coral reef that surrounds a lagoon. *See* ATOLL.   [P.Cl.]

Regolith   The mantle or blanket of unconsolidated or loose rock material that overlies the intact bedrock and nearly everywhere forms the land surface. The regolith may be residual (weathered in place), or it may have been transported to its present site. The undisturbed residual regolith may grade from agricultural soil at the surface, through fresher and coarser weathering products, to solid bedrock several feet or more beneath the surface. The transported regolith includes the alluvium of rivers, sand dunes, glacial deposits, volcanic ash, coastal deposits, and the various mass-wasting deposits that occur on hillslopes. The lunar surface also has a regolith. This layer of fragmental debris is believed to derive from prolonged meteoritic and secondary fragment impact. *See* SOIL; WEATHERING PROCESSES.   [V.R.B.]

Remote sensing   The gathering and recording of information about terrain and ocean surfaces without actual contact with the object or area being investigated. Remote sensing uses the visual, infrared, and microwave portions of the electromagnetic spectrum. Remote sensing is generally conducted by means of remote sensors installed in aircraft and satellites.

Photography. Photography is probably the most useful remote sensing system. Much of the experience gained over the years from photographs of the terrain taken from aircraft is being drawn upon for use in space.

Multispectral photography isolates the reflected energy from a surface in a number of given wavelength bands and records each spectral band separately on film. This technique allows selection of the significant bandwidths in which a given area of terrain displays maximum tonal contrast and, hence, increases the effective spectral resolution of the system over conventional black-and-white or color systems. Because of its spectral selectivity capabilities, the multispectral approach provides a means of collecting a great amount of specific information.

Multispectral imagery. Multispectral scanning systems record the spectral reflectance by photoelectric means (rather than by photochemical means as in multispectral photography) simultaneously in several individual wavelengths within the visual and near-infrared portions of the electromagnetic spectrum.

In satellite applications, optical energy is sensed by an array of detectors simultaneously in four spectral bands from 0.47 to 1.1 micrometers. As the optical sensors for the various frequency bands sweep across the underlying terrain in a plane perpendicular to the flight direction of the satellite, they record energy from individual areas on the ground. The smallest individual area distinguished by the scanner is called a picture element or pixel, and a separate spectral reflectance is recorded in analog or digital form for each pixel. The spectral reflectance values for each pixel can be transmitted electronically to ground receiving stations in near-real time, or stored on magnetic tape in the satellite until it is over a receiving station. When the signal intensities are received on the ground, they can be reconstructed almost instantaneously into the virtual equivalent of conventional aerial photographs.

Infrared. Thermal infrared radiation is mapped by means of infrared scanners similar to multispectral scanners, but in this case radiated energy is recorded generally in the 8–14-μm portion of the electromagnetic spectrum. The imagery provided by an infrared scanning system gives information that is not available from ordinary photography or from multispectral scanners operating in the visual portion of the electromagnetic spectrum.

In the past, thermal infrared images were generally recorded on photographic film. Videotape records are replacing film as the primary recording medium and permit better imagery to be produced and greater versatility in interpretation of data.

Thermal infrared mapping (thermography) from satellite altitudes is proving to be useful for a number of purposes, one of which is the mapping of thermal currents in the ocean. Thermal infrared mapping from aircraft and satellite altitudes has many other uses also, including the mapping of volcanic activity and geothermal sites, location of ground-water discharge into surface and marine waters, and regional pollution monitoring.

Microwave radar. This type of remote sensing utilizes both active and passive sensors. The active sensors such as radar supply their own illumination and record the reflected energy. The passive microwave sensors record the natural radiation. A variety of sensor types are involved. These include imaging radars, radar scatterometers and altimeters, and over-the-horizon radars using large ground-based antenna arrays, as well as passive microwave radiometers and imagers. One of the most significant advantages of these instruments is their all-weather capability, both day and night.

High-frequency (hf) radar. Such radars utilize frequencies in the 3–30-MHz portion of the electromagnetic spectrum (median wavelength of about 20 m) and are thus not within the microwave part of the spectrum. The energy is transmitted by ground-based antennas in either a sky-wave or surface-wave mode. In the sky-wave mode, the energy is refracted by the various ionospheric layers back down to the Earth's surface some 480–1800 mi or 800–3000 km (on a single-hop basis) away from the hf radar antenna site. The incident waves are reflected from such surface features as sea waves.

[P.C.B.]

Restoration ecology A field in the science of conservation that is concerned with the application of ecological principles to restoring degraded, derelict, or fragmented ecosystems. The primary goal of restoration ecology (also known as ecological restoration) is to return a community or ecosystem to a condition similar in ecological structure, function, or both, to that existing prior to site disturbance or degradation.

A reference framework is needed to guide any restoration attempt—that is, to form the basis of the design (for example, desired species composition and density) and monitoring plan (for example, setting milestones and success criteria for restoration projects). Such a reference system is derived from ecological data collected from a suite of similar ecosystems in similar geomorphic settings within an appropriate biogeographic region. Typically, many sites representing a range of conditions (for example, pristine to highly degraded) are sampled, and statistical analyses of these data reveal what is possible given the initial conditions at the restoration site. *See* ECOLOGY, APPLIED.   [P.L.Fi.]

Rhizosphere The soil region subject to the influence of plant roots. It is characterized by a zone of increased microbiological activity and is an example of the relationship of soil microbes to higher plants.

A sharp boundary cannot be drawn between the rhizosphere and the soil unaffected by the plant (edaphosphere). At the root surface the rhizosphere effect is most intense, falling off sharply with increasing distance.

Growth of a plant markedly changes the microbial population of soil within its influence. In the rhizosphere there are more microorganisms than in soil distant from the plant. This increase is most pronounced with bacteria but is evident with other groups. The rhizosphere effect is seen in seedling plants; it increases with the age of the plant and usually reaches a maximum at the stage of greatest vegetative growth. Upon death of the plant the microbial population reverts to the level of the surrounding soil. Leguminous plants support higher rhizosphere populations than nonlegumes. The stimulation of microorganism growth in the rhizosphere results chiefly from the liberation of readily available organic substances by the growing plant.   [A.G.L.]

Rhodochrosite   The mineral form of manganese carbonate. Calcium, iron, magnesium, and zinc have all been reported to replace some of the manganese. The equilibrium replacement of manganese by calcium increases with the temperature of crystallization.

Rhodochrosite occurs more often in massive or columnar form than in distinct crystals. The color ranges from pale pink to brownish pink. Hardness is 3.5–4 on Mohs scale, and specific gravity is 3.70.

Well-known occurrences of rhodochrosite are in Europe, Asia, and South America. In the United States large quantities occur at Butte, Montana. As a source of manganese, rhodochrosite is also important at Chamberlain, South Dakota, and in Aroostook County, Maine. See CARBONATE MINERALS.   [R.I.Ha.]

Rhodonite   A mineral inosilicate with composition $MnSiO_3$. Hardness is 5.5–6 on Mohs scale, and specific gravity is 3.4–3.7. The luster is vitreous and the color is rose red, pink, or brown. Rhodonite is similar in color to rhodochrosite, manganese carbonate, but it may be distinguished by its greater hardness and insolubility in hydrochloric acid. It has been found at Langban, Sweden; near Sverdlovsk in the Ural Mountains; and at Broken Hill, Australia. Fine crystals of a zinc-bearing variety, fowlerite, are found at Franklin, New Jersey. See SILICATE MINERALS.   [C.S.Hu.]

Rhyolite   A very light-colored, aphanitic (not visibly crystalline), volcanic rock that is rich in silica and broadly equivalent to granite in composition. Migration of rhyolitic magma through the Earth's crust, which causes much of the Earth's explosive and hazardous volcanic activity, represents a major process of chemical fractionation by which continental crust grows and evolves. See GRANITE.

Rhyolites are formed by the process of molten silica-rich magma flowing toward the Earth's surface. Small differences in this process, notably those related to the release of gas from the magma at shallow depth, produce extremely diverse structural features. The high silica content gives rhyolitic lava a correspondingly high viscosity; this hinders crystallization and often causes young rhyolite to be a mixture of microcrystalline aggregates and glassy material. Because of the glassy nature of most rhyolites, they are best characterized by chemical analysis. They typically have 70–75 wt % silicon dioxide (SiO_2) and more potassium oxide (K_2O) than sodium oxide (Na_2O). See LAVA; MAGMA; VOLCANIC GLASS; VOLCANO.

Rhyolite is one of the most common volcanic rocks in continental regions; it is virtually absent in the ocean basins. The rock often occurs in large quantities associated with andesite and basalt. It is common in environments ranging from accretionary prisms at continental margins to magmatic arcs related to subduction zones. Rhyolite is also prevalent in extensional regions and hot spots in continental interiors. *See* ANDESITE; BASALT.   [L.W.Y.]

Rift valley   One of the geomorphological expressions between two tectonic plates that are opening relative to each other or sliding past each other. The term originally was used to describe the central graben structures of such classic continental rift zones as the Rhinegraben and the East African Rift, but the definition now encompasses mid-oceanic ridge systems with central valleys such as the Mid-Atlantic Ridge. *See* MID-OCEANIC RIDGE; PLATE TECTONICS.

Continental and oceanic rift valleys are end members in what many consider to be an evolutionary continuum. In the case of continental rift valleys, plate separation is incomplete, and the orientation of the stress field relative to the rift valley can range from nearly orthogonal to subparallel. Strongly oblique relationships are probably the norm. In contrast, oceanic rift valleys mark the place where the trailing edges of two distinctly different plates are separating. The separation is complete, and the spreading is organized and focused, resulting in rift valleys that tend to be oriented orthogonal or suborthogonal to the spreading directions.

The basic cross-sectional form of rift valleys consists of a central graben surrounded by elevated flanks. It is almost universally accepted that the central grabens of continental rift valleys are subsidence features. The crystalline basement floors of some parts of the Tanganyika and Malawi (Nyasa) rift valleys in East Africa lie more than 5 mi (9 km) below elevated flanks. *See* GRABEN.

In continental rift valleys the true cross-sectional form is typically asymmetric, with the rift floors tilted toward the most elevated flank. Most of the subsidence is controlled by one border fault system, and most of the internal faults parallel the dip of the border faults. *See* FAULT AND FAULT STRUCTURES.

Oceanic rift valleys are also distinctly separated into segments by structures known as transform faults. The cross-sectional form of oceanic rift valleys can be markedly asymmetric. It is unlikely that the cross-sectional form of oceanic rift valleys is related genetically to that of continental rift valleys, except in the broadest possible terms. *See* TRANSFORM FAULT.   [B.R.R.]

River   A natural, fresh-water surface stream that has considerable volume compared with its smaller tributaries. The tributaries are known as brooks, creeks, branches, or forks. Rivers are usually the main stems and larger tributaries of the drainage systems that convey surface runoff from the land. Rivers flow from headwater areas of small tributaries to their mouths, where they may discharge into the ocean, a major lake, or a desert basin.

Rivers flowing to the ocean drain about 68% of the Earth's land surface. Regions draining to the sea are termed exoreic, while those draining to interior closed basins are endoreic. Areic regions are those which lack surface streams because of low rainfall or lithoogic conditions.

Sixteen of the largest rivers account for nearly half of the total world river flow of water. The Amazon River alone carries nearly 20% of all the water annually discharged by the world's rivers. Rivers also carry large loads of sediment. The total sediment load for all the world's rivers averages about 22×10^9 tons (20×10^9 metric tons) brought to the sea each year. Sediment loads for individual rivers vary considerably. The Yellow River of northern China is the most prolific transporter of sediment. Draining an agricultural region of easily eroded loess, this river averages about 2×10^9 tons (1.8×10^9 metric tons) of sediment per year, one-tenth of the world average. *See* DEPOSITIONAL SYSTEMS AND ENVIRONMENTS; LOESS.

River discharge varies over a broad range, depending on many climatic and geologic factors. The low flows of the river influence water supply and navigation. The high flows are a concern as threats to life and property. However, floods are also beneficial. The ancient Egyptian civilization was dependent upon the Nile River floods to provide new soil and moisture for crops. Floods are but one attribute of rivers that affect human society. Means of counteracting the vagaries of river flow have concerned engineers for centuries. In modern times many of the world's rivers are managed to conserve the natural flow for release at times required by human activity, to confine flood flows to the channel and to planned areas of floodwater storage, and to maintain water quality at optimum levels. *See* FLOODPLAIN.   [V.R.B.]

River tides   Tides that occur in rivers emptying directly into tidal seas. These fides show three characteristic modifications of ocean tides. (1) The speed at which the tide travels upstream depends on the depth of the channel. (2) The further upstream, the longer the duration of the falling tide and the shorter the duration of the rising tide. (3) The range of the tide decreases with distance upstream. *See* TIDE.

In a river the difference between the depths of water at high and low tides may be relatively large, leading to a marked difference between the speeds at which high and low tides move. The difference in depth between various points on the river also partially explains the second modification, or duration of fall and rise. In addition, the river flow, which may fluctuate widely, helps a failing tide but hinders a rising tide, increasing the difference in duration.

The third modification or decrease in tidal range upstream may be accounted for by loss of energy of the water through friction with the sides and bottom of the channel. Although friction always saps energy from the tide, if the channel becomes constricted within a short distance, the water may be forced into a smaller space, thus producing a larger tidal range.   [B.K.]

Rock   A relatively common aggregate of mineral grains. Some rocks consist essentially of but one mineral species (monomineralic, such as quartzite, composed of quartz); others consist of two or more minerals (polymineralic, such as granite, composed of quartz, feldspar, and biotite). Rock names are not given for those rare combinations of minerals that constitute ore deposits, such as quartz, pyrite, and gold. In the popular sense rock is considered also to denote a compact substance, one with some coherence; but geologically, friable volcanic ash also is a rock. A genetic classification of rocks is shown below.

    Igneous
      Intrusive
        Plutonic (deep)
        Hypabyssal (shallow)
      Extrusive
        Flow
        Pyroclastic (explosive)
      Sedimentary
        Clastic (mechanical or detrital)
        Chemical (crystalline or precipitated)
        Organic (biogenic)
      Metamorphic
        Cataclastic
        Contact metamorphic and pyrometasomatic

Regional metamorphic (dynamothermal)
Hybrid
    Metasomatic
    Migmatitic

Exceptions to the requirement that rocks consist of minerals are obsidian, a volcanic rock consisting of glass; and coal, a sedimentary rock which is a mixture of organic compounds. *See* COAL; IGNEOUS ROCKS; METAMORPHIC ROCKS; OBSIDIAN; ROCK MECHANICS; SEDIMENTARY ROCKS; VOLCANIC GLASS. [E.W.H.]

Rock, electrical properties of The effect of changes in pressure and temperature on electrical properties of rocks. There has been increasing interest in the electrical properties of rocks at depth within the Earth and the Moon. The reason for this interest has been consideration of the use of electrical properties in studying the interior of the Earth and its satellite, particularly to depths of tens or hundreds of kilometers. At such depths pressures and temperatures are very great, and laboratory studies in which these pressures and temperatures are duplicated have been used to predict what the electrical properties at depth actually are. More direct measurements of the electrical properties deep within the Earth have been made by using surface-based electrical surveys of various sorts. An important side aspect of the study of electrical properties has been the observation that, when pressures near the crushing strength are applied to a rock, marked changes in electrical properties occur, probably caused by the development of incipient fractures. Such changes in resistivity might be used in predicting earthquakes, if they can be measured in the ground. *See* GEOELECTRICITY; GEOPHYSICAL EXPLORATION.

Attempts to measure the electrical properties of rocks to depths of tens or hundreds of kilometers in the Earth indicate that the Earth's crust is zoned electrically. The surface zone, with which scientists are most familiar, consists of a sequence of sedimentary rocks, along with fractured crystalline and metamorphic rocks, all of which are moderately good conductors of electricity because they contain relatively large amounts of water in pore spaces and other voids. This zone, which may range in thickness from a kilometer to several tens of kilometers, has conductivities varying from about $\frac{1}{2}$ ohm-m in recent sediments to 1000 ohm-m or more in weathered crystalline rock.

The basement rocks beneath this surface zone are crystalline, igneous, or metamorphic rocks which are much more dense, having little pore space in which water may collect. Since most rock-forming minerals are good insulators at normal temperatures, conduction of electricity in such rocks is determined almost entirely by the water in them. As a result, this part of the Earth's crust is electrically resistant (the resistivity lies in the range 10,000–1,000,000 ohm-m).

At rather moderate depths beneath the surface of the second zone, resistivity begins to decrease with depth. This decrease is considered to be the result of higher temperatures, which almost certainly are present at great depths. High temperatures lead to partial ionization of the molecular structure of minerals composing a rock, and the ions render even the insulating minerals conductive. [G.V.K.]

Rock age determination Finding the age of rocks based on the presence of naturally occurring long-lived radioactive isotopes of several elements in certain minerals and rocks. Measurements of rock ages have enabled geologists to reconstruct the geologic history of the Earth from the time of its formation 4.6×10^9 years ago to the present. Age determinations of rocks from the Moon have also contributed to

Parent-daughter pairs used for dating rocks and minerals

Parent	Daughter	Half-life, 10^9 years
Potassium-40	Argon-40	11.8
Potassium-40	Calcium-40	1.47
Rubidium-87	Strontium-87	48.8
Samarium-147	Neodymium-143	107
Rhenium-187	Osmium-187	43
Thorium-232	Lead-208	14.008
Uranium-235	Lead-207	0.7038
Uranium-238	Lead-206	4.468

knowledge of the history of the Moon, and may someday be used to study the history of Mars and of other bodies within the solar system.

Many rocks and minerals contain radioactive atoms that decay spontaneously to form stable atoms of other elements. Under certain conditions these radiogenic daughter atoms accumulate within the mineral crystals so that the ratio of the daughter atoms divided by the parent atoms increases with time. This ratio can be measured very accurately with a mass spectrometer, and is then used to calculate the age of the rock by means of an equation based on the law of radioactivity. The radioactive atoms used for dating rocks and minerals have very long half-lives, measured in billions of years. They occur in nature only because they decay very slowly. The pairs of parents and daughters used for dating are listed in the table. *See* DATING METHODS.

The rubidium-strontium method is based on rubidium-87, which decays to stable strontium-87 (^{87}Sr) by emitting a beta particle from its nucleus. The abundance of the radiogenic strontium-87 therefore increases with time at a rate that is proportional to the Rb/Sr ratio of the rock or mineral. The method is particularly well suited to the dating of very old rocks such as the ancient gneisses near Godthaab in Greenland, which are almost 3.8×10^9 years old. This method has also been used to date rocks from the Moon and to determine the age of the Earth by analyses of stony meteorites.

The potassium-argon method is based on the assumption that all of the atoms of radiogenic argon-40 that form within a potassium-bearing mineral accumulate within it. This assumption is satisfied only by a few kinds of minerals and rocks, because argon is an inert gas that does not readily form bonds with other atoms. The K-Ar method of dating has been used to establish a chronology of mountain building events in North America beginning about 2.8×10^9 years ago and continuing to the present. In addition, the method has been used to date reversals of the polarity of the Earth's magnetic field during the past 1.3×10^7 years. *See* OROGENY; PALEOMAGNETISM.

The uranium, thorium-lead method is based on uranium and thorium atoms which are radioactive and decay through a series of radioactive daughters to stable atoms of lead (Pb). Minerals that contain both elements can be dated by three separate methods based on the decay of uranium-238 to lead-206, uranium-235 to lead-207, and thorium-232 to lead-208. The three dates agree with each other only when no atoms of uranium, thorium, lead, and of the intermediate daughters have escaped. Only a few minerals satisfy this condition. The most commonly used mineral is zircon ($ZrSiO_4$), in which atoms of uranium and thorium occur by replacing zirconium. *See* LEAD ISOTOPES (GEOCHEMISTRY); RADIOACTIVE MINERALS.

The common-lead method is based on the common ore mineral galena (PbS) which consists of primordial lead that dates from the time of formation of the Earth and varying amounts of radiogenic lead that formed by decay of uranium and thorium in the Earth. The theoretical models required for the interpretation of common lead have provided insight into the early history of the solar system and into the relationship between meteorites and the Earth.

The fission-track method is based on uranium-238 which can decay both by emitting an alpha particle from its nucleus and by spontaneous fission. The number of spontaneous fission tracks per square centimeter is proportional to the concentration of uranium and to the age of the sample. When the uranium content is known, the age of the sample can be calculated. This method is suitable for dating a variety of minerals and both natural and manufactured glass. Its range extends from less than 100 years to hundreds of millions of years. See FISSION TRACK DATING.

The samarium-neodymium method of dating separated minerals or whole-rock specimens is similar to the Rb-Sr method. The Sm-Nd method is even more reliable than the Rb-Sr method of dating rocks and minerals, because samarium and neodymium are less mobile than rubidium and strontium. The isotopic evolution of neodymium in the Earth is described by comparison with stony meteorites.

The rhenium-osmium method is based on the beta decay of naturally occurring rhenium-187 to stable osmium-187. It has been used to date iron meteorites and sulfide ore deposits containing molybdenite.   [G.Fau.]

Rock cleavage   A secondary, planar structure of deformed rocks. A cleavage is penetrative and systematic, as opposed to fractures and shear zones which may occur alone or in widely spaced sets. It is generally better developed in fine grained rocks than in coarse ones. Application of the term derives from the ability to split rocks along the structure.

Simple cleavages are generally parallel to the axial surfaces of folds. This is true for folds formed by uniform flow of the rock mass where primary layering behaves as a passive marker. When rock layers are buckled, the primary layering behaving as mechanical discontinuities, cleavage that develops early may be fanned by subsequent growth of the fold. Cleavage is also associated with faults. See FAULT AND FAULT STRUCTURES; FOLD AND FOLD SYSTEMS.

Continuous (microscopically penetrative) cleavages, as in slates, are the earliest tectonic fabric elements that can be recognized in rocks. Cleavage also develops in rocks of lower deformational and metamorphic grade than slates. In these rocks, the cleavage occurs as discrete surfaces or seams, often coated with a film of clay or carbonaceous material; the cleavage surfaces are separated by zones of undeformed sedimentary rock. The surfaces may be smooth, anastomosing, or stylolitic. See METAMORPHISM; ROCK MECHANICS; SLATE.   [D.B.Bi.]

Rock magnetism   The permanent and induced magnetism of rocks and minerals on scales ranging from the atomic to the global, including applications to magnetic field anomalies and paleomagnetism. Natural compasses, concentrations of magnetite (Fe_3O_4) called lodestones, are one of humankind's oldest devices. W. Gilbert in 1600 discovered that the Earth itself is a giant magnet, and speculated that its magnetism might be due to subterranean lodestone deposits. Observations by B. Brunhes in 1906 that some rocks are magnetized reversely to the present Earth's magnetic field, and by M. Matuyama in 1929 that reversely and normally magnetized rocks correspond to different geological time periods, made it clear that geomagnetism is dynamic, with frequent reversals of north and south poles. Nevertheless, permanent magnetism of rocks remains important because it alone provides a memory of the intensity, direction, and polarity of the Earth's magnetic field in the geological past. From this magnetic record comes much of the evidence for continental drift, sea-floor spreading, and plate tectonics. *See* CONTINENTAL DRIFT; GEOMAGNETISM; PALEOMAGNETISM; PLATE TECTONICS.

The magnetism of rocks arises from the ferromagnetism or ferrimagnetism of a few percent or less of minerals such as magnetite. The magnetic moments of neighboring atoms in such minerals are coupled parallel or antiparallel, creating a spontaneous magnetization M_S. All magnetic memory, including that of computers, permanent magnets, and rocks, is due to the spontaneous and permanent nature of this magnetism. Spontaneous magnetization requires no magnetic field to create it, and cannot be demagnetized.

The magnetism can be randomized on the scale of magnetic mineral grains because different regions of a crystal tend to have their M_S vectors in different directions. These regions are called magnetic domains. Grains so small that they contain only one domain (single-domain grains) are the most powerful and stable paleomagnetic recorders. Larger, multidomain grains can also preserve a paleomagnetic memory, through imbalance in the numbers, sizes, or directions of domains, but this memory is more easily altered by time and changing geological conditions. *See* MAGNETITE; MAGNETOMETER.

[D.J.Du.]

Rock mechanics   Application of the principles of mechanics and geology to quantify the response of rock when it is acted upon by environmental forces, particularly when human-induced factors alter the original ambient conditions. Rock mechanics is an interdisciplinary engineering science that requires interaction between physics, mathematics, and geology, and civil, petroleum, and mining engineering. The present state of knowledge permits only limited correlations between theoretical predictions and empirical results. Therefore, the most useful principles are based upon data obtained from laboratory and in-place measurements and from prototype behavior (behavior of the completed engineering works). Increasing emphasis is upon in-place measurements because rock properties are regarded as site-specific; that is, the properties of the rock system at one site probably will be significantly different from those at another site, even if geologic environments are similar. *See* ENGINEERING GEOLOGY.

Because of the interdisciplinary aspects, there is no standardization of rock mechanics terminology. However, the following terms and definitions are useful.

Environmental factors are the natural factors and human influences that require consideration in engineering problems in rock mechanics. The major natural factors are geology, ambient stresses, and hydrology. The human influences derive from the application of chemical, electrical, mechanical, or thermal energy during construction (or destruction) processes.

The ambient stress field is the distribution and numerical value of the stresses in the environment prior to its disturbance by humans.

The term rock system includes the complete environment that can influence the behavior of that portion of the Earth's crust that will become part of an engineering structure. Generally, all natural environmental factors are included.

A rock element is the coherent, intact piece of rock that is the basic constituent of the rock system and which has physical, mechanical, and petrographic properties that can be described or measured by laboratory tests on each such element. The concepts of rock system and rock element enable the concomitant engineering design to be optimized according to the principles of system engineering.

"Rock failure" occurs when a rock system or element no longer can perform its intended engineering function. Failure may be evidenced by fractures, distortion of shape, or reduction in strength. "Failure mechanism" includes the causes for the manner of rock failure.

[W.R.J.]

Rock varnish   A dark coating on rock surfaces exposed to the atmosphere. Rock varnish is probably the slowest-accumulating sedimentary deposit, growing at only a few micrometers to tens of micrometers per thousand years. Its thickness ranges from less than 5 μm to over 600 μm, and is typically 100 μm or so. Although found in all terrestrial environments, varnish is mostly developed and well preserved in arid to semiarid deserts; thus, another common name is desert varnish. Rock varnish is composed of about 30% manganese and iron oxides, up to 70% clay minerals, and over a dozen trace and rare-earth elements. The building blocks of rock varnish are mostly blown in as airborne dust. Although the mechanism responsible for the formation of rock varnish remains unclear, two hypotheses have been proposed to explain the great enrichment in manganese within varnish (typically over 50 times compared with the adjacent environment such as soils, underlying rock, or dust). The abiotic hypothesis assumes that small changes in pH can concentrate manganese by geochemical processes. The biotic hypothesis suggests that bacteria, and perhaps other microorganisms, concentrate manganese; this is supported by culturing experiments and direct observations of bacterial enhancement of manganese. *See* SEDIMENTARY ROCKS.

Since rock varnish records environmental, especially climatic, events that are regionally or even globally synchronous, varnish microstratigraphy can be used as a tool for age dating. Without radiometric calibration, varnish microstratigraphy itself may be used to estimate relative ages of varnished geomorphic or archeological features in deserts. Once calibrated, varnish microstratigraphy can provide numerical age estimates for geomorphic and archeological features. Specifically, for petroglyphs and geoglyphs, the layering patterns of rock varnish hold the greatest potential for assigning ages. *See* ARCHEOLOGICAL CHRONOLOGY; DATING METHODS; ROCK AGE DETERMINATION.

[T.L.]

Romanechite   A barium-containing manganese oxide mineral with an ideal composition of $(Ba,H_2O)_2(Mn^{4+}, Mn^{3+})_5O_{10}$. Romanechite is a basic member of a group of manganese oxide minerals that are similar in physical appearance and have closely related structures. It is an ore of manganese and occurs in a variety of widespread localities. Its name is derived from its occurrence in Romaneche, Soane-et-Loire, France. *See* MANGANESE OXIDE MINERALS.

Romanechite is black to steel gray in color, opaque, and fine-grained. It occurs mainly as hard crusts that are botryoidal or reniform, but also as unusual samples that are soft and powdery. Many botryoidal samples have a fine-scale banding or layering parallel to their surfaces. Crystals are rare. The reported specific gravity of romanechite ranges from approximately 4.4 to 4.8, and the Mohs hardness ranges from 5 to 6 (less for powdery varieties). Romanechite is commonly intimately intergrown with other manganese oxide minerals, and is difficult to identify without the use of analytical techniques such as x-ray diffraction or electron microscopy.   [S.T.]

Rubellite   The red to red-violet variety of the gem mineral tourmaline. Perhaps the most sought-for of the many colors in which tourmaline occurs, it was named for its resemblance to ruby. The color is thought to be caused by the presence of lithium. It has a hardness of 7–7.5 on Mohs scale, a specific gravity near 3.04, and refractive indices of 1.624 and 1.644. Fine gem-quality material is found in Brazil, Madagascar, Maine, southern California, the Ural Mountains, and elsewhere. *See* GEM; TOURMALINE.

[R.T.L.]

Ruby   The red variety of the mineral corundum, in its finest quality the most valuable of gemstones. Only medium to dark tones of red to slightly violet-red or very slightly orange-red are called ruby; light reds, purples, and other colors are properly called sapphires. In its pure form the mineral corundum, with composition Al_2O_3, is colorless. The rich red of fine-quality ruby is the result of the presence of a minute amount of chromic oxide. The chromium presence permits rubies to be used for lasers producing red light. *See* CORUNDUM; SAPPHIRE.

The finest ruby is the transparent type with a medium tone and a high intensity of slightly violet-red, which has been likened to the color of pigeon's blood. Star rubies do not command comparable prices, but they, too, are in great demand. The ruby was among the first of the gemstones to be duplicated synthetically and the first to be used extensively in jewelry. *See* GEM.   [R.T.L.]

Rutile   The most frequent of the three polymorphs of titania, TiO_2; the two other polymorphs are brookite and anatase.

The mineral occurs as striated tetragonal prisms and needles, commonly repeatedly twinned. The color is deep blood red, reddish brown, to black, rarely violet or yellow. Specific gravity is 4.2, and hardness 6.5 on Mohs scale. Melting point is 1825°C (3317°F).

Rutile occurs as an accessory in many rock types, ranging from plutonic to metamorphic rocks, and even as detrital material in sediments and placers because of its resistance to weathering. Large crystals have been found in some granite pegmatites; in Brazil it often occurs as inclusions in clear quartz crystals (rutilated quartz). Rutile is commonly associated with apatite in high-temperature veins. In sufficient quantities, it is marketed as an ore of titanium.   [P.B.M.]

S

Saline evaporites   Deposits of bedded sedimentary rocks composed of salts precipitated during solar evaporation of surface or near-surface brines derived from seawater or continental waters. Dominant minerals in ancient evaporite beds are anhydrite (along with varying amounts of gypsum) and halite, which make up more than 85% of the total sedimentary evaporite salts. Many other salts make up the remaining 15%; their varying proportions in particular beds can be diagnostic of the original source of the mother brine. *See* ANHYDRITE; GYPSUM; HALITE; SEAWATER; SEDIMENTARY ROCKS.

Today, brines deposit their salts within continental playas or coastal salt lakes and lay down beds a few meters thick and tens of kilometers across. In contrast, ancient, now-buried evaporite beds are often much thicker and wider; they can be up to hundreds of meters thick and hundreds of kilometers wide. Most ancient evaporites were formed by the evaporation of saline waters within hyperarid areas of huge seaways typically located within arid continental interiors. The inflow brines in such seaways were combinations of varying proportions of marine and continental ground waters and surface waters. There are few modern depositional counterparts to these ancient evaporites, and none to those beds laid down when whole oceanic basins dried up, for example, the Mediterranean some 5.5 million years ago. *See* BASIN; DEPOSITIONAL SYSTEMS AND ENVIRONMENTS; MEDITERRANEAN SEA; PLAYA.

Evaporite salts precipitate by the solar concentration of seawater, continental water, or hybrids of the two. The chemical makeup, salinity (35‰), and the proportions of the major ions in modern seawater are near-constant in all the world's oceans, with sodium (Na) and chloride (Cl) as the dominant ions and calcium (Ca) and sulfate (SO_4) ions present in smaller quantities [$Na(Ca)SO_4Cl$ brine]. Halite and gypsum anhydrites have been the major products of seawater evaporation for at least the past 2 billion years, but the proportions of the more saline minerals, such as sylvite/magnesium sulfate ($MgSO_4$) salts, appear to have been more variable.   [J.Wa.]

Salt dome   An upwelling of crystalline rock salt and its aureole of deformed sediments. A salt pillow is an immature salt dome comprising a broad salt swell draped by concordant strata. A salt stock is a more mature, pluglike diapir of salt that has pierced, or appears to have pierced, overlying strata. Most salt stocks are 0.6–6 mi (1–10 km) wide and high. Salt domes are closely related to other salt upwellings, some of which are much larger. Salt canopies, which form by coalescence of salt domes and tongues, can be more than 200 mi (300 km) wide. *See* DIAPIR.

Exploration for oil and gas has revealed salt domes in more than 100 sedimentary basins that contain rock salt layers several hundred meters or more thick. The salt was precipitated from evaporating lakes in rift valleys, intermontaine basins, and especially along divergent continental margins. Salt domes are known in every ocean and continent. *See* BASIN.

Salt domes consist largely of halite (NaCl, common table salt). Other evaporites, such as anhydrite ($CaSO_4$) and gypsum ($CaSO_4 \cdot 2H_2O$), form thinner layers within the rock salt. *See* HALITE; SALINE EVAPORITES.

Salt domes supply industrial commodities, including fuel, minerals, chemical feedstock, and storage caverns. Giant oil or gas fields are associated with salt domes in many basins around the world, especially in the Middle East, North Sea, and South Atlantic regions. Salt domes are also used to store crude oil, natural gas (methane), liquefied petroleum gas, and radioactive or toxic wastes. [M.P.A.J.]

Sand Unconsolidated granular material consisting of mineral, rock, or biological fragments between 63 micrometers and 2 mm in diameter. Finer material is referred to as silt and clay; coarser material is known as gravel. Sand is usually produced primarily by the chemical or mechanical breakdown of older source rocks, but may also be formed by the direct chemical precipitation of mineral grains or by biological processes. Accumulations of sand result from hydrodynamic sorting of sediment during transport and deposition. *See* CLAY MINERALS; DEPOSITIONAL SYSTEMS AND ENVIRONMENTS; GRAVEL; MINERAL; ROCK; SEDIMENTARY ROCKS.

Most sand originates from the chemical and mechanical breakdown, or weathering, of bedrock. Since chemical weathering is most efficient in soils, most sand grains originate within soils. Rocks may also be broken into sand-size fragments by mechanical processes, including diurnal temperature changes, freeze-thaw cycles, wedging by salt crystals or plant roots, and ice gouging beneath glaciers. *See* WEATHERING PROCESSES.

Because sand is largely a residual product left behind by incomplete chemical and mechanical weathering, it is usually enriched in minerals that are resistant to these processes. Quartz not only is extremely resistant to chemical and mechanical weathering but is also one of the most abundant minerals in the Earth's crust. Many sands dominantly consist of quartz. Other common constituents include feldspar, and fragments of igneous or metamorphic rock. Direct chemical precipitation or hydrodynamic processes can result in sand that consists almost entirely of calcite, glauconite, or dense dark-colored minerals such as magnetite and ilmenite. *See* FELDSPAR; QUARTZ.

Although sand and gravel has one of the lowest average per ton values of all mineral commodities, the vast demand makes it among the most economically important of all mineral resources. Sand and gravel is used primarily for construction purposes, mostly as concrete aggregate. Pure quartz sand is used in the production of glass, and some sand is enriched in rare commodities such as ilmenite (a source of titanium) and in gold. [M.J.J.]

Sandstone A clastic sedimentary rock comprising an aggregate of sand-sized (0.06–2.0-mm) fragments of minerals, rocks, or fossils held together by a mineral cement. Sandstone forms when sand is buried under successive layers of sediment. During burial the sand is compacted, and a binding agent such as quartz, calcite, or iron oxide is precipitated from ground water which moves through passageways between grains. Sandstones grade upward in grain size into conglomerates and breccias; they grade downward in size into siltstones and shales. When the proportion of fossil fragments or carbonate grains is greater than 50%, sandstones grade into clastic limestones. *See* BRECCIA; CONGLOMERATE; LIMESTONE; SAND; SHALE.

The basic components of a sandstone are framework grains (sand particles), which supply the rock's strength; matrix or mud-sized particles, which fill some of the space between grains; and crystalline cement. The composition of the framework grains reveals much about the history of the derivation of the sand grains, including the parent rock type and weathering history of the parent rock. Textural attributes of sandstone are the same as those for sand, and they have the same genetic significance. *See* SAND.

[L.J.S.]

Sandstones are classified according to the relative proportion of quartz to other grain types, and according to the ratio of feldspar grains to finely crystalline lithic fragments. Quartz-rich sandstones are commonly called quartz-arenite. Sandstones poor in quartz are commonly called arkose, when feldspar grains are more abundant than lithic fragments, and litharenite (or graywacke) when the reverse is true. Subarkose and sublitharenite (or subgraywacke) refer to analogous sandstones of intermediate quartz content. Sandstones composed dominantly of calcareous grains are called calcarenite, and represent a special variety of limestone. Other sandstones composed exclusively of volcanic debris are called volcanic sandstone, and are gradational, through the interplay of eruptive and erosional processes, to tuff, the fragmental volcanic rocks produced by the disintegration of magma during explosive volcanic eruptions. *See* ARENACEOUS ROCKS; ARKOSE; FELDSPAR; GRAYWACKE; QUARTZ; TUFF.   [W.R.D.]

Because sandstone can possess up to 35% connected pore space, it is the most important reservoir rock in the Earth's crust. In the future sandstone may serve as a reservoir into which hazardous fluids, such as nuclear wastes, are injected for storage.

Sandstone which is easily split (flagstone) and has an attractive color is used as a building stone. Sandstone is also an important source of sand for the glass industry and the construction industry, where it is used as a filler in cement and plaster. Crushed sandstone is used as road fill and railroad ballast. Silica-cemented sandstone is used as firebrick in industrial furnaces. Some of the most extensive deposits of uranium are found in sandstones deposited in ancient stream channels. *See* SEDIMENTARY ROCKS; STONE AND STONE PRODUCTS.   [L.J.S.]

Sapphire   The name given to all gem varieties of the mineral corundum, except those that have medium to dark tones of red that characterize ruby. Although the name sapphire is most commonly associated with the blue variety, there are many other colors of gem corundum to which sapphire is applied correctly; these include yellow, brown, green, pink, orange, purple, colorless, and black. Sapphire has a hardness of 9, a specific gravity near 4.00, and refractive indices of 1.76–1.77. Asterism, the star effect, is the result of reflections from tiny, lustrous, needlelike inclusions of the mineral rutile, plus a domed form of cutting. *See* CORUNDUM; GEM; RUBY; RUTILE.   [R.T.L.]

Sapropel   A term originally defined as an aquatic sediment rich in organic matter that formed under reducing conditions (lack of dissolved oxygen in the water column) in a stagnant water body. Such inferences about water-column dissolved-oxygen contents are not always easy to make for ancient environments. Therefore, the term sapropel or sapropelic mud has been used loosely to describe any discrete black or dark-colored sedimentary layers (>1 cm or 0.4 in. thick) that contain greater than 2 wt % organic carbon. Sapropels may be finely laminated (varved) or homogeneous, and may less commonly exhibit structures indicating reworking or deposition of the sediment by currents. Sapropels largely contain amorphous organic matter derived from planktonic organisms (such as planktonic or benthic algae in lakes or plankton in marine settings). Such organic matter possesses a large hydrogen-to-carbon ratio; therefore, sapropelic sequences are potential petroleum-forming deposits. The enhanced preservation of amorphous organic matter in sapropels may indicate conditions of exceptionally great surface-water productivity, extremely low bottom-water dissolved-oxygen contents, or both. Some sapropels may, however, contain substantial amounts of organic matter derived from land plants. *See* ANOXIC ZONES; MARINE SEDIMENTS; ORGANIC GEOCHEMISTRY; PETROLEUM; VARVE.   [M.A.Ar.]

Satellite meteorology   The branch of meteorological science that uses meteorological sensing elements on satellites to define the past and present state of the atmosphere. Meteorological satellites can measure a wide spectrum of electromagnetic radiation in real time, providing the meteorologist with a supplemental source of data.

Modern satellites are sent aloft with multichannel high-resolution radiometers covering an extensive range of infrared and microwave wavelengths. Radiometers sense cloudy and clear-air atmospheric radiation at various vertical levels, atmospheric moisture content, ground and sea surface temperatures, and ocean winds, and provide visual imagery as well. *See* METEOROLOGICAL SATELLITES.

There are two satellite platforms used for satellite meteorology: geostationary and polar. Geostationary (geo) satellites orbit the Earth at a distance that allows them to make one orbit every 24 hours. By establishing the orbit over the Equator, the satellite appears to remain stationary in the sky. This is important for continuous scanning of a region on the Earth for mesoscale (approximately 10–1000 km horizontal) forecasting.

Polar satellites orbit the Earth in any range of orbital distances with a high inclination angle that causes part of the orbit to fly over polar regions. The orbital distance of 100–200 mi (160–320 km) is selected for meteorological applications, enabling the satellite to fly over a part of the Earth at about the same time every day. With orbital distances of a few hundred miles, the easiest way to visualize the Earth-satellite relationship is to think of a satellite orbiting the Earth pole-to-pole while the Earth rotates independently beneath the orbiting satellite. The advantage of polar platforms is that they eventually fly over most of the Earth. This is important for climate studies since one set of instruments with known properties will view the entire world.

The enormous aerial coverage by satellite sensors bridges many of the observational gaps over the Earth's surface. Satellite data instantaneously give meteorologists up-to-the minute views of current weather phenomena.

Images derived from the visual channels are presented as black and white photographs. The brightness is solely due to the reflected solar light illuminating the Earth. Visible images are useful for determining general cloud patterns and detailed cloud structure. In addition to clouds, visible imagery shows snowcover, which is useful for diagnosing snow amount by observing how fast the snow melts following a storm. Cloud patterns defined by visual imagery can give the meteorologist detailed information about the strength and location of weather systems, which is important for determining storm motion and provides a first guess or forecast as to when a storm will move into a region. *See* CLOUD.

More quantitative information is available from infrared sensors, which measure radiation at longer wavelengths (from infrared to microwave). By analyzing the infrared data, the ground surface, cloud top, and even intermediate clear air temperatures can be determined 24 hours a day. By relating the cloud top temperature in the infrared radiation to an atmospheric temperature profile from balloon data, cloud top height can be estimated. This is a very useful indicator of convective storm intensity since more vigorous convection will generally extend higher in the atmosphere and appear colder.

The advent of geosynchronous satellites allowed the position of cloud elements to be traced over time. These cloud movements can be converted to winds, which can provide an additional source of data in an otherwise unobserved region. These techniques are most valuable for determination of mid- and high-level winds, particularly over tropical ocean areas. Other applications have shown that low-level winds can be determined in more spatially limited environments, such as those near thunderstorms, but those winds become more uncertain when the cloud elements grow vertically into air with a different speed and direction (a sheared environment). *See* WIND.

By using a wide variety of sensors, satellite data provide measurements of phenomena from the largest-scale global heat and energy budgets down to details of individual thunderstorms. Having both polar orbiting and geosynchronous satellites allows coverage over most Earth locations at time intervals from 3 minutes to 3 hours.

The greatest gain with the introduction of weather satellites was in early detection, positioning, and monitoring of the strength of tropical storms (hurricanes, typhoons). Lack of conventional meteorological data over the tropics (particularly the oceanic areas) makes satellite data indispensable for this task. The hurricane is one of the most spectacular satellite images. The exact position, estimates of winds, and qualitative determination of strength are possible with continuous monitoring of satellite imagery in the visible channels. In addition, infrared sensors provide information on cloud top height, important for locating rain bands. Microwave sensors can penetrate the storm to provide an indication of the interior core's relative warmth, closely related to the strength of the hurricane, and sea surface temperature to assess its development potential. *See* HURRICANE; TROPICAL METEOROLOGY.

Most significant weather events experienced by society—heavy rain or snow, severe thunderstorms, or high winds—are organized by systems that have horizontal dimensions of about 60 mi (100 km). These weather systems, known as mesoscale convective systems, often fall between stations of conventional observing networks. Hence, meteorologists might miss them were it not for satellite sensing. *See* HAIL; MESOMETEOROLOGY; METEOROLOGY; PRECIPITATION (METEOROLOGY); STORM; THUNDERSTORM; TORNADO; WEATHER FORECASTING AND PREDICTION.   [D.L.B.; J.A.McG.]

Scapolite An aluminosilicate mineral. It is commonly found as light-colored, translucent tetragonal prisms. Scapolite is normally white, but many other colors are known, including some used as semiprecious gems resembling amethyst and citrine. The mineral has a Mohs hardness of 5–6. The formula of scapolite is $(Na,Ca)_4(Al,Si)_6SI_6O_{24}(Cl,CO_3,SO_4)$. In nature significant amounts of K and SO_4 substitute for Na and CO_3.

Scapolite is a common mineral in metamorphic rocks, particularly in those which contain calcite. It is found in marbles, gneisses, skarns, and schists. Scapolite probably forms about 0.1% of the Earth's upper crust. It is commonly found as inclusions in igneous rocks derived from deep within the Earth's crust, and probably makes up several percent of the lower crust. *See* SILICATE MINERALS.   [D.E.E.]

Scattering layer A layer of organisms in the sea which causes sound to scatter and returns echoes. Recordings by sonic devices of echoes from sound scatterers indicate that the scattering organisms are arranged in approximately horizontal layers in the water, usually well above the bottom. The layers are found in both shallow and deep water.   [J.B.H.]

Scheelite A mineral consisting of calcium tungstate, $CaWO_4$. Scheelite occurs in colorless to white, tetragonal crystals; it may also be massive and granular. Its fracture is uneven, and its luster is vitreous to adamantine. Scheelite has a hardness of 4.5–5 on Mohs scale and a specific gravity of 6.1. Its streak is white. The mineral is transparent and fluoresces bright bluish-white under ultraviolet light.

Scheelite is an important tungsten mineral and occurs in small amounts in vein deposits. The most important scheelite deposit in the United States is near Mill City, Nevada.   [E.C.T.C.]

Schist   Medium- to coarse-grained, mica-bearing metamorphic rock with well-developed foliation (layered structure) termed schistosity. Schist is derived primarily from fine-grained, mica-bearing rocks such as shales and slates. The schistosity is formed by rotation, recrystallization, and new growth of mica; it is deformational in origin. The planar to wavy foliation is defined by the strong preferred orientation of platy minerals, primarily muscovite, biotite, and chlorite. The relatively large grain size of these minerals (up to centimeters) produces the characteristic strong reflection when light shines on the rock. *See* BIOTITE; CHLORITE; MUSCOVITE.

Schists are named by the assemblage of minerals that is most characteristic in the field; for example, a garnet-biotite schist contains porphyroblasts of garnet and a schistosity dominated by biotite. Schists can provide important information on the relationship between metamorphism and deformation. *See* METAMORPHIC ROCKS; METAMORPHISM; PETROFABRIC ANALYSIS.   [B.A.V.D.P.]

Sea breeze   A diurnal, thermally driven circulation in which a surface convergence zone often exists between airstreams having over-water versus over-land histories. The sea breeze is one of the most frequently occurring small-scale (mesoscale) weather systems. It results from the unequal sensible heat flux of the lower atmosphere over adjacent solar-heated land and water masses. Because of the large thermal inertia of a water body, during daytime the air temperature changes little over the water while over land the air mass warms. Occurring during periods of fair skies and generally weak large-scale winds, the sea breeze is recognizable by a wind shift to onshore, generally several hours after sunrise. On many tropical coastlines the sea breeze is an almost daily occurrence. It also occurs with regularity during the warm season along mid-latitude coastlines and even occasionally on Arctic shores. Especially during periods of very light winds, similar though sometimes weaker wind systems occur over the shores of large lakes and even wide rivers and estuaries (lake breezes, river breezes). At night, colder air from the land often will move offshore as a land breeze. Typically the land breeze circulation is much weaker and shallower than its daytime counterpart. *See* ATMOSPHERIC GENERAL CIRCULATION; MESOMETEOROLOGY; METEOROLOGY.

The occurrence and strength of the sea breeze is controlled by a variety of factors, including land-sea surface temperature differences; latitude and day of the year; the synoptic wind and its orientation with respect to the shoreline; the thermal stability of the lower atmosphere; surface solar radiation as affected by haze, smoke, and stratiform and convective cloudiness; and the geometry of the shoreline and the complexity of the surrounding terrain. *See* WIND.   [W.A.Ly.]

Sea-floor imaging   The process whereby mapping technologies are used to produce highly detailed images of the sea floor. High-resolution images of the sea floor are used to locate and manage marine resources such as fisheries and oil and gas reserves, identify offshore faults and the potential for coastal damage due to earthquakes, and map out and monitor marine pollution, in addition to providing information on what processes are affecting the sea floor, where these processes occur, and how they interact. *See* MARINE GEOLOGY.

Side-scan sonar provides a high-resolution view of the sea floor. In general, a side-scan sonar consists of two sonar units attached to the sides of a sled tethered to the back of a ship. Each sonar emits a burst of sound that insonifies a long, narrow corridor of the sea floor extending away from the sled. Sound reflections from the corridor that echo back to the sled are then recorded by the sonar in their arrival sequence, with echoes from points farther away arriving successively later. The sonars repeat this sequence of "talking" and listening every few seconds as the sled is pulled through the water so that consecutive recordings build up a continuous swath of sea-floor reflections, which provide information about the texture of the sea floor. *See* ECHO SOUNDER; UNDERWATER SOUND.

The best technology for mapping sea-floor depths or bathymetry is multibeam sonar. These systems employ a series of sound sources and listening devices that are mounted on the hull of a survey ship. As with side-scan sonar, every few seconds the sound sources emit a burst that insonifies a long, slim strip of the sea floor aligned perpendicular to the ship's direction. The listening devices then begin recording sounds from within a fan of narrow sea-floor corridors that are aligned parallel to the ship and that cross the insonified strip. By running the survey the same way that one mows a lawn, adjacent swaths are collected parallel to one another to produce a complete sea-floor map of an area.

The most accurate and detailed view of the sea floor is provided by direct visual imaging through bottom cameras, submersibles, remotely operated vehicles, or if the waters are not too deep, scuba diving. Because light is scattered and absorbed in waters greater than about 33 ft (10 m) deep, the sea-floor area that bottom cameras can image is no more than a few meters. This limitation has been partly overcome by deep-sea submersibles and remotely operated vehicles, which provide researchers with the opportunity to explore the sea floor close-up for hours to weeks at a time. But even the sea-floor coverage that can be achieved with these devices is greatly restricted relative to side-scan sonar, multibeam sonar, and satellite altimetry.

The technology that provides the broadest perspective but the lowest resolution is satellite altimetry. A laser altimeter is mounted on a satellite and, in combination with land-based radars that track the satellite's altitude, is used to measure variations in sea-surface elevation to within 2 in. (5 cm). Removing elevation changes due to waves and currents, sea-surface height can vary up to 660 ft (200 m). These variations are caused by minute differences in the Earth's gravity field, which in turn result from heterogeneities in the Earth's mass. These heterogeneities are often associated with sea-floor topography. By using a mathematical function that equates sea-surface height to bottom elevations, global areas of the sea floor can be mapped within a matter of weeks. However, this approach has limitations. Sea-floor features less than 6–9 mi (10–15 km) in length are generally not massive enough to deflect the ocean surface, and thus go undetected. Furthermore, sea-floor density also affects the gravity field; and where different-density rocks are found, such as along the margins of continents, the correlation between Earth's gravity field and sea-floor topography breaks down.

[L.F.Pr.]

Sea ice Ice formed by the freezing of seawater. Ice in the sea includes sea ice, river ice, and land ice. Land ice is principally icebergs. River ice is carried into the sea during spring breakup and is important only near river mouths. The greatest part, probably 99% of ice in the sea, is sea ice. *See* ICEBERG.

The freezing point temperature and the temperature of maximum density of seawater vary with salinity. When freezing occurs, small flat plates of pure ice freeze out of solution to form a network which entraps brine in layers of cells. As the temperature decreases more water freezes out of the brine cells, further concentrating the remaining brine so that the freezing point of the brine equals the temperature of the surrounding pure ice structure. The brine is a complex solution of many ions.

The brine cells migrate and change size with changes in temperature and pressure. The general downward migration of brine cells through the ice sheet leads to freshening of the top layers to near zero salinity by late summer. During winter the top surface temperature closely follows the air temperature, whereas the temperature of the underside remains at freezing point, corresponding to the salinity of water in contact.

The sea ice in any locality is commonly a mixture of recently formed ice and old ice which has survived one or more summers. Except in sheltered bays, sea ice is continually in motion because of wind and current.   [W.Ly.]

Sea of Okhotsk   A semienclosed basin adjacent to the North Pacific Ocean, bounded on the north, east, and west by continental Russia, the Kamchatka Peninsula, and northern Japan. On its southeast side the Sea of Okhotsk is connected to the North Pacific via a number of straits and passages through the Kuril Islands. The sea covers a surface area of approximately 590,000 mi^2 (1.5 million km^2), or about 1% of the total area of the Pacific, and has a maximum depth of over 9000 ft (3000 m). Its mean depth is about 2500 ft (830 m). Owing to the cold, wintertime Arctic winds that blow to the southeast, from Russia toward the North Pacific Ocean, the Sea of Okhotsk is partially covered with ice during the winter months, from November through April. See BASIN; PACIFIC OCEAN.

The amount of water exchanged between the Sea of Okhotsk and the North Pacific Ocean is not well known, but it is thought that the waters of the Sea of Okhotsk that do enter the North Pacific may play an important role in the Pacific's large-scale circulation. The reason is the extreme winter conditions over the Sea of Okhotsk: its waters are generally colder and have a lower salinity than the waters at the same density in the North Pacific. See OCEAN CIRCULATION; SEAWATER.   [S.C.Ri.]

Seamount and guyot   A seamount is a mountain that rises from the ocean floor; a submerged flat-topped seamount is termed a guyot. By arbitrary definition, seamounts must be at least 3000 ft (about 900 m) high, but in fact there is a continuum of smaller undersea mounts, down to heights of only about 300 ft (100 m). Some seamounts are high enough temporarily to form oceanic islands, which ultimately subside beneath sea level. There are on the order of 10,000 seamounts in the world ocean, arranged in chains (for example, the Hawaiian chain in the North Pacific) or as isolated features. In some chains, seamounts are packed closely to form ridges. Very large oceanic volcanic constructions, hundreds of kilometers across, are called oceanic plateaus. See MARINE GEOLOGY; OCEANIC ISLANDS; VOLCANO.

Almost all seamounts are the result of submarine volcanism, and most are built within less than about 1 million years. Seamounts are made by extrusion of lavas piped upward in stages from sources within the Earth's mantle to vents on the seafloor. Seamounts provide data on movements of tectonic plates on which they ride, and on the rheology of the underlying lithosphere. The trend of a seamount chain traces the direction of motion of the lithospheric plate over a more or less fixed heat source in the underlying asthenosphere part of the Earth's mantle. See LITHOSPHERE; PLATE TECTONICS.   [E.L.Wi.]

Seawater   An aqueous solution of salts of a rather constant composition of elements whose presence determines the climate and makes life possible on the Earth and which constitutes the oceans, the mediterranean seas, and their embayments. The physical, chemical, biological, and geological events therein are the studies that are grouped as oceanography. Water is most often found in nature as seawater (about 98%). The rest is ice, water vapor, and fresh water. The basic properties of seawater, their distribution, the interchange of properties between sea and atmosphere or land,

Major constituents of seawater (salinity 35 psu)*

Positive ions	Amount, g/kg	Negative ions	Amount, g/kg
Sodium (Na$^+$)	10.752	Chloride (Cl$^-$)	19.345
Magnesium (Mg^{2+})	1.295	Bromide (Br$^-$)	0.066
Potassium (K$^+$)	0.390	Fluoride (F$^-$)	0.0013
Calcium (Ca^{2+})	0.416	Sulfate (SO$_4^-$)	2.701
Strontium (Sr^{2+})	0.013	Bicarbonate (HCO$_3^-$)	0.145
		Boron hydroxide (B(OH)$_3^-$)	0.027

*Water, 965 psu; dissolved materials, 35 psu.

the transmission of energy within the sea, and the geochemical laws governing the composition of seawater and sediments are the fundamentals of oceanography. *See* HYDROSPHERE; OCEANOGRAPHY.

The major chemical constituents of seawater are cations (positive ions) and anions (negative ions) [see table]. In addition, seawater contains the suspended solids, organic substances, and dissolved gases found in all natural waters. A standard salinity of 35 practical salinity units (psu; formerly parts per thousand, or ‰) has been assumed. While salinity does vary appreciably in oceanic waters, the fractional composition of salts is remarkably constant throughout the world's oceans. In addition to the dissolved salts, natural seawater contains particulates in the form of plankton and their detritus, sediments, and dissolved organic matter, all of which lend additional coloration beyond the blue coming from Rayleigh scattering by the water molecules. Almost every known natural substance is found in the ocean, mostly in minute concentrations.   [J.L.Re.]

Seawater fertility   A measure of the potential ability of seawater to support life. Fertility is distinguished from productivity, which is the actual production of living material by various trophic levels of the food web. Fertility is a broader and more general description of the biological activity of a region of the sea, while primary production, secondary production, and so on, is a quantitative description of the biological growth at a specified time and place by a certain trophic level. Primary production that uses recently recycled nutrients such as ammonium, urea, or amino acids is called regenerated production to distinguish it from the new production that is dependent on nitrate being transported by mixing or circulation into the upper layer where primary production occurs. New production is organic matter, in the form of fish or sinking organic matter, that can be exported from the ecosystem without damaging the productive capacity of the system.

The potential of the sea to support growth of living organisms is determined by the fertilizer elements that marine plants need for growth. Fertilizers, or inorganic nutrients as they are called in oceanography, are required only by the first trophic level in the food web, the primary producers; but the supply of inorganic nutrients is a fertility-regulating process whose effect reaches throughout the food web. When there is an abundant supply to the surface layer of the ocean that is taken up by marine plants and converted into organic matter through photosynthesis, the entire food web is enriched, including zooplankton, fish, birds, whales, benthic invertebrates, protozoa, and bacteria. *See* DEEP-SEA FAUNA; FOOD WEB; MARINE FISHERIES.

The elements needed by marine plants for growth are divided into two categories depending on the quantities required: The major nutrient elements that appear to determine variations in ocean fertility are nitrogen, phosphorus, and silicon. The micronutrients are elements required in extremely small, or trace, quantities including essential metals such as iron, manganese, zinc, cobalt, magnesium, and copper, as well as vitamins and specific organic growth factors such as chelators. Knowledge of the fertility consequences of variations in the distribution of micronutrients is incomplete, but consensus among oceanographers is that the overall pattern of ocean fertility is set by the major fertilizer elements—nitrogen, phosphorus, and silicon—and not by micronutrients.

Two types of marine plants carry out primary production in the ocean: microscopic planktonic algae collectively called phytoplankton, and benthic algae and sea grasses attached to hard and soft substrates in shallow coastal waters.

The benthic and planktonic primary producers are a diverse assemblage of plants adapted to exploit a wide variety of marine niches; however, they have in common two basic requirements for the photosynthetic production of new organic matter: light energy and the essential elements of carbon, hydrogen, nitrogen, oxygen, phosphorus, sulfur, and silicon for the synthesis of new organic molecules. These two requirements are the first-order determinants of photosynthetic growth for all marine plants and, hence, for primary productivity everywhere in the ocean.

The regions of the world's oceans differ dramatically in overall fertility. In the richest areas, the water is brown with diatom blooms, fish schools are abundant, birds darken the horizon, and the sediments are fine-grained black mud with a high organic content. In areas of low fertility, the water is blue and clear, fish are rare, and the bottom sediments are well-oxidized carbonate or clay. These extremes exist because the overall pattern of fertility is determined by the processes that transport nutrients to the sunlit upper layer of the ocean where there is energy for photosynthesis. *See* SEAWATER.

[R.T.B.]

Sedimentary rocks   Rocks that accumulate at the surface of the Earth, under ambient temperatures. Together with extruded hot lavas, sedimentary rocks form a thin cover of stratified material (the stratisphere) over the deep-seated igneous and metamorphic rocks that constitute the bulk of the Earth's crust. Sediments cover about three-quarters of the land and of the ocean floor. The thickness of the stratisphere is generally measured in kilometers, and locally reaches about 15 km (50,000 ft). *See* EARTH CRUST; IGNEOUS ROCKS; METAMORPHIC ROCKS; ROCK.

Most sediments accumulate as sand and dust or mud. Being deposited from fluids (air, water) under the influence of gravity, they tend to assume level surfaces (though locally steep slopes may be developed, as in dunes and reefs). Changes in supply of sediment and in depositing agencies change the nature of the deposits from day to day and from millennium to millennium, and commonly interrupt the process altogether. As a result, the accumulated mantle of sediment has a layered structure, divided into beds or strata. Sediments become compacted as waters are squeezed out of them during burial and tectonism, and become cemented as remaining pore space becomes filled by newly growing minerals, mainly calcite or quartz. Bacterial degradation of organic matter, invasion by other fluids, and changes in temperature continue to alter the chemical environment, and lead to alteration of unstable mineral phases. Such processes are included in the term diagenesis. Soft sediment thus becomes converted to rock, but the geologist includes both in the concept of sedimentary rocks. *See* CALCITE; DIAGENESIS; QUARTZ.

When sediments are carried to greater depths or are otherwise subjected to high heat or pressure, growth of new minerals and plastic deformation destroy sedimentary structures and metamorphose the rock. Alternatively, the sediment melts in transition to igneous rock. Thus, sedimentary rocks are recycled through geologic time. Most of the crust under the continents, consisting of igneous and metamorphic rocks, has probably passed through the sedimentary state at some point. Despite such losses, sedimentary rocks have locally survived from very early (Archean) times, nearly 4 billion years ago. *See* ARCHEAN.

Sediments are almost entirely derived from transfer of materials within the Earth's crust. First in importance is gradation, the wearing away of the highlands and the deposition of the products in the low spots: subsiding basins and the oceans. Second is crustal volcanism, which produces large ash falls from explosive volcanoes, and recycles ions to the surface in hot springs. Small amounts are contributed from the mantle underlying the crust: mainly pumice produced when mantle-derived oceanic basalts interact with water. A small fraction of sediment consists of organic matter created by organisms from carbon dioxide and water. Water frozen in the atmosphere transiently covers parts of the stratisphere with ice, while traces of extraterrestrial matter continue to be added from meteorites. *See* COSMIC SPHERULES; WEATHERING PROCESSES.

Though sediments contain such a large range of diverse constituents occurring in a wide variety of mixtures, such mixtures are generally dominated by one or two constituents, and thus may be grouped into a number of classes, each of which can be divided into families.

Detrital sediments are alternately transported and deposited, reeroded, and redeposited on their way to a more permanent resting place, so that their constituents may carry the imprints of a complex history, while the structure of the deposit testifies to the last depositional episode. *See* DEPOSITIONAL SYSTEMS AND ENVIRONMENTS.

Pyroclastic sediments originate from volcanic vents. Submarine eruptions form pumice, or frothy glass, much of which floats widely. The important contributions are great eruptions of glass droplets are ejected into atmosphere and stratosphere to fall as a rain of pumice, sand, and silt, in some cases mixed with crystals. Pyroclastic rocks, largely composed of glass, are readily altered to clay minerals (montmorillonite) in weathering. They produce excellent soils. Beds of montmorillonite (bentonites) are mined for preparation of artificial muds such as those used in well drilling. *See* MONT-MORILLONITE; PUMICE; VOLCANO; VOLCANOLOGY.

Chemical sediments represent the precipitation of materials carried in solution, either by simple chemical precipitation or by the activity of organisms.

Carbonate rocks form about 20% of all sediments. In natural waters, calcium and magnesium are mainly held in solution by virtue of carbon dioxide. In many fresh waters and in the surficial ocean, withdrawal of carbon dioxide—by warming of the water or by the consumption of carbon dioxide in green-plant photosynthesis—leads to supersaturation and to the deposition of calcium carbonate. This normally yields a lime mud of microscopic crystals. Even more important is the secretion of calcium carbonate skeletons, ultimately deposited on ocean floors, by some algae and by a large variety of animals, ranging from microscopic foraminifera to corals and molluscan shells. Carbonate rocks are a major ingredient of portland cement. They are crushed in large quantities for use in road building, agriculture, and smelting, and in the chemical industry. They also furnish building and ornamental stone. Carbonate rocks contain a large share of the world's petroleum resources. *See* ARAGONITE; CALCITE; DOLOMITE; LIMESTONE; OOLITE; STYLOLITES.

Evaporites are formed in bays, estuaries, and lakes of arid regions. On progressive evaporation, seawater first forms deposits of calcium sulfate as gypsum or anhydrite, followed by halite (NaCl) and ultimately potash and magnesium salts. Evaporation of lake water may yield different precipitates such as trona, borax, and silicates. Much of what is sold as table salt is mined from evaporite deposits, as is potash fertilizer. Plaster of paris is produced from gypsum or anhydrite, and the chemical industry relies on evaporite deposits of various types. *See* GYPSUM; HALITE; SALINE EVAPORITES.

Nondetrital siliceous rocks such as silicon dioxide (silica) is second only to carbonate in the dissolved load of most streams. Organisms take up nearly all silica supplied, covering much of the deep-sea floor with radiolarian and diatomaceous ooze. Over geological time spans, diagenetic alteration converts these into dull white opal-ct or quartz porcellanites, or into the solid, waxy-looking mosaics of fine quartz grains known as chert or flint. Diatom ooze is mined for abrasives and filters, as well as for insulation. *See* CHERT; SILICA MINERALS.

Carbonaceous sediments are the result of organic activity, and are of two sorts: the peat-coal series and the kerogens. Peat is used for local fuel in boggy parts of the world. Lignite and bituminous coals continue to be important fuels. *See* COAL; LIGNITE; PEAT.

[A.G.F.]

Sedimentology   The study of natural sediments, both lithified (sedimentary rocks) and unlithified, and of the processes by which they are formed. Sedimentology includes all those processes that give rise to sediment or modify it after deposition: weathering, which breaks up or dissolves preexisting rocks so that sediment may form from them; mechanical transportation; deposition; and diagenesis, which modifies sediment after deposition and burial within a sedimentary basin and converts it into sedimentary rock. Sediments deposited by mechanical processes (gravels, sands, muds) are known as clastic sediments, and those deposited predominantly by chemical or biological processes (limestones, dolomites, rock salt, chert) are known as chemical sediments. *See* SEDIMENTARY ROCKS; WEATHERING PROCESSES.

The raw materials of sedimentation are the products of weathering of previously formed igneous, metamorphic, or sedimentary rocks. In the present geological era, 66% of the continents and almost all of the ocean basins are covered by sedimentary rocks. Therefore, most of the sediment now forming has been derived by recycling previously formed sediment. Identification of the oldest rocks in the Earth's crust, formed more than 3×10^9 years ago, has shown that this process has been going on at least since then. Old sedimentary rocks tend to be eroded away or converted into metamorphic rocks, so that very ancient sedimentary rocks are seen at only a few places on Earth. *See* EARTH CRUST; IGNEOUS ROCKS; METAMORPHIC ROCKS.

Major controls. The major controls on the sedimentary cycle are tectonics, climate, worldwide (eustatic) changes in sea level, the evolution of environments with geological time, and the effect of rare events.

Tectonics are the large-scale motions (both horizontal and vertical) of the Earth's crust. Tectonics are driven by forces within the interior of the Earth but have a large effect on sedimentation. These crustal movements largely determine which areas of the Earth's crust undergo uplift and erosion, thus acting as sources of sediment, and which areas undergo prolonged subsidence, thus acting as sedimentary basins. Rates of uplift may be very high (over 10 m or 33 ft per 1000 years) locally, but probably such rates prevail only for short periods of time. Over millions of years, uplift even in mountainous regions is about 1 m (3.3 ft) per 1000 years, and it is closely balanced by rates of erosion. Rates of erosion, estimated from measured rates of sediment transport in rivers and from various other techniques, range from a few meters per 1000 years in mountainous areas to a few millimeters per 1000 years averaged over entire continents. *See* BASIN; PLATE TECTONICS.

Climate plays a secondary but important role in controlling the rate of weathering and sediment production. The more humid the climate, the higher these rates are. A combination of hot, humid climate and low relief permits extensive chemical weathering, so that a larger percentage of source rocks goes into solution, and the clastic sediment produced consists mainly of those minerals that are chemically inert (such as quartz) or that are produced by weathering itself (clays). Cold climates and high relief favor physical over chemical processes. *See* CLIMATOLOGY.

Tectonics and climate together control the relative level of the sea. In cold periods, water is stored as ice at the poles, which can produce a worldwide (eustatic) lowering of sea level by more than 100 m (330 ft). Changes in sea level, whether local or worldwide, strongly influence sedimentation in shallow seas and along coastlines; sea-level changes also affect sedimentation in rivers by changing the base level below which a stream cannot erode its bed.

One of the major conclusions from the study of ancient sediments has been that the general nature and rates of sedimentation have been essentially unchanged during the last billion years of geological history. However, this conclusion, uniformitarianism, must be qualified to take into account progressive changes in the Earth's environment through geological time, and the operation of rare but locally or even globally important catastrophic events. The most important progressive changes have been in tectonics and atmospheric chemistry early in Precambrian times, and in the nature of life on the Earth, particularly since the beginning of the Cambrian. *See* CAMBRIAN.

Throughout geological time, events that are rare by human standards but common on a geological time scale, such as earthquakes, volcanic eruptions, and storms, produced widespread sediment deposits. There is increasing evidence for a few truly rare but significant events, such as the rapid drying up of large seas (parts of the Mediterranean) and collisions between the Earth and large meteoric or cometary bodies (bolides).

Sediment is moved either by gravity acting on the sediment particles or by the motions of fluids (air, water, flowing ice), which are themselves produced by gravity. Deposition takes place when the rate of sediment movement decreases in the direction of sediment movement; deposition may be so abrupt that an entire moving mass of sediment and fluid comes to a halt (mass deposition, for example, by a debris flow), or so slow that the moving fluid (which may contain only a few parts per thousand of sediment) leaves only a few grains of sediment behind. The settling velocity depends on the density and viscosity of the fluid, as well as on the size, shape, and density of the grains. *See* STREAM TRANSPORT AND DEPOSITION.

Chemical sedimentation. Chemical weathering dissolves rock materials and delivers ions in solution to lakes and the ocean. The concentrations of ions in river and ocean water are quite different, showing that some ions must be removed by sedimentation. Comparison of the modern rate of delivery of ions to the ocean, with their concentration in the oceans, shows that some are removed very rapidly (residence times of only a few thousand years) whereas others, such as chlorine and sodium, are removed very slowly (residence times of hundreds of millions of years).

Biological effects. Many so-called chemical sediments are actually produced by biochemical action. Much is then reworked by waves and currents, so that the chemical sediment shows clastic textures and consists of grains rounded and sorted by transport. Depositional and diagenetic processes, however, are often strongly affected by organic action, no matter what the origin of the sediment. Plants in both terrestrial and marine environments tend to trap sediment, enhancing deposition and slowing erosion.

Sedimentary environments and facies. Sedimentary rocks preserve the main direct evidence about the nature of the surface environments of the ancient Earth and the way they have changed through geological time. Thus, besides trying to understand the basic principles of sedimentation, sedimentologists have studied modern and ancient sediments as records of ancient environments. For this purpose, fossils and primary sedimentary structures are the best guide. These structures are those formed at the time of deposition, as opposed to those formed after deposition by diagenesis, or by deformation. In describing sequences of sedimentary rocks in the field (stratigraphic sections), sedimentologists recognize compositional, structural, and organic aspects of rocks that can be used to distinguish one unit of rocks from another. Such units are known as sedimentary facies, and they can generally be interpreted as having formed in different environments of deposition. Though there are a large number of different sedimentary environments, they can be classified in a number of general classes, and their characteristic facies are known from studies of modern environments. *See* FACIES (GEOLOGY); STRATIGRAPHY; TRACE FOSSILS.   [G.V.M.]

Seiche   A short-period oscillation in an enclosed or semienclosed body of water, analogous to the free oscillation of water in a dish. The initial displacement of water from a level surface can arise from a variety of causes, and the restoring force is gravity, which always tends to maintain a level surface. Once formed, the oscillations are characteristic only of the geometry of the basin itself and may persist for many cycles before decaying under the influence of friction. The term "seiche" appears to have been first used to describe the rhythmic oscillation of the water surface in Lake Geneva, which occasionally exposed large areas of the lake bed that are normally submerged.

Seiches can be generated when the water is subject to changes in wind or atmospheric pressure gradients or, in the case of semi-enclosed basins, by the oscillation of adjacent connected water bodies having a periodicity close to that of the seiche or of one of its harmonics. Other, less frequent causes of seiches include heavy precipitation over a portion of the lake, flood discharge from rivers, seismic disturbances, submarine mudslides or slumps, and tides. The most dramatic seiches have been observed after earthquakes.   [A.Wu.; D.M.F.]

Seismic risk   The probability that social or economic consequences of earthquakes will equal or exceed specified values at a site, at several sites, or in an area, during a specified exposure time.

Although the term seismic risk is occasionally used in a general sense to mean the potential for both the occurrence of natural phenomena and the economic and life loss associated with earthquakes, it is useful to differentiate between the concepts of seismic hazard and seismic risk. Seismic hazard may be defined as any physical phenomena that result either from surface faulting during shallow earthquakes or from the ground shaking resulting from an earthquake and that may produce adverse effects on human activities.

The exposure time is the time period of interest for seismic hazard or risk calculations. In practical applications, the exposure time may be considered to be the design lifetime of a building or the length of time over which the numbers of casualties will be estimated.
                    [S.T.A.]

Seismic stratigraphy   Determination of the nature of sedimentary rocks and their fluid content from analysis of seismic data. Seismic stratigraphy is divided into seismic-sequence (facies) analysis and reflection-character analysis.

In seismic-sequence analysis the first step is to separate seismic-sequence units, also called seismic-facies units. This is usually done by mapping unconformities where they are shown by angularity. Angularity below an unconformity may be produced by erosion at an angle across the former bedding surfaces or by toplap (offlap), and angularity above an unconformity may be produced by onlap or downlap, the latter distinction being based on geometry. The unconformities are then followed along reflections from the points where they cannot be so identified, advantage being taken of the fact that the unconformity reflection is often relatively strong. The procedure often followed is to mark angularities in reflections by small arrows before drawing in the boundaries. *See* UNCONFORMITY.

Seismic-facies units are three-dimensional, and many of the conclusions from them are based on their three-dimensional shape. The appearance on seismic lines in the dip and strike directions is often very different. For example, a fan-shaped unit might show a progradational pattern in the dip direction and discontinuous, overlapping arcuate reflections in the strike direction. *See* FACIES (GEOLOGY).

Reflection-character analysis may be based on information from boreholes which suggests that a particular interval may change nearby in a manner which increases its likelihood to contain hydrocarbon accumulations. Lateral changes in the wave shape of individual reflection events may suggest where the stratigraphic changes or hydrocarbon accumulations may be located.

Where sufficient information is available to develop a reliable model, expected changes are postulated and their effects are calculated and compared to observed seismic data. The procedure is called synthetic seismogram manufacture; it usually involves calculating seismic data based on sonic and density logs from boreholes, sometimes based on a model derived in some other way. The sonic and density data are then changed in the manner expected for a postulated stratigraphic change, and if the synthetic seismogram matches the actual seismic data sufficiently well, it implies that the changes in earth layering are similar to those in the model. *See* GEOPHYSICAL EXPLORATION; SEISMOGRAPHIC INSTRUMENTATION; SEISMOLOGY; STRATIGRAPHY.   [R.E.Sh.]

Seismographic instrumentation   Various devices or systems of devices for measuring movement in the Earth. Ground motion is generally the result of passing seismic waves, gravitational tides, atmospheric processes, and tectonic processes. Seismographic instrumentation typically consists of a sensing element (seismometer), a signal-conditioning element or elements (galvanometer, mechanical or electronic amplifier, filters, analog-to-digital conversion circuitry, telemetry, and so on), and a recording element (analog visible or direct, frequency modulation, or digital magnetic tape or disk). Seismographs are used for earthquake studies, investigations of the Earth's gravity field, nuclear explosion monitoring, petroleum exploration, and industrial vibration measurement.

Seismographic instruments may be required to measure ground motions accurately over a range approaching 12 orders of magnitude, from as small as 10^{-11} m to as large as several meters (a very large earthquake). The instruments may be required to measure frequencies as low as $\sim 10^{-5}$ Hz (the semidiurnal gravitational tides) to as high as $\sim 10^4$ Hz (as observed from acoustic emissions from rock failures in mines). Seismic waves from earthquakes are observed in the bandwidth of $\sim 3 \times 10^{-4}$ Hz (the gravest free oscillations of the Earth) to ~ 200 Hz (a local earthquake). In exploration seismology the frequency range of interest is typically 10–1000 Hz.

The seismometer is the basic sensing element in seismographic instruments, and there are two fundamentally different types: inertial and strain. The inertial seismometer generates an output signal that is proportional to the relative motion between its frame (usually attached to the ground or a point of interest) and an internal inertial reference mass. The strain seismometer (or linear extensometer) generates an output that is proportional to the distance between two points.

A seismoscope is a device that indicates only the occurrence of relatively strong ground shaking and not its time of occurrence or duration. A typical seismoscope inscribes a hodograph of horizontal strong ground motion on a smoked watch glass.

A dilatometer continuously and precisely measures volumetric strain. The quantity measured is the change ΔV in the reference volume V, and the ratio $\Delta V/V$ gives the volumetric strain. Dilatometers are typically installed in a borehole in competent rock (preferably granite) at a depth of 100–300 m (330–1000 ft).

A tiltmeter monitors the relative change in the elevation between two points, usually with respect to a liquid-level surface. The horizontal distance between the reference points may be as little as a few millimeters or as large as several hundred meters.

The gravity meter is just a vertical-component accelerometer, that is, a pendulum sensing ground motion and equipped with a displacement transducer, analogous to the inertial tiltmeter. Gravimeters are widely used in geophysical exploration, in the study of earth tides, and in the recording of very low frequency (0.0003–0.01 Hz) seismic waves from earthquakes.

The complete seismograph produces a record of the properly conditioned signal from the seismometer, along with appropriate timing information. The recording system may be as simple as a mechanical stylus scratching a line on a smoke-covered drum in a portable microearthquake seismograph, or as complex as a multichannel computer-controlled system handling 25,000 24-bit digital words per second in a modern seismic reflection survey for petroleum exploration. The range between these extremes includes many special-purpose seismographs, all designed to record ground motion in a particular application. *See* EARTH TIDES; EARTHQUAKE; GEOPHYSICAL EXPLORATION; SEISMOLOGY.

[T.V.McE.; R.A.U.]

Seismology
The study of the shaking of the Earth's interior caused by natural or artificial sources. Throughout the period in which plate tectonics was advanced and its basic tenets tested and confirmed in the early 1960s, and into the latest phase of inquiry into basic processes, seismology (and particularly seismic imaging) has provided critical observational evidence upon which discoveries have been made and theory has been advanced regarding the structure of the Earth's crust, mantle, and core. *See* PLATE TECTONICS.

Theoretical seismology. A seismic source is an energy conversion process that over a short time (generally less than a minute and usually less than 1–10 s) transforms stored potential energy into elastic kinetic energy. This energy then propagates in the form of seismic waves through the Earth until it is converted into heat by internal (molecular) friction. Large sources, that is, sources that release large amounts of potential energy, can be detected worldwide. Earthquakes above Richter magnitude 5 and explosions above 50 kilotons or so are large enough to be observed globally before the seismic waves dissipate below modern levels of detection. Small charges of dynamite or small earthquakes are detectable at a distance of a few tens to a few hundreds of kilometers, depending on the type of rock between the explosion and the detector. *See* EARTHQUAKE.

Seismic vibrations are recorded by instruments known as seismometers that sense the change in the position of the ground (or water pressure) as seismic waves pass underneath. The record of ground motion as a function of time is a seismogram, which may be in either analog or digital form. Advances in computer technology have made analog recording virtually obsolete: most seismograms are recorded digitally, which makes quantitative analysis much more feasible.

The response of the Earth to a seismic disturbance can be approximated by the equation of motion for a disturbance in a perfectly elastic body. This equation holds regardless of the type of source, and is closely related to the acoustic-wave equation governing the propagation of sound in a fluid. The equation of motion for an isotropic perfectly elastic solid separates into two equations describing the propagation of purely dilatational (volume changing, curl-free) and purely rotational (no volume changing, divergence-free) disturbances. These propagate with wave speeds α and β, respectively. These velocities are also known as the compressional or primary (P) and shear or secondary (S) velocities, and the corresponding waves are called P and S waves. The compressional velocity is always faster than the shear velocity. In the Earth, α can range from a few hundred meters per second in unconsolidated sediments to more than 13.7 km/s (8.2 mi/s) just above the core–mantle boundary. Wave speed β ranges from zero in fluids (ocean, fluid outer core) to about 7.3 km/s (4.4 mi/s) at the core-mantle boundary.

A P wave has no curl and thus only causes the material to undergo a volume change with no other distortion. An S wave has no divergence, thus causing no volume change, but right angles embedded in the material are distorted. Explosions are relatively efficient generators of compressional disturbances, but earthquakes generate both compressional and shear waves. Compressional waves, by virtue of the mechanical stability condition, always arrive before shear waves.

Compressional and shear waves can exist in an elastic body irrespective of its boundaries. For this reason, seismic waves traveling with speed α or β are known as body waves. A third type of wave motion is produced if the elastic material is bounded by a free surface. The free-surface boundary conditions help trap energy near the surface, resulting in a boundary or surface wave. This in turn can be of two types. A Rayleigh wave combines both compressional and shear motion and requires only the presence of a boundary to exist. A Love wave is a pure-shear disturbance that can propagate only in the presence of a change in the elastic properties with depth from the free surface. Both are slower than body waves.

Solutions of the elastic-wave equation in which a wave function of a particular shape propagates with a particular speed are known as traveling waves. An important property of traveling waves is their causality; that is, the wave function has no amplitude before the first predicted arrival of energy. The complete seismic wavefield can be constructed by summing up every possible traveling wave.

Traveling-wave or full-wave theory provides the basis for a very useful theoretical abstraction of elastic-wave propagation in terms of the more common notions of wavefronts and their outwardly directed normals, called rays. Ray theory makes the prediction of certain kinematic quantities such as ray path, travel time, and distance by a simple geometric exercise. Ray theory can be developed in the context of an Earth comprising flat-lying layers of uniform velocities; this is a very useful approximation for most problems in crustal seismology and can be extended to spherical geometry for global studies.

Kinematic equations have been developed to describe what happens to rays as they impinge on the boundaries between layers. The illustration shows a single ray propagating in the stack of horizontal layers that define the model Earth. At each interface, part of the ray's energy is reflected, but a portion also passes through into the layer below. The transmitted portion of the ray is refracted; that is, it changes the angle at which it is propagating. The relationship between the incident angle and the refracted angle is exactly the same as that describing the refraction of light between two media of differing refractive index.

These simple geometric equations can be extended to the computation of amplitudes provided that there are no sharp discontinuities in the velocity as a function of depth. More exact representations of the amplitudes and wave shapes that solve the full-wave equation to varying extents can be constructed with the aid of powerful computers; these methods are collectively known as seismogram synthesis, and the seismograms thus computed are known as synthetics. Synthetics can be computed for elastic or dissipative media that vary in one, two, or three dimensions.

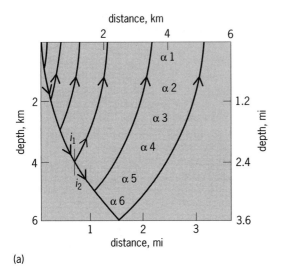

(a)

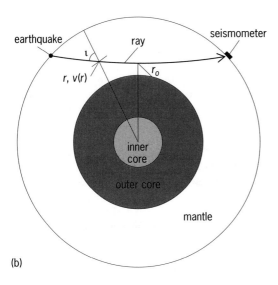

(b)

Seismic ray paths. (*a*) A single ray passing through a multilayered Earth comprising a stack of uniform velocity layers will be reflected from each layer and also be refracted as it passes from one layer into the layer below in a manner that obeys Snell's law. Each ray therefore is considered to give rise to a new system of rays. (*b*) Ray diagram for a cross section of the spherical Earth. At the point labeled $v(r)$, $r =$ radial distance and $v =$ velocity. $r_0 =$ radial distance to the turning point. α is the angle of incidence.

In a typical experiment for crustal imaging, a source of seismic energy is discharged on the surface, and instruments record the disturbance at numerous locations. Many different types of sources have been devised, from simple explosives to mechanical vibrators and devices known as airguns that discharge a "shot" of compressed air. The details of the source-receiver geometry vary with the type of experiment and its objective, but the work always involves collecting a large number of recordings at increasing distance from the source. This seismogram is complex, exhibiting a number of distinct arrivals with a variety of shapes and having amplitudes that change with distance. Although this seismogram clearly does not resemble the structure of the Earth in any sensible way and is therefore not what would normally be thought of as an image, it can be analyzed to recover estimates of those physical properties of the Earth that govern seismic-wave propagation.

Two- and three-dimensional imaging. A volume of the crust can be directly imaged by seismic tomography. In crustal tomography, active sources are used (explosives on land, airguns at sea) so that the source location and shape are already known. Experiments can be constructed in which sources and receivers are distributed in such a way that many rays pass through a particular volume and the tomographic inversion can produce relatively high-resolution images of velocity pertubations in the crust. Crustal tomography uses transmitted rays like those that pass from a surface source through the crust to receivers that are also on the same surface. See GEOPHYSICAL EXPLORATION.

Seismic source imaging. Another imaging problem in global seismology is constructing models of the seismic source. So-called first-motion representations of seismic sources (earthquakes) are the result of measurements made on the very first P waves or S waves arriving at an instrument; therefore they represent the very beginning of the rupture on the fault plane. This is not a problem if the rupture is approximately a point source, but this is true in practice only if the earthquake is quite small or exceptionally simple. An alternative is to examine only longer-period seismic phases, including surface waves, to obtain an estimate of the average point source that smooths over the space and time complexities of a large rupture. This so-called centroid-moment-tensor representation is routinely computed for events with magnitudes greater than about 5.5. Because an estimate for a centroid moment tensor is derived from much more of the seismogram than the first arrivals, it gives a better estimate of the energy content of the earthquake. This estimate, known as the seismic moment, represents the total stress reduction resulting from the earthquake; it is the basis for a new magnitude number M_W. This value is equivalent to the Richter body wave (m_b) or surface-wave magnitude (M_S) at low magnitudes, but it is much more accurate for magnitudes above about 7.5.

Some large events comprise smaller subevents distributed in space and time and contributing to the total rupture and seismic moment. The position and individual rupture characteristics of these subevents can be mapped with remarkable precision, given data of exceptional bandwidth and good geographical distribution. An outstanding problem is whether the location of these subevents is related to stress heterogeneities within the fault zone. These stress heterogeneities are known as barriers or asperities, depending on whether they stop or initiate rupture. See SEISMOGRAPHIC INSTRUMENTATION.

[J.Mu.; A.L.L.]

Sepiolite A complex hydrated magnesium silicate mineral named for its resemblance to cuttlefish bone, alternately named meerschaum (sea foam). The ideal composition, $Mg_8(H_2O)_4(OH)_4Si_{12}O_{30}$, is modified by some additional water of hydration, but is otherwise quite representative. Interlaced disoriented fibers aggregate into a massive stone so porous that it floats on water. These stones are easily carved, take a high polish with wax, and harden when warmed. See CLAY MINERALS; SILICATE MINERALS.

[W.F.B.]

Sequence stratigraphy The study of stratigraphic sequences, defined as stratigraphic units bounded by unconformities. With improvements in the acquisition and processing of reflection-seismic data by petroleum exploration companies in the 1970s came the recognition that unconformity-bounded sequences could be recognized in most sedimentary basins. This was the beginning of an important development, seismic stratigraphy, which also included the use of seismic reflection character to make interpretations about large-scale depositional facies and architecture. See SEISMIC STRATIGRAPHY; UNCONFORMITY.

Underpinning sequence-stratigraphic methods are the following interrelated principles: (1) The volume of sediment accumulating in any part of a sedimentary basin is dependent on the space made available for sediment by changes in sea level or basin-floor elevation. This space is referred to as accommodation. (2) Changes in accommodation tend to be cyclic, and they are accompanied by corresponding changes in sedimentary environment and depositional facies. Thus, a rise in base level typically leads to an increase in accommodation, deepening of the water in the basin, with corresponding changes in facies, and a transgression, with a consequent landward shift in depositional environments and in depositional facies. A fall in base level may lead to exposure and erosion (negative accommodation), with the development of a widespread unconformity. (3) These predictable changes provide the basis for a model of the shape and internal arrangement or architecture of a sequence, including the organization and distribution of sedimentary facies and the internal bedding surfaces that link these facies together. *See* BASIN; DEPOSITIONAL SYSTEMS AND ENVIRONMENTS; FACIES (GEOLOGY).

Clastic-dominated sequences are bounded by unconformities. These surfaces (sequence boundaries) are typically well developed within coastal and shelf sediments, where they form as a result of subaerial exposure and erosion during falling sea level. In deeper-water settings, including the continental slope and base of slope, there may be no corresponding sedimentary break; and sequences may be mapped into such settings only if the unconformity can be correlated to the equivalent conformable surface (the correlative conformity). In some instances, the surface of marine transgression, which develops during the initial rise in sea level from a lowstand, forms a distinctive surface that is close in age to the subaerial unconformity and may be used as the sequence boundary. *See* MARINE GEOLOGY; MARINE SEDIMENTS.

The cycle of rise and fall of sea level may be divided into four segments: lowstand, transgressive, highstand, and falling stage. The deposits that form at each stage are distinctive, and are assigned to systems tracts named for each of these stages.

Carbonate-dominated sequences are derived from carbonate sedimentation which is most active in warm, clear, shallow, shelf seas. During the sea-level cycle, these conditions tend to be met during the highstand phase. Sediment production may be so active, including that of reef development at the platform margin, that it outpaces accommodation generation, leading to deposition on the continental slope. Oversteepened sediment slopes there may be remobilized, triggering sediment gravity flows and transportation into the deep ocean. This process is called highstand shedding.

There are several processes of sequence generation that range from a few tens of thousands of years to hundreds of millions of years for the completion of a cycle of rise and fall of sea level. More than one such process may be in progress at any one time within a basin, with the production of a range of sequence styles nested within or overlayering each other.

High-frequency sequence generation is driven by orbital forcing of climate (the so-called Milankovitch effects), of which glacial eustasy is the best-known outcome. The effects of glacioeustasy have dominated continental-margin sedimentation since the freeze-up of Antarctica in the Oligocene. Regional tectonism—such as the process of thermal subsidence following rifting, and flexural loading in convergent plate settings— develops changes in basement elevation that drive changes in relative sea level. These cycles have durations of a few millions to a few tens of millions of years, and they are confined to individual basins or the flanks of major orogens or plate boundaries. *See* PALEOGEOGRAPHY.

Sequence concepts enable petroleum exploration and development geologists to construct predictive sequence models for stratigraphic units of interest from the limited information typically available from basins undergoing petroleum exploration. These models can guide regional exploration, and can also assist in the construction of production models that reflect the expected partitioning of reservoir-quality facies within individual stratigraphic units. *See* CLIMATE HISTORY; GEOLOGIC TIME SCALE; GEOPHYSICAL EXPLORATION; GLACIOLOGY; PALEOCLIMATOLOGY; STRATIGRAPHY.   [A.D.M.]

Serpentine   The name traditionally applied to three hydrated magnesium silicate minerals, antigorite, chrysotile, and lizardite. All have similar chemical compositions but with three different but closely related layered crystal structures. Serpentine also has been used as a group name for minerals with the same layered structures but with a variety of compositions. The general formula is $M_3T_2O_5(OH)_4$, where M may be magnesium (Mg), ferrous iron (Fe^{2+}), ferric iron (Fe^{3+}), aluminum (Al), nickel (Ni), manganese (Mn), cobalt (Co), chromium (Cr), zinc (Zn), or lithium (Li); and T may be silicon (Si), Al, Fe^{3+}, or boron (B).

Lizardite has a planar structure, with the misfit accommodated by slight adjustments of the atomic positions within the layers. Chrysotile has a cylindrical structure in which the layers are either concentrically or spirally rolled to produce fiber commonly ranging from 15 to 30 nanometers in diameter, and micrometers to centimeters in length. These fibers have great strength and flexibility and are the most abundant and commonly used form of asbestos. Antigorite has a modulated wave structure, with wavelengths generally varying between 3 and 5 nm. *See* ASBESTOS.   [F.J.W.]

Serpentinite   A common rock composed of serpentine minerals; usually formed through the hydration of ultramafic rocks, dunites, and peridotites in a process known as serpentinization. The result is the formation of hydrated magnesium-rich minerals, such as antigorite, chrysotile, or lizardite, commonly with magnetite or, less frequently, brucite. *See* ASBESTOS; DUNITE; PERIDOTITE.

Serpentinites can be distinguished by, and are named for, the dominant serpentine mineral in the rock, that is, antigorite-serpentinite, chrysotile-serpentinite, and lizardite-serpentinite. Lizardite-serpentinites are the most abundant. They have been formed in retrograde terrains and are characterized by the pseudomorphic replacement of the original olivine, pyroxenes, amphiboles, and talc by lizardite with or without magnetite or brucite. Antigorite-serpentinites can form directly from minerals such as olivine, pyroxene, and so forth in retrograde terrains similar to lizardite, but at a high temperature. Chrysotile-serpentinites usually occur only in chrysotile asbestos deposits. The occurrence of serpentinites is widespread, particularly in greenstone belts, mountain chains, and mid-ocean ridges, where they have formed through the serpentinization of ultramafic rocks. *See* MID-OCEANIC RIDGE; SERPENTINE.   [F.J.W.]

Sferics   Electromagnetic radiations produced primarily by lightning strokes from thunderstorms. It is estimated that globally there occur about 2000 thunderstorms at any one time, and that these give rise to about 100 lightning strokes every second. The radiations are short impulses that usually last a few milliseconds, with a frequency content ranging from the low audio well into the gigahertz range. Sferics (short for atmospherics) are easily detected with an ordinary amplitude-modulation (AM) radio tuned to a region between radio stations, especially if there are thunderstorms within a few hundred miles. These sounds or noises have been identified and characterized with specific names, for example, hiss, pop, click, whistler, and dawn chorus. They fall into what is generally known as radio noise. *See* ATMOSPHERIC ELECTRICITY; DUST STORM; LIGHTNING; THUNDERSTORM.

The various types of sferics include terrestrial, magnetospheric, and Earth-ionospheric. Terrestrial sferics includes anthropogenic noise from sources such as automobile ignition, motor brushes, coronas from high-voltage transmission lines, and various high-current switching devices. Dust storms and dust devils have also been observed to produce sferics.

Lightning-generated sferics are sometimes coupled into the magnetosphere, where they are trapped and guided by the Earth's magnetic field. In this mode, the impulse travels in an ionized region. As a result, the frequencies present in the original impulse are separated by dispersion (the higher frequencies travel faster than the lower) and produce the phenomena known as whistlers. *See* MAGNETOSPHERE.

By far the dominant and most readily observed sferics are the lightning-produced impulses that travel in the spherical cavity formed by the ionosphere and the Earth's surface. Lightning currents produce strong radiation in the very low-frequency band, 3–30 kHz, and in the extremely low-frequency band, 6 Hz–3 kHz. *See* IONOSPHERE.

[M.Bro.]

Shale   A class of fine-grained clastic sedimentary rocks with a mean grain size of less than 0.0625 mm (0.0025 in.), including siltstone, mudstone, and claystone. One-half to two-thirds of all sedimentary rocks are shales. *See* SEDIMENTARY ROCKS.

Shale is deposited as mud, that is, small particles of silt and clay. The particles are deposited when fluid turbulence caused by currents or waves is no longer adequate to counteract the force of gravity, or if the water evaporates or infiltrates into the ground. Clay particles often form larger aggregates which settle from suspension more rapidly than individual particles. Silt particles and clay aggregates are often deposited as thin layers less than 10 mm (0.4 in.) thick called laminae. *See* DEPOSITIONAL SYSTEMS AND ENVIRONMENTS.

Mineralogically, most shales are made up of clay minerals, silt-sized quartz and feldspar grains, carbonate cements, accessory minerals such as pyrite and apatite, and amorphous material such as volcanic glass, iron and aluminum oxides, silica, and organic matter. The most common clay minerals in shales are smectite, illite, kaolinite, and chlorite. The type of clay particles deposited is dependent on the mineralogy, climate, and tectonics of the source area. *See* CLAY MINERALS; CHLORITE; ILLITE; KAOLINITE.

Shales are usually classified or described according to the amount of silt, the presence and type of lamination, mineralogy, chemical composition, and color. Variations in these properties are related to the type of environment in which the shale was deposited and to postdepositional changes caused by diagenesis and compaction. *See* DIAGENESIS.

The small size of pores in shale relative to those in sandstone causes shale permeability to be much lower than sand permeability. Although fracturing due to compaction stresses or to tectonic movements can create deviations from this general trend, shales often form permeability barriers to fluid movement; this has important bearing on the occurrence of subsurface water and hydrocarbons. Ground-water aquifers are commonly confined by an underlying low-permeability shale bed or aquiclude, which prevents further downward movement of the water. Hydrocarbon reservoirs are often capped by low-permeability shale which forms an effective seal to prevent hydrocarbons from escaping. *See* AQUIFER.

[J.R.D.]

Siderite   A mineral ($FeCO_3$) with the same space group and hexagonal crystal system as calcite ($CaCO_3$). Siderite has a gray, tan, brown, dark brown, or red color, has rhombohedral cleavages, and occasionally may show rhombohedral crystal terminations. It may display curved crystal faces like dolomite ($CaMg[CO_3]_2$), but more commonly is found as massive, compact, or earthy masses. It has a high specific gravity of 3.94, a medium hardness of 3.5–4, and a high index of refraction, 1.88. *See* CARBONATE MINERALS; DOLOMITE.

Siderite, a widespread mineral in near-surface sediments and ore deposits, occurs in hydrothermal veins, lead-silver ore deposits, sedimentary concretions formed in limestones and sandstones, and Precambrian banded iron formations that precipitated under acidic conditions. Famous localities for siderite are found in Styria (Austria), Westphalia (Germany), Cornwall (Britain), Wawa (Northern Ontario, Canada), Minas Gerais (Brazil), and Llallagua and Potosi (Bolivia). These and other occurrences have provided locally significant quantities of siderite as iron ore. *See* MAGNETITE; METAMOR-PHISM; ORE AND MINERAL DEPOSITS.   [E.J.E.]

Silica minerals   Silica (SiO_2) occurs naturally in at least nine different varieties (polymorphs), which include tridymite, cristobalite, coesite, and stishovite, in addition to high (β) and low (α) quartz. These forms are characterized by distinctive crystallography, optical characteristics, physical properties, pressure-temperature stability ranges, and occurrences.

The crystal structures of all silica polymorphs except stishovite contain silicon atoms surrounded by four oxygens, thus producing tetrahedral coordination polyhedra. Each oxygen is bonded to two silicons, creating an electrically neutral framework. Stishovite differs from the other silica minerals in having silicon atoms surrounded by six oxygens (octahedral coordination.) Ideal high tridymite is composed of sheets of SiO_4 tetrahedra oriented perpendicular to the *c* crystallographic axis (Fig. 1) with adjacent tetrahedra in these sheets pointing in opposite directions. High cristobalite, like tridymite, is composed of parallel sheets of SiO_4 tetrahedra with neighboring tetrahedra pointing in opposite directions. However, the hexagonal rings are distorted and adjacent sheets are rotated 60° with respect to one another, resulting in the geometry shown in Fig. 2. Coesite also contains silicon atoms tetrahedrally coordinated by oxygen. These polyhedra

Fig. 1. Portion of an idealized sheet of tetrahedrally coordinated silicon atoms similar to that found in tridymite and cristobalite. Sharing of apical oxygens (which point in alternate directions) between silicons in adjacent sheets generates a continuous framework. (*After J. J. Papike and M. Cameron, Crystal chemistry of silicate minerals of geophysical interest, Rev. Geophys. Space Phys., 14:37–80; copyright © 1976 by American Geophysical Union*)

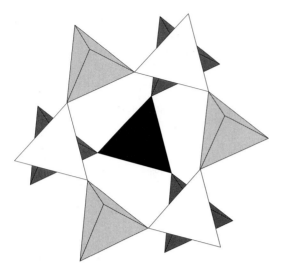

Fig. 2. Portion of cubic high cristobalite illustrating the distortion of tetrahedral sheets, with are oriented parallel to (111), and the 60° rotation of adjacent sheets. (*After J. J. Pa-pike and M. Cameron, Crystal chemistry of silicate minerals of geophysical interest, Rev. Geophys. Space Phys., 14:37–80; copyright © 1976 by American Geophysical Union*)

share corners to form chains composed of four-membered rings. Silicon in stishovite is octahedrally coordinated by oxygen. These coordination polyhedra share edges and corners to form chains of octahedra parallel to the *c* crystallographic axis.

Chemically, all silica polymorphs are ideally 100% SiO_2. However, unlike quartz which commonly contains few impurities, the compositions of tridymite and cristobalite generally deviate significantly from pure silica. This usually occurs because of a coupled substitution in which a trivalent ion such as Al^{3+} or Fe^{3+} substitutes for Si^{4+}, with electrical neutrality being maintained by monovalent or divalent cations occupying interstices In the relatively open structures of these two minerals. *See* COESITE; QUARTZ.

[J.C.D.]

Silicate minerals   All silicates are built of a fundamental structural unit, the so-called SiO_4 tetrahedron. The crystal structure may be based on isolated SiO_4 groups or, since each of the four oxygen ions can bond to either one or two silicon (Si) ions, on SiO_4 groups shared in such a way as to form complex isolated groups or indefinitely extending chains, sheets, or three-dimensional networks. Mixed structures in which more than one type of shared tetrahedra are present also are known.

Silicates are classified according to the nature of the sharing mechanism, as revealed by x-ray diffraction study. The sharing mechanism gives rise to a characteristic ratio of Si to O, but it is possible for oxygen ions that are not bonded to Si to be present in the structure, and sometimes some or all of any aluminum present must be counted as equivalent to Si.

The detailed crystallographic and physical properties of the various silicates are broadly related to the type of silicate framework that they possess. Thus, the phyllosilicates as a group typically have a platy crystal habit, with a cleavage parallel to the plane of layering of the structure, and are optically negative with rather high birefringence. The inosilicates, based on an extended one-dimensional rather than two-dimensional linkage of the SiO_4 tetrahedra, generally form crystals of prismatic habit; if cleavage is present, it will be parallel to the direction of elongation. The tectosilicates commonly are equant in habit, without marked preference for cleavage direction, and tend to have a relatively low birefringence.

Silicate minerals make up the bulk of the outer crust of the Earth and form in a wide range of geologic environments. Many silicates are of economic importance. For discussions of certain silicate mineral groups see AMPHIBOLE; ANDALUSITE; CHLORITE; CHLORITOID; EPIDOTE; FELDSPAR; FELDSPATHOID; GARNET; HUMITE; MICA; OLIVINE; PYROXENE; SCAPOLITE; SERPENTINE; ZEOLITE. [C.Fr.]

Silicate phase equilibria Silicate phase equilibria studies define the conditions of temperature, composition, and pressure at which silicates can stably coexist. Silicate phase equilibria relations are used by geologists, ceramists, and cement manufacturers to explain the variation of composition of silica-bearing minerals, as well as their number and order of appearance in rocks, slags, glasses, and cements. They are also useful to interpret the chemistry of refractories, boiler scale deposits, and welding fluxes.

Silica itself makes up nearly 60% by weight of the Earth's crust. The next most abundant oxides, in decreasing order, are Al_2O_3, CaO, Na_2O, FeO, MgO, K_2O, and Fe_2O_3; all of these occur principally combined with silica as silicates. Free silica and the hundreds of silicate minerals make up nearly 97% of the Earth's crust. The study of silicate phase equilibria was initiated by geologists seeking to apply the phase rule of J. Willard Gibbs to these abundant natural substances. See SILICATE MINERALS.

Dynamic and static methods are used to determine equilibrium in silicate systems. Dynamic methods require large samples of silicates that are difficult to prepare, and require also that equilibrium be reached quickly. Silicates in general are slow to react, and supercooling or superheating of hundreds of degrees before reaction occurs is common. Many silicates react sluggishly at temperatures of 1000°C (1830°F) or higher.

Most silicate phase equilibria are determined by the static method of holding a sample under controlled conditions until equilibrium is attained, then quenching the sample for examination.

Equilibrium is established when the products obtained by heating a sample to a given temperature are identical with the products of cooling a sample to that temperature; no requirement is made regarding the texture, shape, or grain size of the products. Another criterion used in recognizing equilibrium is that no change of the sample can be observed after holding the charge at a given temperature for very long periods of time.

The use of diagrams to present silicate phase equilibria is customary, since such diagrams express quantitatively the amount and composition of each phase present at any bulk composition in the system at any temperature. [D.B.St.]

Siliceous sediment Fine-grained sediment and sedimentary rock dominantly composed of the microscopic remains of the unicellular, silica-secreting plankton diatoms and radiolarians. Minor constituents include extremely small shards of sponge spicules and other microorganisms such as silicoflagellates. Siliceous sedimentary rock sequences are often highly porous and can form excellent petroleum source and reservoir rocks. See SEDIMENTARY ROCKS.

Given their biologic composition, siliceous sediments provide some of the best geologic records of the ancient oceans. Diatoms did not evolve until the late Mesozoic; thus the majority of siliceous rocks older than approximately 150 million years are formed by radiolarians. Geologists map the distribution of ancient siliceous sediments now pushed up onto land by plate tectonic processes, and can thus determine which portions of the ancient seas were biologically productive; this knowledge in turn can give great insight into regions of the Earth's crust that may be economically productive (for example oil-containing regions). The vast oil reserves of coastal California are predominantly found in the Monterey Formation, a highly porous diatomaceous siliceous sedimentary sequence distributed along the western seaboard of the United States. *See* CHALK; CHERT; LIMESTONE.   [R.W.Mu.]

Siliceous sinter   A porous silica deposit formed around hot springs. It is white to light gray and sometimes friable. Geyserite is a variety of siliceous sinter formed around geysers. The siliceous sinters are deposited as the hot subterranean waters cool after issuing at the surface and become supersaturated with silica that was picked up at depth. The sinters are frequently deposited on algae that live in the pools around the hot springs. *See* GEYSER.   [R.Si.]

Sillimanite   A nesosilicate mineral of composition $Al_2O[SiO_4]$, crystallizing in the orthorhombic system. Sillimanite commonly occurs in slender crystals or parallel groups, and is frequently fibrous, hence the synonym fibrolite. There is one perfect cleavage, luster is vitreous, color is brown, pale green, or white, hardness is 6–7 on Mohs scale, and the specific gravity is 3.23.

Sillimanite, andalusite, and kyanite are polymorphs of $Al_2O[SiO_4]$. The three $Al_2O[SiO_4]$ polymorphs are important in assessing the metamorphic grade of the rocks in which they crystallized. *See* ANDALUSITE; KYANITE; SILICATE MINERALS.   [P.B.M.]

Silurian   The third oldest period of the Paleozoic Era, spanning an interval from about 412 to 438 million years before the present. The Silurian system includes all sedimentary rocks deposited and all igneous and metamorphic rocks formed in the Silurian Period. Both the base and top of the Silurian have been designated by international agreement at the first appearances of certain graptolite species in rock sequences at easily examined and well-studied outcrops. *See* GEOLOGIC TIME SCALE.

The most prominent feature of Silurian paleogeography was the immense Gondwana plate. It included much of present-day South America, Africa, the Middle East, Antarctica, Australia, and the Indian subcontinent. During the Silurian, many plates continued the relative northward motion that had commenced during the mid-Ordovician. Plate positions and plate motions as well as topographic features of the plates controlled depositional environments and lithofacies. These, in turn, significantly influenced organismal development and distributions. Silurian Northern Hemisphere plates, other than a portion of Siberia, are not known north of the Northern Hemisphere tropics. Presumably, nearly all of the Northern Hemisphere north of the tropics was ocean throughout the Silurian. *See* CONTINENTS, EVOLUTION OF; PALEOGEOGRAPHY.

Absence of plates bearing continental or shallow shelf marine environments north of about 45° north latitude indicates that ocean circulation in most of the Silurian Northern Hemisphere was zonal. Ocean surface currents in the tropics would have been influenced strongly by the prevailing westerlies. The large size of the Gondwana plate and the presence of land over much of it would have led to development of seasonal monsoon conditions. Surface circulation south of 30° south would have hit the western side of Laurentia and flowed generally northward. *See* PALEOCEANOGRAPHY.

Collision of the Avalonian and Laurentian plates in the latest Ordovician coincides with development of the Southern Hemisphere continental glaciation. Erosion of the land area formed at the Avalon-Laurentian plate collision generated a large volume of coarse to fine-grained siliclastic materials. That part of South America (modern eastern South America) near the South Pole for the early part of the Silurian was the site of as many as four brief glacial episodes. *See* PALEOCLIMATOLOGY.

Both nonvascular and vascular plants continued to develop in land environments following their originations in the early mid-Ordovician. Many of these Silurian plants were mosslike and bryophytelike. Psilophytes assigned to the genus *Cooksonia* were relatively widespread in Late Silurian terrestrial environments. The probable lycopod (club moss) *Baragwanathia* apparently lived in nearshore settings in modern Australia during the latter part of the Silurian. Silurian land life also included probable arthopods and annelid worms. Fecal pellets of wormlike activity have been found as well as remains of centipede-, millepede-, and spiderlike arthropods.

Shallow marine environments in the tropics were scenes of rich growths of algae, mat-forming cyanobacteria, spongelike organisms, sponges, brachiopods, bryozoans, corals, crinoids, and ostracodes. Nearshore marine siliclastic strata bear ostracodes, small clams, and snails and trilobites. Certain nearshore strata bear the remains of horseshoe-crab-like eurypterids.

Fish are prominent in a number of Silurian nearshore and some offshore marine environments. Jawless armored fish include many species of thelodonts that had bodies covered with minute bony scales, heterostracans, and galeaspids that had relatively heavily armored head shields, and anaspids that possessed body armor consisting of scales and small plates. Jawed fish were relatively rare in the Silurian. They were primarily spiny sharks or acanthodians. As well, there are remains of true sharklike fish and fish with interior bony skeletons (osteichthyans) in Late Silurian rocks.   [W.B.N.B.]

Sirocco   A southerly or southeasterly wind current from the Sahara or from the deserts of Saudi Arabia which occurs in advance of cyclones moving eastward through the Mediterranean Sea. The sirocco is most pronounced in the spring, when the deserts are hot and the Mediterranean cyclones are vigorous. It is observed along the southern and eastern coasts of the Mediterranean Sea from Morocco to Syria as a hot, dry wind capable of carrying sand and dust great distances from the desert source. The sirocco is cooled and moistened in crossing the Mediterranean and produces an oppressive, muggy atmosphere when it extends to the southern coast of Europe. *See* AIR MASS; WIND.   [F.S.]

Skarn   A broad range of rock types made up of calc-silicate minerals such as garnet, regardless of their association with ores, that originate by replacement of precursor rocks. It was a term originally coined by miners in reference to rock consisting of coarse-grained, calc-silicate minerals associated with iron ores in central Sweden. Ore deposits that contain skarn are termed skarn deposits; such deposits are the world's premier sources of tungsten. They are also important sources of copper, iron, molybdenum, zinc, and other metals. Skarns also serve as sources of industrial minerals such as graphite, asbestos, and magnesite. *See* ORE AND MINERAL DEPOSITS; SILICATE MINERALS.

Based on mineralogy, three idealized types of skarn are recognized: calcic skarn characterized by calcium- and iron-rich silicates (andradite, hedenbergite, wollastonite); magnesian skarn characterized by calcium- and magnesium-rich silicates (forsterite, diopside, serpentine); and aluminous skarn characterized by aluminum- and magnesium-rich calc-silicates (grossularite, vesuvianite, epidote).   [M.T.E.]

Skutterudite   A mineral with composition $(Co,Ni)As_3$, an ore of cobalt and nickel. Commonly the mineral is massive with metallic luster and tin-white color. The hardness is $5^{1}/_{2}$–6 on Mohs scale and the specific gravity 6.6. Skutterudite is found at Freiberg, Annaberg, and Schneeberg in Germany, and at Cobalt, Ontario.   [C.S.Hu.]

Slate   Any of the deformed fine-grained, mica-rich rocks that are derived primarily from mudstones and shales, containing a well-developed, penetrative foliation that is called slaty cleavage. Slaty cleavage is a secondary fabric element that forms under low-temperature conditions (less than 540°F or 300°C), and imparts to the rock a tendency to split along planes. It is a type of penetrative fabric; that is, the rock can be split into smaller and smaller pieces, down to the size of the individual grains. If there is an obvious spacing between fabric elements (practically, greater than 1 mm), the fabric is called spaced. Slates typically contain clay minerals (for example, smectite), muscovite/illite, chlorite, quartz, and a variety of accessory phases (such as epidote or iron oxides). Under increasing temperature conditions, slate grades into phyllite and schist. *See* ARGILLACEOUS ROCKS; CLAY MINERALS; PHYLLITE; SCHIST; SHALE.

Slaty cleavage is defined by a strong dimensional preferred orientation of clay in a very clay rich, low-grade metamorphic rock, and the resulting rock is a slate. Slaty cleavage tends to be smooth and planar. Coupled with the penetrative nature of slaty cleavage, these characteristics enable slates to split into very thin sheets. This and the durability of the rock are reasons why slates are used in the roofing industry, in the tile industry, and in the construction of pool tables.   [B.A.V.D.P.]

Smithsonite   A naturally occurring rhombohedral zinc carbonate ($ZnCO_3$), with a crystal structure similar to that of calcite ($CaCO_3$). Smithsonite has a hardness on the Mohs scale of $4^1/_2$, has a specific gravity of 4.30–4.45, and exhibits perfect rhombohedral cleavage. *See* CALCITE.

Smithsonite most commonly forms as an alteration product of the mineral sphalerite (ZnS) during supergene enrichment of zinc ores in arid or semiarid environments. It is seldom pure zinc carbonate, commonly containing other divalent metal ions such as manganese (Mn^{2+}), ferrous iron (Fe^{2+}), magnesium (Mg^{2+}), calcium (Ca^{27+}), cadmium (Cd^{2+}), cobalt (Co^{2+}), or lead (Pb^{2+}) substituting for zinc ion. Substitution of other elements for zinc may result in different colors such as blue-green (copper), yellow (cadmium), and pink (cobalt). Smithsonite has been mined as an ore of zinc and has also been used as an ornamental stone. *See* CARBONATE MINERALS.   [J.C.D.]

Smog   The noxious mixture of gases and particles commonly associated with air pollution in urban areas. Harold Antoine des Voeux is credited with coining the term in 1905 to describe the air pollution in British towns. *See* AIR POLLUTION.

The constituents of smog affect the human cardio-respiratory system and pose a health threat. Individuals exposed to smog can experience acute symptoms ranging from eye irritation and shortness of breath to serious asthmatic attacks. Under extreme conditions, smog can cause mortality, especially in the case of the infirm and elderly. Smog can also harm vegetation and likely leads to significant losses in the yields from forests and agricultural crops in affected areas.

The only characteristic of smog that is readily apparent to the unaided observer is the low visibility or haziness that it produces, due to tiny particles suspended within the smog. Observation of the more insidious properties of smog—the concentrations of toxic constituents—requires sensitive analytical instrumentation. Technological advances in these types of instruments, along with the advent of high-speed computers to simulate smog formation, have led to an increasing understanding of smog and its causes.

Smog is an episodic phenomenon because specific meteorological conditions are required for it to accumulate near the ground. These conditions include calm or stagnant winds which limit the horizontal transport of the pollutants from their sources, and a temperature inversion which prevents vertical mixing of the pollutants from the boundary layer into the free troposphere. *See* METEOROLOGY; STRATOSPHERE; TROPOSPHERE.

Smog can be classified into three types: London smog, photochemical smog, and smog from biomass burning.

London smog arises from the by-products of coal burning. These by-products include soot particles and sulfur oxides. During cool damp periods (often in the winter), the soot and sulfur oxides can combine with fog droplets to form a dark acidic fog. As nations switch from coal to cleaner-burning fossil fuels such as oil and gas as well as alternate energy sources such as hydroelectric and nuclear, London smogs cease. *See* ACID RAIN; COAL.

Photochemical smog is a more of a haze than a fog and is produced by chemical reactions in the atmosphere that are triggered by sunlight. A. J. Hagen-Smit first unraveled the chemical mechanism that produces photochemical smog. He irradiated mixtures of volatile organic compounds (VOC) and nitrogen oxides (NO_x) in a reaction chamber. After a few hours, Hagen-Smit observed the appearance of cracks in rubber bands stretched across the chamber. Knowing that ozone (O_3) can harden and crack rubber, Hagen-Smit correctly reasoned that photochemical smog was caused by photochemical reactions involving VOC and NO_x, and that one of the major oxidants produced in this smog was O_3. *See* ATMOSPHERIC CHEMISTRY.

While generally not as dangerous as London smog, photochemical smog contains a number of noxious constituents. Ozone, a strong oxidant that can react with living tissue, is one of these noxious compounds. Another is peroxyacetyl nitrate (PAN), an eye irritant that is produced by reactions between NO_2 and the breakdown products of carbonyls. Particulate matter having diameters of about 10 micrometers or less are of concern because they can penetrate into the human respiratory tract during breathing and have been implicated in a variety of respiratory ailments.

Probably the oldest type of smog known to humankind is produced from the burning of biomass or wood. It combines aspects of both London smog and photochemical smog since the burning of biomass can produce copious quantities of smoke as well as VOC and NO_x. [W.L.Ch.]

Snow   Frozen precipitation resulting from the growth of ice crystals from water vapor in the Earth's atmosphere.

As ice particles fall out in the atmosphere, they melt to raindrops when the air temperature is a few degrees above 32°F (0°C), or accumulate on the ground at colder temperatures. At temperatures above −40°F (−40°C), individual crystals begin growth on icelike aerosols (often clay particles 0.1 micrometer in diameter), or grow from cloud droplets (10 μm in diameter) frozen by similar particles. At lower temperatures, snow crystals grow on cloud droplets frozen by random molecular motion. At temperatures near 25°F (−4°C), crystals sometimes grow on ice fragments produced during soft hail (graupel) growth. Snow crystals often grow in the supersaturated environment provided by a cloud of supercooled droplets; this is known as the Bergeron-Findeisen process for formation of precipitation. When crystals are present in high concentrations (100 particles per liter) they grow in supersaturations lowered by mutual competition for available vapor.

Ice crystals growing under most atmospheric conditions (air pressure down to 0.2 atm or 20 kilopascals and temperatures 32 to −58°F or 0 to −50°C) have a hexagonal crystal structure, consistent with the arrangement of water molecules in the ice lattice, which leads to striking hexagonal shapes during vapor growth. The crystal habit (ratio of growth along and perpendicular to the hexagonal axis) changes dramatically with temperature. Both field and laboratory studies of crystals grown under known or controlled conditions show that the crystals are platelike above 27°F (−3°C) and between 18 and −13°F (−8 and −25°C), and columnlike between 27 and 18°F (−3 and −8°C) and below −13°F (−25°C).

Individual crystals fall in the atmosphere at velocity up to 0.5 m s^{-1} (1.6 ft s^{-1}). As crystals grow, they fall at higher velocity, which leads, in combination with the high moisture availability in a supercooled droplet cloud, to sprouting of the corners to form needle or dendrite skeletal crystals.

Under some conditions crystals aggregate to give snowflakes. This happens for the dendritic crystals that grow near 5°F (-15°C), which readily interlock if they collide with each other, and for all crystals near 32°F (0°C). Snowflakes typically contain several hundred individual crystals.

When snow reaches the ground, changes take place in the crystals. At temperatures near 32°F (0°C) the crystals rapidly lose the delicate structure acquired during growth, sharp edges evaporate, and the crystals take on a rounded shape, some 1–2 mm (0.04–0.08 in.) in diameter. These grains sinter together at their contact points to give snow some structural rigidity. The specific gravity varies from ~0.05 for freshly fallen "powder" snow to ~0.4 for an old snowpack. *See* PRECIPITATION (METEOROLOGY).

[J.Hal.]

Snow gage   An instrument for measuring the amount of water equivalent in snow; more commonly known as a snow sampler. Frequently snow samplers are made of a lightweight seamless aluminum tube consisting of easily coupled lengths. Other snow samplers have been developed from material such as fiber glass and plastic for use in shallow snow, deep dense snow, and so forth.

To obtain a measurement the sampler is pushed vertically through the snow to the ground surface. The sampler, together with its snow core, is withdrawn and weighed. The water equivalent of the snow layer is obtained by subtracting the weight of the sampler from the total. In addition to any error introduced by the scale, there is usually a 6–8% error in the weight of samples taken in this manner.

Automatic devices that permit the remote observation of the water equivalent of the snow have been developed. These devices also permit telemetering the data to a central location, eliminating the need for travel to the snowfields. *See* SNOW; SNOW SURVEYING. [R.T.Be.]

Snow line   A term generally used to refer to the elevation of the lower edge of a snow field. In mountainous areas, it is not truly a line but rather an irregular, commonly patchy border zone, the position of which in any one sector has been determined by the amount of snowfall and ablation. These factors may vary considerably from one part to another. On glacier surfaces the snow line is sometimes referred to as the glacier snow line or névé line (the outer limit of retained winter snow cover on a glacier).

Year-to-year variation in the position of the orographical snow line is great. The mean position over many decades, however, is important as a factor in the development of nivation hollows and protalus ramparts in deglaciated cirque beds. *See* GLACIOLOGY; SNOWFIELD AND NÉVÉ. [M.M.Mi.]

Snow surveying   A technique for providing an inventory of the total amount of snow covering a drainage basin or a given region. Most of the usable water in western North America originates as mountain snowfall that accumulates during the winter and spring and appears several months later as streamflow. Snow surveys were established to provide an estimate of the snow water equivalent (that is, the depth of water produced from melting the snow) for use in predicting the volume of spring runoff. They are also extremely useful for flood forecasting, reservoir regulation, determining hydropower requirements, municipal and irrigation water supplies, agricultural productivity, wildlife survival, and building design, and for assessing transportation and recreation conditions.

Conventional snow surveys are made at designated sites, known as snow courses, at regular intervals each year throughout the winter period. A snow sampler is used to measure the snow depth and water equivalent at a series of points along a snow course. Average depth and water equivalent are calculated for each snow course. Satellite remote sensing and data relay are technologies used to obtain information on snow cover in more remote regions. *See* REMOTE SENSING; SNOW GAGE.   [B.E.Go.]

Snowfield and névé   The term snowfield is usually applied to mountain and glacial regions to refer to an area of snow-covered terrain with definable geographic margins. Where the connotation is very general and without regard to geographical limits, the term snow cover is more appropriate; but glaciology requires more precise terms with respect to snowfield areas.

These terms differentiate according to the physical character and age of the snow cover. Technically, a snowfield can embrace only new or old snow (material from the current accumulation year). Anything older is categorized as firn or ice. The term névé is a descriptive phrase used to refer to consolidated granular snow not yet changed to glacier ice. Because of the need for simple terms, however, it has become acceptable to use the term névé when specifically referring to a geographical area of snowfields on mountain slopes or glaciers (that is, an area covered with perennial "snow" and embracing the entire zone of annually retained accumulation). *See* GLACIOLOGY.   [M.M.Mi.]

Soapstone   A soft talc-rich rock. Soapstones are rocks composed of serpentine, talc, and carbonates (magnesite, dolomite, or calcite). They represent original peridotites which were altered at low temperatures by hydrothermal solutions containing silicon dioxide, carbon dioxide, and other dissolved materials (products of low-grade metasomatism). The whole group of rocks may loosely be referred to as soapstones because of their soft, soapy consistency.   [T.F.W.B.]

Soda niter   A nitrate mineral having chemical composition $NaNO_3$ (sodium nitrate); also known as nitratite, it is by far the most abundant of the nitrate minerals. It sometimes occurs as simple rhombohedral crystals but is usually massive granular. The mineral has a perfect rhombohedral cleavage, conchoidal fracture, and is rather sectile. Its hardness is 1.5 to 2 on Mohs scale, and its specific gravity is 2.266. It has a vitreous luster and is transparent. It is colorless to white, but when tinted by impurities, it is reddish brown, gray, or lemon yellow.

Soda niter is a water-soluble salt found principally as a surface efflorescence in arid regions, or in sheltered places in wetter climates. It is usually associated with niter, nitrocalcite, gypsum, epsomite, mirabilite, and halite. The only large-scale commercial deposits of soda niter in the world occur in a belt roughly 450 mi (725 km) long and 10–50 mi (16–80 km) wide along the eastern slope of the coast ranges in the Atacama, Tarapaca, and Antofagasta Deserts of northern Chile. Chilean nitrate had a monopoly of the world's fertilizer market for many years, but now occupies a subordinate position owing to the development of synthetic processes for nitrogen fixation which permit the production of nitrogen from the air. *See* CALICHE; NITER; NITRATE MINERALS.   [G.Sw.]

Sodalite   A mineral of the feldspathoid group with chemical composition $Na_4Al_3Si_3O_{12}Cl$. Sodalite is usually massive or granular with poor cleavage. The Mohs hardness is 5.5–6.0, and the density 2.3. The luster is vitreous, and the color is usually blue but may also be white, gray, or green. Notable occurrences are at Mount

Vesuvius; Bancroft, Ontario; and on the Kola Peninsula of Russia. *See* FELDSPATHOID; SILICATE MINERALS.   [L.Gr.]

Soil   Finely divided rock-derived material containing an admixture of organic matter and capable of supporting vegetation. Soils are independent natural bodies, each with a unique morphology resulting from a particular combination of climate, living plants and animals, parent rock materials, relief, the ground waters, and age. Soils support plants, occupy large portions of the Earth's surface, and have shape, area, breadth, width, and depth. Soil, as used here, differs in meaning from the term as used by engineers, where the meaning is unconsolidated rock material. *See* PEDOLOGY.

Origin and classification.   Soil covers most of the land surface as a continuum. Each soil grades into the rock material below and into other soils at its margins, where changes occur in relief, ground water, vegetation, kinds of rock, or other factors which influence the development of soils. Soils have horizons, or layers, more or less parallel to the surface and differing from those above and below in one or more properties, such as color, texture, structure, consistency, porosity, and reaction (see illustration). The succession of horizons is called the soil profile.

Soil formation proceeds in stages, but these stages may grade indistinctly from one into another. The first stage is the accumulation of unconsolidated rock fragments, the

Photograph of a soil profile showing horizons. The dark crescent-shaped spots at the soil surface are the result of plowing. The dark horizon is the principal horizon of accumulation of organic matter that has been washed down from the surface. The thin wavy lines were formed in the same manner. 1 in. = 2.5 cm.

Soil orders

Order	Formative element in name	General nature of soils
Alfisols	alf	Gray to brown surface horizons, medium to high base supply, with horizons of clay accumulation; usually moist, but may be dry during summer
Aridisols	id	Pedogenic horizons, low in organic matter, and usually dry
Entisols	ent	Pedogenic horizons lacking
Histosols	ist	Organic (peats and mucks)
Inceptisols	ept	Usually moist, with pedogenic horizons of alteration of parent materials but not of illuviation
Mollisols	oil	Nearly black organic-rich surface horizons and high base supply
Oxisols	ox	Residual accumulations of inactive clays, free oxides, kaolin, and quartz; mostly tropical
Spodosols	od	Accumulations of amorphous materials in subsurface horizons
Ultisols	ult	Usually moist, with horizons of clay accumulation and a low supply of bases
Vertisols	ert	High content of swelling clays and wide deep cracks during some season

parent material. Parent material may be accumulated by deposition of rock fragments moved by glaciers, wind, gravity, or water, or it may accumulate more or less in place from physical and chemical weathering of hard rocks. The second stage is the formation of horizons. This stage may follow or go on simultaneously with the accumulation of parent material. Soil horizons are a result of dominance of one or more processes over others, producing a layer which differs from the layers above and below. *See* WEATHERING PROCESSES.

Systems of soil classification are influenced by concepts prevalent at the time a system is developed. The earliest classifications were based on relative suitability for different crops, such as rice soils, wheat soils, and vineyard soils. Over the years, many systems of classification have been attempted but none has been found markedly superior. Two bases for classification have been tried. One basis has been the presumed genesis of the soil; climate and native vegetation were given major emphasis. The other basis has been the observable or measurable properties of the soil.

The Soil Survey staff of the U.S. Department of Agriculture and the land-grant colleges adopted the current classification scheme in 1965. This system differs from earlier systems in that it may be applied to either cultivated or virgin soils. Previous systems have been based on virgin profiles, and cultivated soils were classified on the presumed characteristics or genesis of the virgin soils. The new system has six categories, based on both physical and chemical properties. These categories are the order, suborder, great group, subgroup, family, and series, in decreasing rank. The orders and the general nature of the included soils are given in the table. The suborder narrows the ranges in soil moisture and temperature regimes, kinds of horizons, and composition, according to which of these is most important. The taxa (classes) in the great group category group soils that have the same kinds of horizons in the same sequence and have similar moisture and temperature regimes. The great groups are divided into subgroups that show the central properties of the great group, intergrade subgroups that show properties of more than one great group, and other subgroups for soils with atypical properties that are not characteristic of any great group.

The families are defined largely on the basis of physical and mineralogic properties of importance to plant growth. The soil series is a group of soils having horizons similar in differentiating characteristics and arrangement in the soil profile, except for texture of the surface portion, and developed in a particular type of parent material.

Surveys. Soil surveys include those researches necessary (1) to determine the important characteristics of soils, (2) to classify them into defined series and other units, (3) to establish and map the boundaries between kinds of soil, and (4) to correlate and predict adaptability of soils to various crops, grasses, and trees; behavior and productivity of soils under different management systems; and yields of adapted crops on soils under defined sets of management practices. Although the primary purpose of soil surveys has been to aid in agricultural interpretations, many other purposes have become important, ranging from suburban planning, rural zoning, and highway location, to tax assessment and location of pipelines and radio transmitters. This has happened because the soil properties important to the growth of plants are also important to its engineering uses.

Two kinds of soil maps are made. The common map is a detailed soil map, on which soil boundaries are plotted from direct observations throughout the surveyed area. Reconnaissance soil maps are made by plotting soil boundaries from observations made at intervals. The maps show soil and other differences that are of significance for present or foreseeable uses. [G.D.S.]

Physical properties. Physical properties of soil have critical importance to growth of plants and to the stability of cultural structures such as roads and buildings. Such properties commonly are considered to be: size and size distribution of primary particles and of secondary particles, or aggregates, and the consequent size, distribution, quantity, and continuity of pores; the relative stability of the soil matrix against disruptive forces, both natural and cultural; color and textural properties, which affect absorption and radiation of energy; and the conductivity of the soil for water, gases, and heat. These usually would be considered as fixed properties of the soil matrix, but actually some are not fixed because of influence of water content. The additional property, water content—and its inverse, gas content—ordinarily is transient and is not thought of as a property in the same way as the others. However, water is an important constituent, despite its transient nature, and the degree to which it occupies the pore space generally dominates the dynamic properties of soil. Additionally, the properties listed above suggest a macroscopic homogeneity for soil which it may not necessarily have. In a broad sense, a soil may consist of layers or horizons of roughly homogeneous soil materials of various types that impart dynamic properties which are highly dependent upon the nature of the layering. Thus, a discussion of dynamic soil properties must include a description of the intrinsic properties of small increments as well as properties it imparts to the system.

From a physical point of view it is primarily the dynamic properties of soil which affect plant growth and the strength of soil beneath roads and buildings. While these depend upon the chemical and mineralogical properties of particles, particle coatings, and other factors discussed above, water content usually is the dominant factor. Water content depends upon flow and retention properties, so that the relationship between water content and retentive forces associated with the matrix becomes a key physical property of a soil. *See* EROSION; GROUND-WATER HYDROLOGY. [W.H.G.]

Soil chemistry   The study of the composition and chemical properties of soil. Soil chemistry involves the detailed investigation of the nature of the solid matter from which soil is constituted and of the chemical processes that occur as a result of the

Table 1. Average percentages of total carbon, total nitrogen, and organic phosphorus in selected soils

Soil	% C	% N	% P
Sand	2.5	.23	.04
Fine sandy loam	3.3	.23	.06
Medium loam	2.3	.22	.05
Clay loam, well drained	4.6	.36	.10
Clay loam, poorly drained	8.0	.43	.05
Peat	46.1	1.32	.03

action of hydrological, geological, and biological agents on the solid matter. Because of the broad diversity among soil components and the complexity of soil chemical processes, the application of a wide variety of concepts and methods employed in the chemistry of aqueous solutions, of amorphous and crystalline solids, and of solid surfaces is required.

Elemental composition. The elemental composition of soil varies over a wide range, permitting only a few general statements to be made. Those soils that contain less than 12–20% organic carbon are termed mineral. All other soils are termed organic. Carbon, oxygen, hydrogen, nitrogen, phosphorus, and sulfur are the most important constituents of organic soils and of soil organic matter in general. Carbon, oxygen, and hydrogen are most abundant; the content of nitrogen is often about one-tenth that of carbon, while the content of phosphorus or sulfur is usually less than one-fifth that of nitrogen (Table 1).

Besides oxygen, the most abundant elements found in mineral soils are silicon, aluminum, and iron. The distribution of chemical elements will vary considerably from soil to soil and, in general, will be different in a specific soil from the distribution of elements in the crustal rocks of the Earth. The most important micro or trace elements in soil are boron, copper, manganese, molybdenum, and zinc, since these elements are essential in the nutrition of green plants. Also important are cobalt, selenium, cadmium, and nickel. The average distribution of trace elements in soil is not greatly different from that in crustal rocks (Table 2).

The elemental composition of soil varies with depth below the surface because of pedochemical weathering. The principal processes of this type that result in the

Table 2. Average amounts of trace elements commonly found in soils and crustal rocks

Trace element	Soil, ppm*	Crustal rocks, ppm
As	6	1.8
B	10	10
Cd	.06	.2
Co	8	25
Cr	100	100
Cu	20	55
Mo	2	1.5
Ni	40	75
Pb	10	13
Se	.2	.05
V	100	135
Zn	50	70

*ppm = parts per million.

removal of chemical elements from a given soil horizon are: (1) soluviation (ordinary dissolution in water), (2) cheluviation (complexation by organic or inorganic ligands), (3) reduction, and (4) suspension. The principal effect of these four processes is the appearance of alluvial horizons in which compounds such as aluminum and iron oxides, aluminosilicates, or calcium carbonate have been precipitated from solution or deposited from suspension. *See* WEATHERING PROCESSES.

Minerals. The minerals in soils are the products of physical, geochemical, and pedochemical weathering. Soil minerals may be either amorphous or crystalline. They may be classified further, approximately, as primary or secondary minerals, depending on whether they are inherited from parent rock or are produced by chemical weathering, respectively.

The bulk of the primary minerals that occur in soil are found in the silicate minerals. Chemical weathering of the silicate minerals is responsible for producing the most important secondary minerals in soil. These are found in the clay fraction and include aluminum and iron hydrous oxides (usually in the form of coatings on other minerals), carbonates, and aluminosilicates. *See* CLAY MINERALS; SILICATE MINERALS.

Ion exchange. A portion of the chemical elements in soil is in the form of cations that are not components of inorganic salts but that can be replaced reversibly by the cations of leaching salt solutions or acids. These cations are said to be exchangeable, and their total quantity is termed the cation exchange capacity (CEC) of the soil. The CEC of a soil generally will vary directly with the amounts of clay and organic matter present and with the distribution of clay minerals.

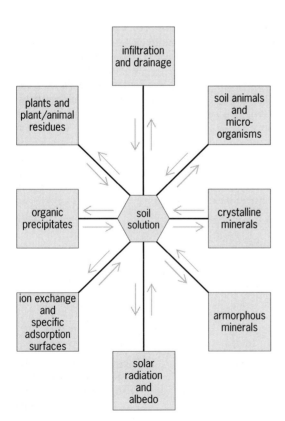

Factors influencing the chemistry of soil solution. (*Modified from J. F. Hodgson, Chemistry of the micronutrients in soil, Adv. Agron., 15:141, 1963*)

The stoichiometric exchange of the anions in soil for those in a leaching salt solution is a phenomenon of relatively small importance in the general scheme of anion reactions with soils. Under acid conditions (pH < 5) the exposed hydroxyl groups at the edges of the structural sheets or on the surfaces of clay-sized particles become protonated and thereby acquire a positive charge. The degree of protonation is a sensitive function of pH, the ionic strength of the leaching solution, and the nature of the clay-sized particle.

Soil solution. The solution in the pore space of soil acquires its chemical properties through time-varying inputs and outputs of matter and energy that are mediated by the several parts of the hydrologic cycle and by processes originating in the biosphere (see illustration). The soil solution thus is a dynamic and open natural water system whose composition reflects the many reactions that can occur simultaneously between an aqueous solution and an assembly of mineral and organic solid phases that varies with both time and space. *See* SOIL.   [G.Sp.]

Soil ecology   The study of the interactions among soil organisms, and between biotic and abiotic aspects of the soil environment. Soil is made up of a multitude of physical, chemical, and biological entities, with many interactions occurring among them. Soil is a variable mixture of broken and weathered minerals and decaying organic matter. Together with the proper amounts of air and water, it supplies, in part, sustenance for plants as well as mechanical support. *See* SOIL.

Abiotic and biotic factors lead to certain chemical changes in the top few decimeters (8–10 in.) of soil. The work of the soil ecologist is made easier by the fact that the surface 10–15 cm (4–6 in.) of the A horizon has the majority of plant roots, microorganisms, and fauna. A majority of the biological-chemical activities occur in this surface layer.

The biological aspects of soil range from major organic inputs, decomposition by primary decomposers (bacteria, fungi, and actinomycetes), and interactions between microorganisms and fauna (secondary decomposers) which feed on them. The detritus decomposition pathway occurs on or within the soil after plant materials (litter, roots, sloughed cells, and soluble compounds) become available through death or senescence. Plant products are used by microorganisms (primary decomposers). These are eaten by the fauna which thus affect flows of nutrients, particularly nitrogen, phosphorus, and sulfur. The immobilization of nutrients into plants or microorganisms and their subsequent mineralization are critical pathways. The labile inorganic pool is the principal one that permits subsequent microorganism and plant existence. Scarcity of some nutrient often limits production. Most importantly, it is the rates of flux into and out of these labile inorganic pools which enable ecosystems to successfully function. *See* ECOLOGY; ECOSYSTEM; SOIL; SYSTEMS ECOLOGY.   [D.C.C.]

Solenoid (meteorology)   In meteorological usage, solenoids are hypothetical tubes formed in space by the intersection of a set of surfaces of constant pressure (isobaric surfaces) and a set of surfaces of constant specific volume of air (isosteric surfaces) or density (isopycnic surfaces). The isobaric and isosteric surfaces are such that the values of pressure and specific volume, respectively, change by one unit from one surface to the next. The state of the atmosphere is said to be barotropic when there are no solenoids, that is, when isobaric and isosteric surfaces coincide. The number of solenoids cut by any plane surface element of unit area is a measure of the torque exerted by the pressure gradient force, tending to accelerate the circulation of air around the boundary of the area. *See* BAROCLINIC FIELD; BAROTROPIC FIELD; ISOPYCNIC.   [F.S.; H.B.B.]

Solid solution Compositional variation of a crystalline substance due to substitution or omission of various atomic constituents within a crystal structure. Solid solutions can be classified as substitutional, interstitial, or omissional. They may also be categorized by the nature of their thermodynamic properties, such as enthalpy, entropy, and free energy (for example ideal, nonideal, and regular solid solutions).

The concept of solid solution can be understood by considering a specific mineral group such as olivine. Although olivine-group minerals may exhibit a range of compositions they all have a similar crystal structure. Thus all of the minerals in the olivine group are isostructural (having a similar crystal structure). Yet within this structural framework, compositions vary considerably. Such chemical variation can be described by defining the nature and extent of the atomic substitutions involved or by describing intermediate compositions in terms of limiting end-member compositions. In addition to substitution within the crystal structure, compositional variation may take place by interstitial substitution or omission solid solution. Interstitial solid solution occurs when ions or atoms occupy a position in a crystal structure that is usually vacant. Omission solid solution occurs when an atom or ion is missing from a specific crystallographic position. *See* OLIVINE.

Solid solution is widespread among minerals. In fact, very few naturally occurring minerals exist as pure end-member substances but exhibit trace to extensive solid solution. The extent of solid solution depends upon the relative sizes of the atoms or ions involved, the charges of the ions, the coexistence and composition of other minerals or liquid (for example, magma), and the temperature and pressure conditions of formation (with temperature having a more pronounced effect).   [J.C.D.]

South America The southernmost of the New World or Western Hemisphere continents, with three-fourths of it lying within the tropics. South America is approximately 4500 mi (7200 km) long and at its greatest width 3000 mi (4800 km). Its area is estimated to be about 7,000,000 mi^2 (18,000,000 km^2). South America has many unique physical features, such as the Earth's longest north-south mountain range (the Andes), highest waterfall (Angel Falls), highest navigable fresh-water lake (Lake Titicaca), and largest expanse of tropical rainforest (Amazonia). The western side of the continent has a deep subduction trench offshore, whereas the eastern continental shelf is more gently sloping and relatively shallow.

Deciduous forest are found in areas where there is seasonal drought and the trees lose their leaves in order to slow transpiration. The lower slopes of the Andes, central Venezuela, and central Brazil are areas where these formations are found. Conifer forests occur in the higher elevations of the Andes and the higher latitudes of Chile and Argentina.

Tropical savannas occupy an extensive range in northern South America through southeastern Venezuela and eastern Colombia. Temperate savannas are found in Paraguay, Uruguay, the Pampas of Argentina, and to the south, Patagonia. Savannas are composed of a combination of grass and tree species. The climate in these areas is often quite hot with high rates of evapotranspiration and a pronounced dry season. Most of the plants and animals of these zones are drought-adapted and fire-adapted. Tall grasses up to 12 ft (3.5 m) are common as are thorny trees of the Acacia (Fabaceae) family. Many birds and mammals are found in these zones, including anteater, armadillo, capybara (the largest rodent on Earth), deer, jaguar, and numerous species of venomous snake, including rattlesnake and bushmaster (*mapanare*).

South America is unique in having a west-coast desert (the Atacama) that extends almost to the Equator, probably receiving less rain than any on Earth, and an east coast desert located poleward from latitude 40°S (the Patagonian).

In Bolivia and Peru the zone from 10,000 to 13,000 ft (3000 to 3900 m), though occasionally to 15,000–16,000 ft (4500 to 4800 m), is known as the *puna*. Here the hot days contrast sharply with the cold nights. Above the *puna*, from timberline to snowline, is the *paramo*, a region of broadleaf herbs and grasses found in the highest elevations of Venezuela, Colombia, and Ecuador. Many of the plant species in these environments are similar to those found at lower elevations; however, they grow closer to the ground in order to conserve heat and moisture.   [D.A.Sa.; C.L.W.]

South Pole   That end of the Earth's rotational axis opposite the North Pole. It is the southernmost point on the Earth and one of the two points through which all meridians pass (the North Pole being the other point). This is the geographic pole and is not the same as the south magnetic pole. The South Pole lies inland from the Ross Sea, within the land mass of Antarctica, at an elevation of about 9200 ft (2800 m).

There is no natural way to determine local Sun time because there is no noon position of the Sun, and shadows point north at all times, there being no other direction from the South Pole. *See* MATHEMATICAL GEOGRAPHY; NORTH POLE.   [V.H.E.]

Southeast Asian waters   All the seas between Asia and Australia and the Pacific and the Indian oceans. They form a geographical and oceanographical unit because of their special structure and position, and make up an area of 3,450,000 mi^2 (8,940,000 km^2), or about 2.5% of the surface of all oceans.

The surface circulation is completely reversed twice a year by the changing monsoon winds. The subsurface circulation carries chiefly the outrunners of the intermediate waters of the Pacific Ocean into these seas. The tides are mostly of the mixed type. Diurnal tides are found in the Java Sea, in the Gulf of Tonkin, and in the Gulf of Thailand. Semidiurnal tides with high amplitudes occur in the Malacca Straits. *See* INDIAN OCEAN; PACIFIC OCEAN.   [K.W.]

The Southeast Asian seas are characterized by the presence of numerous major plate boundaries. In Southeast Asia the plate boundaries are identified, respectively, by young, small ocean basins with their spreading systems and associated high heat flow; deep-sea trenches and their associated earthquake zones and volcanic chains; and major strike-slip faults such as the Philippine Fault (similar to the San Andreas Fault in California). The Southeast Asian seas are thus composed of a mosaic of about 10 small ocean basins whose boundaries are defined mainly by trenches and volcanic arcs. The dimensions of these basins are much smaller than the basins of the major oceans. The major topographic features of the region are believed to represent the surface expression of plate interactions, the scars left behind on the sea floor. *See* PLATE TECTONICS.   [D.E.H.]

Space   Physically, space is that property of the universe associated with extension in three mutually perpendicular directions. Space, from a newtonian point of view, may contain matter, but space exists apart from matter. Through usage, the term space has come to mean generally outer space or the region beyond Earth. Geophysically, space is that portion of the universe beyond the immediate influence of Earth and its atmosphere. From the point of view of flight, space is that region in which a vehicle cannot obtain oxygen for its engines or rely upon an atmospheric gas for support (either by buoyancy or by aerodynamic effects). Astronomically, space is a part of the space-time continuum by which all events are uniquely located. Relativistically, the space and time variables of uniformly moving (inertial) reference systems are connected by the Lorentz transformations. Gravitationally, one characteristic of space is that all bodies undergo the same acceleration in a gravitational field and therefore that inertial forces are equivalent to gravitational forces. Perceptually, space is sensed indirectly by the objects and events within it. Thus, a survey of space is more a survey of its contents.

[S.F.S.]

Sphalerite   A mineral, β-ZnS, also called blende. It is the low-temperature form and more common polymorph of ZnS, Pure β-ZnS on heating inverts to wurtzite, α-ZnS, at 1868°F (1020°C).

The mineral is most commonly in coarse to fine, granular, cleavable masses. The luster is resinous to submetallic; the color is white when pure, but is commonly yellow, brown, or black, darkening with increased percentage of iron. There is perfect dodecahedral cleavage; the hardness is $3^1/_2$ on Mohs scale; specific gravity is 4.1 for pure sphalerite.

Sphalerite is a common and widely distributed mineral. It occurs both in veins and in replacement deposits in limestones. As the chief ore mineral of zinc, sphalerite is mined on every continent. The United States is the largest producer, followed by Canada, Mexico, Russia, Australia, Peru, the Congo River area, and Poland. *See* WURTZITE.

[C.S.Hu.; P.B.M.]

Spilite   An aphanitic (microscopically crystalline) to very-fine-grained igneous rock, with more or less altered appearance, resembling basalt but composed of albite or oligoclase, chlorite, epidote, calcite, and actinolite.

In spite of the highly sodic plagioclase, spilites are generally classed with basalts because of the low silica content (about 50%). They also retain many textural and structural features characteristic of basalt.

Spilites are found most frequently as lava flows and more rarely as small dikes and sills. Spilitic lavas typically show pillow structure, in which the rock appears composed of closely packed, elongated, pillow-shaped masses up to a few feet across. Pillows are typical of subaqueous lava flows. Vesicles, commonly filled with various minerals, may give the rock an amygdaloidal structure. *See* BASALT; IGNEOUS ROCKS.   [W.I.R.]

Spinel   Any of a family of important AB_2O_4 oxide minerals, where A and B represent cations. Spinel minerals are widely distributed in the earth, in meteorites, and in rocks from the Moon. While the ideal spinel formula is $MgAl_2O_4$, some 30 elements, with valences from 1 to 6, are known to substitute in the A or B cation sites, resulting in well over 150 synthetic compounds having the spinel crystal structure. The term spinel is derived from *spina* (Latin, thorn) in reference to its pointed octahedral, crystal habit, and also to its dendritic snowflake form in rapidly chilled high-temperature slags and lavas.

The named spinel minerals that have so far been recorded in nature are oxides that occur as a matrix of A^{2+} versus B^{3+} cations. Three spinel series evolve from the classification: spinel, magnetite, and chromite. In addition to these spinels, there are other oxide, sulfur (thiospinels), silicate, and rare selenide-bearing spinels, all of which have relatively simple end-member compositions. *See* MAGNETITE.

Aluminous spinels are highly refractory, vary from translucent to transparent, and vary from colorless to green, blue, brown, and black. All other oxide and thiospinels are opaque with metallic lusters. Mohs hardness varies from about 4.5 (linnaeite) to about 8 (spinel). Density is approximately 5 g/cm^3 (3 $oz/in.^3$). Magnetite and maghemite are ferrimagnetic, with high saturation magnetizations and Curie temperatures of 580°C (1076°F) and 675°C (1247°F), respectively.

Spinels are widely employed to deduce the evolutionary history of rocks because compositions are extremely sensitive to environmental conditions of formation. Emery, the abrasive, is magnetite + corundum. Chromite is the chief source of chromium. Iron, titanium and vanadium are derived from magnetite, and zinc is extracted from franklinite. Spinel also occurs as a semiprecious gem, and it is widely employed as a mechanically robust ceramic. The compass used in ancient times was a mixture of magnetite and maghemite; the entire fields of rock magnetism and paleomagnetism as well as the recorded history of the Earth's magnetic field hinge on the magnetic properties of these inverse spinels. Maghemite is widely employed in magnetic recording tapes and magnetic colloids. *See* MINERALOGY; PALEOMAGNETISM; PETROLOGY.   [S.E.H.]

Spodumene   The name given to the monoclinic lithium pyroxene $LiAl(SiO_3)_2$. Spodumene commonly occurs as white to yellowish prismatic crystals, often with a "woody" appearance.

Spodumene is usually found as a constituent in certain granitic pegmatites. The emerald-green variety, hiddenite, and a lilac variety, kunzite, are used as precious stones. Spodumene from pegmatites is used as an ore for lithium. *See* PYROXENE.

[G.W.DeV.]

Spring (hydrology)   A place where groundwater discharges upon the land surface because the natural flow of groundwater to the place exceeds the flow from it. Springs are ephemeral, discharging intermittently, or permanent, discharging constantly. Springs are usually at mean annual air temperatures. The less the discharge, the more the temperature reflects seasonal temperatures. Spring water usually originates as rain or snow (meteoric water).

Hot-spring water may differ in composition from meteoric water through exchange between the water and rocks. Common minerals consist of component oxides. Oxygen of minerals has more ^{18}O than meteoric water. Upon exchange, the water is enriched in ^{18}O. Most minerals contain little deuterium, so that slight deuterium changes occur. Some hot-spring waters are acid from the oxidation of hydrogen sulfide to sulfate.

Mineral spring waters have high concentrations of solutes and wide ranges in chemistry and temperatures; hot mineral springs may be classified as hot springs as well as mineral springs. Most mineral springs are high either in sodium chloride or sodium bicarbonate (soda springs) or both; other compositions are found, such as a high percentage of calcium sulfate from the solution of gypsum.

The chemical compositions of spring waters are seldom in chemical equilibrium with the air. Groundwaters whose recharge is through grasslands may contain a thousand times as much CO_2 as would be in equilibrium with air, and those whose recharge is through forests may contain a hundred times as much as would be in equilibrium with air. Sulfate in groundwater may be reduced in the presence of organic matter to H_2S, giving some springs the odor of rotten eggs. *See* GEYSER.   [I.B.]

Squall   A strong wind with sudden onset and more gradual decline, lasting for several minutes. Wind speeds in squalls commonly reach 30–60 mi/h (13–27 m/s), with a succession of brief gusts of 80–100 mi/h (36–45 m/s) in the more violent squalls. Squalls may be local in nature, as with isolated thunderstorms, or may occur over a wide area in the vicinity of a well-developed cyclone, where the squalls locally reinforce already strong winds. Because of their sudden violent onset, and the heavy rain, snow, or hail showers which often accompany them, squalls cause heavy damage to structures and crops and present severe hazards to transportation.

The most common type of squall is the thundersquall or rain squall associated with heavy convective clouds, frequently of the cumulonimbus type. Such a squall usually sets in shortly before onset of the thunderstorm rain, blowing outward from the storm and generally lasting for only a short time. It is formed when cold air, descending in the core of the thunderstorm rain area, reaches the Earth's surface and spreads out. Particularly in desert areas, the thunderstorm rain may largely or wholly evaporate before reaching the ground, and the squall may be dry, often associated with dust storms. *See* DUST STORM; SQUALL LINE; THUNDERSTORM.

Squalls of a different type result from cold air drainage down steep slopes. The force of the squall is derived from gravity and depends on the descending air which is colder and more dense than the air it replaces. So-called fall winds of this kind are common on mountainous coasts of high latitudes, where cold air forms on elevated plateaus and drains down fiords or deep valleys. Violent squalls also characterize the warm foehn winds of the Alps and the similar chinook winds on the eastern slopes of the Rocky Mountains. *See* CHINOOK; WIND.   [C.W.N.; E.Ke.]

Squall line   A line of thunderstorms, near whose advancing edge squalls occur along an extensive front. The thundery region, 12–30 mi (20–50 km) wide and up to 1200 mi (2000 km) long, moves at a typical speed of 30 knots (15 m/s) for 6–12 h or more and sweeps a broad area. In the United States, severe squall lines are most common in spring and early summer when northward incursions of maritime tropical air east of the Rockies interact with polar front cyclones. Ranking next to hurricanes in casualities and damage caused, squall lines also supply most of the beneficial rainfall in some regions. *See* FRONT; SQUALL; THUNDERSTORM.   [C.W.N.]

Stalactites and stalagmites   Stalactites, stalagmites, dripstone, and flow-stone are travertine deposits in limestone caverns, formed by the evaporation of waters bearing calcium carbonate. Stalactites grow down from the roofs of caves and tend to be long and thin, with hollow cores. The water moves down the core and precipitates at the bottom, slowly extending the length while keeping the core open for more water to move down.

Stalagmites grow from the floor up and are commonly found beneath stalactites; they are formed from the evaporation of the same drip of water that forms the stalactite. Stalagmites are thicker and shorter than stalactites and have no central hollow core. *See* CAVE; LIMESTONE.   [R.Si.]

Staurolite   A nesosilicate mineral occurring in metamorphic rocks. The chemical formula of staurolite may be written as $A_4B_4C_{18}D_4T_8O_{40}X_8$, where A = Fe^{2+}, Mg; B = Fe^{2+}, Zn, Co, Mg, Li, Al, Fe^{3+}, Mn^{2+}; C = Al, Fe^{3+}, Cr, V, Ti; D = Al, Mg; T = Si, Al; X = OH, F, O. Staurolite occurs as well-formed, often-twinned, prismatic crystals. It is brown-black, reddish brown, or light brown in color and has a vitreous to dull luster. Light color and dull luster can result from abundant quartz inclusions. There is no cleavage, specific gravity is 3.65–3.75, and hardness is 7–7.5 (Mohs scale).

Typical minerals occurring with staurolite are quartz, micas (muscovite and biotite), garnet (almandine), tourmaline, and kyanite, sillimanite, or andalusite. Staurolite is common where pelitic schists reach medium-grade metamorphism. Examples are the Swiss and Italian Alps (notable at Saint Gotthard, Switzerland), and all the New England states, Virginia, the Carolinas, Georgia, New Mexico, Nevada, and Idaho. *See* METAMORPHISM; SILICATE MINERALS.   [F.C.Ha.]

Stibnite   A mineral with composition Sb_2S_3 (antimony trisulfide), the chief ore of antimony. It crystallizes in slender, prismatic, vertically striated crystals which may be curved or bent. It is often in bladed, granular, or massive aggregates. The hardness is 2 on Mohs scale and the specific gravity 4.5–4.6. The luster is metallic and the color lead-gray to black. It is one of few minerals that fuses easily in the match flame (525°C or 977°F).

Stibnite has been found in various mining districts in Germany, Romania, France, Bolivia, Peru, and Mexico. In the United States the Yellow Pine mine at Stibnite, Idaho, is the largest producer. Other deposits are in Nevada and California. The finest crystals have come from the island of Shikoku, Japan.   [C.S.Hu.]

Stilbite   A mineral belonging to the zeolite family of silicates. It crystallizes in sheaflike aggregates of thin tabular crystals. There is perfect cleavage parallel to the side pinacoid, and here the mineral has a pearly luster; elsewhere the luster is vitreous. The color is usually white but may be brown, red, or yellow. Hardness is $3^1/_2$–4 on Mohs scale; specific gravity is 2.1–2.2. See ZEOLITE.

Stilbite is found in Iceland, India, Scotland, Nova Scotia, and in the United States at Bergen Hill, New Jersey, and the Lake Superior copper district in Michigan.   [C.S.Hu.]

Stishovite   Naturally occurring stishovite, SiO_2, is a mineral formed under very high pressure with the silicon atom in sixfold, or octahedral, coordination instead of the usual fourfold, or tetrahedral, coordination. The presence of stishovite indicates formation pressures in excess of 10^6 lb/in.^{2} (7.5 gigapascals). The possibility of the existence of stishovite at great depths strongly influences the interpretations of geophysicists and solid-state physicists regarding the phase transitions of mineral matter, as well as the interpretation of seismic data in the study of such regions of the interior of the Earth. See SILICA MINERALS.

Stishovite occurs in submicrometer size in very small amounts (less than 1% of the rock) in samples of Coconino sandstones from the Meteor Crater of Arizona, which contains up to 10% of coesite, the other high-pressure polymorph of silica. Because of its extremely fine grain size and because of the sparsity of this mineral in the rock, positive identification of the mineral is possible only by the x-ray diffraction method after chemical concentration. See COESITE.

The specific gravity of stishovite, calculated from the x-ray data, is 4.28, compared with the value of 4.35 for the synthetic material. It is 46% denser than coesite and much denser than other modifications of silica. See RUTILE.   [E.C.T.C.]

Stone and stone products   The term stone is applied to rock that is cut, shaped, broken, crushed, and otherwise physically modified for commercial use. The two main divisions are dimension stone and crushed stone. Other descriptive terms may be used, for example, building stone, roofing stone, or precious stone. See GEM; ROCK.

The term dimension stone is applied to blocks that are cut and milled to specified sizes, shapes, and surface finishes. The principal uses are for building and ornamental applications. Granites, limestones, sandstones, and marbles are widely used; basalts, diabases, and other dark igneous rocks are used less extensively. Soapstone is used to some extent. Rock suitable for use as dimension stone must be obtainable in large, sound blocks, free from incipient cracks, seams, and blemishes, and must be without mineral grains that might cause stains as a result of weathering. It must have an attractive color, and generally a uniform texture.

Slate differs from other dimension stone because it can be split into thin sheets of any desired thickness. Commercial slate must be uniform in quality and texture and reasonably free from knots, streaks, or other imperfections, and have good splitting properties. Roofing slates are important products of most slate quarries. However, the roofing-slate industry has declined considerably because of competition from other types of roofing. Slate is also used for milled products such as blackboards, electrical panels, window and door sills and caps, baseboards, stair treads, and floor tile. *See* SLATE.

Nearly all the principal types of stone—granite, diabase, basalt, limestone, dolomite, sandstone, and marble—may be used as sources of commercial crushed stone; limestone is by far the most important. Crushed stone is made from sound, hard stone, free from surface alteration by weathering. Stone that breaks in chunky, more or less cubical fragments is preferred. Commercial stone should be free from certain deleterious impurities, such as opalescent quartz, and free from clay or silt. Crushed stone is used principally as concrete aggregate, as road stone, or as railway ballast. Other uses for limestone are as a fluxing material to remove impurities from ores smelted in metallurgical furnaces, in the manufacture of alkali chemicals, calcium carbide, glass, paper, paint, and sugar, and for filter beds and for making mineral wool.   [R.L.B.]

Storm   An atmospheric disturbance involving perturbations of the prevailing pressure and wind fields on scales ranging from tornadoes (0.6 mi or 1 km across) to extratropical cyclones (1.2–1900 mi or 2–3000 km across); also, the associated weather (rain storm, blizzard, and the like). Storms influence human activity in such matters as agriculture, transportation, building construction, water impoundment and flood control, and the generation, transmission, and consumption of electric energy. *See* WIND.

The form assumed by a storm depends on the nature of its environment, especially the large-scale flow patterns and the horizontal and vertical variation of temperature; thus the storms most characteristic of a given region vary according to latitude, physiographic features, and season. Extratropical cyclones and anticyclones are the chief disturbances over roughly half the Earth's surface. Their circulations control the embedded smaller-scale storms. Large-scale disturbances of the tropics differ fundamentally from those of extratropical latitudes. *See* HURRICANE; SQUALL; THUNDERSTORM; TORNADO.

Cyclones form mainly in close proximity to the jet stream, that is, in strongly baroclinic regions where there is a large increase of wind with height. Weather patterns in cyclones are highly variable, depending on moisture content and thermodynamic stability of air masses drawn into their circulations. Warm and occluded fronts, east of and extending into the cyclone center, are regions of gradual upgliding motions, with widespread cloud and precipitation but usually no pronounced concentration of stormy conditions. Extensive cloudiness also is often present in the warm sector. Passage of the cold front is marked by a sudden wind shift, often with the onset of gusty conditions, with a pronounced tendency for clearing because of general subsidence behind the front. Showers may be present in the cold air if it is moist and unstable because of heating from the surface. Thunderstorms, with accompanying squalls and heavy rain, are often set off by sudden lifting of warm, moist air at or near the cold front, and these frequently move eastward into the warm sector. *See* CYCLONE; JET STREAM; WEATHER.

Extratropical cyclones alternate with high-pressure systems or anticyclones, whose circulation is generally opposite to that of the cyclone. The circulations of highs are not so intense as in well-developed cyclones, and winds are weak near their centers. In low levels the air spirals outward from a high; descent in upper levels results in warming and drying aloft. Anticyclones fall into two main categories, the warm "subtropical" and the cold "polar" highs.

Between the scales of ordinary air turbulence and of cyclones, there exist a variety of circulations over a middle-scale or mesoscale range, loosely defined as from about one-half up to a few hundred miles. Alternatively, these are sometimes referred to as subsynoptic-scale disturbances because their dimensions are so small that they elude adequate description by the ordinary synoptic network of surface weather stations. Thus their detection often depends upon observation by indirect sensing systems. *See* METEOROLOGICAL SATELLITES; RADAR METEOROLOGY; STORM DETECTION. [C.W.N.]

Storm detection Identifying storm formation, monitoring subsequent storm evolution, and assessing the potential for destruction of life and property through application of various methods and techniques. Doppler radars, satellite-borne instruments, lightning detection networks, and surface observing networks are used to detect the genesis of storms, to diagnose their nature, and to issue warnings when a threat to life and property exists. *See* STORM.

Radar surveillance. Radars emit pulses of electromagnetic radiation that are broadcast in a beam, whose angular resolution is about $1°$ with a range resolution of about 0.5 km (0.3 mi). The radar beam may intercept precipitation particles in a storm that reflect a fraction of the transmitted energy to the transmitter site (generally called reflectivity or the scatter cross section per unit volume). As the transmitter sweeps out a volume by rotating and tilting the transmitting antenna, the reflectivity pattern of the precipitation particles embodied in the storm is defined. Doppler radars also can measure the velocities of precipitation particles along the beam (radial velocity). Reflectivity and velocity patterns of the storm hydrometeors then make it possible to diagnose horizontal and vertical circulations that may arise within the storm, and to estimate the type and severity of weather elements attending the storm, such as rainfall, hail, damaging winds, and tornadoes. *See* PRECIPITATION (METEOROLOGY); RADAR METEOROLOGY; WIND.

Satellite surveillance. Since the early 1960s, meteorological data from satellites have had an increasing impact on storm detection and monitoring. In December 1966 the first geostationary *Applications Technology Satellite (ATS 1)* allowed forecasters in the United States to observe storms in animation. A Geostationary Operational Environmental Satellite (GOES) program was initiated within the National Oceanic and Atmospheric Administration (NOAA) with the launch of *GOES 1* in October 1975. The visible and infrared spin scan radiometer (VISSR) provided imagery, which significantly advanced the ability of meteorologists to detect and observe storms by providing frequent-interval visible and infrared imagery of the Earth surface, cloud cover, and atmospheric moisture patterns.

The first of NOAA's next generation of geostationary satellites, *GOES 8* was launched in the spring of 1994. *GOES 8* introduced improved capabilities to detect and observe storms. The *GOES 8* system includes no conflict between imaging and sounding operation, multispectral imaging with improved resolution and better signal-to-noise in the infrared bands, and more accurate temperature and moisture soundings of the storm environment. The Earth's atmosphere is observed nearly continuously.

Derived-product images showing fog and stratus areas from *GOES 8* are created by combining direct satellite measurements, such as by subtracting brightness temperatures at two different wavelengths. *GOES 8* shows the fog and stratus much more clearly because of its improved resolution. This capability enables forecasters to detect boundaries between rain-cooled areas having fog or low clouds, and clear areas. Such boundaries are frequently associated with future thunderstorm development. The sounder on *GOES 8* is capable of fully supporting routine forecasting operations. This advanced sounding capability consists of better vertical resolution in both temperature and moisture, and improved coverage of soundings in and around cloudy weather systems. *See* CLOUD; FOG; METEOROLOGICAL SATELLITES; SATELLITE METEOROLOGY.

Surface observing systems. Larger convective storm systems such as squall lines and mesoscale convective systems can be detected (but not fully described) by the temperature, moisture, wind, and pressure patterns observed by appropriate surface instrumentation. Automatic observing systems provide frequent data on pressure, temperature, humidity, wind, cloud base, and most precipitation types, intensity, and accumulation. Analyses of these data, combined with improved conceptual models of convective storm systems, enable forecasters to detect and monitor the intense mesoscale fluctuations in pressure and winds that often accompany the passage of convective weather systems such as bow echoes, derechos (strong, straight-line winds), and squall lines. A bow echo is a specific radar reflectivity pattern associated with a line of thunderstorms. The middle portion of the thunderstorm line is observed to move faster than the adjacent portions, causing the line of storms to assume a bowed-out configuration. Other analyses of these mesoscale data fields aid the forecaster in detecting favorable areas for thunderstorm cell regeneration, which may produce slowly moving mesoscale convective storms attended by heavy rains and flash floods. *See* SQUALL LINE; WEATHER OBSERVATIONS.

Cloud-to-ground lightning detectors. Lightning location stations provide forecasters with the location, polarity, peak current, and number of strokes in a flash to ground within seconds of the flash occurrence. Useful applications have emerged with regard to the detection and tracking of thunderstorms, squall lines, other mesoscale convective systems, and the weather activity that accompany these phenomena, such as tornadoes and hail. *See* LIGHTNING; MESOMETEOROLOGY; SFERICS; THUNDERSTORM; WEATHER FORECASTING AND PREDICTION. [C.F.Ch.]

Storm electricity Processes responsible for the separation of positive and negative electric charges in the atmosphere during storms, including the spectacular manifestation of this charge separation: lightning discharges. Cloud electrification is almost invariably associated with convective activity and with the formation of precipitation in the form of liquid water (rain) and ice particles (graupel and hail). The most vigorous convection and active lightning occurs in the summertime, when the energy source for convection, water vapor, is most prevalent. Winter snowstorms can also be strongly electrified, but they produce far less lightning than summer storms. Electrified storm clouds occasionally occur in complete isolation; more commonly they are found in convective clusters or in lines that may extend horizontally for hundreds of kilometers. *See* PRECIPITATION (METEOROLOGY).

Measurements of electric field at the ground and from instrumented balloons within thunderclouds have disclosed an electrostatic structure that appears to be fairly systematic throughout the world. The measurements show that the principal variations in charge occur in the vertical and are affected by the temperature of the cloud. The charge structure within a thundercloud is tripolar, with a region of dominant negative charge sandwiched between an upper region of positive charge and a subsidiary lower region of positive charge. In addition to the charge accumulations described within the cloud, electrical measurements disclose the existence of charge-screening layers at the upper cloud boundary and a layer of positive charge near the Earth's surface beneath the cloud. These secondary charge accumulations arising from charge migration outside the cloud are caused by electrostatic forces of attraction set up by the charges within the cloud.

Large differences of electric potential are associated with the distribution of charge maintained by active thunderclouds. These large differences in potential are maintained by charging currents that result from the motions of air and particles. The charging currents range from milliamperes in small clouds that are not producing lightning to several amperes for large storms with high rates of lightning. *See* CLOUD PHYSICS.

In response to charge separation within a thundercloud, the electric field increases to a value of approximately 10^6 V/m (300,000 V/ft) at which point dielectric breakdown occurs and lightning is initiated. Most lightning extends through the cloud at speeds of 10,000–100,000 m/s (22,000–220,000 mi/h). The peak temperature of lightning, which is a highly ionized plasma, may exceed 30,000 K (54,000°F). The acoustic disturbance caused by the sudden heating of the atmosphere by lightning is thunder. *See* LIGHTNING; THUNDER.

Meteorologists have shown a growing interest in the large-scale display of real-time lightning activity, since lightning is one of the most sensitive indicators of convective activity. Research has expanded into relationships between lightning characteristics and the meteorological evolution of different types of storms. The discovery of the sensitive dependence of local lightning activity on the temperature of surface air has led to research focused on the use of the global electrical circuit as a diagnostic for global temperature change. *See* ATMOSPHERIC ELECTRICITY; STORM DETECTION; THUNDERSTORM.

[E.Wi.]

Storm surge An anomalous rise in water elevations caused by severe storms approaching the coast. A storm surge can be succinctly described as a large wave that moves with the storm that caused it. The surge is intensified in the nearshore, shallower regions where the surface stress caused by the strong onshore winds pile up water against the coast, generating an opposing pressure head in the offshore direction. However, there are so many other forces at play in the dynamics of the storm surge phenomenon, such as bottom friction, Earth's rotation, inertia, and interaction with the coastal geometry, that a simple static model cannot explain all the complexities involved. Scientists and engineers have dedicated many years in the development and application of sophisticated computer models to accurately predict the effects of storm surges.

The intensity and dimension of the storm causing a surge, and thus the severity of the ensuing surge elevations, depend on the origin and atmospheric characteristics of the storm itself. Hurricanes and severe extratropical storms are the cause of most significant surges. In general, hurricanes are more frequent in low to middle latitudes, and extratropical storms are more frequent in middle to high latitudes. *See* HURRICANE; STORM.

[S.R.S.]

Strand line The line that marks the separation of land and water along the margin of a pond, lake, sea, or ocean; also called the shoreline. The strand line is very dynamic. It changes with the tides, storms, and seasons, and as long-term sea-level changes take place. The sediments on the beach respond to these changes, as do the organisms that live in this dynamic environment. On a beach organisms move with the tides, and on a rocky coast they tend to be organized relative to the strand line because of special limitations or adaptations to exposure.

Geologists who study ancient coastal environments commonly try to establish where the strand line might be in the rock strata. This can sometimes be determined by a combination of the nature and geometry of individual laminations in the rock, by identifying sedimentary structures that occur at or near the strand line. The most indicative of these structures are swash marks, which are very thin accumulations of sand grains that mark the landward uprush of a wave on the beach. *See* FACIES (GEOLOGY); PALEOGEOGRAPHY; STRATIGRAPHY.

[R.A.D.]

Stratigraphy A discipline involving the description and interpretation of layered sediments and rocks, and especially their correlation and dating. Correlation is a procedure for determining the relative age of one deposit with respect to another. The term "dating" refers to any technique employed to obtain a numerical age, for example, by making use of the decay of radioactive isotopes found in some minerals in sedimentary rocks or, more commonly, in associated igneous rocks. To a large extent, layered rocks are ones that accumulated through sedimentary processes beneath the sea, within lakes, or by the action of rivers, the wind, or glaciers; but in places such deposits contain significant amounts of volcanic material emplaced as lava flows or as ash ejected from volcanoes during explosive eruptions. See DATING METHODS; IGNEOUS ROCKS; ROCK AGE DETERMINATION; SEDIMENTARY ROCKS.

Sedimentary successions are locally many thousands of meters thick owing to subsidence of the Earth's crust over millions of years. Sedimentary basins therefore provide the best available record of Earth history over nearly 4 billion years. That record includes information about surficial processes and the varying environment at the Earth's surface, and about climate, changing sea level, the history of life, variations in ocean chemistry, and reversals of the Earth's magnetic field. Sediments also provide a record of crustal deformation (folding and faulting) and of large-scale horizontal motions of the Earth's lithospheric plates (continental drift). Stratigraphy applies not only to strata that have remained flat-lying and little altered since their time of deposition, but also to rocks that may have been strongly deformed or recrystallized (metamorphosed) at great depths within the Earth's crust, and subsequently exposed at the Earth's surface as a result of uplift and erosion. As long as original depositional layers can be identified, some form of stratigraphy can be undertaken. See BASIN; CONTINENTAL DRIFT; FAULT AND FAULT STRUCTURES; SEDIMENTOLOGY.

An important idea first articulated by the Danish naturalist Nicolaus Steno in 1669 is that in any succession of strata the oldest layer must have accumulated at the bottom, and successively younger layers above. It is not necessary to rely on the present orientation of layers to determine their relative ages because most sediments and sedimentary rocks contain numerous features, such as current-deposited ripples, minor erosion surfaces, or fossils of organisms in growth position, that have a well-defined polarity with respect to the up direction at the time of deposition (so-called geopetal indicators). This principle of superposition therefore applies equally well to tilted and even overturned strata. Only where a succession is cut by a fault is a simple interpretation of stratigraphic relations not necessarily possible, and in some cases older rocks may overlie younger rocks structurally. See DEPOSITIONAL SYSTEMS AND ENVIRONMENTS.

The very existence of layers with well-defined boundaries implies that the sedimentary record is fundamentally discontinuous. Discontinuities are present in the stratigraphic record at a broad range of scales, from that of a single layer or bed to physical surfaces that can be traced laterally for many hundreds of kilometers. Large-scale surfaces of erosion or nondeposition are known as unconformities, and they can be identified on the basis of both physical and paleontological criteria. See PALEONTOLOGY; UNCONFORMITY.

Most stratal discontinuities possess time-stratigraphic significance because strata below a discontinuity tend to be everywhere older than strata above. To the extent that unconformities can be recognized and traced widely within a sedimentary basin, it is possible to analyze sedimentary rocks in a genetic framework, that is, with reference to the way they accumulated. This is the basis for the modern discipline of sequence stratigraphy, so named because intervals bounded by unconformities have come to be called sequences.

Traditional stratigraphic analysis has focused on variations in the intrinsic character or properties of sediments and rocks—properties such as composition, texture, and included fossils (lithostratigraphy and biostratigraphy)—and on the lateral tracing of distinctive marker beds such as those composed of ash from a single volcanic eruption (tephrostratigraphy). The techniques of magnetostratigraphy and chemostratigraphy are also based on intrinsic characteristics, although these techniques require sophisticated laboratory analysis. Sequence stratigraphy attempts to integrate these approaches in the context of stratal geometry, thereby providing a unifying framework in which to investigate the time relations between sediment and rock bodies as well as to measure their numerical ages (chronostratigraphy and geochronology). Seismic stratigraphy is a variant of the technique of sequence stratigraphy in which unconformities are identified and traced in seismic reflection profiles on the basis of reflection geometry. *See* GEOCHRONOMETRY; SEISMIC STRATIGRAPHY; SEISMOLOGY.

Conventional stratigraphy currently recognizes two kinds of stratigraphic unit: material units, distinguished on the basis of some specified property or properties or physical limits; and temporal or time-related units. A common example of a material unit is the formation, a lithostratigraphic unit defined on the basis of lithic characteristics and position within a stratigraphic succession. Each formation is referred to a section or locality where it is well developed (a type section), and assigned an appropriate geographic name combined with the word formation or a descriptive lithic term such as limestone, sandstone, or shale (for example, Tapeats Sandstone). Some formations are divisible into two or more smaller-scale units called members and beds. In other cases, formations of similar lithic character or related genesis are combined into composite units called groups and supergroups.

Sequence stratigraphy differs from conventional stratigraphy in two important respects. The first is that basic units (sequences) are defined on the basis of bounding unconformities and correlative conformities rather than material characteristics or age. The second is that sequence stratigraphy is fundamentally not a system for stratigraphic classification, but a procedure for determining how sediments accumulate. *See* SEQUENCE STRATIGRAPHY.   [N.C.B.]

Stratosphere   The atmospheric layer that is immediately above the troposphere and contains most of the Earth's ozone. Here temperature increases upward because of absorption of solar ultraviolet light by ozone. Since ozone is created in sunlight from oxygen, a by-product of photosynthesis, the stratosphere exists because of life on Earth. In turn, the ozone layer allows life to thrive by absorbing harmful solar ultraviolet radiation. The mixing ratio of ozone is largest (10 parts per million by volume) near an altitude of 30 km (18 mi) over the Equator. The distribution of ozone is controlled by solar radiation, temperature, wind, reactive trace chemicals, and volcanic aerosols. *See* ATMOSPHERE; TROPOSPHERE.

The heating that results from absorption of ultraviolet radiation by ozone causes temperatures generally to increase from the bottom of the stratosphere (tropopause) to the top (stratopause) near 50 km (30 mi), reaching 280 K (45°F) over the summer pole. This temperature inversion limits vertical mixing, so that air typically spends months to years in the stratosphere. *See* TEMPERATURE INVERSION; TROPOPAUSE.

The lower stratosphere contains a layer of small liquid droplets. Typically less than 1 micrometer in diameter, they are made primarily of sulfuric acid and water. Occasional large volcanic eruptions maintain this aerosol layer by injecting sulfur dioxide into the stratosphere, which is converted to sulfuric acid and incorporated into droplets. Enhanced aerosol amounts from an eruption can last several years. By reflecting sunlight, the aerosol layer can alter the climate at the Earth's surface. By absorbing upwelling infrared radiation from the Earth's surface, the aerosol layer can warm the stratosphere. The aerosols also provide surfaces for a special set of chemical reactions that affect the ozone layer. Liquid droplets and frozen particles generally convert chlorine-bearing compounds to forms that can destroy ozone. They also tend to take up nitric acid and water and to fall slowly, thereby removing nitrogen and water from the stratosphere. The eruption of Mount Pinatubo (Philippines) in June 1991 is believed to have disturbed the Earth system for several years, raising stratospheric temperatures by more than 1 K (1.8°F) and reducing global surface temperatures by about 0.5 K (0.9°F).

Ozone production is balanced by losses due to reactions with chemicals in the nitrogen, chlorine, hydrogen, and bromine families. Reaction rates are governed by temperature, which depends on amounts of radiatively important species such as carbon dioxide. Human activities are increasing the amounts of these molecules and are thereby affecting the ozone layer. Evidence for anthropogenic ozone loss has been found in the Antarctic lower stratosphere. Near polar stratospheric clouds, chlorine and bromine compounds are converted to species that, when the Sun comes up in the southern spring, are broken apart by ultraviolet radiation and rapidly destroy ozone. This sudden loss of ozone is known as the anthropogenic Antarctic ozone hole. *See* STRATOSPHERIC OZONE.   [M.H.H.]

Stratospheric ozone   While ozone is found in trace quantities throughout the atmosphere, the largest concentrations are located in the lower stratosphere in a layer between 9 and 18 mi (15 and 30 km). Atmospheric ozone plays a critical role for the biosphere by absorbing the ultraviolet radiation with wavelength (λ) 240–320 nanometers. This radiation is lethal to simple unicellular organisms (algae, bacteria, protozoa) and to the surface cells of higher plants and animals. It also damages the genetic material of cells and is responsible for sunburn in human skin. The incidence of skin cancer has been statistically correlated with the observed surface intensity of ultraviolet wavelength 290–320 nm, which is not totally absorbed by the ozone layer. *See* STRATOSPHERE.

Ozone also plays an important role in photochemical smog and in the purging of trace species from the lower atmosphere. Furthermore, it heats the upper atmosphere by absorbing solar ultraviolet and visible radiation ($\lambda < 710$ nm) and thermal infrared radiation ($\lambda \simeq 9.6$ micrometers). As a consequence, the temperature increases steadily from about −60°F (220 K) at the tropopause (5–10 mi or 8–16 km altitude) to about 45°F (280 K) at the stratopause (30 mi or 50 km altitude). This ozone heating provides the major energy source for driving the circulation of the upper stratosphere and mesosphere. *See* ATMOSPHERIC GENERAL CIRCULATION; TROPOPAUSE.

Above about 19 mi (30 km), molecular oxygen (O_2) is dissociated to free oxygen atoms (O) during the daytime by ultraviolet photons, ($h\nu$), as shown in reaction (1). The oxygen atoms produced then form ozone (O_3) by reaction (2), where M is an arbitrary

$$O_2 + h\nu \rightarrow O + O \qquad \lambda < 242 \text{ nm} \tag{1}$$

$$O + O_2 + M \rightarrow O_3 + M \tag{2}$$

molecule required to conserve energy and momentum in the reaction. Ozone has a short lifetime during the day because of photodissociation, as shown in reaction (3).

$$O_3 + h\nu \rightarrow O_2 + O \qquad \lambda < 710 \text{ nm} \tag{3}$$

However, except above 54 mi (90 km), where O_2 begins to become a minor component of the atmosphere, reaction (3) does not lead to a net destruction of ozone. Instead the O is almost exclusively converted back to O_3 by reaction (2). If the odd oxygen concentration is defined as the sum of the O_3 and O concentrations, then odd oxygen is produced by reaction (1). It can be seen that reactions (2) and (3) do not affect the odd oxygen concentrations but merely define the ratio of O to O_3. Because the rate

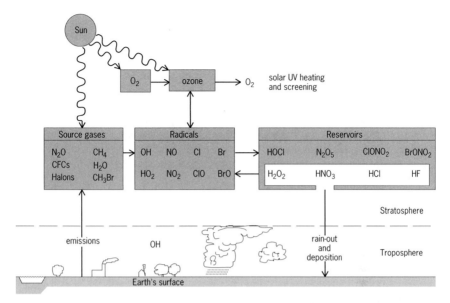

Principal chemical cycles in the stratosphere. The destruction of ozone is affected by the presence of radicals which are produced by photolysis or oxidation of source gases. Chemical reservoirs are relatively stable but are removed from the stratosphere by transport toward the troposphere and rain-out.

of reaction (2) decreases with altitude while that for reaction (3) increases, most of the odd oxygen below 36 mi (60 km) is in the form of O_3 while above 36 mi (60 km) it is in the form of O. The reaction that is responsible for a small fraction of the odd oxygen removal rate is shown as reaction (4). A significant fraction of the removal is

$$O + O_3 \rightarrow O_2 + O_2 \tag{4}$$

caused by the presence of chemical radicals [such as nitric oxide (NO), chlorine (Cl), bromine (Br), hydrogen (H), or hydroxyl (OH)], which serve to catalyze reaction (4) (see illustration).

The discovery in the mid-1980s of an ozone hole over Antarctica, which could not be explained by the classic theory of ozone and had not been predicted by earlier chemical models, led to many speculations concerning the causes of this event, which can be observed each year in September and October. As suggested by experimental and observational evidence, heterogeneous reactions on the surface of liquid or solid particles that produce Cl_2, HOCl, and $ClNO_2$ gas, and the subsequent rapid photolysis of these molecules, produces chlorine radicals (Cl, ClO) which in turn lead to the destruction of ozone in the lower stratosphere by a catalytic cycle [reactions (5)–(7)].

$$Cl + O_3 \rightarrow ClO + O_2 \tag{5}$$

$$ClO + ClO \rightarrow Cl_2O_2 \tag{6}$$

$$Cl_2O_2 + h\nu \rightarrow 2Cl + O_2 \tag{7}$$

Solar radiation is needed for these processes to occur.

Sites on which the reactions producing Cl_2, HOCl, and $ClNO_2$ can occur are provided by the surface of ice crystals in polar stratospheric clouds (PSCs). These clouds are formed between 8 and 14 mi (12 and 22 km) when the temperature drops below approximately −123°F (187 K). Other types of particles are observed at temperatures above the frost point of −123°F (187 K). These particles provide additional surface area for these reactions to occur. Clouds are observed at high latitudes in winter. Because the winter temperatures are typically 20–30°F (10–15 K) colder in the Antarctic than in the Arctic, their frequency of occurrence is highest in the Southern Hemisphere. Thus, the formation of the springtime ozone hole over Antarctica is explained by the activation of chlorine and the catalytic destruction of O_3 which takes place during September, when the polar regions are sunlit but the air is still cold and isolated from midlatitude air by a strong polar vortex. Satellite observations made since the 1970s suggest that total ozone in the Arctic has been abnormally low during the 1990s, probably in relation to the exceptionally cold winter tempratures in the Arctic lower stratosphere recorded during that decade.   [G.P.B.; R.G.Pr.]

Stream gaging   The measurement of water discharge in streams. Discharge is the rate of movement of the stream's water volume. It is the product of water velocity times cross-sectional area of the stream channel. Several techniques have been developed for measuring stream discharge; selection of the gaging method usually depends on the size of the stream. The most accurate methods for measuring stream discharge make use of in-stream structures through which the water can be routed, such as flumes and weirs.

A flume is a constructed channel that constricts the flow through a control section, the exact dimensions of which are known. Through careful hydraulic design and calibration by laboratory experiments, stream discharge through a flume can be determined by simply measuring the water depth (stage) in the inlet or constricted sections. Appropriate formulas relate stage to discharge for the type of flume used.

A weir is used in conjunction with a dam in the streambed. The weir itself is usually a steel plate attached to the dam that has a triangular, rectangular, or trapezoidal notch over which the water flows. Hydraulic design and experimentation has led to calibration curves and appropriate formulas for many different weir designs. To calculate stream discharge through a weir, only the water stage in the reservoir created by the dam needs to be measured. Stream discharge can be calculated by using the appropriate formula that relates stage to discharge for the type of weir used. *See* HYDROLOGY; SURFACE WATER.
                                                                        [T.C.Wi.]

Stream transport and deposition   The sediment debris load of streams is a natural corollary to the degradation of the landscape by weathering and erosion. Eroded material reaches stream channels through rills and minor tributaries, being carried by the transporting power of running water and by mass movement, that is, by slippage, slides, or creep. The size represented may vary from clay to boulders. At any place in the stream system the material furnished from places upstream either is carried away or, if there is insufficient transporting ability, is accumulated as a depositional feature. The accumulation of deposited debris tends toward increased ease of movement, and this tends eventually to bring into balance the transporting ability of the stream and the debris load to be transported.   [L.B.L.]

Stromatolite   A laminated, microbial structure in carbonate rocks (limestone and dolomite). Stromatolites are the oldest macroscopic evidence of life on Earth, at least 2.5 billion years old, and they are still forming in the seas. During the 1.5 billion years of Earth history before marine invertebrates appeared, stromatolites were the most obvious evidence of life, and they occur sporadically throughout the remainder of the geologic record. In Missouri and Africa, stromatolite reefs have major accumulations of lead, zinc, or copper; and in Montana, New Mexico, and Oman, stromatolites occur within oil and gas reservoirs. For geologists, the shapes of stromatolites are useful indications of their environmental conditions, and variations in form and microstructure of the laminations may be age-diagnostic in those most ancient sedimentary rocks that lack invertebrate fossils. *See* DOLOMITE; LIMESTONE; REEF.

Stromatolites are readily recognizable in outcrops by their characteristic convex-upward laminated structure. Individual, crescent-shaped laminations, which are generally about a millimeter thick, are grouped together to produce an enormous range of shapes and sizes.

The tiny, filamentous cyanobacteria (blue-green algae) that make present-day stromatolites, and similar filaments associated with the oldest stromatolites known, are considered one of the most successful organisms on Earth. Living stromatolites in the Bahamas and Western Australia possess laminations that record the episodic trapping and binding of sediment particles by the microbial mat. In the modern oceans, stromatolites develop almost exclusively in extreme marine conditions that exclude or deter browsing invertebrates and fish from destroying the microbial mats and inhibit colonization by competing algae.   [R.Gi.]

Strontianite   The mineral form of strontium carbonate, usually with some calcium replacing strontium. It characteristically occurs in veins with barite or celestite or as masses in certain sedimentary rocks. Strontianite is normally prismatic, but it may also be massive. It may be colorless or gray with yellow, green, or brownish tints. The hardness is $3^{1}/_{2}$ on Mohs scale, and the specific gravity of 3.76. It occurs at Strontian, Scotland, and in Germany, Austria, Mexico, and India and, in the United States, in the Strontium Hills of Calfornia. *See* CARBONATE MINERALS.   [R.I.Ha.]

Structural geology   The branch of geology that deals with study and interpretation of deformation of the Earth's crust. Deformation brings about changes in size (dilation), shape (distortion), position (translation), or orientation (rotation). Evidence for the changes caused by deformation are commonly implanted into geologic bodies in the form of recognizable structures, such as faults and joints, folds and cleavage, and foliation and lineation. The geologic record of structures and structural relations is best developed and most complicated in mountain belts, the most intensely deformed parts of the Earth's crust. *See* MOUNTAIN SYSTEMS.

The discipline of structural geology harnesses three interrelated strategies of analysis: descriptive analysis, kinematic analysis, and dynamic analysis. Descriptive analysis is concerned with recognizing and describing structures and measuring their orientations. Kinematic analysis focuses on interpreting the deformational movements responsible for the structures. Dynamic analysis interprets deformational movements in terms of forces, stresses, and mechanics. The ultimate goal of these interdependent approaches is to interpret the physical evolution of crustal structures, that is, tectonic analysis. A major emphasis in modern structural geology is strain analysis, the quantitative analysis of changes in size and shape of geologic bodies, regardless of scale.

There are many significant practical applications of structural geology. An understanding of the descriptive and geometric properties of folds and faults, as well as mechanisms of folding and faulting, is of vital interest to exploration geologists in the petroleum industry. Ore deposits commonly are structurally controlled, or structurally disturbed, and for these reasons detailed structural geologic mapping is an essential component of mining exploration. Other applications of structural geology include the evaluation of proposals for the disposal of radioactive waste in the subsurface, and the targeting of safe sites for dams, hospitals, and the like in regions marked by active faulting. *See* FAULT AND FAULT STRUCTURES; FOLD AND FOLD SYSTEMS; ORE AND MINERAL DEPOSITS; PETROLEUM GEOLOGY.   [G.H.D.]

Structural petrology   The study of the structural aspects of rocks, as distinct from the purely chemical and mineralogical studies that are generally emphasized in other branches of petrology. The term was originally used synonymously with petro-fabric analysis, but is sometimes restricted to denote the analysis of only microscopic structural and textural features. *See* PETROFABRIC ANALYSIS; PETROGRAPHY.   [J.M.Ch.]

Stylolites   Irregular surfaces occurring in certain rocks, mostly parallel to bedding planes, in which small toothlike projections on one side of the surface fit into cavities of like shape on the other side (see illustration). Stylolites are most common in limestones

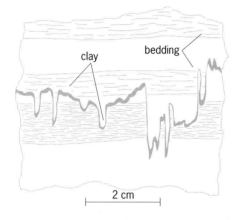

Stylolite in limestone.

and dolomites but are also present in many other kinds of rock, including sandstones, gypsum beds, and cherts. *See* DOLOMITE; LIMESTONE; SEDIMENTARY ROCKS.   [R.Si.]

Subduction zones   Regions where portions of the Earth's tectonic plates are diving beneath other plates, into the Earth's interior. Subduction zones are defined by deep oceanic trenches, lines of volcanoes parallel to the trenches, and zones of large earthquakes that extend from the trenches landward.

Plate tectonic theory recognizes that the Earth's surface is composed of a mosaic of interacting lithospheric plates, with the lithosphere consisting of the crust (continental or oceanic) and associated underlying mantle, for a total thickness of about 100 km (60 mi). Oceanic lithosphere is created by sea-floor spreading at mid-ocean ridges (divergent, or accretionary, plate boundaries) and destroyed at subduction zones (at convergent, or destructive, plate boundaries). At subduction zones, the oceanic litho-sphere dives beneath another plate, which may be either oceanic or continental. Part of the material on the subducted plate is recycled back to the surface (by being scraped off the subducting plate and accreted to the overriding plate, or by melting and rising as magma), and the remainder is mixed back into the Earth's deeper mantle. This process balances the creation of lithosphere that occurs at the mid-ocean ridge system. The convergence of two plates occurs at rates of 1–10 cm/yr (0.4–4 in./yr) or 10–100 km (6–60 mi) per million years (see illustration).

During subduction, stress and phase changes in the upper part of the cold descending plate produce large earthquakes in the upper portion of the plate, in a narrow band

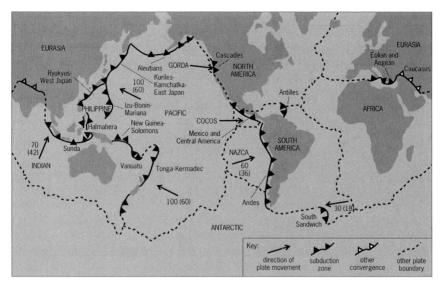

Principal subduction zones. Plate names are in all-capitals. Numbers at arrows indicate velocity of plate movement in kilometers (miles) per million years. (*After J. Gill, Orogenic Andesites and Plate Tectonics, Springer-Verlag, 1981*)

called the Wadati-Benioff zone that can extend as deep as 700 km (420 mi). The plate is heated as it descends, and the resulting release of water leads to melting of the overlying mantle. This melt rises to produce the linear volcanic chains that are one of the most striking features of subduction zones. *See* LITHOSPHERE.

Subduction zones can be divided in two ways, based either on the nature of the crust in the overriding plate or on the age of the subducting plate. The first classification yields two broad categories: those beneath an oceanic plate, as in the Mariana or Tonga trenches, and those beneath a continental plate, as along the west coast of South America (see illustration). The first type is known as an intraoceanic convergent margin, and the second is known as an Andean-type convergent margin. *See* MID-OCEANIC RIDGE; PLATE TECTONICS.

Active volcanoes are highly visible features of subduction zones. The volcanoes that have developed above subduction zones in East Asia, Australasia, and the western Americas surround the Pacific Ocean in a so-called Ring of Fire. At intraoceanic convergent margins, volcanoes may be the only component above sea level, leading to the name "island arc." The more general term "volcanic arc" refers to volcanoes built on either oceanic or continental crust. *See* VOLCANO.

An eventual consequence of subduction is orogeny, or mountain building. Subduction zones are constantly building new crust by the production of volcanic material or the accretion of oceanic sediments. However, the development of the greatest mountain ranges—the Alps or the Himalayas—occurs not during "normal" subduction but during the death of a subduction zone, when it becomes clogged with a large continent or volcanic arc. *See* OROGENY. [R.J.Ste.; S.H.B.]

Submarine canyon A steep-sided valley developed on the sea floor of the continental slope. Submarine canyons serve as major conduits for sediment transport from land and the continental shelf to the deep sea. Modern canyons are relatively narrow, deeply incised, steeply walled, often sinuous valleys with predominantly

V-shaped cross sections. Most canyons originate near the continental-shelf break, and generally extend to the base of the continental slope. Canyons often occur off the mouths of large rivers, such as the Hudson River or the Mississippi River, although many other canyons, such as the Bering Canyon in the southern Bering Sea, are developed along structural trends. *See* CONTINENTAL MARGIN.

Modern submarine canyons vary considerably in their dimensions. The average lengths of canyons has been estimated to be about 34 mi (55 km); although the Bering Canyon is more than 680 mi (1100 km) long and is the world's longest submarine canyon. The shortest canyons are those off the Hawaiian Islands, and average about 6 mi (10 km) in length. Submarine canyons are characterized by relatively steep gradients. The average slope of canyon floors is 309 ft/mi (58 m/km). In general, shorter canyons tend to have higher gradients. For example, shorter canyons of the Hawaiian group have an average gradient of 766 ft/mi (144 m/km), whereas the Bering Canyon has a slope of only 42 ft/mi (7.9 m/km).

In comparison to modern canyons, dimensions of ancient canyons are considerably smaller. Deposits of ancient canyons are good hydrocarbon reservoirs. This is because submarine canyons and channels are often filled with sand that has the potential to hold oil and gas. Examples of hydrocarbon-bearing canyon-channel reservoirs are present in the subsurface in California, Louisiana, and Texas.

Physical and biological processes that are common in submarine canyons are mass wasting, turbidity currents, bottom currents, and bioerosion, mass wasting and turbidity currents being the more important. Mass wasting is a general term used for the failure, dislodgement, and downslope movement of sediment under the influence of gravity. Common examples of mass wasting are slides, slumps, and debris flows. Major slumping events can lead to formation of submarine canyons. The Mississippi Canyon in the Gulf of Mexico is believed to have been formed by retrogressive slumping during the late Pleistocene fall in sea level and has been partially infilled during the Holocene rise in sea level. Turbidity currents are one of the most important erosional and depositional processes in submarine canyons. There is considerable evidence to suggest that turbidity currents flow at velocities of 11–110 in./s (28–280 cm/s) in submarine canyons. Therefore, turbidity currents play a major role in the erosion of canyons. *See* DEPOSITIONAL SYSTEMS AND ENVIRONMENTS; MARINE GEOLOGY; MARINE SEDIMENTS; MASS WASTING; TURBIDITY CURRENT. [G.Sh.]

Sulfide phase equilibria   The chemistry of the sulfides is rather simple inasmuch as most sulfide minerals involve only two major elements, although some involve three, and a few four. Understanding of the phase relations among these minerals is important both to the geologist, whose task it is to locate and exploit ore deposits, and to the metallurgist, whose task it is to extract the metals from the ores for industrial use.

Sulfide ore deposits are the most important sources of numerous metals such as lead, zinc, copper, nickel, cobalt, and molybdenum. In addition, sulfide ores provide substantial amounts of noble metals such as platinum, gold, and silver, and of other industrially important elements such as cadmium, rhenium, and selenium. Although iron sulfides usually are the most common minerals in such deposits, most commercial iron is mined from iron oxide ores and most sulfur from elemental sulfur deposits.

Some of the more common sulfide minerals are listed in the table. The first eight sulfide minerals listed may be plotted in the ternary system copper-iron-sulfur. Similarly, the first two and the last two minerals may be plotted in the ternary system iron-nickel-sulfur. Thus, by studying in detail the phase equilibria in two ternary systems a great deal of information can be obtained about many of the common sulfides occurring in ore deposits. However, before the ternary systems can be explored in a

Common sulfide minerals

Mineral name	Chemical formula
Pyrite	FeS_2
Pyrrhotites (three or more varieties)	$Fe_{1-x}S$
Covellite	CuS
Digenite	Cu_9S_5
Chalcocite	Cu_2S
Bornite	Cu_5FeS_4
Chalcopyrite	$CuFeS_2$
Cubanite	$CuFe_2S_3$
Galena	PbS
Sphalerite and wurtzite	ZnS
Metacinnabar and cinnabar	HgS
Argentite and acanthite	Ag_2S
Molybdenite	MoS_2
Millerite	NiS
Pentlandite	$(Fe,Ni)_9S_8$

systematic way, the binary systems bounding the ternaries have to be studied in detail. Similarly, it is necessary that the four bounding ternary systems be fully understood before a quaternary system can be systematically investigated. Thus, it is seen that before a systematic study of, for example, the immensely important quaternary system copper-iron-nickel-sulfur can proceed, much preliminary information is required. The prerequisite data include complete knowledge of the four ternary systems Cu-Fe-S, Cu-Fe-Ni, Fe-Ni-S, and Cu-Ni-S. In turn, the phase relations in these ternary systems cannot be systematically studied before the six binary systems Fe-Cu, Fe-Ni, Cu-Ni, Fe-S, Cu-S, and Ni-S have been thoroughly explored.

The enormous differences in the vapor pressures over the different phases occurring in sulfide systems add complications to the diagrammatic representation. For instance, in the Fe-S system the vapor pressure over pure iron is about 10^{-25} atm (10^{-20} pascal) at 450°C (840°F), whereas that over pure sulfur is a little more than 1 atm (10^5 pascals) at the same temperature. A complete diagrammatic representation of the relations in such a system, therefore, requires coordinates for composition and temperature, as well as for pressure. In a two-component system, such as the Fe-S system, such a representation is feasible because only three coordinates are necessary. However, in ternary (where such diagrams involve four-dimensional space) and in multicomponent systems, this type of diagrammatic representation is not possible. For this reason it is customary to use composition and temperature coordinates only for the diagrammatic representation of sulfide systems. The relations as shown in such diagrams in reality represent a projection from composition-temperature-pressure space onto a two-dimensional composition-temperature plane or onto a three-dimensional prism, depending upon whether the system contains two or three components.

Pyrite or pyrrhotite, or both, occur almost ubiquitously, not only in ore deposits but in nearly all kinds of rocks. Of the binary systems mentioned above, therefore, the iron-sulfur system is of the most importance to the economic geologist. *See* PYRITE.

[G.Kul.]

Sun dog   A bright spot of light that sometimes appears on either side of the Sun, the same distance above the horizon as the Sun, and separated from it by an angle of about 22° (see illustration). For higher Sun elevations, the angle increases slightly. These spots are known by many common names: sun dogs, mock suns, false suns, or the 22° parhelia. They usually show a red edge on the side closest to the Sun. On

Sun dogs on either side of the Sun. Also visible are the 22° halo and the parhelic circle. (*Courtesy of Robert Greenler*)

some occasions the entire spectrum of colors can be spread out in the sun-dog spot but, commonly, the red edge is followed by an orange or yellow band that merges into a diffuse white region. The effects result from the refraction by sunlight through small, flat, hexagonal-shaped ice crystals falling through the air such that their flat faces are oriented nearly horizontally. *See* HALO; METEOROLOGICAL OPTICS.   [R.Gr.]

Supercontinent   The six major continents today are Africa, Antarctica, Australia, Eurasia, North America, and South America. Prior to the formation of the Atlantic, Indian, and Southern ocean basins over the past 180 million years by the process known as sea-floor spreading, the continents were assembled in one supercontinent called Pangea (literally "all Earth"). Pangea came together by the collision, about 300 million years ago (Ma), of two smaller masses of continental rock, Laurasia and Gondwanaland. Laurasia comprised the combined continents of ancient North America (known as Laurentia), Europe, and Asia. Africa, Antarctica, Australia, India, and South America made up Gondwanaland (this name comes from a region in southern India). The term "supercontinent" is also applied to Laurasia and Gondwanaland; hence it is used in referring to a continental mass significantly bigger than any of today's continents. A supercontinent may therefore incorporate almost all of the Earth's continental rocks, as did Pangea, but that is not implied by the word. *See* CONTINENT; CONTINENTS, EVOLUTION OF.

Laurasia, Gondwanaland, and Pangea are the earliest supercontinental entities whose former existence can be proven. Evidence of older rifted continental margins, for example surrounding Laurentia and on the Pacific margins of South America, Antarctica, and Australia, point to the existence of older supercontinents. The hypothetical Rodinia (literally "the mother of all continents") may have existed 800–1000 Ma, and Pannotia (meaning "the all-southern supercontinent") fleetingly around 550 Ma. Both are believed to have included most of the Earth's continental material. There may have been still earlier supercontinents, because large-scale continents, at least the size of southern Africa or Western Australia, existed as early as 2500 Ma at the end of Archean times. *See* ARCHEAN; CONTINENTAL MARGIN.

The amalgamation and fragmentation of supercontinents are the largest-scale manifestation of tectonic forces within the Earth. The cause of such events is highly controversial. *See* PLATE TECTONICS.   [I.W.O.D.]

Surface water   A term commonly used to designate the water flowing in stream channels. The term is sometimes used in a broader sense as opposed to "subsurface water." In this sense, surface water includes water in lakes, marshes, glaciers, and reservoirs as well as that flowing in streams. In the broadest sense, surface water is all the water on the surface of the Earth and thus includes the water of the oceans. Subsurface water includes water in the root zone of the soil and ground water flowing or stored in the rock mantle of the Earth. Subsurface water differs from surface water in the mechanics of its movement as well as in its location. Surface and subsurface water are two stages of the movement of the Earth's water through the hydrologic cycle. The world's ocean and atmospheric moisture are two other main stages of the grand water cycle of the Earth. *See* HYDROLOGY.

The table gives estimates of the amounts of water in various parts of the hydrologic cycle and their detention periods. It may be noted that surface water on the continents is but a small part of the world's water and that the bulk of that is in fresh-water lakes. However, the detention period is also short. This means that the surface-water part, and especially the water in the streams, is rapidly discharged and replenished. That is why surface water, as well as the shallower ground water, is called a renewable resource. Water that has a detention period of more than a generation is not renewed within sufficient time to be so considered. *See* GROUND-WATER HYDROLOGY; RIVER.

Precipitation that reaches the Earth is subdivided by processes of evaporation and infiltration into various routes of subsequent travel. Evaporation from wet land surfaces and from vegetation returns some of the water to the atmosphere immediately. Precipitation that falls at rates less than the local rate of infiltration enters the soil. Some of the infiltrated water is retained in the soil, sustaining plant life, and some reaches

Distribution of the world's supply of water

Location	Volume of water, 10^9 acre-ft*	Percentage total	Detention period, years
World's oceans	1,060,000	97.39	5,000
Surface water on the continents			
Glaciers and polar			
ice caps	20,000	1.83	2,000
Fresh-water lakes	100	0.0093	100
Saline lakes and			
inland seas	68	0.0063	50
Average in stream			
channels	0.25	0.00002	0.05
Total surface water	20,200		700 av
Subsurface water on the continents			
Root zone of the soil	10	0.00094	0.25
Ground water above			
Ground water above			
2500 ft	3,700	0.339	5
Ground water below			
2500 ft	4,600	0.425	100
Total subsurface			
water	8,300		
Atmospheric water	115	0.0011	0.03
Total world water			
(rounded)	1,088,000	100	3,000

*109 acre-ft $= 1.233 \times 10^8$ ha · m $= 1.233 \times 10^{12}$ m³.
†2500 ft $= 750$ m.

the ground water. The precipitation that exceeds the capacity of the soil to absorb water flows overland in the direction of the steepest slope and concentrates in rills and minor channels. During storms most of the water in surface streams is derived from that portion of the precipitation which fails to infiltrate the soil. *See* PRECIPITATION (METEOROLOGY).

The distinction between surface and subsurface water, though useful, should not obscure the fact that water on the surface and water underground is physically connected through pores, cracks, and joints in rock and soil material. In many areas, particularly in humid regions, surface water in stream channels is the visible part of a reservoir, which is partly underground; the water surface of a river is the visible extension of the surface of the ground water.   [L.B.L.]

Syenite   A phaneritic (visibly crystalline) plutonic rock with granular texture composed largely of alkali feldspar (orthoclase, microcline, usually perthitic) with subordinate plagioclase (oligoclase) and dark-colored (mafic) minerals (biotite, amphibole, and pyroxene). If sodic plagioclase (oligoclase or andesine) exceeds the quantity of alkali feldspar, the rock is called monzonite. Monzonites are generally light to medium gray, but syenites are found in a wide variety of colors (gray, green, pink, red), some of which make the material ideal for use as ornamental stone. Syenite is an uncommon plutonic rock and usually occurs in relatively small bodies (dikes, sills, stocks, and small irregular plutons). *See* IGNEOUS ROCKS.   [C.A.C.]

Syncline   In its simplest form, a geologic structure marked by the folding of originally horizontal rock layers into a systematically curved, concave upward profile geometry (illus. *a*). A syncline is convex in the direction of the oldest beds in the folded sequence, concave in the direction of the youngest beds. Although typically upright, a syncline may be overturned, recumbent, or upside down (illus. *d*). Synclines occur in all sizes, from microscopic to regional. Profile forms may be curved smoothly (illus. *a*) to sharply angular (illus. *b*). Fold tightness of a syncline, as measured by the angle at which the limbs of the syncline join, may be so gentle that the fold is barely discernible, to so tight that the limbs are virtually parallel to one another (illus. *c*). The orientation of the axis of folding is horizontal to shallowly plunging, but synclines may plunge as steeply as vertical.

Synclines are products of the layer-parallel compression that arises commonly during mountain building. The final profile form of the fold reflects the mechanical properties

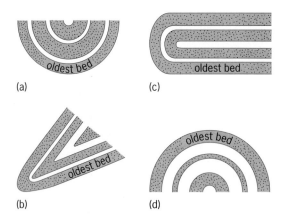

(a)

(c)

(b)

(d)

Varieties of synclines as seen in profile view. (*a*) Upright syncline with smoothly curved limbs. (*b*) Overturned, sharply angular syncline with planar limbs. (*c*) Recumbent, isoclinal syncline with parallel limbs. (*d*) Upside-down syncline, sometimes called an antiformal syncline.

of the rock sequence under the temperature-pressure conditions of folding, and the percentage of shortening required by the deformation. *See* ANTICLINE; FOLD AND FOLD SYSTEMS. [G.H.D.]

Systems ecology   The analysis of how ecosystem function is determined by the components of an ecosystem and how those components cycle, retain, or exchange energy and nutrients. Systems ecology typically involves the application of computer models that track the flow of energy and materials and predict the responses of systems to perturbations that range from fires to climate change to species extinctions. Systems ecology is closely related to mathematical ecology, with the major difference stemming from systems ecology's focus on energy and nutrient flow and its borrowing of ideas from engineering. Systems ecology is one of the few theoretical tools that can simultaneously examine a system from the level of individuals all the way up to the level of ecosystem dynamics. It is an especially valuable approach for investigating systems so large and complicated that experiments are impossible, and even observations of the entire system are impractical. In these overwhelming settings, the only approach is to break down the research into measurements of components and then assemble a system model that pieces together all components. An important contribution of ecosystem science is the recognition that there are critical ecosystem services such as cleansing of water, recycling of waste materials, production of food and fiber, and mitigation of pestilence and plagues. *See* ECOLOGICAL COMMUNITIES; ECOLOGY; ECOSYSTEM; GLOBAL CLIMATE CHANGE; THEORETICAL ECOLOGY. [P.M.Ka.]

T

Taconite The name given to the siliceous iron formation from which the high-grade iron ores of the Lake Superior district have been derived. It consists chiefly of fine-grained silica mixed with magnetite and hematite. As the richer iron ores approach exhaustion in the United States, taconite becomes more important as a source of iron.

[C.S.Hu.]

Talc A hydrated magnesium layer silicate (phyllosilicate) with composition close to $Mg_3Si_4O_{10}(OH)_2$. Talc commonly is white, but it may appear pale green or grayish depending on the amount of minor impurities. Talc has a greasy feel and pearly luster, and has been used as one of the hardness standards for rock-forming minerals with the value 1 on Mohs scale. Because talc is soft, it can be scratched by fingernails.

Talc frequently occurs in magnesium-rich metamorphosed serpentinites and siliceous dolomites. Talc-rich rocks include massive soapstones, massive steatite, and foliated talc schists.

Talc is a good insulating material. It has been commonly used in industry as a raw material for ceramics, paints, plastics, cosmetics, papers, rubber, and many other applications. *See* SILICATE MINERALS.

[J.G.L.]

Tantalite A mineral with compositon $(Fe,Mn)Ta_2O_6$, an oxide of iron, manganese, and tantalum. Niobium substitutes for tantalum in all proportions; a complete series extends to columbite $(Fe,Mn)Nb_2O_6$. Pure tantalite is rare. Iron and manganese vary considerably in their relative proportions. Tantalite crystallizes in the orthorhombic system and is common in short prismatic crystals. The hardness is 6 on Mohs scale, and the specific gravity 7.95 (pure tantalite). The luster is submetallic and the color iron black. Tantalite is the principal ore of tantalum. It is found chiefly in granite pegmatites and as a detrital mineral, in some places in important amounts, having weathered from such rocks. The chief producing areas are the Republic of the Congo and Nigeria. *See* COLUMBITE.

[C.S.Hu.]

Tektite A member of one of several groups of objects that are composed almost entirely of natural glass formed from the melting and rapid cooling of terrestrial rocks by the energy accompanying impacts of large extraterrestrial bodies. Tektites are dark brown to green, show laminar to highly contorted flow structure on weathered surfaces and in thin slices, are brittle with excellent conchoidal fracture, and occur in large masses but are mostly small to microscopic in size.

Five major groups of tektites are known: North American, 34,000,000 years old; Czechoslovakian (moldavites), 15,000,000 years; Ivory Coast, 1,300,000 years old; Russian (irgizites) 1,100,000 years old; and (5) Australasian, 700,000 years old. The North American, Ivory Coast, and Australasian tektites also occur as microtektites in oceanic sediment cores near the areas of their land occurrences. In the land occurrences, virtually all of the tektites are found mixed with surface gravels and recent sediments that are younger than their formation ages.

The chemical compositions of tektites differ from those of ordinary terrestrial rocks principally in that they contain less water and have a greater ratio of ferrous to ferric iron, both of which are almost certainly a result of their very-high-temperature history.

[E.A.K.]

Temperature inversion   The increase of air temperature with height; an atmospheric layer in which the upper portion is warmer than the lower. Such an increase is opposite, or inverse, to the usual decrease of temperature with height, or lapse rate, in the troposphere. However, above the tropopause, temperature increases with height throughout the stratosphere, decreases in the mesosphere, and increases again in the thermosphere. Thus inversion conditions prevail throughout much of the atmosphere much or all of the time, and are not unusual or abnormal. *See* AIR TEMPERATURE; ATMOSPHERE.

Inversions are created by radiative cooling of a lower layer, by subsidence heating of an upper layer, or by advection of warm air over cooler air or of cool air under warmer air. Outgoing radiation, especially at night, cools the Earth's surface, which in turn cools the lowermost air layers, creating a nocturnal surface inversion a few inches to several hundred feet thick.

Inversions effectively suppress vertical air movement, so that smokes and other atmospheric contaminants cannot rise out of the lower layer of air. California smog is trapped under an extensive subsidence inversion; surface radiation inversions, intensified by warm air advection aloft, can create serious pollution problems in valleys throughout the world; radiation and subsidence inversions, when horizontal air motion is sluggish, create widespread pollution potential, especially in autumn over North America and Europe. *See* AIR POLLUTION; SMOG.   [A.Cou.]

Terrain areas   Subdivisions of the continental surfaces distinguished from one another on the basis of the form, roughness, and surface composition of the land. The pattern of landform differences is strongly reflected in the arrangement of such other features of the natural environment as climate, soils, and vegetation. These regional associations must be carefully considered in planning of activities as diverse as agriculture, transportation, city development, and military operations.

Eight classes of terrain are distinguished on the basis of steepness of slopes, local relief (the maximum local differences in elevation), cross-sectional form of valleys and divides, and nature of the surface material. Approximate definitions of terms used and percentage figures indicating the fraction of the world's land area occupied by each class are as follows; (1) flat plains: nearly level land, slight relief, 4%; (2) rolling and irregular plains: mostly gently sloping, low relief, 30%; (3) tablelands: upland plains broken at intervals by deep valleys or escarpments, moderate to high relief, 5%; (4) plains with hills or mountains: plains surmounted at intervals by hills or mountains of limited extent, 15%; (5) hills: mostly moderate to steeply sloping land of low to moderate relief, 8%; (6) low mountains: mostly steeply sloping, high relief, 14%; (7) high mountains: mostly steeply sloping, very high relief, 13%; and (8) ice caps: surface material, glacier ice, 11%.   [E.H.Ha.]

Terrestrial coordinate system   The perpendicular intersection of two curves or two lines, one relatively horizontal and the other relatively vertical, is the basis for finding and describing terrestrial location. The Earth's graticule, consisting of an imaginary grid of east-to-west-bearing lines of latitude and north-to-south bearing lines of longitude, is derived from the Earth's shape and rotation, and is rooted in spherical geometry. Plane coordinate systems, equivalent to horizontal X and vertical Y coordinates, are based upon cartesian geometry and differ from the graticule in that they have no natural origin or beginning for their grids.

The Earth, which is essentially a sphere, rotates about an axis that defines the geographic North and South poles. The poles serve as the reference points on which the system of latitude and longitude is based (see illustration).

Latitude is arc distance (angular difference) from the Equator and is defined by a system of parallels, or lines that run east to west, each fully encompassing the Earth. The Equator is the parallel that bisects the Earth into the Northern and Southern hemispheres, and lies a constant 90° arc distance from both poles. As the only parallel to bisect the Earth, the Equator is considered a great circle. All other parallels are small circles (do not bisect the Earth), and are labeled by their arc distance north or south from the Equator and by the hemisphere in which they fall. Parallels are numbered from 0° at the Equator to 90° at the poles. For example, 42°S describes the parallel 42 degrees arc distance from the Equator in the Southern Hemisphere. For increased

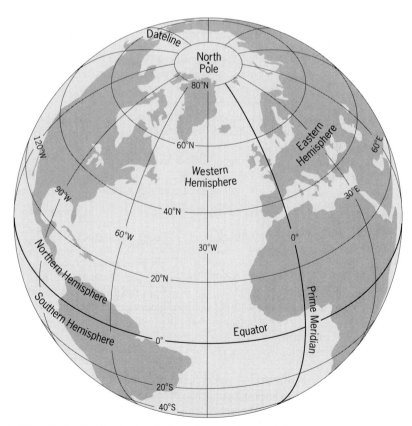

Earth's graticule. Meridians of longitude run from north to south, but are measured east or west of the Prime Meridian. Parallels run from east to west, but are measured north or south of the Equator.

location precision, degrees of latitude and longitude are further subdivided into minutes ($1° = 60'$) and seconds ($1' = 60''$). *See* EQUATOR; GREAT CIRCLE, TERRESTRIAL.

Longitude is defined by a set of imaginary curves extending between the two poles, spanning the Earth. These curves, called meridians, always point to true geographical north (or south) and converge at the poles. In the present-day system of longitude, meridians are numbered by degrees east or west of the beginning meridian, called the Prime Meridian or the Greenwich Meridian, which passes through the Royal Observatory in Greenwich, England. The Prime Meridian was assigned a longitude of $0°$.

Since the Earth is fundamentally a sphere, its circumference describes a circle containing $360°$, the arc distance through which the Earth rotates in 24 hours. The arc distance from the Prime Meridian describes the location of any meridian (see illustration). The $180°$ meridian is commonly referred to as the International Dateline. Together, the Prime Meridian and the International Dateline describe a great circle that bisects the Earth, as do all other meridian circles. The west half of the Earth, located between the Prime Meridian and the International Dateline, comprises the Western Hemisphere, and the east half on the opposite side forms the Eastern Hemisphere. Meridians within the Western Hemisphere are labeled with a W, and meridians within the Eastern Hemisphere are labeled with an E. A complete description of longitude includes an angular measurement and a hemispheric label. For example, $78°$W is the meridian $78°$ west of the Prime Meridian. Neither the $0°$ meridian (Prime) nor the $180°$ meridian (Dateline) is given a hemispheric suffix because they divide the two hemispheres, and therefore do not belong to either one.

Coordinate system alternatives to the graticule evolved in the early twentieth century because of the complexity of using spherical geometry in determining latitude, longitude, and direction. Plane (two-dimensional) or cartesian coordinate systems presume that a relatively nonspherical Earth exists in smaller areas. Plane coordinates are superimposed upon these small areas, with coordinates being determined by the equivalent of a grid composed of a number of parallel vertical lines (X) and a complementary set of parallel horizontal lines (Y).

The State Plane Coordinate system (SPC) is used only in the United States and partitions each state into zones. Each zone has its own coordinate system. The number of zones designated in each state is determined by the size of the state. Zone boundaries follow either meridians or parallels depending on the shape of the state. All measurements are made in feet.

The Universal Transverse Mercator (UTM) system is a worldwide coordinate system in which locations are expressed using metric units. The basis for the UTM system is the Universal Transverse Mercator map projection. This projection becomes vastly distorted in polar areas above $80°$, and for this reason the UTM system is confined to extend from $84°$N to $80°$S. The UTM system partitions the Earth into 60 north-south elongated zones, each having a width of $6°$ of longitude. *See* MAP PROJECTIONS.

A number of other coordinate systems are in use today. Foremost among these are the U.S. Public Land Survey System, the Universal Polar Stereographic (UPS) system, and the World Geographic Reference (GEOREF) system.   [S.Lav.]

Terrestrial ecosystem   A community of organisms and their environment that occurs on the land masses of continents and islands. Terrestrial ecosystems are distinguished from aquatic ecosystems by the lower availability of water and the consequent importance of water as a limiting factor. Terrestrial ecosystems are characterized by greater temperature fluctuations on both a diurnal and seasonal basis than occur in aquatic ecosystems in similar climates. The availability of light is greater in terrestrial ecosystems than in aquatic ecosystems because the atmosphere is more transparent than water. Gases are more available in terrestrial ecosystems than in aquatic ecosystems. Those gases include carbon dioxide that serves as a substrate for photosynthesis, oxygen that serves as a substrate in aerobic respiration, and nitrogen that serves as a substrate for nitrogen fixation. Terrestrial environments are segmented into a subterranean portion from which most water and ions are obtained, and an atmospheric portion from which gases are obtained and where the physical energy of light is transformed into the organic energy of carbon-carbon bonds through the process of photosynthesis.

Terrestrial ecosystems occupy 55,660,000 mi^2 (144,150,000 km^2), or 28.2%, of Earth's surface. Although they are comparatively recent in the history of life (the first terrestrial organisms appeared in the Silurian Period, about 425 million years ago) and occupy a much smaller portion of Earth's surface than marine ecosystems, terrestrial ecosystems have been a major site of adaptive radiation of both plants and animals. Major plant taxa in terrestrial ecosystems are members of the division Magnoliophyta (flowering plants), of which there are about 275,000 species, and the division Pinophyta (conifers), of which there are about 500 species. Members of the division Bryophyta (mosses and liverworts), of which there are about 24,000 species, are also important in some terrestrial ecosystems. Major animal taxa in terrestrial ecosystems include the classes Insecta (insects) with about 900,000 species, Aves (birds) with 8500 species, and Mammalia (mammals) with approximately 4100 species.

Organisms in terrestrial ecosystems have adaptations that allow them to obtain water when the entire body is no longer bathed in that fluid, means of transporting the water from limited sites of acquisition to the rest of the body, and means of preventing the evaporation of water from body surfaces. They also have traits that provide body support in the atmosphere, a much less buoyant medium than water, and other traits that render them capable of withstanding the extremes of temperature, wind, and humidity that characterize terrestrial ecosystems. Finally, the organisms in terrestrial ecosystems have evolved many methods of transporting gametes in environments where fluid flow is much less effective as a transport medium.

The organisms in terrestrial ecosystems are integrated into a functional unit by specific, dynamic relationships due to the coupled processes of energy and chemical flow. Those relationships can be summarized by schematic diagrams of trophic webs, which place organisms according to their feeding relationships. The base of the food web is occupied by green plants, which are the only organisms capable of utilizing the energy of the Sun and inorganic nutrients obtained from the soil to produce organic molecules. Terrestrial food webs can be broken into two segments based on the status of the plant material that enters them. Grazing food webs are associated with the consumption of living plant material by herbivores. Detritus food webs are associated with the consumption of dead plant material by detritivores. The relative importance of those two types of food webs varies considerably in different types of terrestrial ecosystems. Grazing food webs are more important in grasslands, where over half of net primary productivity may be consumed by herbivores. Detritus food webs are more important in forests, where less than 5% of net primary productivity may be consumed by herbivores. *See* FOOD WEB; SOIL ECOLOGY.

There is one type of extensive terrestrial ecosystem due solely to human activities and eight types that are natural ecosystems. Those natural ecosystems reflect the variation of precipitation and temperature over Earth's surface. The smallest land areas are occupied by tundra and temperate grassland ecosystems, and the largest land area is occupied by tropical forest. The most productive ecosystems are temperate and tropical forests, and the least productive are deserts and tundras. Cultivated lands, which together with grasslands and savannas utilized for grazing are referred to as agroecosystems, are of intermediate extent and productivity. Because of both their areal extent and their high average productivity, tropical forests are the most productive of all terrestrial ecosystems, contributing 45% of total estimated net primary productivity on land. See DESERT; ECOLOGICAL COMMUNITIES; ECOSYSTEM.   [S.J.McN.]

Terrestrial radiation   Electromagnetic radiation emitted from the Earth and its atmosphere. Terrestrial radiation, also called thermal infrared radiation or outgoing longwave radiation, is determined by the temperature and composition of the Earth's atmosphere and surface. The temperature structure of the Earth and the atmosphere is a result of numerous physical, chemical, and dynamic processes. In a one-dimensional context, the temperature structure is determined by the balance between radiative and convective processes.

The Earth's surface emits electromagnetic radiation according to the laws that govern a blackbody or a graybody. A blackbody absorbs the maximum radiation and at the same time emits that same amount of radiation so that thermodynamic equilibrium is achieved as to define a uniform temperature. A graybody is characterized by incomplete absorption and emission and is said to have emissivity less than unity. The thermal infrared emissivities from water and land surfaces are normally between 90 and 95%. It is usually assumed that the Earth's surfaces are approximately black in the analysis of infrared radiative transfer. Exceptions include snow and some sand surfaces whose emissivities are wavelength-dependent and could be less than 90%. Absorption and emission of radiation by atmospheric molecules are more complex and require a fundamental understanding of quantum mechanics. See ATMOSPHERE; HEAT BALANCE, TERRESTRIAL ATMOSPHERIC.

The radiant energy emitted from a number of temperatures covering the Earth and the atmosphere is measured as a function of wavenumber and wavelength. This energy is called Planck intensity (or radiance), and the units that are commonly used are denoted as watt per square meter per solid angle per wavenumber ($W/m^2 \cdot sr \cdot cm^{-1}$). Terrestrial radiation originating from the Earth-atmosphere-ocean system, as well as solar radiation reflected and scattered back to space, is measured on a daily basis by meteorological satellites. Instruments on meteorological satellites measure visible, ultraviolet, infrared, and microwave radiation. See METEOROLOGICAL SATELLITES.

Each spectral region provides meteorologists and other Earth system scientists with information about atmospheric ozone, water vapor, temperature, aerosols, clouds, precipitation, lightning, and many other parameters. Measuring atmospheric radiation allows the detection of sea and land temperature, snow and ice cover, and winds at the surface of the ocean. By tracking the movement of clouds and other atmospheric features, such as aerosols and water vapor, it is possible to obtain estimates of winds above the surface. See SATELLITE METEOROLOGY.   [K.N.L.; T.H.V.H.]

Terrestrial water   The total inventory of water on the Earth. Water is unevenly distributed over the Earth's surface in oceans, rivers, and lakes. In addition, the world's water is distributed throughout the atmosphere and also occurs as soil moisture, groundwater, ice caps, and glaciers. See ATMOSPHERE; GLACIOLOGY; GROUND-WATER HYDROLOGY; HYDROLOGY; LAKE; SURFACE WATER.   [R.L.N.]

Tertiary   The older major subdivision (period) of the Cenozoic Era, extending from the Cretaceous (top of the Mesozoic Era) to the beginning of the Quaternary (younger Cenozoic Period). The term Tertiary corresponds to all the rocks and fossils formed during this period. Typical sedimentary rocks include widespread limestones, sandstones, mudstones, marls, and conglomerates deposited in both marine and terrestrial environments; igneous rocks include extrusive and intrusive volcanics as well as rocks formed deep in the Earth's crust (plutonic). *See* CRETACEOUS; FOSSIL; ROCK.

The Tertiary Period is characterized by a rapid expansion and diversification of marine and terrestrial life. In the marine realm, a major radiation of oceanic microplankton occurred following the terminal Cretaceous extinction events. This had its counterpart on land in the rapid diversification of multituberculates, marsupials, and insectivores—holdovers from the Mesozoic—and primates, rodents, and carnivores, among others, in the ecologic space vacated by the demise of the dinosaurs and other terrestrial forms. Shrubs and grasses and other flowering plants diversified in the middle Tertiary, as did marine mammals such as cetaceans (whales), which returned to the sea in the Eocene Epoch. The pinnipeds (walruses, sea lions, and seals) are derived from land carnivores, or fissipeds, and originated in the Neogene temperate waters of the North Atlantic and North Pacific. Indeed, the great diversification on land and in the sea of birds and, particularly, mammals has led to the informal designation of the Tertiary as the Age of Mammals in textbooks on historical geology.

The modern configuration of continents and oceans developed during the Cenozoic Era as a result of the continuing process known as plate tectonics. Mountain-building events (orogenies) and uplifts of large segments of the Earth's crust (epeirogenies) alternated with fluctuating transgressions and regressions of the seas over land. The middle to late Tertiary Alpine-Himalayan orogeny and the late Tertiary Cascadian orogeny led to the east-west and north-south mountain ranges, respectively, which are located in Eurasia and western North America. *See* CORDILLERAN BELT; MOUNTAIN SYSTEMS; OROGENY; PLATE TECTONICS.   [W.A.Ber.]

Tetrahedrite   A mineral having the composition $(Cu,Fe,Zn,Ag)_{12}Sb_4S_{13}$. It is massive or granular. Its hardness is $3^1/_2-4$ on Mohs scale and the specific gravity varies from 4.6 to 5.1, depending on the composition. The luster is metallic and the color grayish black; thus, in some mining localities, this mineral is called gray copper.

Tetrahedrite is a widely distributed mineral, usually found in silver and copper veins. In some places it has sufficient silver to be a valuable ore, as at Freiberg, Germany, and silver mines of Peru, Bolivia, and Mexico. It is found in silver and copper mines in the western United States.   [C.S.Hu.]

Theoretical ecology   The use of models to explain patterns, suggest experiments, or make predictions in ecology. Because ecological systems are idiosyncratic, extremely complex, and variable, ecological theory faces special challenges. Unlike physics or genetics, which use fundamental laws of gravity or of inheritance, ecology has no widely accepted first-principle laws. Instead, different theories must be invoked for different questions, and the theoretical approaches are enormously varied. A central problem in ecological theory is determining what type of model to use and what to leave out of a model. The traditional approaches have relied on analytical models based on differential or difference equations; but recently the use of computer simulation has greatly increased. *See* ECOLOGY; ECOLOGY, APPLIED.

The nature of ecological theory varies depending on the level of ecological organization on which the theory focuses. The primary levels of ecological organization are (1) physiological and biomechanical, (2) evolutionary (especially applied to behavior), (3) population, and (4) community.

At the physiological and biomechanical level, the goals of ecological theory are to understand why particular structures are present and how they work. The approaches of fluid dynamics and even civil engineering have been applied to understanding the structures of organisms, ranging from structures that allow marine organisms to feed, to physical constraints on the stems of plants.

At the behavioral evolutionary level, the goals of ecological theory are to explain and predict the different choices that individual organisms make. Underlying much of this theory is an assumption of optimality: the theories assume that evolution produces an optimal behavior, and they attempt to determine the characteristics of the optimal behavior so it can be compared with observed behavior. One area with well-developed theory is foraging behavior (where and how animals choose to feed). Another example is the use of game theory to understand the evolution of behaviors that are apparently not optimal for an individual but may instead be better for a group.

The population level has the longest history of ecological theory and perhaps the broadest application. The simplest models of single-species populations ignore differences among individuals and assume that the birth rates and death rates are proportional to the number of individuals in the population. If this is the case, the rate of growth is exponential, a result that goes back at least as far as Malthus's work in the 1700s. As Malthus recognized, this result produces a dilemma: exponential growth cannot continue unabated. Thus, one of the central goals of population ecology theory is to determine the forces and ecological factors that prevent exponential growth and to understand the consequences for the dynamics of ecological populations. *See* POPULATION ECOLOGY.

Modifications and extensions of theoretical approaches like the logistic model (which uses differential equations to explain the stability of populations) have also been used to guide the management of renewable natural resources. Here, the most basic concept is that of the maximum sustainable yield, which is the greatest level of harvest at which a population can continue to persist.

The primary goal of ecological theory at the community level is to understand diversity at local and regional scales. Recent work has emphasized that a great deal of diversity in communities may depend on trade-offs. For example, a trade-off between competitive prowess and colonization ability is capable of explaining why so many plants persist in North American prairies. Another major concept in community theory is the role of disturbance. Understanding how disturbances (such as fires, hurricanes, or wind storms) impacts communities is crucial because humans typically alter disturbance. *See* BIODIVERSITY; ECOLOGICAL COMMUNITIES.   [A.Has.]

Thermal ecology   The examination of the independent and interactive biotic and abiotic components of naturally heated environments. Geothermal habitats are present from sea level to the tops of volcanoes and occur as fumaroles, geysers, and hot springs. Hot springs typically possess source pools with overflow, or thermal, streams (rheotherms) or without such streams (limnotherms). Hot spring habitats have existed since life began on Earth, permitting the gradual introduction and evolution of species and communities adapted to each other and to high temperatures. Other geothermal habitats do not have distinct communities.

Hot-spring pools and streams, typified by temperatures higher than the mean annual temperature of the air at the same locality and by benthic mats of various colors, are found on all continents except Antarctica. They are located in regions of geologic activity where meteoric water circulates deep enough to become heated. The greatest densities occur in Yellowstone National Park (Northwest United States), Iceland, and New Zealand. Source waters range from 40°C (104°F) to boiling (around 100°C or 212°F depending on elevation), and may even be superheated at the point of emergence. Few hot springs have pH 5–6; most are either acidic (pH 2–4) or alkaline (pH 7–9).   [C.Wi.]

Thermal wind   The difference in the geostrophic wind between two heights in the atmosphere over a given position on Earth. It approximates the variation of the actual winds with height for large-scale and slowly changing motions of the atmosphere. Such structure in the wind field is of fundamental importance to the description of the atmosphere and to processes causing its day-to-day changes. The thermal wind embodies a basic relationship between vertical fluctuations of the horizontal wind and horizontal temperature gradients in the atmosphere. This relationship arises from the combination of the geostrophic wind law, the hydrostatic equation, and the gas law.

The geostrophic wind law applies directly to steady, straight, and unaccelerated horizontal motion and is a good approximation for large-scale and slowly changing motions in the atmosphere. The hydrostatic equation combined with the gas law relates the atmospheric pressure and temperature fields. The relationship is accurate for most atmospheric situations but not for small scale and rapidly changing conditions such as in turbulence and thunderstorms. The equation gives the change of pressure in the vertical direction as a function of pressure and temperature. The key conclusion is that at a given level in the atmosphere the pressure change (decrease) with height is more rapid in cold air than in warm air. *See* ATMOSPHERE; GEOSTROPHIC WIND; TROPOSPHERE.

[D.D.H.]

Thermosphere   A rarefied portion of the atmosphere, lying in a spherical shell between 50 and 300 mi (80 and 500 km) above the Earth's surface, where the temperature increases dramatically with altitude. The thermosphere responds to the variable outputs of the Sun, the ultraviolet radiation at wavelengths less than 200 nanometers, and the solar wind plasma that flows outward from the Sun and interacts with the Earth's geomagnetic field. This interaction energizes the plasma, accelerates charged particles into the thermosphere, and produces the aurora borealis and aurora australis, which are nearly circular-shaped regions of luminosity that surround the magnetic north and south poles respectively. Embedded within the thermosphere is the ionosphere, a weakly ionized plasma. *See* IONOSPHERE; MAGNETOSPHERE.

In the thermosphere, these molecular species are subjected to intense solar ultraviolet radiation and photodissociation that gradually turns the molecular species into the atomic species oxygen, nitrogen, and hydrogen. Up to above 60 mi (100 km), atmospheric turbulence keeps the atmosphere well mixed, with the molecular concentrations dominating in the lower atmosphere. Above 60 mi, solar ultraviolet radiation most strongly dissociates molecular oxygen, and there is less mixing from atmospheric turbulence. The result is a transition area where molecular diffusion dominates and atmospheric species settle according to their molecular and atomic weights. Above 60 mi, atomic oxygen is the dominant species. *See* ATMOSPHERE.

About 60% of the solar ultraviolet energy absorbed in the thermosphere and ionosphere heats the ambient neutral gas and ionospheric plasma; 20% is radiated out of the thermosphere as airglow from excited atoms and molecules; and 20% is stored as chemical energy of the dissociated oxygen and nitrogen molecules, which is released later when recombination of the atomic species occurs. Most of the neutral gas heating that establishes the basic temperature structure of the thermosphere is derived from excess energy released by the products of ion-neutral and neutral chemical reactions occurring in the thermosphere and ionosphere. *See* AIRGLOW.

The average vertical temperature profile is determined by a balance of local solar heating by the downward conduction of molecular thermal product to the region of minimum temperature near 50 mi (80 km). For heat to be conducted downward within the thermosphere, the temperature of the thermosphere must increase with altitude. The global mean temperature increases from about 200 K (−100°F) near 50 mi to

700–1400 K (800–2100°F) above 180 mi (300 km), depending upon the intensity of solar ultraviolet radiation reaching the Earth. Above 180 mi, molecular thermal conduction occurs so fast that vertical temperature differences are largely eliminated; the isothermal temperature in the upper thermosphere is called the exosphere temperature.

As the Earth rotates, absorption of solar energy in the thermosphere undergoes a daily variation. Dayside heating causes the atmosphere to expand, and the loss of heat at night causes it to contract. This heating pattern creates pressure differences that drive a global circulation, transporting heat from the warm dayside to the cool nightside.   [R.G.R.]

Thorianite   A radioactive mineral with the idealized composition ThO_2 (thorium dioxide) and isostructural with uraninite (pitchblende). Rare earths and uranium are often present in variable amounts, together with small amounts of radiogenic lead. Thorianite usually occurs as worn cubic crystals. The hardness is about 7 on Mohs scale, and the specific gravity is 9.7–9.8. The color is brownish black to reddish brown, and the luster usually is resinous. Thorianite is a primary mineral found chiefly in pegmatites. It is best known as a detrital mineral. It has been obtained commercially from detrital deposits and pegmatites in Madagascar and Sri Lanka. See PEGMATITE; RADIOACTIVE MINERALS; URANINITE.   [C.Fr.]

Thorite   A mineral, thorium silicate. The idealized chemical formula of thorite is $ThSiO_4$. All natural material departs widely from this composition owing to the partial substitution of uranium, rare earths, calcium, and iron for thorium. The specific gravity ranges between about 4.3 and 5.4. The hardness on Mohs scale is about $4^1/_2$. The color commonly is brownish yellow to brownish black and black.

Vein deposits containing thorite occur in Colorado, Idaho, and Montana. A vein deposit of monazite containing thorium is mined at Steenkampskraal near Van Rhynsdorp, Cape Province, South Africa. See METAMICT STATE; RADIOACTIVE MINERALS; SILICATE MINERALS.   [C.Fr.]

Thunder   The acoustic radiation produced by thermal lightning channel processes. The lightning return stroke is a high surge of electric current that has a very short duration, depositing approximately 95% of its electrical energy during the first 20 microseconds. Spectroscopic studies have shown that the lightning channel is heated to temperatures in the 20,000–30,000 K (36,000–54,000°F) range by this process. The pressure of the hot channel exceeds 10 atm ($>10^6$ pascals). The hot, high-pressure channel expands supersonically and forms a shock wave as it pushes against the surrounding air. Because of the momentum gained in expanding, the shock wave overshoots, causing the pressure in the core of the channel to go below atmospheric pressure temporarily. The outward-propagating wave separates from the core of the channel, forming an N-shaped wave that eventually decays into an acoustic wavelet. See STORM ELECTRICITY.

The sound that is eventually heard or detected, thunder, is the sum of many individual acoustic pulses, each a remnant of a shock wave, that have propagated to the point of observation from the generating channel segments. The first sounds arrive from the nearest part of the lightning channel and the last sounds from the most distant parts.

The higher the source of the sound, the farther it can be heard. Frequently, the thunder that is heard originates in the cloud and not in the visible channel. On some occasions, the observer may hear no thunder at all; this is more frequent at night when lightning can be seen over long distances and thunder can be heard only over a limited range (~10 km or 6 mi). *See* LIGHTNING; THUNDERSTORM. [A.A.F.]

Thunderstorm A convective storm accompanied by lightning and thunder and a variety of weather such as locally heavy rainshowers, hail, high winds, sudden temperature changes, and occasionally tornadoes. The characteristic cloud is the cumulonimbus or thunderhead, a towering cloud, generally with an anvil-shaped top. A host of accessory clouds, some attached and some detached from the main cloud, are often observed in conjunction with cumulonimbus. *See* LIGHTNING; THUNDER.

Thunderstorms are manifestations of convective overturning of deep layers in the atmosphere and occur in environments in which the decrease of temperature with height (lapse rate) is sufficiently large to be conditionally unstable and the air at low levels is moist. In such an atmosphere, a rising air parcel, given sufficient lift, becomes saturated and cools less rapidly than it would if it remained unsaturated because the released latent heat of condensation partly counteracts the expansional cooling. The rising parcel reaches levels where it is warmer (by perhaps as much as $18°F$ or $10°C$ over continents) and less dense than its surroundings, and buoyancy forces accelerate the parcel upward. The rising parcel is decelerated and its vertical ascent arrested at altitudes where the lapse rate is stable, and the parcel becomes denser than its environment. The forecasting of thunderstorms thus hinges on the identification of regions where the lapse rate is unstable, low-level air parcels contain adequate moisture, and surface heating or uplift of the air is expected to be sufficient to initiate convection. *See* CONVECTIVE INSTABILITY; FRONT.

Thunderstorms are most frequent in the tropics, and rare poleward of 60° latitude. Thunderstorms are most common during late afternoon because of the diurnal influence of surface heating.

Radar is used to detect thunderstorms at ranges up to 250 mi (400 km) from the observing site. Much of present-day knowledge of thunderstorm structure has been deduced from radar studies, supplemented by visual observations from the ground and satellites, and in-place measurements from aircraft, surface observing stations, and weather balloons. *See* METEOROLOGICAL INSTRUMENTATION; RADAR METEOROLOGY; SATELLITE METEOROLOGY.

Thunderstorms are considered severe when they produce winds greater than 58 mi/h (26 m/s or 50 knots), hail larger than 3/4 in. (19 mm) in diameter, or tornadoes. While thunderstorms are generally beneficial because of their needed rains (except for occasional flash floods), severe storms have the capacity of inflicting utter devastation over narrow swaths of the countryside. Severe storms are most frequently supercells which form in environments with high convective instability and moderate-to-large vertical wind shears. The supercell may be an isolated storm or part of a squall line. *See* HAIL; SQUALL; SQUALL LINE; TORNADO. [R.D.J.]

Tidal bore A part of a tidal rise in a river which is so rapid that water advances as a wall often several feet high. The phenomenon is favored by a substantial tidal range and a channel which shoals and narrows rapidly upstream, but the conditions are so critical that it is not common. Although the bore is a very striking feature, the tide continues to rise after the passage of the bore. Bores may be eliminated by changing channel depth or shape. *See* RIVER TIDES; TIDE.

In North America three bores have been observed: at the head of the Bay of Fundy (see illustration), at the head of the Gulf of California, and at the head of Cook Inlet, Alaska. The largest known bore occurs in the Tsientang Kiang, China. At spring tides

Tidal bore of the Petitcodiac River, Bay of Fundy, New Brunswick, Canada. Rise of water is about 4 ft (1.2 m). (*New Brunswick Travel Bureau*)

this bore is a wall of water 15 ft (4.5 m) high moving upstream at 25 ft/s (7.5 m/s).

[B.K.]

Tidal datum   A reference elevation of the sea surface from which vertical measurements are made, such as depths of the ocean and heights of the land. The intersection of the elevation of a tidal datum with the sloping shore forms a line used as a horizontal boundary. In turn, this line is also a reference from which horizontal measurements are made for the construction of additional coastal and marine boundaries.

Since the sea surface moves up and down from infinitely small amounts to hundreds of feet over periods of less than a second to millions of years, it is necessary to stop the vertical motion in order to have a practical reference. This is accomplished by hydraulic filtering, numerical averaging, and segment definition of the record obtained from a tide gage affixed to the adjacent shore. Waves of periods up through wind waves are effectively damped by a restricting hole in the measurement well. Recorded hourly heights are averaged to determine the mean of the higher (or only) high tide of each tidal day (24.84 h), all the high tides, all the hourly heights, all the low tides, and the lower (or only) low tide. The length of the averaging segment is a specific 19 year, which averages all the tidal cycles through the regression of the Moon's nodes and the metonic cycle. [The metonic cycle is a time period of 235 lunar months (19 years); after this period the phases of the Moon occur on the same days of the same months.] But most of all, the 19-year segment is meaningful in terms of measurement capability, averaging meteorological events, and for engineering and legal interests. However, the 19-year segment must be specified and updated because of sea-level changes occurring over decades. The present tidal datum epoch is 1983 through 2001.

Tidal datums are legal entities. Because of variations in gravity, semistationary meteorological conditions, semipermanent ocean currents, changes in tidal characteristics, ocean density differences, and so forth, the sea surface (at any datum elevation) does not conform to a mathematically defined spheroid. *See* GEODESY; TIDE.   [S.D.H.]

Tidalites   Sediments of varied composition deposited by tidal processes in subtidal, intertidal, and supratidal environments. Tides along modern coastlines produce flood and ebb currents which advance and retreat between high- and low-water levels and also operate in subtidal environments. These currents are largely responsible for the formation of tidalites. *See* TIDE.

The tidal flat, including the supratidal and intertidal zones, is the best understood of the tidal environments. In these zones both terrigenous and nonterrigenous tidal flats occur, the former often incised by tidal channels.

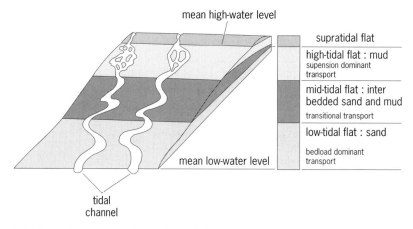

Tidal-flat sedimentation model for the North Sea coasts of Germany and the Netherlands. (*After G. de V. Klein, Clastic Tidal Facies, Continuing Educational Publication Company, 1977*)

Terrigenous tidal flats consist of an intertidal zone that is subdivided into a low, mid, and high tidal flat (see illustration). Each of the specific tidal processes operate to produce sediments with characteristic textures and sedimentary structures. On extensive tidal flats, fine-grained sediments accumulate from suspension near high-water line, and coarser sandy sediments are deposited by bed-load processes around low-water level. On mid flats, interbedded sands and muds are developed under conditions of alternating bed-load and suspension sedimentation (see illustration). The supratidal flat receives limited quantities of marine sediment, and instead comprises salt marshes and mangrove swamps in humid climates, with sabkhas, consisting of evaporite minerals such as gypsum and halite, commonly developed under arid conditions.

Nonterrigenous tidal flats do not receive terrigenous clastic sediment. Instead, sediment which accumulates on these tidal flats consists predominantly of calcium carbonate produced within the basin. Various organisms which live mainly below low-water level form calcium carbonate in their life cycle, and upon death this hard framework breaks down to produce clay, silt, and sand-size carbonate particles. These are available for reworking both within the subtidal zone and onto the tidal flats.

Sedimentation below low-water level and under the influence of tides is occurring today in shallow shelf seas. The best understood is the North Sea, where sand transport by tidal currents is taking place down to depths of 150 ft (50 m). Echo sounding has shown that the sands form linear bodies up to 90 ft (30 m) high. They are asymmetric in cross section and, as revealed by seismic profiling, are structured internally by giant cross-strata.

The recognition of tidalites in the geological record requires evidence of a marine environment, ebb and flood current flow, and exposure of the depositional surface during low-water stage for intertidalites. *See* DEPOSITIONAL SYSTEMS AND ENVIRONMENTS; SEDIMENTOLOGY. [K.A.Eri.]

Tide   Stresses exerted in a body by the gravitational action of another, and related phenomena resulting from these stresses. Every body in the universe raises tides, to some extent, on every other. This article deals only with tides on the Earth, since these are fundamentally the same as tides on all bodies, and more specifically with variations of sea level, whatever their origin.

The tide-generating forces arise from the gravitational action of Sun and Moon, the effect of the Moon being about twice as effective as that of the Sun in producing tides. The tidal effects of all other bodies on the Earth are negligible. The tidal forces act to generate stresses in all parts of the Earth and give rise to relative movements of the matter of the solid Earth, ocean, and atmosphere. In the ocean the tidal forces act to generate alternating tidal currents and displacements of the sea surface.

If the Moon attracted every point within the Earth with equal force, there would be no tide. It is the small difference in direction and magnitude of the lunar attractive force, from one point of the Earth's mass to another, which gives rise to the tidal stresses. The tide-generating force is proportional to the mass of the disturbing body (Moon) and to the inverse cube of its distance. This inverse cube law accounts for the fact that the Moon is 2.17 times as important, insofar as tides are concerned, as the Sun, although the latter's direct gravitational pull on the Earth, which is governed by an inverse-square law, is about 180 times the Moon's pull.

At most places in the ocean and along the coasts, sea level rises and falls in a regular manner. The highest level usually occurs twice in any lunar day, the times bearing a constant relationship with the Moon's meridional passage. The time between the Moon's meridional passage and the next high tide is called the lunitidal interval. The difference in level between successive high and low tides, called the range of the tide, is generally greatest near the time of full or new Moon, and smaller near the times of quadrature. The range of the tide usually exhibits a secondary variation, being greater near the time of perigee (when the Moon is closest to the Earth) and smaller at apogee (when the Moon is farthest away).

The above situation is observed at places where the tide is predominantly semidiurnal. At many other places, it is observed that one of the two maxima in any lunar day is higher than the other. This effect is known as the diurnal inequality and represents the presence of an appreciable diurnal variation. At these places, the tide is said to be of the "mixed" type. At a few places, the diurnal tide actually predominates, there generally being only one high and low tide during the lunar day.

The range of the ocean tide varies between wide limits. The highest range is encountered in the Bay of Fundy, where values exceeding 50 ft (15 m) have been observed. In some places in the Mediterranean, South Pacific, and Arctic, the tidal range never exceeds 2 ft (0.6 m).

Owing to the rotation of the Earth, there is a gyroscopic, or Coriolis, force acting perpendicularly to the motion of any water particle in motion. In the Northern Hemisphere this force is to the right of the current vector. The horizontal, or tractive, component of the tidal force generally rotates in the clockwise sense in the Northern Hemisphere. As a result of both these influences the tidal currents in the open ocean generally rotate in the clockwise sense in the Northern Hemisphere, and in the counterclockwise sense in the Southern Hemisphere. *See* CORIOLIS ACCELERATION; EARTH TIDES.   [G.W.G.]

Till   The generic term for sediment deposited directly from glacier ice. Till is characteristically nonsorted and nonstratified and is deposited by lodgement or melt-out beneath a glacier or by melt-out on the surface of a glacier. The texture of till varies greatly and all tills are characterized by a wide range, of particle sizes. Till contains a variety of rock and mineral fragments which reflect the source material over which the glacier flowed. The particles in the deposit usually show a preferred orientation related to the nature and direction of the ice flow. The overall character of the fill reflects the source material, position and distance of transport, nature and position of deposition, and postdepositional changes.   [W.H.J.]

Titanite   A calcium, titanium silicate, $CaTiOSiO_4$, of high titanium content. Titanite is also known as sphene. Titanite is an orthosilicate (nesosilicate) with a hardness of $5–5\frac{1}{2}$ on the Mohs scale, a distinct cleavage, a specific gravity of 3.4–3.55, and an adamantine to resinous luster. It commonly occurs as distinct wedge-shaped crystals that are usually brown in hand specimens. Titanite may also be gray, green, yellow, or black.

Titanite is a common accessory mineral in many igneous and metamorphic rocks. It may be the principal titanium-bearing silicate mineral. It occurs in abundance in the Magnet Cove, igneous complex in Arkansas and in the intrusive alkalic-rocks of the Kola Penninsula, Russia.

The composition of titanite may diverge from pure $CaTiSiO_4$ because of a variety of chemical substitutions. Calcium ions (Ca^{2+}) can be partially replaced by strontium ions (Sr^{2+}) and rare-earth ions such as thorium (Th^{4+}) and uranium (U^{4+}). Because titanite commonly contains radioactive elements, it has been used for both uranium-lead and fission track methods of dating. *See* DATING METHODS; IGNEOUS ROCKS; METAMORPHIC ROCKS; SILICATE MINERALS.   [J.C.D.]

Todorokite   A hydrated manganese oxide mineral containing calcium, barium, potassium, sodium, and sometimes magnesium, general formula $(Na,Ca,K,Ba,Sr)_{0.3–0.7}$ $(Mn,Mg,Al)_6O_{12} \cdot 3.2–4.5\ H_2O$. Todorokite is a major constituent of manganese nodules, which occur in large quantities ($>10^{12}$ tons) on the ocean floors. It has been shown to host some of the copper, nickel, and cobalt that occur in some manganese nodules in quantities of up to several weight percent. First described in 1934, todorokite is named for its occurrence at the Todoroki mine, Hokkaido, Japan. Terrestrial todorokite has also been found in deposits in Cuba, Portugal, Austria, France, United States, Brazil, and South Africa.

Todorokite is black to brown and is commonly very fine grained. It occurs as massive samples or as fibrous aggregates. Its hardness on the Mohs scale is low (1.5–2.5), and its reported specific gravity ranges from approximately 3.1 to 3.8.

The basic todorokite structure consists of triple chains of manganese-oxygen octahedra that are linked at roughly right angles to form a tunnel structure. This [3 × 3] structure can accommodate large cations and water in the tunnel sites. *See* MANGANESE NODULES.   [S.T.]

Topaz   A mineral best known for its use as a gemstone. Crystals are usually colorless but may be red, yellow, green, blue, or brown. The wine-yellow variety is the one usually cut and most highly prized as a gem. Corundum of similar color sometimes goes under the name of Oriental topaz. Citrine, a yellow variety of quartz, is the most common substitute and may be sold as quartz topaz.

Topaz is a nesosilicate with chemical composition $Al_2SiO_4(F,OH)_2$. The mineral crystallizes in the orthorhombic system and is commonly found in well-developed prismatic crystals with pyramidal terminations. It has a perfect basal cleavage which enables it to be distinguished from minerals otherwise similar in appearance. Hardness is 8 on Mohs scale; specific gravity is 3.4–3.6.

Fine yellow and blue crystals have come from Siberia and much of the wine-yellow gem material from Minas Gerais, Brazil. In the United States topaz has been found near Florissant, Colorado; in Thomas Range, Utah; in San Diego County, California; and near Topsham, Maine. *See* GEM; SILICATE MINERALS.   [C.S.Hu.]

Topographic surveying and mapping   The measurement of surface features and configuration of an area or a region, and the graphic expression of those features. Surveying is the art and science of measurement of points on, above, or under the surface of the Earth. Topographic maps show the natural and cultural features of a piece of land. The natural features include configuration (relief), hydrography, and vegetation. The cultural features include roads, buildings, bridges, political boundaries, and the sectional breakdown of the land. Topographic maps are used by a wide variety of people, such as engineers designing a new road; backpackers finding their way into remote areas; scientists describing soil or vegetation types, wildlife habitat, or hydrology; and military personnel planning field operations. *See* CARTOGRAPHY; MAP PROJECTIONS.

Topographic maps that show natural and cultural features only in plan view are called planimetric maps, while maps that show relief are called hypsometric maps. Contour lines join points along a line of the same elevation across the ground. Contours show not only the elevation of the ground but also the geomorphic shape of features. *See* CONTOUR.

A digital terrain model (DTM) is a computer-generated grid laid over the topographic information, which can then be rotated, tilted, and vertically exaggerated to give a three-dimensional view of the ground from different perspectives, including oblique representations. This technology is an excellent presentation tool: it utilizes the advances that have been made in computer mapping and drafting software.

Prior to starting collection of data, a network of known horizontal and vertical control points must be established. The network also allows measurements made from several different locations in the same coordinate system to fit together into the same reference datum (the basis for the coordinate system). Field methods that are used to make measurements include ground surveys, geographic positioning systems, and hydrographic surveys. Photogrammetry and remote sensing techniques involve the use of photography to obtain reliable measurements. Photographs can be taken from airplanes, helicopters, and even satellites; thus the term remote sensing is applied to this technology. *See* HYDROGRAPHY; REMOTE SENSING.   [C.Br.]

Torbanite   A variety of coat that resembles a carbonaceous shale in outward appearance. It is fine-grained, brown to black, and tough, and breaks with a conchoidal or subconchoidal fracture. Torbanite is synonymous with boghead coal and is related to cannel coal. It is derived from colonial algae identified with the modern species of *Botryococcus braunii* and antecedent forms. High-assay torbanite yields paraffinic oil, whereas low-assay material yields asphaltic oil. *See* COAL.   [I.A.B.]

Tornado   A violently rotating, tall, narrow column of air (vortex), typically about 300 ft (100 m) in diameter, that extends to the ground from a cumulonimbus cloud. The vast majority of tornadoes rotate cyclonically (counterclockwise in the Northern Hemisphere). Of all atmospheric storms, tornadoes are the most violent. *See* CLOUD; CYCLONE.

Tornadoes are made visible by a generally sharp-edged, funnel-shaped cloud pendant from the cloud base, and a swirling cloud of dust and debris rising from the ground (see illustration). The funnel consists of small waterdroplets that form as moist air entering the tornado's partial vacuum expands and cools. The condensation funnel may not extend all the way to the ground and may be obscured by dust. Many condensation funnels exist aloft without tangible signs that the vortex is in contact with the ground; these are known as funnel clouds. Tornado funnels assume various forms: a slender smooth rope, a cone (often truncated by the ground), a thick turbulent black cloud on the ground, or multiple funnels (vortices) that revolve around the axis of the overall tornado.

The Cordell, Oklahoma, tornado of May 22, 1981, in its decay stage. (*National Severe Storms Laboratory/University of Mississippi Tornado Intercept Project*)

Many tornadoes evolve as follows: The tornado begins outside the precipitation region as a dust whirl on the ground and a short funnel pendant from a wall cloud on the southwest side of the thunderstorm; it intensifies as the funnel lengthens downward, and attains its greatest power as the funnel reaches its greatest width and is almost vertical; then it shrinks and becomes more tilted, and finally becomes contorted and ropelike as it decays. A downdraft and curtain of rain and large hail gradually spiral from the northeast cyclonically around the tornado, which often ends its life in rain. *See* HAIL; PRECIPITATION (METEOROLOGY); THUNDERSTORM.

Most tornadoes and practically all violent ones develop from a larger-scale circulation, the mesocyclone, which is 2–6 mi (3–9 km) in diameter and forms in a particularly virulent variety of thunderstorm, the supercell. The mesocyclone forms first at midaltitudes of the storm and in time develops at low levels and may extend to high altitudes as well. The tornado forms on the southwest side (Northern Hemisphere) of the storm's main updraft, close to the downdraft, after the development of the mesocyclone at low levels. Some supercells develop up to six mesocyclones and tornadoes repeatedly over great distances at roughly 45-min intervals. Tornadoes associated with supercells are generally of the stronger variety and have larger parent cyclones. Hurricanes during and after landfall may spawn numerous tornadoes from small supercells located in their rainbands. *See* HURRICANE.

Tornadoes are classified as weak, strong, or violent, or from F0 to F5 on the Fujita (F) scale of damage intensity. Sixty-two percent of tornadoes are weak (F0 to F1). These tornadoes have maximum windspeeds less than about 50 m/s (110 mi/h) and inflict only minor damage, such as peeling back roofs, overturning mobile homes, and pushing cars into ditches. Thirty-six percent of tornadoes are strong (F2 to F3) with maximum windspeeds estimated to be 50–90 m/s (110–200 mi/h). Strong tornadoes extensively damage the roofs and walls of houses but leave some walls partially standing. They demolish mobile homes, and lift and throw cars. The remaining 2% are violent (F4 to F5), with windspeeds in excess of about 90 m/s (200 mi/h). They level houses to their foundations, strew heavy debris over hundreds of yards, and make missiles out of heavy objects such as roof sections, vehicles, utility poles, and large, nearly empty storage tanks. *See* WIND.

Tornadoes occur most often at latitudes between 20° and 60°, and they are relatively frequent in the United States, Russia, Europe, Japan, India, South Africa, Argentina, New Zealand, and parts of Australia. Violent tornadoes are confined mainly to the United States, east of the Rocky Mountains.

Essentially, there are five atmospheric conditions that set the stage for wide-spread tornado development: (1) a surface-based layer, at least 3000 ft (1 km) deep, of warm, moist air, overlain by dry air at midlevels; (2) an inversion separating the two layers, preventing deep convection until the potential for explosive overturning is established; (3) rapid decrease of temperature with height above the inversion; (4) a combination of mechanisms, such as surface heating and lifting of the air mass by a front or upper-level disturbance, to eliminate the inversion locally; (5) pronounced vertical wind shear (variation of the horizontal wind with height). Specifically, storm-relative winds in the lowest 6000 ft (2 km) should exceed 20 knots (10 m/s) and veer (turn anticyclonically) with height at a rate of more than 10°/1000 ft (30°/km). Such conditions are prevalent in the vicinity of the jet stream and the low-level jet.

The first three conditions above indicate that the atmosphere is in a highly metastable state. There is a strong potential for thunderstorms with intense updrafts and down-drafts. The fourth condition is the existence of a trigger to release the instability and initiate the thunderstorms. The fifth is the ingredient for updraft rotation. *See* AIR MASS; FRONT; JET STREAM; TEMPERATURE INVERSION.   [R.D.J.]

Tourmaline   A cyclosilicate mineral family with (BO_3) triangular groups and a complex chemical composition. The general formula can be written $XY_3Al_6(OH)_4(BO_3)_3(Si_6O_{18})$, in which X = Na, Ca, and Y = Al, Fe^{3+}, Li, Mg, Mn^{2+}. The more common tourmalines are dravite (X,Y = Na,Mg), schorl (X,Y = Na,Fe), uvite (X,Y = Ca,Mg), and elbaite (X,Y = Na,Li). Fluorine commonly substitutes in the hydroxyl position. Tourmaline is a hard ($7\frac{1}{2}$ on Mohs scale), varicolored mineral which can be an important semiprecious gemstone. *See* GEM; SILICATE MINERALS.

The tourmaline crystal is polar; thus it is piezoelectric; that is, if pressure is exerted at one end, opposite electrical charges will occur at opposite poles. It is also pyroelectric, with the electrical charges developed at the ends of the polar axis on a change in temperature. Because of its piezoelectric property, tourmaline can be cut into gages to measure transient pressures. *See* PYROELECTRICITY.   [P.B.M.]

Trace fossils   Fossilized evidence of animal behavior, also known as ichnofossils, biogenic sedimentary structures, bioerosion structures, or lebensspuren. The fossils include burrows, trails, and trackways created by animals in unconsolidated sediment (see illustration), as well as borings, gnawings, raspings, and scrapings excavated by organisms in harder materials, such as rock, shell, bone, or wood. Some workers also consider coprolites (fossilized feces), regurgitation pellets, burrow excavation pellets, rhizoliths (plant root penetration structures), and algal stromatolites to be trace fossils. *See* STROMATOLITE.

Trace fossils are important in paleontology and paleoecology, because they provide information about the presence of unpreserved soft-bodied members of the original

Agrichnial farming traces (burrows produced in order to farm or trap food inside the sediment) of unknown organisms, including a double-spiral tunnel (*Spirorhaphe*) and a meshlike network of tunnels (*Paleodictyon*). Tertiary, Austria. (*Photograph by W. Häntzschel*)

communities, life habits of fossil organisms, evolution of certain behavior patterns through geologic time, and biostratigraphy of otherwise unfossiliferous deposits. Trace fossils also are useful in sedimentology and paleoenvironmental studies, because they are sedimentary structures that are preserved in place and are very rarely reworked and transported, as body fossils of animals and plants commonly are. This fact allows trace fossils to be regarded as reliable indicators of original conditions in the sedimentary environment. The production of trace fossils involves disruption of original stratification and sometimes results in alteration of sediment texture or composition. *See* FOSSIL; PALEONTOLOGY; SEDIMENTOLOGY.

Trace fossils occur in sedimentary deposits of all ages from the late Precambrian to the Recent. Host rocks include limestone, sandstone, siltstone, shale, coal, and other sedimentary rocks. These deposits represent sedimentation in a broad spectrum of settings, ranging from subaerial (such as eolian dunes and soil horizons) to subaqueous (such as rivers, lakes, swamps, tidal flats, beaches, continental shelves, and the deep-sea floor). *See* DEPOSITIONAL SYSTEMS AND ENVIRONMENTS.

Organisms may produce fossilizable traces on the sediment surface (epigenic structures) or within the sediment (endogenic structures). Trace fossils may be preserved in full three-dimensional relief (either wholly contained within a rock or weathered out as a separate piece) or in partial relief (either as a depression or as a raised structure on a bedding plane). Simply because a trace fossil is preserved on a bedding plane does not indicate that it originally was an epigenic trace. Diagenetic alteration of sediment commonly enhances the preservation of trace fossils by differential cementation or selective mineralization. In some cases, trace fossils have been preferentially replaced by chert, dolomite, pyrite, glauconite, apatite, siderite, or other minerals. *See* DIAGENESIS.

The study of trace fossils is known as ichnology. The prefix "ichno-" (as in ichnofossil and ichnotaxonomy) and the suffix "-ichnia" (as in epichnia and hypichnia) commonly are employed to designate subjects relating to trace fossils. The suffix "-ichnus" commonly is attached to the ichnogenus name of many trace fossils (as in *Dimorphichnus* and *Teichichnus*).   [A.A.E.]

Trachyte   A light-colored, aphanitic (very finely crystalline) rock of volcanic origin, composed largely of alkali feldspar with minor amounts of dark-colored (mafic) minerals (biotite, hornblende, or pyroxene). If sodic plagioclase (oligoclase or andesine) exceeds the quantity of alkali feldspar, the rock is called latite. Trachyte and latite are chemically equivalent to syenite and monzonite, respectively. *See* SYENITE.

Streaked, banded, and fluidal structures due to flowage of the solidifying lava are commonly visible in many trachytes and may be detected by a parallel arrangement of tabular feldspar phenocrysts. A distinctive microscopic feature is trachytic texture in which the tiny, lath-shaped sanidine crystals of the rock matrix are in parallel arrangement and closely packed.

Trachyte is not an abundant rock, but it is widespread. It occurs as flows, tuffs, or small intrusives (dikes and sills). It may be associated with alkali rhyolite, latite, or phonolite. *See* IGNEOUS ROCKS; MAGMA; SPILITE.   [C.A.C.]

Transform fault   One of the three fundamental types of boundaries between the mobile lithospheric plates that cover the surface of the Earth. Whereas spreading centers mark sites where crust is created between diverging plates, and subduction zones are where crust is destroyed between convergent plates, transform faults separate plates that are sliding past each other with neither creation nor destruction of crust. The primary tectonic feature of all transform faults is a strike-slip fault zone, a generally vertical fracture parallel to the relative motion between the two plates that it separates. Strike-slip fault zones are described as right-lateral if the far side is moving right relative to the near side or left-lateral if it is moving to the left. Not all such fault zones are plate-bounding transform faults. Small-scale strike-slip faulting is a common secondary feature of many subduction zones, especially where plate convergence is oblique, and of some spreading centers, especially those with propagating rifts; it also occurs locally deep in plate interiors. The distinguishing characteristic of a transform fault is that both ends extend to a junction with another type of plate boundary. At these junctions the divergent or convergent motion along the other boundaries is transformed into purely lateral slip. *See* EARTH CRUST; PLATE TECTONICS; SUBDUCTION ZONES.

Transform faults are most readily classified by the types of plate boundary intersected at their ends, the variety of lithosphere (oceanic or continental) they separate, and by whether they are isolated or are part of a multifault system. The common oceanic type is the ridge-ridge transform, linking two literally offset axes of a spreading center. Also common are transform faults that link the end of a spreading center to a triple junction, the meeting place of three plates and three plate boundaries. *See* LITHOSPHERE; MID-OCEANIC RIDGE.

Other types are long trench-trench transforms at the northern and southern margins of the Caribbean plate, and the combined San Andreas/Gulf of California transform, which separates the North American and Pacific plates for 1500 mi (2400 km) between triple junctions at Cape Mendocino (California) and the mouth of the Gulf of California. Strike-slip faulting in the Gulf of California (and on the northern Caribbean plate boundary) occurs along several parallel zones linked by short spreading centers, and the overall structure is more properly called a transform fault system; similar fault patterns are found at many ridge-ridge transforms. Along a few strike-slip fault zones, lithospheric plates slide quietly and almost continuously past each other by the process called aseismic creep. Much more often, frictional resistance to the sliding in the brittle crust causes the accumulation of shear stresses that are episodically or periodically relieved by sudden shifts of crustal blocks, creating earthquakes. The largest lateral shifts (slips) of the ground surface along major continental transform faults have been associated with some of the largest earthquakes on record; in 1906 the Pacific plate alongside 270 mi (450 km) of the San Andreas Fault suddenly moved an average of 15 ft (4.5 m) northwest relative to the North American plate on the other side, and the resulting magnitude-8.2 earthquake destroyed much of San Francisco. The average slip in this single event was equivalent to about 150–250 years of Pacific–North American plate motion. *See* EARTHQUAKE; FAULT AND FAULT STRUCTURES; SEISMOLOGY. [P.Lo.]

Travertine A rather dense, banded limestone, sometimes moderately porous, that is formed either by evaporation about springs, as is tufa, or in caves, as stalactites, stalagmites, or dripstone. Where travertine or tufa (calcareous sinter) is deposited by hot springs, it may be the result of the loss of carbon dioxide from the waters as pressure is released upon emerging at the surface; the release of carbon dioxide lowers the solubility of calcium carbonate and it precipitates. High rates of evaporation in hot-spring pools also lead to supersaturation. Travertine formed in caves is simply the result of complete evaporation of waters containing mainly calcium carbonate. *See* LIMESTONE; STALACTITES AND STALAGMITES; TUFA. [R.Si.]

Tremolite The name given to magnesium-rich monoclinic calcium amphibole $Ca_2Mg_5Si_8O_{22}(OH)_2$ The mineral is white to gray, but colorless in thin section. Unlike other end-member compositions of the calcium amphibole group, very pure tremolite is found in nature. Substitution of Fe for Mg is common, but pure ferrotremolite, $Ca_2Fe_5Si_8O_{22}(OH)_2$ is rare. Intermediate compositions between tremolite and ferrotremolite are referred to as actinolites, and are green in color and encompass a large number of naturally occurring calcium amphiboles. *See* AMPHIBOLE. [B.L.D.]

Triassic The oldest period of the Mesozoic Era. The Triassic encompasses a time frame between about 248 and 213 million years ago (m.y.a.) that was named for the threefold division of rocks at its type locality in central Germany, where continental redbeds and evaporites are separated by a marine limestone. *See* MESOZOIC.

The Triassic Period uniquely embraces both the final consolidation of Pangaea and the initial breakup of the landmass, which in the Middle Jurassic led to the origin of the Central Atlantic ocean basin. The Triassic thus marks the beginning of a new cycle of ocean-basin opening, through continental extension, and oceanic closing through subducting oceanic lithospheres along continental margins. *See* LITHOSPHERE; PALEOGEOGRAPHY; PLATE TECTONICS.

The most important tectonic event in the Mesozoic Era was the rifting of the Pangaea craton, which began in the Late Triassic, culminating in the Middle Jurassic with the formation of the Central Atlantic ocean basin. Rifting began in the Tethys region in the Early Triassic, and progressed from western Europe and the Mediterranean into the Central Atlantic off Morocco and eastern North America by the Late Triassic. As crustal extension continued throughout the Triassic, the Tethys seaway spread farther westward and inland. By that time, rifting and sea-floor spreading extended into the Gulf of Mexico, separating North and South America.

Continental rift basins, passive continental margins, and ocean basins form in response to divergent stresses that extend the crust. Crustal extension, as it pertains to the Atlantic, embraces a major tectonic cycle marked by Late Triassic–Early Jurassic rifting and Middle Jurassic to Recent (Holocene) drifting. The rift stage, involving heating and stretching of the crust, was accompanied by uplift, faulting, basaltic igneous activity, and rapid filling of deep elongate rift basins. The drift stage, involving the slow cooling of the lithosphere over a broad region, was accompanied by thermal subsidence with concomitant marine transgression of the newly formed plate margin. The transition from rifting to drifting, accompanied by sea-floor spreading, is recorded by the postrift unconformity. *See* CONTINENTAL DRIFT; CONTINENTAL MARGIN; HOLOCENE; UNCONFORMITY.

Permian-to-Triassic consolidation of Pangaea in western North America led to the Sonoma orogeny (mountain building), which resulted from overthrusting and suturing of successive island-arc and microcontinent terranes to the western edge of the North American Plate. However, toward the end of the Triassic Period, as crustal extension was occurring in the Central Atlantic region, the plate moved westward, overriding the Pacific Plate along a reversed subduction zone. This created for the remainder of the Mesozoic Era an Andean-type plate edge with a subducting sea floor and associated deep-sea trench and magmatic arc. These effects can be studied in the Cordilleran mountain belt. *See* CORDILLERAN BELT; OROGENY.

As these epicontinental seas regressed westward, nonmarine fluvial, lacustrine, and windblown sands were deposited on the craton. Today many of these red, purple, ash-gray, and chocolate-colored beds are some of the most spectacular and colorful scenery in the American West. For example, the Painted Desert of Arizona, which is known for its petrified logs of conifer trees, was developed in the Chinle Formation. *See* CRATON.

Triassic faunas are distinguished from earlier ones by newly evolved groups of plants and animals. In marine communities, molluscan stocks proliferated vigorously. Bivalves diversified greatly and took over most of the niches previously occupied by brachiopods; ammonites proliferated rapidly from a few Permian survivors. The scleractinian (modern) corals appeared, as did the shell-crushing placodont reptiles and the ichthyosaurs. In continental faunas, various groups of reptiles appeared, including crocodiles and crocodilelike forms, the mammallike reptiles, and the first true mammals, as well as dinosaurs.

Triassic land plants contain survivors of many Paleozoic stocks, but the gymnosperms became dominant and cycads appeared. *See* PALEOZOIC.   [W.Man.]

Tropic of Cancer   The parallel of latitude about $23^1/_2{}^\circ$ (23.45°) north of the Equator. The importance of this line lies in the fact that its degree of angle from the Equator is the same as the inclination of the Earth's axis from the vertical to the plane of the ecliptic. Because of this inclination of the axis and the revolution of the Earth in its orbit, the vertical overhead rays of the Sun may progress as far north as $23^1/_2{}^\circ$. At no place north of the Tropic of Cancer will the Sun, at noon, be 90° overhead.

On June 21, the summer solstice (Northern Hemisphere), the Sun is vertical above the Tropic of Cancer. On this same day the Sun is 47° above the horizon at noon at the Arctic Circle, and at the Tropic of Capricorn, only 43° above the horizon. The Tropic of Cancer is the northern boundary of the equatorial zone called the tropics, which lies between the Tropic of Cancer and Tropic of Capricorn. *See* MATHEMATICAL GEOGRAPHY.

[V.H.E.]

Tropic of Capricorn   The parallel of latitude approximately $23^1/_2{}^\circ$ (23.45°) south of the Equator. It was named for the constellation Capricornus (the goat), for astronomical reasons which no longer prevail.

Because the Earth, in its revolution around the Sun, has its axis inclined $23^1/_2{}^\circ$ from the vertical to the plane of the ecliptic, the Tropic of Capricorn marks the southern limit of the zenithal position of the Sun. Thus, on December 22 (Southern Hemisphere summer, but northern winter solstice) the Sun, at noon, is 90° above the horizon.

The Tropic of Capricorn is the southern boundary of the equatorial zone referred to as the tropics, which lies between the Tropic of Capricorn and the Tropic of Cancer. *See* MATHEMATICAL GEOGRAPHY; TROPIC OF CANCER.   [V.H.E.]

Tropical meteorology   The study of atmospheric structure and behavior in the areas astride the Equator, roughly between 30° north and south latitude. The weather and climate of the tropics involve phenomena such as trade winds, hurricanes, intertropical convergence zones, jet streams, monsoons, and the El Niño Southern Oscillation. More energy is received from the Sun over the tropical latitudes than is lost to outer space (infrared radiation). The reverse is true at higher latitudes, poleward of 30°. The excess energy from the tropics is transported by winds to the higher latitudes, largely by vertical circulations that span roughly 30° in latitudinal extent. These circulations are known as Hadley cells.

For the most part, the oceanic tropics (the islands) experience very little change of day-to-day weather except when severe events occur. Tropical weather can be more adverse during the summer seasons of the respective hemispheres. The near equatorial belt between 5°S and 5°N is nearly always free from hurricanes and typhoons: the active belt lies outside this region over the tropics. The land areas experience considerable heating of the Earth's surface, and the summer-to-winter contrasts are somewhat larger there. For instance, the land areas of northern India experience air temperatures as high as 108°F (42°C) in the summer (near the Earth's surface), while in the winter season the temperatures remain 72°F (22°C) for many days. The diurnal range of temperature is also quite large over land areas on clear days during the summer (32°F or 18°C) as compared to winter (18°F or 10°C).

The steady northeast surface winds over the oceans of the Northern Hemisphere between 5° and 20°N and southeast winds over the corresponding latitudes of the southern oceans constitute the trade winds. Trade winds have intensities of around 5–10 knots (2.5–5 m/s). They are the equatorial branches of the anticyclonic circulation (known as the subtropical high pressure). The steadiness of wind direction is quite high in the trades. *See* WIND.

Hurricanes are also known as typhoons in the west Pacific and tropical cyclones in Indian Ocean and south Pacific. If the wind speed exceeds 65 knots (33 m/s) in a tropical storm, the storm is labeled a hurricane. A hurricane usually forms over the tropical oceans, north or south of 5° latitude from the Equator. *See* HURRICANE.

Intertropical convergence zones are located usually between 5 and 10°N latitude. They are usually oriented west to east and contain cloud clusters with rainfall of the order of 1.2–2 in. (30–50 mm) per day. The trade winds of the two hemispheres supply moisture to this precipitating system. *See* CLOUD PHYSICS; PRECIPITATION (METEOROLOGY).

A number of fast-moving air currents, known as jets, are important elements of the tropical general circulation. With speeds in excess of 30 knots (15 m/s), they are found over several regions of the troposphere. *See* ATMOSPHERIC GENERAL CIRCULATION; JET STREAM; TROPOSPHERE.

Basically the entire landmass from the west coast of Africa to Asia and extending to the date line experiences a phenomenon known as the monsoon. Monsoon circulations are driven by differential heating between relatively cold oceans and relatively warm landmasses. *See* MONSOON METEOROLOGY.

Every 2–6 years the eastern equatorial Pacific Ocean experiences a rise in sea surface temperature of about 5–9°F (3–5°). This phenomenon is known as El Niño, which is part of a larger cycle referred to as the El Niño Southern Oscillation (ENSO). The other extreme in the cycle is referred to as La Niña. El Niño has been known to affect global-scale weather. *See* EL NIÑO; MARITIME METEOROLOGY.   [T.N.Kr.]

Tropopause   The boundary between the troposphere and the stratosphere in the atmosphere. The tropopause is broadly defined as the lowest level above which the lapse rate (decrease) of temperature with height becomes less than 5.8°F mi^{-1}

($2°C$ km^{-1}). In low latitudes the tropical tropopause is at a height of 9.3–11 mi at about $-135°F$ (15–17 km at about 180 K), and the polar tropopause between tropics and poles is at about 6.2 mi at about $-63°F$ (10 km at about 220 K). There is a well-marked "tropopause gap" or break where the tropical and polar tropopauses overlap at 30–40° latitude. The break is in the region of the subtropical jet stream and is of major importance for the transfer of air and tracers (humidity, ozone, radioactivity) between stratosphere and troposphere. The height of the tropopause varies seasonally and also daily with the weather systems, being higher and colder over anticyclones than over depressions. *See* AIR TEMPERATURE; ATMOSPHERE; STRATOSPHERE; TROPOSPHERE.   [R.J.Mu.]

Troposphere   The lowest major layer of the atmosphere. The troposphere extends from the Earth's surface to a height of 6–10 mi (10–16 km), the base of the stratosphere. It contains about four-fifths of the mass of the whole atmosphere. *See* ATMOSPHERE.

On the average, the temperature decreases steadily with height throughout this layer, with a lapse rate of about $19°F$ mi^{-1} ($6.5°C$ km^{-1}), although shallow inversions (temperature increases with height) and greater lapse rates occur, particularly in the boundary layer near the Earth's surface. Appreciable water-vapor contents and clouds are almost entirely confined to the troposphere. Hence it is the seat of all important weather processes and the region where interchange by evaporation and precipitation (rain, snow, and so forth) of water substance between the surface and the atmosphere takes place. *See* CLIMATOLOGY; CLOUD PHYSICS; METEOROLOGY; WEATHER.   [R.J.Mu.]

Tsunami   A set of ocean waves caused by any large, abrupt disturbance of the sea surface. If the disturbance is close to the coastline, tsunamis can demolish local coastal communities within minutes. A very large disturbance can both cause local devastation and export tsunami destruction thousands of miles away. Since 1850, tsunamis have been responsible for the loss of over 120,000 lives and billions of dollars of damage to coastal structures and habitats. Methods for predicting when and where the next tsunami will strike have not been developed; but once the tsunami is generated, forecasting its arrival and impact is possible through wave theory and measurement technology. *See* OCEAN WAVES.

Tsunamis are most commonly generated by earthquakes in marine and coastal regions. Major tsunamis are produced by large (greater than 7 on the Richter scale), shallow-focus (<30-km or 19-mi depth in the Earth) earthquakes associated with the movement of oceanic and continental plates. They frequently occur in the Pacific, where dense oceanic plates slide under the lighter continental plates. When these plates fracture, they cause a vertical movement of the sea floor that allows a quick and efficient transfer of energy from the solid earth to the ocean. The resulting tsunami propagates as a set of waves whose energy is concentrated at wavelengths corresponding to the earth movements ($\sim$100 km or 60 mi), at wave heights determined by vertical displacement ($\sim$1 m or 3 ft), and at wave directions determined by the adjacent coastline geometry. Because each earthquake is unique, every tsunami has unique wavelengths, wave heights, and directionality. From a warning perspective, this makes the problem of forecasting tsunamis in real time daunting. *See* EARTHQUAKE; PLATE TECTONICS.

Other large-scale disturbances of the sea surface that can generate tsunamis are explosive volcanoes and asteroid impacts. The eruption of the volcano Krakatoa in the East Indies on August 27, 1883, produced a 30-m (100-ft) tsunami that killed over 36,000 people. *See* VOLCANO.   [E.N.B.]

Tufa   A spongy, porous limestone formed by precipitation from evaporating spring and river waters; also known as calcareous sinter. Calcium carbonate commonly precipitates from supersaturated waters on the leaves and stems of plants growing around the springs and pools and preserves some of their plant structures. Tufa tends to be fragile and friable. *See* LIMESTONE; TRAVERTINE.   [R.Si.]

Tuff   Fragmental volcanic products from explosive eruptions that are consolidated, cemented, or otherwise hardened to form solid rock. In strict scientific usage, the term "tuff" refers to consolidated volcanic ash, which by definition consists of fragments smaller than 2 mm. However, the term is also used for many pyroclastic rocks composed of fragments coarser than ash and even for pyroclastic material that has undergone limited posteruption reworking. If the thickness, temperature, and gas content of a tuff-forming pyroclastic flow are sufficiently high, the constituent fragments can become compacted and fused to form welded tuff. The term "tuff" is also used in the naming of several related types of small volcanic edifices formed by hydrovolcanic eruptions, triggered by the explosive interaction of hot magma or lava with water. *See* IGNIMBRITE; PYROCLASTIC ROCKS; VOLCANO.   [R.Ti.]

Turbidite   A bed of sediment or sedimentary rock that was deposited from a turbidity current. The term turbidite is fundamentally genetic and interpretive in nature, rather than being a descriptive term (like common rock names). Turbidites are clastic sedimentary rocks, but they may be composed of silicic grains (quartz, feldspar, rock fragments) and therefore be a type of sandstone, or they may be composed of carbonate grains and therefore be a type of limestone. A geologist's description of a rock as a turbidite is actually an expression of an opinion that the rock was deposited by a turbidity current, rather than being a description of a particular type of rock. *See* TURBIDITY CURRENT.

No single feature of a deposit is sufficient to identify it as a turbidite and not all turbidites are marine. There are well-documented examples of modern turbidity currents and turbidites described from lakes. Although probably most turbidites were originally deposited in water of considerable depth (hundreds to thousands of meters), it is generally difficult to be specific about estimating the depth of deposition. The most that can be said is that (in most cases) there is no sign of sedimentary structures formed by the action of waves. *See* DEPOSITIONAL SYSTEMS AND ENVIRONMENTS; SEDIMENTARY ROCKS; SEDIMENTOLOGY.   [G.V.M.]

Turbidity current   A flow of water laden with sediment that moves downslope in an otherwise still body of water. The driving force of a turbidity current is obtained from the sediment, which renders the turbid water heavier than the clear water above. Turbidity currents occur in oceans, lakes, and reservoirs. They may be triggered by the direct inflow of turbid water, by wave action, by subaqueous slumps, or by anthropogenic activities such as dumping of mining tailings and dredging operations.

Turbidity currents are characterized by a well-defined front, also known as head, followed by a thinner layer known as the body of the current. They are members of a larger class of stratified flows known as gravity or density currents. Sediment can be entrained from or deposited on the bed, thus changing the total amount of sediment in suspension. A turbidity current must generate enough turbulence to hold its sediment in suspension. Under certain conditions, a turbidity current might erode its bed, pick up sediment, become heavier, accelerate, and pick up even more sediment, increasing its driving force in a self-reinforcing cycle akin to the formation of a snow avalanche. *See* DEPOSITIONAL SYSTEMS AND ENVIRONMENTS.

Turbidity currents constitute a major mechanism for the transport of fluvial, littoral, and shelf sediments onto the ocean floor. These flows are considered to be responsible for the scouring of submarine and sublacustrine canyons. These canyons are often of massive proportions and rival the Grand Canyon in scale. Below the mouths of most canyons, turbidity currents form vast depositional fans that have many of the features of alluvial fans built by rivers and constitute major hydrocarbon reservoirs. The sedimentary deposits created by turbidity currents, known as turbidites, are a major constituent of the geological record. *See* MARINE GEOLOGY; MARINE SEDIMENTS; SUBMARINE CANYON; TURBIDITE.   [M.H.Ga.]

Turquoise   A mineral of composition $CuAl_6(PO_4)_4(OH)_8 \cdot 5H_2O$ in which considerable ferrous ion (Fe^{2+}) may substitute for copper. Ferric ion (Fe^{3+}) may also substitute for part or all of the aluminum (Al), forming a complete chemical series from turquoise to chalcosiderite [$CuFe_6(PO_4)_4(OH)_8 \cdot 5H_2O$]. Turquoise with a strong sky-blue or bluish-green to apple green color is easily recognized, and such material is commonly used as a gem. Some variscite, of composition $AlPO_4 \cdot 2H_2O$ with minor chemical substitutions of Fe^{3+} and or chromium ion (Cr^{3+}) for aluminum and with a soft, clear green color, may be marketed as green turquoise. *See* GEM.

Most turquoise is massive, dense, and cryptocrystalline to fine-granular. It commonly occurs as veinlets or crusts and in stalactitic or concretionary shapes. It has a hardness on the Mohs scale of about 5 to 6 and a vitreous to waxy luster. The distinctive light blue coloration of much turquoise is the result of the presence of cuprous ion (Cu^{2+}); limited substitution of the copper by Fe^{2+} produces greenish colors.

Turquoise is a secondary mineral, generally formed in arid regions by the interaction of surface waters with high-alumina igneous or sedimentary rocks. It occurs most commonly as small veins and stringers traversing more or less decomposed volcanic rocks. Since the times of antiquity, turquoise of very fine quality has been produced from a deposit in Persia (now Iran) near Nishapur. It occurs also in Siberia, Turkistan, China, the Sinai Peninsula, Germany, and France.

The southwestern United States has been a major source of turquoise, especially the states of Nevada, Arizona, New Mexico, and Colorado. Extensive deposits in the Los Cerillos Mountains, near Santa Fe, New Mexico, were mined very early by Native Americans and were a major early source of gem turquoise. However, much of the gem-quality turquoise has been depleted in the Southwest.   [C.K.]

U

Unconformity   In the stratigraphic sequence of the Earth's crust, a surface of erosion that cuts the underlying rocks and is overlain by sedimentary strata. The unconformity represents an interval of geologic time, called the hiatus, during which no deposition occurred and erosion removed preexisting rock. The result is a gap, in some cases encompassing millions of years, in the stratigraphic record. *See* STRATIGRAPHY.

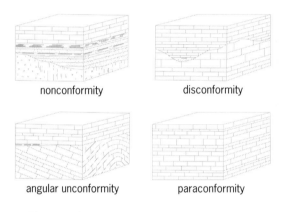

nonconformity   disconformity

angular unconformity   paraconformity

Four types of unconformity. (After C.O. Dunbar and J. Rodgers, Principles of Stratigraphy, Wiley, 1957)

There are four kinds of unconformable relations (see illustration): *Nonconformity*—underlying rocks are not stratified, such as massive crystalline rocks formed deep in the Earth. *Angular unconformity*—underlying rocks are stratified but were deformed before being eroded, resulting in angular discordance; this was the first type to be recognized; the term unconformity was originally used to describe the geometric relationship between the underlying and overlying bedding planes. *Disconformity*—underlying strata are undeformed and parallel to overlying strata, but separated by an evident erosion surface. *Paraconformity*—strata are parallel and the erosion surface is subtle, or even indistinguishable from a simple bedding plane.   [C.W.By.]

Underwater photography   The techniques involved in using photographic equipment underwater. By far the greatest percentage of underwater photography is done within sport-diving limits in the tropical oceans.

Underwater photographers are faced with specific technical challenges. Water is 600 times denser than air and is predominantly blue in color. Depth affects light and creates physiological considerations for the photographer. As a result, underwater photography requires an understanding of certain principles of light beneath the sea.

As in all photography, consideration of the variables of light transmission is crucial to underwater photography. When sunlight strikes the surface of the sea, its quality and quantity change in several ways. As light travels from air to a denser medium, such as water, the light rays are bent (refracted); one result is magnification of underwater objects by one-third as compared to viewing them in air. The magnification effect must be considered when estimating distances underwater, which is critical for both focus and exposure. Light is absorbed when it propagates through water. Variables affecting the level of light penetration include the time of day (affects the angle at which the sunlight strikes the surface of the water); cloud cover; clarity of the water; depth (light is increasingly absorbed with increasing depth); and surface conditions (if the sea is choppy, more light will be reflected off the surface and less light transmitted to the underwater scene).

Depth affects not only the quantity of light but also the quality of light. Once light passes from air to water, different wavelengths of its spectrum are absorbed as a function of the color of the water and depth. Even in the clearest tropical sea, water serves as a powerful cyan (blue-green) filter. Natural full-spectrum photographs can be taken only with available light in very shallow depths. In ideal daylight conditions and clear ocean water, photographic film fails to record red at about 15 ft (4.5 m) in depth. Orange disappears at 30 ft (9 m), yellow at 60 ft (18 m), green at 80 ft (24 m), and at greater depth only blue and black are recorded on film. To restore color, underwater photographers must use artificial light. *See* SEAWATER.

The water column between photographer and subject degrades both the resolution of the image and the transmission of artificial light (necessary to restore color). Therefore, the most effective underwater photos are taken as close as possible to the subject, thereby creating the need for a variety of optical tools to capture subjects of various sizes within this narrow distance limitation.

There are two types of underwater cameras—amphibious and housed. Amphibious cameras may be used either underwater or topside, although some lenses are for underwater use only (known as water contact lenses). A housed camera is a conventional above-water camera that has been protected from the damaging effects of seawater by a waterproof enclosure. The amphibious camera is protected by a series of O-rings, primarily located at the lens mount, film loading door, shutter release, and other places where controls are necessary. The O-rings make the system not only resistant to leaks but also impervious to dust or inclement weather when used above water.   [S.Fri.]

Deep-sea underwater photography—approximately 150 ft (35 m)—requires the design and use of special camera and lighting equipment. Watertight cases are required for both camera and light source, and they must be able to withstand the pressure generated by the sea. For each 33 ft (10 m) of depth, approximately one additional atmosphere ($\sim 10^2$ kilopascals) of pressure is exerted. At the greatest ocean depths, about 40,000 ft (12,000 m), a case must be able to withstand 17,600 lb/in.^{2} (1200 kg/cm^2). The windows for the lens and electrical seals must also be designed for such pressure to prevent water intrusion.

Auxiliary lighting is required, since daylight is absorbed in both intensity and hue. The camera must be positioned and triggered to render the desired photograph, and the great depths preclude a free-swimming human operator. Operation is often from a cable via sonar sensing equipment or from deep-diving underwater vehicles. Bottom-sensing switches can operate deep-sea cameras for photographing the sea floor, and remotely operated vehicles (ROVs) can incorporate both video and still cameras. *See* UNDERWATER VEHICLE.

When an observer descends to great depths in a diving vehicle, the camera can assist in documentation by recording what is seen. Furthermore, the visual data will assist in accurate description of the observed phenomena. Elapsed-time photography with a motion picture camera in the sea is important in studying sedimentation deposits caused by tides, currents, and storms. Similarly, the observation of biological activity taken with the elapsed-time camera and then speeded up for viewing may reveal processes that cannot ordinarily be observed. *See* DIVING; UNDERWATER TELEVISION.   [H.E.E.; S.Fri.]

Underwater sound   The production, propagation, reflection, scattering, and reception of sound in seawater. The sea covers approximately 75% of the Earth's surface. In terms of exploration, visible observation of the sea is limited due to the high attenuation of light, and radar has very poor penetrability into salt water. Because of the extraordinary properties that sound has in the sea, and because of some of the inherent characteristics of the sea, acoustics is the principal means by which the sea has been explored. *See* OCEAN.

Absorption. Sound has a remarkably low loss of energy in seawater, and it is that property above all others that allows it to be used in research and other application. Absorption is the loss of energy due to internal causes, such as viscosity. Over the frequency range from about 100 Hz (cycles per second) to 100 kHz, absorption is dominated by the reactions of two molecules, magnesium sulfate ($MgSO_4$) and boric acid [$B(OH)_3$]. These molecules are normally in equilibrium with their ionic constituents. The pressure variation caused by an acoustic wave changes the ionic balance and, during the passage of the pressure-varying acoustic field, it cannot return to the same equilibrium, and energy is given up. This is called chemical relaxation. At about 65 kHz magnesium sulfate dominates absorption, and boric acid is important near 1 kHz. *See* SEAWATER.

Sound speed. The speed of sound in seawater and its dependence on the parameters of the sea, such as temperature, salinity, and density, have an enormous effect on acoustics in the sea. Generally the environmental parameter that dominates acoustic processes in oceans is the temperature, because it varies both spatially and temporally. Solar heating of the upper ocean has one of the most important effects on sound propagation. As the temperature of the upper ocean increases, so does the sound speed. Winds mix the upper layer, giving rise to a layer of water of approximately constant temperature, below which is a region called the thermocline. Below that, most seawater reaches a constant temperature. All these layers depend on the season and the geographical location, and there is considerable local variation, depending on winds, cloud cover, atmospheric stability, and so on. Shallow water is even more variable due to tides, fresh-water mixing, and interactions with the sea floor. Major ocean currents, such as the Gulf Stream and Kuroshio, have major effects on acoustics. The cold and warm eddies that are spun off from these currents are present in abundance and significantly affect acoustic propagation. *See* GULF STREAM; KUROSHIO; OCEANOGRAPHY.

Pressure waves. The science of underwater sound is the study of pressure waves in the sea over the frequency range from a few hertz to a few megahertz. The International System (SI) units are the pascal (Pa) for pressure (equal to one newton per square meter) and the watt per square meter (W/m^2) for sound intensity (the flow of energy through a unit area normal to the direction of wave propagation). In acoustics, it is more convenient to refer to pressures, which are usually much smaller than a pascal, and the consequent intensities with a different reference, the decibel. Intensity in decibels (dB) is ten times the logarithm to the base ten of the measured intensity divided by a reference intensity.

Wave propagation. The mathematical equation that sound obeys is known as the wave equation. Its derivation is based on the mathematical statements of Newton's second law for fluids (the Navier-Stokes equation), the equation of continuity (which essentially states that when a fluid is compressed, its mass is conserved), and a law of compression, relating a change of volume to a change in pressure. By the mathematical manipulation of these three equations, and the assumption that only very small physical changes in the fluid are taking place, it is possible to obtain a single differential equation that connects the acoustic pressure changes in time to those in space by a single quantity, the square of the sound speed (c), which is usually a slowly varying function of both space and time.

Knowing the sound speed as a function of space and time allows for the investigation of the spatial and temporal properties of sound, at least in principle. The mathematics used to find solutions to the wave equation are the same as those that are used in other fields of physics, such as optics, radar, and seismics.

In addition to knowing the speed of sound, it is necessary to know the location and nature of the sources of sound, the location and features of the sea surface, the depth to the sea floor, and, in many applications, the physical structure of the sea floor. It is not possible to know the sound speed throughout the water column or know the boundaries exactly. Thus the solutions to the wave equation are never exact representations of nature, but estimates, with an accuracy that depends on both the quality of the knowledge of the environment and the degree to which the mathematical or numerical solutions to the wave equation represent the actual physical situation.

Ambient noise. A consequence of the remarkable transmission of sound is that unwanted sounds are transmitted just as efficiently. One of the ultimate limitations to the use of underwater sound is the ability to detect a signal above the noise. In the ocean, there are four distinct categories of ambient sound: biological, oceanographic physical processes, seismic, and anthropogenic.

Scattering and reverberation. The other source of unwanted sound is reverberation. Sound that is transmitted inevitably finds something to scatter from in the water column, at the sea surface, or at the sea floor. The scatter is usually in all directions, and some of it will return to the system that processes the return signals. Sources of scattering in the water column are fish, particulates, and physical inhomogeneities. The sea surface is, under normal sea conditions, agitated by winds and has the characteristic roughness associated with the prevailing atmospheric conditions. Rough surfaces scatter sound with scattering strengths that depend on the roughness, the acoustic frequency (or wavelength), and the direction of the signal. The scattering is highly time-dependent, and needs to be studied with an appropriate statistical approach. The sea floor has inherent roughness and is usually inhomogeneous, both properties causing scatter. Although scatter degrades the performance of sonars, the characteristics of the return can be determined to enable its cancellation through signal processing or array design. Scattering can also be used to study the sea surface, the sea floor, fish types and distribution, and inhomogeneities in the water column. *See* SCATTERING LAYER. [R.R.G.]

Underwater television

Any type of electronic camera that is located underwater in order to collect and display images. It must be packaged in a waterproof housing. The underwater camera may be packaged with its own recording device, or it can be attached to a television that is located on a ship, in a laboratory, or at a remote site. In the latter case, the images by the camera are real-time images. An underwater television may be used for sport, ocean exploration, industrial applications, or military purposes. Common imaged subjects are animals, coral reefs, underwater shipwrecks, and underwater structures such as piers, bridges, and offshore oil platforms.

During the daytime and at minimal depths, underwater television can be used to view objects illuminated with natural sunlight. For nighttime viewing or at very deep depths, artificial lights must be used. Light in the ocean is reduced in intensity quite severely: even in the clearest natural waters, a beam of blue or green light will be reduced in intensity by approximately 67% every 230 ft (70 m). Light that is propagating through an aqueous medium such as seawater or lake water will also be selectively reduced in intensity, based on wavelength. In most situations, the extreme reds and blues will be most severely attenuated, with the region of highest clarity being the yellow to blue-green wavelengths.

Modern advances in television cameras have opened up a host of applications for underwater viewing, such as of standard-video, charge-injection-device, silicon-intensified-target, or charge-coupled-device cameras. These cameras can be adapted for high frame rate, low light level, or underwater color imaging. Computer modeling of underwater images can be used to predict the performance of an underwater imaging system in terms of range of viewing and quality of images as a function of the clarity of the water. *See* UNDERWATER PHOTOGRAPHY.   [J.S.J.]

Underwater vehicle   A submersible work platform designed to be operated either remotely or directly. Underwater vehicles are grouped into three categories: deep submersible vehicles (DSVs), remotely operated vehicles (ROVs), and autonomous underwater vehicles (AUVs). There are also hybrid vehicles which combine two or three categories on board a single platform. Within each category of submersible there are specially adaped vehicles for specific work tasks. These can be purpose-built or modifications of standard submersibles.

There are five types of DSVs: one-atmosphere untethered vehicles; one-atmosphere tethered vehicles, including observation/work bells; atmospheric diving suits; diver lock-out vehicles; and wet submersibles. While they differ mainly in configuration, source of power, and number of crew members, all carry a crew at 1-atm (10^2-kilopascal) pressure within a dry chamber. An exception is the wet submersible, where the crew is exposed to full depth pressure. The purpose of the DSV is to put the trained mind and eye to work inside the ocean. The earliest submersibles had very small viewing ports fitted into thick-walled steel hulls. In the mid-1960s, experimental work began on use of massive plastics (acrylics) as pressure hull materials. Today, submersibles with depth capabilities to 3300 ft (1000 m) are being manufactured with pressure hulls made entirely of acrylic. Essentially the hull is now one huge window.

The first ROVs were developed in the late 1950s for naval use. By the mid-1970s, they were used in the civil sector. The rapid acceptance of these submersibles is due to their relatively low cost and the fact that they do not put human life at risk when undertaking hazardous missions. However, their most important attribute is that they are less complex. By virtue of their surface-connecting umbilical cable, they can operate almost indefinitely since there is no human inside requiring life support and no batteries to be recharged. There are four types of ROV: tethered free-swimming vehicles, towed vehicles, bottom reliant vehicles, and structurally reliant vehicles.

AUVs are crewless and untethered submersibles which operate independent of direct human control. Their operations are controlled by a preprogrammed, on-board computer. They were first developed in the military where applications include such tasks as minefield location and mapping, minefield installation, submarine decoys, and covert intelligence collection. Civilian tasks include site monitoring, basic oceanographic data gathering, under-ice mapping, offshore structure and pipeline inspection, and bottom mapping. These submersibles are particularly useful where long-duration measurements and observations are to be made and where human presence is not required.

AUVs span a wide range of sizes and capabilities, related to their intended missions. Each is a mobile instrumentation platform with propulsion, sensors, and on-board intelligence designed to complete sampling tasks autonomously. At the large end of the scale, transport-class platforms in the order of 10 m (33 ft) length and 10 metric ton (11 tons) weight in air have been designed for missions requiring long endurance, high speed, large payloads, or high-power sensors. At the small end of the scale, network-class platforms in the order of 1 m (3.3 ft) length and 100 kg (220 lb) weight in air address missions requiring portability, multiple platforms, adaptive spatial sampling, and sustained presence in a specific region. Vehicles can also be categorized in terms of propulsion method (propeller-driven or buoyancy-driven) or in terms of their maximum operating depth.

Hybrid vehicles are those that combine crewed vehicles, remotely operated vehicles, and divers. For example, the hybrid *DUPLUS II* can operate either as a tethered free-swimming ROV or as a 1-atm tethered crewed vehicle. This evolved to provide capability for remotely conducting those tasks for which human skills are not needed, and then to put the human at the place where those skills are required. Other hybrid examples include ROVs that can be controlled remotely from the surface or at the work site by a diver performing maintenance and repair tasks.   [D.Wal.; T.B.C.]

Upper-atmosphere dynamics   The motion of the atmosphere above 50 km (30 mi). The predominant dynamical phenomena of the upper atmosphere are quite different from those encountered in the lower atmosphere. Among those encountered in the lower atmosphere are cyclones, anticyclones, tropical hurricanes, thunderstorms and shower clouds, tornadoes, and dust devils. Even the largest of these phenomena do not penetrate far into the upper atmosphere. Above an altitude of about 50 km (30 mi), the predominant dynamical phenomena are internal gravity waves, tides, sound waves (including infrasonic), turbulence, and large-scale circulation.

Except under meteorological conditions characterized by convection, the atmosphere is stable against small vertical displacements of small air parcels; this results from buoyancy forces that tend to restore displaced air parcels to their original levels. An air parcel therefore tends to oscillate around its undisturbed position at a frequency known as the Brunt-Vaisala frequency ω_B. If pressure waves are generated in the atmosphere with frequencies much greater than ω_B, they propagate as sound waves. For frequencies much less than ω_B, the waves propagate as internal gravity waves; in this case, the restoring forces for the wave motion are provided primarily by buoyancy (that is, gravity) rather than by compression.

Tides are internal gravity waves of particular frequencies. The term tidal usually implies that the exciting force is gravitational attraction by the Moon or Sun. However, it is conventional in the case of atmospheric tides to include also those waves that are excited by solar heating. One is therefore concerned with three separate excitation functions—lunar gravitation, solar gravitation, and solar heating.

Tidal wind patterns in the upper atmosphere generate electrical currents in the ionosphere through a dynamo action. These in turn give rise to diurnal variations in the geomagnetic field that can be observed at the Earth's surface. *See* ATMOSPHERIC TIDES; GEOMAGNETIC VARIATIONS; IONOSPHERE.

Sound waves generated in the lower atmosphere may propagate upward; to maintain continuity of energy flow, the waves might be expected to grow in relative amplitude as they move into the more rarefied upper atmosphere. However, higher temperatures in the upper atmosphere refract most of the energy back toward the Earth's surface, giving rise to the phenomenon known as anomalous propagation. This involves the redirection of upward-moving sound waves back to the surface beyond the point where the source can be heard by waves propagating along the surface. Infrasonic waves with periods from 20 to 80 s have been observed occasionally with detectors at the Earth's surface in connection with auroral activity.

There is clear visual evidence of turbulence in the upper atmosphere; this evidence is obtained by examination of vapor trails released from rockets or of long-persisting meteor trails. The source of the turbulence is not clear. The atmosphere is thermo-dynamically stable against vertical displacements throughout the region above the troposphere, and work has to be done against buoyancy forces in order to produce and maintain turbulence. The only apparent source of energy is internal gravity waves, either tidal or of random period.

There are prevailing patterns of atmospheric circulation in the upper atmosphere, but they are very different from those that occur in the lower atmosphere, which are associated with weather systems and have complicated structures resulting from growth of instabilities. The upper atmospheric large-scale wind systems are mainly diurnal in nature and global in scale.

The main heat source that is responsible for the upper atmospheric circulation is ul-traviolet radiation from the Sun, radiation that is mainly absorbed at altitudes between 100 and 200 km (60 and 120 mi). The atmosphere is not a good infrared radiator in this altitude region, so the temperature rises rapidly with altitude, providing a temper-ature gradient of such a magnitude that molecular conduction transfers the absorbed heat downward to altitudes below 100 km (60 mi) where the atmosphere does have the capability of radiating the energy back to space. Above about 300 km (180 mi), the temperature becomes roughly constant with altitude because very little energy is absorbed there (the gas is exceedingly rarefied) and the thermal conductivity is good enough under these circumstances to virtually eliminate vertical temperature gradients.

The region of rising temperature above 80 km (48 mi) is known as the thermosphere. The exosphere is that region of the atmosphere that is so rarefied that for many purposes collisions between molecules can be neglected; it is roughly the region above 500 km (300 mi).

At high latitudes, the ionosphere moves in response to electric fields imposed as a consequence of interactions between the Earth's magnetic field and the solar wind. The imposed electric field causes the ionosphere to drift in a generally antisunward direction over the polar caps (regions at higher latitudes than the auroral zone, or magnetic latitudes greater than about 68°), with a return circulation (that is, generally sunward in direction) just outside the polar caps. *See* ATMOSPHERIC GENERAL CIRCULATION; WIND.

Ultraviolet radiation of suitable wavelengths can photodissociate atmospheric molecules—something of great importance in the upper atmosphere. It is even impor-tant in the stratosphere, where ozone is formed as a result of absorption by molecular oxygen of ultraviolet radiation, the important wavelengths being below 242 nanome-ters. *See* STRATOSPHERE.   [F.S.J.]

Upwelling   The phenomenon or process involving the ascending motion of water in the ocean. Vertical motions are an integral part of ocean circulation, but they are a thousand to a million times smaller than the horizontal currents. Vertical motions are inhibited by the density stratification of the ocean because with increasing depth, as the temperature decreases, the density increases, and energy must be expended to displace water vertically. The ocean is also stratified in other properties; for example, nutrient concentration generally increases with depth. Thus even weak vertical flow may cause a significant effect by advecting nutrients to a new level. *See* GEOPHYSICAL FLUID DYNAMICS.

There are two important upwelling processes. One is the slow upwelling of cold abyssal water, occurring over large areas of the world ocean, to compensate for the formation and sinking of this deep water in limited polar regions. The other is the upwelling of subsurface water into the euphotic (sunlit) zone to compensate for a horizontal divergence of the flow in the surface layer, usually caused by winds. *See* OCEAN CIRCULATION.   [R.L.Sm.]

Uraninite   The chief ore mineral of uranium. Uraninite has the idealized chemical composition UO_2, uranium dioxide. Thorium and rare earths, chiefly cerium, are usually present in variable and sometimes large amounts. Lead always is present by radioactive decay of the thorium and uranium present. Complete solid-solution series extend between UO_2, ThO_2 (thorianite), and CeO_2 (cerianite). *See* RADIOACTIVE MINERALS; THORIANITE.

The color of uraninite is black, grading to brownish black and dark brown in the more highly oxidized material. The luster of fresh material is steel-gray. The hardness is $5^1/_2-6$ on Mohs scale. The specific gravity of pure UO_2 is 10.9, but that of most natural material is 9.7−7.5.   [C.Fr.]

Urban climatology   The branch of climatology concerned with urban areas. These locales produce significant changes in the surface of the Earth and the quality of the air. In turn, surface climate in the vicinity of urban sites is altered. The era of urbanization on a worldwide scale has been accompanied by unintentional, measurable changes in city climate. *See* CLIMATOLOGY.

The process of urbanization changes the physical surroundings and induces alterations in the energy, moisture, and motion regime near the surface. Most of these alterations may be traced to causal factors such as air pollution; anthropogenic heat; surface waterproofing; thermal properties of the surface materials; and morphology of the surface and its specific three-dimensional geometry—building spacing, height, orientation, vegetative layering, and the overall dimensions and geography of these elements. Other factors that must be considered are relief, nearness to water bodies, size of the city, population density, and land-use distributions.

In general, cities are warmer than their surroundings, as documented over a century ago. They are islands or spots on the broader, more rural surrounding land. Thus, cities produce a heat island effect on the spatial distribution of temperatures. The timing of a maximum heat island is followed by a lag shortly after sundown, as urban surfaces, which absorbed and stored daytime heat, retain heat and affect the overlying air. Meantime, rural areas cool at a rapid rate.

A number of energy processes are altered to create warming, and various features lead to those alterations. City size, the morphology of the city, land-use configuration, and the geographic setting (such as relief, elevation, and regional climate) dictate the intensity of the heat island, its geographic extent, its orientation, and its persistence through time. Individual causes for heat island formation are related to city geometry, air pollution, surface materials, and anthropogenic heat emission. There are two atmosphere layers in an urban environment, besides the planetary boundary layer outside and extending well above the city: (1) The urban boundary layer is due to the spatially integrated heat and moisture exchanges between the city and its overlying air. (2) The surface of the city corresponds to the level of the urban canopy layer. Fluxes across this plane comprise those from individual units, such as roofs, canyon tops, trees, lawns, and roads, integrated over larger land-use divisions (for example, suburbs).   [A.J.B.]

V

Van Allen radiation The high-energy, charged particles that are trapped into orbits by the geomagnetic field, forming radiation belts that surround the Earth. The belts consist primarily of electrons and protons and extend from a few hundred kilometers above the Earth to a distance of about $8 R_e$ (R_e = radius of Earth = 6371 km = 3959 mi). James Van Allen and coworkers discovered them in 1958 using radiation detectors carried on satellites *Explorer 1* and *3*, and they are often referred to as the Van Allen belts.

A charged particle under the influence of the geomagnetic field follows a trajectory that can be conveniently described as a superposition of three separate motions. The first motion, produced by the magnetic force acting at right angles to both the particle velocity and the magnetic field, is a rapid spiral about magnetic field lines. As the spiraling particle moves along the field line toward either the North Pole or South Pole, the increase in magnetic field strength causes the particle to be reflected so that it bounces between the Earth's two hemispheres. Superimposed on the spiral and bounce motions is a slow east-west drift; electrons drift eastward and protons or heavier ions drift westward. (The resulting current, called the ring current, acts to decrease the strength of the Earth's (surface) northward magnetic field at low latitudes.) Thus, individual trapped particles move completely around the Earth in a complicated pattern, their motion being constrained to lie on magnetic shells.

The spatial structure of trapped radiation shows two maxima: an inner radiation belt centered at about $1.5R_e$ and an outer belt centered at 4–$5R_e$. In the inner radiation belt, the most penetrating particles are protons with energies extending to several hundred megaelectronvolts (MeV). However, the flux of high-energy protons decreases rapidly with increasing distance from the Earth and becomes insignificant beyond $4R_e$. Electrons and low-energy protons with energies up to a few MeV occur throughout the stable trapping region. The electron energies extend to several MeV, and in the outer radiation belt electrons are the most penetrating component.

The intensity, energy spectrum, and spatial distribution of particles within the radiation belts vary with time. The most dramatic variations are associated with magnetic storms. The changes are most pronounced for particles in the outer belt where the magnetic variations are the largest. During a magnetic storm, the electron flux in the outer belt may increase by an order of magnitude or more.

It is believed that most of the very high energy protons (>50 MeV) in the inner belt result from the spontaneous decay of high-energy neutrons, produced by collisions of cosmic rays with atmospheric atoms. However, the vast majority of the radiation belt populations are ions and electrons that originate from either the atmosphere or the solar wind and are accelerated by processes only partly understood.

The radiation belts are just one feature of the space plasma environment. For example, plasma from the Sun is continually impinging on the Earth's magnetic field, resulting in a "cavity" known as the magnetosphere.

Because the conditions that lead to the formation of Earth's radiation belts are so general, it is believed that any planet or moon that has a large enough magnetic field will also have radiation belts. Jupiter, Saturn, Uranus, and Neptune have strong magnetic fields and very large, intense radiation belts analogous to those of the Earth.

The term "space weather" describes the conditions in space that affect Earth and its technological systems. It is a consequence of the behavior of the Sun, and the interaction of the solar wind with the Earth's magnetic field. The fluxes of electrons and protons trapped in the radiation belts can injure both personnel and equipment on board spacecraft if the vehicle is exposed to these energetic particles for a sufficiently long time. Because of the spatial structure of the belts, the degree of damage will be strongly dependent on the position in space and hence on the orbit of the vehicle. In the inner belt region, high-energy protons can pass through several centimeters of aluminum structure and injure components in the interior of the spacecraft. In most other regions of the radiation belts, the trapped particles are less penetrating, and damage is confined to exposed equipment such as solar cells.   [P.Ril.; M.W.]

Varve   Any of a variety of distinct sediment laminations or beds deposited within the span of a single year. They are formed commonly in saline or fresh-water lakes, but examples from marine environments are known as well. Usually, varves occur in repetitive series and thus comprise vertical sequences of annual cyclic deposits. Varves range in thickness from less than a millimeter (0.04 in.) to over a meter (3 ft), but typically are a few millimeters or centimeters thick.

The classic varves are found in glacial lake sediments formed during the Pleistocene ice ages. These glacial varves occur typically as couplets of light-colored silt or sand and dark clay. The relatively coarse silt-sand layers are formed during the warm summer months when meltwater inflows and sediment yields to the lake are large. During winter when meltwater inflow is greatly reduced or stopped, the fine-grained clay settles slowly to the lake bottom to deposit the fine, dark winter layer of the varve. Similar varved sediments are found commonly in modern glacier-fed lakes that undergo large seasonal variations in inflow. See PLEISTOCENE.

Varves can be used as tools for correlation as well as for chronological reconstructions. In addition to their use in dating sedimentary deposits, varves have been used to investigate sedimentation rates, cyclic deposition, climate variations, glacial histories, and as standards of comparison for other dating techniques. Varves have been identified occasionally in ancient sedimentary rocks. See DATING METHODS; GLACIAL EPOCH; MARINE SEDIMENTS; SEDIMENTARY ROCKS; SEDIMENTOLOGY.   [N.D.S.]

Vegetation and ecosystem mapping   The graphic portrayal of spatial distributions of vegetation, ecosystems, or their characteristics. Vegetation is one of the most conspicuous and characteristic features of the landscape and has long been a convenient way to distinguish different regions; maps of ecosystems and biomes have been mainly vegetation maps. As pressure on the Earth's natural resources grows and as natural ecosystems are increasingly disturbed, degraded, and in some cases replaced completely, the mapping of vegetation and ecosystems, at all scales and by various methods, has become more important. See BIOME; ECOSYSTEM.

Three approaches have arisen for mapping general vegetation patterns, (1) based on vegetation structure or gross physiognomy, (2) based on correlated environmental patterns, and (3) based on important floristic taxa. The environmental approach provides the least information about the actual vegetation but succeeds in covering regions where the vegetation is poorly understood. Most modern classification systems use a combination of physiognomic and floristic characters.

Mapping has expanded to involve other aspects of vegetation and ecosystems as well as new methodologies for map production. Functional processes such as primary production, decomposition rates, and climatic correlates (such as evapotranspiration) have been estimated for enough sites so that world maps can be generated. Structural aspects of ecosystems, such as total standing biomass or potential litter accumulations, are also being estimated and mapped. Quantitative maps of these processes or accumulations can be analyzed geographically to provide first estimates of important aspects of world biogeochemical budgets and resource potentials.

Computer-produced maps, using Geographic Information Systems (GIS), often coupled directly with predictive models, remote-sensing capabilities, and other techniques, have also revolutionized vegetation and ecosystem mapping. This gives scientists a powerful tool for modeling and predicting the outcome from global climate change, in that feedback from the world's vegetation can be accounted for. Before computer technology exploded in the early 1980s, the spatial scale and related resolution or grain of vegetation and ecosystem mapping was limited by the static nature of hard-copy maps. The advent of GIS technology enabled the analysis of digital maps at any spatial scale, and the only limitation was the resolution at which the data were originally mapped. In addition, GIS software is used for sophisticated spatial analyses on maps, and this was virtually impossible before. *See* CARTOGRAPHY; CLIMATE MODELING; GEOGRAPHIC INFORMATION SYSTEMS; REMOTE SENSING.   [B.E.F.; E.O.B.]

Vermiculite   A group of minerals common in some soils and clays and belonging to the family of minerals called layer silicates. Species within the vermiculite group are denoted as either dioctahedral vermiculite or trioctahedral vermiculite, with two or three octahedral cation sites occupied per formula unit, respectively. Trioctahedral vermiculite is a 2:1 layer silicate with a fundamental unit similar to that of mica. An octahedral sheet forms the basis of the layer and is sandwiched between two opposing tetrahedral sheets. *See* MICA.

Perfect basal cleavage develops by the layerlike structure. Density varies but is near 2.4 g/cm^3; hardness on Mohs scale is near 1.5; and luster is pearly. Thin sheets deform easily and may be yellow to brown.

Heat-treated and expanded vermiculite is used as an insulator in construction. Mixed with plaster and cement, vermiculite is used to make lightweight versions of these materials. Vermiculite is also useful as an absorbent for some environmentally hazardous liquids. *See* CLAY MINERALS; SILICATE MINERALS.   [S.Gu.]

Vesuvianite   A sorosilicate mineral of complex composition crystallizing in the tetragonal system; also known by the name idocrase. Crystals, frequently well formed, are usually prismatic with pyramidal terminations. It commonly occurs in columnar aggregates but may be granular or massive. The luster is vitreous to resinous; the color is usually green or brown but may be yellow, blue, or red. Hardness is $6\frac{1}{2}$ on Mohs scale; specific gravity is 3.35–3.45.

The composition of vesuvianite is expressed by the formula $Ca_{10}Al_4(Mg,Fe)_2$-$Si_9O_{34}(OH)_4$. Magnesium and ferrous iron are present in varying amounts, and boron or fluorine is found in some varieties. Beryllium has been reported in small amounts.

Vesuvianite is found characteristically in crystalline limestones resulting from contact metamorphism. It is there associated with other contact minerals such as garnet, diopside, wollastonite, and tourmaline. Noted localities are Zermatt, Switzerland; Christiansand, Norway; River Vilui, Siberia; and Chiapas, Mexico. In the United States it is found in Sanford, Maine; Franklin, New Jersey; Amity, New York; and at many contact metamorphic deposits in western states. A compact green variety resembling jade is found in California and is called californite. *See* SILICATE MINERALS.   [C.S.Hu.]

Vivianite   A mineral of the vivianite group, other important members of which are annabergite and erythrite.

Vivianite is a hydrated ferrous phosphate, $Fe_3(PO_4)_2 \cdot 8H_2O$; usually ferric iron is present as the result of oxidation. It crystallizes in the monoclinic system, with crystals generally prismatic. Vivianite also occurs in earthy form and as globular and encrusting masses of fibrous structure. Crystals are colorless and transparent when fresh. Oxidation changes the color progressively to pale blue, greenish blue, dark blue, or bluish black.

[W.R.Lo.]

Volcanic glass   A natural glass formed by rapid cooling of magma. Magmas typically comprise crystals and bubbles of gas within a silicate liquid. On slow cooling, the liquid portion of the magma usually crystallizes, but if cooling is sufficiently rapid, it may convert to glass—an amorphous, metastable solid that lacks the long-range microscopic order characteristic of crystalline solids. *See* LAVA; MAGMA.

Silica-rich, rhyolitic magmas frequently quench to glass during explosive eruptions and make up the bulk of the solid material in many pyroclastic deposits (usually as shards, pumice lumps, and other fragments); but they also can erupt quiescently to form massive glassy rocks (known as obsidian, the most common source of volcanic glass on land) even in the slowly cooled interiors of flows tens of meters thick. In contrast, more basic, basaltic glasses (sometimes known as tachylite) are less common and rarely form in more than small quantities unless rapidly cooled in a volcanic eruption. Pele's hair is an example of basaltic glass formed in this way. *See* BASALT; OBSIDIAN; RHYOLITE.

[E.M.St.]

Volcano   A mountain or hill, generally steep-sided, formed by accumulation of magma (molten rock with associated gas and crystals) erupted through openings or volcanic vents in the Earth's crust; the term volcano also refers to the vent itself. During the evolution of a long-lived volcano, a permanent shift in the locus of principal vent activity can produce a satellitic volcanic accumulation as large as or larger than the parent volcano, in effect forming a new volcano on the flanks of the old.

Planetary exploration has revealed dramatic evidence of volcanoes and their products on the Earth's Moon, Mars, Mercury, Venus, and the moons of Jupiter, Neptune, and Uranus on a scale much more vast than on Earth. However, only the products and landforms of terrestrial volcanic activity are described here. *See* VOLCANOLOGY.

Volcanic vents. Volcanic vents, channelways for magma to ascend toward the surface, can be grouped into two general types: fissure and central (pipelike). Magma consolidating below the surface in fissures or pipes forms a variety of igneous bodies, but magma breaking the surface produces fissure or pipe eruptions. Fissures, most of them less than 10 ft (3 m) wide, may form in the summit region of a volcano, on its flanks, or near its base; central vents tend to be restricted to the summit area of a volcano. For some volcanoes or volcanic regions, swarms of fissure vents are clustered in swaths called rift zones.

Volcanic products. Magma erupted onto the Earth's surface is called lava. If the lava is chilled and solidifies quickly, it forms volcanic glass; slower rates of chilling result in greater crystallization before complete solidification. Lava may accrete near the vent to form various minor structures or may pour out in streams called lava flows, which may travel many tens of miles from the vents. During more violent eruption, lava torn into fragments and hurled into the air is called pyroclastic (fire-broken materials). *See* LAVA; MAGMA; PYROCLASTIC ROCKS; VOLCANIC GLASS.

Volcanic gases. Violent volcanic explosions may throw dust and aerosols high into the stratosphere, where it may drift across the surface of the globe for many thousands of miles. Most of the solid particles in the volcanic cloud settle out within a few days, and nearly all settle out within a few weeks, but the gaseous aerosols (principally sulfuric acid droplets) may remain suspended in the stratosphere for several years. Such stratospheric clouds of volcanic aerosols, if sufficiently voluminous and long-lived, can have an impact on global climate. *See* ACID RAIN; AIR POLLUTION.

In general, water vapor is the most abundant constituent in volcanic gases; the water is mostly of meteoric (atmospheric) origin, but in some volcanoes can have a significant magmatic or juvenile component. Excluding water vapor, the most abundant gases are the various species of carbon, sulfur, hydrogen, chlorine, and fluorine.

Mudflows are common on steep-side volcanoes where poorly indurated or non-welded pyroclastic material is abundant. Probably by far the most common cause, however, is simply heavy rain saturating a thick cover of loose, unstable pyroclastic material on the steep slope of the volcano, transforming the material into a mobile, water-saturated "mud," which can rush downslope at a speed as great as 50–55 mi (80–90 km) per hour. Such a dense, fast-moving mass can be highly destructive, sweeping up everything loose in its path.

Volcanic landforms. Much of the Earth's solid surface, on land and below the sea, has been shaped by volcanic activity. Landscape features of volcanic origin may be either positive (constructional) forms, the result of accumulation of volcanic materials, or negative forms, the result of the lack of accumulation or collapse.

Not all volcanoes show a graceful, symmetrical cone shape, such as that exemplified by Mount Fuji, Japan. Most volcanoes, especially those near tectonic plate boundaries, are more irregular, though of grossly conical shape. Such volcanoes, called stratovolcanoes or composite volcanoes, typically erupt explosively and are composed dominantly of andesitic, relatively viscous and short lava flows, interlayered with beds of ash and cinder that thin away from the principal vents. Volcanoes constructed primarily of fluid basaltic lava flows, which may spread great distances from the vents, typically are gentle-sloped, broadly upward convex structures. Such shield volcanoes, classic examples of which are Mauna Loa volcano, Hawaii, tend to form in oceanic intraplate regions and are associated with hot-spot volcanism. The shape and size of a volcano can vary widely between the simple forms of composite and shield volcanoes, depending on magma viscosity, eruptive style (explosive versus nonexplosive), migration of vent locations, duration and complexity of eruptive history, and posteruption modifications.

Some of the largest volcanic edifices are not shaped like the composite or shield volcanoes. In certain regions of the world, voluminous extrusions of very fluid basaltic lava from dispersed fissure swarms have built broad, nearly flat-topped accumulations. These voluminous outpourings of lava are known as flood basalts or plateau basalts. *See* BASALT.

Submarine volcanism. Deep submarine volcanism occurs along the spreading ridges that zigzag for thousands of miles across the ocean floor, and it is exposed above sea level only in Iceland. Because of the logistical difficulties in making direct observations posed by the great ocean depths, no deep submarine volcanic activity has been actually observed during eruption. However, evidence that deep-sea eruptions are happening is clearly indicated by (1) seismic and acoustic monitoring networks; (2) the presence of deep-ocean floor hydrothermal vents; (3) episodic hydrothermal discharges, measured and mapped as thermal and geochemical anomalies in the ocean water; and (4) the detection of new lava flows in certain segments of the oceanic ridge system. *See* HYDROTHERMAL VENT; MID-OCEANIC RIDGE.

Volcanic eruptions in shallow water are very similar in character to those on land but, on average, are probably somewhat more explosive, owing to heating of water and resultant violent generation of supercritical steam. Much of the ocean basin appears to be floored by basaltic lava. *See* OCEANIC ISLANDS.

Fumaroles and hot springs. Vents at which volcanic gases issue without lava or after the eruption are known as fumaroles. They are found on active volcanoes during and between eruptions and on dormant volcanoes, persisting long after the volcano itself has become inactive. Fumaroles grade into hot springs and geysers. The water of most, if not all, hot springs is predominantly of meteoric origin, and is not water liberated from magma. Some hot springs are of volcanic origin and the water may contain volcanic gases. *See* GEYSER.

Distribution of volcanoes. Over 500 active volcanoes are known on the Earth, mostly along or near the boundaries of the dozen or so lithospheric plates that compose the Earth's solid surface. Lithospheric plates show three distinct types of boundaries: divergent or spreading margins—adjacent plates are pulling apart; convergent margins (subduction zones)—plates are moving toward each other and one is being destroyed; and transform margins—one plate is sliding horizontally past another. All these types of plate motion are well demonstrated in the Circum-Pacific region, in which many active volcanoes form the so-called Ring of Fire. Some volcanoes, however, are not associated with plate boundaries, and many of these so-called intraplate volcanoes form roughly linear chains in the interior parts of the oceanic plates, for example, the Hawaiian-Emperor, Austral, Society, and Line archipelagoes in the Pacific Basin. Intraplate volcanism also has resulted in voluminous outpourings of fluid lava to form extensive plateau basalts, or of more viscous and siliceous pyroclastic products to form ash flow plains.   [R.I.T.]

Volcanology   The scientific study of volcanic phenomena, especially the processes, products, and hazards associated with active or potentially active volcanoes. It focuses on eruptive activity that has occurred within the past 10,000 years of the Earth's history, particularly eruptions during recorded history. Strictly speaking, it emphasizes the surface eruption of magmas and related gases, and the structures, deposits, and other effects produced thereby. Broadly speaking, however, volcanology includes all studies germane to the generation, storage, and transport of magma, because the surface eruption of magma represents the culmination of diverse physicochemical processes at depth. This article considers the activity of erupting volcanoes and the nature of erupting lavas. For a discussion of the distribution of volcanoes and the surface structures and deposits produced by them, *see* PLATE TECTONICS; VOLCANO.

On average, about 50 to 60 volcanoes worldwide are active each year. About half of these constitute continuing activity that began the previous year, and the remainder are new eruptions. Analysis of historic records indicates that eruptions comparable in size to that of Mount St. Helens or El Chichón tend to occur about once or twice per decade, and larger eruptions such as Pinatubo about once per one or two centuries. On a global basis, eruptions the size of that at Nevado del Ruiz in November 1985 are orders of magnitude more frequent.

Modern volcanology perhaps began with the founding of well-instrumented observations at Asama Volcano (Japan) in 1911 and at Kilauea Volcano (Hawaii) in 1912. The Hawaiian Volcano Observatory, located on Kilauea's caldera rim, began to

Generalized relationships between magma composition, relative viscosity, and common eruptive characteristics

Magma composition	Relative viscosity	Common eruptive characteristics
Basaltic	Fluidal	Lava fountains, flows, and pools
Andesitic	Less fluidal	Lava flows, explosive ejecta, ashfalls, and pyroclastic flows
Dacitic-rhyolitic	Viscous	Explosive ejecta, ashfalls, pyroclastic flows, and lava domes

conduct systematic and continuous monitoring of seismic activity preceding, accompanying, and following eruptions, as well as other geological, geophysical, and geochemical observations and investigations.

The eruptive characteristics, products, and resulting landforms of a volcano are determined predominantly by the composition and physical properties of the magmas involved in the volcanic processes (see table). Formed by partial melting of existing solid rock in the Earth's lower crust or upper mantle, the discrete blebs of magma consist of liquid rock (silicate melt) and dissolved gases. Driven by buoyancy, the magma blebs, which are lighter than the surrounding rock, coalesce as they rise toward the surface to form larger masses. *See* IGNEOUS ROCKS; LITHOSPHERE; MAGMA.

Magma consists of three phases: liquid, solid, and gas. Volcanic gases generally are predominantly water; other gases include various compounds of carbon, sulfur, hydrogen, chlorine, and fluorine. All volcanic gases also contain minor amounts of nitrogen, argon, and other inert gases, largely the result of atmospheric contamination at or near the surface.

Temperatures of erupting magmas have been measured in lava flows and lakes, pyroclastic deposits, and volcanic vents by means of infrared sensors, optical pyrometers, and thermocouples. Reasonably good and consistent measurements have been obtained for basaltic magmas erupted from Kilauea and Mauna Loa volcanoes, Hawaii, and a few other volcanoes. Measured temperatures typically range between 2100 and 2200°F (1150 and 1200°C), and many measurements in cooling Hawaiian lava lakes indicate that the basalt becomes completely solid at about 1800°F (980°C). *See* GEOLOGIC THERMOMETRY.

The character of a volcanic eruption is determined largely by the viscosity of the liquid phase of the erupting magma and the abundance and condition of the gas it contains. Viscosity is in turn affected by such factors as the chemical composition and temperature of the liquid, the load of suspended solid crystals and xenoliths, the abundance of gas, and the degree of vesiculation. The subsequent violent expansion during eruption shreds the frothy liquid into tiny fragments, generating explosive showers of volcanic ash and dust, accompanied by some larger blocks (volcanic "bombs"); or it may produce an outpouring of a fluidized slurry of gas, semisolid bits of magma froth, and entrained blocks to form high-velocity pyroclastic flows, surges, and glowing avalanches. *See* PYROCLASTIC ROCKS.

Types of eruptions customarily are designated by the name of a volcano or volcanic area that is characterized by that sort of activity, even though all volcanoes show different modes of eruptive activity on occasion and even at different times during a single eruption.

Eruptions of the most fluid lava, in which relatively small amounts of gas escape freely with little explosion, are designated Hawaiian eruptions. Most of the lava is extruded as successive, thin flows that travel many miles from their vents. An occasional feature of Hawaiian activity is the lava lake, a pool of liquid lava with convectional circulation that occupies a preexisting shallow depression or pit crater. *See* LAVA.

Strombolian eruptions are somewhat more explosive eruptions of lava, with greater viscosity, and produce a larger proportion of pyroclastic material. Many of the volcanic bombs and lapilli assume rounded or drawn-out forms during flight, but commonly are sufficiently solid to retain these shapes on impact.

Generally still more explosive are the vulcanian type of eruptions. Angular blocks of viscous or solid lava are hurled out, commonly accompanied by voluminous clouds of ash but with little or no lava flow.

Peléean eruptions are characterized by the heaping up of viscous lava over and around the vent to form a steep-sided hill or volcanic dome. Explosions, or collapses of portions of the dome, may result in glowing avalanches (nuées ardentes).

Plinian eruptions are paroxysmal eruptions of great violence—named after Pliny the Elder, who was killed in A.D. 79 while observing the eruption of Vesuvius—and are characterized by voluminous explosive ejections of pumice and by ash flows. The copious expulsion of viscous siliceous magma commonly is accompanied by collapse of the summit of the volcano, forming a caldera, or by collapse of the broader region, forming a volcano-tectonic depression. *See* CALDERA.

A major component of the science of volcanology is the systematic and, preferably, continuous monitoring of active and potentially active volcanoes. Scientific observations and measurements—of the visible and invisible changes in a volcano and its surroundings—between eruptions are as important, perhaps even more crucial, than during eruptions. Measurable phenomena important in volcano monitoring include earthquakes; ground movements; variations in gas compositions; and deviations in local gravity, electrical, and magnetic fields. These phenomena reflect pressure and stresses induced by subsurface magma movements and or pressurization of the hydrothermal envelope surrounding the magma reservoir. The monitoring of volcanic seismicity and ground deformations before, during, and following eruptions has provided the most useful and reliable information. *See* EARTHQUAKE; SEISMOLOGY.

Volcanoes are in effect windows into the Earth's interior; thus research in volcanology, in contributing to an improved understanding of volcanic phenomena, provides special insights into the chemical and physical processes operative at depth. However, volcanology also serves an immediate role in the mitigation of volcanic and related hydrologic hazards (mudflows, floods, and so on). Progress toward hazards mitigation can best be advanced by a combined approach. One aspect is the preparation of comprehensive volcanic hazards assessments of all active and potentially active volcanoes, including a volcanic risk map for use by government officials in regional and local land-use planning to avoid high-density development in high-risk areas. The other component involves improvement of predictive capability by upgrading volcano-monitoring methods and facilities to adequately study more of the most dangerous volcanoes. An improved capability for eruption forecasts and predictions would permit timely warnings of impending activity, and give emergency-response officials more lead time for preparation of contingency plans and orderly evacuation, if necessary.   [R.I.T.]

Water table   The upper surface of the zone of saturation in permeable rocks not confined by impermeable rocks. It may also be defined as the surface underground at which the water is at atmospheric pressure. Saturated rock may extend a little above this level, but the water in it is held up above the water table by capillarity and is under less than atmospheric pressure; therefore, it is the lower part of the capillary fringe and is not free to flow into a well by gravity. Below the water table, water is free to move under the influence of gravity.

The position of the water table is shown by the level at which water stands in wells penetrating an unconfined water-bearing formation. Where a well penetrates only impermeable material, there is no water table and the well is dry. But if the well passes through impermeable rock into water-bearing material whose hydrostatic head is higher than the level of the bottom of the impermeable rock, water will rise approximately to the level it would have assumed if the whole column of rock penetrated had been permeable. This is called artesian water. *See* ARTESIAN SYSTEMS; GROUND-WATER HYDROLOGY.

[A.N.S./R.K.L.]

Waterspout   An intense columnar vortex (not necessarily containing a funnel-shaped cloud) of small horizontal extent, over water. Typical visible vortex diameters are of the order of 33 ft (10 m), but a few large waterspouts may exceed 330 ft (100 m) across. In the case of Florida waterspouts, only rarely does the visible funnel extend from parent cloudbase to sea surface. Like the tornado, most of the visible funnel is condensate. Therefore, the extension of the funnel cloud downward depends upon the distribution of ambient water vapor, ambient temperature, and pressure drop due to the vortex circulation strength. These vortices are most frequently observed during the warm season in the oceanic tropics and subtropics.

All waterspouts undergo a regular life cycle composed of five discrete but overlapping stages. (1) The dark-spot stage signifies a complete vortex column extending from cloud-base to sea surface. (2) The spiral-pattern stage is characterized by development of alternating dark- and light-colored bands spiraling around the dark spot on the sea surface. (3) The spray ring (incipient spray vortex) stage is characterized by a concentrated spray ring around the dark spot, with a lengthening funnel cloud above. (4) The mature waterspout stage (see illustration) is characterized by a spray vortex of maximum intensity and organization. (5) The decay stage occurs when the waterspout dissipates (often abruptly).

Waterspouts and tornadoes are qualitatively similar, differing only in certain quantitative aspects: tornadoes are usually more intense, move faster, and have longer lifetimes—especially maxi-tornadoes. Tornadoes are associated with intense, baroclinic (frontal), synoptic-scale disturbances with attendant strong vertical wind shear, while

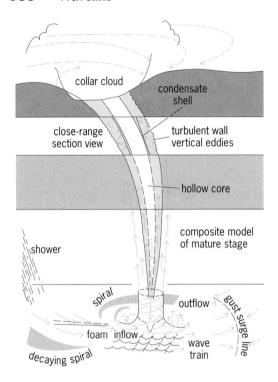

collar cloud

condensate shell

close-range section view

turbulent wall vertical eddies

hollow core

composite model of mature stage

shower

spiral

outflow

gust surge line

foam inflow

wave train

decaying spiral

Composite schematic model of a mature waterspout. For scaling reference, the maximum funnel diameters in this stage, just below the collar cloud, range from 10 to 460 ft (3 to 140 m).

waterspouts are associated with weak, quasibarotropic disturbances (weak thermal gradients) and consequent weak vertical wind shear. *See* TORNADO; WIND.   [J.H.G.]

Wavellite   A hydrated phosphate of aluminum mineral with composition $Al_3(OH)_3(PO_4)_2 \cdot 5H_2O$, in which small amounts of fluorine and iron may substitute for the hydroxyl group (OH) and aluminum (Al), respectively. Wavellite crystallizes in the orthorhombic system. The crystals are stout to long prismatic, but are rare. Wavellite commonly occurs as globular aggregates of fibrous structure and as encrusting and stalactitic masses. Wavellite crystals range in color from colorless and white to different shades of blue, green, yellow, brown, and black. Wavellite is found at many places in Europe and North America; it is also abundant in tin veins at Llallagua, Bolivia. *See* PHOSPHATE MINERALS.   [W.R.Lo.]

Weather   The state of the atmosphere, as determined by the simultaneous occurrence of several meteorological phenomena at a geographical locality or over broad areas of the Earth. When such a collection of weather elements is part of an interrelated physical structure of the atmosphere, it is termed a weather system, and includes phenomena at all elevations above the ground. More popularly, weather refers to a certain state of the atmosphere as it affects humans' activities on the Earth's surface. In this sense, it is often taken to include such related phenomena as waves at sea and floods on land.

A weather element is any individual physical feature of the atmosphere. At a given locality, at least seven such elements may be observed at any one time. These are clouds, precipitation, temperature, humidity, wind, pressure, and visibility. Each of these principal elements is divided into many subtypes. *See* WEATHER MAP.

The various forms of precipitation are included by international agreement among the hydrometeors, which comprise all the visible features in the atmosphere, besides clouds, that are due to water in its various forms. For convenience in processing weather data and information, this definition is made to include some phenomena not due to water, such as dust and smoke. Some of the more common hydrometeors include rain, snow, fog, hail, dew, and frost. See PRECIPITATION (METEOROLOGY).

Certain optical and electrical phenomena have long been observed among the weather elements. These include lightning, aurora, solar or lunar corona, and halo. See AIR MASS; ATMOSPHERE; AURORA; CLOUD; FRONT; LIGHTNING; METEOROLOGY; STORM; WEATHER OBSERVATIONS; WIND.   [P.F.C.]

Weather forecasting and prediction   Processes for formulating and disseminating information about future weather conditions based upon the collection and analysis of meteorological observations. Weather forecasts may be classified according to the space and time scale of the predicted phenomena. Atmospheric fluctuations with a length of less than 100 m (330 ft) and a period of less than 100 s are considered to be turbulent. The study of atmospheric turbulence is called micrometeorology; it is of importance for understanding the diffusion of air pollutants and other aspects of the climate near the ground. Standard meteorological observations are made with sampling techniques that filter out the influence of turbulence. Common terminology distinguishes among three classes of phenomena with a scale that is larger than the turbulent microscale: the mesoscale, synoptic scale, and planetary scale.

The mesoscale includes all moist convection phenomena, ranging from individual cloud cells up to the convective cloud complexes associated with prefrontal squall lines, tropical storms, and the intertropical convergence zone. Also included among mesoscale phenomena are the sea breeze, mountain valley circulations, and the detailed structure of frontal inversions. Most mesoscale phenomena have time periods less than 12 h. The prediction of mesoscale phenomena is an area of active research. Most forecasting methods depend upon empirical rules or the short-range extrapolation of current observations, particularly those provided by radar and geostationary satellites. Forecasts are usually couched in probabilistic terms to reflect the sporadic character of the phenomena. Since many mesoscale phenomena pose serious threats to life and property, it is the practice to issue advisories of potential occurrence significantly in advance. These "watch" advisories encourage the public to attain a degree of readiness appropriate to the potential hazard. Once the phenomenon is considered to be imminent, the advisory is changed to a "warning," with the expectation that the public will take immediate action to prevent the loss of life. See MESOMETEOROLOGY; SQUALL.

The next-largest scale of weather events is called the synoptic scale, because the network of meteorological stations making simultaneous, or synoptic, observations serves to define the phenomena. The migratory storm systems of the extratropics are synoptic-scale events, as are the undulating wind currents of the upper-air circulation which accompany the storms. The storms are associated with barometric minima, variously called lows, depressions, or cyclones. The synoptic method of forecasting consists of the simultaneous collection of weather observations, and the plotting and analysis of these data on geographical maps. An experienced analyst, having studied several of these maps in chronological succession, can follow the movement and intensification of weather systems and forecast their positions. This forecasting technique requires the regular and frequent use of large networks of data. See WEATHER MAP.

Planetary-scale phenomena are persistent, quasistationary perturbations of the global circulation of the air with horizontal dimensions comparable to the radius of the

Earth. These dominant features of the general circulation appear to be correlated with the major orographic features of the globe and with the latent and sensible heat sources provided by the oceans. They tend to control the paths followed by the synoptic-scale storms, and to draw upon the synoptic transients for an additional source of heat and momentum. *See* ATMOSPHERE; METEOROLOGICAL INSTRUMENTATION; WEATHER OBSERVATIONS.

Numerical weather prediction is the prediction of weather phenomena by the numerical solution of the equations governing the motion and changes of condition of the atmosphere. Numerical weather prediction techniques, in addition to being applied to short-range weather prediction, are used in such research studies as air-pollutant transport and the effects of greenhouse gases on global climate change. *See* AIR POLLUTION; GREENHOUSE EFFECT; JET STREAM; UPPER-ATMOSPHERE DYNAMICS.   [J.P.G.; J.R.G.]

The first operational numerical weather prediction model consisted of only one layer, and therefore it could model only the temporal variation of the mean vertical structure of the atmosphere. Computers now permit the development of multilevel (usually about 10–20) models that could resolve the vertical variation of the wind, temperature, and moisture. These multilevel models predict the fundamental meteorological variables for large scales of motion. Global models with horizontal resolutions as fine as 125 mi (200 km) are being used by weather services in several countries. Global numerical weather prediction models require the most powerful computers to complete a 10-day forecast in a reasonable amount of time.

Research models similar to global models could be applied for climate studies by running for much longer time periods. The extension of numerical predictions to long time intervals (many years) requires a more accurate numerical representation of the energy transfer and turbulent dissipative processes within the atmosphere and at the air-earth boundary, as well as greatly augmented computing-machine speeds and capacities.

Long-term simulations of climate models have yielded simulations of mean circulations that strongly resemble those of the atmosphere. These simulations have been useful in explaining the principal features of the Earth's climate, even though it is impossible to predict the daily fluctuations of weather for extended periods. Climate models have also been used successfully to explain paleoclimatic variations, and are being applied to predict future changes in the climate induced by changes in the atmospheric composition or characteristics of the Earth's surface due to human activities. *See* CLIMATE HISTORY; CLIMATE MODIFICATION; PALEOCLIMATOLOGY.   [R.A.An.]

Surface meteorological observations are routinely collected from a vast continental data network, with the majority of these observations obtained from the middle latitudes of both hemispheres. Commercial ships of opportunity, military vessels, and moored and drifting buoys provide similar in-place measurements from oceanic regions. Information on winds, pressure, temperature, and moisture throughout the troposphere and into the stratosphere is routinely collected from (1) balloon-borne instrumentation packages (radiosonde observations) and commercial and military aircraft which sample the free atmosphere directly; (2) ground-based remote-sensing instrumentation such as wind profilers (vertically pointing Doppler radars), the National Weather Service Doppler radar network, and lidars; and (3) special sensors deployed on board polar orbiting or geostationary satellites. The remotely sensed observations obtained from meteorological satellites have been especially helpful in providing crucial measurements of areally and vertically averaged temperature, moisture, and winds in data-sparse (mostly oceanic) regions of the world. Such measurements are necessary to accommodate modern numerical weather prediction practices and to enable forecasters to continuously monitor global storm (such as hurricane) activity. *See* METEOROLOGICAL INSTRUMENTATION; RADAR METEOROLOGY.

Forecast products and forecast skill are classified as longer term (greater than 2 weeks) and shorter term. These varying skill levels reflect the fact that existing numerical prediction models such as the medium-range forecast have become very good at making large-scale circulation and temperature forecasts, but are less successful in making weather forecasts. An example is the prediction of precipitation amount and type given the occurrence of precipitation and convection. Each of these forecasts is progressively more difficult because of the increasing importance of mesoscale processes to the overall skill of the forecast. *See* Precipitation (meteorology).   [L.F.B.]

Nowcasting is a form of very short range weather forecasting. The term nowcasting is sometimes used loosely to refer to any area-specific forecast for the period up to 12 h ahead that is based on very detailed observational data. However, nowcasting should probably be defined more restrictively as the detailed description of the current weather along with forecasts obtained by extrapolation up to about 2 h ahead. Useful extrapolation forecasts can be obtained for longer periods in many situations, but in some weather situations the accuracy of extrapolation forecasts diminishes quickly with time as a result of the development or decay of the weather systems. *See* Weather.

[K.A.B.]

Forecasts of time averages of atmospheric variables, for example, sea surface temperature, where the lead time for the prediction is more than 2 weeks, are termed long-range or extended-range climate predictions. Extended-range predictions of monthly and seasonal average temperature and precipitation are known as climate outlooks. The accuracy of long-range outlooks has always been modest because the predictions must encompass a large number of possible outcomes, while the observed single event against which the outlook is verified includes the noise created by the specific synoptic disturbances that actually occur and that are unpredictable on monthly and seasonal time scales. According to some estimates of potential predictability, the noise is generally larger than the signal in middle latitudes.   [E.A.O'L.]

Weather map   A map or a series of maps that is used to depict the evolution and life cycle of atmospheric phenomena at selected times at the surface and in the free atmosphere. Weather maps are used for the analysis and display of in-place observational measurements and computer-generated analysis and forecast fields derived from weather and climate prediction models by research and operational meteorologists, government research laboratories, and commercial firms. Similar analyses derived from sophisticated computer forecast models are displayed in map form for forecast periods of 10–14 days in advance to provide guidance for human weather forecasters. *See* Meteorological instrumentation; Weather observations.

Rapid advances in computer technology and visualization techniques, as well as the continued explosive growth of the Internet distribution of global weather observations, satellite and radar imagery, and model analysis and forecast fields, have revolutionized how weather, climate, and forecast data and information can be conveyed to both the general public and sophisticated users in the public and commercial sectors. People and organizations with access to the Internet can access weather and climate information in a variety of digital or map forms in support of a wind range of professional and personal activities. *See* Climatology; Meteorological satellites; Meteorology; Radar meteorology; Weather forecasting and prediction.   [L.F.B.]

Weather modification   Human influence on the weather and, ultimately, climate. This can be either intentional, as with cloud seeding to clear fog from airports or to increase precipitation, or unintentional, as with air pollution, which increases aerosol concentrations and reduces sunlight. Weather is considered to be the day-to-day variations of the environment—temperature, cloudiness, relative humidity, wind-speed, visibility, and precipitation. Climate, on the other hand, reflects the average and extremes of these variables, changing on a seasonal basis. Weather change may lead to climate change, which is assessed over a period of years. See AIR POLLUTION; CLIMATE HISTORY; CLOUD PHYSICS.

Specific processes of weather modification are as follows: (1) Change of precipitation intensity and distribution result from changes in the colloidal stability of clouds. For example, seeding of supercooled water clouds with dry ice (solid carbon dioxide, CO_2) or silver iodide (AgI) leads to ice crystal growth and fall-out; layer clouds may dissipate, convective clouds may grow. (2) Radiation change results from changes of aerosol or clouds (deliberately with a smoke screen, or unintentionally with air pollution from combustion), from changes in the gaseous constituents of the atmosphere (as with carbon dioxide from fossil fuel combustion), and from changes in the ability of surfaces to reflect or scatter back sunlight (as replacing farmland by houses.) (3) Change of wind regime results from change in surface roughness and heat input, for example, replacing forests with farmland. See PRECIPITATION (METEOROLOGY).   [J.Hal.]

Weather observations   The measuring, recording, and transmitting of data of the variable elements of weather. In the United States the National Weather Service (NWS), a division of the National Oceanic and Atmospheric Administration (NOAA), has as one of its primary responsibilities the acquisition of meteorological information. The data are sent by various communication methods to the National Meteorological Center.

At the Center, the raw data are fed into large computers that are programmed to plot, analyze, and process the data and also to make prognostic weather charts. The processed data and the forecast guidance are then distributed by special National Weather Service systems and conventional telecommunications to field offices, other government agencies, and private meteorologists. They in turn prepare forecasts and warnings based on both processed and raw data. See WEATHER MAP.

A wide variety of meteorological data are required to satisfy the needs of meteorologists, climatologists, and users in marine activities, forestry, agriculture, aviation, and other fields. This has led to a dual surface-observation program: the Synoptic Weather Program and the Basic Observations Program. See AERONAUTICAL METEOROLOGY; AGRICULTURAL METEOROLOGY; INDUSTRIAL METEOROLOGY.

The Synoptic Weather Program is designed to assist in the preparation of forecasts and to provide data for international exchange. Worldwide surface observations are taken at standard times [0000, 0600, 1200, and 1800 Universal Time Coordinated (UTC)] and sent in synoptic code.

The Basic Observations Program routinely provides meteorological data every hour. Special observations are taken at any intervening time to report significant weather events or changes. Observation sites are located primarily at airports; a few are in urban centers. At these sites, human observers report the weather elements.

Present weather consists of a number of hydrometers, such as liquid or frozen precipitation, fog, thunderstorms, showers, and tornadoes, and of lithometers, such as haze, dust, smog, dust devils, and blowing sand. The amount of cloudiness is also reported. See FOG; METEOROLOGICAL OPTICS; PRECIPITATION (METEOROLOGY); SMOG; THUNDERSTORM; TORNADO.

Pressure measurements are read from either a mercury or precision aneroid barometer located at the station. A microbarograph provides a continuous record of the pressure, from which changes in specific intervals of time are reported. Pressure changes are frequently quite helpful in short-range prediction of weather events. See AIR PRESSURE.

Temperature and humidity are measured by a hygrothermometer, located near the center of the runway complex at many airport stations. The readings are transmitted to the observation site. The temperature dial indicator is equipped with pointers to determine maximum and minimum temperature extremes. *See* HUMIDITY.

Wind speed and direction measurements are telemetered into most airport stations. The equipment, consisting of an anemometer and a wind vane, is located near the center of the runway complex at participating airports; elsewhere it is placed in an unsheltered area. *See* WIND MEASUREMENT.

Various types of clouds and their heights are reported. The lowest height of opaque clouds covering half or more of the sky is known as the ceiling, and is normally measured by a ceilometer at first-order stations. *See* CLOUD.

Upper-air observations have been made by the National Weather Service with radiosondes. The radiosonde is a small, expendable instrument package that is suspended below a 6-ft-diameter (2-m) balloon filled with hydrogen or helium. As the radiosonde is carried aloft, sensors on it measure profiles of pressure, temperature, and relative humidity. By tracking the position of the radiosonde in flight with a radio direction finder or radio navigation system, such as Loran or the Global Positioning System (GPS), information on wind speed and direction aloft is also obtained.

Understanding and accurately predicting changes in the atmosphere requires adequate observations of the upper atmosphere. Radiosonde observations, plus routine aircraft reports, radar, and satellite observations, provide meteorologists with a three-dimensional picture of the atmosphere. *See* METEOROLOGICAL INSTRUMENTATION; WEATHER OBSERVATIONS; WIND MEASUREMENT.

Weather radars distributed throughout the United States are used to observe precipitation within a radius of about 250 nmi (460 km), and associated wind fields (utilizing the Doppler principle) within about 125 nmi (230 km). The primary component of this set of weather radars is known as NEXRAD (Next Generation Weather Radar). These radars provide information on rainfall intensity, likelihood of tornadoes or severe thunderstorms, projected paths of individual storms (both ambient and within-storm wind fields), and heights of storms for short-range (up to 3 h) forecasts and warnings. *See* RADAR METEOROLOGY.

Geostationary weather satellites near 22,000 mi (36,000 km) above the Earth transmit pictures depicting the cloud cover over vast expanses of the hemisphere. Using still photographs and animated images, the meteorologist can determine, among other things, areas of potentially severe weather and the motion of clouds and fog. In addition, the satellite does an outstanding job of tracking hurricanes over the ocean where few other observations are taken. *See* HURRICANE; METEOROLOGICAL SATELLITES.

Ground-based lightning detection systems detect the electromagnetic wave that emanates from the lightning path as the lightning strikes the ground. Lightning information has proven to be operationally valuable to a wide variety of users and as a supplement to other observing systems, particularly radar and satellites. *See* LIGHTNING; MESOMETEOROLOGY; METEOROLOGY; WEATHER FORECASTING AND PREDICTION.   [F.S.Z.; R.L.L.]

Weathering processes   The response of geologic materials to the environment (physical, chemical, and biological) at or near the Earth's surface. This response typically results in a reduction in size of the weathering materials; some may become as tiny as ions in solution.

The agents and energies that activate weathering processes and the products resulting therefrom have been classified traditionally as physical and chemical in type. In classic physical weathering, rock materials are broken by action of mechanical forces into smaller fragments without change in chemical composition, whereas in chemical weathering the process is characterized by change in chemical composition. In practice, however, the two processes commonly overlap.

Specific agents of weathering may be recognized and correlated with the types of effects they produce. Important agents of weathering are water in all surface occurrences (rain, soil and ground water, streams, and ocean); the atmosphere (H_2O, O_2, CO_2, wind); temperature (ambient and changing, especially at the freezing point of water); insolation (on large bare surfaces); ice (in soil and glaciers); gravity; plants (bacteria and macroforms); animals (micro and macro, including humans). Human modifications of otherwise geologic weathering that have increased exponentially during recent centuries include construction, tillage, lumbering, use of fire, chemically active industry (fumes, liquid, and solid effluents), and manipulation of geologic water systems.

Products of physical weathering include jointed (horizontal and vertical) rock masses, disintegrated granules, frost-riven soil and surface rock, and rock and soil flows. Products of chemical weathering include the soil, and the clays used in making ceramic structural products, whitewares, refractories, various fillers and coating of paper, portland cement, absorbents, and vanadium. These are the relatively insoluble products of weathering; characteristically they occur in clays, siltstones, and shales. Sand-size particles resulting from both physical and chemical weathering may accumulate as sandstones.

After precipitation, the relatively soluble products of chemical weathering give rise to products and rocks such as limestone, gypsum, rock salt, silica, and phosphate and potassium compounds useful as fertilizers.   [W.D.K.]

West Indies   An archipelago, including the Bahamas, the Greater Antilles (including Cuba, Jamaica, Hispaniola—the Dominican Republic and Haiti—and Puerto Rico), the Lesser Antilles, and other islands, curving 2500 mi (4000 km) from Yucatan Peninsula and southeastern Florida to northern Venezuela and enclosing the Caribbean Sea. Situated between latitude $10°$ and $27°N$ and longitude $59°$ and $85°W$, in the zone of the northeast trade winds, the West Indies have a subtropical and predominantly oceanic climate, with even warmth and steady breezes. Temperatures vary little from season to season, ranging from means of 80–85°F (27–29°C) in July to 70–78°F (21–26°C) in January at sea level. Freezing is unknown, and the hottest temperatures rarely exceed 90°F (32°C). Precipitation ranges from a low of 25–50 in. (64–127 cm) a year on low-lying islands and drier coasts up to 300 in. (7.6 m) on the highest peaks, which are almost perpetually cloud-capped. At lower elevations, rainfall is erratic from year to year and from season to season, but reaches a maximum in the summer and fall, when the northeast trades are replaced by light, variable winds. This is also the season of hurricanes, destructive tropical cyclones which sweep west and northwest across the Caribbean, sparing only the southernmost islands. The winter months are generally dry, and there is frequently a shorter dry season in July or August. *See* TROPICAL METEOROLOGY.

The West Indian flora is chiefly derived from Central and South America, but there are a number of endemic species, notably palms; many mainland plants failed to colonize the islands, which are floristically poor. The effect of isolation and small size is evident in the meager character of West Indian fauna. Animal species are limited; there are few mammals and no large ones, except for domesticated animals and, especially on Hispaniola, feral cattle, goats, pigs, and horses.   [D.Low.]

Wetlands   Ecosystems that form transitional areas between terrestrial and aquatic components of a landscape. Typically they are shallow-water to intermittently flooded ecosystems, which results in their unique combination of hydrology, soils, and vegetation. Examples of wetlands include swamps, fresh- and salt-water marshes, bogs, fens, playas, vernal pools and ponds, floodplains, organic and mineral soil flats, and tundra. As transitional elements in the landscape, wetlands often develop at the interface between drier uplands such as forests and farmlands, and deep-water aquatic systems such as lakes, rivers, estuaries, and oceans. Thus, wetland ecosystems are characterized by the presence of water that flows over, ponds on the surface of, or saturates the soil for at least some portion of the year.

Wetland soils can be either mineral (composed of varying percentages of sand, silt, or clay) or organic (containing 12–20% organic matter). Through their texture, structure, and landscape position, soils control the rate of water movement into and through the soil profile (the vertical succession of soil layers). Retention of water and organic carbon in the soil environment controls biogeochemical reactions that facilitate the functioning of wetland soils. *See* BIOGEOCHEMISTRY.

Vegetated wetlands are dominated by plant species, called hydrophytes, that are adapted to live in water or under saturated soil conditions. Adaptations that allow plants to survive in a water-logged environment include morphological features, such as pneumatophores (the "knees," or exposed roots, of the bald cypress), buttressed tree trunks, shallow root systems, floating leaves, hypertrophied lenticels, inflated plant parts, and adventitious roots. Physiological adaptations also allow plants to survive in a wetland environment. These include the ability of plants to transfer oxygen from the root system into the soil immediately surrounding the root (rhizosphere oxidation); the reduction or elimination of ethanol accumulation due to low concentrations of alcohol dehydrogenase; and the ability to concentrate malate (a nontoxic metabolite) instead of ethanol in the root system.

Wetlands differ with respect to their origin, position in the landscape, and hydrologic and biotic characteristics. For example, work has focused on the hydrology as well as the geomorphic position of wetlands in the landscape. This hydrogeomorphic approach recognizes and uses the fundamental physical properties that define wetland ecosystems to distinguish among classes of wetlands that occur in riverine, depressional, estuarine or lake fringe, mineral or organic soil flats, and slope environments.

The extent of wetlands in the world is estimated to be $2–3 \times 10^6 \, \text{mi}^2$ ($5–8 \times 10^6 \, \text{km}^2$), or about 4–6% of the Earth's land surface. Wetlands are found on every continent except Antarctica and in every clime from the tropics to the frozen tundra. Rice paddies, which comprise another $500,000–600,000 \, \text{mi}^2$ ($1.3–1.5 \times 10^6 \, \text{km}^2$), can be considered as a type of domesticated wetland of great value to human societies worldwide. *See* BOG; PLAYA.

Wetlands are often an extremely productive part of the landscape. They support a rich variety of waterfowl and aquatic organisms, and represent one of the highest levels of species diversity and richness of any ecosystem. Wetlands are an extremely important habitat for rare and endangered species.

Wetlands often serve as natural filters for human and naturally generated nutrients, organic materials, and contaminants. The ability to retain, process, or transform these substances is called assimilative capacity, and is strongly related to wetland soil texture and vegetation. The assimilative capacity of wetlands has led to many projects that use wetland ecosystems for wastewater treatment and for improving water quality. Wetlands also have been shown to prevent downstream flooding and, in some cases, to prevent ground-water depletion as well as to protect shorelines from storm damage. The best wetland management practices enhance the natural processes of wetlands by maintaining conditions as close to the natural hydrology of the wetland as possible. *See* GROUND-WATER HYDROLOGY.

The world's wetlands are becoming a threatened landscape. Loss of wetlands world-wide currently is estimated at 50%. Wetland loss results primarily from habitat destruction, alteration of wetland hydrology, and landscape fragmentation. Global warming may soon be added to this list, although the exact loss of coastal wetlands due to sea-level rise is not well documented. Worldwide, destruction of wetland ecosystems primarily has been through the conversion of wetlands to agricultural land.

Hydrologic modifications that destroy, alter, and degrade wetland systems include the construction of dams and water diversions, ground-water extraction, and the artificial manipulation of the amount, timing, and periodicity of water delivery. The primary impact of landscape fragmentation on wetland ecosystems is the disruption and degradation of wildlife migratory corridors, reducing the connectivity of wildlife habitats and rendering wetland habitats too small, too degraded, or otherwise irreversibly altered to support the critical life stages of plants and animals.

The heavy losses of wetlands in the world, coupled with the recognized values of these systems, have led to a number of policy initiatives at both the national and international levels.

Wetland restoration usually refers to the rehabilitation of degraded or hydrologically altered wetlands, often involving the reestablishment of vegetation. Wetland enhancement generally refers to the targeted restoration of one or a set of ecosystem functions over others, for example, the focused restoration of a breeding habitat for rare, threatened, or endangered amphibians. Wetland creation refers to the construction of wetlands where they did not exist before. Created wetlands are also called constructed or artificial wetlands. Restoring, enhancing, or creating a wetland requires a comprehensive understanding of hydrology and ecology, as well as engineering skills. *See* ECOSYSTEM; ESTUARINE OCEANOGRAPHY; HYDROLOGY; RESTORATION ECOLOGY.

[W.J.M.; P.L.Fi.; L.C.Le.; S.R.St.]

Willemite A rare nesosilicate mineral, composition Zn_2SiO_4, crystallizing in the hexagonal system. It is usually massive or granular with a vitreous luster; crystals are rare. The mineral may be variously colored, most commonly green, red, or brown. Hardness is $5\frac{1}{2}$ on Mohs scale; specific gravity is 3.9–4.2. Willemite forms a valuable ore of zinc at Franklin, New Jersey. At this famous zinc deposit Willemite fluoresces a yellow-green. *See* SILICATE MINERALS.   [C.S.Hu.]

Wind The motion of air relative to the Earth's surface. The term usually refers to horizontal air motion, as distinguished from vertical motion, and to air motion averaged over a chosen period of 1–3 min. Micrometeorological circulations (air motion over periods of the order of a few seconds) and others small enough in extent to be obscured by this averaging are thereby eliminated.

The direct effects of wind near the surface of the Earth are manifested by soil erosion, the character of vegetation, damage to structures, and the production of waves on water surfaces. At higher levels wind directly affects aircraft, missile and rocket operations, and dispersion of industrial pollutants, radioactive products of nuclear explosions, dust, volcanic debris, and other material. Directly or indirectly, wind is responsible for the production and transport of clouds and precipitation and for the transport of cold and warm air masses from one region to another. *See* ATMOSPHERIC GENERAL CIRCULATION; WIND MEASUREMENT.

Cyclonic and anticyclonic circulation are each a portion of the pattern of airflow within which the streamlines (which indicate the pattern of wind direction at any instant) are curved so as to indicate rotation of air about some central point of the cyclone or anticyclone. The rotation is considered cyclonic if it is in the same sense as the rotation of the surface of the Earth about the local vertical, and is considered anti-cyclonic if in the opposite sense. Thus, in a cyclonic circulation, the streamlines indicate counterclockwise (clockwise for anticylonic) rotation of air about a central point on the Northern Hemisphere or clockwise (counterclockwise for anticyclonic) rotation about a point on the Southern Hemisphere. When the streamlines close completely about the central point, the pattern is denoted respectively a cyclone or an anticyclone. Since the gradient wind represents a good approximation to the actual wind, the center of a cyclone tends strongly to be a point of minimum atmospheric pressure on a horizontal surface. Thus the terms cyclone, low-pressure area, or low are often used to denote essentially the same phenomenon. *See* GRADIENT WIND.

Convergent or divergent patterns are said to occur in areas in which the (horizontal) wind flow and distribution of air density is such as to produce a net accumulation or depletion, respectively, of mass of air. The horizontal mass divergence or convergence is intimately related to the vertical component of motion. For example, since local temporal rates of change of air density are relatively small, there must be a net vertical export of mass from a volume in which horizontal mass convergence is taking place. Only thus can the total mass of air within the volume remain approximately constant.

The horizontal mass divergence or convergence is closely related to the circulation. In a convergent wind pattern the circulation of the air tends to become more cyclonic; in a divergent wind pattern the circulation of the air tends to become more anticyclonic. A convergent surface wind field is typical of fronts. As the warm and cold currents impinge at the front, the warm air tends to rise over the cold air, producing the typical frontal band of cloudiness and precipitation. *See* FRONT.

Zonal surface winds patterns result from a longitudinal averaging of the surface circulation. This averaging typically reveals a zone of weak variable winds near the Equator (the doldrums) flanked by northeasterly trade winds in the Northern Hemisphere and southeasterly trade winds in the Southern Hemisphere, extending poleward in each instance to about latitude $30°$. The doldrum belt, particularly at places and times at which it is so narrow that the trade winds from the two hemispheres impinge upon it quite sharply, is designated the intertropical convergence zone, or ITCZ. The resulting convergent wind field is associated with abundant cloudiness and locally heavy rainfall. *See* MONSOON METEOROLOGY.

Local winds commonly represent modifications by local topography of a circulation of large scale. They are often capricious and violent in nature and are sometimes characterized by extremely low relative humidity. Examples are the mistral which blows down the Rhone Valley in the south of France, the bora which blows down the gorges leading to the coast of the Adriatic Sea, the foehn winds which blow down the Alpine valleys, the williwaws which are characteristic of the fiords of the Alaskan coast and the Aleutian Islands, and the chinook which is observed on the eastern slopes of the Rocky Mountains. *See* CHINOOK.   [F.S.; H.B.B.]

Wind measurement   The determination of three parameters: the size of an air sample, its speed, and its direction of motion. Air movement or wind is a vector that is specified by speed and direction; meteorological convention indicates wind direction is the direction from which the wind blows (for example, a southeast wind blows toward the northwest). Anemometers measure wind speed, while wind vanes indicate direction. On average, the wind blows horizontally over flat terrain; however, gusts, thermals, cloud outflows, and many other conditions have associated with them significant short-term vertical wind components. While research wind instruments typically measure both horizontal and vertical air movement, operational and personal wind sensors measure only the horizontal component.

There are many types of wind measurement instruments. In situ devices measure characteristics of air in contact with the instrument; often they are referred to as immersion sensors because they are immersed in the fluid (air) they measure. Remote wind sensors make measurements without physical contact with the portion of the atmosphere measured. Active remote sensors emit electromagnetic (for example, light or radio waves) or sound waves into the atmosphere and measure the amount and nature of the electromagnetic or acoustic power returned from the atmosphere. *See* METEOROLOGICAL INSTRUMENTATION; METEOROLOGICAL RADAR; WIND.   [W.F.D.]

Wind rose   A diagram in which statistical information concerning the direction and speed of the wind at a particular location may be conveniently summarized. In the standard wind rose a line segment is drawn in each of perhaps eight compass directions from a common origin (see illustration). The length of a particular segment

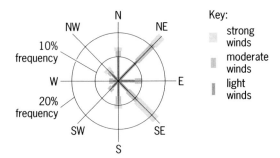

Standard wind rose.

is proportional to the frequency with which winds blow from that direction. Parts of a given segment are given various thicknesses, indicating frequencies of occurrence of various classes of wind speed from the given direction. *See* WIND MEASUREMENT.   [F.S.]

Wind stress   The drag or tangential force per unit area exerted on the surface of the Earth by the adjacent layer of moving air. Erosion of ground surfaces and the production of waves on water surfaces are manifestations of wind stress. Surface wind stress determines the exchange of momentum between the Earth and the atmosphere and exerts a strong influence on the typical variation of wind through the lowest kilometer of the atmosphere. Estimated values of the surface wind stress range up to several dynes per square centimeter (0.1 pascal), depending on the nature of the surface and the character of the adjacent airflow. *See* METEOROLOGY.

Significant stresses arise within the lower atmosphere because of the strong shear of the wind between the slowly moving air near the ground and the more rapidly moving air a kilometer above and because of the turbulent nature of the airflow in this region. The turbulent eddies referred to here have characteristic dimensions ranging up to a few hundreds of meters.

Wind pressure is the force exerted by the wind per unit area of solid surface exposed normal to the wind direction and is also known as dynamic pressure. In contrast to shearing stresses, the wind pressure arises from the difference in pressure between the windward and lee sides of the exposed surface. Wind pressure thus represents a substantial force when the wind speed is high. *See* WIND.   [F.S.]

The drag or tangential force of the wind on the sea is expressed in units of dynes per square centimeter or micronewtons per square meter but is normally taken to represent the mean drag over an undefined area, perhaps several kilometers square, containing many waves. It is usually related to an appropriate time and space average of the wind near the sea surface (at 10 m above the mean level, for example).

The drag coefficient over the sea is an important quantity in both meteorology and oceanography since it relates the wind speed to the drag, which generates ocean waves, drives the ocean currents, and sets the scale of the atmospheric turbulence that transfers water vapor and heat from the ocean to the atmosphere to provide the energy for clouds and weather systems. The drag coefficient of the sea surface depends on the wave field and on the turbulent structure of the flow in the air and the water. Present knowledge of the complicated fluid mechanics involved is not sufficient to allow theoretical calculation of it. *See* ATMOSPHERIC GENERAL CIRCULATION; MARITIME METEOROLOGY; OCEAN CIRCULATION; OCEAN WAVES.

There is substantial agreement that the drag of the wind on the sea is small relative to that of a fixed soil surface with the same geometry. It is largely independent of the fetch and so seems to depends less on the larger waves than on the short waves and ripples. Surface-active agents, which affect the shortest waves, may therefore be important.

[H.C.]

Witherite   The mineral form of barium carbonate. Witherite has orthorhombic symmetry and the aragonite structure type. Crystals, often twinned, may appear hexagonal in outline. It may be white or gray with yellow, brown, or green tints. Its hardness is 3.5 and its specific gravity 4.3.

Witherite may be found in veins with barite and galena. It is found in many places in Europe, and large crystals occur at Rosiclare, Illinois. *See* CARBONATE MINERALS.

[R.I.Ha.]

Wolframite   A mineral with composition $(Fe,Mn)WO_4$, intermediate between ferberite and huebnerite, which form a complete solid solution series. Wolframite occurs commonly in short, brownish-black, monoclinic, prismatic, bladed crystals. It is probably the most important tungsten mineral. China is the major producer of wolframite. Tungsten minerals of the wolframite series occur in many areas of the western United States; the major producing district is Boulder and northern Gilpin counties in Colorado. *See* HUEBNERITE.

[E.C.T.C.]

Wollastonite   A mineral inosilicate with composition $CaSiO_3$. Commonly it is massive, or in cleavable to fibrous aggregates. Hardness is $5-5^1/_2$ on Mohs scale; specific gravity is 2.85. On the cleavages the luster is pearly or silky; the color is white to gray.

Wollastonite is found in large masses in the Black Forest of Germany; Brittany, France; Chiapas, Mexico; and Willsboro, New York, where it is mined as a ceramic material. *See* SILICATE MINERALS.

[C.S.H.]

Wulfenite   A mineral consisting of lead molybdate, $PbMoO_4$. Wulfenite occurs commonly in yellow, orange, red, and grayish-white crystals, with a luster from adamantine to resinous. Wulfenite may also be massive or granular. Its fracture is uneven. Its hardness is 2.7–3 on Mohs scale and its specific gravity 6.5–7. Its streak is white.

Wulfenite is found in numerous localities in the western and southwestern United States. Brilliant orange tabular wulfenite crystals up to 2 in. (5 cm) in size have been found from the Red Cloud and Hamburg mines in Yuma County, Arizona.

[E.C.T.C.]

Wurtzilite   A black, infusible carbonaceous substance occurring in Uinta County, Utah. It is insoluble in carbon disulfide, has a density of about 1.05, and consists of

79–80% carbon, 10.5–12.5% hydrogen, 4–6% sulfur, 1.8–2.2% nitrogen, and some oxygen. Wurtzilite is derived from shale beds deposited near the close of Eocene (Green River) time. The material was introduced into the calcareous shale beds as a fluid after which it polymerized to form nodules or veins. *See* ASPHALT AND ASPHALTITE; IMPSONITE.

[I.A.B.]

Wurtzite   A mineral with composition ZnS (zinc sulfide). It exists most commonly in fibrous or columnar aggregates and banded crusts with resinous luster and brownish-black color. Hardness is $3^1/_2$ on Mohs scale and its specific gravity 4.0. Zinc sulfide is dimorphous, crystallizing as both the high-temperature form wurtzite, α-ZnS, and the low-temperature form sphalerite, β-ZnS. Wurtzite is the rarer and at room temperature the less stable form and, unlike sphalerite, is rarely mined as an ore of zinc. Wurtzite usually contains iron and cadmium, and a complete solid solution series extends to greenockite, CdS. *See* GREENOCKITE; SPHALERITE.   [C.S.Hu.]

X, Z

Xenolith A rock fragment enclosed in another rock, and of varying degrees of foreignness. Cognate xenoliths, for example, are pieces of rock that are genetically related to the host rock that contains them, such as pieces of a border zone in the interior of the same body. Included blocks of unrelated rocks are more deserving of the xenolith label. Such foreign rocks help establish the once molten condition of invading magma capable of incorporating and mixing an assemblage of unrelated rock inclusions. *See* LAVA; MAGMA.

Xenoliths tend to react with the enclosing magma, so that their constituent minerals become like those in equilibrium with the melt. Reaction is rarely complete, however. Even completely equilibrated xenoliths may be conspicuous because the equilibration process does not require either the texture or the proportions of the minerals in the xenolith to match those in the enclosing rock. *See* PLUTON.

Xenoliths may be angular to round, millimeters to meters in diameter, aligned or haphazard, and sharply or gradationally bounded. Xenoliths are present in most bodies of igneous rock. *See* IGNEOUS ROCKS.   [A.T.A.]

Zeolite Any mineral belonging to the zeolite family of minerals and synthetic compounds characterized by an aluminosilicate tetrahedral framework, ion-exchangeable large cations, and loosely held water molecules permitting reversible dehydration. The general formula can be expressed as $X_y^{1+,2+}Al_x^{3+}Si_{1-x}^{4+}O_2 \cdot nH_2O$. The amount of large cations (X) present is conditioned by the aluminum/silicon (Al:Si) ratio and the formal charge of these large cations. Typical large cations are the alkalies and alkaline earths such as sodium (Na^+), potassium (K^+), calcium (Ca^{2+}), strontium (Sr^{2+}), and barium (Ba^{2+}). The large cations, coordinated by framework oxygens and water molecules, reside in large cavities in the crystal structure; these cavities and channels may even permit the selective passage of organic molecules. Thus, zeolites are extensively studied from theoretical and technical standpoints because of their potential and actual use as molecular sieves, catalysts, and water softeners. *See* SILICATE MINERALS.

Zeolites are low-temperature and low-pressure minerals and commonly occur as late minerals in amygdaloidal basalts, as devitrification products, as authigenic minerals in sandstones and other sediments, and as alteration products of feldspars and nepheline. Phillipsite and laumontite occur extensively in sediments on the ocean floor. Stilbite, heulandite, analcime, chabazite, and scolecite are common as large crystals in vesicles and cavities in various basalts.

Zeolites are usually white, but often may be colored pink, brown, red, yellow, or green by inclusions; the hardness is moderate (3–5) and the specific gravity low (2.0–2.5) because of their rather open framework structures.   [P.B.M.]

The porous, yet crystalline nature of zeolites has been exploited commercially in three main areas—as sorbents, as cation-exchange materials, and as catalysts.

Most aluminosilicate zeolites have an extremely high (yet reversible) affinity for water and are widely used as desiccants. The discrimination between molecules on the basis of their size as compared to the zeolite pore size is the basis of several extremely important separation processes in a procedure known as molecular sieving.

The major commodity use of zeolites takes advantage of their cation-exchange ability and leads to their use in low-phosphate detergents. Mineral zeolites have found utility in agricultural and wastewater treatment applications, where they either ion-exchange harmful metal ions out of the stream being treated or absorb ammonia by reacting, as their acidic form, to produce the absorbed ammonium ion.

The most important and expanding area of application for zeolites is as heterogeneous catalysts. By combining the properties of excellent thermal stability ($>800°C$ or $1500°F$), a pore size of molecular dimensions, and the ready introduction of a wide assortment of cations via ion exchange, numerous very selective catalysts can be prepared.   [N.Her.]

Zincite   A mineral with composition ZnO (zinc oxide). It crystallizes in the hexagonal system with a wurtzite-type structure. Thus its principal axis is polar and different forms appear at top and bottom of crystals. Such crystals are rare and the mineral is usually massive. Its hardness is 4 and its specific gravity 5.6. The mineral has a subadamantine luster and a deep-red to orange-yellow color. Zincite is rare except at the zinc deposits at Franklin and Sterling Hill, N.J. There, associated with franklinite and willemite, it is mined as a valuable ore of zinc. *See* FRANKLINITE; WILLEMITE.   [C.S.Hu.]

Zircon   A mineral with the idealized composition $ZrSiO_4$, one of the chief sources of the element zirconium. Structurally, zircon is a nesosilicate, with isolated SiO_4 groups. *See* SILICATE MINERALS.

Zircon often occurs as well-formed crystals. The color is variable, usually brown to reddish brown, but also colorless, pale yellowish, green, or blue. The transparent colorless or tinted varieties are popular gemstones. Hardness is $7^1/_2$ on Mohs scale; specific gravity is 4.7, decreasing in metamict types.

Because of its chemical and physical stability, zircon resists weathering and accumulates in residual deposits and in beach and river sands, from which it has been obtained commercially in Florida and in India, Brazil, and other countries. *See* HEAVY MINERALS.
                                                                        [C.Fr.]

Zooarcheology   Zooarcheology or archeozoology is the study of animal remains from archeological sites. Most such remains derive from people's meals. In other words, zooarcheology is essentially the study of ancient garbage, mainly bones and teeth of mammals, as well as birds, fish, mollusks, and even insects. Zooarcheology helps to provide a more complete picture of people's environment and way of life, especially their economy, as reflected in the relationship between people and animals 100,000 or even just 100 years ago. It is a multidisciplinary endeavor requiring knowledge of anatomy and biometry as well as an appreciation of the archeological questions that need to be addressed. Unlike most paleontological collections, zooarcheological collections are usually well dated and comprise large numbers of bones. They provide excellent opportunities to study microevolution. However, much remains to be explored in this relatively new science. *See* ARCHEOLOGY.   [S.J.M.D.]

1
Appendix

2
Contributors

3
Index

BIBLIOGRAPHIES

CLIMATOLOGY AND METEOROLOGY

Ahrens, C.D., *Meteorology Today*, 7th ed., 2002.
Berger, A.L., S.H. Schneider, and J.C. Duplessy (eds.), *Climate and Geo-Sciences*, 1989.
Bryant, E., *Climate Process and Change*, 1997.
Danielson, E.W., J. Levin, and E. Abrams, *Meteorology*, 2d ed., 2002.
Finlayson-Pitts, B.J., and J.N. Pitts, Jr., *Chemistry of the Upper and Lower Atmosphere*, 2000.
Glantz, M.H. (ed.), *The Role of Regional Organizations in Context of Climate Change*, 1993.
Holton, J.R., *An Introduction to Dynamic Meteorology*, 3d ed., 1992.
Hornak, K.A., *Dictionary of Meteorology and Climatology, English/Spanish*, 1997.
Liou, K.N., *An Introduction to Atmospheric Radiation*, 2d ed., 2002.
Lutgens, F., E. Tarbuck, and D. Tasa, *The Atmosphere: An Introduction to Meteorology*, 9th ed., 2003.
Oliver, J.E., and J.J. Hidore, *Climatology: An Atmospheric Science*, 2d ed., 2001.
Pruppacher, H.R., and J.D. Klett, *Microphysics of Clouds and Precipitation*, 1997.
Scorer, R.S., *Dynamics of Meteorology and Climate*, 1997.
Wang, P.K., *Ice Microdynamics*, 2002.

Journals:
Bulletin of the American Meteorological Society, American Meteorological Society, monthly.
Journal of the Atmospheric Sciences, American Meteorological Society, semimonthly.
Journal of Climate, American Meteorological Society, monthly.
Journal of Applied Meteorology, American Meteorological Society, monthly.

ECOLOGY

Bolen, E.G., and W.L. Robinson, *Wildlife Ecology and Management*, 5th ed., 2002.
Conner, J.K., and D.L. Hartl, *A Primer of Ecological Genetics*, 2004.
Gotelli, N.J., *A Primer of Ecology*, 3d ed., 2001.
Molles, Jr., M.C., *Ecology: Concepts and Applications*, 2d ed., 2002.
Pianka, E.R., *Evolutionary Ecology*, 6th ed., 1999.
Ricklefs, R.E., and G. Miller, *Ecology*, 4th ed., 1999.
Roughgarden, J., et al. (eds.), *Perspectives in Ecological Theory*, 1989.
Smith, R.L., and T. Smith, *Ecology and Field Biology*, 6th ed., 2001.
Walker, L.R., and R. del Moral, *Primary Succession and Ecosystem Rehabilitation* (Cambridge Studies in Ecology), 2003.
Westman, W.E., *Ecology Impact, Assessment and Environmental Planning*, 1985.

Journals:
Annual Review of Ecology and Systematics, annually.
Ecological Monographs, Ecological Society of America, quarterly.
Ecology, Ecological Society of America, 6 issues per year.
Journal of Environmental Sciences, Institute of Environmental Sciences, bimonthly.

ENVIRONMENTAL CHEMISTRY

Armour, M., *Hazardous Laboratory Chemicals Disposals Guide*, 3d spiral ed., 2003.
Burke, R., *Hazardous Materials Chemistry for Emergency Responders*, 2d ed., 2002.
Furr, K.A., *CRC Handbook of Laboratory Safety*, 5th ed., 2000.

Hathaway, G.J., N.H. Proctor, and J.P. Hughes, *Proctor and Hughes' Chemical Hazards of the Workplace*, 4th ed., 1996.

Lund, H.F. (ed.), *The McGraw-Hill Recycling Handbook*, 1993.

Manahan, S.E., *Environmental Chemistry*, 8th ed., 2004.

Parkin, G., *Chemistry for Environmental Engineering and Science*, 2002.

Pawlowski, L., et al. (eds.), *Chemistry for Protection of the Environment*, 1998.

Quigley, D.R., *The Essential Pocket Book of Emergency Chemical Management*, spiral ed., 1996.

Journals:

Environmental Science and Technology, American Chemical Society, semimonthly.

Journal of Environmental Engineering, American Society of Civil Engineers, monthly.

ENVIRONMENTAL ENGINEERING

Banham, R., *The Architecture of the Well-Tempered Environment*, 2d rev. ed., 1984.

Barrett, G.W., and R. Rosenberg (eds.), *Stress Effects on Natural Ecosystems*, 1982.

Holmes, J.R., *Practical Waste Management*, 1990.

Kiely, G., *Environmental Engineering*, 1996.

Lin, S.D., *Handbook of Environmental Engineering Calculations*, 2000.

Linaweaver, F.P. (ed.), *Environmental Engineering*, 1992.

Revelle, C.S., E.E. Whitlatch, and J.R. Wright, *Civil Engineering Systems*, 2d ed., 2003.

Salvato, J.A., N.L. Nemerow, and F.J. Agardy, *Environmental Engineering and Sanitation*, 5th ed., 2003.

White, I.D., D. Mottershead, and S.J. Harrison, *Environmental Systems: An Introductory Text*, 2d ed., 1993.

Journals:

The Diplomate, American Academy of Environmental Engineers, quarterly.

Environmental Engineering Science, monthly.

Journal of Environmental Engineering, monthly.

Journal of Environmental Sciences, Institute of Environmental Engineers, bimonthly.

GEOCHEMISTRY

Brownlow, A.H., *Geochemistry*, 2d ed., 1995.

Chester, R., *Marine Geochemistry*, 2d ed., 2003.

Engel, M., and S. Mako (eds.), *Organic Chemistry, Principles and Applications* (Topics in Geobiology), 1993.

Faure, G., *Principles and Applications of Inorganic Geochemistry*, 2d ed., 1997.

Hoefs, J., *Stable Isotope Geochemistry*, 5th ed., 2004.

Journals:

Chemical Geology, Elsevier, monthly.

Geochimica et Cosmochimica Acta, Elsevier, semimonthly.

GEOGRAPHY

Gaile, G.L., and C.J. Willmott (eds.), *Geography in America at the Dawn of the 21st Century*, 2004.

Gould, P., *Becoming a Geographer (Space, Place, and Society)*, 2000.

McKnight, T.L., and D. Hess, *Physical Geography: A Landscape Appreciation*, 8th ed., 2004.

Muehrcke, P.C., *Map Use: Reading, Analysis and Interpretation*, 1998.

Robinson, A.H., et al., *Elements of Cartography*, 1995.

Strahler, A.H., and A.N. Strahler, *Introducing Physical Geography*, 3d ed., 2002.

Wieden, F.T., *Physical Geography*, 1995.

Journals:

Focus, American Geographical Society, quarterly.
Geographical Review, American Geographical Society, quarterly.
The Professional Geographer, Association of American Geographers, quarterly.

GEOLOGY

Ager, D.V., *Nature of the Stratigraphical Record*, 3d ed., 1993.
Condie, K.C., and R.E. Sloon, *Origin and Evolution of Earth: Principles of Historical Geology*, 1998.
Hallam, A., *Great Geological Controversies*, 2d ed., 1990.
Lutgens, F.K., E.J. Tarbuck, and D. Tasa, *Essentials of Geology*, 8th ed., 2002.
Montgomery, C.W., *Physical Geology*, 3d ed., 1992.
Press, F., and R. Siever, *Understanding Earth*, 3d ed., 2000.
Skinner, B.J., and S.C. Porter, *Dynamic Earth: An Introduction to Physical Geology*, 5th ed., 2003.
Tarbuck, E.J., F.K. Lutgens, and D. Tasa, *The Earth: An Introduction to Physical Geology*, 8th ed., 2004.

Journals:

Annual Reviews of Earth and Planetary Sciences, annually.
Episodes, International Union of Geological Sciences, quarterly.
Geotimes, American Geological Institute, monthly.
GSA Bulletin, Geological Society of America, monthly.
Journal of Geology, University of Chicago Press, bimonthly.

GEOMORPHOLOGY

Bloom, A.L., *Geomorphology,: A Systematic Analysis of Late Cenozoic Landforms*, 3d ed., 1997.
Easterbrook, D.J., *Surface Processes and Landforms*, 2d ed., 1998.
Goudie, A., *The Changing Earth: Geomorphological Processes and Time*, 1995.
Leopold, L.B., M.G. Wolmon, and J.P. Miller, *Fluvial Process in Geomorphology*, 1995.
Thomas, D.S. (ed.), *Arid Zone Geomorphology: Process, Form and Change in Drylands*, 2d ed., 1999.
Walker, H.J., and W.E. Grabau (eds.), *The Evolution of Geomorphology: A Nation-by-Nation Summary of Development*, 1993.

Journals:

Earth Surface Processes and Landforms, Wiley, bimonthly.

GEOPHYSICS

Bertotti, B., and P. Farinella, *Physics of the Earth and the Solar System, Dynamics, Evolution, Space, Navigation, Space-time Structure* (Geophysics and Astrophysics Monographs), 1990.
Fowler, C.M., *The Solid Earth: An Introduction to Global Geophysics*, 2d ed., 2004.
Shearer, P., *Introduction to Seismology*, 1999.
Telford, W.M., L.P. Geldart, and R.E. Sheriff, *Applied Geophysics*, 2d ed., 1990.
Turcotte, D.L., and G. Schubert, *Geodynamics*, 2d ed., 2001.

Journals:

Eos, American Geophysical Union, weekly.
Journal of Geophysical Research (issued in three parts), American Geophysical Union, monthly.
Reviews in Geophysics, American Geophysical Union, quarterly.

HYDROLOGY

Barcelona, M., *Contamination of Groundwater: Prevention Assessment, Restoration* (Pollution Technology Review), 1990.
Black, P.E., *Watershed Hydrology*, 2d ed., 1996.
Dunne, T., and L.B. Leopold, *Water in Environmental Planning*, 1995.
Gupta, R.S.S., *Hydrology and Hydraulic Systems*, 2001.
Linsley, R.K., et al., *Hydrology for Engineers*, 3d ed. (McGraw-Hill Series in Water Resources & Environmental Engineering), 1982.
Ponce, V.M., *Engineering Hydrology: Principles and Practices*, 1989.
Wilson, E.M., *Engineering Hydrology*, 4th ed., 1990.

Journals:

Hydrological Sciences Journal, International Association of Hydrological Sciences (U.K.), quarterly.
Journal of Hydrology, Elsevier, monthly.

MINERALOGY

Klein, C., *Manual of Mineral Science*, 22d ed., 2001.
Howie, R.A., J. Zussman, and W.A. Deer, *An Introduction to the Rock-Forming Minerals*, 2d ed., 1992.
Reviews in Mineralogy and Geochemistry *(Series 1974–2004)*, P.H. Ribbe (series editor), Mineralogical Society of America.

Journals:

American Mineralogist, Mineralogical Society of America, monthly.
Mineralogical Magazine, Mineralogical Society of Great Britain, monthly.
Physics and Chemistry of Minerals, Springer-Verlag, monthly.

OCEANOGRAPHY

Colling, A., *Ocean Circulation*, 2d ed., 2001.
Garrison, T.S., *Oceanography: An Invitation to Marine Science*, 5th ed., 2005.
Knauss, J.A., *An Introduction to Physical Oceanography*, 2d ed., 1996.
Lalli, C.M., *Biological Oceanography*, 2d ed., 1997.
Pilson, M.E.Q., *An Introduction to the Chemistry of the Sea*, 1998.
Thurman, H.V., and A.P. Trujillo, *Essentials of Oceanography*, 8th ed., 2004.

Journals:

Journal of Physical Oceanography, American Meteorological Society, monthly.
Oceanus, Woods Hole Oceanographic Institute, semiannually.

PALEONTOLOGY

Benton, M.J., and J. Sibbock, *Vertebrate Paleontology*, 2000.
Briggs, D., D. Erwin, and F.J. Collier, *The Fossils of the Burgess Shale*, 1995.
Briggs, D.E.G., and P.R. Crowther, *Palaeobiology II*, 2002.
Carroll, R., *Vertebrate Paleontology and Evolution*, 1987.
Clack, J.A., *Gaining Ground: The Origin and Early Evolution of Tetrapods*, 2002.
Gould, S.J., *The Book of Life: An Illustrated History of the Evolution of Life on Earth*, 2d ed., 2001.
Jackson, J.B.C., S. Lidgard, and A.H. Cheetham, *Evolutionary Patterns: Growth, Form, and Tempo in the Fossil Record*, 2001.
Knoll, A.H., *Life on a Young Planet: The First Three Million Years of Evolution on Earth*, 2003.

Lowell, D., and T. Rowe, *The Mistaken Extinction: Dinosaur Evolution and the Origin of Birds*, 1997.
Minkoff, E.C., E.H. Colbert, and M. Morales, *Colbert's Evolution of the Vertebrates*, 2001.
Prothero, D.R., *Bringing Fossils to Life: An Introduction to Paleobiology*, 2003.
Rudwick, M.J., *The Meaning of Fossils: Episodes in the History of Paleontology*, 2d ed., 1985.
Simpson, G.G., *Life in the Past: An Introduction to Paleontology*, 2003.
Vermeij, G.J., *Evolution and Escalation: An Ecological History of Life*, 1987.
Willis, K., and J. McElwain, *The Evolution of Plants*, 2002.

Journals:

Journal of Paleontology, Society of Economic Paleontologists and Mineralogists, bimonthly.
Journal of Vertebrate Paleontology, Society of Vertebrate Paleontology, quarterly.
Palaeontology, Palaeontological Association, bimonthly.
Paleobiology, Paleontology Society, quarterly.

PETROLOGY

Best, M.G., *Igneous and Metamorphic Petrology*, 2d ed., 2002.
Greensmith, J.T., *Petrology of the Sedimentary Rocks*, 7th ed., 1988.
Hyundman, D.W., *Petrology of Igneous and Metamorphic Rocks*, 2d ed., 1985.
Krumbein, W.C., and F.J. Pettijohn, *Manual of Sedimentary Petrography*, reprinted, 1990.
McBirney, A.R., *Igneous Petrology*, 2d ed., 1992.
Philpotts, A.R., *Principles of Igneous and Metamorphic Petrology*, 1990.
Wilson, M., *Igneous Petrogenesis*, 1989.
Yardley, B.W.D., *Introduction to Metamorphic Petrology*, 1989.

Journals:

Contributions to Mineralogy and Petrology, Springer-Verlag, monthly.
Journal of Petrology (England), quarterly.

SOIL

Bowles, J.E., *Engineering Properties of Soils and Their Measurement*, 4th ed., 1992.
Brady, N.C., and R.R. Weil, *The Nature and Properties of Soils*, 13th ed., 2001.
Crunkilton, J.R., et al., *The Earth and Agriscience*, 1995.
Daniels, R.B., and R.D. Hammer, *Soil Geomorphology*, 1992.
Dowdy, R.H. (ed.), *Chemistry in the Soil Environment*, 1981.
Foster, A.B., and D.A. Bosworth, *Approved Practices in Soil Conservation*, 5th ed., 1982.
Foth, H.D., *Fundamentals of Soil Science*, 8th ed., 1990.
Hanks, R.J., *Applied Soil Physics: Soil Water and Temperature Applications*, 2d ed., 1992.
Morgan, R.P.C., *Soil Erosion and Conservation*, 2d ed., 1995.
Plaster, E.J., *Soil Science and Management*, 3d ed., 1997.

Journals:

Agronomy Journal, American Society of Agronomy, bimonthly.

Equivalents of commonly used units for the U.S. Customary System and the metric system

1 inch = 2.5 centimeters (25 millimeters)	1 centimeter = 0.4 inch	1 inch = 0.083 foot
1 foot = 0.3 meter (30 centimeters)	1 meter = 3.3 feet	1 foot = 0.33 yard (12 inches)
1 yard = 0.9 meter	1 meter = 1.1 yards	1 yard = 3 feet (36 inches)
1 mile = 1.6 kilometers	1 kilometer = 0.62 mile	1 mile = 5280 feet (1760 yards)
1 acre = 0.4 hectare	1 hectare = 2.47 acres	
1 acre = 4047 square meters	1 square meter = 0.00025 acre	
1 gallon = 3.8 liters	1 liter = 1.06 quarts = 0.26 gallon	1 quart = 0.25 gallon (32 ounces; 2 pints)
1 fluid ounce = 29.6 milliliters	1 milliliter = 0.034 fluid ounce	1 pint = 0.125 gallon (16 ounces)
32 fluid ounces = 946.4 milliliters		1 gallon = 4 quarts (8 pints)
1 quart = 0.95 liter	1 gram = 0.035 ounce	1 ounce = 0.0625 pound
1 ounce = 28.35 grams	1 kilogram = 2.2 pounds	1 pound = 16 ounces
1 pound = 0.45 kilogram	1 kilogram = 1.1×10^{-3} ton	1 ton = 2000 pounds
1 ton = 907.18 kilograms		

$$°F = (1.8 \times °C) + 32 \qquad °C = (°F - 32) \div 1.8$$

Conversion factors for the U.S. Customary System, metric system, and International System

A. Units of length

Units		cm	m	$in.$	ft	yd	mi
1 cm	= 1		0.01	0.3937008	0.03280840	0.01093613	6.213712×10^{-6}
1 m	= 100.		1	39.37008	3.280840	1.093613	6.213712×10^{-4}
1 in.	= 2.54		0.0254	1	$0.08333333\ldots$	$0.02777777\ldots$	1.578283×10^{-5}
1 ft	= 30.48		0.3048	12.	1	$0.3333333\ldots$	$1.893939\ldots \times 10^{-4}$
1 yd	= 91.44		0.9144	36.	3.	1	$5.681818\ldots \times 10^{-4}$
1 mi	= 1.609344×10^{5}		1.609344×10^{3}	6.336×10^{4}	5280.	1760.	1

B. Units of area

Units		cm^2	m^2	$in.^2$	ft^2	yd^2	mi^2
1 cm^2	= 1		10^{-4}	0.1550003	1.076391×10^{-3}	1.195990×10^{-4}	3.861022×10^{-11}
1 m^2	= 10^{4}		1	1550.003	10.76391	1.195990	3.861022×10^{-7}
1 in.^{2}	= 6.4516		6.4516×10^{-4}	1	$6.944444\ldots \times 10^{-3}$	7.716049×10^{-4}	2.490977×10^{-10}
1 ft^2	= 929.0304		0.09290304	144.	1	$0.1111111\ldots$	3.587007×10^{-8}
1 yd^2	= 8361.273		0.8361273	1296.	9.	1	3.228306×10^{-7}
1 mi^2	= 2.589988×10^{10}		2.589988×10^{6}	4.014490×10^{9}	2.78784×10^{7}	3.0976×10^{6}	1

Conversion factors for the U.S. Customary System, metric system, and International System (*cont.*)

C. Units of volume

Units	m^3	cm^3	liter	$in.^3$	ft^3	qt	gal
1 m^3	= 1	10^6	10^3	6.102374×10^4	35.31467	1.056688×10^3	264.1721
1 cm^3	= 10^{-6}	1	10^{-3}	0.06102374	3.531467×10^{-5}	1.056688×10^{-3}	2.641721×10^{-4}
1 liter	= 10^{-3}	1000.	1	61.02374	0.03531467	1.056688	0.2641721
1 $in.^3$	= 1.638706×10^{-5}	16.38706	0.01638706	1	5.787037×10^{-4}	0.01731602	4.329004×10^{-3}
1 ft^3	= 2.831685×10^{-2}	28316.85	28.31685	1728.	1	2.992208	7.480520
1 qt	= 9.463529×10^{-4}	946.3529	0.9463529	57.75	0.03342014	1	0.25
1 gal (U.S.)	= 3.785412×10^{-3}	3785.412	3.785412	231.	0.1336806	4.	1

D. Units of mass

Units	g	kg	oz	lb	metric ton	ton
1 g	= 1	10^{-3}	0.03527396	2.204623×10^{-3}	10^{-6}	1.102311×10^{-6}
1 kg	= 1000.	1	35.27396	2.204623	10^{-3}	1.102311×10^{-3}
1 oz (avdp)	= 28.34952	0.02834952	1	0.0625	2.834952×10^{-5}	3.125×10^{-5}
1 lb (avdp)	= 453.5924	0.4535924	16.	1	4.535924×10^{-4}	$5. \times 10^{-4}$
1 metric ton	= 10^6	1000.	35273.96	2204.623	1	1.102311
1 ton	= 907184.7	907.1847	32000.	2000.	0.9071847	1

E. Units of density

Units	$g \cdot cm^{-3}$	$g \cdot L^{-1}, kg \cdot m^{-3}$	$oz \cdot in^{-3}$	$lb \cdot in^{-3}$	$lb \cdot ft^{-3}$	$lb \cdot gal^{-1}$
1 $g \cdot cm^{-3}$ =	1	1000.	0.5780365	0.03612728	62.42795	8.345403
1 $g \cdot L^{-1}, kg \cdot m^{-3}$ =	10^{-3}	1	5.780365×10^{-4}	3.612728×10^{-5}	0.06242795	8.345403×10^{-3}
1 $oz \cdot in^{-3}$ =	1.729994	1729.994	1	0.0625	108.	14.4375
1 $lb \cdot in^{-3}$ =	27.67991	27679.91	16.	1	1728.	231.
1 $lb \cdot ft^{-3}$ =	0.01601847	16.01847	9.259259×10^{-3}	5.787037×10^{-4}	1	0.1336806
1 $lb \cdot gal^{-1}$ =	0.1198264	119.8264	4.749536×10^{-3}	4.329004×10^{-3}	7.480519	1

F. Units of pressure

Units	$Pa, N \cdot m^{-2}$	$dyn \cdot cm^{-2}$	bar	atm	$kgf \cdot cm^{-2}$	$mmHg$ (torr)	$in. Hg$	$lbf \cdot in^{-2}$
1 Pa, 1 $N \cdot m^{-2}$ =	1	10	10^{-5}	9.869233×10^{-6}	1.019716×10^{-5}	7.500617×10^{-3}	2.952999×10^{-4}	1.450377×10^{-4}
1 $dyn \cdot cm^{-2}$ =	0.1	1	10^{-6}	9.869233×10^{-7}	1.019716×10^{-6}	7.500617×10^{-4}	2.952999×10^{-5}	1.450377×10^{-5}
1 bar =	10^{5}	10^{6}	1.	0.9869233	1.019716	750.0617	29.52999	14.50377
1 atm =	101325	1013250	1.01325	1	1.033227	760.	29.92126	14.69595
1 $kgf \cdot cm^{-2}$ =	98066.5	980665	0.980665	0.9678411	1	735.5592	28.95903	14.22334
1 $mmHg$ (torr) =	133.3224	1333.224	1.333224×10^{-3}	1.315789×10^{-3}	1.359510×10^{-3}	1	0.03937008	0.0193368
1 $in. Hg$ =	3386.388	33863.88	0.03386388	0.03342105	0.03453155	25.4	1	0.4911541
1 $lbf \cdot in^{-2}$ =	6894.757	68947.57	0.06894757	0.06804596	0.07030696	51.71493	2.036021	1

Conversion factors for the U.S. Customary System, metric system, and International System (cont.)

G. Units of energy

Units	g mass (energy equiv)	J	eV	cal	cal_{IT}	Btu_{IT}	kWh	hp-h	ft-lbf	$ft^3 \cdot lbf \cdot in.^{-2}$	liter-atm
1 g mass (energy equiv) =	1	8.987552×10^{13}	5.609589×10^{32}	2.148076×10^{13}	2.146640×10^{13}	8.518555×10^{10}	2.496542×10^{7}	3.347918×10^{7}	6.628878×10^{13}	4.603388×10^{11}	8.870024×10^{11}
1 J =	1.112650×10^{-14}	1	6.241510×10^{18}	0.2390057	0.2388459	9.478172×10^{-4}	2.777777×10^{-7}	3.725062×10^{-7}	0.7375622	5.121960×10^{-3}	9.869233×10^{-3}
1 eV =	1.782662×10^{-33}	1.602176×10^{-19}	1	3.829293×10^{-20}	3.826733×10^{-20}	1.518570×10^{-22}	4.450490×10^{-26}	5.968206×10^{-26}	1.181705×10^{-19}	8.206283×10^{-22}	1.581225×10^{-21}
1 cal =	4.655328×10^{-14}	4.184	2.611448×10^{19}	1	0.9993312	3.965667×10^{-3}	1.1622222×10^{-6}	1.558562×10^{-6}	3.085960	2.143028×10^{-2}	0.04129287
1 cal_{IT} =	4.658443×10^{-14}	4.1868	2.613195×10^{19}	1.000669	1	3.968321×10^{-3}	1.163×10^{-6}	1.559609×10^{-6}	3.088025	2.144462×10^{-2}	0.04132050
1 Btu_{IT} =	1.173908×10^{-11}	1055.056	6.585141×10^{21}	252.1644	251.9958	1	2.930711×10^{-4}	3.930148×10^{-4}	778.1693	5.403953	10.41259
1 kWh =	4.005540×10^{-8}	3600000.	2.246944×10^{25}	860420.7	859845.2	3412.142	1	1.341022	2655224.	18349.06	35529.24
1 hp-h =	2.986931×10^{-8}	2384519.	1.675545×10^{25}	641615.6	641186.5	2544.33	0.7456998	1	1980000.	13750.	26494.15
1 ft-lbf =	1.508551×10^{-14}	1.355818	8.462351×10^{18}	0.3240483	0.3238315	1.285067×10^{-3}	3.766161×10^{-7}	$5.050505\ldots \times 10^{-7}$	1	$6.944444\ldots \times 10^{-3}$	0.01338088
1 $ft^2 \cdot lbf \cdot in.^{-2}$ =	2.172313×10^{-12}	195.2378	1.218579×10^{21}	46.66295.	46.63174	0.1850497	5.423272×10^{-5}	$7.272727\ldots \times 10^{-5}$	144.	1	1.926847
1 liter-atm =	1.127393×10^{-12}	101.325	6.324210×10^{20}	24.21726	24.20106	0.09603757	2.814583×10^{-5}	3.774419×10^{-5}	74.73349	0.5189825	1

Geologic column and scale of time

Eon	Era	Period	Epoch	Dates (10^4 years before present)
Phanerozoic	Cenozoic	Quaternary	Holocene	0.01
			Pleistocene	1.8
		Tertiary	Pliocene	5
			Miocene	23
			Oligocene	38
			Eocene	54
			Paleocene	65
	Mesozoic	Cretaceous		144
		Jurassic		208
		Triassic		245
	Paleozoic	Permian		286
		Pennsylvanian		325
		Mississippian		360
		Devonian		410
		Silurian		440
		Ordovician		505
		Cambrian		544
Proterozoic*	No subdivisions in wide use			2500
Archean*	No subdivisions in wide use			3800
Hadean	No subdivisions			4500

*Proterozoic plus Archean also called Precambrian.

Compositions of important rock types in the earth's crust and the average continental crust

Composition	Anorthosite	Peridotite	Oceanic basalt	Andesite	Dacite	Granodiorite	Granite	Graywacke	Sandy shale	Continental crust upper 9 mi (15 km)
Chemical					*Weight, %*					
SiO_2	54.0	44.0	50.0	60.0	65.5	66.0	70.5	64.0	65.5	66.0
TiO_2	.8	.2	1.5	.8	.3	.5	.3	.5	.5	.5
Al_2O_3	24.0	2.5	15.5	17.5	15.0	15.5	14.6	14.5	14.0	15.5
Fe_2O_3	.8	1.0	1.5	3.0	.8	2.0	1.6	1.5	3.5	2.0
FeO	2.5	8.0	8.0	3.2	2.5	2.6	1.8	3.5	2.0	3.0
MgO	1.5	40.0	7.0	2.8	2.0	2.0	.8	2.2	1.7	2.0
CaO	10.0	2.5	10.5	6.0	3.7	4.0	2.0	2.6	2.5	4.2
Na_2O	4.5	.1	2.9	3.5	3.8	3.6	3.5	3.2	1.5	3.5
K_2O	.1	.02	.25	3.0	2.4	2.8	4.3	2.0	4.0	3.0
Mineralogical					*Approximate volume*					
Olivine	—	*	†	—	—	—	—	—	—	—
Fe, T, Mg oxides	†	†	†	—	—	—	—	—	—	†
Pyroxene	†	†	*	•	†	†	—	—	—	—
Amphibole	—	—	—	•	•	•	†	•	†	†
Plagioclase	*	—	•	•	•	•	•	†	†	•
K-feldspar	—	—	—	†	†	•	•	•	•	•
Micas	—	—	—	—	•	†	•	•	•	•
Quartz	—	—	—	—	†	•	•	•	•	•
Chlorites	—	—	—	—	—	—	—	•	•	—
Clay minerals	—	—	—	—	—	—	—	†	•	—

*Major constituent.
†Subordinate mineral.

Elemental composition of earth's crust based on igneous and sedimentary rock

Element	Weight %	Atomic %	Volume %
Oxygen	46.71	60.5	94.24
Silicon	27.69	20.5	0.51
Titanium	0.62	0.3	0.03
Aluminum	8.07	6.2	0.44
Iron	5.05	1.9	0.37
Magnesium	2.08	1.8	0.28
Calcium	3.65	1.9	1.04
Sodium	2.75	2.5	1.21
Potassium	2.58	1.4	1.88
Hydrogen	0.14	3.0	

Physical properties of some common rocks

Rock	Specific gravity	Porosity, %	Compressive strength, $lb/in.^2$	Tensile strength, $lb/in.^2$
Igneous				
Granite	2.67	1	30,000–50,000	500–1000
Basalt	2.75	1	25,000–30,000	
Sedimentary				
Sandstone	2.1–2.5	5–30	5,000–15,000	100–200
Shale	1.9–2.4	7–25	5,000–10,000	
Limestone	2.2–2.5	2–20	2,000–20,000	400–850
Metamorphic				
Marble	2.5–2.8	0.5–2	10,000–30,000	700–1000
Quartzite	2.5–2.6	1–2	15,000–40,000	
Slate	2.6–2.8	0.5–5	15,000–30,000	

Simplified classification of major igneous rocks on the basis of composition and texture

SiO₂-rich (acidic) → → → → → → → → → → SiO₂-poor (basic)

	Light colored	Gray	Dark colored		Black
Mineral composition:	Quartz, potash feldspar, biotite	Potash feldspar, biotite, or amphibole	Sodic plagioclase, hornblende, or augite	Augite, olivine, hypersthene, calcic plagioclase	Olivine, enstatite, augite
INTRUSIVE					
Medium-grained	Granite	Syenite	Diorite	Gabbro	Peridotite
EXTRUSIVE					
Fine-grained to aphanitic	Rhyolite	Trachyte	Andesite	Basalt	
	← Felsite →				
Porphyritic	Rhyolite porphyry	Trachyte porphyry	Andesite porphyry	Basalt porphyry	
Glassy	Obsidian				
Vesicular	Pumice			Scoria	
Fragmental	Tuff and agglomerate of each type				

Approximate concentration of ore elements in earth's crust and in ores

Element	In average igneous rocks, %	In ores, %
Iron	5.0	50
Copper	0.007	0.5–5
Zinc	0.013	1.3–13
Lead	0.0016	1.6–16
Tin	0.004	0.01*–1
Silver	0.00001	0.05
Gold	0.0000005	0.0000015*–0.01
Uranium	0.0002	0.2
Tungsten	0.003	0.5
Molybdenum	0.001	0.6

*Placer deposits.

Some historical volcanic eruptions

Volcano	Year	Estimated casualties	Principal causes of death
Merapi (Indonesia)	1006	>1000	Explosions
Kelut (Indonesia)	1586	10,000	Lahars (mudflows)
Vesuvius (Italy)	1631	18,000	Lava flows, mudflows
Etna (Italy)	1669	10,000	Lava flows, explosions
Merapi (Indonesia)	1672	>300	Nuées ardentes, lahars
Awu (Indonesia)	1711	3,200	Lahars
Papandayan (Indonesia)	1772	2,957	Explosions
Laki (Iceland)	1783	10,000	Lava flows, volcanic gas, starvation*
Asama (Japan)	1783	1,151	Lava flows, lahars
Unzen (Japan)	1792	15,000	Lahars, tsunami
Mayon (Philippines)	1814	1,200	Nuées ardentes, lava flows
Tambora (Indonesia)	1815	92,000	Starvation*
Galunggung (Indonesia)	1822	4,000	Lahars
Awu (Indonesia)	1856	2,800	Lahars
Krakatau (Indonsia)	1883	36,000	Tsunami
Awu (Indonesia)	1892	1,500	Nuées ardentes, lahars
Mont Pelée, Martinique (West Indies)	1902	36,000	Nuées ardentes
Soufrière, St. Vincent (West Indies)	1902	1,565	Nuées ardentes
Taal (Philippines)	1911	1,332	Explosions
Kelut (Indonesia)	1919	5,000	Lahars
Lamington (Papua New Guinea)	1951	3,000	Nuées ardentes, explosions
Merapi (Indonesia)	1951	1,300	Lahars
Agung (Indonesia)	1963	3,800	Nuées ardentes, lahars
Taal (Philippines)	1965	350	Explosions
Mount St. Helens (United States)	1980	57	Lateral blast, mudflows
El Chichón (Mexico)	1982	>2,000	Explosions, nuées ardentes
Nevado del Ruiz (Colombia)	1985	>25,000	Mudflows
Unzen (Japan)	1991	41	Nuées ardentes
Pinatubo (Philippines)	1991	>300	Nuées ardentes, mudflows, ash fall (roof collapse)
Merapi (Indonesia)	1994	>41	Nuées ardentes from dome collapse
Soufrière Hills, Montserrat (West Indies)	1997	19	Nuées ardentes

*Deaths directly attributable to the destruction or reduction of food crops, livestock, agricultural lands, pasturage, and other disruptions of food chain.

Types of volcanic structure

Name	Characteristics
Shield	Low height, broad area; formed by successive fluid flows accumulating around a single, central vent
Cinder cone	Cone of moderate size with apex truncated; circular in plan, gently sloping sides; composed of pyroclastic particles, usually poorly consolidated
Spatter cone	Small steep-sided cone with well-defined crater composed of pyroclastic particles, well consolidated (agglomerate)
Composite cone	Composed of interlayered flows and pyroclastics; flows from sides (flank flows) common, as are radial dike swarms; slightly concave in profile, with central crater
Caldera	Basins of great size but relatively shallow; formed by explosive decapitation of stratocones, by collapse into underlying magma chamber, or both
Plug dome	Domal piles of viscous (usually rhyolitic) lava, growing by subsurface accretion and accompanied by outer fragmentation
Cryptovolcanic structures	Circular areas of highly fractured rocks in regions generally free of other structural disturbances; believed to have formed either by subsurface explosions or by sinking of cylindrical rock masses over magma chambers

World's estimated water supply

Location	Surface area, mi^2 (km^2)	Water volume, mi^3 (km^3)	Percentage of total water
Surface water			
Fresh-water lakes	330,000 (855,000)	30,000 (130,0000)	0.009
Saline lakes and inland seas	270,000 (700,000)	25,000 (104,000)	0.008
Average in stream channels	—	300 (1300)	0.0001
Subsurface water			
Vadose water (includes soil moisture)		16,000 (67,000)	0.005
Groundwater within depth of a half mile	50,000,000 (130,000,000)	1,000,000 (4,200,000)	0.31
Groundwater, deep-lying		1,000,000 (4,200,000)	0.31
Other water locations			
Ice caps and glaciers	6,900,000 (18,000,000)	7,000,000 (29,000,000)	2.15
Atmosphere (at sea level)	197,000,000 (510,000,000)	3,100 (12,900)	0.001
World ocean	139,500,000 (361,300,000)	317,000,000 (1,321,000,000)	97.2
TOTALS (rounded)		326,000,000 (1,360,000,000)	100

Dimensions of some major lakes

Lake	Area, mi² (km²)	Volume (approx.), 10³ acre-ft (10⁶ m³)	Shoreline, mi (km)	Depth	
				Av., ft (m)	Max., ft (m)
Caspian Sea	168,500 (436,412)	71,300 (87,947)	3,730 (6,003)	675 (207)	3,080 (939)
Superior	32,200 (83,398)	9,700 (11,965)	1,860 (2,993)	475 (145)	1,000 (305)
Victoria	26,200 (67,858)	2,180 (2,689)	2,130 (3,428)		
Aral Sea	26,233 (67,943)*	775 (956)			
Huron	23,010 (59,596)	3,720 (4,589)	1,680 (2,704)		
Michigan	22,400 (58,016)	4,660 (5,748)			870 (256)
Baikal	13,300 (34,447)*	18,700 (23,066)		2,300 (701)	5,000 (1,524)
Tanganyika	12,700 (32,893)	8,100 (9,991)			4,700 (1,433)
Great Bear	11,490 (29,759)*		1,300 (2,092)		
Great Slave	11,170 (28,930)*		1,365 (2,197)		
Nyasa	11,000 (28,490)	6,800 (8,388)		900 (274)	2,310 (704)
Erie	9,940 (25,744)	436 (538)			
Winnipeg	9,390 (24,320)*		1,180 (1,899)		
Ontario	7,540 (19,529)	1,390 (1,714)			
Balkash	7,115 (18,428)				
Ladoga	7,000 (18,130)	745 (919)			
Chad	6,500 (16,835)*				
Maracaibo	4,000 (10,360)*				
Eyre	3,700 (9,583)*				
Onega	3,764 (9,749)	264 (326)			
Rudolf	3,475 (9,000)*				
Nicaragua	3,089 (8,000)	87 (107)			
Athabaska	3,085 (7,990)				
Titicaca	3,200 (8,288)	575 (709)			
Reindeer	2,445 (6,332)				

*Area fluctuates.

Characteristics of some of the world's major rivers

River	Average discharge, ft³/s (m³/s)	Drainage area, 10³ mi² (10³ km²)	Average annual sediment load, 10³ tons (10³ metric tons)	Length, mi (km)
Amazon	6,390,000 (181,000)	2770 (7180)	990,000 (900,000)	3899 (6275)
Congo	1,400,000 (39,620)	1420 (3690)	71,300 (64,680)	2901 (4670)
Orinoco	800,000 (22,640)	571 (1480)	93,130 (86,490)	1600 (2570)
Yangtze	770,000 (21,790)	749 (1940)	610,000 (550,000)	3100 (4990)
Brahmaputra	706,000 (19,980)	361 (935)	880,000 (800,000)	1700 (2700)
Mississippi-Missouri	630,000 (17,830)	1240 (3220)	379,000 (344,000)	3890 (6260)
Yenisei	614,000 (17,380)	1000 (2590)	11,600 (10,520)	3550 (5710)
Lena	546,671 (15,480)	1170 (3030)	—	2900 (4600)
Mekong	530,000 (15,000)	350 (910)	206,850 (187,650)	2600 (4180)
Parana	526,000 (14,890)	1200 (3100)	90,000 (81,650)	2450 (3940)
St. Lawrence	500,000 (14,150)	564 (1460)	4,000 (3,630)	2150 (3460)
Ganges	497,600 (14,090)	451 (1170)	1,800,000 (1,600,000)	1640 (2640)
Irrawaddy	478,900 (13,560)	140 (370)	364,070 (330,280)	1400 (2300)
Ob	440,700 (12,480)	1000 (2590)	15,700 (14,240)	2800 (4500)
Volga	350,000 (9,900)	591 (1530)	20,780 (18,840)	2320 (3740)
Amur	338,000 (9,570)	788 (2040)	—	2900 (4670)

Cloud classification based on air motion and associated physical characteristics

Kind of motion	Kind of cloud	Name	Characteristic precipitation
Widespread slow ascent, associated with cyclones (stable atmosphere)	Thick layers	Cirrus, later becoming: cirrostratus altostratus altocumulus nimbostratus	Snow trails Prolonged moderate rain or snow
Convection, due to passage over warm surface (unstable atmosphere)	Small heap cloud	Cumulus	None
	Shower- and thundercloud	Cumulonimbus	Intense showers of rain or hail
Irregular stirring causing cooling during passage over cold surface (stable atmosphere)	Shallow low layer clouds, fogs	Stratus Stratocumulus	None, or slight drizzle or snow

The 100 highest mountain peaks

Mountain peak	Range	Location	Height Feet	Height Meters
Everest	Himalayas	Nepal-China	29,028	8,848
K2 (Godwin Austen)	Karakoram	Kashmir	28,250	8,611
Kanchenjunga	Himalayas	Nepal-India	28,208	8,598
Lhotse I	Himalayas	Nepal-China	27,923	8,511
Makalu I	Himalayas	Nepal-China	27,824	8,481
Lhatose II	Himalayas	Nepal-China	27,560	8,400
Dhaulagri	Himalayas	Nepal	26,810	8,172
Manaslu I	Himalayas	Nepal	26,760	8,156
Cho Oyu	Himalayas	Nepal-China	26,750	8,153
Nanga Parbat	Himalayas	Kashmir	26,660	8,126
Annapurna	Himalayas	Nepal	26,504	8,078
Gasherbrum	Karakoram	Kashmir	26,470	8,068
Broad	Karakoram	Kashmir	26,400	8,047
Gosainthan	Himalayas	China	26,287	8,012
Annapura II	Himalayas	Nepal	26,041	7,937
Gyachung Kang	Himalayas	Nepal-China	25,910	7,897
Disteghil Sar	Karakoram	Kashmir	25,858	7,882
Hinalchuli	Himalayas	Nepal	25,801	7,864
Nuptse	Himalayas	Nepal-China	25,726	7,841
Masherbrum	Karakoram	Kashmir	25,660	7,821
Nandi Devi	Himalayas	India	25,645	7,817
Rakaposhi	Karakoram	Kashmir	25,550	7,788
Kanjut Sar	Karakoram	Kashmir	25,461	7,761
Kamet	Himalayas	India-China	25,447	7,756
Namcha Barwa	Himalayas	China	25,445	7,756
Gurla Manghata	Himalayas	China	25,355	7,728
Ulugh Mustagh	Kunlun	China	25,340	7,724
Kungur	Mustagh Ata	China	25,325	7,719
Tirich Mir	Hindu Kush	Pakistan	25,230	7,690
Saser Kangri	Karakoram	Kashmir	25,172	7,672

Name	Range	Country		
Makalu II	Himalayas	Nepal-China	25,120	7,657
Conggashan	Daxue Shan	China	24,900	7,590
Kula Kangri	Himalayas	Bhutan-China	24,784	7,554
Chang-tzu	Himalayas	Nepal-China	24,780	7,533
Muztagh Ata	Muztagh Ata	China	24,757	7,546
Skyang Kangri	Himalayas	Kashmir	24,750	7,544
Ismail Samani Peak	Pamirs	Tajikistan	24,590	7,495
Jongsong Peak	Himalayas	Nepal-India	24,472	7,459
Pobeda Peak	Tian Shan	Kyrgyzstan	24,406	7,439
Sia Kangri	Himalayas	Kashmir	24,350	7,422
Haramosh Peak	Karakoram	Kashmir	24,270	7,397
Istoro Nal	Hindu Kush	Pakistan	24,240	7,388
Tent Peak	Himalayas	Nepal-India	24,165	7,365
Chomo Lhari	Himalayas	Bhutan-China	24,040	7,327
Chamlang	Himalayas	Nepal	24,012	7,319
Kabru	Himalayas	Nepal-India	24,002	7,316
Alung Gangri	Himalayas	China	24,000	7,315
Baltoro Knagri	Himalayas	Kashmir	23,990	7,312
Muztagh Ata	Kunlun	China	23,890	7,282
Mana	Himalayas	India	23,860	7,273
Baruntse	Himalayas	Nepal	23,688	7,220
Nepal Peak	Himalayas	Nepal-India	23,500	7,163
Amne Machin	Kunlun	China	23,490	7,160
Gauri Sankar	Himalayas	Nepal-China	23,440	7,145
Badrinath	Himalayas	India	23,420	7,138
Nunkun	Himalayas	Kashmir	23,410	7,135
Lenin Peak	Pamirs	Tajikistan/Kyrgyzstan	23,405	7,134
Pyramid	Himalayas	Nepal-India	23,400	7,132
Api	Himalayas	Nepal	23,399	7,132
Pauhunri	Himalayas	India-China	23,385	7,128
Trisul	Himalayas	India	23,360	7,120
Korzhenevski Peak	Pamirs	Tajikistan	23,310	7,105
Kangto	Himalayas	India-China	23,260	7,090
Nyainqentanglha	Nyainqentanglha Shan	China	23,255	7,088
Trisuli	Himalayas	India	23,210	7,074
Dunagiri	Himalayas	India	23,184	7,066
Revolution Peak	Pamirs	Tajikistan	22,880	6,974
Aconcagua	Andes	Argentina	22,834	6,960
Ojos del Salado	Andes	Argentina-Chile	22,572	6,880

The 100 highest mountain peaks (cont.)

Mountain peak	Range	Location	Height	
			Feet	Meters
Bonete	Andes	Argentina	22,546	6,872
Tupungato	Andes	Argentina-Chile	22,310	6,800
Moscow Peak	Pamirs	Tajikistan	22,260	6,785
Pissis	Andes	Argentina	22,241	6,779
Mercedario	Andes	Argentina	22,211	6,770
Huascaran	Andes	Peru	22,205	6,768
Llullaillaco	Andes	Argentina-Chile	22,057	6,723
El Libertador	Andes	Argentina	22,047	6,720
Cachi	Andes	Argentina	22,047	6,720
Kailas	Himalayas	China	22,027	6,714
Incahusai	Andes	Argentina-Chile	21,720	6,620
Yerupaja	Andes	Peru	21,709	6,617
Kurumda	Pamirs	Tajikistan	21,686	6,610
Galan	Andes	Argentina	21,654	6,600
El Muerto	Andes	Argentina-Chile	21,457	6,540
Sajama	Andes	Bolivia	21,391	6,520
Nacimiento	Andes	Argentina	21,302	6,493
Illimani	Andes	Bolivia	21,201	6,462
Coropuna	Andes	Peru	21,083	6,426
Laudo	Andes	Argentina	20,997	6,400
Ancohuma	Andes	Bolivia	20,958	6,388
Ausangate	Andes	Peru	20,945	6,384
Toro	Andes	Argentina-Chile	20,932	6,380
Illampu	Andes	Bolivia	20,873	6,362
Tres Cruces	Andes	Argentina-Chile	20,853	6,356
Huandoy	Andes	Peru	20,852	6,356
Parinacota	Andes	Bolivia-Chile	20,768	6,330
Tortolas	Andes	Argentina-Chile	20,745	6,323
Ampato	Andes	Peru	20,702	6,310
El Condor	Andes	Argentina	20,669	6,300
Salcantay	Andes	Peru	20,574	6,271

BIOGRAPHICAL LISTING

Agricola, Georgius, real name Georg Bauer (1494–1555), German physician and mineralogist. Known as the father of systematic mineralogy.

Aitken, John (1839–1919), Scottish physicist. Studied dust particles in the atmosphere, known as Aitken nuclei.

Anaximander (611–547 B.C.), Greek astronomer and mathematician. Reputed inventor of geographical maps; formulated the concept of the universe as infinite (apeiron).

Appleton, Edward Victor (1892–1965), English physicist. Demonstrated the existence of the ionosphere and discovered its region known as the Appleton layer; contributed to the development of radar; Nobel Prize, 1947.

Barnett, Samuel Jackson (1873–1956), American physicist. Discovered Barnett effect and used it to measure the gyromagnetic ratio of ferromagnetic materials; gave experimental proof of existence of ionosphere.

Beaufort, Francis (1774–1857), English hydrographer. Devised the scale of wind velocity.

Becquerel, Antoine César (1788–1878), French physicist. Pioneer in electrochemistry; first to extract metals from ore by electrolysis.

Benioff, Hugo (1899–1968), American geophysicist. Investigated earthquakes, particularly through instrumental seismology.

Bjerknes, Vilhelm Fremann Doren (1862–1951), Norwegian physicist. Research on electric waves; originated the polar-front theory in meteorology.

Bouguer, Pierre (1698–1758), French geodesist, hydrographer, and physicist. Laid foundations of photometry; discovered Bouguer-Lambert law of light intensity.

Bowen, Norman Levi (1887–1956), Canadian-born American geologist. Studied physical chemistry of geological processes and phase equilibria of silicates; discovered significance of reaction principle in petrogenesis.

Buffon, Georges Louis Leclerc, Comte de (1707–1788), French naturalist. Compiled *Histoire Naturelle*, a monumental work on natural history.

Buys-Ballot, Christoph Hendrik Didericus (1817–1890), Dutch meteorologist. Devised a system of storm signals; formulated Buys-Ballot's law for determination of wind direction.

Cavendish, Henry (1731–1810), English physicist and chemist. Determined the density of the Earth and the composition of the atmosphere; studied properties of carbon dioxide and hydrogen.

Celsius, Anders (1701–1744), Swedish astronomer. Constructed the thermometer using the Celsius (centigrade) scale.

Chamberlin, Thomas Chrowder (1843–1928), American geologist. Studied the fundamental geology of the solar system.

Chapman, Sydney (1888–1970), English mathematician and physicist. Discovered (independently of D. Enskog) gaseous thermal diffusion; studied the daily variations of the geomagnetic field and magnetic storms.

Clarke, Alexander Ross (1828–1914), British geodesist. Worked on the triangulation of the British Isles; proposed Clarke ellipsoids as geodetic standards.

Cronstedt, Axel Fredrik, Baron (1722–1765), Swedish mineralogist. Discovered nickel; developed a chemical classification system for minerals.

Darwin, Charles Robert (1809–1882), English naturalist. Proposed far-reaching theory of evolution of species and theory of natural selection in his *Origin of Species*.

Darwin, George Howard (1845–1912), English mathematician and astronomer. Applied detailed dynamical analysis to cosmological and geological problems.

Doppler, Christian Johann (1803–1853), Austrian physicist and mathematician. Formulated Doppler's principle, relating the frequency of wave motion to velocity; described the Doppler effect.

Elsasser, Walter Maurice (1904–1991), German-born American geophysicist. Formulated the dynamo theory of the Earth's permanent terrestrial magnetic force.

Eötvös, Roland, Baron (1848–1919), Hungarian physicist. Research on gravitation and terrestrial magnetism; formulated a law which relates surface tension to temperature of liquids; designed the Eötvös torsion balance.

Eratosthenes (3d century B.C.), Greek astronomer. Suggested an extra day in the calendar every fourth year; made a determination of the size of the Earth; measured obliquity of the ecliptic.

Ewing, William Maurice (1906–1974), American geophysicist. Made fundamental contributions to seismology, geodesy, oceanography, and submarine geology.

Fahrenheit, Gabriel Daniel (1686–1736), German physicist. Constructed thermometers; invented the Fahrenheit temperature scale.

Flamsteed, John (1646–1719), English astronomer. Made a trustworthy catalog of stars; invented conical projection in mapmaking.

Forbush, Scott Ellsworth (1904–1984), American geophysicist. Discovered the worldwide decrease in cosmic-ray intensity associated with some magnetic storms.

Foucault, Jean Bernard Léon (1819–1868), French physicist. Accurately determined the velocity of light; constructed the Foucault pendulum and the Foucault prism; determined experimentally the rotation of the Earth.

Franklin, Benjamin (1706–1790), American physicist, oceanographer, meteorologist, and inventor. Formulated a theory of general electrical "action"; introduced principle of conservation of charge; showed that lightning is an electrical phenomenon; invented lighting rod.

Friedel, Charles (1832–1899), French chemist and mineralogist. With J. M. Crafts, described the Friedel-Crafts reaction; work on artificial production of minerals; studied crystals, ketones, and aldehydes.

Geitel, Hans Friedrich (1855–1923), German experimental physicist. With J. Elster, studied atmospheric electricity, radioactivity, and photoelectricity, and invented photocell.

Gunter, Edmund (1581–1626), English mathematician and astronomer. Invented Gunter's chain used in surveying, and the logarithmic scale (Gunter's scale) which is the principle of the slide rule.

Hagen, Gotthilf Heinrich Ludwig (1797–1884), German hydraulic engineer. Discovered Hagen-Poiseuille law (governing nonturbulent flow of fluids through circular tubes) independently of J. L. M. Poiseuille; directed construction of dikes, harbor installations, and dune fortifications.

Heaviside, Oliver (1850–1925), English physicist. Proposed the Heaviside layer in the upper atmosphere.

Humboldt, Friedrich Heinrich Alexander, Baron von (1769–1859), German naturalist. Founder of physical geography; made scientific explorations of South America and Central Asia.

Kennelly, Arthur Edwin (1861–1939), American electrical engineer. Discovered the ionized layer in the atmosphere, independently of O. Heaviside; proposed the theory of alternating currents.

Knudsen, Martin Hans Christian (1871–1949), Danish physicist and hydrographer. Studied flow and diffusion of gases at low pressure; developed Knudsen cell and Knudsen gage; developed methods to measure the properties of seawater.

Lacaille, Nicolas Louis de (1713–1762), French astronomer and geodesist, Determined positions of nearly 10,000 stars in southern skies; measured lunar and solar parallax; showed that Earth has equatorial bulge.

Libby, Willard Frank (1908–1980), American chemist. Developed the method of radiocarbon dating; Nobel Prize, 1960.

Lyell, Charles (1797–1875), British geologist. Wrote *Principles of Geology*, refuting catastrophic theory of geological changes.

Mercator, Gerhardus, real name Gerhard Kremer (1512–1594), Flemish geographer. Created a chart of the world (Mercator projection); made surveying instruments.

Miller, William Hallowes (1801–1880), British crystallographer and mineralogist. Introduced Miller indices for identifying crystallographic planes.

Mohorovičić, Andrija (1857–1936), Yugoslav meteorologist and seismologist. Discovered Mohorovičić seismic discontinuity.

Mohs, Friedrich (1773–1839), German mineralogist. Developed Mohs scale of hardness.

Molina, Mario J. (1943–), American chemist. Demonstrated that chemically inert chlorofluorocarbon (CFC) could be transported up to the ozone layer and could react with ultraviolet light and deplete the ozone layer; Nobel Prize, 1995.

Murchison, Roderick Impey (1792–1871), British geologist. Studied the order of rock formations in Great Britain; with A. Sedgwick, differentiated the Silurian and Devonian.

Piccard, Auguste (1884–1962), Swiss physicist. Conducted a data-collecting exploration of the stratosphere in an airtight gondola of a balloon; constructed and tested a bathysphere for deep-sea exploration.

Pomeranchuk, Isaak Yakolevich (1913–1966), Soviet physicist. Showed that energy of cosmic-ray electrons reaching the atmosphere is limited by their radiation in Earth's magnetic field; proved the Pomeranchuk theorem for scattering cross sections.

Porro, Ignazio (1801–1875), Italian topographer, geodesist, and physicist. Invented optical surveying instruments, Porro prism erecting system, and modern prism binoculars.

Ptolemy (2d century), Greco-Egyptian astronomer, geographer, and geometer at Alexandria. Proposed the Ptolemaic system, with the Earth as the center of the universe.

Rossby, Carl Gustaf Arvid (1898–1957), Swedish-born American meteorologist. Formulated theories of large-scale air movements; derived the Rossby formula, relating speed of propagation of perturbations to airflow and wavelengths of perturbations; devised the Rossby diagram, used to plot air mass properties.

Sedgwick, Adam (1785–1873), English geologist. With R. I. Murchison, established the Devonian system.

Störmer, Carl Fredrik Mülertz (1874–1957), Norwegian mathematician and geophysicist. Studied atmospheric phenomena; discovered the Stormer cone concerning cosmic rays.

Teisserenc de Bort, Léon Philippe (1855–1913), French meteorologist. Discovered the stratosphere.

Torricelli, Evangelista (1608–1647), Italian physicist. Invented the mercury barometer.

Van Allen, James Alfred (1914–), American physicist. Discovered that the Earth is circled by two high-energy radiation belts, leading to major revisions in concepts of the Earth's atmosphere and magnetic field.

Wegener, Alfred Lothar (1880–1930), German geologist. Presented the idea of continental drift.

Contributor Initials

A

A.A.E. A. A. Ekdale
A.A.F. Arthur A. Few
A.A.L. Andrew A. Lacis
A.B.W. A. B. Watts
A.Cou. Arnold Court
A.C.G. Arch C. Gerlach
A.C.L. Alcinda C. Lewis
A.D.H. Arthur D. Hasler
A.D.M. Andrew D. Miall
A.F.H. A. F. Hagner
A.F.T. Alex Francis Trendall
A.G.F. Alfred G. Fischer
A.G.L. Allan G. Lochhead
A.Ha. A. Hallam: Jurassic
A.Has. Alan Hastings
A.H.Ro. Arthur H. Robinson
A.J.Ba. Alain J. Baronnet
A.K.H. Ardith K. Hansel
A.L.G. Arnold L. Gordon
A.L.L. Art Lerner-Lam
A.M.D. Andrew M. Davis
A.M.G. Alan M. Gaines
A.M.K. Albert M. Kudo
A.Mor. Andrew Morton
A.Na. Alexandra Navrotsky
A.N.H. Alex N. Halliday
A.N.S. Albert N. Sayre
A.P.R. Andrew P. Roberts
A.R.P. Allison R. Palmer
A.T.A. Alfred T. Anderson, Jr.
A.Wu. Alfred Wüest
A.W.H.D. Antoni W. H. Damman

B

B.A.M. Brian A. Maurer
B.A.T. Brian A. Tinsley
B.A.V.D.P. Ben A. van der Pluijm
B.C.C. Bryan C. Chakoumakos
B.C.R. Benoit Cushman-Roisin
B.C.Sc. B. Charlotte Schreiber
B.E.F. Blake E. Feist
B.E.Go. Barry E. Goodison
B.F.H. Benjamin F. Howell, Jr.
B.Hi. Bruce Hicks
B.J.C. Bernard J. Conkley
B.J.M. Basil J. Mason
B.J.S. Brian J. Skinner
B.J.Wo. B. J. Wood
B.J.Wu. Bernhardt J. Wuensch
B.K. Blair Kinsman
B.L.B. Blaine L. Blad
B.L.D. Barry L. Doolan
B.L.I. B. Lynn Ingram
B.Rus. Brian Rust
B.R.L. Bruce R. Lipin

B.R.R. Brian R. Rust
B.U.H. Bilal U. Haq
B.V. Bernard Vonnegut
B.Y.Ta. Bernard Y. Tao
B.Z.B. Barbara Zakrewska Borowieka

C

C.An. Christian Andreasen
C.A.C. Carleton A. Chapman
C.A.R. Charles A. Ross
C.A.Sa. Charles A. Sandberg
C.Br. Chris Brod
C.C. Charles Collinson
C.Fr. Clifford Frondel
C.F.Ch. Charles F. Chappell
C.G.J. Clive G. Jones
C.H.Mo. Clyde H. Moore
C.H.S. Christopher H. Scholz
C.K. Cornelis Klein
C.K.S. Charles K. Shearer
C.L.Ch. Charles L. Christ
C.L.W. C. Langdon. White
C.Pa. Camille Parmesan
C.R.Sc. Christopher R. Scotese
C.Sc. Charlotte Schreiber
C.Si. Carol Simpson
C.S.H. C. S. Hammen
C.S.Hu. Cornelius S. Hurlbut, Jr.
C.V.C. Charles V. Crittenden
C.Wi. Conrad Wickstrom
C.W.Be. Curt W. Beck
C.W.Bu. Charles W. Burnham
C.W.By. Charles W. Byers
C.W.N. Chester W. Newton

D

D.A.Sa. Deborah A. Salazar
D.B.Bi. David B. Bieler
D.B.Lo. D. B. Loope
D.B.St. David B. Stewart
D.C.C. David C. Coleman
D.C.F. Derek C. Ford
D.DiB. David DiBiase
D.D.E. Dennis D. Eberl
D.D.Fo. Dennis D. Focht
D.D.H. David D. Houghton
D.Eb. Dennis Eberl
D.E.Br. Don E. Brownlee, II
D.E.E. David E. Ellis
D.E.H. Dennis E. Hayes
D.Fo. Derek Ford
D.F.Ow. Denis F. Owen
D.Ge. Dennis Genito
D.H.T. D. H. Tarling
D.H.Th. David Hurst Thomas
D.J.D. Durk J. Doornbos

D.J.Dep. Donald J. DePaulo
D.J.Du. David J. Dunlop
D.J.E. Don J. Easterbrook
D.J.J. Daniel J. Jones
D.J.Ov. D. J. Over
D.L.B. Daniel L. Birkenheuer
D.L.Ha. Dennis L. Hartmann
D.L.I. Douglas L. Inman
D.L.J. Donald Lee Johnson
D.Lam. Dennis Lamb
D.Low. David Lowenthal
D.M.F. David M. Farmer
D.M.Fo. David M. Fountain
D.R.P. Donald R. Peacor
D.R.V. David R. Veblen
D.R.W. David R. Wones
D.Sim. Daniel Simberloff
D.Sk. David Skelly
D.S.Ba. Daniel S. Barker
D.T.G. Dana T. Griffen
D.W.Mo. David W. Mohr
D.W.P. Darrell W. Pepper

E

E.A.A. Edward A. Ackerman
E.A.K. Elbert A. King
E.C.T.C. Edward C. T. Chao
E.E.W. E. Eugene Weaver
E.F.McB. Earle F. McBride
E.H.Ha. Edwin H. Hammond
E.I. Earl Ingerson
E.J.Ba. Eric J. Barron
E.J.E. Eric J. Essene
E.Ke. Edwin Kessler
E.K.B. Elizabeth K. Berner
E.L.Cl. Elmond L. Claridge
E.L.Wi. Edward L. Winterer
E.M.P. E. M. Parmentier
E.M.R. Eugene M. Rasmusson
E.M.St. Edward M. Stolper
E.N.B. Eddie N. Bernard
E.O.B. Elgene O. Box
E.P.K. E. Philip Krider
E.R.R. Elmar R. Reiter
E.S.B. Elso S. Barghoorn
E.S.S. Edward S. Sarachik
E.Wi. Earle Williams
E.W.H. E. William Heinrich

F

F.B. Fred Barker
F.B.Go. Frank B. Golley
F.B.Sh. Frederick B. Shuman
F.B.V.H. Franklyn B. Van Houten
F.C.Ha. Frank C. Hawthorne
F.Go. Fraser Goff

F.G.Et. Frank G. Ethridge
F.H.L. Fritz H. Laves
F.H.Lu. Frank H. Ludlam
F.J.Sc. Francis J. Schmidlin
F.J.W. Fred J. Weibell
F.L.M. Florence Lansana Margai
F.M.W. Floyd M. Wahl
F.R.B. F. R. Boyd
F.S. Frederick Sanders
F.S.B. Francis H. Brown
F.S.J. Francis S. Johnson
F.S.Z. Frederick S. Zbar
F.V.B. Ferdinando V. Boero

G

G.An. Giuseppe Angrisano
G.A.D. Gregory A. Davis
G.A.Da. Gilles A. Daigle
G.A.M. Gordon A. Macdonald
G.A.Ma. George A. Maul
G.Bo. Gerard Bond
G.B.L. Glen B. Lesins
G.D.K. Garry D. Karner
G.D.S. Guy D. Smith
G.E.Li. Gene E. Likens
G.Fau. Gunter Faure
G.F.H. G. F. Herzog
G.F.P. George F. Pinder
G.H.C. Gilbert H. Cady
G.H.D. George H. Davis
G.J. Giles Johnson
G.J.R. Gerald J. Romick
G.Kul. Gunnar Kullerud
G.K.Cz. Gerald K. Czamanske
G.M.Y. Grant M. Young
G.P.B. Guy P. Brasseur
G.R. George Rapp, Jr.
G.R.G. G. Ronald Gray
G.R.N. Gerald R. North
G.Sh. G. Shanmugam
G.Sp. Garrison Sposito
G.Sw. George Switzer
G.T. Geza Teleki
G.V.K. George V. Keller
G.V.M. Gerard V. Middleton
G.We. Gunter Weller
G.W.Br. George W. Brindley
G.W.DeV. George W. DeVore
G.W.G. George W. Gokel
G.W.G. Gordon W. Groves

H

H.Bl. Harvey Blatt
H.B.B. Howard B. Bluestein
H.B.J. Howard B. Jensen
H.B.Pi. Harry B. Pionke
H.Co. Helen Couclelis
H.C.Wi. Kurd C. Willett
H.D.D. H. D. Dooley
H.D.G.M. Henri D. Grissino-Mayer
H.D.O. Harold D. Orville
H.E.L. H. E. Landsberg

H.Gr. H. Granicher
H.H.D. Heinz H. Damberger
H.H.Mu. Haydn H. Murray
H.J.B.D. Henry J. B. Dick
H.J.Wi. Herold J. Wiens
H.Mor. Helmut Moritz
H.N.P. Henry N. Pollack
H.O.A.M. Henry O. A. Meyer
H.S.L. Harry S. Ladd
H.T.R. H. Thomas Rossby
H.W.J. Holger W. Jannasch

I

I.A.B. Irving A. Breger
I.B. Ivan Barnes
I.W.O.D. Ian W. O. Dalziel

J

J.A.K. John A. Knauss
J.A.McG. John A. McGinley
J.A.R. John A. Robbins
J.A.Smi. Jerome A. Smith
J.A.Sp. J. Alexander Speer
J.B.D. John B. Dunning, Jr.
J.B.H. Jacques B. Hadler
J.B.Jo. Jeremy B. Jones
J.Cai. Joseph Cain
J.C.Co. Jonathan C. Comer
J.C.D. John C. Drake
J.D.MacD. J. D. MacDougall
J.Ev. John Everetts, Jr.
J.E.K. Jessica Elzea Kogel
J.E.K. John E. Kutzbach
J.E.O. John E. Oliver
J.E.S. James E. Schindler
J.Fo. Joseph Ford
J.F.Ba. J. F. Bash
J.F.Hu. John F. Hubert
J.F.P. James F. Price
J.Ga. John Gamble
J.G.B. James B. Blencoe
J.G.J. James G. Johnson
J.Hal. John Hallett
J.H.G. Joseph H. Golden
J.H.L. Jean H. Langenheim
J.H.Sh. John H. Shergold
J.H.Z. John H. Zifcak
J.J.Gr. J. J. Griffin
J.J.Lo. J. J. Love
J.J.Se. J. John Sepkoski, Jr.
J.K.W. Jean K. Whelan
J.Li. Juhn Liou
J.Ly. John Lyman
J.L.Ba. Jeffrey L. Bada
J.L.K. J. Laurence Kulp
J.L.McH. J. L. McHugh
J.L.Re. Joseph L. Reid
J.L.Sta. J. L. Stanford
J.L.Wi. James L. Wilson
J.M. Jonathan Marks
J.Marr. John Marr
J.Mu. John Mutter
J.M.Ch. John M. Christie
J.M.Hu. John M. Hughes

J.M.Wa. John M. Wallace
J.N. Jerome Namias
J.P.G. J. P. Gerrity
J.P.Gr. John P. Grotzinger
J.Ra. John Rakovan
J.R.B. James R. Boles
J.R.C. Joseph R. Curray
J.R.Co. James R. Cochran
J.R.D. James R. Daniel
J.R.F. J. R. Fulks
J.R.G. John R. Gyakum
J.R.H. James R. Holton
J.R.Har. Jay R. Harman
J.R.He. James R. Hein
J.Sim. Joanne Simpson
J.Sup. John Suppe
J.S.C. John S. Campbell
J.S.Fe. Jay S. Fein
J.S.H. Jeffrey S. Hanor
J.T.D. J. Thomas Dutro, Jr.
J.T.K. John T. Kuo
J.T.Le. J. T. Lee
J.T.P. Judith Totman Parrish
J.V.C. J. V. Chernosky
J.Wa. John Warren
J.W.Gei. John W. Geissman
J.W.M. James W. Murray
J.W.S. John Ward Smith

K

K.A. Knut Aagaard
K.A.Em. Kerry A. Emanuel
K.A.Eri. Kenneth A. Eriksson
K.Bu. Kevin Burke
K.B.C. Kenneth B. Cumberland
K.C.C. Kent C. Condie
K.G. K. Grasshoff
K.Ho. Kip Hodges
K.M.S. Kaye M. Shedlock
K.N.L. K. N. Liou
K.R.D. K. R. Dyer
K.Si. Kerry Sieh
K.T.C. Kang-tsung Chang
K.W. Klaus Wyrtki
K.W.F. Karl W. Flessa

L

L.B.L. Luna B. Leopold
L.B.S. Louis B. Slichter
L.C.Le. Lyndon C. Lee
L.E.Mo. Leonard E. Mortenson
L.F.B. Lance F. Bosart
L.F.Pr. Lincoln F. Pratson
L.Gr. Lawrence Grossman
L.J.E. L. J. Ehrenberger
L.Ke. Ljuba Kerschofer
L.L.Y.C. Luke L. Y. Chang

M

M.A.A. Michael A. Adewumi
M.A.Ar. Michael A. Arthur
M.A.Ma. Michelle A. Marvier

W

W.A.Ber.   W. A. Berggren
W.A.Ly.   Walter A. Lyons
W.B.A.   Walter B. Ayers
W.B.N.B.   William B. N. Berry
W.C.P.   Walter C. Pitman, III
W.C.Wo.   William C. Wonders
W.D.K.   Walter D. Keller
W.D.Mo.   Walter D. Mooney
W.E.De.   Walter E. Dean
W.F.B.   William F. Bradley
W.F.D.   Walter F. Dabberdt

W.G.McG.   William G.
   McGinnies
W.H.B.   Wolfgang H. Berger
W.H.Ca.   Wallace H. Campbell
W.H.J.   W. Hilton Johnson
W.I.R.   William Ingersoll
   Rose, Jr.
W.J.M.   William J. Mitsch
W.Ly.   Waldo Lyon
W.L.Ch.   William L. Chameides
W.L.Fi.   William L. Fisher
W.Man.   Warren Manspeizer
W.R.J.   William R. Judd

W.R.Lo.   Wayne R. Lowell
W.R.V.S.   W. Randall Van
   Schmus
W.Sp.   William Spackman
W.Stu.   Wilson Sturges
W.Sw.   William Swider
W.Swe.   Will Swearingen
W.S.Ba.   W. Scott Baldridge

Y, Z

Y.O.   Yoshi Ogasawara
Z.S.   Zdenek Sekera

Contributor Affiliations

This list below comprises all contributors to the Encyclopedia. This list may be used in conjunction with the previous section to fully identify the contributor of each article.

A

Aagaard, Prof. Knut. Department of Oceanography, University of Washington.

Ackerman, Dr. Edward A. Deceased; formerly, Carnegie Institution of Washington.

Adewumi, Prof. Michael A. Petroleum and Natural Gas Engineering, Pennsylvania State University.

Akasofu, S.-I. Director, International Arctic Research Center, University of Alaska, Fairbanks.

Alexander, Dr. Vera. Institute of Marine Science, University of Alaska.

Algermissen, Dr. S. T. Aurora, Colorado.

Anderson, Dr. Alfred T., Jr. Department of Geophysical Sciences, University of Chicago.

Andreasen, Dr. Christian. Bureau Hydrographique International, Monaco.

Angrisano, Rear Admiral Guiseppe. Bureau Hydrographique International, Monaco.

Armstrong, Dr. Richard L. Institute of Arctic and Alpine Research, University of Colorado.

Arthur, Dr. Michael A. Graduate School of Oceanography, University of Rhode Island.

Ayers, Dr. Walter B. Bureau of Economic Geology, University of Texas, Austin.

B

Bachinski, Dr. Sharon W. Department of Geology, University of New Brunswick, Fredericton, Canada.

Bada, Dr. Jeffrey L. Scripps Institution of Oceanography, University of California, San Diego-La Jolla.

Badgley, Dr. Peter C. Earth Science Division, Office of the Naval Reserve, Arlington, Virginia.

Bailey, Dr. S. W. Department of Geology, University of Wisconsin.

Baker, Dr. Paul A. Department of Geology, Duke University.

Baker, Dr. Victor R. Department of Geosciences, University of Arizona.

Baldridge, Dr. W. Scott. Earth and Space Science Division, Los Alamos National Laboratory.

Barber, Dr. Richard T. Duke University Marine Laboratory.

Barghoorn, Prof. Elso S. Department of Biology and Botanical Museum, Harvard University.

Barker, Prof. Daniel S. Department of Geological Sciences, University of Texas, Austin.

Barker, Dr. Fred. Geological Survey, U.S. Department of the Interior, Denver, Colorado.

Barnes, Ivan. Water Resources Division, Geological Survey, U.S. Department of the Interior, Menlo Park, California.

Barron, Prof. Eric. Director, Earth System Science Center, Pennsylvania State University, University Park.

Barronet, Dr. Alain. Centre de Recherche sur les Méchanismes de la Croissance Cristalline, Marseille, France.

Barth, Prof. T. F. W. Deceased; formerly, Geologic Museum, Oslo.

Barton, Dr. Mark D. Department of Earth and Space Sciences, University of California, Los Angeles.

Bash, Dr. Jack. University–National Oceanographic System Office, University of Rhode Island, Saunderstown.

Bates, Dr. Robert L. Department of Geology, Ohio State University.

Bauer, Dr. Simon H. Department of Chemistry, Cornell University.

Beaumont, Robert T. Director of Meteorological Operations, E. G. G., Inc., Boulder, Colorado.

Beck, Prof. Curt W. Department of Chemistry, Amber Research Laboratory, Vassar College, Poughkeepsie, New York.

Becker, Dr. Philippe. AT&T Bell Laboratories, Murray Hill, New Jersey.

Bell, Robin E. Lamont-Doherty Geological Observatory of Columbia University, Palisades, New York.

Berger, Prof. Wolfgang H. Geological Research Division, Scripps Institution of Oceanography, La Jolla, California.

Berggren, Dr. W. A. Woods Hole Oceanographic Institution, Woods Hole, Massachusetts.

Bernard, Dr. E. N. National Oceanic and Atmospheric Administration, Pacific Marine Environmental Laboratory, Seattle, Washington.

Berner, Dr. Elizabeth K. Department of Geology and Geophysics, Yale University.

Berner, Dr. Robert A. Department of Geology and Geophysics, Yale University.

Berry, Dr. William B. N. Museum of Paleontology, University of California, Berkeley.

Bieler, David B. Department of Geology and Geophysics, Centenary College of Louisiana, Shreveport.

Birkenheuer, Dr. Daniel L. Lakewood, Colorado.

Black, Dr. Robert F. Department of Geology, University of Connecticut.

Blad, Dr. Blaine L. Agricultural Meteorology Section, University of Nebraska.

Blatt, Dr. Harvey. Hebrew University of Jerusalem, Givat Ram, Jerusalem, Israel.

Blencoe, Dr. James G. Chemistry Division, Oak Ridge National Laboratory.

Bluestein, Prof. Howard B. Department of Meteorology, University of Oklahoma.

Boero, Dr. Ferdinando. Professor of Zoology, Università di Lecce, Dipartmento di Biologia, Stazione di Biologia Marina, Lecce, Italy.

Boles, Prof. James R. Department of Geological Sciences, University of California, Santa Barbara.

Bond, Dr. Gerard. Lamont-Doherty Geological Observatory, Palisades, New York.

Borowiecki, Prof. Barbara Z. Department of Geography, University of Wisconsin, Milwaukee.

Bosart, Prof. Lance F. Department of Atmospheric Sciences, State University of New York, Albany.

Box, Dr. Elgene O. Department of Geography, University of Georgia.

Boyd, Dr. Richard H. Department of Chemical Engineering, University of Utah.

Bradley, Prof. William F. Department of Chemical Engineering, University of Texas, Austin.

Brassell, Prof. Simon C. Professor of Geological Sciences, Biogeochemical Laboratories, Indiana University, Bloomington.

Brasseur, Dr. Guy P. Director, Atmospheric Chemistry Division, National Center for Atmospheric Research, Boulder, Colorado.

Breger, Dr. Irving A. Deceased; formerly, Office of Energy Resources, U.S. Geological Survey, Reston, Virginia.

Brindley, Dr. George W. Department of Geosciences, Pennsylvania State University.

Brod, Dr. Chris. Glen Canyon Environmental Studies, Bureau of Reclamation, Flagstaff, Arizona.

Brook, Dr. Marx. Department of Physics, New Mexico Institute of Mining and Technology, Socorro.

Brownlee, Dr. Donald E., II. Department of Astronomy, University of Washington, Seattle.

Burke, Prof. Kevin C. Department of Geosciences, University of Houston.

Burnham, Prof. Charles W. Department of Earth and Planetary Sciences, Harvard University.

Burns, Roger G. Department of Earth Science, Massachusetts Institute of Technology.

Byers, Dr. Charles W. Department of Geology, University of Wisconsin.

C

Cady, Dr. Gilbert H. Deceased; formerly, Consulting Coal Geologist, Urbana, Illinois.

Cain, Dr. Joseph C. Department of Geology, Florida State University.

Caine, Nel. Institute of Arctic and Alpine Research, University of Colorado at Boulder.

Campbell, Dr. John S. Department of Biological Sciences, University of Lethbridge, Alberta, Canada.

Campbell, Dr. Wallace H. U.S. Geological Survey, Denver, Colorado.

Carbone, Dr. Richard E. Director, Atmospheric Technology Division, National Center for Atmospheric Research, Boulder, Colorado.

Chakoumakos, Dr. Bryan C. Department of Geology, Northrup Hall, University of New Mexico.

Chameides, Prof. William L. School of Earth and Atmospheric Sciences, Georgia Institute of Technology, Atlanta.

Chang, Prof. Kang-tsung. Department of Geography, University of Idaho, Moscow.

Chang, Dr. Luke L. Y. Department of Geology, University of Maryland, College Park.

Chao, Dr. Edward C. T. Geological Survey, U.S. Department of the Interior, Reston, Virginia.

Chapman, Dr. Carleton A. Department of Geology, University of Illinois.

Chappell, Dr. Charles F. University Corporation for Atmospheric Research, COMET, Boulder, Colorado.

Chernosky, Prof. J. V. Department of Geological Science, University of Maine.

Christ, Dr. Charles L. Physicist, U.S. Geological Survey.

Christie, Dr. John M. Department of Geology, University of California, Los Angeles.

Christie-Blick, Dr. Nicholas. Professor of Geology, Lamont-Doherty Earth Observatory, Columbia University, Palisades, New York.

Clapp, Philip F. National Weather Service, National Oceanic and Atmospheric Administration, Washington, D.C.

Claridge, Dr. Elmond L. Retired; formerly, Director of Graduate Program in Petroleum Engineering, Chemical Engineering Department, University of Houston Central Campus, Houston.

Cloud, Dr. Preston E., Jr. Department of Geology, University of California, Santa Barbara.

Cobine, Dr. James D. Deceased; formerly, Professor and Senior Scientist, Atmospheric Sciences Research Center, State University of New York, Albany.

Coleman, Dr. David C. Department of Entomology, University of Georgia.

Collinson, Dr. Charles. Department of Geology, University of Illinois.

Comer, Prof. Jonathan. Deparment of Geography, Oklahoma State University, Stillwater.

Condie, Dr. Kent C. Department of Earth and Environmental Sciences, New Mexico Institute of Mining and Technolgy, Socorro.

Conway Morris, Dr. Simon. Department of Earth Sciences, The Open University, Milton Keynes, England.

Couclelis, Prof. Helen. Department of Geography, University of California, Santa Barbara.

Court, Dr. Arnold. Retired; formerly, Department of Climatology, California State University.

Crittenden, Dr. Charles V. Geographer, Economic Development Administration, U.S. Department of Commerce.

Cumberland, Prof. Kenneth B. Professor of Geography, University of Auckland, New Zealand.

Curray, Dr. Joseph R. Scripps Institution of Oceanography, University of California, La Jolla.

Cushman-Roisin, Dr. Benoit. Thayer School of Engineering, Dartmouth College, Hanover, New Hampshire.

Czamanske, Dr. Gerald K. U.S. Geological Survey, Menlo Park, California.

D

Dabberdt, Dr. Walter F. Atmospheric Technology Division, National Center for Atmospheric Research, Boulder, Colorado.

Dahl, Dr. Peter S. Department of Geology, Kent State University, Kent, Ohio.

Daigle, Dr. Gilles A. National Research Council, Institute for Microstructural Sciences, Ottawa, Canada.

Dalziel, Dr. Ian W. O. Department of Geological Sciences, University of Texas at Austin.

Damberger, Dr. Heinz H. Illinois State Geological Survey, Coal Section, Champaign.

Damman, Prof. Antoni W. H. Department of Biology, University of Connecticut.

Daniel, James R. Department of Biochemistry, Purdue University.

Davies-Jones, Dr. Robert P. Meteorologist, National Severe Storms Laboratory, U.S. Department of Commerce, Norman, Oklahoma.

Davis, Dr. Andrew M. Enrico Fermi Institute, University of Chicago, Illinois.

Davis, Dr. George H. Department of Geological Sciences, University of Arizona.

Davis, Prof. Gregory A. Department of Geological Sciences, University of Southern California.

Davis, Dr. Richard A., Jr. Department of Geology, University of South Florida.

Davis, Dr. Simon J. M. Ancient Monuments Laboratory, Saville Row, London, England.

Deacon, Margaret. Department of Oceanography, University of Southampton, United Kingdom.

Dean, Dr. Walter E. U.S. Geological Survey, Denver Federal Center, Denver, Colorado.

Deland, Dr. Raymond J. Retired; formerly, Department of Meteorology and Oceanography, New York University.

DePaolo, Dr. Donald J. Department of Geology and Geophysics, University of California, Berkeley.

DeVore, Dr. George W. Department of Geology, Florida State University.

DiBiase, Dr. David. Department of Geography, Pennsylvania State University, University Park.

Dick, Dr. Henry J. B. Senior Scientist, Department of Geology and Geophysics, Woods Hole Oceanographic Institution, Woods Hole, Massachusetts.

Dickinson, Prof. Robert E. Department of Atmospheric Physics, University of Arizona, Tucson.

Dill, Dr. Robert F. Ocean Mining Administration, U.S. Department of the Interior.

Dinsmore, Robertson P. International Ice Patrol, Woods Hole Oceanographic Institution, Woods Hole, Massachusetts.

Dodds, Robert H. Deceased; formerly, Personnel Manager, Gibbs and Hill, Inc., New York.

Donovan, Dr. Nowell. Department of Geology, Texas Christian University, Forth Worth.

Doolan, Barry L. Department of Geology, Memorial University of Newfoundland, St. John's, Canada.

Dooley, H. D. Marine Laboratory, Department of Agriculture and Fisheries, Aberdeen, Scotland.

Doornbos, Dr. Durk J. Deceased; formerly, Institute for Geophysics, University of Oslo, Norway.

Dott, Dr. Robert H., Jr. Department of Geology, University of Wisconsin.

Drake, Prof. John C. Department of Geology, University of Vermont.

Dunlop, Prof. David J. Department of Physics, University of Toronto, Ontario, Canada.

Dunning, Dr. John B., Jr. Department of Forestry and Natural Resources, Purdue University.

Dutro, Dr. J. Thomas, Jr. Museum of Natural History, U.S. Department of the Interior, Washington, D.C.

Dyer, Dr. K. R. Institute of Oceanographic Sciences, Somerset, England.

E

Easterbrook, Don J. Department of Geology, Western Washington University, Bellingham.

Eberl, Dr. Dennis D. U.S. Geological Survey, Denver Federal Center, Denver, Colorado.

Einaudi, Prof. Marco T. Department of Applied Earth Science, Stanford University.

Eisenbud, Dr. Merril. Institute of Environmental Medicine, New York University Medical Center.

Ekdale, Dr. Allan A. Department of Geology and Geophysics, College of Mines and Earth Sciences, University of Utah.

Ellis, Dr. David E. Research and Development Department, Continental Oil Company, Ponco City, Oklahoma.

Emanuel, Dr. Kerry Andrew. Department of Earth, Atmospheric and Planetary Sciences, Massachusetts Institute of Technology.

Emerson, Dr. Steven. Department of Oceanography, University of Washington.

English, Prof. Van H. Department of Geography, Dartmouth College.

Eriksson, Dr. Kenneth A. Department of Geology, University of Texas, Richardson.

Essene, Prof. Eric J. Department of Geological Sciences, University of Michigan, Ann Arbor.

Ethridge, Dr. Frank G. Department of Earth Resources, Colorado State University.

Everetts, Dr. John Jr. Professor of Architectural Engineering, Pennsylvania State University.

Ewing, Prof. Rodney. Department of Earth and Planetary Sciences, University of New Mexico, Albuquerque.

F

Farmer, Dr. David M. Department of the Environment, Pacific Region, Institute of Ocean Sciences, Sidney, British Columbia, Canada.

Faure, Dr. Gunter. Department of Geological Sciences, Ohio State University.

Fein, Dr. Jay S. Division of Atmospheric Sciences, National Science Foundation, Washington, D.C.

Feist, Dr. Blake E. Northwest Fisheries Science Center, Environmental Conservation Division–Watershed Program, Seattle, Washington.

Fiedler, Dr. Peggy L. Piedmont, California.

Fischer, Prof. Alfred G. San Pedro, California.

Fisher, Dr. Robert L. Geological Research Division, Scripps Institution of Oceanography, La Jolla, California.

Fisher, Dr. William L. Department of Geological Science, University of Texas, Austin.

Fleischer, Dr. R. L. Physical Science Branch, General Physics Laboratory, General Electric Company, Schenectady, New York.

Flessa, Dr. Karl Walter. Division of Earth Sciences, National Science Foundation, Washington, D.C.

Flint, Dr. Richard F. Department of Geology and Geophysics, Yale University.

Focht, Prof. Dennis D. Department of Soil and Environmental Sciences, University of California, Riverside.

Ford, Prof. Derek C. Department of Geography, McMaster University, Hamilton, Ontario, Canada.

Fountain, Dr. David M. Department of Geology and Geophysics, University of Wyoming.

Frondel, Prof. Clifford. Retired; Department of Geological Sciences, Harvard University.

Frost, Dr. Thomas M. Center for Limnology, University of Wisconsin.

Fulks, J. R. Retired; formerly, National Weather Service, Chicago, Illinois.

G

Gaines, Dr. Alan M. National Science Foundation.

Gamble, Dr. John. Department of Geology, Victoria University, Wellington, New Zealand.

Garcia, Dr. John. Department of Psychology, University of California, Los Angeles.

Genito, Dennis. Pasture Systems and Watershed Management Research Laboratory, U.S. Department of Agriculture–Agricultural Research Service, University Park, Pennsylvania.

Gerlach, Dr. Arch C. Chief Geographer, Geological Survey, U.S. Department of the Interior.

Gerrity, Dr. Joseph P. National Meteorological Center, National Oceanic and Atmospheric Administration, Camp Springs, Maryland.

Ginsburg, Dr. Robert N. Rosenstiel School of Marine and Atmospheric Science, University of Miami.

Glantz, Dr. Michael. National Center for Atmospheric Research, Boulder, Colorado.

Goff, Dr. Fraser. Los Alamos National Laboratories, Los Alamos, New Mexico.

Goldfarb, Dr. Marjorie S. School of Oceanography, University of Washington, Seattle.

Goldich, Dr. Samuel S. Department of Geology, Colorado School of Mines, Golden.

Golley, Dr. Frank B. Institute of Ecology, Athens, Georgia.

Goodison, Dr. Barry E. Superintendent, Hydrometeorological Impact and Development Section, Canadian Climate Centre, Atmospheric Environment Service, Downsview, Ontario.

Goodman, Dr. Ralph R. Naval Research Laboratory, Stennis Space Center, Mississippi.

Gordon, Dr. Arnold L. Lamont-Doherty Geological Observatory, Palisades, New York.

Gotelli, Dr. Nick. Department of Biology, University of Vermont, Burlington.

Gould, Dr. Stephen J. Professor of Geology, Museum of Comparative Zoology, Harvard University.

Granicher, Prof. H. Laboratory of Solid State Physics, Swiss Federal Institute of Technology, Zurich.

Grasshoff, Dr. K. Institut für Meereskunde an der Universitat Kiel, Germany.

Gray, G. Ronald. Director, Syncrude Canada Ltd., Edmonton, Alberta, Canada.

Greenler, Dr. Robert. Department of Physics, University of Wisconsin.

Griffen, Prof. Dana T. Department of Geology, Brigham Young University, Provo, Utah.

Griffin, Dr. James. Graduate School of Oceanography, University of Rhode Island, Charleston.

Grim, Dr. Ralph E. Department of Geology, University of Illinois, Urbana.

Grissino-Mayer, Dr. Henri D. Department of Physics, Astronomy, and Geosciences, Valdosta State University, Valdosta, Georgia.

Grossman, Dr. Lawrence. Department of Geophysical Sciences, University of Chicago.

Grotzinger, Dr. John. Department of Earth, Atmosphere, and Space Sciences, Massachusetts Institute of Technology.

Groves, Dr. Gordon W. Institute de Geofisica, Torre de Ciencias, Ciudad Universitaria, Mexico.

Guggenheim, Prof. Stephen. Department of Geological Sciences, University of Illinois, Chicago.

Gyakum, Dr. John R. Department of Meteorology, McGill University, Montreal, Quebec, Canada.

H

Hadler, Jacques B. Webb Institute of Naval Architecture, Glen Cove, New York.

Hagan, Dr. Maura. National Center for Atmospheric Research, Boulder, Colorado.

Haggerty, Dr. Stephen E. Department of Geology and Geography, University of Massachusetts.

Halbouty, Dr. Michael T. Consulting Geologist and Petroleum Engineer, Houston, Texas.

Hallam, Dr. Anthony. Department of Geology, University of Birmingham, England.

Hallet, Dr. John. Director, Atmospheric Ice Physics Laboratory, Desert Research Institute, Atmospheric Sciences Center, University of Nevada.

Halliday, Dr. Alex. Department of Geological Sciences, University of Michigan, Ann Arbor.

Hammen, Dr. C. S. Department of Zoology, University of Rhode Island.

Hammond, Prof. Edwin H. Department of Geography. University of Tennessee.

Haney, Dr. Robert L. Department of Meteorology, Naval Postgraduate School, Monterey, California.

Hanna, Dr. Steven R. Sigma Research Corporation, Concord, Massachusetts.

Hanor, Jeffrey S. Department of Geology, Louisiana State University.

Hansel, Dr. Ardith K. Illinois State Geological Survey, Champaign.

Haq, Dr. Bilal U. Department of Marine Geology and Geophysics, National Science Foundation, Washington, D.C.

Harker, Dr. Robert I. Department of Geology, University of Pennsylvania.

Harman, Prof. Jay R. Department of Geography, Michigan State University, East Lansing.

Hartmann, Dr. Dennis L. Department of Atmospheric Sciences, University of Washington, Seattle.

Hasler, Dr. Arthur D. Laboratory of Limnology, University of Wisconsin.

Hastings, Dr. Alan. Division of Environmental Studies, University of California at Davis.

Hawthorne, Dr. Frank C. Department of Geological Sciences, University of Manitoba, Winnipeg, Canada.

Hayes, Dr. Dennis E. Lamont-Doherty Geological Observatory, Columbia University, Palisades, New York.

Heaney, Dr. Peter J. Department of Geological and Geophysical Sciences, Princeton University.

Heckel, Dr. P. H. Department of Geology, University of Iowa.

Hein, Dr. James R. U.S. Geological Survey, Branch of Pacific Marine Geology, Menlo Park, California.

Heinrich, Dr. E. William. Department of Geology and Mineralogy, University of Michigan.

Helman, Marc L. Department of Geological Sciences, University of Durham, England.

Herring, Dr. Thomas A. Department of Earth, Atmospheric, and Planetary Sciences, Massachusetts Institute of Technology, Cambridge.

Herzog, Prof. Gregory F. Department of Chemistry, Rutgers University, Piscataway, New Jersey.

Hicks, Dr. Bruce B. Atmospheric Turbulence and Diffusion Division, National Oceanic and Atmospheric Administration, Oak Ridge, Tennessee.

Hicks, Steacy D. Sterling, Virginia.

Hodges, Prof. Kip. Department of Earth, Atmospheric, and Planetary Sciences, Massachusetts Institute of Technology, Cambridge.

Hoffman, Dr. Paul. Department of Earth and Planetary Sciences, Harvard University.

Hoffman, Dr. Paul. Department of Microbiology and Immunology, Faculty of Medicine, Dalhousie University, Halifax, Nova Scotia, Canada.

Holdaway, Dr. M. J. Department of Geological Sciences, Southern Methodist University.

Holland, Dr. Timothy J. B. Department of Mineralogy and Petrology, University of Cambridge, England.

Holton, Prof. James R. Department of Atmospheric Sciences, University of Washington.

Horsnail, Dr. R. F. Amax Exploration, Inc., Denver, Colorado.

Houghton, Dr. David D. Department of Meteorology, University of Wisconsin, Madison.

Howell, Dr. Benjamin F., Jr. Department of Geosciences, Pennsylvania State University.

Hubert, Prof. John F. Department of Geosciences, University of Massachusetts, Amherst.

Hughes, Dr. John M. Associate Dean, College of Arts and Science, Department of Geology, Miami University, Oxford, Ohio.

Hurlbut, Prof. Cornelius S., Jr. Professor of Mineralogy, Department of Geological Sciences, Harvard University.

I

Ingersoll, Dr. Raymond V. University of California at Los Angeles, Department of Earth and Space Science.

Ingerson, Dr. Earl. Department of Geology, University of Texas, Austin.

Ingram, Dr. B. Lynn. Department of Geology and Geophysics, University of California, Berkeley.

Inman, Dr. Douglas L. Professor of Oceanography, Scripps Institution of Oceanography, La Jolla, California.

J

Jackson, Dr. M. P. A. Senior Research Scientist, Bureau of Economic Geology, University of Texas, Austin.

Jahns, Dr. Richard H. Department of Earth Sciences, Stanford University.

Jannasch, Dr. Holger W. Senior Scientist, Woods Hole Oceanographic Institution, Woods Hole, Massachusetts.

Jensen, Dr. Howard B. Research Supervisor, Laramie Energy Research Center, Energy Research and Development Administration, Wyoming.

Johnson, Dr. Donald Lee. Department of Geography, University of Illinois, Urbana.

Johnson, Dr. Francis S. Acting President, University of Texas, Dallas.

Johnson, Dr. Giles. School of Biological Sciences, University of Manchester, United Kingdom.

Johnson, Dr. James G. Department of Geology, Oregon State University.

Johnson, Dr. W. Hilton. Department of Geology, University of Illinois, Urbana-Champaign.

Jones, Dr. Clive G. Institute of Ecosystem Studies, New York Botanical Garden, Millbrook, New York.

Jones, Dr. Daniel J. Professor and Chairman, Department of Earth Sciences, California State College, Bakersfield.

Jones, Dr. Jeremy B. Department of Biological Sciences, University of Nevada, Las Vegas.

Judd, Dr. William R. School of Civil Engineering, Purdue University.

K

Kareiva, Prof. Peter. Department of Zoology, University of Washington, Seattle.

Karner, Dr. Garry D. Lamont-Doherty Earth Observatory, Palisades, New York.

Keller, George V. Department of Geophysics, Colorado School of Mines.

Keller, Prof. Walter D. Department of Geology, University of Missouri.

Kelley, Prof. Michael C. Engineering and Theory Center, Cornell University.

Kessler, Prof. Edwin. Director, National Severe Storms Laboratory, Norman, Oklahoma.

King, Dr. Elbert A. Department of Geology, University of Houston.

Kinsman, Prof. Blair. College of Marine Studies, University of Delaware.

Klein, Cornelius. Department of Geology, University of New Mexico.

Kogel, Dr. Jessica Elzea. Thiele Kaolin Company, Sanders–ville, Georgia.

Kominz, Michelle. Lamont-Doherty Geological Observatory, Palisades, New York.

Krider, Prof. E. Philip. Director, Institute of Atmospheric Physics, University of Arizona.

Krishnamurti, Dr. T. N. Department of Meteorology, Florida State University.

Kudo, Prof. Albert M. Department of Earth and Planetary Sciences, University of New Mexico, Albuquerque.

Kullerud, Dr. Gunnar. Department of Geosciences, Purdue University.

Kulp, J. Laurence. President, Teledyne Isotopes, Westwood, New Jersey.

Kuo, Prof. John T. Department of Mining Engineering, Columbia University.

Kutzbach, Prof. John E. Department of Meteorology, University of Wisconsin.

L

Labotka, Dr. Theodore C. Department of Geological Sciences, College of Liberal Arts, University of Tennessee.

Lacis, Dr. Andrew A. NASA Goddard Institute for Space Studies, New York.

Ladd, Dr. Harry S. (Retired) National Museum, Smithsonian Institution.

Lamb, Prof. Dennis. Department of Meteorology, Pennsylvania State University.

Landsberg, Dr. H. E. Deceased; formerly, Institute for Physical Science and Technology, University of Maryland.

Langenheim, Prof. Jean H. Division of Natural Sciences, University of California, Santa Cruz.

Laves, Dr. Friz H. Deceased; formerly, Eidg. Technische Hoschschule, Institut fiir Kristallographie und Petrographie, Zurich, Switzerland.

Lavin, Prof. Stephen. Department of Geography, University of Nebraska, Lincoln.

Lavoie, Dr. Ronald L. Chief, Program Requirements and Development Division, U.S. Department of Commerce, National Oceanic and Atmospheric Administration, Silver Spring, Maryland.

Lee, Dr. J. T. Cooperative Institute for Mesoscale Meterological Studies, National Oceanic and Atmospheric Administration, University of Oklahoma.

Lee, Lyndon C. L. C. Lee & Associates, Inc., Seattle, Washington.

Leopold, Dr. Luna B. Department of Geology and Geophysics, University of California, Berkeley.

Lerner-Lam, Dr. Art. Lamont-Doherty Geological Observatory of Columbia University, Palisades, New York.

Lesins, Dr. Glen. Department of Oceanography, Dalhousie University, Halifax, Nova Scotia, Canada.

Lewis, Dr. Alcinda C. Institute of Ecosystem Studies, New York Botanical Garden, Millbrook.

Liddicoat, Richard T., Jr. Gemological Institute of America, Los Angeles, California.

Likens, Prof. Gene E. Director, Institute of Ecosystem Studies, The New York Botanical Garden, Millbrook, New York.

Linde, Dr. Ronald K. President, Envirodyne, Inc., Los Angeles, California.

Lindzen, Prof. Richard S. Center for Meteorology and Physical Oceanography, Massachusetts Institute of Technology.

Linsley, Prof. Ray K. Department of Civil Engineering, Stanford University.

Liou, Juhn G. Department of Geology, Stanford University.

Liou, Prof. Kuo-Nan. Department Chair, Department of Atmospheric Sciences, University of California, Los Angeles.

Lipin, Bruce R. U.S. Department of the Interior, Reston, Virginia.

Liu, Dr. Tanzhuo. Lamont-Doherty Earth Observatory, Columbia University, Palisades, New York.

Lochhead, Dr. Allan G. Microbiology Research Institute, Canada Department of Agriculture, Ottawa, Ontario.

Lonsdale, Dr. Peter. Scripps Institution of Oceanography, La Jolla, California.

Loope, Dr. David B. Department of Geology, University of Nebraska Medical Center.

Love, Dr. Jeffrey J. Institute of Geophysics and Planetary Physics, University of California, San Diego.

Lowell, Prof. Wayne R. Department of Geology, Indiana University.

Lowenthal, Dr. David. Research Associate, American Geographical Society, New York.

Ludlam, Prof. Frank H. Deceased; formerly, Department of Meteorology, Imperial College, London, England.

Lyman, Dr. John. Department of Oceanography, University of North Carolina.

Lyon, Dr. Waldo. Arctic Submarine Research Laboratory, Naval Undersea Warfare Center, San Diego, California.

Lyons, Prof. Walter. Consultant in Telecommunications, Flushing, New York.

M

Macdonald, Dr. Gordon A. Deceased; formerly, Institute of Geophysics, University of Hawaii.

Macdoughall, Prof. J. Douglas. Geological Research Division, Scripps Institution of Oceanography, La Jolla, California.

Manspeizer, Prof. Warren. Department of Geological Sciences, Faculty of Arts and Sciences, Rutgers University.

Margai, Prof. Florence Lansana. Department of Geography, State University of New York at Binghamton.

Mariño, Prof. Miguel A. Hydrology Program, University of California, Davis.

Marks, Dr. Jonathan. Department of Sociology and Anthropology, University of North Carolina at Charlotte.

Marshall, Dr. Norman B. Retired; formerly, Senior Principal Scientific Officer, British Museum of Natural History, London, England.

Maruyama, Prof. Shige. Department of Earth and Planetary Sciences, Tokyo Institute of Technology, Tokyo, Japan.

Marvier, Dr. Michelle A. Department of Biology, Santa Clara University, Santa Clara, California.

Mason, Dr. Basil J. Program Director, Center for Environmental Technology, Imperial College of Science and Technology, London, England.

Maul, Prof. George A. Director, Division of Marine and Environmental Systems, Florida Institute of Technology, Melbourne.

Maurer, Dr. Brian A. Department of Fisheries and Wildlife, Michigan State University, East Lansing.

McBride, Dr. Earle F. Department of Geology, University of Texas, Austin.

McEvilly, Dr. Thomas V. Department of Geology, University of California, Berkeley.

McGinley, Dr. John A. Chief, Forecast Research Group, National Oceanic and Atmospheric Administration, Environmental Research Laboratory, Boulder, Colorado.

McGinnies, Dr. William G. Professor of Dendrochronology and Arid Land Ecologist, Office of Arid Land Studies, University of Arizona.

McHugh, Dr. J. L. Marine Sciences Research Center, State University of New York, Stony Brook.

McKnight, Prof. Tom. Department of Geography, University of Caifornia, Los Angeles, California.

McNaughton, Prof. Samuel J. Department of Biology, Syracuse University.

McWilliams, Dr. Michael O. Department of Geophysics, School of Earth Sciences, Stanford University.

Meyer, Prof. Henry O. A. Department of Earth and Atmospheric Sciences, Purdue University.

Miall, Dr. Andrew D. Department of Geology, University of Toronto, Ontario, Canada.

Middleton, Dr. Gerard V. Department of Geology, McMaster University, Hamilton, Ontario, Canada.

Miller, Prof. Maynard M. Department of Geology, Michigan State University; Director, Foundation for Glacial and Environmental Research, Seattle, Washington.

Mitsch, Prof. William J. Graduate Program Environmental Science, School of Natural Resources, Ohio State University, Columbus.

Mitterer, Dr. Richard M. Department of Geosciences, University of Texas, Dallas.

Miyawaki, Dr. Ritsuro. Department of Geology, National Science Museum, Tokyo, Japan.

Mohr, Prof. David W. Department of Geology, Texas A & M University.

Moore, Prof. Clyde H., Jr. Director, Applied Carbonate Research Program, Department of Geology, Agricultural and Mechanical College, Louisiana State University.

Moore, Dr. Paul B. Department of the Geophysical Sciences, University of Chicago.

Moritz, Prof. Helmut. Technische Universität Graz, Abteilung für Physikalische Geodäsie, Graz, Austria.

Morse, Dr. Robert W. Associate Director and Dean of Oceanographic Studies, Woods Hole Oceanographic Institute, Woods Hole, Massachusetts.

Mortenson, Leonard E. Department of Biological Sciences, Purdue University.

Morton, Dr. Andrew. Department of Geology and Petroleum Geology, University of Aberdeen, Kings College, Aberdeen, United Kingdom.

Murgatroyd, Dr. R. J. Meteorological Office, Bracknell, England.

Murray, Dr. Haydn H. Executive Vice President, Georgia Kaolin Company, Elizabeth, New Jersey.

Murray, Dr. James W. School of Oceanography, University of Washington, Seattle.

Murray, Dr. Royce W. Department of Chemistry, University of North Carolina.

Mutter, Dr. John. Lamont-Doherty Geological Observatory, Columbia University, Multichannel Seismics Group, Palisades, New York.

N

Nace, Dr. Raymond L. Geological Survey, U.S. Department of the Interior, Raleigh, North Carolina.

Namias, Jerome. Climate Research Group, Scripps Institution of Oceanography, La Jolla, California.

Navrotsky, Dr. Alexandra. Department of Geological and Geophysical Sciences, Princeton University.

Newton, Dr. Chester W. National Center for Atmospheric Research, Boulder, Colorado.

Newton, Dr. R. C. Department of the Geophysical Sciences, University of Chicago.

Nolan, Dr. Robert P. Environmental Sciences Laboratory, Brooklyn College, Brooklyn, New York.

North, Dr. Gerald R. Director, Climate System Research Program, Department of Meteorology, Texas A&M University.

Nyman, Dr. Matthew W. Senior Research Associate, Department of Earth and Planetary Sciences, University of New Mexico, Albuquerque.

O

Oliver, Prof. John E. Department of Geography, Geology, and Anthropology, Indiana State University, Terre Haute, Indiana.

Orville, Dr. Harold D. Institute for Atmospheric Science, South Dakota School of Mines, Rapid City, South Dakota.

Osberg, Prof. Philip H. Department of Geological Sciences, University of Maine.

Over, Dr. Jeff. Department of Geological Sciences, State University of New York at Geneseo.

Owen, Dr. Denis F. Department of Biology, Oxford Polytechnic.

P

Palmer, Dr. Allison R. Department of Earth and Space Sciences, State University of New York, Stony Brook.

Parmentier, Prof. E. Mark. Department of Geological Sciences, Brown University, Providence, Rhode Island.

Parmesan, Dr. Camille. National Center for Ecological Analysis and Synthesis, Santa Barbara, California.

Parrish, Dr. Judith Totman. Department of Geosciences, University of Arizona.

Peacor, Dr. Donald R. Department of Geology, University of Michigan.

Pepper, Prof. Darrell W. Department of Mechanical Engineering, University of Nevada, Las Vegas.

Philander, Prof. S. George. Program in Atmospheric and Oceanic Sciences, Princeton University.

Pimm, Dr. Stuart. Department of Zoology, University of Tennessee, Knoxville.

Pinder, Dr. George F. Department of Civil and Geological Engineering, Princeton University.

Pinker, Prof. Rachel T. Department of Meteorology, University of Maryland.

Pionke, Dr. Harry B. Pasture Systems and Watershed Management Research Laboratory, U.S. Department of Agriculture–Agricultural Research Service, University Park, Pennsylvania.

Pitman, Waker C., III. Lamont-Doherty Geological Observatory, Palisades, New York.

Pollack, Dr. Henry N. Department of Geological Sciences, University of Michigan.

Praston, Dr. Lincoln F. Research Scientist, Institute of Arctic and Alpine Research, Department of Geology, University of Colorado, Boulder.

Price, Dr. James F. Department of Physical Oceanography, Woods Hole Oceanographic Institution, Massachusetts.

Prinn, Prof. Ronald G. Department of Earth, Atmosphere and Planetary Sciences, Massachusetts Institute of Technology, Cambridge.

R

Radok, Dr. Uwe. Environmental Research (CIRES), University of Colorado.

Rakovan, Prof. John. Department of Geology, Miami University, Oxford, Ohio.

Randerson, Dr. Peter. Department of Applied Biology, University of Wales Institute of Science and Technology, Cardiff.

Rapp, Prof. George, Jr. Dean, College of Letters and Science, University of Minnesota.

Rasmusson, Dr. Eugene M. Geophysical Fluid Dynamics Laboratory, Environmental Science Services Administration, Princeton, New Jersey.

Rauber, Prof. Robert M. Department of Atmospheric Science, University of Illinois at Urbana-Champaign, Urbana.

Reeder, Dr. Richard J. Geosciences Program, Department of Earth and Space Sciences, State University of New York, Stony Brook.

Reid, Joseph L. Scripps Institution of Oceanography, La Jolla, California.

Reiter, Dr. Elmar R. Department of Atmospheric Science, Colorado State University.

Ribbe, Dr. Paul H. Department of Geological Sciences, Virginia Polytechnic Institute.

Richardson, Dr. Philip. Department of Physical Oceanography, Woods Hole Oceanographic Institution, Woods Hole, Massachusetts.

Riley, Dr. Pete. Science Applications International Corporation, San Diego, California.

Riser, Dr. Stephen C. School of Oceanography, University of Washington, Seattle.

Risser, Dr. Paul G. Vice President for Research, University of New Mexico.

Robbins, Dr. John A. Great Lakes Environmental Research Laboratory, National Oceanic and Atmospheric Administration, Ann Arbor, Michigan.

Roberts, Dr. Andrew. Department of Oceanography, University of Southampton, Southampton Oceanography Centre, United Kingdom.

Robinson, Prof. Arthur H. Department of Geography, University of Wisconsin.

Romick, Dr. Gerald J. Applied Physics Laboratory, Hopkins University, Laurel, Maryland.

Rose, Dr. William Ingersoll, Jr. Department of Geology and Geological Engineering, Michigan Technological University.

Ross, Dr. Charles A. Department of Geology, Western Washington State College.

Ross, Dr. Malcolm. Washington, D.C.

Rossby, Prof. H. Thomas. Graduate School of Oceanography, University of Rhode Island, Kingston.

Rust, Dr. Brian R. Department of Geology, University of Ottawa, Ontario, Canada.

Rutledge, Dr. Steven A. Department of Atmospheric Science, Colorado State University, Fort Collins.

S

Saito, Dr. Tsunemasa. Lamont-Doherty Geological Observatory, Palisades, New York.

Salazar, Prof. Deborah A. Department of Geography, Oklahoma State University, Stillwater.

Samson, Prof. Perry. Department of Atmospheric, Oceanic, and Space Sciences, University of Michigan, Ann Arbor.

Sandberg, Dr. Charles. U.S. Geological Survey, Denver, Colorado.

Sandberg, Dr. Philip A. Department of Geology, University of Illinois at Urbana-Champaign.

Sanders, Dr. Frederick. Department of Meteorology, Massachusetts Institute of Technology.

Sarachik, Prof. Edward S. Department of Atmospheric Sciences, University of Washington, Seattle.

Sayre, Dr. Albert N. Deceased; formerly, Consulting Groundwater Geologist, Behre Dolbear and Company.

Schindler, Dr. James E. Department of Biological Sciences, College of Sciences, Clemson University.

Schmid, Dr. Rudolf. Department of Botany, University of California, Berkeley.

Schmidlin, Dr. Francis. National Aeronautics and Space Administration, Observational Science Branch, Laboratory for Oceans, Goddard Space Flight Center, Wallops Flights Facility, Wallops Island, Virginia.

Schmitt, Dr. Raymond. Woods Hole Oceanographic Institution, Woods Hole, Massachusetts.

Schnabel, Dr. Ronald. Pasture Systems and Watershed Management Research Laboratory, U.S. Department of Agriculture–Agricultural Research Service, University Park, Pennsylvania.

Scholle, Prof. Peter A. Department of Geological Sciences, Southern Methodist University, Dallas, Texas.

Scholz, Christopher H. Lamont-Doherty Geological Observatory, Palisades, New York.

Schreiber, Prof. B. Charlotte. Department of Earth and Environmental Sciences, Queens College, New York.

Schultz, Robert M. General Manager, William Langer Jewel Bearing Plant, Bulova Watch Company, Inc., Rolla, North Dakota.

Scotese, Dr. Christopher R. Department of Geology, University of Texas, Arlington.

Sekera, Prof. Zdenek. Deceased; formerly, Department of Meteorology, Institute for Geophysics and Planetary Physics, University of California, Los Angeles.

Selby, Prof. Michael J. Department of Earth Science, University of Waikato, New Zealand.

Sepkoski, Dr. J. John, Jr. Department of Geophysical Sciences, Henry Hinds Laboratory, University of Chicago.

Shanmugam, Dr. G. Mobil Research and Development Corporation, Research Department, Dallas Research Laboratory, Dallas, Texas.

Shearer, Dr. Charles K. Institute of Meteorites, Department of Earth and Planetary Sciences, University of New Mexico, Albuquerque.

Shedlock, Dr. Kaye M. U.S. Geological Survey, Administrative Officer, Denver, Colorado.

Shergold, Dr. John. H. Chairman, International Subcommission on Cambrian Stratigraphy, Masseret, France.

Sheriff, Dr. Robert E. Department of Geology, University of Houston.

Shuman, Dr. Frederick B. Director, National Meteorological Center, National Oceanic and Atmospheric Administration, Washington, D.C.

Sieh, Prof. Kerry. Seismological Laboratory, California Institute of Technology, Pasadena.

Siever, Dr. Raymond. Department of Geological Sciences, Harvard University.

Signorini, Dr. Sergio R. Division of Ocean Sciences, National Science Foundation, Arlington, Virginia.

Simberloff, Prof. Daniel. Department of Biological Science, Florida State University.

Simon, Dr. Steven. Department of Geophysical Science, University of Chicago, Illinois.

Simonson, Dr. Roy W. Director (retired), Soil Classification and Correlation, U.S. Department of Agriculture, Hyattsville, Maryland.

Simpson, Dr. Carol. Department of Earth and Planetary Sciences, Johns Hopkins University.

Simpson, Dr. Joanne. Head, Severe Storms Branch, Goddard Space Flight Center, NASA, Greenbelt, Maryland.

Simpson, Prof. Robert H. Retired; formerly, Director, Experimental Meteorology Laboratory, National Weather Service, Miami, Florida.

Singer, Dr. S. F. Formerly, Deputy Assistant Secretary, U.S. Department of the Interior.

Skelly, Dr. David. School of Forestry and Environmental Studies, Yale University, New Haven, Connecticut.

Skinner, Prof. Brian J. Department of Geology and Geophysics, Yale University, New Haven, Connecticut.

Slichter, Prof. Louis B. Deceased; formerly, Institute of Geophysics, University of California, Los Angeles.

Smith, Dr. Guy D. Deceased; formerly, Soil Conservation Service, U.S. Department of Agriculture.

Smith, Dr. Jerome A. Marine Physics Laboratory, Scripps Institution of Oceanography, University of California, La Jolla.

Smith, Dr. John Ward. Deceased; formerly, Research Supervisor, Laramie Energy Research Center, Energy Research and Development Administration, Laramie, Wyoming.

Smith, Prof. Norman D. Department of Geosciences, University of Nebraska, Lincoln.

Smith, Dr. Robert L. College of Oceanic and Atmospheric Sciences, Oregon State University, Corvallis.

Spackman, Dr. William. Department of Geology, Pennsylvania State University.

Speer, Dr. J. Alexander. Department of Marine, Earth, and Atmospheric Sciences, School of Physical and Mathematical Sciences, North Carolina State University.

Sposito, Dr. Garrison. Department of Plant and Soil Biology, University of California, Berkeley.

Stage, Dr. Steven A. Baton Rouge, Louisiana.

Stanford, Prof. John L. Department of Physics and Astronomy, Iowa State University, Ames.

Stern, Dr. Robert J. Department of Geosciences, University of Texas, Richardson.

Stewart, Dr. David B. Experimental Geochemistry and Mineralogy Branch, U.S. Geological Survey, Department of the Interior.

Stewart, Scott R. L. C. Lee & Associates, Inc., Seattle, Washington.

Stolper, Dr. Edward M. Division of Geological and Planetary Science, California Institute of Technology.

Sturges, Dr. Wilson. Department of Oceanography, Florida State University.

Suppe, John. Department of Geological Sciences, Princeton University.

Swearingen, Dr. Will. NASA-Montana State University TechLink, Bozeman, Montana.

Swider, Dr. William, Jr. Space Physics Division, Air Force Geophysics Laboratory, Hanscom Air Force Base, Bedford, Massachusetts.

Switzer, Dr. George. Curator, Department of Mineral Sciences, National Museum of Natural History, Smithsonian Institution.

T

Tao, Dr. B. Y. School of Agricultural and Biological Engineering, Purdue University, West Lafayette, Indiana.

Tarling, Dr. D. H. Department of Geophysics and Planetary Physics, University of Newcastle, England.

Taylor, Dr. R. E. Associate Professor and Director, Radiocarbon Laboratory, University of California, Riverside.

Taylor, Dr. Stuart Ross. Department of Geology, Australian National University, Canberra, Australia.

Taylor, Prof. Thomas N. Department of Biology, University of Kansas, Lawrence.

Teleki, Dr. Geza. Deceased; formerly, Department of Geology, George Washington University.

Thomas, Dr. David Hurst. American Museum of Natural History, New York.

Tinsley, Dr. Brian A. Center for Space Science, Department of Physics, University of Texas, Dallas.

Toy, Dr. T. J. Department of Geography, University of Denver, Colorado.

Tracy, Robert J. Department of Geological Sciences, Harvard University.

Trendall, Dr. Alec. Adjunct Professor, Department of Applied Physics, Curtin University of Technology, Perth, Western Australia.
Turner, Dr. Shirley. National Institute of Standards and Technology, U.S. Department of Commerce, Gaithersburg, Maryland.

U,V

Uhrhammer, Dr. Robert A. Seismological Laboratory, University of California, Berkeley.
Van Andel, Dr. Tjeerd H. Department of Geology, Stanford University.
Van der Pluijm, Prof. Ben. Department of Geological Sciences, University of Michigan, Ann Arbor.
Van Houten, Prof. Franklyn B. Department of Geology, Princeton University.
Van Schmus, Prof. W. Randall. Department of Geology, University of Kansas.
Veblen, Dr. David R. Department of Earth and Planetary Sciences, Johns Hopkins University.
Vogt, Dr. Peter. Department of the Navy, Naval Research Laboratory, Washington, D.C.
Vonnegut, Dr. Bernard. Retired; formerly, Atmospheric Sciences Research Center, State University of New York, Albany.

W

Wahl, Prof. Floyd M. Professor and Chairman, Department of Geology, University of Florida.
Walker, Dr. R. M. Department of Physics, Washington University.
Wallace, Dr. John M. Department of Atmospheric Sciences, University of Washington, Seattle.
Walt, Dr. Martin. Physical Sciences Laboratory, Lockheed Missiles and Space Company, Palo Alto, California.
Ward, Jeffrey. American Radio, Newington, Connecticut.
Warren, Dr. John J. K. Resources Pty. Ltd., Mitcham, Australia.
Watts, Dr. Anthony B. Department of Earth Sciences, University of Oxford, England.
Weaver, Dr. E. Eugene. Research Scientist, Product Development Group, Ford Motor Company, Dearborn, Michigan.
Weibell, Dr. Fred J. Biomedical Engineering Society, Culver City, California.
Weller, Prof. Gunter. Global Change and Arctic Systems, University of Alaska, Fairbanks.
Wetzel, Dr. Richard. College of William and Mary, Virginia Institute of Marine Science, School of Marine Science, Gloucester Point, Virginia.
Whelan, Dr. Jean K. Senior Research Specialist, Department of Chemistry, Woods Hole Oceanographic Institution, Woods Hole, Massachusetts.
White, Dr. C. Langdon. Professor of Geography, Stanford University.
White, Dr. Sidney E. Department of Geology and Mineralogy, Ohio State University.

Wickstrom, Dr. Conrad. Department of Biological Sciences, Kent State University.
Wiens, Prof. Herold J. Deceased; formerly, Department of Geography, Yale University.
Wikle, Prof. Thomas A. Professor and Head, Department of Geography, Oklahoma State University, Stillwater.
Wilkening, Dr. M. Department of Physics, New Mexico Institute of Mining and Technology.
Willett, Prof. Kurd C. Department of Meteorology, Massachusetts Institute of Technology.
Williams, Dr. Earle R. Department of Earth, Atmospheric and Planetary Sciences, Massachusetts Institute of Technology, Center for Meteorology and Physical Oceanography.
Wilson, Dr. James L. Department of Geology and Mineralogy, University of Michigan.
Winter, Dr. Thomas C. U.S. Geological Survey, Department of the Interior, Denver, Colorado.
Winterer, Dr. Edward L. Geological Research Division, Scripps Institution of Oceanography, La Jolla, California.
Wonders, Dr. William C. Department of Geography, University of Alberta, Canada.
Wones, Dr. David R. Department of Geological Sciences, Virginia Polytechnic Institute and State University.
Wood, Dr. B. J. Department of the Geophysical Sciences, University of Chicago.
Wuensch, Dr. Bernhardt J. Department of Metallurgy and Materials Science, Massachusetts Institute of Technology.
Wurtele, Prof. Morgan G. Department of Atmospheric Sciences, University of California, Los Angeles.
Wyrtki, Dr. Klaus. Department of Oceanography, University of Hawaii, Manoa.
Wüest, Dr. Alfred Johny. Applied Aquatic Ecology (APEC), Kastanienbaum, Switzerland.

Y

Yagi, Prof. Takehiko. Institute of Solid State Physics, University of Tokyo, Japan.
Young, Prof. Grant M. Department of Earth Sciences, University of Western Ontario, London, Ontario, Canada.
Yungul, Dr. Sulhi H. Chevron Resources Company, San Francisco, California.

Z

Zbar, Dr. Frederick. Chevy Chase, Maryland.
Zhang, Dr. Ruyuan. Department of Geological and Environmental Sciences, Stanford University, Stanford, California.
Zifcak, John H. The Foxboro Company, Foxboro, Massachusetts.
Zumberge, Dr. Mark A. Institute of Geophysics and Planetary Physics, Scripps Institution of Oceanography, La Jolla, California.

Index

The asterisk indicates page numbers of an article title.